// ANNALS OF
THE NEW YORK ACADEMY
OF SCIENCES

Volume 806

EDITORIAL STAFF

Executive Editor
BILL M. BOLAND

Managing Editor
JUSTINE CULLINAN

Associate Editor
STEPHANIE J. BLUDAU

The New York Academy of Sciences
2 East 63rd Street
New York, New York 10021

PAPERS ON GENERAL TOPOLOGY AND APPLICATIONS

ELEVENTH SUMMER CONFERENCE AT THE UNIVERSITY OF SOUTHERN MAINE

ANNALS OF THE NEW YORK ACADEMY OF SCIENCES
Volume 806

PAPERS ON GENERAL TOPOLOGY AND APPLICATIONS

ELEVENTH SUMMER CONFERENCE AT THE UNIVERSITY OF SOUTHERN MAINE

Edited by Susan Andima, Robert C. Flagg, Gerald Itzkowitz, Yung Kong, Ralph Kopperman, and Prabudh Misra

The New York Academy of Sciences
New York, New York
1996

♾ *The paper used in this publication meets the minimum requirements of American National Standard for Information Sciences—Permanence of Paper for Printed Library Materials, ANSI Z39.48-1984.*

Cover: *The art on the cover of the softcover edition of this book is part of a piece entitled "Chaos on a Torus (T16)" generated by Dr. James Yorke, who has graciously given permission for its use.*

Library of Congress Cataloging-in-Publication Data

Papers on general topology and applications : eleventh summer conference at the University of Southern Maine / edited by Susan Andima . . . [et al.].
p. cm. — (Annals of the New York Academy of Sciences; ISSN 0077-8923 ; v. 806).
Papers presented at the Eleventh Summer Conference on General Topology and Applications, held Aug. 10–13, 1995, at the University of Southern Maine.
Includes bibliographical references and index.
ISBN 1-57331-090-5 (cloth : alk. paper). — ISBN 1-57331-091-3 (pbk. : alk. paper)
1. Topology—Congresses. I. Andima, Susan, 1943– . II. Summer Conference on General Topology and Applications (11th : 1995 : University of Southern Maine) III. Series.
Q11.N5 vol. 806
[QA611.A1]
500 s—dc21 96-51642
[514] CIP

GYAT / PCP
Printed in the United States of America
ISBN 1-57331-090-5 (cloth)
ISBN 1-57331-091-3 (paper)

ANNALS OF THE NEW YORK ACADEMY OF SCIENCES

Volume 806
December 31, 1996

PAPERS ON GENERAL TOPOLOGY AND APPLICATIONS[a]

ELEVENTH SUMMER CONFERENCE AT THE UNIVERSITY OF SOUTHERN MAINE

Editors
SUSAN ANDIMA, ROBERT C. FLAGG, GERALD ITZKOWITZ,
YUNG KONG, RALPH KOPPERMAN, AND PRABUDH MISRA

Conference Organizers
ROBERT FLAGG (CHAIR)
DAVID BRIGGS, SAM SHORE, AND MERLE GUAY

Advisory Committee
BEVERLY BRECHNER, ROBERT C. FLAGG, ELISE GRABNER,
GARY GRUENHAGE, MEL HENRIKSEN, NEIL HINDMAN,
GERALD ITZKOWITZ, JUDY KENNEDY, YUNG KONG,
RALPH KOPPERMAN (CHAIR), RONNIE LEVY, JAN VAN MILL,
LARRY NARICI, ANDRZEJ SZYMANSKI, MURAT TUNCALI,
JERRY VAUGHAN, STEVE WATSON, SCOTT WILLIAMS
AND RICHARD WILSON

CONTENTS

[a]The papers in this volume were presented at the Eleventh Summer Conference on General Topology and Applications held at the University of Southern Maine, Gorham, Maine on August 10–13, 1995.

Financial assistance was received from:

- BARUCH COLLEGE—CUNY
- THE CITY COLLEGE—CUNY
- THE COLLEGE OF STATEN ISLAND—CUNY
- DELFT UNIVERSITY OF TECHNOLOGY, DELFT, THE NETHERLANDS
- LONG ISLAND UNIVERSITY—C.W. POST CAMPUS
- THE NEW YORK ACADEMY OF SCIENCES (MATHEMATICS SECTION)
- NIPISSING UNIVERSITY, NORTH BAY, ONTARIO
- QUEENS COLLEGE—CUNY

Preface

DAVID BRIGGS, ROBERT C. FLAGG, MERLE GUAY
University of Southern Maine
Portland, Maine 04103

AND

SAM SHORE
University of New Hampshire
Durham, New Hampshire 03824

The 11th Summer Conference on General Topology and Applications was held August 10–13 at the University of Southern Maine. There were three special sessions: *Generalized Metrics and Metrization* organized by Gary Gruenhage of Auburn University, Ralph Kopperman of the City College of New York, and Sam Shore of the University of New Hampshire; *Set Theoretic Topology* organized by Jerry Vaughan of the University of North Carolina and James Baumgartner of Dartmouth; and *Topology in Computer Science* organized by Jaco deBakker of the Centre for Mathematics and Computer Science, Amsterdam, Bob Flagg of the University of Southern Maine, Yung Kong of Queens College, and Jan Rutten of the Centre for Mathematics and Computer Science, Amsterdam. There were also two workshops: *Set Theory and Logic in Topology* conducted by Alan Dow of York University and *Topology in Computer Science* conducted by Mike Mislove of Tulane University.

The conference had over 140 participants with nearly 100 contributed talks at the general and special sessions. There were, in addition, six invited lectures:

- Zoltan Balogh's on programming second and third order properties of topological spaces;
- Peter Collins' on metrization theory and the theory of generalized metric spaces;
- Horst Herrlich's on compactness and the axiom of choice;
- Jimmie Lawson's on the interface between topology and domain theory;
- Stephen Watson's on topological methods in general relativity and statistics; and
- Ta Sun Wu's on maximal almost periodic groups and a theorem of Glicksberg.

The organizers of the Conference are grateful to the University of Southern Maine for hosting the conference and to the New York Academy of Science for publishing the proceedings.

We appreciate the support we received from the University of Southern Maine and the University of New Hampshire. Dr. Richard Stebbins, Dean of Arts and Sciences, University of Southern Maine, generously helped defray some of the costs of the conference. Elizabeth Morin, Conference Coordinator, suggested the location of the conference and made sure that all locations were prepared and available when needed. The Mathematics and Statistics Chairman, Joel Irish, provided much necessary support and encouragement. We would also especially like to thank Rosemary Austin, the Department of Mathematics and Statistics administrative assistant, for her hard work and support throughout.

The student assistants Anissa Bacon and Barrett Stetson from the University of Southern Maine and Mike Cullinane, John Griffin, Bill Hillery, David Kao, and Neil Portnoy from the University of New Hampshire kindly helped with transportation at the conference. John Griffin deserves special thanks for providing rides to and from the Portland Jetport at the most inconvenient times.

It would not be possible to publish this volume without the diligence and timeliness of the referees to whom we particularly express our warm thanks. Finally, we must acknowledge with gratitude the professional help extended by the editorial staff of the New York Academy of Sciences. Stephanie J. Bludau shepherded these papers through the press with grace and good humor, while Bill Boland, Justine Cullinan, and Sheila Kane arranged for publication of this volume as part of the Annals series of the Academy.

Note on the Constructible Sets of a Topological Space

JEAN-PAUL ALLOUCHE[a]
CNRS, LRI
Bâtiment 490
F-91405 Orsay Cedex
France

INTRODUCTION

Some twenty years ago, I tried to "compute," in a purely formal way, as many relations as possible involving the classical symbols of general topology, like $\overline{A}$ for the closure of the set A, $\overset{\circ}{A}$ for the interior of A and all usual well-known other notations. In particular, I obtained a characterization of the sets in the Boolean algebra generated by closed and open sets, that I never published. Recently, R. Mneimné, writing a book on *Group Actions* (see [9]), told me that the *constructible sets* (i.e., the sets in the Boolean algebra generated by the open and the closed subsets of a topological space), are a useful tool in classical Algebraic Geometry. For example, Chevalley proved the orbits of an algebraic affine group operating algebraically on an algebraic affine variety are locally closed (for the Zariski topology) by first proving the image of a polynomial map between algebraic varieties is a constructible set (see Cartan-Chevalley [3], or Borel [2]; see also Hartshorne [6, p. 94] or Humphreys [8, 4.4, p. 33]). For other occurrences of locally closed sets one can read Godement [4, p. 138], Herrlich and Hůsek [7], or the very recent paper by Arhangel'skiĭ and Collins [1].

Having had the opportunity at the *11th Summer Conference in Topology* to present the characterization of the constructible sets given below, I not only learned this was apparently not known *stricto sensu*, but also, that this result could be compared to a general characterization of "locally nice spaces" given by A.H. Stone in [10]. Recently, A.H. Stone also pointed out that R. Gurevič asked in [5] whether a result similar to the first part of our theorem is in the literature. In this paper, we first give our characterization, with an autonomous proof, then we show how our result also can be deduced from Stone's. Note, however, that our result is "constructive" and that we also

Mathematics Subject Classification: 54D99, 04A99.
Key words and phrases: constructible set, locally closed set, Boolean algebra.
[a]E-mail: allouche@lri.lri.fr.

give a characterization of the sets which are the union of a constructible set and of a closed set.

1. CHARACTERIZATION OF CONSTRUCTIBLE SETS

Let us recall that the constructible sets in a topological space E are the sets in the Boolean algebra generated by the closed and the open sets of this topological space, i.e., the smallest subset of $\mathcal{P}(E)$ containing the open and closed sets and stable under finite union and complementation — hence, also under finite intersection.

It is not hard to prove that these sets are exactly the finite unions of locally closed sets: a subset A of a topological space E is said to be locally closed if it can be written $A = \omega \cap F$, where ω is an open subset of E, and F a closed subset of E. Furthermore, if A is a locally closed subset of E, one can write it in a "canonical" way as:

$$A = \overline{A} \cap \overset{\circ}{\overbrace{(A \cup \complement \mathring{A})}} .$$

We briefly recall the proof of this result: plainly it suffices to prove that, if the set A is locally closed, then the set $A \cup \complement \mathring{A}$ is open, or equivalently, the complementary set $\overline{A} \cap \complement A$ is closed. Let $A = \omega \cap F$ with ω open and F closed, then

$$\overline{A} \cap \complement A = \overline{\omega \cap F} \cap (\complement\omega \cup \complement F) = (\overline{\omega \cap F} \cap \complement\omega) \cup (\overline{\omega \cap F} \cap \complement F)$$

$$= \overline{\omega \cap F} \cap \complement\omega$$

(as $\overline{\omega \cap F} \cap \complement F \subset \overline{\omega} \cap \overline{F} \cap \complement F = \overline{\omega} \cap F \cap \complement F = \varnothing$). Hence, as $\complement\omega$ is closed, the set $\overline{A} \cap \complement A$ is closed. Furthermore, the reciprocal is clearly true: if $\overline{A} \cap \complement A$ is closed — which is equivalent to $A \cup \complement \mathring{A}$ is open — then A is locally closed and indeed $A = \overline{A} \cap (A \cup \complement \mathring{A})$.

DEFINITION: For any subset A of the topological space E let

$$\check{A} = \overline{A} \cap \complement A,$$

(hence,

$$\overset{\vee\vee}{A} \text{ stands for } \overset{\vee}{\overbrace{(\check{A})}} \text{, and } \overset{n\vee}{A}$$

is obtained from A by iterating n times the operator $\vee$.

PROPOSITION 1:
(i) The set A is closed if and only if the set $\check{A}$ is empty.

(ii) The set A is locally closed if and only if $\check{A}$ is closed, which in turn is equivalent to

$$\check{\check{A}} = \varnothing.$$

(iii) For any set A one has $A = (A \setminus \check{\check{A}}) \cup \check{\check{A}}$.

(iv) For any set A, one has:

$$A \setminus \check{\check{A}} = A \cap \overbrace{(A \cup \overset{\circ}{\complement A})}^{\circ} = \overline{A} \cap \overbrace{(A \cup \overset{\circ}{\complement A})}^{\circ} = \overline{A} \setminus \overline{\check{A}}\ .$$

The set $A \setminus \check{\check{A}} = \overline{A} \setminus \overline{\check{A}}$ is thus locally closed for any set A; it is denoted in [9] by $\mathcal{B}(A)$ (note that a set A is thus locally closed if and only if $A = \mathcal{B}(A)$).

(v) For any integer n and any set A, one has

$$\mathcal{B}(\overset{n\vee}{A}) \subset \overline{\mathcal{B}(A)}\ .$$

(vi) For any integer n one has:

$$\overset{(2n+1)\vee}{A} = \overline{\overset{(2n)\vee}{A}} \cap \complement A \quad \text{and} \quad \overset{(2n+2)\vee}{A} = \overline{\overset{(2n+1)\vee}{A}} \cap A.$$

Proof: (i) The first part is trivial.

(ii) As recalled above, A is locally closed if and only if $\overline{A} \cap \complement A = \check{A}$ is closed, and from the first part, this happens if and only if

$$\check{\check{A}} = \varnothing.$$

(iii) The third assertion is easy to prove:

$$\check{\check{A}} = \overline{\check{A}} \cap \complement(\check{A}) = \overline{\overline{A} \cap \complement A} \cap \complement(\overline{A} \cap \complement A) = \overline{\overline{A} \cap \complement A} \cap (\complement\overline{A} \cup A)$$

$$= \overline{\overline{A} \cap \complement A} \cap A \text{ as } \overline{\overline{A} \cap \complement A} \text{ is included in } \overline{A},$$

hence, $\check{\check{A}}$ is a subset of A.

(iv) To prove the fourth assertion, one writes:

$$A \setminus \check{\check{A}} = A \cap \complement(\overline{\overline{A} \cap \complement A} \cap A) = A \cap (\complement(\overline{\overline{A} \cap \complement A}) \cup \complement A) = A \cap \complement(\overline{\overline{A} \cap \complement A})$$

$$= A \cap \overbrace{(A \cup \overset{\circ}{\complement A})}^{\circ} \subset \overline{A} \cap \overbrace{(A \cup \overset{\circ}{\complement A})}^{\circ}.$$

To prove the reverse inclusion it suffices to write:

$$\bar{A} \cap \overbrace{(A \cup \complement \mathring{A})}^{\circ} \subset \bar{A} \cap (A \cup \complement \mathring{A}) = (\bar{A} \cap A) \cup (\bar{A} \cap \complement \mathring{A}) = A \cup \varnothing = A.$$

(v) To prove assertion (v), it suffices to prove it for $n = 1$, because then, if it is true for an integer n and for every set A, one has

$$\mathcal{B}(\overset{(n+1)\vee}{A}) \subset \overline{\mathcal{B}(\overset{n\vee}{A})} \subset \overline{\overline{\mathcal{B}(A)}} = \overline{\mathcal{B}(A)}\,.$$

Let

$$\mathcal{B}(\check{A}) = \overline{\check{A}} \setminus \overline{\check{\check{A}}} \subset \bar{A} \setminus \overline{\check{\check{A}}}\,.$$

But $A = \mathcal{B}(A) \cup \check{\check{A}}$. Hence,

$$\bar{A} \setminus \overline{\check{\check{A}}} = \overline{\left(\mathcal{B}(A) \cup \check{\check{A}}\right)} \setminus \overline{\check{\check{A}}} = \left(\overline{\mathcal{B}(A)} \cup \overline{\check{\check{A}}}\right) \setminus \overline{\check{\check{A}}} \subset \overline{\mathcal{B}(A)}\,.$$

(vi) The last claim is proved by induction on n. Define $R_n = A$ if n is odd, and $R_n = \complement A$ if n is even. We shall prove by induction on n that:

$$\overset{(n+1)\vee}{A} = \overline{\overset{n\vee}{A}} \cap R_n.$$

The result is clear for $n = 1$ from the above computation:

$$\check{\check{A}} = \overline{\bar{A} \cap \complement A} \cap A = \overline{\check{A}} \cap A.$$

Then if the property is true for n, one has:

$$\overset{(n+2)\vee}{A} = \overline{\overset{(n+1)\vee}{A}} \cap \complement(\overset{(n+1)\vee}{A}) = (\overline{\overline{\overset{n\vee}{A}} \cap R_n}) \cap \complement(\overline{\overset{n\vee}{A}} \cap R_n) =$$

$$\overline{\overline{\overset{n\vee}{A}} \cap R_n} \cap (\complement \overline{\overset{n\vee}{A}} \cup \complement R_n)$$

$$= \overline{\overline{\overset{n\vee}{A}} \cap R_n} \cap \complement R_n = \overline{\overset{(n+1)\vee}{A}} \cap \complement R_n = \overline{\overset{(n+1)\vee}{A}} \cap R_{n+1}. \quad \square$$

REMARK 1: The sequence $(\overline{\overset{n\vee}{A}})_n$ is thus decreasing.

We now give a characterization of the constructible sets in a topological space in terms of the operator $\vee$. For a constructible set A, this characterization also gives the *smallest integer n* such that A is the union of n locally closed sets.

THEOREM: (a) The set A is the union of n locally closed sets if and only if

$$\overset{2n\vee}{A} = \emptyset.$$

Hence, the smallest such n is the smallest integer n such that $\overset{2n\vee}{A} = \emptyset$. Furthermore, in this case, the set A admits the following "canonical" decomposition into n disjoint locally closed sets:

$$A = (\overline{A}\setminus\overline{\overset{\vee}{A}}) \cup \left(\overline{\overset{\vee\vee}{A}} \setminus \overline{\overset{\vee\vee\vee}{A}}\right) \cup \cdots \cup \left(\overline{\overset{(2n-2)\vee}{A}} \setminus \overline{\overset{(2n-1)\vee}{A}}\right).$$

(b) The set A is the union of n locally closed sets and of a closed set if and only if

$$\overset{(2n+1)\vee}{A} = \emptyset.$$

Hence, the smallest such n is the smallest integer n such that

$$\overset{(2n+1)\vee}{A} = \emptyset.$$

Proof: (a) One first notices from Proposition 1 above that, for any set A and any integer n, one has (iterating assertion (iii)):

$$A = (A\setminus\overset{\vee\vee}{A}) \cup (\overset{\vee\vee}{A}\setminus\overset{4\vee}{A}) \cdots \cup \left(\overset{(2n-2)\vee}{A}\setminus\overset{2n\vee}{A}\right) \cup \overset{(2n\vee)}{A}.$$

Hence, if $\overset{2n\vee}{A} = \emptyset$, one gets:

$$A = (A\setminus\overset{\vee\vee}{A}) \cup (\overset{\vee\vee}{A}\setminus\overset{4\vee}{A}) \cdots \cup \left(\overset{(2n-2)\vee}{A}\setminus\overset{2n\vee}{A}\right),$$

which is the union of n locally closed sets from part (iv) of Proposition 1; this also gives the "canonical" decomposition.

(b) Suppose now that $A = X_1 \cup X_2 \cup \cdots \cup X_n$, where each X_i is a locally closed set. One proves easily by induction on r and using the last part of Proposition 1 that:

$$\overline{\overset{2r\vee}{A}} = \bigcup_{i_1, i_2, \ldots, i_{r+1}} \left(\overline{\overline{\overline{\overline{X_{i_1}} \cap CA \cap X_{i_2}} \cap CA \cap \cdots \cap X_{i_r}} \cap CA \cap X_{i_{r+1}}}\right)$$

$$= \bigcup_{i_1, i_2, \ldots, i_{r+1}} F_{i_1, \ldots, i_{r+1}}.$$

Hence, for $r = n$,

$$\overset{\overline{2n\vee}}{A} = \bigcup_{i_1, i_2, \ldots, i_{n+1}} \left(\overline{\overline{\overline{\overline{\overline{\overline{\overline{X_{i_1}} \cap \complement A} \cap X_{i_2}} \cap \complement A} \cap \cdots \cap X_{i_n}} \cap \complement A} \cap X_{i_{n+1}}}\right)$$

$$= \bigcup_{i_1, i_2, \ldots, i_{n+1}} F_{i_1, \ldots, i_{n+1}}.$$

Let us prove that every term in this union is empty. One first notices that for every choice of $k < l$ one has:

$$F_{i_1, i_2, \ldots, i_{n+1}} \subset \overline{\overline{\overline{X_{i_k}} \cap \complement A} \cap X_{i_l}} \subset \overline{\overline{\overline{X_{i_k}} \cap \complement X_{i_k}} \cap X_{i_l}} = \overline{\overline{X_{i_k}} \cap \complement X_{i_k} \cap X_{i_l}}$$

(remember that $A = \cup_j X_j$, and that each X_k is locally closed, which allows us to apply part (ii) of Proposition 1).

Now every $(n+1)$-uple $i_1, \ldots, i_{n+1}$ cannot consist of different integers (as they all take their values in $[1, n]$), hence there exist two indices, say i_k and i_l, with $k < l$, such that: $i_k = i_l = i$. Hence,

$$F_{i_1, \ldots, i_{n+1}} = \overline{\overline{\overline{X_{i_k}} \cap \complement X_{i_k}} \cap X_{i_l}} = \overline{\overline{X_i} \cap \complement X_i \cap X_i} = \varnothing.$$

The last part of the theorem is proved by imitating the proof above, details are left to the reader. ❑

REMARK 2: A result from this theorem is that, if A is the union of $n+1$ locally closed sets, then

$$\overset{\vee\vee}{A} \text{ is the union of } n \text{ locally closed sets.}$$

I did not find a simple direct proof of this result (which would — of course — imply the theorem by induction).

To play a little bit with the notions defined above, we want to give a "computational" proof of a result given in [9] (see also [2] in the Noetherian case).

PROPOSITION 2: (a) Let A be a subset of a topological space. There exists a locally closed set X such that $X \subset A \subset \overline{X}$ if and only if one has $A \subset \overline{\mathcal{B}(A)}$, (using notations from part (iv) of Proposition 1, see [9]).

Then $\mathcal{B}(A)$ satisfies the same conditions as X, contains X and hence is the largest locally closed set included in A, and whose closure contains A.

(b) The set of A's such that there exists a locally closed set X, with $X \subset A \subset \overline{X}$ is stable under finite union, and contains the constructible sets.

Proof: (a) As $\mathcal{B}(A)$ is a locally closed set included in A, the inclusion $A \subset \overline{\mathcal{B}(A)}$ implies the existence of a locally closed set X (indeed $\mathcal{B}(A)$ itself), such that $X \subset A \subset \overline{X}$. Reciprocally, suppose the existence of such a set X. Then, taking closures in the inclusions $X \subset A \subset \overline{X}$, one gets $\overline{X} = \overline{A}$, or equivalently $\complement^{\circ} A = \complement^{\circ} X$. Hence (using the fact that X is locally closed, which implies $\overline{X} \cap \complement X$ closed, or equivalently $X \cup \complement^{\circ} X$ open),

$$\mathcal{B}(A) = A \cap \left(\overset{\circ}{\overbrace{A \cup \complement^{\circ} A}} \right) \supset X \cap \left(\overset{\circ}{\overbrace{X \cup \complement^{\circ} X}} \right) = X \cap (X \cup \complement^{\circ} X) = X.$$

This implies $A \subset \overline{X} \subset \overline{\mathcal{B}(A)}$.

(b) To prove the second part of Proposition 2, it suffices clearly to prove that the set of subsets A for which there exists a locally closed set X with $X \subset A \subset \overline{X}$ is stable under finite unions. Let A and B be two such sets, let X and Y be two locally closed sets such that $X \subset A \subset \overline{X}$ and $Y \subset B \subset \overline{Y}$, then $(X \cup Y) \subset (A \cup B) \subset (\overline{X} \cup \overline{Y}) = \overline{(X \cup Y)}$. Suppose we find a locally closed set Z with $Z \subset (X \cup Y) \subset \overline{Z}$, then $\overline{Z} = \overline{X \cup Y}$ and Z satisfies: $Z \subset (A \cup B) \subset \overline{Z}$.

Now define $C = X \cup Y$. As

$$C = \mathcal{B}(C) \cup \check{\check{C}},$$

in order to prove that $C \subset \overline{\mathcal{B}(C)}$, it suffices to prove that

$$\check{\check{C}} \subset \overline{\mathcal{B}(C)}.$$

But C is the union of two locally closed sets, hence $\check{\check{C}}$ is locally closed. Hence,

$$\check{\check{C}} = \mathcal{B}(\check{\check{C}}) \subset \overline{\mathcal{B}(C)},$$

from part (v) of Proposition 1. ❑

2. EXAMPLES OF APPLICATIONS

We give two examples of applications, the proofs of which are left to the reader (who might also try to find proofs which do not use the notions above).

EXAMPLE 1: Prove that the set of rational numbers $\mathbb{Q}$ is not a constructible subset of the set of real numbers $\mathbb{R}$ for the usual topology.

EXAMPLE 2: Let $\mathcal{M}_n(\mathbb{C})$ be the set of all squares matrices of size n with complex coefficients, equipped with the usual topology, or with the Zariski topology. Note that the subset of matrices of given rank is a locally closed set (for both topologies). Now let $\mathcal{P}$ be the subset of the matrices of even rank. It is straightforward that $\mathcal{P}$ is the union of $\left[\frac{n}{2}\right] + 1$ locally closed sets (for both topologies). Prove that $\mathcal{P}$ is not the union of $\left[\frac{n}{2}\right]$ (or less) locally closed sets.

3. ANOTHER PROOF OF THE PREVIOUS CHARACTERIZATION OF CONSTRUCTIBLE SETS

After we explained the above result at the *11th Summer Conference in Topology,* A.H. Stone told us this might be related to his paper [10]. This paper explains how, for certain "nice" properties of a topological space, one can first characterize the spaces which are finite unions of subspaces each having locally the "nice" property and second give the smallest number of such subspaces. His results apply to locally closed spaces and, after simplifying Stone's notations, his result reads in this case: *let A be any subset of a topological space, define $H(A)$ to be the largest open locally closed subset of A and let $K(A)$ be defined by $K(A)=A\setminus H(A)$. Define two sequences of subsets H_i and K_i by:*

$$H_1 = H(A), \quad K_1 = K(A), \quad H_{i+1} = H(K_i), \quad K_{i+1} = K_i \setminus H_{i+1}.$$

Then A is the union of a finite number of locally closed sets if and only if there exists an integer n such that $K_n = \varnothing$. Furthermore, the smallest such n is the smallest number of locally closed sets whose union is equal to A.

In this statement, "the largest open locally closed subset of A" means of course "the largest open (in A) locally closed subset of A." Hence, to show our result can be deduced from Stone's, it suffices to show that, given any subset A of a topological space,

$$H(A) = A \setminus \check{\check{A}}, \quad \text{i.e.,}\ K(A) = \check{\check{A}},$$

$$K_i = K^{(i)}(A).$$

The second assertion is an easy induction. The first one is our next result. Before proving it, we note that this does not give (directly) the result in the second part of our Theorem. We also note that our statement on constructible sets is ... constructive. As alluded to in the introduction, one might develop a kind of "formal computation" in topological spaces: to speak in a rather vague way, the basic operations would be the finite set-theoretic operations (say, finite intersection and complementation, hence finite union) on one hand and topological operations on the other hand. One would like to use the smallest possible number of topological operations and to deduce the other ones from these "primitive" ones. For example, $\mathring{A}$ the interior of A, cannot in general be computed by using finitely many operations from set theory, even by allowing to "recognize" an open set. Hence, it is reasonable to take the operation $A \longrightarrow \mathring{A}$ as a "primitive" one. Of course, the operations

$$A \longrightarrow \overline{A} \text{ and } A \longrightarrow \check{A} = \overline{A} \cap \complement A$$

are then deduced easily from the "primitive" operations. Hence our criterion for constructible set is "constructive" in this sense, which is not immediately the case with Stone's result. Of course our next proposition unifies both points of view.

PROPOSITION 3: The largest open (in A) locally closed subset of the set A in the topological space E is the set

$$\mathcal{B}(A) = A \backslash \check{\check{A}} = \overline{A} \backslash \overline{\check{A}}$$

introduced in Proposition 1(iv).

Proof: We first recall that, for any set A, the set $\mathcal{B}(A)$ is a locally closed subset of the set A as [Proposition 1(iv)]

$$\mathcal{B}(A) = A \backslash \check{\check{A}} = A \cap \overset{\circ}{\overbrace{(A \cup \complement \mathring{A})}} = \overline{A} \cap \overset{\circ}{\overbrace{(A \cup \complement \mathring{A})}} = \overline{A} \backslash \overline{\check{A}}\,.$$

Now let B be a subset of A both open in A and locally closed. As B is open in A it can be written $B = A \cap \theta$ where θ is an open set. Let us show that B is then included in

$$\mathcal{B}(A) = A \backslash \check{\check{A}}.$$

It suffices to show that

$$B \cap \check{\check{A}} = \varnothing,$$

which is equivalent to

$$B \cap \overline{\overline{A} \cap \complement A} = \varnothing.$$

Let

$$\begin{aligned} B \cap \overline{\overline{A} \cap \complement A} = A \cap \theta \cap \overline{\overline{A} \cap \complement A} \quad & = B \cap A \cap \theta \cap \overline{\overline{A} \cap \complement A} \\ \subset B \cap A \cap \overline{\theta \cap \overline{A} \cap \complement A} \quad & \subset B \cap A \cap \overline{\overline{\theta \cap \overline{A}} \cap \complement A} \end{aligned}$$

$$= B \cap A \cap \overline{\overline{B} \cap \complement A} \qquad \subset B \cap A \cap \overline{\overline{B} \cap \complement B}.$$

But B is locally closed, hence $\overline{B} \cap \complement B$ is closed and equal to $\overline{\overline{B} \cap \complement B}$. Finally, the set

$$B \cap A \cap \overline{\overline{B} \cap \complement B} \text{ is equal to } B \cap A \cap \overline{B} \cap \complement B,$$

hence empty, and our proposition is proved.

ACKNOWLEDGMENTS

We thank R. Mneimné and G. Skordev for interesting discussions about these questions, the organizers of the *11th Summer Conference in Topology* where this paper has been presented, and A.H. Stone for having brought to our attention references [10] and [5].

REFERENCES

1. ARHANGEL'SKIĬ, A.V. & P.J. COLLINS. 1995. On submaximal spaces. Topology and its Applications. **64:** 219–241.
2. BOREL, A. 1991. Linear Algebraic Groups. 2nd edit. Springer Verlag. New York.
3. CARTAN, H. & C. CHEVALLEY. 1955/56. Géométrie Algébrique. Séminaire Cartan-Chevalley, Secrétariat Math., Paris.
4. GODEMENT, R. 1964. Topologie algébrique et théorie des faisceaux. Hermann.
5. GUREVIČ, R. 1980. Notices of the A.M.S. **27:** 278.
6. HARTSHORNE, R. 1977. Algebraic Geometry. Springer Verlag. New York.
7. HERRLICH, V. & M. HUŠEK. 1992. Categorical topology. *In:* Recent Progress in General Topology (M. Hušek & J. van Mill, Eds.). Elsevier Science Publisher. New York. 369–403
8. HUMPHREYS, J.E. 1975. Linear Algebraic Groups. Springer Verlag. New York.
9. MNEIMNÉ, R. Éléments de Géométrie — Actions de Groupes. To appear.
10. STONE, A.H. 1991. Finite union of locally nice spaces. Topology and its Applications. **41:** 57–64

A Metric Approach to Control Flow Semantics

J.W. DE BAKKER[a,c] AND E.P. DE VINK[b,d]

[a]*Department of Software Technology*
CWI, P.O.Box 94079
NL–1090 GB Amsterdam, The Netherlands
and
[b]*Faculty of Mathematics and Computer Science*
Vrije Universiteit
De Boelelaan 1081a
NL–1081–HV Amsterdam, The Netherlands

ABSTRACT: The authors' monograph "Control Flow Semantics" (MIT Press 1996) gives an extensive exposition of comparative programming language semantics using techniques from metric topology. In the book Banach's fixed-point theorem for complete metric spaces plays a prominent role in the construction and comparison of semantical models. Here we present the basic idea of exploiting Banach's theorem. The approach is illustrated with the definition of an operational and a denotational model for an abstract programming language with parallelism.

Note: The work reported here is not novel, but intended to provide an introduction of the metric approach to programming language semantics for a nonspecialist audience.

1. INTRODUCTION

Traditionally in the field of programming language semantics one distinguishes between operational semantics, which captures the computational intuition, and denotational semantics, which provides a technically convenient starting point for reasoning about programs. In the development of either type of model for a programming language featuring parallelism, the aspect of nondeterminism as well as the issue of infinite behavior should be dealt with. Indeterminacy enters the scene, since parallel components can proceed independent of each other. As a result, different computations may be observed from distinct runs of the same program. Also, for some components of a parallel system, e.g., a clock or the departmental file server, nontermination may be the ideal behavior. So, *a priori*, both finite and infinite computations should be considered as equally desirable.

Mathematics Subject Classification: 54E50, 68Q10, 68Q55.

Key words phrases: complete metric spaces, operational and denotational semantics, transition systems, concurrency.

[c]E-mail: jaco@cwi.nl.

[d]E-mail: vink@cs.vu.nl.

When concentrating on the flow of control, i.e., on the sequences of actions, instead of on the data flow, i.e., the cummulative effect of the actions, metric topology fits in very naturally: Intuitively, the set of behaviors of a program is the limit of finite sets, since a computational device is only capable of performing finitely many actions in finite time. Thus, to a program one may associate a *compact* set of computations. Below, for a metric space of words, the role of the hyperspace of compacta will be illustrated. Since both finite and infinite sequences are considered first class citizens, the principle of natural induction is not an appropriate proof principle. Instead one wants a technique applicable to infinite computations as well. It turns out that the well-known Banach fixed-point theorem serves as a very attractive alternative.

The importance of the Banach-theorem is threefold: for a contraction on a complete metric space there *exists* a fixed point, which is *unique* and, moreover, it can be obtained as *limit* from the iteration sequence starting from any point. All three aspects are exploited in the semantical modeling in this paper. For the construction of semantical definitions the existence of fixed points is used. The characterization of the fixed point as limit of iterations proves useful in deriving properties of semantical mappings obtained by a fixed point definition. Furthermore, when comparing operational and denotational semantics (i.e., when justifying the denotational model with respect to the operational one), the uniqueness of the fixed point is pivotal, and it is here that we find the major advantage of using the structure of complete metric spaces.

Further, we discuss the methodology for a single abstract programming language with parallelism. It should be noted that the metric approach is applicable to a wide range of programming constructs. One of the aims of the present paper is to encourage the reader to consult, e.g., the textbook [6] where, based on the principles discussed here, operational and denotational models for the control flow of a broad spectrum of programming concepts are developed and compared.

2. MATHEMATICAL PRELIMINARIES

In this section we collect some basic notation, definitions, and results. In particular we present the metric space $\mathcal{A}^\infty$ of finite and infinite words, completeness of the hyperspace of nonempty compact subsets of $\mathcal{A}^\infty$, and the so-called Lifting Lemma.

DEFINITION 2.1: Let $\mathbb{N}$ be the set of natural numbers including 0. Let $\mathbb{N}_{\geq 1}$ be the set of positive natural numbers. A (possibly empty) subset N of $\mathbb{N}_{\geq 1}$ is called an initial segment if $\forall n,m \in \mathbb{N}_{\geq 1}$: $n \in N \wedge m < n \Rightarrow m \in N$.

(a) Let $\mathcal{A}$ be a nonempty set. Define the collection $\mathcal{A}^\infty$ of *words* over $\mathcal{A}$

by $\mathcal{A}^\infty = \{w : N \to \mathcal{A} \mid N \subseteq \mathbb{N}_{\geq 1}$ an initial segment$\}$. For $n \in \mathbb{N}$ and $w \in \mathcal{A}^\infty$, $w : N \to \mathcal{A}$ say, the word $w[n]$ is the restriction $w \restriction N \cap \{1, \ldots, n\}$.

(b) The Baire-distance $d_B : \mathcal{A}^\infty \times \mathcal{A}^\infty \to [0,1]$ (cf. [1]) is given by

$$d_B(w, v) = 2^{-\sup\{n \in \mathbb{N} \mid w[n] = v[n]\}}$$

for $w,v \in \mathcal{A}^\infty$. By convention $2^{-\infty} = 0$.

We usually write $w = a_1a_2 \cdots a_n$ if $w : \{1, \ldots, n\} \to \mathcal{A}$ is such that $w(i) = a_i$ $(1 \leq i \leq n)$ and similarly $w = a_1a_2 \cdots$ if $w : \mathbb{N}_{\geq 1} \to \mathcal{A}$ satisfies $w(i) = a_i$ $(1 \leq i < \infty)$. In the first case w is called a *finite* word of length n, in the second case w is called *infinite*. $\mathcal{A}^*$ and $\mathcal{A}^\omega$ will be the subsets of finite and infinite words, respectively. So $\mathcal{A}^\infty = \mathcal{A}^* \cup \mathcal{A}^\omega$. We refer to $w[n]$ as the prefix of w of length (at most) n. We distinguish $\epsilon : \varnothing \to \mathcal{A}$. It is called the empty word. Note $w[0] = \epsilon$ for each $w \in \mathcal{A}^\infty$. Typically we have:

$d_B(abc, abe) = \frac{1}{4}$, $d_B(abc, ab) = \frac{1}{4}$, $d_B(abc, acb) = \frac{1}{2}$, $d_B(abc, abc) = 0$

$d_B(\epsilon, w) = 1$ for $w \in \mathcal{A}^\infty \setminus \{\varepsilon\}$, since $w[1] \neq \epsilon$ for nonempty words w

$d_B(a^n, a^\omega) = 2^{-n}$, $n \geq 0$ (employing the abbreviation a^n for the word of n consecutive a's and a^ω for the infinite word of a's).

A nonempty word w has a first letter a, say, and a possibly empty remainder w'. We write $w = a \cdot w'$ in such a situation. A basic property of the Baire distance is $d_B(a \cdot w, a \cdot v) = \frac{1}{2} d_B(w, v)$ for arbitrary $a \in \mathcal{A}$, $w, v \in \mathcal{A}^\infty$. The following result is well-known.

THEOREM 2.2: $(\mathcal{A}^\infty, d_B)$ is a complete metric space.

Next we consider the hyperspace of nonempty compact subsets of $\mathcal{A}^\infty$.

DEFINITION 2.3:

(a) Let $a \cdot W = \{a \cdot w \mid w \in W\}$ for $a \in \mathcal{A}$, $W \subseteq \mathcal{A}^\infty$.

(b) The Hausdorff-distance $d_H : \mathcal{P}(\mathcal{A}^\infty) \times \mathcal{P}(\mathcal{A}^\infty) \to [0,1]$ (cf. [9]) is given by $d_H(W, V) =$

$$\max\{\sup\{d_B(w,V) \mid w \in W\}, \sup\{d_B(v,W) \mid v \in V\}\}$$

for all $W,V \subseteq \mathcal{A}^\infty$, where $d_B(x,Y) = \inf\{d_B(x,y) \mid y \in Y\}$.

We have

$$d_H\left(\textstyle\bigcup_{i \in I} W_i, \bigcup_{i \in I} V_i\right) \leq \sup\{d_H(W_i, V_i) \mid i \in I\} \tag{2.1}$$

for arbitrary index set I and sequences $(W_i)_{i \in I}$, $(V_i)_{i \in I}$ in $\mathcal{P}(\mathcal{A}^\infty)$. Moreover, for $W,V \subseteq \mathcal{A}^\infty$ and $a \in \mathcal{A}$ we have

$$d_H(a \cdot W, a \cdot V) = \tfrac{1}{2} d_H(W, V) \tag{2.2}$$

provided W,V are *nonempty*.

THEOREM 2.4: Put $\mathcal{P}_{nco}(\mathcal{A}^\infty) = \{W \subseteq \mathcal{A}^\infty \mid W \text{ nonempty and compact}\}$. Then $(\mathcal{P}_{nco}(\mathcal{A}^\infty), d_H)$ is a complete metric space.

Proof: By Theorem 2.2 and a result due to Kuratowski [10].

DEFINITION 2.5: For a 1-bounded metric space (M,d) and any nonempty set X we use $X \to M$ to denote the function space of all mappings from X to M endowed with the distance d_F of uniform convergence, i.e.,

$$d_F(f, g) = \sup\{d(f(x), g(x)) \mid x \in X\}$$

for any $f,g: X \to M$.

An elementary result states that the metric space $(X \to M, d_F)$ is complete whenever (M,d) is.

A final mathematical prerequisite is the so-called Lifting Lemma. It is a useful technical result exploited below in Section 4.

DEFINITION 2.6: For metric spaces (M_i, d_i) and real numbers $\alpha_i \in [0, \infty)$, $0 \le i \le k$, we say that $f: M_1 \times \cdots \times M_k \to M_0$ is $(\alpha_1, \ldots, \alpha_k)$-Lipschitz if

$$d_0(f(x_1, \ldots, x_k), f(y_1, \ldots, y_k)) \le \max\{\alpha_i \cdot d_i(x_i, y_i) \mid 1 \le i \le k\}$$

for all $(x_1, \ldots, x_k), (y_1, \ldots, y_k) \in M_1 \times \cdots \times M_k$.

The function f is nonexpansive in case $\alpha_i = 1$ for all i $(1 \le i \le k)$ and f is contractive if $\alpha_i < 1$ for all i $(1 \le i \le k)$.

LEMMA 2.7: (Lifting Lemma) Suppose $f: \mathcal{A}^\infty \times \cdots \times \mathcal{A}^\infty \to \mathcal{P}_{nco}(\mathcal{A}^\infty)$ is $(\alpha_1, \ldots, \alpha_k)$-Lipschitz. Then $F: \mathcal{P}_{nco}(\mathcal{A}^\infty) \times \cdots \times \mathcal{P}_{nco}(\mathcal{A}^\infty) \to \mathcal{P}_{nco}(\mathcal{A}^\infty)$ given by

$$F(W_1, \ldots, W_k) = \bigcup \{f(w_1, \ldots, w_k) \mid \forall i, 1 \le i \le k: w_i \in W_i\}$$

is well-defined and also $(\alpha_1, \ldots, \alpha_k)$-Lipschitz.

Proof: See [6, p.95–96]. ❑

3. SYNTAX AND OPERATIONAL SEMANTICS

In this section we introduce the abstract programming language $\mathcal{L}$ and define its operational semantics $\mathfrak{O}$. After discussing the syntax for $\mathcal{L}$, we present an automaton that can "execute" programs for $\mathcal{L}$. The operational meaning of a program then will be defined in terms of the computations of the automaton.

DEFINITION 3.1:

(a) Fix sets $\mathcal{A}$ and $\mathcal{X}$ of atomic syntactic objects, referred to as *actions* and *procedure variables*, respectively.

(b) Let the set $\mathcal{S}$ of syntactic objects called *statements* be built up from $\mathcal{A}$, $\mathcal{X}$ and the symbols (,),;,+,|| such that

$$\mathcal{A}, \mathcal{X} \subseteq \mathcal{S} \quad \text{and} \quad s_1, s_2 \in \mathcal{S} \ \text{ implies } (s_1;s_2), (s_1+s_2), (s_1 \,||\, s_2) \in \mathcal{S}.$$

(c) Let the subset $\mathcal{G}$ of $\mathcal{S}$ of *guarded statements* be the least set such that

$$\mathcal{A} \subseteq \mathcal{G},\ g \in \mathcal{G}, s \in \mathcal{S} \text{ implies } (g;s) \in \mathcal{G} \text{ and}$$
$$g_1, g_2 \in \mathcal{G} \text{ implies } (g_1+g_2), (g_1 \,||\, g_2) \in \mathcal{G}.$$

(d) Let the set $\mathcal{D}$ of *declarations* consist of mappings from finite subsets of $\mathcal{X}$ to $\mathcal{G}$.

(e) Let the set $\mathcal{L}$, also referred to as the *language* $\mathcal{L}$ with *programs* as elements, consist of the pairs (D,s) in $\mathcal{D} \times \mathcal{S}$ such that each procedure variable occurring in D or s is in the domain of D.

The sets $\mathcal{A}$ and $\mathcal{X}$ are supposed to be taken nonempty, disjoint, and may be finite or infinite. They are considered primitive for $\mathcal{S}$. Typically, metavariables a, x, s, g, D (possibly with sub- or superscripts) are used to range over $\mathcal{A}$, $\mathcal{X}$, $\mathcal{S}$, $\mathcal{G}$, $\mathcal{D}$, respectively.

Our actions a from $\mathcal{A}$ can be thought of as abstract representations of constructs like assignments, tests, etc., that occur in concrete programming languages as **Pascal** or **C++**. Since we want to concentrate on the *flow of control*, i.e., the sequences in which actions take place and not on their effect, it suffices here to leave them unspecified.

A similar remark applies to $\mathcal{X}$. For procedure calls x we omit details concerning procedure arguments or parameters. Also, for declarations we avoid cluttering up the syntactic definition by too many details describing their precise form. Necessary for our purpose is that for any procedure variable occurring in a program (D,s), either in the declaration part D or in the program body s, there is a unique (guarded) statement associated to it by D.

That guarded statements — rather than general statements — are used as definitions of procedures will be crucial for the techniques explained below. A statement is by definition either an action a, a procedure variable x, a *sequential composition* $(s_1;s_2)$, an *alternative composition* or choice (s_1+s_2), or a *parallel composition* or merge $(s_1 \,||\, s_2)$. A guarded statement is either an action a, a sequential composition $(g;s)$ of a guarded statement g followed by an arbitrary statement s, the alternative composition (g_1+g_2) or the parallel composition $(g_1 \,||\, g_2)$ of two guarded statements.

The effect of the definition is that in a guarded statement, hence, in each definition $D(x)$ of a procedure variable x in a declaration D, no procedure variable (either equal to or different from x) occurs "at the front" of $D(x)$. So, e.g., self-referential declarations D with $D(x) = x$ and $D(x) = (a + (x;b))$ are excluded, but also less harmful D's with simultaneously $D(x) = y$ and $D(y) = a$. For notational convenience, usually outermost parentheses are dropped from statements and programs. Also, we consider the symbol ";" to bind tighter than "+", which in turn binds tighter than "||", and write, e.g., $a;x + b$ instead of $(a;x) + b$ or $((a;x) + b)$ where $a,b \in \mathcal{A}$ and $x,y \in \mathcal{X}$.

In order to capture the computational intuition of the language $\mathcal{L}$ we will construct an abstract automaton or virtual machine. This automaton will be able to run the programs in $\mathcal{L}$, making moves from one configuration to another while some action can be observed. Sometimes more transitions are possible from a configuration (the automaton is nondeterministic), some configurations do not admit a transition (the computation of the automaton then terminates). The transitions of the automaton are given by the transition relation. This relation consists of triples: a source configuration, a label or observation of the transition, and a target configuration. The labels of the transitions will be used in the formulation of the operational semantics below.

DEFINITION 3.2:

(a) Let E be a new symbol. The set $\mathcal{R}$ of *resumptions* is defined by $\mathcal{R} = S \cup \{E\}$. $\mathcal{R}$ is ranged over by the metavariable r. The set C of *configurations* is then given by $C = \mathcal{D} \times \mathcal{R}$. Typically, a configuration is denoted by a pair (D,r).

(b) The ternary relation $\rightarrow \subseteq C \times \mathcal{A} \times C$, called the transition relation for $\mathcal{L}$, is defined as follows: Let $r_1 \xrightarrow{a}_D r_2$ be short for $((D,r_1), a, (D,r_2))$. Then "$\rightarrow$" is the smallest relation such that

$$a \xrightarrow{a}_D \mathrm{E} \qquad \text{(Act)}$$

$$D(x) \xrightarrow{a}_D r \text{ implies } x \xrightarrow{a}_D r \qquad \text{(Rec)}$$

$$s_1 \xrightarrow{a}_D s' \text{ implies } s_1;s_2 \xrightarrow{a}_D s';s_2 \qquad \text{(Seq 1)}$$

$$s_1 \xrightarrow{a}_D \mathrm{E} \text{ implies } s_1;s_2 \xrightarrow{a}_D s_2 \qquad \text{(Seq 2)}$$

$$s_1 \xrightarrow{a}_D r' \text{ implies } s_1 + s_2 \xrightarrow{a}_D r' \qquad \text{(Choice 1)}$$

$$s_2 \xrightarrow{a}_D r'' \text{ implies } s_1 + s_2 \xrightarrow{a}_D r'' \qquad \text{(Choice 2)}$$

$$s_1 \xrightarrow{a}_D s' \text{ implies } s_1 \parallel s_2 \xrightarrow{a}_D s' \parallel s_2 \qquad \text{(Par 1)}$$

$$s_1 \xrightarrow{a}_D \mathrm{E} \text{ implies } s_1 \parallel s_2 \xrightarrow{a}_D s_2 \qquad \text{(Par 2)}$$

$$s_2 \xrightarrow{a}_D s'' \text{ implies } s_1 \parallel s_2 \xrightarrow{a}_D s_1 \parallel s'' \qquad \text{(Par 3)}$$

$$s_2 \xrightarrow{a}_D \mathrm{E} \text{ implies } s_1 \parallel s_2 \xrightarrow{a}_D s_1 \qquad \text{(Par 4)}$$

EXAMPLES: Suppose $D(x) = a;x + b$, then $x \xrightarrow{a}_D x$ (since $a \xrightarrow{a}_D \mathrm{E}$ by (Act), so $a;x \xrightarrow{a}_D x$ by (Seq 2), hence $a;x + b \xrightarrow{a}_D x$ by (Choice 1) and $x \xrightarrow{a}_D x$ by (Rec)) and $x \xrightarrow{b}_D \mathrm{E}$ (since $b \xrightarrow{b}_D \mathrm{E}$ by (Act), so $a;x + b \xrightarrow{b}_D \mathrm{E}$ by (Choice 2) and $x \xrightarrow{b}_D \mathrm{E}$ by (Rec)).

A resumption is either a statement (which encodes the things still to be done), or the symbol E (which denotes *termination*). In general the computation of the virtual machine *resumes* with r after a transition $s \xrightarrow{a}_D r$.

Configurations are pairs (D,r). They are the states of the abstract automaton for $\mathcal{L}$. Some configurations are final states, viz., those of the format (D,E), most are not, viz. the programs (D,s), i.e., from these configurations there are further transitions is possible.

In the setting of the language $\mathcal{L}$, the bindings between procedure variables and their associated bodies do not change. A computation of the virtual machine starting in state (D,s) will visit only configurations of the form (D,r). This justifies the use of the notation "$\rightarrow_D$".

We next discuss the clauses (Act) to (Par 4) of the transition relation. The only thing the automaton can do in a configuration (D,a) is to perform the action a and move to the terminated configuration (D,E). This is reflected by the clause (Act). For a configuration of the form (D,x), according to the clause (Rec), the machine looks up the body $D(x)$ of x, and checks the possible transitions for $(D,D(x))$. The same transitions then are available for (D,x). In a sequential composition $s_1;s_2$ the activity of s_1 precedes that of s_2. Therefore, the machine inspects the transitions for (D,s_1). If (D,s_1) admits a transition to (D,s') with label a, then also for $(D,s_1;s_2)$ a transition with label a is possible but now to $(D,s';s_2)$. The remainder s' of s_1 still has be executed before the machine processes the statement s_2. In case the execution of s_1 terminates, i.e., the machine makes an a-transition to E, the configuration $(D,s_1;s_2)$ will make an a-transition to the configuration (D,s_2) where the machine can start elaborating s_2. So the transitions added in virtue of the clauses (Seq 1) and (Seq 2) are in essence based on the possible transitions for the s_1-part.

This asymmetry is not present in the clauses for "+" and "||". If (D,s_1) can do a step, the same step can be performed by (D,s_1+s_2), and also, if (D,s_2) can do a step the same step is available to (D,s_1+s_2). So for the configuration (D,s_1+s_2) both the transitions of (D,s_1) and (D,s_2) are available. Similarly, for $(D,s_1 \parallel s_2)$, if for (D,s_1) there is a step to (D,s') or (D,E) then there is a step to $(D,s' \parallel s_2)$ or (D,s_2) in accordance with (Par 1) and (Par 2), respectively. A symmetric explanation applies to the cases (Par 3) and (Par 4).

A useful technical notion is the *weight* $wgt(D,s)$ of a program. Many properties in our setting cannot be shown using structural induction whereas induction on the weight is indeed appropriate. For example, one can show by weight induction that, with respect to some declaration D, for each s there is at least one transition but only finitely many (i.e., the set $\{(a,r) \mid s \xrightarrow{a}_D r\}$ is nonempty and finite).

DEFINITION 3.3: The weight function $wgt\colon \mathcal{L} \rightarrow \mathbb{N}$ is given by

$$wgt(D,a) = 1$$

$$wgt(D,x) = wgt(D,D(x)) + 1$$

$$wgt(D,s_1;s_2) = wgt(D,s_1) + 1$$

$$wgt(D,s_1 * s_2) = wgt(D,s_1) + wgt(D,s_2) + 1 \text{ with } * \in \{+,||\}.$$

Well-definedness of *wgt* can be checked first by structural induction for guarded statements in $\mathcal{G}$ and then again by structural induction for arbitrary statements in $\mathcal{S}$. (It is here that our set-up with guarded statements is crucial.) Note that according to the definition of *wgt*, the sources of the right-hand sides of the clauses (Rec) to (Par 4) have strictly more weight than the sources of the corresponding left-hand sides. In particular, by definition, $wgt(D,x) > wgt(D,D(x))$. Also note that generally the weight of (D,s_2) can not be compared to the weight of $(D,s_1;s_2)$; it may be more, equal, or less. In the denotational model an additional property will be needed to compensate for this (cf. Lemmas 4.3b and 4.5b).

The transition relation "$\rightarrow$" determines the single steps that the abstract automaton for $\mathcal{L}$ can perform. *Computations* performed by the virtual machine are those sequences of transitions that are "connected" and, moreover, are either finite and end with a terminating configuration, or are infinite. Now, the operational semantics, the basic computational model for $\mathcal{L}$, collects all the possible computation sequences for the program in which the abstract machine is supposed to start.

DEFINITION 3.4: The operational semantics $\mathcal{O}: \mathcal{L} \rightarrow \mathcal{P}(\mathcal{A}^{\infty})$ is given by the following clauses

(1) $a_1 \cdots a_n \in \mathcal{O}(D,s) \Leftrightarrow$

$$\exists r_0, \ldots, r_n : s = r_0 \wedge \forall i, 1 \leq i \leq n : r_{i-1} \xrightarrow{a_i}_D r_i \wedge r_n = E$$

(2) $a_1 a_2 \cdots \in \mathcal{O}(D,s) \Leftrightarrow \exists r_0, r_1, \ldots : s = r_0 \wedge \forall i, 1 \leq i < \infty : r_{i-1} \xrightarrow{a_i}_D r_i$

(where $a_1 \cdots a_n \in \mathcal{A}^*$ and $a_1 a_2 \cdots \in \mathcal{A}^{\omega}$). Define $\mathcal{O}': C \rightarrow \mathcal{P}(\mathcal{A}^{\infty})$ by $\mathcal{O}'(D,E) = \{\epsilon\}$ and $\mathcal{O}'(D,s) = \mathcal{O}(D,s)$.

EXAMPLES: Suppose $D(x) = a;x + b$, $D(y) = b;y$, $D(z) = c;z$. Elaborating the transitions for $(D,a;b+c)$, (D,x), $(D,y||c)$, and $(D,y||z)$ yields $\mathcal{O}(D,a;b||c) = \{abc, acb, cab\}$, $\mathcal{O}(D,x) = \{a^n b \mid n \geq 0\} \cup \{a^{\omega}\}$, $\mathcal{O}(D,y||c) = \{b^n c b^{\omega} \mid n \geq 0\} \cup \{b^{\omega}\}$ and $\mathcal{O}(D,y||z) = \{b,c\}^{\omega}$.

The operational semantics $\mathcal{O}$ will serve as a touchstone for other models of the language $\mathcal{L}$, in particular for the denotational semantics D developed in the next section. However, Definition 3.4 is not well suited to support a comparison with other meaning functions. We therefore, relying on Banach's fixed-point theorem, formulate an equivalent of Definition 3.4 that does facilitate this. The auxiliary mapping $\mathcal{O}'$ will be characterized as the *unique fixed point* of a contraction Φ on a complete metric space *Sem*.

DEFINITION 3.5: Put $Sem = Conf \rightarrow \mathcal{P}_{nco}(\mathcal{A}^{\infty})$ and define $\Phi: Sem \rightarrow Sem$ by

$$\Phi(S)(D,\text{E}) = \{\epsilon\}$$

$$\Phi(S)(D,s) = \bigcup \{a \cdot S(D,r) \mid s \xrightarrow{a}_D r\}$$

for $S \in Sem$, $D \in \mathcal{D}$, $s \in S$.

Note that functions $S \in Sem$ also take terminating configurations (D,E) as an argument. Well-definedness of Φ is based on the fact that the transition relation "$\rightarrow$" is *finitely branching*, i.e., that $\{(a,r) \in \mathcal{A} \times \mathcal{R} \mid s \xrightarrow{a}_D r\}$ is finite for each program (D,s). So the right-hand side in the second clause is a finite union of compacta.

THEOREM 3.6: Let Φ be as in Definition 3.5. Then Φ is a contraction and $\mathfrak{O}' = fix(\Phi)$.

Proof: See [6, pp.102–105, 527–528]. ❑

Using the fixed-point characterization of (the extension $\mathfrak{O}'$ of) $\mathfrak{O}$ we may actually prove that the operational meaning, e.g., of (D,x) as given above is indeed $\{a^n b \mid n \geq 0\} \cup \{a^\omega\}$ as follows: We have $x \xrightarrow{a}_D x$ and $x \xrightarrow{b}_D \text{E}$ as only transitions for (D,x). Therefore, by virtue of Theorem 3.6, $\mathfrak{O}(D,x) = a \cdot \mathfrak{O}(D,x) \cup \{b\}$. Putting $W = \{a^n b \mid n \geq 0\} \cup \{a^\omega\}$, we also have $W = \{a^n b \mid n \geq 0\} \cup \{a^\omega\} = a \cdot (\{a^n b \mid n \geq 0\} \cup \{a^\omega\}) \cup \{b\} = a \cdot W \cup \{b\}$. Hence, $d_H (\mathfrak{O}(D,x))\, W \leq \max\{d_H(a \cdot \mathfrak{O}(D,x), a \cdot W), d_H\{b\}, \{b\}\} = \frac{1}{2} d_H(\mathfrak{O}(D,x), W)$ by Equations (2.1) and (2.2). So $d_H(\mathfrak{O}(D,x), W) = 0$. Since both $\mathfrak{O}(D,x)$ (by Theorem 3.6) and W are closed, it follows that $\mathfrak{O}(D,x) = W = \{a^n b \mid n \geq 0\} \cup \{a^\omega\}$.

4. DENOTATIONAL SEMANTICS AND EQUIVALENCE OF $\mathfrak{O}$ AND $\mathfrak{D}$

The mathematical model $\mathfrak{D}$ for $\mathcal{L}$ that we develop (cf. Definition 4.6) below features two typical properties:

- Programs will be represented or "*denoted*" by elements of some mathematical structure $\mathbb{D}$, in our case a complete metric space. Then $\mathfrak{D}: \mathcal{L} \rightarrow \mathbb{D}$.
- The meaning of a syntactic construction will be given in terms of the meanings of its constituent parts. This is the *compositionality* principle. So, for $* \in \{;, +, \|\}$ we will have $\mathfrak{D}(D, s_1 * s_2) = \mathfrak{D}(D,s_1) * \mathfrak{D}(D,s_2)$ with at the right-hand side mappings $*: \mathbb{D}^2 \rightarrow \mathbb{D}$ still to be defined. These are the *semantical* operators ";", "+", and "||" on $\mathbb{D}$, and are the counterparts of the *syntactical* operators of $\mathcal{L}$.

Once $\mathbb{D}$ and the semantical operators have been chosen, the two conditions above determine almost all of the denotational semantics $\mathfrak{D}$, leaving only unsettled the cases for an action a and a procedure variable x.

As denotational domain $\mathbb{D}$ we take the nonempty and compact subsets of $\mathcal{A}^\infty$ (compact with respect to the Baire metric on $\mathcal{A}^\infty$). Taking nonempty compact sets does not only support the Hausdorff distance but more importantly paves the way for an application of the Lifting Lemma 2.7 for the construction of the semantical operators.

DEFINITION 4.1: The denotational domain $\mathbb{D}$ is defined as the collection $\{W \subseteq \mathcal{A}^\infty \mid W \text{ nonempty and compact}\}$ endowed with the Hausdorff distance d_H induced by d_B.

The semantical operation "+" can be given directly, viz. as set-union. The mappings ";" and "||" on $\mathbb{D}$ will be developed in two stages. First we define ";" and "||" on $\mathcal{A}^\infty$ (delivering elements of $\mathbb{D}$) and establish appropriate Lipschitz properties. Application of the Lifting Lemma 2.7 then yields operations ";" and "||" on $\mathbb{D}$ itself.

Following the computational interpretation of ";" we should have $\epsilon \,;w_2 = w_2$ and $(a \cdot w')\,;\, w_2 = a \cdot (w'\,; w_2)$: If the first component w_1 is the empty word then the sequential composition coincides with w_2. In case w_1 is nonempty, w_1 is of the form $a \cdot w'$ and $w_1\,;\,w_2$ starts with a followed by the sequential composition of w' and w_2. Though intuitively convincing the self-referential equations above cannot qualify as a definition of ";" on $\mathcal{A}^\infty$.

Similarly, introducing an auxiliary function "$⌊⌊$" called *leftmerge*, we expect for "||" that $w_1 \,||\, w_2 = w_1 ⌊⌊\, w_2 \cup w_2 ⌊⌊\, w_1$, $\epsilon ⌊⌊\, w_2 = \{w_2\}$ and $(a \cdot w') ⌊⌊\, w_2 = a \cdot (w' \,||\, w_2)$: A parallel composition of w_1 and w_2 can equally well start with the first element of w_1 or that of w_2 (if available). The mapping "$⌊⌊$" expresses that the left component should yield the first element, while the remainder should be merged. For example, $ab \,||\, c$ should yield $\{abc, acb, cab\}$: the action c can come after, in between or before the actions a and b. (Below, the "$⌊⌊$" will prove to be useful in the comparison of the operational and denotational model.)

Again, as for the mapping ";", the equations are circular ("||" depends on "$⌊⌊$", "$⌊⌊$" depends on "||"). They cannot serve as a definition. Our way out of this self-referentiality is the introduction of higher-order contractions $\Omega_;$ and $\Omega_{||}$, i.e., contractions on a suitable complete metric space of mappings. Their fixed points — unique by Banach's theorem — will be our ";" and "||", respectively.

DEFINITION 4.2: Put $Op = \mathcal{A}^\infty \times \mathcal{A}^\infty \to \mathbb{D}$. Define $\Omega_;, \Omega_{||} : Op \to Op$ such that, for $\phi \in Op$,

$$\Omega_;(\phi)(\epsilon, w_2) = \{w_2\}$$

$$\Omega_;(\phi)(a \cdot w', w_2) = a \cdot \phi(w', w_2)$$

$$\Omega_{||}(\phi)(w_1, w_2) = \Omega_;(\phi)(w_1, w_2) \cup \Omega_;(\phi)(w_2, w_1).$$

Then $;, ||, ⌊⌊ : \mathcal{A}^\infty \times \mathcal{A}^\infty \to \mathbb{D}$ are given by $; = fix(\Omega_;)$, $|| = fix(\Omega_{||})$, $⌊⌊ = \Omega_;(||)$.

EXAMPLES: We have $ab;cde = \{abcde\}$, for $ab;cde = a \cdot (b;cde) = a \cdot (b \cdot (\epsilon\,;cde)) = a \cdot (b \cdot \{cde\}) = a \cdot \{bcde\} = \{abcde\}$. Also $a^\omega\,;cde = \{a^\omega\}$. This can be seen as follows: Since $a^\omega = a \cdot a^\omega$ it holds that $a^\omega;cde = a \cdot (a^\omega;cde)$ and $\{a^\omega\} = a \cdot \{a^\omega\}$. Therefore, $d_H(a^\omega;cde, \{a^\omega\}) = d_H(a \cdot (a^\omega\,; cde), a \cdot \{a^\omega\}) = \frac{1}{2} d_H(a^\omega;cde, \{a^\omega\})$ by Equation (2.2). Thus $d_H(a^\omega;cde, \{a^\omega\}) = 0$ and $a^\omega;cde = \{a^\omega\}$.

For example, $ab \,||\, c = \{abc, acb, cab\}$, $b^\omega \,||\, c = \{b^n cb^\omega \mid n \geq 0\} \cup \{b^\omega\}$ and $b^\omega \,||\, c^\omega = \{b,c\}^\omega$. We elaborate the case $b^\omega \,||\, c$. Exploiting already the equality $\epsilon \,||\, b^\omega = \{b^\omega\}$ to be discussed in a minute, we find that $b^\omega \,||\, c = b^\omega \,\underline{\|}\, c \cup c \,\underline{\|}\, b^\omega = b \cdot (b^\omega \,||\, c) \cup c \cdot (\epsilon \,||\, b^\omega) = b \cdot (b^\omega \,||\, c) \cup c \cdot \{b^\omega\} = b \cdot (b^\omega \,||\, c) \cup \{cb^\omega\}$. Put $W = \{b^n cb^\omega \mid n \geq 0\} \cup \{b^\omega\}$. We then have $W = b \cdot (\{b^n cb^\omega \mid n \geq 0\} \cup \{b^\omega\}) \cup \{cb^\omega\} = b \cdot W \cup \{cb^\omega\}$. Hence, $d_H(\{b^\omega \,||\, c, W) = d_H(b \cdot (b^\omega \,||\, c) \cup \{cb^\omega\}, b \cdot W \cup \{cb^\omega\}) \leq \max\{d_H(b \cdot (b^\omega \,||\, c), b \cdot W), d_H(\{cb^\omega\}, \{cb^\omega\}) = \frac{1}{2} d_H(b^\omega \,||\, c, W)$. So $b^\omega \,||\, c = W = \{b^n cb^\omega \mid n \geq 0\} \cup \{b^\omega\}$.

The proof of the equality $\epsilon \,||\, w = \{w\}$ for $w \in \mathcal{A}^\infty$ may serve as an illustration of what one may call the "$\varepsilon \leq \frac{1}{2}\varepsilon$" argument. Since $\mathcal{A}^\infty$ also includes *infinite* words w, we can not reason by induction on the length of a word. The "$\varepsilon \leq \frac{1}{2}\varepsilon$" technique will serve as a substitute.

Put $\varepsilon = \sup\{d_H(\epsilon \,||\, w, \{w\}) \mid w \in \mathcal{A}^\infty\}$. We will argue that $\varepsilon \leq \frac{1}{2}\varepsilon$, so $\varepsilon = 0$, hence, $d_H(\epsilon \,||\, w, \{w\}) = 0$ and $\epsilon \,||\, w = \{w\}$ for all $w \in \mathcal{A}^\infty$ follows. We distinguish for this two cases: one for the empty string ϵ, a second for the nonempty strings. Straightforward calculation shows $\epsilon \,||\, \epsilon = \{\epsilon\}$, hence, $d_H(\epsilon \,||\, \epsilon, \{\epsilon\}) \leq \frac{1}{2}\varepsilon$. For nonempty $w \in \mathcal{A}^\infty$ we have $w = a \cdot w'$ for suitable $a \in \mathcal{A}$, $w' \in \mathcal{A}^\infty$. So, $\epsilon \,||\, w = \epsilon \,\underline{\|}\, w \cup (a \cdot w') \,\underline{\|}\, \epsilon = \{w\} \cup a \cdot (\epsilon \,||\, w')$ by symmetry of "$||$". Therefore, $d_H(\epsilon \,||\, w, \{w\} = d_H(\{w\} \cup a \cdot (\epsilon \,||\, w'), \{w\} \cup \{a \cdot w'\}) \leq \max\{d_H(\{w\}, \{w\}), d_H(\{a \cdot (\epsilon \,||\, w'), a \cdot \{w'\})\} = \frac{1}{2} d_H(\epsilon \,||\, w', \{w'\}) \leq \frac{1}{2}\varepsilon$ (by definition of ε). We conclude $d_H(\epsilon \,||\, w, \{w\} \leq \frac{1}{2}\varepsilon$ for all $w \in \mathcal{A}^\infty$. Taking the supremum over $\mathcal{A}^\infty$ then yields $\varepsilon \leq \frac{1}{2}\varepsilon$ and establishes the desired result.

For technical reasons, viz. contractivity of the higher-order transformation Ψ below, we need nonexpansiveness of the semantical operators involved (and a slightly stronger condition for ";"). Having already given ";" and "$||$" as fixed points of contractions on a suitable complete metric space we can push the application of Banach's fixed-point theorem a little further: ";" and "$||$" can be obtained as the limits of iteration sequences starting from an arbitrary point. This characterization will be exploited in the following lemma.

LEMMA 4.3:

(a) ";" and "$||$" are nonexpansive.

(b) $d_H(w_1;w_2, v_1;v_2) \leq \max\{d_B(w_1, v_1), \frac{1}{2} d_B(w_2, v_2)\}$ for $w_1,w_2,v_1,v_2 \in \mathcal{A}^\infty$ provided $w_1,v_1 \neq \epsilon$.

Proof: Let Op be as in Definition 4.2. Put $Op' = \{\phi \in Op \mid \phi$ nonexpansive}. Note Op' is a nonempty closed subspace of Op. So, if we can prove

$$\Omega_{;}(\phi), \Omega_{||}(\phi) \in Op' \quad \text{for all } \phi \in Op', \tag{4.3}$$

we have, by Banach's fixed-point theorem, ; = $\lim_n \Omega_{;}{}^n(\phi_0)$, || = $\lim_n \Omega_{||}{}^n(\phi_0)$ $\in Op'$ with $\phi_0 \in Op'$ arbitrary. By the very definition of Op' we then conclude that ";" and "||" are nonexpansive. Verification of Equation (4.3) requires a straightforward but detailed case analysis omitted here (see [6, pp.50–51,155]). Part (b) can be shown by a variation on the argument given for part (a). ❑

The Lipschitz-conditions given by the lemma complete the first stage of the construction of the semantical operators. Next we will appeal to the Lifting Lemma for both the definition of ";" and "||" and for establishing nonexpansiveness/contractivity properties. In addition we will introduce "∪" as the denotational counterpart of the syntactic "+". We also define the leftmerge "$\lfloor\!\lfloor$" on $\mathbb{D}$, since it will be of help in the comparison of $\mathfrak{C}$ and $\mathfrak{D}$ below.

DEFINITION 4.4: Define ;, || , $\lfloor\!\lfloor$: $\mathbb{D} \times \mathbb{D} \to \mathbb{D}$ as the liftings of ;, ||, $\lfloor\!\lfloor$ on $\mathcal{A}^\infty$, respectively. That is, for all $W, V \in \mathbb{D}$,

$$W * V = \bigcup \{w * v \mid w \in W, v \in V\} \quad \text{with } * \in \{;, ||, \lfloor\!\lfloor\}$$

where "∗" on the right-hand side is as given in Definition 4.2. Define furthermore +: $\mathbb{D} \times \mathbb{D} \to \mathbb{D}$ by $W + V = W \cup V$.

Well-definedness and nonexpansiveness of "+" is obvious. For ";", "||", and "$\lfloor\!\lfloor$" this is provided by Lemma 4.5. Moreover, the slightly stronger result for ";" on $\mathcal{A}^\infty$ remains valid in the setting of $\mathbb{D}$.

LEMMA 4.5:
(a) The mappings +, ;, ||, $\lfloor\!\lfloor$: $\mathbb{D}^2 \to \mathbb{D}$ are well-defined and nonexpansive.
(b) For $W_1, W_2, V_1, V_2 \in \mathbb{D}$, $d_H\,(W_1; W_2, V_1; V_2) \le \max\{\{d_H(W_1, V_1), \frac{1}{2} d_H(W_2, V_2)\}$ if $\epsilon \notin W_1, V_1$.

Proof: For part (b) we consider ";" as the lifting of (the restriction of) ";" from $(\mathcal{A}^\infty \backslash \epsilon) \times \mathcal{A}^\infty$ to $\mathbb{D}$. ❑

From the definitions of ";", "||", and "$\lfloor\!\lfloor$" on $\mathcal{A}^\infty$ and the construction of ";", "||", and "$\lfloor\!\lfloor$" on $\mathbb{D}$ the following properties can be checked straightforwardly

$$(a \cdot W)\ ;\ V = a \cdot (W; V) \tag{4.4}$$

$$W \,||\, V = W \lfloor\!\lfloor V \cup V \lfloor\!\lfloor W \tag{4.5}$$

$$(a \cdot W) \lfloor\!\lfloor V = a \cdot (W \,||\, V) \tag{4.6}$$

where $W, V \in \mathbb{D}$ and $a \in \mathcal{A}$. ❑

After having ascertained that we have the appropriate semantic operators available, we are now in a position to define $\mathfrak{D}$. We first supply the defining equations for $\mathfrak{D}$, and then prove that these equations can be satisfied uniquely by another argument in terms of a contractive higher-order mapping.

DEFINITION 4.6: $\mathfrak{D}: \mathcal{L} \to \mathbb{D}$ is given as the (unique) function satisfying

$$\mathfrak{D}(D,a) = \{a\}$$

$$\mathfrak{D}(D,x) = \mathfrak{D}(D,D(x))$$

$$\mathfrak{D}(D,s_1;s_2) = \mathfrak{D}(D,s_1)\ ;\ \mathfrak{D}(D,s_2)$$

$$\mathfrak{D}(D,s_1+s_2) = \mathfrak{D}(D,s_1) + \mathfrak{D}(D,s_2)$$

$$\mathfrak{D}(D,s_1 \,||\, s_2) = \mathfrak{D}(D,s_1) \,||\, \mathfrak{D}(D,s_2).$$

Here the operators ";", "+", and "||" on the right-hand side of the equations are as in Definition 4.4.

Examples:

(1) $\mathfrak{D}(D,a;(b+c)) = \mathfrak{D}(D,a);\mathfrak{D}(D,b+c) = \{a\};(\mathfrak{D}(D,b) + \mathfrak{D}(D,c)\ )\} = \{a\};\ (\{b\} + \{c\}) = \{a\};\ \{b,c\} = \{a;b,a;c\} = \{ab,ac\}$.

(2) Suppose $D(x) = a;x + b$. We then have $\mathfrak{D}(D,x) = \mathfrak{D}(D,a;x+b) = \mathfrak{D}(D,a;x) + \mathfrak{D}(D,b) = \{a\};\mathfrak{D}(D,x)) + \{b\} = \{(a \cdot \mathfrak{D}(D,x)) \cup \{b\}$, and, as before, it follows that $d_H(\mathfrak{D}(D,x), \{a^n b \mid n \geq 0\} \cup \{a^\omega\}) = 0$ and $\mathfrak{D}(D,x) = \{a^n b \mid n \geq 0\} \cup \{a^\omega\}$.

Note that the definition of $\mathfrak{D}$ is *not* by structural induction. In the second clause $\mathfrak{D}(D,x)$ is defined as $\mathfrak{D}(D,D(x))$, but elements of $\mathcal{X}$ are considered primitive for S and hence $\mathfrak{D}(D,D(x))$ is not a syntactically simpler program than $\mathfrak{D}(D,x)$. Also, the definition of $\mathfrak{D}$ cannot be justified by induction on the weight $wgt(D,s)$ of a program (as given in Definition 3.3). In the third clause $\mathfrak{D}(D,s_1;s_2)$ is defined both in terms of $\mathfrak{D}(D,s_1)$ and $\mathfrak{D}(D,s_2)$ whereas only $wgt(D,s_1) < wgt(D,s_1;s_2)$ is guaranteed. In the next lemma Definition 4.6 is justified by the introduction of a higher-order transformation Ψ. (Ψ is referred to as higher-order since it takes functions as its arguments.) Ψ will be shown to be contractive and $\mathfrak{D}$ will be its unique fixed point.

The definition of Ψ is by induction on the weight of a program. Therefore, special precaution has to be taken in the clause for the sequential composition: below, $\Psi(S)(D,s_1;s_2)$ will not depend on $\Psi(S)(D,s_2)$ (but on $S(D,s_2)$ instead). In order to compensate for this in the proof of the contractivity of Ψ, we rely on the extra Lipschitz-properties of the semantical operator ";" as established already in Lemma 4.3(b) and 4.5(b). It is exactly this point where the various ingredients of our set-up (weight, Lifting Lemma and Banach's fixed-point theorem) come together.

LEMMA 4.7: Put $Sem = \mathcal{L} \rightarrow \mathbb{D}$, and let Ψ: $Sem \rightarrow Sem$ be given by

$$\Psi(S)(D,a) = \{a\}$$

$$\Psi(S)(D,x) = \Psi(S)(D,D(x))$$

$$\Psi(S)(D,s_1;s_2) = \Psi(S)(D,s_1);\ S(D,s_2)$$

$$\Psi(S)(D,s_1+s_2) = \Psi(S)(D,s_1) + \Psi(S)(D,s_2)$$

$$\Psi(S)(D,s_1 \parallel s_2) = \Psi(S)(D,s_1) \parallel \Psi(S)(D,s_2)$$

for all $S \in Sem$. Then Ψ is contractive.

Proof: By definition of d_F on Sem it suffices to show, for arbitrary S_1,S_2 in Sem, $d_H(\Psi(S_1)(D,s), \Psi(S_2)(D,s)) \leq \frac{1}{2} d_F(S_1,S_2)$ by induction on $wgt(D,s)$. The cases for a and x are straightforward. For the remaining cases we have

$$
\begin{aligned}
&d_H(\Psi(S_1)(D,s_1;s_2), \Psi(S_2)(D,s_1;s_2))\\
&\quad = d_H(\Psi(S_1)(D,s_1);\ S_1(D,s_2), \Psi(S_2)(D,s_1);\ S_1(D,s_2))\\
&\quad \leq \text{[Lemma 4.3(b)]}\\
&\quad \max\{d_H(\Psi(S_1)(D,s_1), \Psi(S_2)(D,s_1))\ ,\ \tfrac{1}{2} d_H(S_1(D,s_2)), S_2(D,s_2))\}\\
&\quad \leq \text{[ind. hyp. for } (D,s_1)\text{, def. } d_F(S_1, S_2)]\ \tfrac{1}{2} d_F(S_1, S_2)
\end{aligned}
$$

and, for $* \in \{+, \parallel\}$,

$$
\begin{aligned}
&d_H(\Psi(S_1)(D,s_1 * s_2), \Psi(S_2)(D,s_1 * s_2))\\
&\quad = d_H(\Psi(S_1)(D,s_1) * \Psi(S_1)(D,s_2), \Psi(S_2)(D,s_1) * \Psi(S_2)(D,s_2))\\
&\quad \leq \text{[Lemma 4.5(a)]} \max\{d_H(\Psi(S_1)(D,s_i), \Psi(S_2)(D,s_i)) \mid i = 1, 2\}\\
&\quad \leq \text{[ind. hyp. for } (D,s_i),\ i = 1, 2]\ \tfrac{1}{2} d_F(S_1, S_2). \quad \square
\end{aligned}
$$

Obviously from the definition of Ψ we have for a map from $\mathcal{L}$ to $\mathbb{D}$ that being a fixed point of Ψ on the one hand, and satisfying the equation for D on the other, are equivalent. By the lemma and Banach's fixed-point theorem Ψ has precisely one fixed point. Thus $\mathfrak{D} = fix(\Psi)$ is well-defined.

Having presented both an operational and a denotational model for $\mathcal{L}$, the question about their relationship should be addressed next. We argue that the two semantics coincide exploiting once again Banach's fixed-point theorem. According to Theorem 3.6, the mapping $\mathfrak{O}'$, an extension of $\mathfrak{O}$ to the set of configurations *Conf*, can be characterized as the fixed point of a higher-order contraction Φ. By extending the denotational semantics $\mathfrak{D}$ to a mapping $\mathfrak{D}'$ on *Conf* and showing that also D' is a fixed point of Φ we obtain, by uniqueness of fixed points, $\mathfrak{O}' = \mathfrak{D}'$. From this $\mathfrak{O} = \mathfrak{D}$ follows directly.

THEOREM 4.8:

(a) Define $\mathcal{O}'\colon Conf \rightarrow \mathcal{P}_{nco}(\mathcal{A}^{\infty})$ by

$$\mathcal{O}'(D,s) = \mathcal{O}(D,s) \quad \text{and} \quad \mathcal{O}'(D,\mathrm{E}) = \{\epsilon\}.$$

Then $\Phi(\mathcal{O}') = \mathcal{O}'$.

(b) $\mathcal{O}(D,s) = \mathcal{O}(D,s)$ for all $(D,s) \in \mathcal{L}$.

Proof: (a) By the definitions, $\Phi(\mathcal{O}')(D,\mathrm{E}) = \mathcal{O}'(D,\mathrm{E})$. Therefore it suffices to check $\Phi(\mathcal{O}')(D,s) = \mathcal{O}'(D,s)$ by induction on $wgt(D,s)$. We restrict the presentation here to the most difficult subcase $(D, s_1 \| s_2)$ (referring for more detail on the remaining cases to [6, pp. 55 & 110]). Identify $\mathrm{E} \| s$ and $s \| \mathrm{E}$ with s. We then have

$$\begin{aligned}
&\Phi(\mathcal{D}')(D,s_1 \| s_2)\\
&\quad = [\text{def. } \Phi] \bigcup \{a \cdot \mathcal{D}'(D,r) \mid s_1 \| s_2 \xrightarrow{a}_D r\}\\
&\quad = [\text{rules (Par 1) to (Par 4)}] \bigcup \{a \cdot \mathcal{D}'(D,r_1 \| s_2) \mid s_1 \xrightarrow{a}_D r_1\} \cup\\
&\qquad\quad \bigcup \{a \cdot \mathcal{D}'(D,s_1 \| r_2) \mid s_2 \xrightarrow{a}_D r_2\}\\
&\quad = [\text{def. } \mathcal{D}'] \bigcup \{a \cdot (\mathcal{D}'(D,r_1) \| \mathcal{D}'(D,s_2)) \mid s_1 \xrightarrow{a}_D r_1\} \cup\\
&\qquad\quad \bigcup \{a \cdot (\mathcal{D}'(D,s_1) \| \mathcal{D}'(D,r_2)) \mid s_2 \xrightarrow{a}_D r_2\}\\
&\quad = [\text{prop. 4.6}] \left(\bigcup \{a \cdot \mathcal{D}'(D,r_1) \mid s_1 \xrightarrow{a}_D r_1\}\right) \mathbin{⫫} \mathcal{D}'(D,s_2) \cup\\
&\qquad\quad \left(\bigcup \{a \cdot \mathcal{D}'(D,r_2) \mid s_2 \xrightarrow{a}_D r_2\}\right) \mathbin{⫫} \mathcal{D}'(D,s_1)\\
&\quad = [\text{def. } \Phi]\ \Phi(\mathcal{D}')(D,s_1) \mathbin{⫫} \mathcal{D}'(D,s_2) \cup \Phi(\mathcal{D}')(D,s_2) \mathbin{⫫} \mathcal{D}'(D,s_1)\\
&\quad = [\text{ind. hyp.}]\ \mathcal{D}'(D,s_1) \mathbin{⫫} \mathcal{D}'(D,s_2) \cup \mathcal{D}'(D,s_2) \mathbin{⫫} \mathcal{D}'(D,s_1)\\
&\quad = [\text{prop. 4.5}]\ \mathcal{D}'(D,s_1) \| \mathcal{D}'(D,s_2)\\
&\quad = [\text{def. } \mathcal{D}']\ \mathcal{D}'(D,s_1 \| s_2).
\end{aligned}$$

(b) By part (a) and Theorem 3.6 we have $\mathcal{O}' = \mathcal{D}'$. From this part (b) follows by the definitions of $\mathcal{O}'$ and $\mathcal{D}'$. ❑

5. CONCLUDING REMARKS

In the above we have sketched for an abstract parallel programming language the metric approach to programming language semantics. Especially we have highlighted the use of Banach's fixed-point theorem in the construction of semantical definitions. Only a general flavor of the main ideas could be given here. Much more information can be found in the monograph

[6] where for 27 languages operational and denotational models are developed.

More widely adopted than the metric approach advocated here, is the order-theoretical approach to the development of denotational models. There, complete partial orders (cpo) and generalizations thereof play a similar role as our complete metric spaces, while Banach's theorem is replaced by the classical Knaster-Tarski theorem that characterizes the *least* fixed point of a continuous function on a cpo. The connections between the cpo-oriented semantics and topology were nicely surveyed by Mike Mislove in a tutorial series of the conference in Portland, Maine (cf. [12]). Recent developments based on the work of Lawvere [11], e.g., [8], [13], and [2] (also in these proceedings), contribute to the reconciliation of the two approaches in the framework of generalized metrics.

Another impression of the interplay of topology and metric semantics is the work on observation frames in [3] (presented at the Topology 1995 Conference) situated in locale theory. Recently, the ideas developed in that paper were applied fruitfully in the field of logic of programs, see [7]. As a final further reference on metric topology and semantics, more specifically operational semantics, we mention [4] and the paper [5], also in these proceedings, on metrically labeled transition systems.

ACKNOWLEDGMENT

Special thanks go to Henno Brandsma for his comments on earlier versions of this paper.

REFERENCES

1. Baire, R. 1909. Sur la représentation des fonctions discontinues. Acta Mathematica. **32:** 97–176.
2. Bonsangue, M.M., F. van Breugel, & J.J.M.M. Rutten. 1995. Alexandroff and Scott topologies for generalized ultrametric spaces. Technical Report IR–394.Vrije Universiteit, Amsterdam.
3. Bonsangue, M.M. 1996. Topological Dualities in Semantics. Ph.D. Thesis, Vrije Universiteit, Amsterdam.
4. van Breugel, F. Topological Models in Comparative Semantics. Birkhauser Verlag. To appear.
5. van Breugel, F. 1995. A theory of metric labelled transition systems. Technical Report SOCS–95.6. McGill Univeristy, Montreal.
6. de Bakker, J.W. & E.P. de Vink. 1996. Control Flow Semantics. The MIT Press. Cambridge, Massachusetts.
7. Bonsangue, M., E.P. de Vink, & J.N. Kok. 1995. Metric predicate transformers: Towards a notion of refinement for concurrency. *In:* Proc. CONCUR'95. Lecture Notes in Computer Science. **962:** 363–377. .

8. FLAGG, B. & R. KOPPERMAN. Continuity spaces: reconciling domains and metric spaces. Theoretical Computer Science. To appear.
9. HAUSDORFF, F. 1914. Grundzüge der Mengenlehre. Leipzig.
10. KURATOWSKI, K. 1956. Sur une méthode de métrisation compléte des certains espaces d'ensembles compacts. Fundamenta Mathematicae. **43:** 114–138.
11. LAWVERE, F.W. 1973. Metric spaces, generalized logic, and closed categories. Rendiconti del Seminario Matematico e Fisico di Milano. **43:** 135–166.
12. MISLOVE, M.W. 1995. A topological approach to theoretical computer science. Draft handed out at the 11*th* Summer Conference on General Topology and Applications, Portland, Maine.
13. WAGNER, K.R. 1994. Solving recursive domain equations with enriched categories. Ph.D. Thesis. Carnegie Mellon University.

IP* Sets in Product Spaces

VITALY BERGELSON[a,c] AND NEIL HINDMAN[b,c]

[a]*Department of Mathematics*
The Ohio State University
Columbus, Ohio 43210
and
[b]*Department of Mathematics*
Howard University
Washington, DC 20059

ABSTRACT: An IP* set in a semigroup $(S,\cdot)$ is a set which meets every set of the form $FP(\langle x_n\rangle_{n=1}^{\infty}) = \{\prod_{n\in F} x_n\colon F$ is a finite nonempty subset of $\mathbb{N}\}$, where the products are taken in increasing order of indices. We show here, using the Stone-Čech compactification of the product space $S_1 \times S_2 \times \ldots \times S_\ell$, that if each S_i is commutative, then whenever C is an IP* set in $S_1 \times S_2 \times \ldots \times S_\ell$ and for each $i \in \{1, 2, \ldots, \ell\}$, $\langle x_{i,n}\rangle_{n=1}^{\infty}$ is a sequence in S_i, C contains Cartesian products of arbitrarily large finite substructures of $FP(\langle x_{1,n}\rangle_{n=1}^{\infty}) \times FP(\langle x_{2,n}\rangle_{n=1}^{\infty}) \times \ldots \times FP(\langle x_{\ell,n}\rangle_{n=1}^{\infty})$. (The notion of "substructure" is made precise in Definition 2.4.) We show further that C need not contain any product of infinite substructures and that the commutativity hypothesis may not be omitted. Similar results apply to arbitrary finite products of semigroups. By way of contrast, we show in Theorem 2.3 that a much stronger conclusion holds for some cell of any finite partition of $S_1 \times S_2 \times \ldots \times S_\ell$ without even any commutativity assumptions.

1. INTRODUCTION

In a semigroup $(S,\cdot)$, we write $\Pi_{n\in F}\, x_n$ for the product written in increasing order of indices. (Thus $\Pi_{n\in\{1,5,7\}}\, x_n = x_1 \cdot x_5 \cdot x_7$.) We further write $FP(\langle x_n\rangle_{n=1}^{\infty}) = \{\Pi_{n\in F}\, x_n\colon F \in \mathcal{P}_f(\mathbb{N})\}$ where $\mathcal{P}_f(\mathbb{N}) = \{A\colon A$ is a finite nonempty subset of $\mathbb{N}\}$ and $\mathbb{N}$ is the set of positive integers. (Similarly, if the operation of the semigroup is denoted by +, we write $FS(\langle x_n\rangle_{n=1}^{\infty}) = \{\Sigma_{n\in F}\, x_n\colon F \in \mathcal{P}_f(\mathbb{N})\}$.) Loosely following Furstenberg [5], we say that a set $A \subseteq S$ is an IP set if and only if there is a sequence $\langle x_n\rangle_{n=1}^{\infty} \in S$ with $FP(\langle x_n\rangle_{n=1}^{\infty}) \subseteq A$. A set $C \subseteq S$ is then an IP* set if and only if $C \cap A \neq \emptyset$ for every IP set A (equivalently if and only if $C \cap FP(\langle x_n\rangle_{n=1}^{\infty}) \neq \emptyset$ for every sequence $\langle x_n\rangle_{n=1}^{\infty} \in S$).

Now IP* sets not only must meet every IP set; a much stronger statement is true. This statement involves the notion of a *product subsystem* which we pause now to define.

Mathematics Subject Classifications: Primary 05D10, 54B10; Secondary 22A30.
Key words and phrases: IP* sets, idempotents, product spaces, finite products.
[c]The authors acknowledge support received from the National Science Foundation via grants DMS 9401093 and DMS 9424421, respectively.

DEFINITION 1.1: Let $(S,\cdot)$ be a semigroup and let $\langle y_n\rangle_{n=1}^{\infty}$ be a sequence in S. The sequence $\langle x_n\rangle_{n=1}^{\infty}$ is a *product subsystem* of $\langle y_n\rangle_{n=1}^{\infty}$ if and only if there is a sequence $\langle H_n\rangle_{n=1}^{\infty}$ in $\mathcal{P}_f(\mathbb{N})$ such that:

(a) for each $n \in \mathbb{N}$, $\max H_n < \min H_{n+1}$ and

(b) for each $n \in \mathbb{N}$, $x_n = \Pi_{t \in H_n} y_t$.

Note that if $\langle x_n\rangle_{n=1}^{\infty}$ is a product subsystem of $\langle y_n\rangle_{n=1}^{\infty}$, then $FP(\langle x_n\rangle_{n=1}^{\infty}) \subseteq FP(\langle y_n\rangle_{n=1}^{\infty})$. (If requirement (a) of the definition is replaced by the requirement that $H_n \cap H_m = \varnothing$ when $n \neq m$, this conclusion may fail if S is not commutative. For example, if $x_1 = y_1 \cdot y_3$ and $x_2 = y_2$, then $x_1 \cdot x_2$ need not be in $FP(\langle y_n\rangle_{n=1}^{\infty})$.) In the event the operation in S is denoted by +, we change products to sums and refer to a *sum subsystem.*

The much stronger statement to which we referred is that given any IP* set $C \subseteq S$ and any sequence $\langle y_n\rangle_{n=1}^{\infty}$ in S, there is a product subsystem $\langle x_n\rangle_{n=1}^{\infty}$ of $\langle y_n\rangle_{n=1}^{\infty}$ such that $FP(\langle x_n\rangle_{n=1}^{\infty}) \subseteq C$. This statement will be proved as Corollary 1.7 below. (It is well known, but we don't have a convenient reference for it.)

It was shown in [2] that IP* sets may be large in unexpected ways. For example [2, Theorem 2.6] an IP* set in $(\mathbb{N}, +)$ must contain an infinite sequence with all of its sums (as expected) *and* its products as well. We investigate in this paper the extent to which IP* sets in product semigroups must be large in terms of sequences in the coordinates. One may ask for example, given semigroups S_1 and S_2, an IP* set C in $S_1 \times S_2$, and sequences $\langle w_n\rangle_{n=1}^{\infty}$ in S_1 and $\langle z_n\rangle_{n=1}^{\infty}$ in S_2, whether there must be product subsystems $\langle x_n\rangle_{n=1}^{\infty}$ of $\langle w_n\rangle_{n=1}^{\infty}$ and $\langle z_n\rangle_{n=1}^{\infty}$ of $\langle z_n\rangle_{n=1}^{\infty}$ with $FP(\langle x_n\rangle_{n=1}^{\infty}) \times FP(\langle y_n\rangle_{n=1}^{\infty}) \subseteq C$.

We give a strong negative answer to this question in Section 2. One does find, however, that if S_1 and S_2 are commutative semigroups, then any IP* set in $S_1 \times S_2$ must contain the product of arbitrarily large finite substructures of $\langle w_n\rangle_{n=1}^{\infty}$ and $\langle z_n\rangle_{n=1}^{\infty}$ and a similar statement applies to any finite product of semigroups (Theorem 2.6).

We also see a surprising turning of the tables. One is accustomed to finding properties that must hold for some cell of any finite partition and automatically expecting at least such a conclusion for IP* sets. For example, as we have already mentioned, while one cell of a partition of $\mathbb{N}$ must contain $FS(\langle x_n\rangle_{n=1}^{\infty})$ for some sequence $\langle x_n\rangle_{n=1}^{\infty}$, any IP* set must contain $FS(\langle x_n\rangle_{n=1}^{\infty}) \cup FP(\langle x_n\rangle_{n=1}^{\infty})$. Similarly, we will see below that given any discrete semigroup S and any idempotent p in βS, while trivially p must belong to the closure of some cell of any partition of S, p must belong also to the closure of *every* IP* set.

By contrast we will see in Theorems 2.2 and 2.7 that one can guarantee a much stronger conclusion for one cell of any partition of the product of two semigroups than can be guaranteed for IP* sets in the same product.

A special semigroup is of significant interest for these problems, namely the semigroup $(\mathcal{P}_f(\mathbb{N}), \cup)$ because a version of Theorem 2.6 for this semigroup is sufficient to imply the validity of Theorem 2.6 for all semigroups. We present this derivation in Section 3.

Our proofs utilize the algebraic structure of the Stone-Čech compactification βS of a discrete semigroup $(S,\cdot)$. We take βS to be the set of all ultrafilters on S, identifying the principal ultrafilters with the points of S. We also denote by $\cdot$ the operation on βS making $(\beta S, \cdot)$ a right topological semigroup with S contained in its topological center. That is, for all $p \in \beta S$, the function ρ_p: $\beta S \longrightarrow \beta S$ defined by $\rho_p(q) = q \cdot p$ is continuous and for all $x \in S$, the function λ_x: $\beta S \longrightarrow \beta S$ defined by $\lambda_x(q) = x \cdot q$ is continuous. The reader is referred to [6] and [7] for an elementary introduction to this operation, with the caution that there $(\beta S, \cdot)$ is taken to be left rather than right topological. (We have made the switch to conform to majority usage, at least among our collaborators.) The basic fact characterizing the right continuous operation on βS is, given $p,q \in \beta S$ and $A \subseteq S$, $A \in p \cdot q$ if and only if $\{x \in S: x^{-1}A \in q\} \in p$ where $x^{-1}A = \{y \in S: x \cdot y \in A\}$. In the event that the operation is denoted by +, the characterization above becomes $A \in p + q$ if and only if $\{x \in S: -x + A \in q\} \in p$ where $-x + A = \{y \in S: x + y \in A\}$.

We will use only a few basic facts about $(\beta S, \cdot)$ which we present here.

THEOREM 1.2: Any compact right topological semigroup has an idempotent.

Proof: [4, Corollary 2.10]. ❑

The proof of the following theorem is the Galvin-Glazer proof of the Finite Sum Theorem, which was never published by them, though it has appeared in surveys. Those places where it has appeared (to our knowledge), however, use left continuity so that the resulting products are in decreasing order of indices. To minimize confusion, we present the proof here.

THEOREM 1.3: Let $(S, \cdot)$ be a semigroup, let p be an idempotent in βS, and let $A \in p$. There is a sequence $\langle x_n\rangle_{n=1}^{\infty}$ in S such that $FP(\langle x_n\rangle_{n=1}^{\infty}) \subseteq A$.

Proof: Let $A_1 = A$ and let $B_1 = \{x \in S: x^{-1}A_1 \in p\}$. Since $A_1 \in p = p \cdot p$, $B_1 \in p$. Pick $x_1 \in B_1 \cap A_1$, let $A_2 = A_1 \cap (x_1{}^{-1}A_1)$, and note that $A_2 \in p$. Inductively, given $A_n \in p$, let $B_n = \{x \in S: x^{-1}A_n \in p\}$. Since $A_n \in p = p \cdot p$, $B_n \in p$. Pick $x_n \in B_n \cap A_n$ and let $A_{n+1} = A_n \cap (x_n{}^{-1}A_n)$.

To see, for example, why $x_2 \cdot x_4 \cdot x_5 \cdot x_7 \in A$, note that $x_7 \in A_7 \subseteq A_6 \subseteq x_5{}^{-1}A_5$ so that $x_5 \cdot x_7 \in A_5 \subseteq x_4{}^{-1}A_4$. Thus $x_4 \cdot x_5 \cdot x_7 \in A_4 \subseteq A_3 \subseteq x_2{}^{-1}A_2$ so that $x_2 \cdot x_4 \cdot x_5 \cdot x_7 \in A_2 \subseteq A_1 = A$.

More formally, we show by induction on $|F|$ that if $F \in \mathcal{P}_f(\mathbb{N})$ and $m = \min F$ then $\Pi_{n \in F}\, x_n \in A_m$. If $|F|=1$, then $\Pi_{n \in F}\, x_n = x_m \in A_m$. Assume $|F| > 1$, let $G = F \setminus \{m\}$, and let $k = \min G$. Note that since $k > m$, $A_k \subseteq A_{m+1}$. Then by the induction hypothesis, $\Pi_{n \in G}\, x_n \in A_k \subseteq A_{m+1} \subseteq x_m{}^{-1}A_m$ so $\Pi_{n \in F}\, x_n = x_m \cdot \Pi_{n \in G}\, x_n \in A_m$. ❑

Theorem 1.4: Let S be a semigroup and let $\langle x_n \rangle_{n=1}^{\infty}$ be a sequence in S. Then $\bigcap_{m=1}^{\infty} clFP(\langle x_n \rangle_{n=m}^{\infty}$ is a subsemigroup of βS. In particular, there is an idempotent p in βS such that for each $m \in \mathbb{N}$, $FP(\langle x_n \rangle_{n=m}^{\infty}) \in p$.

Proof: Let $T = \bigcap_{m=1}^{\infty} clFP(\langle x_n \rangle_{n=m}^{\infty})$. Then T is the intersection of a nested collection of closed nonempty subsets of βS so $T \neq \emptyset$. To see that T is a semigroup, let $p, q \in T$ be given and let $m \in \mathbb{N}$. To see that $FP(\langle x_n \rangle_{n=m}^{\infty}) \in p \cdot q$ we show that $FP(\langle x_n \rangle_{n=m}^{\infty}) \subseteq \{s \in S: s^{-1}FP(\langle x_n \rangle_{n=m}^{\infty}) \in q\}$. (Since $FP(\langle x_n \rangle_{n=m}^{\infty}) \in p$, this will suffice.) To this end, let $s \in FP(\langle x_n \rangle_{n=m}^{\infty})$ be given and pick $F \in \mathcal{P}_f(\mathbb{N})$ with $\min F \geq m$ such that $s = \Pi_{n \in F} x_n$. Let $k = \max F + 1$. Then $FP(\langle x_n \rangle_{n=k}^{\infty}) \in q$ so it suffices to show that $FP(\langle x_n \rangle_{n=k}^{\infty}) \subseteq s^{-1}FP(\langle x_n \rangle_{n=m}^{\infty})$. So, let $t \in FP(\langle x_n \rangle_{n=k}^{\infty})$ be given and pick $G \in \mathcal{P}_f(\mathbb{N})$ with $\min G \geq k$ such that $t = \Pi_{n \in G} x_n$. Then $\max F < \min G$ so $st = \Pi_{n \in F \cup G} x_n \in FP(\langle x_n \rangle_{n=m}^{\infty})$.

For the "in particular" conclusion note that by Theorem 1.2, T has an idempotent. $\quad\square$

The following simple characterization of IP* sets also provides some explanation of the origin of "IP."

THEOREM 1.5: Let $(S, \cdot)$ be a semigroup and let $C \subseteq S$. Then C is an IP* set if and only if for every idempotent $p \in \beta S$, $C \in p$.

Proof: Necessity. Let $p \cdot p = p \in \beta S$ and suppose $C \neq p$. Then $S \backslash C \in p$ so by Theorem 1.3 there is a sequence $\langle x_n \rangle_{n=1}^{\infty}$ with $FP(\langle x_n \rangle_{n=1}^{\infty}) \subseteq S \backslash C$, a contradiction.

Sufficiency. Let $\langle x_n \rangle_{n=1}^{\infty}$ be a sequence in S and pick by Theorem 1.4 an idempotent p with $FP(\langle x_n \rangle_{n=1}^{\infty}) \in p$. Then $C \cap FP(\langle x_n \rangle_{n=1}^{\infty}) \in p$ so $C \cap FP(\langle x_n \rangle_{n=1}^{\infty}) \neq \emptyset$. $\quad\square$

THEOREM 1.6: Let $(S, \cdot)$ be a semigroup, let $\langle y_n \rangle_{n=1}^{\infty}$ be a sequence in S, and let p be an idempotent in $\bigcap_{m=1}^{\infty} cl(FP(\langle y_n \rangle_{n=m}^{\infty}))$. If $A \in p$, then there is a product subsystem $\langle x_n \rangle_{n=1}^{\infty}$ of $\langle y_n \rangle_{n=1}^{\infty}$ with $FP(\langle x_n \rangle_{n=1}^{\infty}) \subseteq A$.

Proof: Pick by Theorem 1.4 an idempotent $p \in \beta S$ such that for each $m \in \mathbb{N}$, $FP(\langle x_n \rangle_{n=m}^{\infty}) \in p$. (The rest of the proof is now a modification of the proof of Theorem 1.3.)

Let $C_1 = A$, let $B_1 = \{y \in S: y^{-1}C_1 \in p\}$, and pick $y_1 \in B_1 \cap C_1$. Pick $H_1 \in \mathcal{P}_f(\mathbb{N})$ such that $y_1 = \Pi_{t \in H_1} x_t$. Let $k_1 = \max H_1 + 1$ and let $C_2 = C_1 \cap y_1^{-1}C_1 \cap FP(\langle x_t \rangle_{t=k_1}^{\infty})$.

Inductively, given $n > 1$ and $k_{n-1} \in \mathbb{N}$ and $C_n \in p$ with $C_n \subseteq FP(\langle x_t \rangle_{t=k_{n-1}}^{\infty})$, let $B_n = \{y \in S: y^{-1}C_n \in p\}$. Pick $y_n \in B_n \cap C_n$ and pick $H_n \in \mathcal{P}_f(\mathbb{N})$ with $\max H_n \geq k_{n-1}$ such that $y_n = \Pi_{t \in H_n} x_t$. Let $k_n = \max H_n + 1$ and let $C_{n+1} = C_n \cap y_n^{-1}C_n \cap FP(\langle x_t \rangle_{t=k_n}^{\infty})$.

Then one immediately sees that $\langle y_n \rangle_{n=1}^{\infty}$ is a product subsystem of $\langle x_n \rangle_{n=1}^{\infty}$. Just as in the proof of Theorem 1.3 one sees that $FP(\langle y_n \rangle_{n=1}^{\infty}) \subseteq A$. $\quad\square$

The following corollary will be used in the proof of Theorem 2.9.

COROLLARY 1.7: Given any IP* set $C \subseteq S$ and any sequence $\langle y_n \rangle_{n=1}^{\infty}$ in S, there is a product subsystem $\langle x_n \rangle_{n=1}^{\infty}$ of $\langle y_n \rangle_{n=1}^{\infty}$ such that $FP(\langle x_n \rangle_{n=1}^{\infty}) \subseteq C$.

Proof: Let $T = \bigcap_{m=1}^{\infty} c\ell FP(\langle y_n \rangle_{n=m}^{\infty})$. By Theorem 1.4 T is a compact subsemigroup of βS, so pick by Theorem 1.2 an idempotent $p \in T$. By Theorem 1.5 $C \in p$, so by Theorem 1.6 there is a product subsystem $\langle x_n \rangle_{n=1}^{\infty}$ of $\langle y_n \rangle_{n=1}^{\infty}$ such that $FP(\langle x_n \rangle_{n=1}^{\infty}) \subseteq C$. ❑

THEOREM 1.8: Let X be a set and let $\mathcal{G} \subseteq \mathcal{P}(X)$. The following statements are equivalent.

(a) For each $r \in \mathbb{N}$, if $X = \bigcup_{i=1}^{r} A_i$, there exist $i \in \{1, 2, \ldots, r\}$ and $G \in \mathcal{G}$ with $G \subseteq A_i$.

(b) There is an ultrafilter p on X such that for each $A \in p$, there exists $G \in \mathcal{G}$ with $G \subseteq A$.

Proof: [7, Theorem 6.7]. ❑

We write $\mathbb{N}$ for the set of positive integers and ω for the set of nonnegative integers.

2. IP* SETS IN PRODUCTS

We show in this section that IP* sets in any finite product of commutative semigroups contain products of arbitrarily large finite subsystems of sequences in the coordinates. We first modify the notion of product subsystems (Definition 1.1) to apply to finite sequences. (Requirement (a) below is of course a triviality, but see the discussion following Definition 1.1 for the fact that requirement (b) can be important.)

DEFINITION 2.1: Let $\langle y_n \rangle_{n=1}^{\infty}$ be a sequence in a semigroup S and let $k,m \in \mathbb{N}$. Then $\langle x_n \rangle_{n=1}^{m}$ is a *product subsystem* of $\langle y_n \rangle_{n=k}^{\infty}$ if and only if there exists a sequence $\langle H_n \rangle_{n=1}^{m}$ in $\mathcal{P}_f(\mathbb{N})$ such that

(a) $\min H_1 \geq k$,

(b) $\max H_n < \min H_{n+1}$ for each $n \in \{1, 2, \ldots, m-1\}$, and

(c) $x_n = \prod_{t \in H_n} y_t$ for each $n \in \{1, 2, \ldots, m\}$.

We establish, using an argument that we have used before [1], a purely Ramsey theoretic property in the product of arbitrary semigroups which is of interest in its own right, independent of IP* sets. It is similar in flavor to the Milliken-Taylor theorem [8, 9]. We first present the two dimensional version, partly because it is simpler, but also to present a stronger conclusion than will be needed here which will contrast sharply with Theorem 2.7 below.

THEOREM 2.2: Let S_1 and S_2 be semigroups and let $\langle y_{1,n} \rangle_{n=1}^{\infty}$ and $\langle y_{2,n} \rangle_{n=1}^{\infty}$ be sequences in S_1 and S_2, respectively. Let $k,r \in \mathbb{N}$ and let $S_1 \times S_2$

$= \bigcup_{j=1}^{r} C_j$. There exist $j \in \{1, 2, \ldots, r\}$ and a product subsystem $\langle x_{1,n}\rangle_{n=1}^{\infty}$ of $\langle y_{1,n}\rangle_{n=k}^{\infty}$ such that for each $m \in \mathbb{N}$ there is a product subsystem $\langle x_{2,n}\rangle_{n=1}^{\infty}$ of $\langle y_{2,n}\rangle_{n=k}^{\infty}$ such that

$$FP(\langle x_{1,n}\rangle_{n=1}^{m}) \times FP(\langle x_{2,n}\rangle_{n=1}^{\infty}) \subseteq C_i.$$

Proof: Given $i \in \{1, 2\}$, pick by Theorem 1.4 an idempotent

$$p_i \in \bigcap_{k=1}^{\infty} c\ell_{\beta S_i} FP(\langle y_{i,n}\rangle_{n=k}^{\infty}).$$

For $x \in S_1$ and $j \in \{1, 2, \ldots, r\}$, define $B_2(x, j) = \{y \in S_2\colon (x, y) \in C_j\}$, and for $j \in \{1, 2, \ldots, r\}$, let $B_1(j) = \{x \in S_1\colon B_2(x, j) \in p_2\}$.

Now for each $x \in S_1$, $S_2 = \bigcup_{j=1}^{r} B_2(x, j)$ so there exists $j \in \{1, 2, \ldots, r\}$ such that $B_2(x, j) \in p_2$. Consequently, $S_i = \bigcup_{j=1}^{r} B_1(j)$ so pick $j \in \{1, 2, \ldots, r\}$ such that $B_1(j) \in p_1$. Pick by Theorem 1.6 a product subsystem $\langle x_{1,n}\rangle_{n=1}^{\infty}$ of $\langle y_{1,n}\rangle_{n=k}^{\infty}$ such that $FP(\langle x_{1,n}\rangle_{n=1}^{\infty}) \subseteq B_1(j)$. Let $m \in \mathbb{N}$. Then $FP(\langle x_{1,n}\rangle_{n=1}^{m}) \subseteq B_1(j)$ so

$$\bigcap\{B_2(a, j)\colon a \in FP(\langle x_{1,n}\rangle_{n=1}^{m})\} \in p_2.$$

Pick by Theorem 1.6 a product subsystem $\langle x_{2,n}\rangle_{n=1}^{\infty}$ of $\langle y_{2,n}\rangle_{n=k}^{\infty}$ such that

$$FP(\langle x_{2,n}\rangle_{n=1}^{\infty}) \subseteq \bigcap\{B_2(a, j)\colon a \in FP(\langle x_{1,n}\rangle_{n=1}^{m})\}.$$

Then $FP(\langle x_{1,n}\rangle_{n=1}^{m}) \times FP(\langle x_{2,n}\rangle_{n=1}^{\infty}) \subseteq C_i$. ❑

It is a consequence of Theorem 2.7 below that one cannot get a product of full infinite subsystems, even if the cells of the partition $\{C_1, C_2, \ldots, C_r\}$ are required to be symmetric. (Recall that a subset C of a Cartesian product is *symmetric* provided $(a, b) \in C$ implies $(b, a) \in C$.)

At first glance it is not clear how to generalize the proof of Theorem 2.2 to higher dimensions, because the two coordinates are treated so differently. It turns out that intermediate coordinates are treated more like the first than the last. One can in fact replace the last coordinate $FP(\langle x_{\ell,n}\rangle_{n=1}^{m})$ in the following lemma by $FP(\langle x_{\ell,n}\rangle_{n=1}^{\infty})$.

THEOREM 2.3: Let $\ell \in \mathbb{N}$ and for each $i \in \{1, 2, \ldots, \ell\}$ let S_i be a semigroup and let $\langle y_{i,n}\rangle_{n=1}^{\infty}$ be a sequence in S_i. Let $m,k,r \in \mathbb{N}$ and let $\times_{i=1}^{\ell} S_i = \bigcup_{j=1}^{r} C_j$. There exists $j \in \{1, 2, \ldots, r\}$ and for each $i \in \{1, 2, \ldots, \ell\}$, there exists a product subsystem $\langle x_{i,n}\rangle_{n=1}^{m}$ of $\langle y_{i,n}\rangle_{n=k}^{\infty}$ such that

$$\times_{i=1}^{\ell} FP(\langle x_{i,n}\rangle_{n=1}^{m}) \subseteq C_j.$$

Proof: Given $i \in \{1, 2, \ldots, \ell\}$, pick by Theorem 1.4 an idempotent

$$p_i \in \bigcap_{k=1}^{\infty} c\ell_{\beta S_i} FP(\langle y_{i,n}\rangle_{n=k}^{\infty}).$$

For $(x_1, x_2, \ldots, x_{\ell-1}) \in \times_{i=1}^{\ell-1} S_i$ and $j \in \{1, 2, \ldots, r\}$, let

$$B_\ell(x_1, x_2, \ldots, x_{\ell-1}, j) = \{y \in S_\ell\colon (x_1, x_2, \ldots, x_{\ell-1}, y) \in C_j\}.$$

Now, given $t \in \{2, 3, \ldots, \ell - 1\}$, assume that $B_{t+1}(x_1, x_2, \ldots, x_t, j)$ has been defined for each $(x_1, x_2, \ldots, x_t) \in \times_{i=1}^{t} S_i$ and each $j \in \{1, 2, \ldots, r\}$. Given $(x_1, x_2, \ldots, x_{t-1}) \in \times_{i=1}^{t-1} S$ and $j \in \{1, 2, \ldots, r\}$, let

$$B_t(x_1, x_2, \ldots, x_{t-1}, j) = \{y \in S_t : B_{t+1}(x_1, x_2, \ldots, x_{t-1}, y, j) \in p_{t+1}\}.$$

Finally, given that $B_2(x,j)$ has been defined for each $x \in S_1$ and each $j \in \{1, 2, \ldots, r\}$, let $B_1(j) = \{x \in S_1 : B_2(x, j) \in p_2\}$.

We show by downward induction on t that for each $t \in \{2, 3, \ldots, \ell\}$ and each $(x_1, x_2, \ldots, x_{t-1}) \in \times_{i=1}^{t-1} S_i$, $S_t = \bigcup_{j=1}^{r} B_t(x_1, x_2, \ldots, x_{t-1}, j)$. This is trivially true for $t = \ell$. Assume $t \in \{2, 3, \ldots, \ell-1\}$ and the statement is true for $t + 1$. Let $(x_1, x_2, \ldots, x_{t-1}) \in \times_{i=1}^{t-1} S_i$. Given $y \in S_t$, one has that

$$S_{t+1} = \bigcup_{j=1}^{r} B_{t+1}(x_1, x_2, \ldots, x_{t-1}, y, j)$$

so one may pick $j \in \{1, 2, \ldots, r\}$ such that $B_{t+1}(x_1, x_2, \ldots, x_{t-1}, y, j) \in p_{t+1}$. Then $y \in B_t(x_1, x_2, \ldots, x_{t-1}, j)$.

Since for each $x \in S_1$, $S_2 = \bigcup_{j=1}^{r} B_2(x, j)$, one sees similarly that $S_1 = \bigcup_{j=1}^{r} B_1(j)$. Pick $j \in \{1, 2, \ldots, r\}$ such that $B_1(j) \in p_1$.

Pick by Theorem 1.6 a product subsystem $\langle x_{1,n} \rangle_{n=1}^{\infty}$ of $\langle y_{1,n} \rangle_{n=1}^{\infty}$ such that $FP(\langle x_{1,n} \rangle_{n=1}^{\infty}) \subseteq B_1(j)$. Let

$$D_2 = \bigcap \{B_2(a, j) : a \in FP(\langle x_{1,n} \rangle_{n=1}^{m})\}.$$

Since $FP(\langle x_{1,n} \rangle_{n=1}^{m})$ is finite, we have $D_2 \in p_2$, so pick by Theorem 1.6 a product subsystem $\langle x_{2,n} \rangle_{n=1}^{\infty}$ of $\langle y_{2,n} \rangle_{n=1}^{\infty}$ such that $FP(\langle x_{2,n} \rangle_{n=1}^{\infty} \subseteq D_2$.

Let $t \in \{2, 3, \ldots, \ell-1\}$ and assume $\langle x_{t,n} \rangle_{n=1}^{\infty}$ has been chosen. Let

$$D_{t+1} = \bigcap \{B_{t+1}(a_1, a_2, \ldots, a_t, j) : (a_1, a_2, \ldots, a_t) \in \times_{i=1}^{t} FP(\langle x_{i,n} \rangle_{n=1}^{m})\}.$$

Then $D_{t+1} \in p_{t+1}$ so pick by Theorem 1.6 a product subsystem $\langle x_{t+1,n} \rangle_{n=1}^{\infty}$ of $\langle y_{t+1,n} \rangle_{n=1}^{\infty}$ such that $FP(\langle x_{t+1,n} \rangle_{n=1}^{\infty} \subseteq D_{t+1}$.

Then $\times_{i=1}^{\ell} FP(\langle x_{t,n} \rangle_{n=1}^{m}) \subseteq C_j$, as required. ❑

In contrast with Theorem 2.3, where the partition conclusion applied to products of arbitrary semigroups, we now restrict ourselves to products of commutative semigroups. We will see in Theorem 2.8 that without this restriction one is not guaranteed products of any subsystems at all.

DEFINITION 2.4: Let S be a semigroup, let $\langle y_n \rangle_{n=1}^{\infty}$ be a sequence in S, and let $m \in \mathbb{N}$. The sequence $\langle x_n \rangle_{n=1}^{m}$ is a *weak product subsystem* of $\langle y_n \rangle_{n=1}^{\infty}$ if and only if there exists a sequence $\langle H_n \rangle_{n=1}^{m}$ in $\mathcal{P}_f(\mathbb{N})$ such that $H_n \cap H_k = \emptyset$ when $n \neq k$ in $\{1, 2, \ldots, m\}$ and $x_n = \prod_{t \in H_n} y_t$ for each $n \in \{1, 2, \ldots, m\}$.

Recall by way of contrast, that in a *product subsystem* one requires that $\max H_n < \min H_{n+1}$.

LEMMA 2.5: Let $\ell \in \mathbb{N}$ and for each $i \in \{1, 2, \ldots, \ell\}$, let S_i be a commutative semigroup and let $\langle y_{i,n}\rangle_{n=1}^{\infty}$ be a sequence in S_i. Let

$$\mathcal{L} = \{p \in \beta(\times_{i=1}^{\ell} S_i)\text{: for each } A \in p \text{ and each } m, k \in \mathbb{N} \text{ there exist for each } i \in \{1, 2, \ldots, \ell\} \text{ a weak product subsystem } \langle x_{i,n}\rangle_{n=1}^{m} \text{ of } \langle y_{i,n}\rangle_{n=k}^{\infty} \text{ such that } \times_{i=1}^{\ell} FP(\langle x_{i,n}\rangle_{n=1}^{m}) \subseteq A\}.$$

Then $\mathcal{L}$ is a compact subsemigroup of $\beta(\times_{i=1}^{\ell} \mathbb{N})$.

Proof: Since any product subsystem is also a weak product subsystem, we have by Theorems 2.3 and 1.8 that $\mathcal{L} \neq \emptyset$. Since $\mathcal{L}$ is defined as the set of ultrafilters all of whose members satisfy a given property, $\mathcal{L}$ is closed, hence compact. To see that $\mathcal{L}$ is a semigroup, let $p,q \in \mathcal{L}$, let $A \in p \cdot q$ and let $m, k \in \mathbb{N}$. Then $\{\vec{a} \in \times_{i=1}^{\ell} S_i : \vec{a}^{-1}A \in q\} \in p$ (where, recall, $a^{-1}A = \{\vec{b}: \vec{a} \cdot \vec{b} \in A\}$) so choose for each $i \in \{1, 2, \ldots, \ell\}$ a weak product subsystem $\langle x_{i,n}\rangle_{n=1}^{m}$ of $\langle y_{i,n}\rangle_{n=k}^{\infty}$ such that

$$\times_{i=1}^{\ell} FP(\langle x_{i,n}\rangle_{n=1}^{m}) \subseteq \{\vec{a} \in \times_{i=1}^{\ell} S_i : \vec{a}^{-1}A \in q\}.$$

Given $i \in \{1, 2, \ldots, \ell\}$ and $n \in \{1, 2, \ldots, m\}$, pick $H_{i,n} \in \mathcal{P}_f(\mathbb{N})$ with $\min H_{i,n} \geq k$ such that $x_{i,n} = \Pi_{t \in H_{i,n}} y_{i,t}$ and if $1 \leq n < s \leq m$, then $H_{i,n} \cap H_{i,s} = \emptyset$.

Let $r = \max(\bigcup_{i=1}^{\ell} \bigcup_{n=1}^{m} H_{i,n}) + 1$ and let

$$B = \bigcap\{\vec{a}^{-1}A : \vec{a} \in \times_{i=1}^{\ell} FP(\langle x_{i,n}\rangle_{n=1}^{m})\}.$$

Then $B \in q$ so choose for each $i \in \{1, 2, \ldots, \ell\}$ a weak product subsystem $\langle z_{i,n}\rangle_{n=1}^{m}$ of $\langle y_{i,n}\rangle_{n=r}^{\infty}$ such that $\times_{i=1}^{\ell} FP(\langle x_{i,n}\rangle_{n=1}^{m}) \subseteq B$. Given $i \in \{1, 2, \ldots, \ell\}$ and $n \in \{1, 2, \ldots, m\}$, pick $K_{i,n} \in \mathcal{P}_f(\mathbb{N})$ with $\min K_{i,n} \geq r$ such that $z_{i,n} = \Pi_{t \in K_{i,n}} y_{i,t}$ and if $1 \leq n < s \leq m$, then $K_{i,n} \cap K_{i,s} = \emptyset$.

For $i \in \{1, 2, \ldots, \ell\}$ and $n \in \{1, 2, \ldots, m\}$, let $L_{i,n} = H_{i,n} \cup K_{i,n}$. Then $\Pi_{t \in L_{i,n}} y_{i,t} = \Pi_{t \in H_{i,n}} y_{i,t} \cdot \Pi_{t \in K_{i,n}} y_{i,t} = x_{i,t} \cdot z_{i,t}$ and if $1 \leq n < s \leq m$, then $L_{i,n} \cap L_{i,s} = \emptyset$. Thus for each $i \in \{1, 2, \ldots, \ell\}$, $\langle x_{i,t} \cdot z_{i,t}\rangle_{n=1}^{m}$ is a weak product subsystem of $\langle y_{i,n}\rangle_{n=k}^{\infty}$.

Finally, we claim that

$$\times_{i=1}^{\ell} FP(\langle x_{i,n} \cdot z_{i,n}\rangle_{n=1}^{m}) \subseteq A.$$

To this end let $\vec{c} \in \times_{i=1}^{\ell} FP(\langle x_{i,n} \cdot z_{i,n}\rangle_{n=1}^{m})$ be given. For each $i \in \{1, 2, \ldots, \ell\}$, pick $F_i \subseteq \{1, 2, \ldots, m\}$ such that $c_i = \Pi_{t \in F_i} x_{i,t} \cdot z_{i,t})$ and let $a_i = \Pi_{t \in F_i} x_{i,t}$ and $b_i = \Pi_{t \in F_i} z_{i,t}$. Then $\vec{b} \in \times_{i=1}^{\ell} FP(\langle z_{i,n}\rangle_{n=1}^{m})$ so $\vec{b} \in B$. Since $\vec{a} \in \times_{i=1}^{\ell} FP(\langle x_{i,n}\rangle_{n=1}^{m})$ one has that $\vec{b} \in \vec{a}^{-1}A$ so that $\vec{a} \cdot \vec{b} \in A$. Since each S_i is commutative, we have for each $i \in \{1, 2, \ldots, \ell\}$ that $\Pi_{t \in F_i}(x_{i,t} \cdot z_{i,t}) = \Pi_{t \in F_i} x_{i,t}) \cdot (\Pi_{t \in F_i} z_{i,t})$ so that $\vec{c} = \vec{a} \cdot \vec{b}$ as required. ❑

THEOREM 2.6: Let $\ell \in \mathbb{N}$ and for each $i \in \{1, 2, \ldots, \ell\}$, let S_i be a commutative semigroup and let $\langle y_{i,n}\rangle_{n=1}^{\infty}$ be a sequence in S_i. Let C be an IP* set in $\times_{i=1}^{\ell} S_i$ and let $m \in \mathbb{N}$. Then for each $i \in \{1, 2, \ldots, \ell\}$ there is a weak product subsystem $\langle x_{i,n}\rangle_{n=1}^{m}$ of $\langle y_{i,n}\rangle_{n=k}^{\infty}$ such that $\times_{i=1}^{\ell} FP(\langle x_{i,n}\rangle_{n=1}^{m}) \subseteq C$.

Proof: Let $\mathcal{L}$ be as in Lemma 2.5. Then $\mathcal{L}$ is a compact subsemigroup of $\times_{i=1}^{\ell} S_i$ so by Theorem 1.2 there is an idempotent $p \in \mathcal{L}$. By Theorem 1.5, $C \in p$. Thus by the definition of $\mathcal{L}$, for each $i \in \{1, 2, \ldots, \ell\}$ there is a weak product subsystem $\langle x_{i,n}\rangle_{n=1}^{m}$ of $\langle y_{i,n}\rangle_{n=k}^{\infty}$ such that $\times_{i=1}^{\ell} FP(\langle x_{i,n}\rangle_{n=1}^{m}) \subseteq C$. ❑

Three natural questions are raised by Theorem 2.6. (1) Can one obtain infinite weak product subsystems (defined in the obvious fashion) such that $\times_{i=1}^{\ell} FP(\langle x_{i,n}\rangle_{n=1}^{\infty}) \subseteq C$? (2) Can one replace "weak product subsystems" with "product subsystems?" (3) Can one omit the requirement that the semigroups S_i be commutative? We answer all three of these questions in the negative.

The first two questions are answered in Theorem 2.7 using the semigroup $(\mathbb{N}, +)$. Since the operation is addition we refer to "sum subsystems" rather than "product subsystems."

THEOREM 2.7: There is an IP* set C in $\mathbb{N} \times \mathbb{N}$ such that:

(a) There do not exist $z \in \mathbb{N}$ and a sequence $\langle x_n\rangle_{n=1}^{\infty}$ in $\mathbb{N}$ such that either $\{z\} \times FS(\langle x_n\rangle_{n=1}^{\infty}) \subseteq C$ or $FS(\langle x_n\rangle_{n=1}^{\infty}) \times \{z\} \subseteq C$. (In particular there do not exist infinite weak sum subsystems $\langle x_{1,n}\rangle_{n=1}^{\infty}$ and $\langle x_{2,n}\rangle_{n=1}^{\infty}$ of $\langle 2^n\rangle_{n=1}^{\infty}$ with $FS(\langle x_{1,n}\rangle_{n=1}^{\infty}) \times FS(\langle x_{2,n}\rangle_{n=1}^{\infty}) \subseteq C$).

(b) There do not exist sum subsystems $\langle x_n\rangle_{n=1}^{2}$ and $\langle y_n\rangle_{n=1}^{2}$ of $\langle 2^n\rangle_{n=1}^{\infty}$ with $\{x_1, x_2\} \times \{y_1, y_2\} \subseteq C$.

Proof: Let

$$C = (\mathbb{N} \times \mathbb{N}) \backslash \{(\Sigma_{n \in F} 2^n, \Sigma_{n \in G} 2^n) : F, G \in \mathcal{P}(\omega) \text{ and } \max F < \min G \text{ or } \max G < \min F\}.$$

To see that C is an IP* set in $\mathbb{N} \times \mathbb{N}$, suppose instead that we have a sequence $\langle (x_n, y_n)\rangle_{n=1}^{\infty}$ in $\mathbb{N} \times \mathbb{N}$ with

$$FS(\langle (x_n, y_n)\rangle_{n=1}^{\infty}) \subseteq \{(\Sigma_{n \in F} 2^n, \Sigma_{n \in G} 2^n): F, G \in \mathcal{P}(\omega) \text{ and } \max F < \min G \text{ or } \max G < \min F\}.$$

Pick F_1 and G_1 in $\mathcal{P}(\omega)$ such that $x_1 = \Sigma_{t \in F_1} 2^t$ and $y_1 = \Sigma_{t \in G_1} 2^t$. Let $k = \max(F_1 \cup G_1)$. Choose $H \in \mathcal{P}_f(\mathbb{N})$ such that $\min H > 1$, $2^{k+1} | \Sigma_{n \in H} x_n$, and $2^{k+1} | \Sigma_{n \in H} y_n$. (Consider congruence classes mod 2^{k+1} to see that one can do this.) Pick $F', G' \in \mathcal{P}(\omega)$ such that $\Sigma_{n \in H} x_n = \Sigma_{n \in F'} 2^n$ and $\Sigma_{n \in H} y_n = \Sigma_{n \in G'} 2^n$. Then $k + 1 \leq \min(F' \cup G')$, so $x_1 + \Sigma_{n \in H} x_n = \Sigma_{n \in F_1 \cup F'} 2^n$ and $y_1 + \Sigma_{n \in H} y_n = \Sigma_{n \in G_1 \cup G'} 2^n$. Also

$$\Sigma_{n \in \{1\} \cup H} (x_n, y_n) = (x_1 + \Sigma_{n \in H} x_n, y_1 + \Sigma_{n \in H} y_n) \in \{(\Sigma_{n \in F} 2^n, \Sigma_{n \in G} 2^n): F, G \in \mathcal{P}(\omega) \text{ and } \max F < \min G \text{ or } \max G < \min F\}.$$

Thus $k+1 \le \max(F_1 \cup F') < \min(G_1 \cup G') \le k$ or $k+1 \le \max(G_1 \cup G') < \min(F_1 \cup F') \le k$, a contradiction.

To establish (a), suppose that one has $z \in \mathbb{N}$ and a sequence $\langle x_n \rangle_{n=1}^{\infty}$ in $\mathbb{N}$ such that either $\{z\} \times FS(\langle x_n \rangle_{n=1}^{\infty}) \subseteq C$ or $FS(\langle x_n \rangle_{n=1}^{\infty}) \times \{z\} \subseteq C$ and assume without loss of generality that $\{z\} \times FS(\langle x_n \rangle_{n=1}^{\infty}) \subseteq C$. Pick $F \in \mathcal{P}(\omega)$ such that $z = \Sigma_{t \in F} 2^t$ and let $k = \max F$. Pick $H \in \mathcal{P}_f(\mathbb{N})$ such that $2^{k+1} | \Sigma_{n \in H} x_n$. Pick $G \in \mathcal{P}(\omega)$ such that $\Sigma_{n \in H} x_n = \Sigma_{t \in G} 2^t$. Then $\max F < \min G$ so $(z, \Sigma_{n \in H} x_n) \notin C$.

To establish (b) suppose that one has sum subsystems $\langle x_n \rangle_{n=1}^{2}$ and $\langle y_n \rangle_{n=1}^{2}$ of $\langle 2^n \rangle_{n=1}^{\infty}$ with $\{x_1, x_2\} \times \{y_1, y_2\} \subseteq C$. Pick $F_1, G_1, F_2, G_2 \in \mathcal{P}(\omega)$ such that $x_1 = \Sigma_{n \in F_1} 2^n$, $x_2 = \Sigma_{n \in F_2} 2^n$, $y_1 = \Sigma_{n \in G_1} 2^n$, $y_2 = \Sigma_{n \in G_2} 2^n$, $\max F_1 < \min F_2$, and $\max G_1 < \min G_2$. Without loss of generality, $\max F_1 \ge \max G_1$. But then we have $\max G_1 \le \max F_1 < \min F_2$ so $(x_2, y_1) \notin C$, a contradiction. ❑

A striking contrast is provided by Theorems 2.2 and 2.7. It is easy to divide most semigroups into two classes, neither of which is an IP* set. Consequently, it is not too surprising when one finds a property that must be satisfied by an IP* set in a semigroup S (such as containing a sequence with all of its sums and products when $S = \mathbb{N}$) which need not be satisfied by any cell of a partition of S. In this case we have a property, namely that of containing $FS(\langle x_{1,n} \rangle_{n=1}^{m}) \times FS(\langle x_{2,n} \rangle_{n=1}^{\infty})$ for some sum subsystems of any given sequences, which must be satisfied by some cell of a partition of $\mathbb{N} \times \mathbb{N}$, but need not be satisfied by IP* sets in $\mathbb{N} \times \mathbb{N}$.

The third question raised by Theorem 2.6 is answered with an example very similar to that used in the proof of Theorem 2.7.

THEOREM 2.8: Let S be the free semigroup on the alphabet $\{y_1, y_2, y_3, \ldots\}$. There is an IP* set C in $S \times S$ such that there do not exist weak product subsystems $\langle x_n \rangle_{n=1}^{2}$ and $\langle w_n \rangle_{n=1}^{2}$ of $\langle y_n \rangle_{n=1}^{\infty}$ such that $\{x_1, x_2\} \times \{w_1, w_2\} \subseteq C$.

Proof: Let

$$C = (S \times S) \setminus \{(\Pi_{n \in F} y_n, \Pi_{n \in G} y_n) : F, G \in \mathcal{P}_f(\mathbb{N}) \text{ and } \max F < \min G \text{ or } \max G < \min F\}.$$

To see that C is an IP* set, suppose one has a sequence $\langle (x_n, w_n) \rangle_{n=1}^{\infty}$ with

$$FP(\langle (x_n, w_n) \rangle_{n=1}^{\infty}) \subseteq \{(\Pi_{n \in F} y_n, \Pi_{n \in G} y_n) : F, G \in \mathcal{P}_f(\mathbb{N}) \text{ and } \max F < \min G \text{ or } \max G < \min F\}.$$

Given any $i < j$ in $\mathbb{N}$, pick $F_i, F_j, G_i, G_j, F_{i,j}, G_{i,j} \in \mathcal{P}_f(\mathbb{N})$ such that $x_i = \Pi_{n \in F_i} y_n$, $x_j = \Pi_{n \in F_j} y_n$, $w_i = \Pi_{n \in G_i} y_n$, $w_j = \Pi_{n \in G_j} y_n$, $x_i x_j = \Pi_{n \in F_{i,j}} y_n$, and $w_i w_j = \Pi_{n \in F_i} y_n$. Since $x_i x_j = \Pi_{n \in F_{i,j}} y_n$, we have that $\max F_i < \min F_j$ and $F_{i,j} = F_i \cup F_j$ and similarly $\max G_i < \min G_j$ and $G_{i,j} = G_i \cup G_j$. Then we may pick $j \in \mathbb{N}$ such that $\max(F_1 \cup G_1) < \min(F_j \cup G_j)$. Then

$$(x_1 x_j, w_1 w_j) \notin \{(\Pi_{n\in F} y_n, \Pi_{n\in G} y_n): F, G \in \mathcal{P}_f(\mathbb{N}) \text{ and } \max F < \min G \text{ or } \max G < \min F\},$$

a contradiction.

Now suppose we have weak product subsystems $\langle x_n\rangle_{n=1}^2$ and $\langle w_n\rangle_{n=1}^2$ of $\langle y_n\rangle_{n=1}^\infty$ such that $\{x_1, x_2\} \times \{w_1, w_2\} \subseteq C$. Pick $F_1, F_2, G_1, G_2 \in \mathcal{P}_f(\mathbb{N})$ such that $x_1 = \Pi_{n\in F_1} y_n$, $x_2 = \Pi_{n\in F_2} y_n$, $w_1 = \Pi_{n\in G_1} y_n$, and $w_2 = \Pi_{n\in G_2} y_n$. Since $x_1 x_2 \in FP(\langle y_n\rangle_{n=1}^\infty)$, we must have $\max F_1 < \min F_2$ and similarly $\max G_1 < \min G_2$. Without loss of generality, $\max F_1 \geq \max G_1$. But then we have $\max G_1 \leq \max F_1 < \min F_2$ so $(x_2, w_1) \notin C$, a contradiction. ❑

We know that there are certain IP* sets, namely sets of returns in a dynamical system, which satisfy stronger conclusions than arbitrary IP* sets. (See, for example, [3].) It is possible to show that if C is such a dynamically defined IP* set and S has an identity, one can choose infinite product subsystems $\langle x_{1,n}\rangle_{n=1}^\infty$ and $\langle x_{2,n}\rangle_{n=1}^\infty$ of given $\langle y_{i,n}\rangle_{n=1}^\infty$ *uniformly* with $FP(\langle x_{1,n}\rangle_{n=1}^\infty) \times FP(\langle x_{2,n}\rangle_{n=1}^\infty) \subseteq C$. That is, one has $\langle H_n\rangle_{n=1}^\infty$ in $\mathcal{P}_f(\mathbb{N})$ such that for each n and each $i \in \{1, 2\}$, $x_{i,n} = \Pi_{t\in H_n} y_{i,t}$.

We do not know if we can impose such uniformity on the weak product subsystems guaranteed by Theorem 2.6. We do have the following simple result establishing a certain amount of uniformity for arbitrary semigroups. It has an obvious generalization to any finite dimension, but for the sake of simplicity, we restrict our attention to two dimensions.

THEOREM 2.9: Let S_1 and S_2 be semigroups and let C be an IP* set in $S_1 \times S_2$. Let $\ell, m \in \mathbb{N}$ and let $\{\langle w_{i,n}\rangle_{n=1}^\infty : i \in \{1, 2, \ldots, \ell\}\}$ be a set of sequences in S_1 and let $\{\langle z_{j,n}\rangle_{n=1}^\infty : j \in \{1, 2, \ldots, m\}\}$ be a set of sequences in S_2. Then there is a sequence $\langle H_n\rangle_{n=1}^\infty$ in $\mathcal{P}_f(\mathbb{N})$ such that:

(a) for each $n \in \mathbb{N}$, $\max H_n < \min H_{n+1}$ and

(b) if for each $n \in \mathbb{N}$, each $i \in \{1, 2, \ldots, \ell\}$ and each $j \in \{1, 2, \ldots, m\}$, $x_{i,n} = \Pi_{t\in H_n} w_{i,t}$ and $y_{j,n} = \Pi_{t\in H_n} z_{j,t}$, then for each $i \in \{1, 2, \ldots, \ell\}$ and each $j \in \{1, 2, \ldots, m\}$, $FP(\langle (x_{i,n}, y_{j,n})\rangle_{n=1}^\infty) \subseteq C$.

Proof: Enumerate $\{1, 2, \ldots, \ell\} \times \{1, 2, \ldots, m\}$ as $\langle (i(k), j(k))\rangle_{k=1}^{\ell m}$. Let for each $n \in \mathbb{N}$, $a_{1,n} = (w_{i(1),n}, z_{j(1),n})$. Pick by Corollary 1.7 a sequence $\langle H_{1,n}\rangle_{n=1}^\infty$ such that if $b_{1,n} = \Pi_{t\in H_{1,n}} a_{1,n}$, then $FP(\langle b_{1,n}\rangle_{n=1}^\infty) \subseteq C$. Let

$$a_{2,n} = (\Pi_{t\in H_{1,n}} w_{i(2),t}, \Pi_{t\in H_{1,n}} z_{j(2),t})$$

Inductively, given $a_{k,n} = (\Pi_{t\in H_{k-1,n}} w_{i(k),t}, \Pi_{t\in H_{k-1,n}} z_{j(k),t})$ pick by Corollary 1.7 a sequence $\langle F_{k,n}\rangle_{n=1}^\infty$ in $\mathcal{P}_f(\mathbb{N})$ such that if $b_{k,n} = \Pi_{t\in F_{k,n}} a_{k,n}$, then $FP(\langle b_{k,n}\rangle_{n=1}^\infty) \subseteq C$. For each n let $H_{k,n} = \bigcup_{t\in F_{k,n}} H_{k-1,t}$ and let (if $k < \ell m$)

$$a_{k+1,n} = (\Pi_{t\in H_{k,n}} w_{i(k+1),t}, \Pi_{t\in H_{k,n}} z_{j(k+1),}).$$

The induction being complete, let $H_n = H_{\ell m,n}$ for each n. For each $n \in \mathbb{N}$, each $i \in \{1, 2, \ldots, \ell\}$ and each $j \in \{1, 2, \ldots, m\}$, let $x_{i,n} = \Pi_{t\in H_n} w_{i,t}$ and $y_{j,n}$

$= \Pi_{t \in H_n} z_{j,t}$. Then, given $i \in \{1, 2, \ldots, l\}$ and $j \in \{1, 2, \ldots, m\}$, pick $k \in \mathbb{N}$ such that $(i, j) = (i(k), j(k))$. Then $FP(\langle(x_{i,n}, y_{j,n})\rangle_{n=1}^{\infty} \subseteq FP(\langle b_{k,n}\rangle_{n=1}^{\infty}) \subseteq C$. ❑

3. AN ALTERNATE DERIVATION OF FINITE SUBSTRUCTURES

We show here that a version of Theorem 2.6 for the semigroup $(\mathcal{P}_f(\mathbb{N}), \cup)$ suffices to derive Theorem 2.6 in its entirety. Since the operation in this semigroup is $\cup$, we write $FU(\langle H_n\rangle_{n=1}^{m}) = \{\bigcup_{n \in F} H_n: \varnothing \neq F \subseteq \{1, 2, \ldots, m\}\}$. Similarly, given a sequence $\langle(H_{1,n}, H_{2,n})\rangle_{n=1}^{\infty}$ in $\mathcal{P}_f(\mathbb{N}) \times \mathcal{P}_f(\mathbb{N})$ we write $FU(\langle(\mathrm{F}_{1,n}, \mathrm{F}_{2,n})\rangle_{n=1}^{m}) = \{(\bigcup_{n \in F} H_{1,n}, \bigcup_{n \in F} H_{2,n}): \varnothing \neq F \subseteq \{1, 2, \ldots, m\}\}$.

For the semigroup $(\mathcal{P}_f(\mathbb{N}), \cup)$, Theorem 2.6 is completely trivial because the only IP* set in $\mathcal{P}_f(\mathbb{N}) \times \mathcal{P}_f(\mathbb{N})$ is $\mathcal{P}_f(\mathbb{N}) \times \mathcal{P}_f(\mathbb{N})$ itself. (This is because every element of $\mathcal{P}_f(\mathbb{N})$ is an idempotent.) We need a weaker notion of IP* set in order to obtain useful results. We will say that a sequence $\langle H_{i,n}\rangle_{n=1}^{\infty}$ is a *disjoint sequence* provided $H_{i,n} \cap H_{i,m} = \varnothing$ whenever $n \neq m$.

DEFINITION 3.1: Let $l \in \mathbb{N}$. A set $C \subseteq \times_{i=1}^{l} \mathcal{P}_f(\mathbb{N})$ is a *weak IP* set* if and only if for any disjoint sequences $\langle H_{1,n}\rangle_{n=1}^{\infty}, \langle H_{2,n}\rangle_{n=1}^{\infty}, \ldots, \langle H_{l,n}\rangle_{n=1}^{\infty}$ in $\mathcal{P}_f(\mathbb{N})$, one has $C \cap FU(\langle(H_{1,n}, H_{2,n}, \ldots, H_{l,n})\rangle_{n=1}^{\infty}) \neq \varnothing$.

We will also need to concern ourselves with a restricted subsemigroup of $\beta(\times_{i=1}^{l} \mathcal{P}_f(\mathbb{N}))$.

DEFINITION 3.2: Let $l \in \mathbb{N}$. Then $I_l = \{p \in \beta(\times_{i=1}^{l} \mathcal{P}_f(\mathbb{N}))$: for each $n \in \mathbb{N}$, $\{(H_1, H_2, \ldots, H_l): (\bigcup_{i=1}^{l} H_i) \cap \{1, 2, \ldots, n\} = \varnothing\} \in p\}$.

It is routine to show that I_l is a subsemigroup of $\beta(\times_{i=1}^{l} \mathcal{P}_f(\mathbb{N}))$. We modify Theorem 1.5.

THEOREM 3.3: Let $l \in \mathbb{N}$, let $C \subseteq \times_{i=1}^{l} \mathcal{P}_f(\mathbb{N})$ be a weak IP* set, and let p be an idempotent in I_l. Then $C \in p$.

Proof: As in the proof of Theorem 1.6 one shows that given any $A \in p$ there exist disjoint sequences $\langle H_{1,n}\rangle_{n=1}^{\infty}, \langle H_{2,n}\rangle_{n=1}^{\infty}, \ldots, \langle H_{l,n}\rangle_{n=1}^{\infty}$ in $\mathcal{P}_f(\mathbb{N})$, such that

$$FU(\langle(H_{1,n}, H_{2,n}, \ldots, H_{l,n})\rangle_{n=1}^{\infty}) \subseteq A.$$

Consequently, one cannot have $\times_{i=1}^{l} \mathcal{P}_f(\mathbb{N}) \setminus C \in p$. ❑

One can define a *strong IP set* in $\times_{i=1}^{l} \mathcal{P}_f(\mathbb{N})$ by requiring that this set contain $FU(\langle(H_{1,n}, H_{2,n}, \ldots, H_{l,n})\rangle_{n=1}^{\infty}$ where each $\langle H_{i,n}\rangle_{n=1}^{\infty}$ is a disjoint sequence. One can then show in a fashion similar to the proof of Corollary 1.7 that any weak IP* set meets any strong IP set along a strong IP set.

Next we modify Lemma 2.5.

LEMMA 3.4: Let $l \in \mathbb{N}$ and for each $i \in \{1, 2, \ldots, l\}$ and each $n \in \mathbb{N}$, let

$Y_{i,n} = \{n\}$. Let

$$\mathcal{M} = \{p \in I_\ell\text{: for each } A \in p \text{ and each } m,k \in \mathbb{N}$$

there exist for each $i \in \{1, 2, \ldots, \ell\}$ a weak product subsystem $\langle H_{i,n}\rangle_{n=1}^m$ of $\langle Y_{i,n}\rangle_{n=k}^\infty$

$$\text{such that } \times_{i=1}^\ell FU(\langle H_{i,n}\rangle_{n=1}^m) \subseteq A\}.$$

Then $\mathcal{M}$ is a compact subsemigroup of $\beta(\times_{i=1}^\ell \mathbb{N})$.

Proof: Let $\mathcal{L}$ be as in Lemma 2.5. Then $\mathcal{M} = \mathcal{L} \cap I_\ell$. Since both $\mathcal{L}$ and I_ℓ are semigroups, it suffices to show that $\mathcal{M} \cap \mathcal{L} \neq \emptyset$.

For each $n \in \mathbb{N}$, let $A_n = \{(H_1, H_2, \ldots, H_\ell) \in \times_{i=1}^\ell \mathcal{P}_f(\mathbb{N})\text{: } (\bigcup_{i=1}^\ell H_i) \cap \{1, 2, \ldots, n\} = \emptyset\}$. Let $\mathcal{B} = \{C \subseteq \times_{i=1}^\ell \mathcal{P}_f(\mathbb{N})$: for each $m,k \in \mathbb{N}$ and each choice of a weak product subsystem $\langle H_{i,n}\rangle_{n=1}^m$ of $\langle Y_{i,n}\rangle_{n=k}^\infty$, $\times_{i=1}^\ell FU(\langle H_{i,n}\rangle_{n=1}^m) \cap C \neq \emptyset\}$. Now if $p \in \beta(\times_{i=1}^\ell \mathcal{P}_f(\mathbb{N}))$ and $\{A_k\text{: } k \in \mathbb{N}\} \cup \mathcal{B} \subseteq p$, then $p \in \mathcal{L} \cap I_\ell$ (since, given any $A \in p$, $\times_{i=1}^\ell \mathcal{P}_f(\mathbb{N})\backslash A \notin \mathcal{B}$). Thus it suffices to show that $\{A_k\text{: } k \in \mathbb{N}\} \cup \mathcal{B}$ has the finite intersection property.

To see this, suppose instead we have some $k \in \mathbb{N}$ and some $C_1, C_2, \ldots, C_r \in \mathcal{B}$ such that $A_k \cap \bigcap_{j=1}^r C_j = \emptyset$. Then $(\times_{i=1}^\ell \mathcal{P}_f(\mathbb{N})\backslash A_k) \cup \bigcup_{j=1}^r (\times_{i=1}^\ell \mathcal{P}_f(\mathbb{N})\backslash C_j)$ so by Theorem 2.3, some one of these sets contains $\times_{i=1}^\ell FU(\langle H_{i,n}\rangle_{n=1}^m)$ for some choice of weak product subsystems $\langle H_{i,n}\rangle_{n=1}^m$ of $\langle Y_{i,n}\rangle_{n=k}^\infty$, which is impossible. ❑

THEOREM 3.5: Let $\ell \in \mathbb{N}$, let $C \subseteq \times_{i=1}^\ell \mathcal{P}_f(\mathbb{N})$ be a weak IP* set, and let $m \in \mathbb{N}$. Then for each $i \in \{1, 2, \ldots, \ell\}$ there is a disjoint sequence $\langle H_{i,n}\rangle_{n=1}^m$ such that $\times_{i=1}^\ell FU(\langle H_{i,n}\rangle_{n=1}^m) \subseteq C$.

Proof: Let $\mathcal{M}$ be as in Lemma 3.4. Then $\mathcal{M}$ is a compact subsemigroup of I_ℓ so pick an idempotent $p \in \mathcal{M}$. By Theorem 3.3 $C \in p$ so by the definition of $\mathcal{M}$ we have for each $i \in \{1, 2, \ldots, \ell\}$ some disjoint sequence $\langle H_{i,n}\rangle_{n=1}^m$ such that $\times_{i=1}^\ell FU(\langle H_{i,n}\rangle_{n=1}^m) \subseteq C$. ❑

Finally, we show how Theorem 3.5 suffices to yield Theorem 2.6 for any semigroups. (We reprint its formulation for the convenience of the reader.)

THEOREM 2.6: Let $\ell \in \mathbb{N}$ and for each i $\in \{1, 2, \ldots, \ell\}$, let S_i be a commutative semigroup and let $\langle y_{i,n}\rangle_{n=1}^\infty$ be a sequence in S_i. Let C be an IP* set in $\times_{i=1}^\ell S_i$ and let $m \in \mathbb{N}$. Then for each $i \in \{1, 2, \ldots, \ell\}$ there is a weak product subsystem $\langle x_{i,n}\rangle_{n=1}^m$ of $\langle y_{i,n}\rangle_{n=k}^\infty$ such that $\times_{i=1}^\ell FP(\langle x_{i,n}\rangle_{n=1}^m) \subseteq C$.

Proof: Let

$$D = \{(H_1, H_2, \ldots, H_\ell) \in \times_{i=1}^\ell \mathcal{P}_f(\mathbb{N})\text{: } (\Pi_{t\in H_1} y_{1,t}, \Pi_{t\in H_2} y_{2,t}, \ldots, \Pi_{t\in H_\ell} y_{\ell,t}) \in C\}.$$

Then it is easy to see that D is a weak IP* set. So pick for each $i \in \{1, 2, \ldots, \ell\}$ a disjoint sequence $\langle H_{i,n}\rangle_{n=1}^m$ as guaranteed by Theorem 3.5. For each $n \in \{1, 2, \ldots, m\}$ and each $i \in \{1, 2, \ldots, \ell\}$, let $x_{i,n} = \Pi_{t\in H_{i,n}} y_{i,t}$. Then $\times_{i=1}^\ell FP(\langle x_{i,n}\rangle_{n=1}^m) \subseteq C$. ❑

ACKNOWLEDGMENT

The authors thank Paul Milnes for some helpful conversations.

REFERENCES

1. BERGELSON, V. & N. HINDMAN. 1989. Ultrafilters and multidimensional Ramsey theorems. Combinatorica. **9:** 1–7.
2. BERGELSON, V. & N. HINDMAN. 1994. IP*-sets and central sets. Combinatorica. **14:** 269–277.
3. BERGELSON, V., N. HINDMAN & B. KRA. 1996. Iterated spectra of numbers — elementary, dynamical, and algebraic approaches. Trans. Amer. Math. Soc. **348:** 893–912.
4. ELLIS, R. 1969. Lectures on Topological Dynamics. Benjamin. New York.
5. FURSTENBERG, H. 1981. Recurrence in Ergodic Theory and Combinatorial Number Theory. Princeton Univ. Press. Princeton, NJ.
6. HINDMAN, N. 1979. Ultrafilters and combinatorial number theory. *In:* Number Theory Carbondale 1979 (M. Nathanson, Ed.). Lecture Notes in Math. **751:** 119–184.
7. HINDMAN, N. 1989. Ultrafilters and Ramsey Theory — an update. *In:* Set Theory and its Applications (J. Steprāns & S. Watson, Eds.). Lecture Notes in Math. **1401:** 97–118.
8. MILLIKEN, K. 1975. Ramsey's Theorem with sums or unions. J. Comb. Theory, Series A. **18:** 276–290.
9. TAYLOR, A. 1976. A canonical partition relation for finite subsets of ω. J. Comb. Theory, Series A. **21:** 137–146.

Irresolvable Products

AMER BEŠLAGIĆ AND RONNIE LEVY

Department of Mathematics
George Mason University
Fairfax, Virginia 22030

INTRODUCTION

A space is *resolvable* if it has complementary dense subsets. A space which is not resolvable is *irresolvable*. If n is a positive integer, a space X is *exactly n-resolvable* if X has n pairwise disjoint dense subsets but not $n + 1$ pairwise disjoint dense subsets. Thus, an irresolvable space is an exactly 1-irresolvable space. Any space having an isolated point is irresolvable, so we are interested only in *crowded* spaces, that is, spaces having no isolated points. (Notice that if $n \geq 2$, then an exactly n-resolvable spaces is automatically crowded.) The question we deal with is whether there are crowded spaces X and Y such that the product $X \times Y$ is exactly n-resolvable. Specifically, we show that the consistency of the existence of crowded spaces X and Y such that $X \times Y$ is irresolvable is equivalent to the consistency of the existence of a measurable cardinal. This result answers a question posed in [1]. We also show that the consistency of the existence of a measurable cardinal implies the consistency of the existence for each positive integer n of spaces X and Y such that $X \times Y$ is exactly n-resolvable.

1. BASIC CONSTRUCTIONS

There are essentially two known methods of producing irresolvable crowded regular spaces. Hewitt's method [3] is to start with a crowded space $\langle X, \tau \rangle$ and to use Zorn's lemma to find a topology μ which is maximal in a partial order of crowded topologies on X which strengthen τ. If τ is Hausdorff, and μ is maximal in the partial order of all crowded topologies which strengthen τ, then μ is also Hausdorff. If τ is regular, μ may be taken to be maximal in the partial order of regular crowded topologies on X which strengthen τ. In either case, $\langle X, \mu \rangle$ is an irresolvable space (see [3] or [1]). Note that if $\langle X, \tau \rangle$ is countable, the construction yields a countable crowded irresolvable space, which can be taken to be regular.

Mathematics Subject Classification: 54D15, 54D10.
Key words and phrases: irresolvable space, exactly n-irresolvable space.

A second method of constructing irresolvable spaces was used in [5] to get irresolvable spaces with certain properties (to be discussed in Section 3), and was used by van Douwen [2] to get exactly n-resolvable spaces for $n \geq 1$. For this construction, we need a definition. A collection C of infinite subsets of ω is *independent* if given any two finite subsets $\mathcal{F}_1$ and $\mathcal{F}_2$ of C such that $\mathcal{F}_1 \cap \mathcal{F}_2 = \varnothing$, then $|\bigcap \mathcal{F}_1 \cap \bigcap\{X \backslash F : F \in \mathcal{F}_2\}| = \omega$. It is known that there are independent families of cardinal 2^{ω} and that every independent family is contained in a *maximal independent family*, that is, an independent family which is not a proper subset of any independent family. Given an infinite independent family C, it is easy to modify the elements of C to get an independent family C' such that if p and q are distinct elements of ω, there is a $C \in C'$ such that $|\{p, q\} \cap C| = 1$. We assume that any infinite independent families we consider have this separation property.

Suppose that I is an infinite maximal independent family. Let X be ω with the topology having for a subbase the set $I \cup \{\omega \backslash I : I \in I\}$. Then the separation property of I assures that X is Hausdorff, and since each subbasic set is clearly closed as well as open, X is regular. A subset D of X is dense in X if and only if D has nonempty intersection with each set of the form $\bigcap \mathcal{F}_1 \cap \bigcap \{X \backslash F : F \in \mathcal{F}_2\}$ where $\mathcal{F}_1$ and $\mathcal{F}_2$ are disjoint finite subsets of I. Thus if both D and $X \backslash D$ were dense in X, $I \cup \{D\}$ would be an independent family which properly contained I contradicting the maximality of I. Therefore, X is irresolvable. This construction was modified in [2] to give 2^n-irresolvable spaces by removing n elements from the maximal independent family I and generating a topology with the remaining elements. We have use for this latter result in Section 3.

THEOREM 1.1: (van Douwen [2]) For each positive integer n there exists a countable crowded Tychonoff space which is exactly n-resolvable.

If n is a positive integer a crowded space is *hereditarily exactly n-resolvable* if every crowded subset is exactly n-resolvable. A crowded space is *open-hereditarily exactly n-resolvable* if every open subset is exactly n-resolvable. Notice that every hereditarily irresolvable space is open-hereditarily irresolvable. We have use of the following result which appears in [2] for the case $n = 1$. We include a sketch of the proof for the convenience of the reader.

THEOREM 1.2: Suppose n is a positive integer. Every crowded exactly n-resolvable space contains a nonempty open subset which is hereditarily exactly n-resolvable.

Proof: If X is a crowded irresolvable space, let $\mathcal{M}$ be a collection of pairwise disjoint subsets of X maximal with respect to the property that each element of $\mathcal{M}$ is the union of $n + 1$ pairwise disjoint dense subsets. Let $U = X \backslash Cl_X \cup \mathcal{M}$. Then U is easily seen to have the required properties. ❑

One technique gives an example of a nonregular T_1 irresolvable product assuming the existence of a measurable cardinal. This technique is due to Malyhin in [6]. Malyhin observed that if $\mathcal{U}$ is a free ultrafilter on a set S, then $\mathcal{U} \cup \{\varnothing\}$ is a base for an irresolvable crowded topology on S. He then notes that if $\mathcal{U}$ is a countably complete free ultrafilter on a set S (of measurable cardinal) and $\mathcal{V}$ is a countably incomplete ultrafilter on ω, then $\{U \times V: U \in \mathcal{U}, V \in \mathcal{V}\}$ is a base for a free ultrafilter on $S \times \omega$.

2. RESOLVABLE PRODUCTS

Our goal is to find two regular crowded spaces X and Y such that $X \times Y$ is irresolvable. It will be helpful to have some idea of where not to look. In this section, we look at some types of spaces which are necessarily resolvable.

The next proposition and corollary tell us something about where to look for spaces whose products are irresolvable. In particular, we can see that the mere fact that a space is irresolvable does not mean that its square is irresolvable — simply take a countable, crowded irresolvable space.

PROPOSITION 2.1: Suppose α is an ordinal and X_0 and X_1 are spaces such that for $k \in 2$, $X_k = \bigcup \{C_k^\xi: \xi \in \alpha\}$, where each set C_k^ξ is nowhere dense in X_k and $C_k^\xi \subseteq C_k^\eta$ for $\xi < \eta$. Then $X_0 \times X_1$ is resolvable.

Proof: Let $A = \bigcup \{C_0^\xi \times (X_1 \backslash C_1^\xi): \xi < \alpha\}$, and let $B = \bigcup \{(X_0 \backslash C_0^\xi) \times C_1^\xi: \xi < \alpha\}$. Since A and B are disjoint, we only need show that each of these sets is dense in $X_0 \times X_1$. Let $U \times V$ be a nonempty open subset of $X_0 \times X_1$ and choose $(p, q) \in U \times V$. Let ξ and η be such that $p \in C_0^\xi$ and $q \in C_1^\eta$. Since each set C_1^τ is nowhere dense in V, there is $\tau > \xi$ such that $V \backslash C_1^\tau \neq \varnothing$. If $y \in V \backslash C_1^\tau$, then $(p, y) \in (U \times V) \cap B$. Therefore, B is dense in $X \times Y$. By symmetry, A is dense in $X \times Y$. ❑

COROLLARY 2.2: Suppose that each of the spaces X and Y is first category in itself. Then $X \times Y$ is resolvable.

Proof: Since in any space the union of finitely many nowhere sets is nowhere dense, any first category space is the union of an increasing ω-chain of nowhere dense subsets. Now we may apply Proposition 2.1 to get the result. ❑

PROPOSITION 2.3: Suppose that for each $n \in \omega$ the space X_n has at least two points. Then the product $\prod_{n \in \omega} X_n$ is resolvable.

Proof: The σ-product based on points which differ in every coordinate are disjoint dense subsets of the full product. ❑

The next proposition tells us that there is no crowded space Z such that $Z \times Z$ is hereditarily irresolvable.

PROPOSITION 2.4: If Z is any crowded space, then there is a subspace X of Z such that $X \times X$ is resolvable.

Proof: Let $\kappa = \min\{\xi$: There is a crowded subset of Z which is the union of ξ nowhere dense subsets.$\}$. Let X be a crowded subset of Z which is the union of κ nowhere dense subsets, say $X = \cup \{N_\alpha : \alpha < \kappa\}$ where the sets N_α are nowhere dense and pairwise disjoint. Let $C_\alpha = \cup \{N_\beta : \beta \leq \alpha\}$. Then by the minimality of κ, each set C_α is nowhere dense, so by Proposition 2.1, $X \times X$ is resolvable. ❑

We now prove that it cannot be shown in ZFC that there are two crowded spaces whose product is irresolvable. It fact, a large cardinal is required to find such spaces.

PROPOSITION 2.5: Suppose that there are crowded spaces X_0 and X_1 such that $X_0 \times X_1$ is irresolvable. Then it is consistent that there is a measurable cardinal.

Proof: We may assume without loss of generality that $X_0 \times X_1$ is hereditarily irresolvable, since $X_0 \times X_1$ contains a nonempty hereditarily irresolvable open subset which, in turn, contains a nonempty open set which is a product. It is shown in [5] that the existence of a space each of whose subsets satisfies the Baire Category Theorem is equiconsistent with the existence of a measurable cardinal. Therefore, if the existence of a measurable cardinal is not consistent, each of the spaces X_k contains a subset C_k which is first category in itself. But then $C_0 \times C_1$ is resolvable by Corollary 2.2, contradicting the assumption that $X_0 \times X_1$ is hereditarily irresolovable. ❑

3. IRRESOLVABLE PRODUCTS

In this section, we discuss how to get two crowded spaces whose product is exactly n-resolvable for n a positive integer. To get such spaces, we have to assume the consistency of the existence of a measurable cardinal. In finding two crowded spaces whose product is irresolvable, we choose one of them to be countable. Then it follows from Corollary 2.2 that the other space must be second category in itself. It is easy to see that every second category space has a nonempty open subset which is a Baire space. Therefore, if there is a crowded space whose product with some countable space is irresolvable, then the union of all open Baire subsets is a Baire space with this property. In fact, we show in Lemma 3.2 that a strong version of the Baire property gives an example. We start with a simple lemma.

LEMMA 3.1: Every somewhere dense subset of an open-hereditarily irresolvable space has nonempty interior in the space.

Proof: Suppose that Z is open-hereditarily irresolvable, and let S be a somewhere dense subset of Z. Let $U = Int_Z\, Cl_Z\, S$ ($\neq \varnothing$). Then $U \cap S$ is dense in U. If $U \cap S$ contained no open subset of Z, then $U \backslash S$ would also be dense in U, contradicting the fact that U is irresolvable. ❑

Recall that if κ is a cardinal, a space X is κ -Baire if every family of $<\kappa$ open dense subsets of X has intersection which is dense in X. The next lemma is the key to finding spaces whose product is exactly n-resolvable.

LEMMA 3.2: Let $\mathbf{c} = 2^{\omega}$. Suppose that Y is a countable open-hereditarily exactly n-resolvable space and X is an open-hereditarily irresolvable space which is $\mathbf{c}^{+}$-Baire. Then $X \times Y$ is exactly n-resolvable.

Proof: Let X and Y be as in the statement of the lemma. Since Y is the union of n pairwise disjoint dense subsets, it is clear that $X \times Y$ is also the union of n pairwise disjoint dense subsets. We must show that $X \times Y$ is not the union of n+1 pairwise disjoint dense subsets. Suppose that $X \times Y = \bigcup_{i=0}^{n} A_i$ where the A_i's are pairwise disjoint. We show that there is an index $i \leq n$ such that A_i is not dense in $X \times Y$. For each $x \in X$ and each $i \leq n$ let $A_i(x) = \mathrm{Int}_{\{x\}\times Y}\, Cl_{\{x\}\times Y} A_i \cap (\{x\} \times Y)$. Since no topological space is the union of finitely many nowhere dense subsets, for each $x \in X$ there is an $i_x \leq n$ such that $A_{i_x}(x) \neq \varnothing$. Since Y is open-hereditarily exactly n-resolvable, $A_{i_x}(x)$ contains an open subset $B_{i_x}(x)$ of $\{x\} \times Y$ which intersects at most n of the sets $A_k \cap (\{x\} \times Y)$. For each nonempty open subset V of Y and for each $i \leq n$, let $B_i(V) = \{x \in X\colon i_x = i \text{ and } B_i(x) = V\}$. Then $X = \bigcup_{i=0}^{n} B_i(V)$. Since Y is countable, it has $\mathbf{c}$ nonempty open subsets, and since by assumption X is $\mathbf{c}^{+}$-Baire, there is a nonempty open subset V of Y and an index $k \leq n$ such that $B_k(V)$ is somewhere dense. By Lemma 3.1, $B_k(V)$ contains a nonempty open subset W. For $j \leq n$, let $W_j = \{x \in W\colon B_{i_x}(x)$ does not intersect $A_j \cap \{x\} \times Y\}$. Since $W = \bigcup_{j=0}^{n} W_j$ and since n is finite, there is an index j_0 such that W_{j_0} is somewhere dense in W and, therefore, by Lemma 3.1, contains a nonempty open subset U. Then $(U \times V) \cap A_{j_0} = \varnothing$ so A_{j_0} is not dense in $X \times Y$. ❑

By Lemma 3.2, if we can find a crowded space X which is open-hereditarily irresolvable and $\mathbf{c}^{+}$-Baire, then the space $X \times Y$ will be exactly n-resolvable if Y is any hereditarily exactly n-resolvable crowded countable space. We know from Theorem 1.1 and Theorem 1.2 that such a space Y exists. It is proved in [5] that if the existence of a measurable cardinal is consistent, then so is the existence of a space X with the required properties. For the sake of completeness, we show how to construct such a space, given the consistency of the existence of a measurable cardinal.

If κ is a cardinal, a subset $S \subseteq \kappa$ is a κ-*independent* family if whenever S_1 and S_2 are disjoint subsets of S each of which has size less than κ, $(\bigcap S_1) \cap \bigcap\{\kappa\backslash S\colon S \in S_2\} \neq \varnothing$. If the intersection has size κ for every S_1 and S_1, the κ-independent family is *uniform*. A κ-independent family is *maximal* if it is not a proper subset of any κ-independent family. As in the case of independent families on ω, given any κ-independent family $\mathcal{S}$ of size at least κ, it is easy to construct such a family with the property that if p and q are distinct elements of κ, there is a set $S \in \mathcal{S}$ such that $\{p,q\} \cap S$ has exactly one element. We assume that all independent families have this property. Kunen proved in [4] that the consistency of the existence of a measurable cardinal implies the consistency of a model of GCH in which there is a strongly inaccessible cardinal κ and a maximal κ-independent family $\mathcal{S} \subseteq \mathcal{P}(\kappa)$ such that $|\mathcal{S}| \geq \kappa$. (The maximal κ-independent family he constructs is in fact uniform.) In such a model, let Z be κ with the topology τ generated by $\{(\bigcap \mathcal{S}_1) \cap \bigcap\{\kappa \backslash S : S \in \mathcal{S}_2\} : \mathcal{S}_1$ and $\mathcal{S}_2$ are disjoint subsets of $\mathcal{S}$ of size less than $\kappa\}$. Notice that every basic open set is nonempty. Our assumption that every two elements of κ can be separated by an element of $\mathcal{S}$ assures that Z is Hausdorff. The complement of each basic open set is a union of elements of $\mathcal{S}$ sets which are complements of sets in $\mathcal{S}$ and is therefore closed. Hence, Z is regular and zero-dimensional, and therefore Tyconoff.

PROPOSITION 3.3: Z is $\mathbf{c}^+$-Baire.

Proof: We first show that if $\{B_\alpha: \alpha < \mathbf{c}\}$ is a chain of basic open sets, then $\bigcap\{B_\alpha: \alpha < \mathbf{c}\}$ is itself a basic open set. To see this, suppose that $B_\alpha = (\bigcap \mathcal{S}_1^\alpha) \cap \bigcap\{\kappa\backslash S: S \in \mathcal{S}_2^\alpha\}$ where for each $\alpha < \mathbf{c}$, the families $\mathcal{S}_1^\alpha$ and $\mathcal{S}_2^\alpha$ are disjoint subsets of $\mathcal{S}$. Let $\mathcal{E}_1 = \bigcup\{\mathcal{S}_1^\alpha: \alpha < \mathbf{c}\}$ and let $\mathcal{E}_2 = \bigcup\{\mathcal{S}_2^\alpha: \alpha < \mathbf{c}\}$. Since each of the sets $\mathcal{E}_1$ and $\mathcal{E}_2$ has size less than κ, we need only show that $\mathcal{E}_1$ and $\mathcal{E}_2$ are disjoint. First notice that if $\alpha < \beta$, then $\mathcal{S}_i^\alpha \subseteq \mathcal{S}_i^\beta$ for $i = 1, 2$. For assume not. Fix $S \in \mathcal{S}_i^\alpha \backslash \mathcal{S}_i^\beta$. Then $B_\beta \cap (\kappa\backslash S) \neq \varnothing$ and $B_\alpha \cap (\kappa\backslash S) = \varnothing$, so B_β is not a subset of B_α. Now if $\mathcal{E}_1$ and $\mathcal{E}_2$ were not disjoint, since $\{B_\alpha: \alpha < \mathbf{c}\}$ is a chain, there would be an index α such that $S \in \mathcal{S}_1^\alpha \cap \mathcal{S}_2^\alpha$.

To complete the proof that Z is $\mathbf{c}^+$-Baire, suppose that $\{D_\alpha: \alpha < \mathbf{c}\}$ is a family of dense open subsets of Z. Let U be a basic open set. Let B_0 be a basic open subset of $U \cap D_0$. Suppose that B_α is defined for $\alpha < \beta$ so that each B_α is a basic open set which is contained in $U \cap \bigcap\{D_\gamma: \gamma \leq \alpha\}$ and so that $\{B_\alpha: \alpha < \beta\}$ forms a chain. By the first paragraph of the proof, $\bigcap\{B_\alpha: \alpha < \beta\}$ is a basic open set so there is a basic open set $B_\beta \subseteq U \cap D_\beta \cap (\bigcap\{B_\alpha: \alpha < \beta\})$. By the first paragraph again, $\bigcap\{B_\alpha: \alpha < \mathbf{c}\}$ is nonempty. Since $\bigcap\{B_\alpha: \alpha < \mathbf{c}\} \subseteq U \cap (\bigcap\{D_\alpha: \alpha < \mathbf{c}\})$, we are done. ❑

COROLLARY 3.4: Suppose n is a positive integer. If the existence of a measurable cardinal is consistent, then it is consistent that there are crowded Tychonoff spaces X and Y such that $X \times Y$ is exactly n-resolvable.

Proof: Let Z be as defined above. We first observe that if D is dense in Z, then D intersects every basic open set, so if $X\backslash D$ intersected every basic open set, the family $\mathcal{S} \cup \{D\}$ would be κ-independent. Since $\mathcal{S}$ is maximal, this would mean that $D \in \mathcal{S}$. But no element of $\mathcal{S}$ is dense in Z because the complement of every element of S is a nonempty open set. Thus Z is irresolvable. (These observations also appear in [5].) By Theorem 1.2, Z has a nonempty open subset X which is hereditarily irresolvable. Then X is $\mathbf{c}^+$-Baire, so if Y is any countable, hereditarily exactly n-resolvable crowded space, Lemma 3.2 implies that $X \times Y$ is exactly n-resolvable. ❑

If W is the discrete union of the spaces X and Y of Corollary 3.4, then W is a crowded space whose square is irresolvable. Under the conditions of Corollary 3.4, there are crowded spaces U and V such that $U \times V$ is hereditarily irresolvable — use Theorem 1.2 to find a basic open hereditarily irresolvable subset $U \times V$ of $X \times Y$. These observations can be contrasted to Proposition 2.4.

The next corollary combines Proposition 2.5with the $n = 1$ case of Corollary 3.4. It is worth noting what the corollary does not say. It does not assert that in any model in which there is a measurable cardinal, there are crowded Tychnoff spaces X_0 and X_1 such that $X_0 \times X_1$ is irresolvable.

COROLLARY 3.5: The existence of crowded Tychonoff spaces X_0 and X_1 such that $X_0 \times X_1$ is irresolvable is equiconsistent with the existence of a measurable cardinal.

ACKNOWLEDGMENT

The authors thank the referee for many helpful suggestions.

REFERENCES

1. COMFORT, W.W. & S. GARCÍA-FERREIRA. Resolvability: a Selective Survey and Some New Results. *In:* Proceedings, December 1994, International Conference on Set-theoretic Topology and Its Applications held in Matsuyama, Japan. To appear.
2. VAN DOUWEN, E.K. 1993. Applications of maximal topologies. Topol. Appl. **51:** 125–139.
3. HEWITT, E. 1943. A problem of set-theoretic topology. Duke J. Math. **10:** 309–333.
4. KUNEN, K. 1983. Maximal σ-independent families. Fund. Math. **117:** 75–80.
5. KUNEN, K., A. SZYMAŃSKI, & F. TALL. 1986. Baire irresolvable spaces and ideal theory. Acta Math. Silesiana. **14:** 98–107.
6. MALYHIN, V.I. 1973. Products of ultrafilters and irresolvable spaces. Math. USSR Sbornik. **19:** 105–115.

Alexandroff and Scott Topologies for Generalized Metric Spaces

M.M. BONSANGUE,[a,d] F. VAN BREUGEL,[b,e] AND J.J.M.M. RUTTEN[c,f]

[a]*Department of Computer Science*
Vrije Universiteit Amsterdam
De Boelelaan 1081a
1081 HV Amsterdam, The Netherlands

[b]*Dipartimento di Informatica*
Università di Pisa
Corso Italia 40
56125 Pisa, Italy

and

[c]*CWI*
P.O. Box 94079
1090 AB Amsterdam, The Netherlands

ABSTRACT: Generalized metric spaces are a common generalization of preorders and ordinary metric spaces. Every generalized metric space can be isometrically embedded in a complete function space by means of a metric version of the categorical *Yoneda embedding*. This simple fact gives naturally rise to: 1. a topology for generalized metric spaces which for arbitrary preorders corresponds to the Alexandroff topology and for ordinary metric spaces reduces to the ε-ball topology; 2. a topology for algebraic generalized metric spaces generalizing both the Scott topology for algebraic complete partial orders and the ε-ball topology for metric spaces.

1. INTRODUCTION

Partial orders and metric spaces play a central role in the semantics of programming languages (see, e.g., [21] and [3]). Parts of their theory have been developed because of semantic necessity (see, e.g., [18] and [1]). Generalized metric spaces provide a common framework for the study of both preorders and ordinary metric spaces. A *generalized metric space* (gms for

Mathematics Subject Classification: 68Q10, 68Q55.
Key words and phrases: generalized metric, preorder, metric, Alexandroff topology, Scott topology, ε-ball topology, Yoneda embedding.
[d]E-mail: marcello@cs.vu.nl.
[e]E-mail: franck@di.unipi.it.
[f]E-mail: jan.rutten@cwi.nl.

short) consists of a set X together with a distance function $X(-, -)\colon X \times X \to [0, \infty]$ satisfying, for all x, y, and z in X,

1. $X(x, x) = 0$ and
2. $X(x, z) \leq X(x, y) + X(y, z)$.

Clearly every ordinary metric space is a gms. A preorder $\leq$ on X can be represented by the gms X with

$$X(x, y) = \begin{cases} 0 & \text{if } x \leq y \\ \infty & \text{if } x \not\leq y, \end{cases}$$

for x and y in X. Reflexivity and transitivity of $\leq$ imply 1. and 2., respectively. By a slight abuse of language, any gms stemming from a preorder in this way will itself be called a preorder.

In this paper we propose two topologies for gms's. The first one is a generalized Alexandroff topology. For preorders it coincides with the Alexandroff topology while for metric spaces it corresponds to the ε-ball topology. The second one is a generalized Scott topology. For algebraic complete partial orders it corresponds to the Scott topology, while for metric spaces it coincides with the ε-ball topology. Both topologies are defined in two ways: by specifying the open sets and by a closure operator. These two alternative definitions are shown to coincide.

Our definition of the generalized Alexandroff topology in terms of open sets is similar to the ones given by Smyth [15], [16] and Flagg and Kopperman [5]. A definition of a generalized Scott topology in terms of open sets similar to ours is briefly mentioned by Smyth in [15].

The definition of the generalized Alexandroff topology in terms of a closure operator already appears in [10], [11], [9]. New is the definition of the generalized Scott topology in terms of a closure operator. Both closure operators are defined by means of an adjunction between preorders. In defining these adjunctions we use the fact — first observed by Lawvere [10] — that, intuitively, one may identify elements x of a gms X with a description of the distances between any element y in X and x. Formally, this description is a function mapping every y in X to the distance $X(y, x)$. These functions from X to $[0, \infty]$ can be interpreted as fuzzy subsets of X. The value a function ϕ assigns to an element y in X is thought of as a measure for the extent to which y is an element of ϕ. This fact corresponds to a generalized metric version of the categorical Yoneda lemma [22]. The corresponding Yoneda embedding isometrically embeds a gms X into the gms of fuzzy subsets of X. By comparing the fuzzy subsets of X with the ordinary subsets of X we obtain an adjunction. This adjunction gives rise to the closure operator defining the generalized Alexandroff topology. Similarly, an algebraic gms X can be isometrically embedded into the gms of fuzzy subsets of its basis B. By comparing the fuzzy subsets of B with the subsets of X we obtain another adjunction inducing the closure operator defining the generalized Scott topology.

Like the ordinary Scott topology for complete partial orders, the generalized Scott topology encodes all information about order, convergence, and continuity (cf. [16]). The generalized Alexandroff topology only encodes the information about order, just like the ordinary Alexandroff topology for preorders (cf. [15], [5]).

The paper is organized as follows. Section 2 and 4 give some basic definitions and facts on gms's. The Yoneda lemma and the generalized Alexandroff topology are discussed in Section 3, while the generalized Scott topology is presented in Section 5. Finally, in Section 6 some related work is discussed.

2. GENERALIZED METRIC SPACES

In this section and Section 4 some basic facts and definitions on gms's are presented. This section is concluded with a table containing the preorder and ordinary metric notions corresponding to the notions introduced below.

An important example of a gms is the set of (extended) real numbers $[0, \infty]$ with the distance function defined, for r and s in $[0, \infty]$, by

$$[0, \infty](r, s) = \begin{cases} 0 & \text{if } r \geq s \\ s - r & \text{if } r < s. \end{cases}$$

This gms is a *quasimetric space* (qms for short): besides the axioms 1. and 2. of the introduction it also satisfies, for all x and y in X, if $X(x, y) = 0$ and $X(y, x) = 0$ then $x = y$. The gms $[0, \infty]$ has the following fundamental property. For all r, s, t in $[0, \infty]$,

$$r + s \geq t \quad \text{if and only if } r \geq [0, \infty](s, t). \tag{1}$$

The above equation expresses that the category with the elements in $[0, \infty]$ as objects and the relation $\geq$ defining the morphisms is a closed category with + as tensor. Many properties of gms's derive from this categorical structure on $[0, \infty]$.

The gms *opposite* to a gms X, denoted by X^{op}, is the set X with the distance function defined, for x and y in X, by

$$X^{op}(x, y) = X(y, x).$$

Let X and Y be gms's. A function $f: X \to Y$ is *nonexpansive* if, for all x and y in X,

$$Y(f(x), f(y)) \leq X(x, y).$$

If the above inequality always is an equality then f is said to be *isometric*. The set of nonexpansive functions from X to Y, denoted by Y^X, together with the distance function defined, for f and g in Y^X, by

$$Y^X(f, g) = \sup_{x \in X} Y(f(x), g(x)),$$

is a gms.

As we have seen in the introduction, a preorder can be viewed as a gms. Conversely, a gms gives rise to a preorder. The *underlying preorder* of a gms X is defined, for x and y in X, by

$$x \leq_X y \quad \text{if and only if } X(x, y) = 0.$$

The preorder and metric notions corresponding to the ones introduced above are listed below.

gms	preorder	metric space
qms	partial order	metric space
opposite	opposite	identity
nonexpansive	monotone	nonexpansive
isometric	order equivalence	isometric
underlying preorder	preorder	identity relation

3. A GENERALIZED ALEXANDROFF TOPOLOGY

We present a generalized Alexandroff topology for gms's. The following lemma turns out to be of great importance for the definitions of topologies for gms's as we shall see below and in Section 5. It is the $[0, \infty]$-enriched version of the famous Yoneda lemma [22], [8] from category theory.

For a gms X, let $\hat{X}$ denote the nonexpansive function space

$$\hat{X} = [0, \infty]^{X^{op}}.$$

An element ϕ in $\hat{X}$ can be interpreted as a fuzzy subset of X. The value that ϕ assigns to an element x in X is thought of as a measure for the extent to which x is an element of ϕ. The smaller this number, the more x should be viewed as an element of ϕ. Every gms can be mapped into the gms $\hat{X}$ of its fuzzy subsets by the *Yoneda embedding* $\mathsf{y}: X \to \hat{X}$ which is defined, for x in X, by

$$\mathsf{y}(x) = \lambda y \in X.\, X(y, x).$$

Note that the nonexpansiveness of $\mathsf{y}(x)$ is an immediate consequence of condition 2. of the introduction and (1).

LEMMA 3.1: (Yoneda) Let X be a gms. For all x in X and ϕ in $\hat{X}$,

$$\hat{X}(\mathsf{y}(x), \phi) = \phi(x).$$

Proof: Let x be in X and ϕ be in $\hat{X}$.

$$\begin{aligned}\phi(x) &= [0, \infty](X(x, x), \phi(x)) \\ &\leq \sup_{y \in X} [0, \infty](X(y, x), \phi(y)) \\ &= \hat{X}(\mathsf{y}(x), \phi)\end{aligned}$$

Because ϕ is nonexpansive, for all y in X,

$$[0, \infty](\phi(x), \phi(y)) \leq X^{op}(x, y) = X(y, x) = \mathsf{y}(x)(y).$$

According to (1), this is equivalent to

$$[0, \infty](\mathsf{y}(x)(y), \phi(y)) \leq \phi(x).$$

Consequently, $\hat{X}(\mathsf{y}(x), \phi) \leq \phi(x)$. ❑

The following corollary is immediate.

COROLLARY 3.2: Let X be a gms. The Yoneda embedding $\mathsf{y}: X \rightarrow \hat{X}$ is isometric.

The closure operator defining the generalized Alexandroff topology for a gms X is obtained by comparing the fuzzy subsets of X with the ordinary subsets of X. Given a fuzzy subset ϕ in $\hat{X}$, by taking only its real elements, i.e., the elements x in X for which $\phi(x) = 0$, we obtain its *extension*

$$\mathsf{e}_A(\phi) = \{x \in X \mid \phi(x) = 0\},$$

where the subscript A stands for Alexandroff. Note that

$$\begin{aligned}\mathsf{e}_A(\phi) &= \{x \in X \mid \phi(x) = 0\} \\ &= \{x \in X \mid \hat{X}(\mathsf{y}(x), \phi) = 0\} \text{ [Yoneda lemma 3.1]} \\ &= \{x \in X \mid \mathsf{y}(x) \leq_{\hat{X}} \phi\}.\end{aligned}$$

Any subset V in $\mathcal{P}(X)$ defines a fuzzy subset $\mathsf{k}_A(V)$ in $\hat{X}$ which is referred to as the *character* of the subset V. It is defined by

$$\mathsf{k}_A(V) = \lambda x \in X. \inf_{v \in V} X(x, v).$$

The closer an element x in X is to the subset V, the more x should be viewed as an element of the character of V. Note that

$$\mathsf{k}_A(V) = \lambda x \in X. \inf_{v \in V} \mathsf{y}(v)(x).$$

The functions $\mathsf{e}_A : \hat{X} \to \mathcal{P}(X)$ and $\mathsf{k}_A : \mathcal{P}(X) \to \hat{X}$ can be nicely related by considering $\hat{X}$ with the underlying preorder $\leq_{\hat{X}}$ and $\mathcal{P}(X)$ ordered by subset inclusion.

PROPOSITION 3.3: Let X be a gms. The functions $\mathsf{e}_A : (\hat{X}, \leq_{\hat{X}}) \to (\mathcal{P}(X), \subseteq)$ and $\mathsf{k}_A : (\mathcal{P}(X), \subseteq) \to (\hat{X}, \leq_{\hat{X}})$ are monotone. Moreover, k_A is left adjoint to e_A.

Proof: Monotonicity of e_A and k_A follows directly from their definitions. We will hence concentrate on the second part of the proposition by proving, for all V in $\mathcal{P}(X)$ and ϕ in $\hat{X}$, $V \subseteq \mathsf{e}_A(\mathsf{k}_A(V))$ and $\mathsf{k}_A(\mathsf{e}_A(\phi)) \leq_{\hat{X}} \phi$, which is equivalent to k_A being left adjoint to e_A (cf. Theorem 0.3.6 of [7]). Because, for all V in $\mathcal{P}(X)$ and v in V, $\mathsf{y}(v) \leq_{\hat{X}} \mathsf{k}_A(V)$, we have that

$$V \subseteq \{x \in X \mid \mathsf{y}(x) \leq_{\hat{X}} \mathsf{k}_A(V)\} = \mathsf{e}_A(\mathsf{k}_A(V)).$$

Furthermore, for ϕ in $\hat{X}$ and x in X,

$$\begin{aligned} \mathsf{k}_A(\mathsf{e}_A(\phi))(x) &= \inf\{X(x, y) \mid y \in X \wedge \mathsf{y}(y) \leq_{\hat{X}} \phi\} \\ &= \inf\{\mathsf{y}(y)(x) \mid y \in X \wedge \forall z \in X\colon \mathsf{y}(y)(z) \geq \phi(z)\} \\ &\geq \inf\{\mathsf{y}(y)(x) \mid y \in X \wedge \mathsf{y}(y)(x) \geq \phi(x)\} \\ &\geq \phi(x). \end{aligned}$$

Consequently, $\mathsf{k}_A(\mathsf{e}_A(\phi)) \leq_{\hat{X}} \phi$. (Note that the ordering underlying $[0, \infty]$ is the opposite of the usual ordering.) ❑

The above fundamental adjunction relates the character of subsets and the extension of fuzzy subsets and is often referred to as the *comprehension schema* [10], [9]. As with any adjoint pair between preorders, the composition $\mathsf{e}_A \circ \mathsf{k}_A$ is a closure operator on X (cf. Theorem 0.3.6 of [7]). It satisfies, for V in $\mathcal{P}(X)$,

$$\begin{aligned} (\mathsf{e}_A \circ \mathsf{k}_A)(V) &= \{x \in X \mid \mathsf{k}_A(V)(x) = 0\} \\ &= \{x \in X \mid \hat{X}(\mathsf{y}(x), \mathsf{k}_A(V)) = 0\} \quad \text{[Yoneda lemma 3.1]} \\ &= \{x \in X \mid \forall y \in X\colon [0, \infty](\mathsf{y}(x)(y), \mathsf{k}_A(V)(y)) = 0\} \\ &= \{x \in X \mid \forall y \in X\colon \mathsf{y}(x)(y) \geq \mathsf{k}_A(V)(y)\} \\ &= \{x \in X \mid \forall y \in X\colon \forall \varepsilon > 0\colon X(y, x) < \varepsilon \Rightarrow \exists v \in V\colon X(y, v) < \varepsilon\}. \quad (2) \end{aligned}$$

By using the above characterization (2) we can prove the following theorem.

THEOREM 3.4: Let X be a gms. The closure operator $\mathsf{e}_A \circ \mathsf{k}_A$ on X is topological.

Proof: It is an immediate consequence of (2) that $(\mathsf{e}_A \circ \mathsf{k}_A)(\emptyset) = \emptyset$. Because $\mathsf{e}_A \circ \mathsf{k}_A$ is a closure operator, for V, W in $\mathcal{P}(X)$,

$$(\mathsf{e}_A \circ \mathsf{k}_A)(V) \cup (\mathsf{e}_A \circ \mathsf{k}_A)(W) \subseteq (\mathsf{e}_A \circ \mathsf{k}_A)(V \cup W).$$

For the reverse inclusion, let x be in $(e_A \circ k_A)(V \cup W)$. Suppose x is not in $(e_A \circ k_A)(V)$. We will show that x is in $(e_A \circ k_A)(W)$. Let y_W be in X and $\varepsilon_W > 0$ with $X(y_W, x) < \varepsilon_W$. We should find a w in W with $X(y_W, w) < \varepsilon_W$. Because x is not in $(e_A \circ k_A)(V)$, there exist a y_V in X and an $\varepsilon_V > 0$ such that

$$X(y_V, x) < \varepsilon_V \wedge \forall v \in V\colon X(y_V, v) \geq \varepsilon_V. \tag{3}$$

Let $\varepsilon = \min\{\varepsilon_V - X(y_V, x), \varepsilon_W - X(y_W, x)\}$. Because x is in $(e_A \circ k_A)(V \cup W)$ and $X(x, x) < \varepsilon$, there exists a w in $V \cup W$ with $X(x, w) < \varepsilon$. The assumption that w is in V contradicts (3), because

$$X(y_V, w) \leq X(y_V, x) + X(x, w) < \varepsilon_V.$$

Thus w is in W. Furthermore,

$$X(y_W, w) \leq X(y_W, x) + X(x,w) < \varepsilon_W. \quad \square$$

The above theorem implies that the closure operator $e_A \circ k_A$ induces a topology on X. In Theorem 3.5 below, it is proved equivalent to the following generalized ε-ball topology. For x in X and $\varepsilon > 0$, we define the *generalized ε-ball* of x by

$$B_\varepsilon(x) = \{y \in X \mid X(x, y) < \varepsilon\}.$$

A subset V of a gms X is *generalized Alexandroff open* (gA-open for short) if, for all x in V,

$$\exists \varepsilon > 0\colon B_\varepsilon(x) \subseteq V.$$

For instance, for every x in X, the generalized ε-ball $B_\varepsilon(x)$ is gA-open. The set of all gA-open subsets of X is denoted by $\mathcal{O}_{gA}$. One can easily verify that $\mathcal{O}_{gA}$ is a topology on X with $\{B_\varepsilon(x) \mid \varepsilon > 0 \wedge x \in X\}$ as basis.

For ordinary metric spaces, the above introduced generalized ε-balls are as usual, while for preorders they are upper-closed sets: if X is a preorder then

$$\begin{aligned} B_\varepsilon(x) &= \{y \in X \mid X(x, y) < \varepsilon\} \\ &= \{y \in X \mid X(x, y) = 0\} \\ &= \{y \in X \mid x \leq_X y\}. \end{aligned} \tag{4}$$

Consequently, the generalized Alexandroff topology restricted to metric spaces is the ε-ball topology, while restricted to preorders it is the ordinary Alexandroff topology.

For V in $\mathcal{P}(X)$, we write $cl_A(V)$ for the closure of V in the generalized Alexandroff topology.

THEOREM 3.5: Let X be a gms. For all V in $\mathcal{P}(X)$, $cl_A(V) = (e_A \circ k_A)(V)$.

Proof: For every topology O on X, the induced topological closure operator cl on X satisfies, for all V in $\mathcal{P}(X)$, $cl(V) = V \cup V^d$, where V^d is the derived set, the set of all accumulation points, of V. It follows from this fact and characterization (2) of $\mathsf{e}_A \circ \mathsf{k}_A$ that it is sufficient to prove

$$V \cup V^d = \{x \in X \mid \forall y \in X\colon \forall \varepsilon > 0\colon X(y, x) < \varepsilon \Rightarrow \exists v \in V\colon X(y, v) < \varepsilon\}. \quad (5)$$

From the definition of accumulation point and the fact that the set of all generalized ε-balls is a basis for the generalized Alexandroff topology, it follows that, for every x in X,

$$\begin{aligned} x \in V^d &\Leftrightarrow \forall W \in O_{gA}\colon x \in W \Rightarrow W \cap (V\backslash\{x\}) \neq \varnothing \\ &\Leftrightarrow \forall y \in X\colon \forall \varepsilon > 0\colon x \in B_\varepsilon(y) \Rightarrow B_\varepsilon(y) \cap (V\backslash\{x\}) \neq \varnothing \\ &\Leftrightarrow \forall y \in X\colon \forall \varepsilon > 0\colon X(y, x) < \varepsilon \Rightarrow \exists v \in (V\backslash\{x\})\colon X(y, v) < \varepsilon. \quad (4) \end{aligned}$$

Therefore, (5) holds. ❑

Every topology O on X induces a preorder on X called the *specialization preorder*: for all x and y in X, $x \leq_O y$ if and only if, for all V in O, if x is in V then y is in V. The specialization preorder on a gms X induced by its generalized Alexandroff topology coincides with the preorder underlying X.

PROPOSITION 3.6: Let X be a gms. For all x and y in X, $x \leq_{O_{gA}} y$ if and only if $x \leq_X y$.

Proof: For any gA-open set V, if x is in V and $X(x, y) = 0$ then y is in V. From this observation the implication from right to left is clear. For the converse, suppose $x \leq_{O_{gA}} y$. Then, for every $\varepsilon > 0$, x is in $B_\varepsilon(x)$ implies y is in $B_\varepsilon(x)$, because generalized ε-balls are gA-open sets. Hence $X(x, y) < \varepsilon$. Since ε was arbitrary, $X(x, y) = 0$, that is $x \leq_X y$. ❑

The above proposition tells us that the underlying preorder of a gms can be reconstructed from its generalized Alexandroff topology.

Note that the specialization preorder $\leq_{O_{gA}}$ is a partial order — this is equivalent to the generalized Alexandroff topology being T_0 — if and only if X is a qms.

4. CAUCHY SEQUENCES, LIMITS, AND COMPLETENESS

Some further basic facts and definitions on gms's are presented. Like Section 2, this section is concluded with a table containing the preorder and ordinary metric notions corresponding to the notions introduced below.

A sequence $(x_n)_n$ in a gms X is *forward-Cauchy* if

$$\forall \varepsilon > 0: \exists N: \forall n \geq m \geq N: X(x_m, x_n) \leq \varepsilon.$$

Since our distance functions need not be symmetric, the following variation exists. A sequence $(x_n)_n$ is *backward-Cauchy* if

$$\forall \varepsilon > 0: \exists N: \forall n \geq m \geq N: X(x_n, x_m) \leq \varepsilon.$$

The *forward-limit* of a forward-Cauchy sequence $(r_n)_n$ in $[0, \infty]$ is given by

$$\underrightarrow{\lim}_n r_n = \sup_n \inf_{k \geq n} r_k.$$

Dually, the *backward-limit* of a backward-Cauchy sequence $(r_n)_n$ in $[0, \infty]$ is defined by

$$\underleftarrow{\lim}_n r_n = \inf_n \sup_{k \geq n} r_k.$$

Forward-limits and backward-limits in $[0, \infty]$ are related as follows. For all forward-Cauchy sequences $(r_n)_n$ in $[0, \infty]$ and r in $[0, \infty]$,

$$[0, \infty](\underrightarrow{\lim}_n r_n, r) = \underleftarrow{\lim}_n [0, \infty](r_n, r). \tag{6}$$

Forward-limits in an arbitrary gms can now be defined in terms of backward-limits in $[0, \infty]$. An element x is a *forward-limit* of a forward-Cauchy sequence $(x_n)_n$ in a gms X, denoted by $x \in \underrightarrow{\lim}_n x_n$, if, for all y in X,

$$X(x, y) = \underleftarrow{\lim}_n X(x_n, y).$$

Note that if $(x_n)_n$ is a forward-Cauchy sequence in X, then, for all y in X, $(X(x_n, y))_n$ is a backward-Cauchy sequence in $[0, \infty]$ because of (1). Our earlier definition of the forward-limit of forward-Cauchy sequences in $[0, \infty]$ is consistent with this definition for arbitrary gms's because of (6). Similarly one can define backward-limits in an arbitrary gms. Since these will not play a role in this paper, their definition is omitted. For simplicity, we shall use *Cauchy* instead of forward-Cauchy and *limit* instead of forward-limit. Note that Cauchy sequences may have more than one limit. Let $(x_n)_n$ be a Cauchy sequence in a gms X and x be in X, with $x \in \underrightarrow{\lim}_n x_n$. For all y in X,

$$y \in \underrightarrow{\lim}_n x_n \text{ if and only if } X(x, y) = 0 \text{ and } X(y, x) = 0. \tag{7}$$

Consequently, limits are unique in qms's. In that case we sometimes write $x = \underrightarrow{\lim}_n x_n$.

A gms X is *complete* if every Cauchy sequence in X has a limit. For example, $[0, \infty]$ is complete. Let X and Y be gms's. If Y is complete then Y^X is also complete (cf. [14, Theorem 6.5]). Consequently, for every gms X, the function space $\hat{X}$ is complete.[1] Limits of the complete function space Y^X are taken pointwise. For all Cauchy sequences $(f_n)_n$ in Y^X and f in Y^X,

$$f \in \underrightarrow{\lim}_n f_n \quad \text{if and only if, for all } x \text{ in } X,\ f(x) \in \underrightarrow{\lim}_n (f_n(x)). \tag{8}$$

Let X and Y be gms's. A nonexpansive function $f: X \to Y$ is *continuous* if it preserves limits: for all Cauchy sequences $(x_n)_n$ in X and x in X, with $x \in \underrightarrow{\lim}_n x_n$, $f(x) \in \underrightarrow{\lim}_n f(x_n)$.

The preordered notion finite can be generalized as follows. An element x in a gms X is *finite* if the function $\lambda y \in X \,.\, X(x, y)$ from X to $[0, \infty]$ is continuous. In order to conclude that x is finite in X, it suffices to prove that, for all Cauchy sequences $(y_n)_n$ in X and y in X, with $y \in \underrightarrow{\lim}_n y_n$, $\underrightarrow{\lim}_n X(x, y_n) \leq X(x, y)$. For example, for all x in X, one can show that

$$\mathsf{y}(x) \text{ is finite in } \hat{X} \tag{9}$$

(cf. [2, Lemma 4.3]).

A subset B of finite elements of a gms X is a *basis* for X if every element in X is a limit of a Cauchy sequence in B. A gms is *algebraic* if there exists a basis. Such a basis is generally not unique.

Below we give the table with the corresponding preorder and metric notions.

gms	preorder	metric space
forward-Cauchy	eventually increasing	Cauchy
backward-Cauchy	eventually decreasing	Cauchy
forward-limit	eventually minimal upperbound	limit
backward-limit	eventually maximal lowerbound	limit
complete	ω-complete	complete
continuous	ω-continuous	continuous
finite	finite	arbitrary
basis	basis	dense subset
algebraic	algebraic	arbitrary

5. A GENERALIZED SCOTT TOPOLOGY

In the Scott topology of a complete partial order X least upperbounds of increasing sequences in X are topological limits. Also, in the ε-ball topology of an ordinary metric space, X limits of Cauchy sequences in X are topolog-

[1]As a consequence, the Yoneda embedding y isometrically embeds a gms X into the complete gms $\hat{X}$. One can define the completion of X as the smallest complete subspace of $\hat{X}$ containing the y-image of X. For details we refer the reader to [2].

ical limits. However, a similar result does not hold for our generalized Alexandroff topology. For example, for complete partial orders the generalized Alexandroff topology coincides with the ordinary Alexandroff topology, for which this result does not hold in general. The Scott topology is the coarsest topology refining the Alexandroff topology with this property (cf.[7], [12], [17]). Also for gms's a suitable refinement of the generalized Alexandroff topology exists.

A key step towards the definition of the generalized Scott topology for algebraic gms's is the following restriction of the Yoneda embedding.

LEMMA 5.1: Let X be a gms. If B is a basis for X, then the function $\mathsf{y}_B: X \to \hat{B}$ defined, for x in X, by

$$\mathsf{y}_B(x) = \lambda b \in B.\ X(b, x)$$

is isometric and continuous.

Proof: According to Corollary 3.2, y is isometric. Consequently, y_B is nonexpansive. According to (8), to prove that y_B is continuous it suffices to show that, for all Cauchy sequences $(x_n)_n$ in X and x in X, with $x \in \underrightarrow{\lim}_n x_n$, and b in B,

$$\mathsf{y}_B(x)(b) \in \underrightarrow{\lim}_n \mathsf{y}_B(x_n)(b).$$

This follows immediately from the fact that b is finite in X. The function y_B is isometric, because, for x and y in X, since B is a basis there exist Cauchy sequences $(b_m)_m$ and $(c_n)_n$ in B such that $x \in \underrightarrow{\lim}_m b_m$ and $y \in \underrightarrow{\lim}_n c_n$, and

$$\begin{aligned}
\hat{B}(\mathsf{y}_B(x), \mathsf{y}_B(y)) &= \underleftarrow{\lim}_m B(\mathsf{y}_B(b_m), \mathsf{y}_B(y)) && [x \in \underrightarrow{\lim}_m b_m,\ \mathsf{y}_B \text{ is continuous}] \\
&= \underleftarrow{\lim}_m \underrightarrow{\lim}_n \hat{B}(\mathsf{y}_B(b_m), \mathsf{y}_B(c_n)) && [y \in \underrightarrow{\lim}_n c_n,\ \mathsf{y}_B \text{ is continuous}, \\
& && \mathsf{y}_B(b_m) \text{ is finite in } \hat{B} \text{ according to (9)}] \\
&= \underleftarrow{\lim}_m \underrightarrow{\lim}_n B(b_m, c_n) && [\text{Corollary 3.2}] \\
&= \underleftarrow{\lim}_m \underrightarrow{\lim}_n X(b_m, c_n) \\
&= \underleftarrow{\lim}_m X(b_m, y) && [y \in \underrightarrow{\lim}_n c_n,\ b_m \text{ is finite in } X] \\
&= X(x, y). && [x \in \underrightarrow{\lim}_m b_m] \quad \square
\end{aligned}$$

The converse of the above lemma holds as well (cf. [2, Theorem 5.6]).

The closure operator defining the generalized Scott topology for an algebraic gms X with basis B is obtained by comparing the fuzzy subsets of B, rather than the fuzzy subsets of X as we have done in Section 3, with the ordinary subsets of X. The *extension* function $\mathsf{e}_S: \hat{B} \to \mathcal{P}(X)$ is defined, for ϕ in $\hat{B}$, by

$$\mathsf{e}_S(\phi) = \{x \in X \mid \mathsf{y}_B(x) \leq_{\hat{B}} \phi\}$$

and the *character* function $\mathsf{k}_S: \mathcal{P}(X) \to \hat{B}$ is defined, for V in $\mathcal{P}(X)$, by

$$\mathsf{k}_S(V) = \lambda b \in B.\ \inf_{v \in V} \mathsf{y}_B(v)(b).$$

As in Proposition 3.3, the functions e_S: $(\hat{B}, \leq_{\hat{B}}) \rightarrow (\mathcal{P}, (X), \subseteq)$ and k_S: $(\mathcal{P}(X), \subseteq) \rightarrow (\hat{B}, \leq_{\hat{B}})$ are monotone and k_S is left adjoint to e_S. Thus, $\mathsf{e}_S \circ \mathsf{k}_S$ is a closure operator on X. Since a basis is generally not unique, one might think that the definition of the closure operator $\mathsf{e}_S \circ \mathsf{k}_S$ depends on the choice of the basis. That this is not the case is a consequence of Theorem 5.6 below. In a way similar to (2), this closure operator can be characterized, for V in $\mathcal{P}(X)$, by

$$(\mathsf{e}_S \circ \mathsf{k}_S)(V) = \{x \in X \mid \forall b \in B: \forall \varepsilon > 0: X(b, x) < \varepsilon \Rightarrow \exists v \in V: X(b, v) < \varepsilon\}. \quad (10)$$

Also this closure operator is topological.

THEOREM 5.2: Let X be an algebraic gms. The closure operator $\mathsf{e}_S \circ \mathsf{k}_S$ on X is topological.

Proof: This theorem is proved along the same lines as Theorem 3.4, but one needs the following additional observation. If B is a basis for X then, for any b_V and b_W in B, $\varepsilon_V, \varepsilon_W > 0$, and x in X, such that $X(b_V, x) < \varepsilon_V$ and $X(b_W, x) < \varepsilon_W$, there exists a b in B such that $X(b_V, b) < \varepsilon_V$, $X(b_W, b) < \varepsilon_W$, and $X(b, x) < \varepsilon$, where $\varepsilon = \min\{\varepsilon_V - X(b_V, b), \varepsilon_W - X(b_W, b)\}$. This fact can be proved as follows. Because X is an algebraic gms with B as basis, there exists a Cauchy sequence $(b_n)_n$ in B with $x \in \underrightarrow{\lim}_n b_n$. Because

$$\begin{aligned} \varepsilon_V &> X(b_V, x) \\ &= \underrightarrow{\lim}_n X(b_V, b_n) \quad [x \in \underrightarrow{\lim}_n b_n, b_V \text{ is finite in } X], \end{aligned}$$

there exists an N_V such that, for all $n \geq N_V$, $X(b_V, b_n) < \varepsilon_V$. Similarly, there exists an N_W such that, for all $n \geq N_W$, $X(b_W, b_n) < \varepsilon_W$. Since

$$\begin{aligned} 0 &= X(x, x) \\ &= \underleftarrow{\lim}_n X(b_n, x) \quad [x \in \underrightarrow{\lim}_n b_n], \end{aligned}$$

there exists an N such that, for all $n \geq N$, $X(b_n, x) < \varepsilon$. The element $b_{\max\{N_V, N_W, N\}}$ in B is the one we were looking for. ❑

Thus, the operator $\mathsf{e}_S \circ \mathsf{k}_S$ induces a topology on X. In the case that X is a preorder with basis B, for every V in $\mathcal{P}(X)$,

$$(\mathsf{e}_S \circ \mathsf{k}_S)(V) = \{x \in X \mid \forall b \in B: b \leq_X x \Rightarrow \exists v \in V: b \leq_X v\},$$

which we recognize as the closure operator induced by the ordinary Scott topology.

Next, an alternative definition of this topology is given by specifying the open sets (this time starting from a gms X). In Theorem 5.6 below, it will be shown that the two definitions coincide whenever X is algebraic.

A subset V of a gms X is *generalized Scott open* (gS-open for short) if, for all Cauchy sequences $(x_n)_n$ in X and x in V, with $x \in \underrightarrow{\lim}_n x_n$,

$$\exists \varepsilon > 0 : \exists N : \forall n \geq N : B_\varepsilon(x_n) \subseteq V.$$

The following proposition gives an example of gS-open sets.

PROPOSITION 5.3: Let X be a gms. An element b in X is finite if and only if, for all $\varepsilon > 0$, the set $B_\varepsilon(b)$ is gS-open.

Proof: Let b be finite in X and $\varepsilon > 0$. We have to show that the generalized ε-ball $B_\varepsilon(b)$ is gS-open. Let $(x_n)_n$ be a Cauchy sequence in X and x be in $B_\varepsilon(b)$, with $x \in \underrightarrow{\lim}_n x_n$. It suffices to prove that

$$\exists \delta > 0 : \exists N : \forall n \geq N : X(b, x_n) < \varepsilon - \delta. \tag{11}$$

Because x is in $B_\varepsilon(b)$, we have that $\exists \delta > 0 : X(b, x) < \varepsilon - \delta$. Since

$$\begin{aligned} \varepsilon - \delta &> X(b, x) \\ &= \underrightarrow{\lim}_n X(b, x_n) \quad [x \in \underrightarrow{\lim}_n x_n,\ b \text{ is finite in } X] \end{aligned}$$

and the sequence $(X(b, x_n))_n$ is Cauchy, we can conclude (11).

Conversely, assume that, for all $\varepsilon > 0$, the set $B_\varepsilon(b)$ is gS-open. Let $(x_n)_n$ be a Cauchy sequence in X and x be in X, with $x \in \underrightarrow{\lim}_n x_n$. Then

$$\forall \varepsilon > 0 : x \in B_{X(b,x)+\varepsilon}(b).$$

Because the set $B_{X(b,x)+\varepsilon}(b)$ is gS-open,

$$\forall \varepsilon > 0 : \exists \delta > 0 : \exists N : \forall n \geq N : B_\delta(x_n) \subseteq B_{X(b,x)+\varepsilon}(b).$$

Hence, $\underrightarrow{\lim}_n X(b, x_n) \leq X(b, x)$. ❑

The set of all gS-open subsets of X is denoted by O_{gS}. This collection forms indeed a topology.

PROPOSITION 5.4: Let X be a gms. O_{gS} is a topology on X. If X is algebraic with basis B, then the set $\{B_\varepsilon(b) \mid \varepsilon > 0 \wedge b \in B\}$ forms a basis for O_{gS}.

Proof: One can easily verify that O_{gS} is closed under finite intersections and arbitrary unions. We will only prove that, for an algebraic gms X with basis B, every gS-open set V in $\mathcal{P}(X)$ is the union of generalized ε-balls of finite elements. Let x be in V. Since X is algebraic, there is a Cauchy sequence $(b_n)_n$ in B with $x \in \underrightarrow{\lim}_n b_n$. Because V is gS-open and $x \in \underrightarrow{\lim}_n b_n$,

$$\exists \varepsilon_x > 0 : \exists N_x : \forall n \geq N_x : B_{\varepsilon_x}(b_n) \subseteq V \wedge x \in B_{\varepsilon_x}(b_n).$$

Therefore, $V \subseteq \bigcup_{x \in V} B_{\varepsilon_x}(b_{N_x})$. Since the other inclusion trivially holds we have that the collection of all generalized ε-balls of finite elements forms a basis for O_{gS}. ❑

Note that every gS-open set is gA-open, because every element x in a gms X is a limit of the constant Cauchy sequence $(x)_n$. Therefore, the generalized Scott topology refines the generalized Alexandroff topology.

Any ordinary metric space X is an algebraic gms in which all elements are finite. Therefore, by the previous proposition, the basic open sets of the generalized Scott topology are all the generalized ε-balls. Hence, for ordinary metric spaces the generalized Scott topology coincides with the standard ε-ball topology.

For a preorder,[2] a subset is gS-open precisely when it is Scott open as is shown in the following proposition. Consequently, the generalized Scott topology restricted to preorders is the ordinary Scott topology.

PROPOSITION 5.5: Let X be a preorder. A set V in $\mathcal{P}(X)$ is gS-open if and only if:

1. for all x, y in X, if x is in V and $x \leq_X y$ then y is in V, and
2. for all sequences $(x_n)_n$ in X satisfying, for all n, $x_n \leq_X x_{n+1}$, and x in V, with $x \in \underrightarrow{\lim}_n x_n$,

$$\exists N : x_N \in V.$$

Proof: Assume the set V in $\mathcal{P}(X)$ is gS-open. Let x, y be in X with x in V and $x \leq_X y$. Because V is gS-open, and hence gA-open,

$$\exists \varepsilon > 0 : B_\varepsilon(x) \subseteq V.$$

Since $x \leq_X y$, we have that y is in $B_\varepsilon(x)$ and consequently y is in V. Let $(x_n)_n$ be a sequence in X satisfying, for all n, $x_n \leq_X x_{n+1}$, and x be in V, with $x \in \underrightarrow{\lim}_n x_n$. Clearly, the sequence $(x_n)_n$ is Cauchy. Because V is gS-open,

$$\exists \varepsilon > 0 : \exists N : \forall n \geq N : B_\varepsilon(x_n) \subseteq V.$$

Hence, x_N is in V.

For the converse, assume 1. and 2. Let $(x_n)_n$ be a Cauchy sequence in X and x be in V, with $x \in \underrightarrow{\lim}_n x_n$. Because X is a preorder, there exists an N such that, for all n, $x_{N+n} \leq_X x_{N+n+1}$. One can easily verify that $x \in \underrightarrow{\lim}_n x_{N+n}$. According to 2.,

$$\exists M : x_{N+M} \in V.$$

From 1. we can conclude that, for all $m \geq M$, x_{N+m} is in V. Again using 1. and (4) we can deduce that, for all $m \geq M$, $B_1(x_{N+m}) \subseteq V$. ❑

[2]Although the Scott topology is usually only defined for complete partial orders, the construction also produces a topology for preorders.

To reconcile the Scott topology for preorders with the ε-ball topology for metric spaces, one could define topologies on gms's which are finer than our generalized Scott topology. For example, call a subset V of a gms X *naive generalized Scott* open (ngS-open for short) if V is generalized Alexandroff open and, for all Cauchy sequences $(x_n)_n$ in X and x in V, with $x \in \underrightarrow{\lim}_n x_n$,

$$\exists N : \forall n \geq N : x_n \in V.$$

Evidently, every gS-open set is ngS-open. Next we show that the naive generalized Scott topology is strictly finer than our generalized Scott topology and that the second part of Proposition 5.4 does not hold for the naive generalized Scott topology. Consider the set

$$X = \{x_1, x_2, \ldots,\} \cup \{x^1, x^2, \ldots\} \cup \{x\},$$

with the distance function defined by the following diagram.

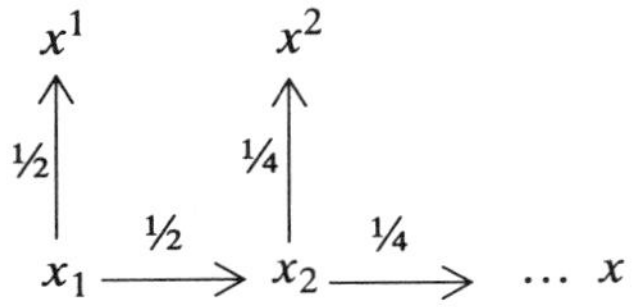

If there is no path from y to z then $X(y, z) = 1$. Otherwise, $X(y, z)$ is the maximum of the labels of the path from y to z. For example, $X(x_2, x) = ¼$ and $X(x, x_2) = 1$. One can easily verify that, for all n, both x_n and x^n are finite in X and that $x \in \underrightarrow{\lim}_n x_n$. Consequently, X is an algebraic gms with basis $X \backslash \{x\}$. Consider now the set

$$V = \{x_1, x_2, \ldots\} \cup \{x\}.$$

Obviously, the set V is ngS-open. However, it is not gS-open as can be proved as follows. By Proposition 5.4, the set

$$\{B_\varepsilon(b) \mid \varepsilon > 0 \wedge b \in X \backslash \{x\}\}$$

forms a basis for the generalized Scott topology of X. Towards a contradiction, assume that V is gS-open. Since x is in V and the above set is a basis, there exists an $\varepsilon > 0$ and a b in $X \backslash \{x\}$ such that x is in $B_\varepsilon(b) \subseteq V$. Because b is in $B_\varepsilon(b) \subseteq V$, we have that $b = x_n$ for some n. Hence $X(x_n, x) < \varepsilon$. By definition, $X(x_n, x) = 2^{-n}$, thus $2^{-n} < \varepsilon$. Since $X(x_n, x^n) = 2^{-n}$, we have that x^n is in V, a contradiction. The above not only proves that V is not gS-open, but it also shows that the set $\{B_\varepsilon(b) \mid \varepsilon > 0 \wedge b \in B\}$ cannot be a basis for the naive generalized Scott topology.

As already announced, we show that the definitions of the generalized Scott topology defined in terms of the closure operator $\mathsf{e}_S \circ \mathsf{k}_S$ and the one defined in terms of open sets coincide. For V in $\mathcal{P}(X)$, we write $cl_S(V)$ for the closure of V in the generalized Scott topology defined in terms of open sets.

THEOREM 5.6: Let X be an algebraic gms. For all V in $\mathcal{P}(X)$, $cl_S(V) = (\mathsf{e}_S \circ \mathsf{k}_S)(V)$.

Proof: This theorem can be proved along the same lines as Theorem 3.5. It follows from characterization (10) of $\mathsf{e}_S \circ \mathsf{k}_S$ and the fact that the generalized ε-balls of finite elements form a basis for the generalized Scott topology. ❑

Since the definition of the closure operator cl_S does not use the basis, the above theorem implies that the choice of the basis is irrelevant for the definition of the closure operator $\mathsf{e}_S \circ \mathsf{k}_S$.

A subset V of a gms X is *generalized Scott closed* (gS-closed for short) if its complement $X \backslash V$ is gS-open. This is equivalent to the following condition. For all Cauchy sequences $(x_n)_n$ in X and x in X, with $x \in \underrightarrow{\lim}_n x_n$,

$$(\forall \varepsilon > 0 : \forall N : \exists n \geq N : \exists y \in V : X(x_n, y) < \varepsilon) \Rightarrow x \in V.$$

Note that if V is a gS-closed set and $(x_n)_n$ is a Cauchy sequence in V, then all its limits should belong to V. Consequently, if V is a subset of X and $(x_n)_n$ is a Cauchy sequence in V, then all its limits belong to $cl_S(V)$. This implies that if B is basis for X then $X \subseteq cl_S(B)$. The converse inclusion is obvious. Hence, the basis B of an algebraic gms X is dense in X.

The following lemma, due to Flagg and Sünderhauf, gives an example of gS-closed sets.

LEMMA 5.7: Let X be a gms. For all x in X and $\delta \geq 0$, the set $\overline{B}_\delta^{op}(x) = \{y \in X \mid X(y, x) \leq \delta\}$ is gS-closed.

Proof: Let $(z_n)_n$ be a Cauchy sequence in X and z be in X, with $z \in \underrightarrow{\lim}_n z_n$, and

$$\forall \varepsilon > 0 : \forall N : \exists n \geq N : \exists y \in \overline{B}_\delta^{op}(x) : X(z_n, y) < \varepsilon.$$

Then

$$\forall \varepsilon > 0 : \forall N : \exists n \geq N : X(z_n, x) < \varepsilon + \delta.$$

Because the sequence $(z_n)_n$ is Cauchy,

$$\forall \varepsilon > 0 : \exists N : \forall n \geq N : X(z_n, x) < \varepsilon + \delta.$$

Consequently, $\underleftarrow{\lim}_n X(z_n, x) \leq \delta$, and hence $X(z, x) \leq \delta$. ❑

Like the generalized Alexandroff topology, the generalized Scott topology provides us all information about the underlying preorder.

PROPOSITION 5.8: Let X be a gms. For all x and y in X, $x \leq_{O_{gS}} y$ if and only if $x \leq_X y$.

Proof: For any gS-open set V, if x is in V and $X(x, y) = 0$, then y is in V. From this observation, the implication from right to left is clear. For the converse, suppose $X(x, y) \neq 0$. Then x is in $X \backslash B_0^{op}(y)$ but y is not in $X \backslash B_0^{op}(y)$. Since, by Lemma 5.7, the set $X \backslash B_0^{op}(y)$ is gS-open it follows that $x \not\leq_{O_{gS}} y$. ❑

An element x is a *topological limit* of a sequence $(x_n)_n$ in a topology O, denoted by $x \in \lim_{O,n} x_n$, if, for all V in O with x in V,

$$\exists N : \forall n \geq N : x_n \in V.$$

The generalized Scott topology also encodes all information about convergence.

PROPOSITION 5.9: Let X be a gms. For all Cauchy sequences $(x_n)_n$ in X and x in X, with $x \in \underrightarrow{\lim}_n x_n$, and y in X, $y \in \lim_{O_{gS},n} x_n$ if and only if $y \leq_{O_{gS}} x$.

Proof: Clearly $x \in \lim_{O_{gS},n} x_n$, and hence $y \leq_{O_{gS}} x$ implies $y \in \lim_{O_{gS},n} x_n$. For the converse, let $y \in \lim_{O_{gS},n} x_n$. Assume $y \not\leq_{O_{gS}} x$. According to Proposition 5.8, there is a $\delta > 0$ such that $X(y, x) \not\leq \delta$. Hence, y is in $X \backslash \overline{B}_\delta^{op}(x)$, which is a gS-open set by Lemma 5.7. Since $y \in \lim_{O_{gS},n} x_n$,

$$\exists N : \forall n \geq N : x_n \in X \backslash \overline{B}_\delta^{op}(x).$$

But

$$\begin{aligned} 0 &= X(x, x) \\ &= \underleftarrow{\lim}_n X(x_n, x) \quad [x \in \underrightarrow{\lim}_n x_n] \end{aligned}$$

so

$$\exists M : \forall m \geq M : X(x_m, x) \leq \delta.$$

This gives a contradiction. Therefore, $y \leq_{O_{gS}} x$. ❑

From the above proposition we can conclude that in a gms every Cauchy sequence topologically converges to its metric limits. However, not every topologically convergent sequence is Cauchy. For example, provide the set $X = \{1, 2, \ldots, \omega\}$ with the distance function

$$X(x, y) = \begin{cases} 0 & \text{if } x = y \\ 1/n & \text{if } x = \omega \text{ and } y = n \\ 1 & \text{otherwise.} \end{cases}$$

Then X is an algebraic complete qms with X itself as basis, since there are no nontrivial Cauchy sequences. The sequence $(n)_n$ topologically converges to ω but is not Cauchy.

Continuity is also encoded by the generalized Scott topology.

PROPOSITION 5.10: Let X and Y be complete gms's. A nonexpansive function $f\colon X \to Y$ is metrically continuous if and only if it is topologically continuous.

Proof: Let $f\colon X \to Y$ be a nonexpansive and metrically continuous function and let V in $\mathcal{P}(Y)$ be gS-open. We need to prove that $f^{-1}(V)$ in $\mathcal{P}(X)$ is gS-open in order to conclude that f is topologically continuous. Indeed, for any Cauchy sequence $(x_n)_n$ in X and x in X, with $x \in \underrightarrow{\lim}_n x_n$, we have

$$\begin{aligned} x \in f^{-1}(V) &\Rightarrow f(x) \in V \\ &\Rightarrow \exists \varepsilon > 0 : \exists N : \forall n \geq N : B_\varepsilon(f(x_n)) \subseteq V \\ &\qquad [f \text{ is metrically continuous, } V \text{ is gS-open}] \\ &\Rightarrow \exists \varepsilon > 0 : \exists N : \forall n \geq N : B_\varepsilon(x_n) \subseteq f^{-1}(V) \\ &\qquad [f \text{ is nonexpansive}] \end{aligned}$$

For the converse, assume $f\colon X \to Y$ is topologically continuous and let $(x_n)_n$ be a Cauchy sequence in X and x be in X, with $x \in \underrightarrow{\lim}_n x_n$. Let $y \in \underrightarrow{\lim}_n f(x_n)$. According to (7) it suffices to prove that $Y(y, f(x)) = 0$ and $Y(f(x), y) = 0$. We have that

$$\begin{aligned} Y(y, f(x)) &= \underleftarrow{\lim}_n Y(f(x_n), f(x)) \; [y \in \underrightarrow{\lim}_n f(x_n)] \\ &\leq \underleftarrow{\lim}_n X(x_n, x) \; [f \text{ is nonexpansive}] \\ &= X(x, x) \; [x \in \underrightarrow{\lim}_n x_n] \\ &= 0. \end{aligned}$$

According to Proposition 5.9, $x \in \lim_{O_{gS},n} x_n$. Because f is continuous, $f(x) \in \lim_{O_{gS},n} f(x_n)$. Using Proposition 5.9 again, we can conclude that $f(x) \leq_{O_{gS}} y$. By Proposition 5.8, $Y(f(x), y) = 0$. ❑

6. RELATED WORK

In this paper we have presented two topologies for gms's. The main contribution of our paper is the reconciliation of the enriched categorical approach of Lawvere [10], [11] (cf.[9], [19], [20]) and the topological approach of Smyth [15], [16] (cf. [5]). The present paper continues the work of Rutten [14] and is part of [2]. In the latter paper, besides the topologies presented in this paper, completion and powerdomains for generalized ultrametric spaces are also defined by means of the Yoneda embedding.

The basic definitions and facts on ordered spaces, metric spaces and topology, and gms's are taken from [7], [13], and [4], [17], and [16], [19], [14], respectively.

Smyth [16], and Flagg and Kopperman [5] have represented algebraic complete partial orders by another gms than the one given in the introduction. The distance function they use is in general not two-valued. In that case, the generalized Alexandroff topology reconciles the Scott topology for algebraic complete partial orders and the ε-ball topology for metric spaces. This approach is simpler than ours, since much of the standard theorems for ordinary metric spaces can be adapted. The price to be paid for this simplicity is that most of their results only hold for a restricted class of spaces: they have to be spectral.

Wagner [20] has also presented a generalized Scott topology. Although for complete partial orders his topology corresponds to the Scott topology, for metric spaces it does not coincide with the ε-ball topology.

Recently, Flagg and Sünderhauf [6] have proved that our generalized Scott topology of an algebraic complete qms arises as the sobrification of its basis taken with the generalized Alexandroff topology.

ACKNOWLEDGMENTS

The authors are grateful to Jaco de Bakker, Paul Gastin, Pietro di Giannantonio, Bart Jacobs, Mike Mislove, Maurice Nivat, Bill Rounds, Erik de Vink, Kim Wagner, and the referee for suggestions, comments, and discussion. We wish to thank Bob Flagg and Philipp Sünderhauf for their careful reading of an earlier version of the paper: Lemma 5.7 is due to them as well as an improvement of Proposition 5.8, Proposition 5.9, and Proposition 5.10.

REFERENCES

1. AMERICA, P. & J.J.M.M. RUTTEN. 1989. Solving Reflexive Domain Equations in a Category of Complete Metric Spaces. J. Computer and System Sciences. **39:** (3) 343–375.
2. BONSANGUE, M.M., F. VAN BREUGEL, & J.J.M.M. RUTTEN. 1995. Generalized Ultrametric Spaces: completion, topology, and powerdomains via the Yoneda embedding. Report CS-R9560, CWI, Amsterdam. Available at **ftp.cwi.nl** as **pub/CWIreports/AP/CS-R9560.ps.Z.**
3. DE BAKKER, J.W. & E.P. DE VINK. 1996. Control Flow Semantics. Foundations of Computing Series. The MIT Press. Cambridge, Massachusetts.
4. ENGELKING, R. 1989. General Topology, Sigma Series in Pure Mathematics, Vol. 6 (revised and completed edition). Heldermann Verlag, Berlin.
5. FLAGG, B. & R. KOPPERMAN. Continuity Spaces: reconciling domains and metric spaces. November 1994. Theoretical Computer Science. To appear.

6. FLAGG, B. & P. SÜNDERHAUF. 1996. The Essence of Ideal Completion in Quantitative Form. March.
7. GIERZ, G., K.H. HOFMANN, K. KEIMEL, J.D. LAWSON, M. MISLOVE, & D.S. SCOTT. 1980. A Compendium of Continuous Lattices. Springer-Verlag. Berlin.
8. KELLY, G.M. 1982. Basic Concepts of Enriched Category Theory. London Mathematical Society Lecture Notes. **64**. Cambridge University Press. Cambridge.
9. KENT, R.E. 1987. The Metric Closure Powerspace Construction. *In:* Proceedings of the 3rd Workshop on Mathematical Foundations of Programming Language Semantics, New Orleans, April 1987. (M. Main, A. Melton, M. Mislove, & D. Schmidt, Eds.). Lecture Notes in Computer Science. **298:** 173–199. Springer-Verlag. Berlin.
10. LAWVER, F.W. 1973. Metric Spaces, Generalized Logic, and Closed Categories. Rendiconti del Seminario Matematico e Fisico di Milano. **43:** 135–166.
11. LAWVER, F.W. 1986. Taking Categories Seriously. Revista Colombiana de Matemáticas. **20:** (3/4) 147–178.
12. MELTON, A. 1989. Topological Spaces for Cpos. *In:* Categorical Methods in Computer Science: with aspects from topology (H. Ehrig, H. Herrlich, H.-J. Kreowski, & G. Preuß, Eds.).Lecture Notes in Computer Science. **393:** 302–314. Springer-Verlag, Berlin.
13. PLOTKIN, G.D. 1983. Domains.
14. RUTTEN, J.J.M.M. 1995. Elements of Generalized Ultrametric Domain Theory. Report CS-R9507. CWI, Amsterdam. Available at **ftp.cwi.nl** as **pub/CWIreports/AP/CS-R9507.ps.Z.** Theoretical Computer Science. To appear.
15. SMYTH, M.B. 1987. Quasi Uniformities: reconciling domains with metric spaces. *In:* Proceedings of the 3rd Workshop on Mathematical Foundations of Programming Language Semantics, New Orleans, April 1987 (M. Main, A. Melton, M. Mislove, & D. Schmidt, Eds.). Lecture Notes in Computer Science. **298:** 236–253. Springer-Verlag. New York.
16. SMYTH, M.B. 1991. Totally Bounded Spaces and Compact Ordered Spaces as Domains of Computation. *In:* Topology and Category Theory in Computer Science (G.M. Reed, A.W. Roscoe, & R.F. Wachter, Eds.). Clarendon Press. Oxford. 207–229.
17. SMYTH, M.B. 1992. Topology. *In:* Handbook of Logic in Computer Science (S. Abramsky, Dov M. Gabbay, & T.S.E. Maibaum, Eds.). **1:** 641–761. Oxford University Press. Oxford.
18. SMYTH, M.B. & G.D. PLOTKIN. 1982. The Category-Theoretic Solution of Recursive Domain Equations. SIAM Journal of Computation. **11:** (4) 761–783.
19. WAGNER, K.R. 1994. Solving Recursive Domain Equations with Enriched Categories. PhD thesis, Carnegie Mellon University. Pittsburgh, Pensylvania.
20. WAGNER, K.R. Liminf Convergence in ω-Categories. Report 371, University of Cambridge Computer Laboratory, Cambridge, June 1995. Theoretical Computer Science. To appear.
21. WINSKEL, G. 1993. The Formal Semantics of Programming Languages: an introduction. Foundations of Computing Series. The MIT Press. Cambridge, Massachusetts.
22. YONEDA, N. 1954. On the Homology Theory of Modules. J. Faculty of Sciences, University of Tokyo, Section I. **7:** (2) 193–227.

A Theory of Metric Labelled Transition Systems

FRANCK VAN BREUGEL[a]

Università di Pisa
Dipartimento di Informatica
Corso Italia 40, 56125 Pisa, Italy

ABSTRACT: Labelled transition systems are useful for giving semantics to programming languages. Kok and Rutten have developed some theory to prove semantic models defined by means of labelled transition systems equal to other semantic models. Metric labelled transition systems are labelled transition systems with the configurations and actions endowed with metrics. The additional metric structure allows us to generalize the theory developed by Kok and Rutten.

INTRODUCTION

The classical result due to Banach [3] that a *contractive function* from a nonempty complete *metric space* to itself has a *unique fixed point* plays an important role in the theory of *metric semantics* for programming languages. Metric spaces and Banach's theorem were first employed by Nivat [20] to give semantics to recursive program schemes. Inspired by the work of Nivat, De Bakker and Zucker [13] gave semantics to concurrent languages by means of metric spaces. The metric spaces they used were defined as solutions of recursive domain equations. By means of Banach's theorem, America and Rutten [1] proved that a particular class of domain equations has unique solutions. Banach's theorem has also been used to prove semantic models to be equal. Kok and Rutten [17] applied a proof principle which we baptize the *unique fixed point proof principle*. By means of this proof principle, elements of a metric space can be proved equal. First, one introduces a function from the metric space to itself, and shows that the function is a contraction. Then one shows that the elements to be proved equal are each a fixed point of the contraction. To apply this proof principle to prove semantic models equal, the models should be elements of a metric space. Furthermore, a contractive function from the metric space to itself with the semantic

Mathematics Subject Classification: 68Q55.
Key words and phrases: labelled transition system, finitely branching, metric labelled transition system, compactly branching, complete metric space, Banach's theorem, unique fixed point proof principle, operational semantics, semantics transformation.
[a]E-mail: franck@di.unipi.it.

models as fixed point is needed. Kok and Rutten developed some theory to prove *operational* semantic models defined by means of *labelled transition systems* a la Plotkin [21] equal to other semantic models — in particular to denotational semantic models — by uniqueness of fixed point. For numerous applications of their theory we refer the reader to De Bakker and De Vink's textbook [12]. Their results are applicable only to operational semantic models induced by *finitely branching* labelled transition systems. Although most programming languages can be modelled operationally by means of a finitely branching labelled transition system, there are languages which cannot. For example, the real-time language ACPrρ introduced by Baeten and Bergstra in [4] gives rise to infinite branching.

In this paper we generalize the theory developed by Kok and Rutten. In the generalized setting we are able to deal with the above mentioned real-time language. To generalize the results we supply the labelled transition systems with some additional metric structure. These enriched labelled transition systems we call *metric labelled transition systems.* The additional metric structure enables us to generalize from finitely branching labelled transition systems to *compactly branching* metric labelled transition systems. All results proved by Kok and Rutten for finitely branching labelled transition systems are generalized to compactly branching metric labelled transition systems. This amounts to a theory of metric labelled transition systems.

1. METRIC LABELLED TRANSITION SYSTEMS

A metric labelled transition system is a labelled transition system with some additional structure. The structure is added by endowing the sets of configurations and actions with 1-bounded complete metrics.

DEFINITION 1.1: A *metric labelled transition system* is a triple $(C, A, \rightarrow)$ consisting of:

(a) a 1-bounded metric space of *configurations* C,
(b) a 1-bounded complete metric space of *actions* A, and
(c) a *transition relation* $\rightarrow \subseteq C \times A \times C$.

Instead of $(c, a, c') \in \rightarrow$ we write $c \xrightarrow{a} c'$. Most of the time we only present the transition relation of a metric labelled transition system. Frequently we depict (the transition relation of) a metric labelled transition system by a directed graph. The nodes are labelled by configurations and the edges are labelled by actions.

EXAMPLE 1.2: The labelled transition system

$(\{0, 0.5, 1\}, [0, 1], \{(0, a, 0.5) \mid a \in [0, 1]\} \cup \{(0, a, 1) \mid a \in [0, 1]\} \cup \{(1, 1, 1)\})$

is presented by

$$\begin{cases} 0 \xrightarrow{a} 0.5 & \text{for } a \in [0, 1] \\ 0 \xrightarrow{a} 1 & \text{for } a \in [0, 1] \\ 1 \xrightarrow{1} 1 & \end{cases}$$

and depicted by

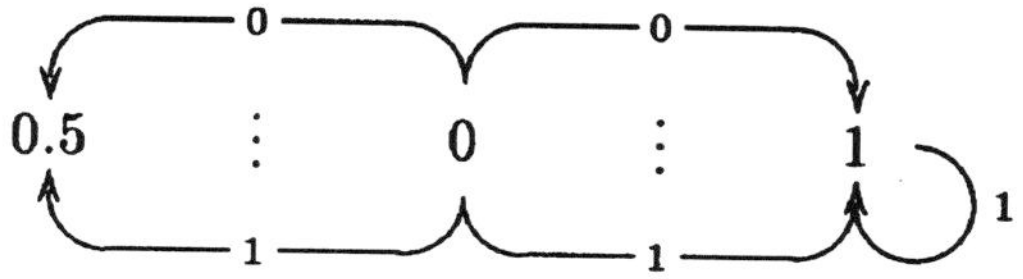

By endowing the set of configurations and the set of actions both with the Euclidean metric we obtain a metric labelled transition system.

If $c \xrightarrow{a} c'$ then we say that there exists a transition from c to c' labelled by a. If there exists a transition from c then we call c a *nonterminal* configuration and write $c \rightarrow$. Otherwise we call c a *terminal* configuration and write $c \not\rightarrow$.

EXAMPLE 1.3: In Example 1.2 the configurations 0 and 1 are nonterminal and the configuration 0.5 is terminal.

A labelled transition system is called *finitely branching* if every configuration has only finitely many outgoing transitions. Because we have a metric on the sets of configurations and actions (and hence, on the Cartesian product of these sets), finitely branching can be generalized to compactly branching: for each configuration, its set of outgoing transitions is compact.

DEFINITION 1.4: A metric labelled transition system $(C, A, \rightarrow)$ is called *compactly branching* if, for all $c \in C$, the set

$$\mathcal{CB}(c) = \{(a, c') \mid c \xrightarrow{a} c'\}$$

is compact.

If we endow the configurations and the actions of a finitely branching labelled transition system both with an arbitrary 1-bounded (complete) metric, then we obtain a compactly branching metric labelled transition system. A compactly branching metric labelled transition system is in general not finitely branching.

EXAMPLE 1.5: The metric labelled transition system introduced in Example 1.2 is not finitely branching but is compactly branching. If, in this example, we endow the actions with the discrete metric, the metric labelled transition system so obtained is no longer compactly branching.

For a compactly branching metric labelled transition system we introduce the condition of transitioning being nonexpansive. To formulate this condition we provide the compact sets[1] of outgoing transitions of the configurations, elements of $\mathcal{P}_c(A \times C)$, with a metric. The set of action-configuration pairs is endowed with the metric obtained from the metric on the actions and the metric on the configurations multiplied by a ½, and the resulting metric space is denoted by $A \times \frac{1}{2} \cdot C$. As we will see below the introduction of the $\frac{1}{2} \cdot$ gives rise to a less restrictive condition. The (compact) sets of these pairs are endowed with the Hausdorff metric [15].

DEFINITION 1.6: A compactly branching metric labelled transition system $(C, A, \rightarrow)$ is called *nonexpansive* if the function

$$\mathcal{CB}: C \rightarrow \mathcal{P}_c(A \times \tfrac{1}{2} \cdot C)$$

defined by

$$\mathcal{CB}(c) = \{(a, c') \mid c \xrightarrow{a} c'\}$$

is nonexpansive.

EXAMPLE 1.7: The compactly branching metric labelled transition system of Example 1.2 is not nonexpansive, because

$$\begin{aligned} d(\mathcal{CB}(0.5), \mathcal{CB}(1)) &= d(\emptyset, \{(1, 1)\}) \\ &= 1 \\ &\not\leq 0.5 \\ &= d(0.5, 1). \end{aligned}$$

By adding the transition

$$0.5 \xrightarrow{1} 0.5$$

we obtain the compactly branching metric labelled transition system

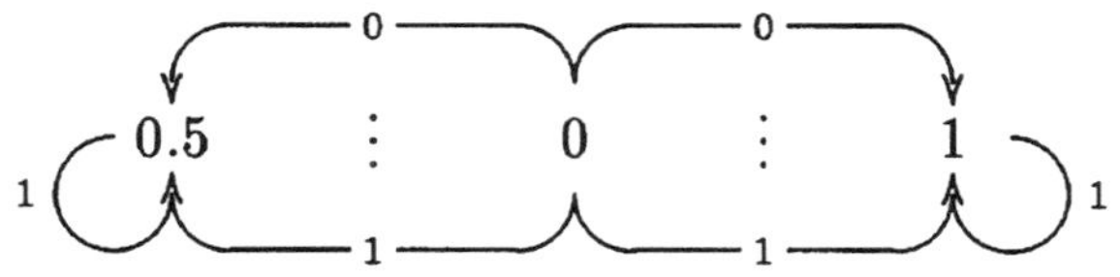

which is nonexpansive.

The $\frac{1}{2} \cdot$ in the above definition does not change the compactness condi-

[1] $\mathcal{P}_c$ denotes the set of compact subsets.

tion. By leaving out the ½ · we obtain a more restrictive nonexpansiveness condition.

EXAMPLE 1.8: The labelled transition system

$$\begin{cases} 0.25 \xrightarrow{0} 0 \\ 0.75 \xrightarrow{0} 1 \end{cases}$$

depicted by

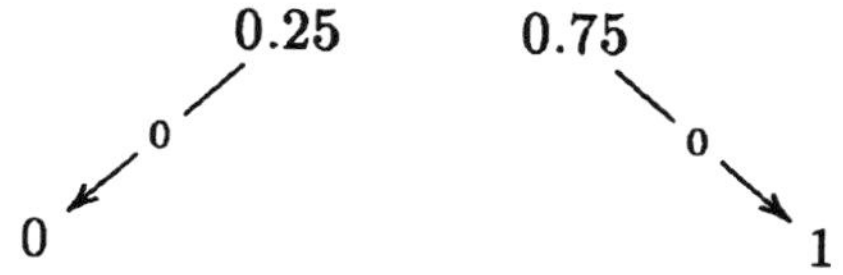

with the set of configurations endowed with the Euclidean metric, is (compactly branching and) nonexpansive, since

$$\begin{aligned} d(\mathcal{CB}(0.25), \mathcal{CB}(0.75)) &= d(\{(0, 0)\}, \{(0, 1)\}) \\ &= 0.5 \\ &= d(0.25, 0.75). \end{aligned}$$

If we leave out the ½ · we have that

$$\begin{aligned} d(\mathcal{CB}(0.25), \mathcal{CB}(0.75)) &= 1 \\ &\not\leq 0.5 \\ &= d(0.25, 0.75). \end{aligned}$$

In some applications the more restricted nonexpansiveness condition is not satisfied, whereas the one given in Definition 1.6 is (cf. [10, Chapter 9]).

A finitely branching labelled transition system with the configurations endowed with an arbitrary 1-bounded metric and the actions endowed with the discrete metric is (compactly branching and) nonexpansive. Consequently we have generalized finitely branching to compactly branching and nonexpansive.

2. OPERATIONAL SEMANTICS

The operational semantics induced by a metric labelled transition system is a function assigning to each configuration a nonempty set[2] of finite and

[2] $\mathcal{P}_n$ denotes the set of nonempty subsets.

infinite action sequences.[3] This assignment is driven by the transition relation of the metric labelled transition system.

DEFINITION 2.1: An *operational semantics* induced by a metric labelled transition system $(C, A, \rightarrow)$ is a function $\mathcal{O}: C \rightarrow \mathcal{P}_n(A^\infty)$ defined by

$$\mathcal{O}(c) = \{a_1a_2\cdots a_n \mid c = c_1 \xrightarrow{a_1} c_2 \xrightarrow{a_2} \cdots \xrightarrow{a_n} c_{n+1} \not\rightarrow\} \cup$$
$$\{a_1a_2\cdots \mid c = c_1 \xrightarrow{a_1} c_2 \xrightarrow{a_2} \cdots\}.$$

In the above definition we use

$$c = c_1 \xrightarrow{a_1} c_2 \xrightarrow{a_2} \cdots \xrightarrow{a_n} c_{n+1} \not\rightarrow$$

as an abbreviation for

$$c = c_1 \wedge \forall 1 \leq m \leq n: c_m \xrightarrow{a_m} c_{m+1} \wedge c_{n+1} \not\rightarrow$$

and

$$c = c_1 \xrightarrow{a_1} c_2 \xrightarrow{a_2} \cdots$$

as an abbreviation for

$$c = c_1 \wedge \forall m \geq 1: c_m \xrightarrow{a_m} c_{m+1}.$$

A sequence $a_1a_2\cdots a_n$ is an element of the operational semantics of the configuration c if there exists a transition sequence from c to some terminal configuration labelled by $a_1a_2\cdots a_n$. If there exists an infinite transition sequence from c labelled by $a_1a_2\cdots$ then the infinite sequence $a_1a_2\cdots$ is an element of the operational semantics of c. Consequently, the operational semantics of a terminal configuration is a singleton set consisting of the empty sequence ε. Note that each configuration is mapped to a nonempty set.

EXAMPLE 2.2: The metric labelled transition system of Example 1.2 induces the operational semantics $\mathcal{O}$ defined by

$$\begin{aligned} \mathcal{O}(0) &= [0, 1] \cdot \{1^\omega\} \cup [0, 1] \\ \mathcal{O}(0.5) &= \{\varepsilon\} \\ \mathcal{O}(1) &= \{1^\omega\}. \end{aligned}$$

To prove an operational semantics equal to another semantics by means of the unique fixed point proof principle, the operational semantics should be an element of a metric space. To turn the space $C \rightarrow \mathcal{P}_n(A^\infty)$ into a metric space we first endow the set of finite and infinite action sequences A^∞ with the following metric.

[3] A^∞ denotes the set of finite and infinite action sequences.

DEFINITION 2.3: The metric $d_{A^\infty}: A^\infty \times A^\infty \to [0, 1]$ is defined by $d_{A^\infty}(\sigma_1, \sigma_2) =$

$$\begin{cases} 0 & \text{if } \sigma_1 = \sigma_2 \\ \sup(\{2^{-n+1} \cdot d_A(\sigma_1(n), \sigma_2(n)) \mid 1 \le n \le |\sigma_1|\} \cup \{2^{-|\sigma_1|}\}) & \text{if } |\sigma_1| < |\sigma_2| \\ \sup(\{2^{-n+1} \cdot d_A(\sigma_1(n), \sigma_2(n)) \mid 1 \le n \le |\sigma_2|\} \cup \{2^{-|\sigma_2|}\}) & \text{if } |\sigma_1| > |\sigma_2| \\ \sup\{2^{-n+1} \cdot d_A(\sigma_1(n), \sigma_2(n)) \mid 1 \le n \le |\sigma_1|\} & \text{otherwise} \end{cases}$$

where $|\sigma_i|$ denotes the length of the sequence σ_i and $\sigma_i(n)$ denotes the n-th element of σ_i.

In case we endow the action set A with the discrete metric we obtain the usual Baire-like metric [2]:

$$d_{A^\infty}(\sigma_1, \sigma_2) = \begin{cases} 0 & \text{if } \sigma_1 = \sigma_2 \\ 2^{-n} & \text{otherwise,} \end{cases}$$

where n is the length of the longest common prefix of σ_1 and σ_2.

Second, we endow the (nonempty) sets of action sequences with the Hausdorff metric induced by the above introduced 1-bounded metric on action sequences. In this way we obtain a pseudometric space rather than a metric space. The restriction to nonempty and *compact* sets[4] of action sequences gives rise to a metric space. Finally, the functions from configurations to (nonempty and compact) sets of action sequences are endowed with the supremum of the pointwise distances. We restrict our attention to operational semantic models that belong to this metric space.

DEFINITION 2.4: An operational semantics $\mathcal{O}: C \to \mathcal{P}_n(A^\infty)$ is called *compact* if $\mathcal{O} \in C \to \mathcal{P}_{nc}(A^\infty)$.

EXAMPLE 2.5: The operational semantics presented in Example 2.2 is compact if the action set [0, 1] is endowed with the Euclidean metric. If, in this example, we endow the action set [0, 1] with the discrete metric, the operational semantics so obtained is not compact.

Not every metric labelled transition system induces a compact operational semantics. If we restrict ourselves to compactly branching and nonexpansive metric labelled transition systems then we obtain compact operational semantic models. Without the additional nonexpansive condition we do in general not obtain compact operational semantic models.

EXAMPLE 2.6: The compactly branching metric labelled transition system

[4] $\mathcal{P}_{nc}$ denotes the set of nonempty and compact subsets.

$$\begin{cases} 0 \xrightarrow{0} 0 \\ 0 \xrightarrow{1/n} 1/n \text{ for } n \in \mathbb{N} \end{cases}$$

depicted by

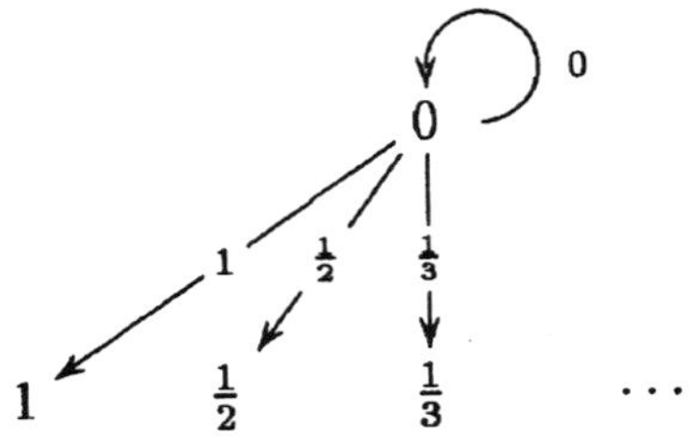

with the set of configurations and the set of actions both endowed with the Euclidean metric, does not induce a compact operational semantics. Note that the function $C\mathcal{B}$ is not nonexpansive.

Next we prove that a compactly branching and nonexpansive metric labelled transition system induces a compact operational semantics. To prove this we first prove two additional propositions. In the first proposition we demonstrate that the nonterminal and terminal configurations of a compactly branching and nonexpansive metric labelled transition system are distance 1 apart.

PROPOSITION 2.7: The nonterminal and terminal configurations of a compactly branching and nonexpansive metric labelled transition system are distance 1 apart.

Proof: For a nonterminal configuration c, $C\mathcal{B}(c) \neq \emptyset$ and for a terminal configuration c', $C\mathcal{B}(c') = \emptyset$. Since the metric labelled transition system is nonexpansive,

$$1 = d(C\mathcal{B}(c), C\mathcal{B}(c')) \leq d(c, c'). \quad \square$$

In the second proposition we show that, for a compactly branching and nonexpansive metric labelled transition system, for all configurations c and natural numbers n, the set of transition sequences starting from the configuration c and truncated at length n is compact.

PROPOSITION 2.8: Let $(C, A, \rightarrow)$ be a compactly branching and nonexpansive metric labelled transition system. For all $c \in C$ and $n \in \mathbb{N}$, the set

$$C\mathcal{B}^n(c) = \{(a_1, c_2, a_2, \ldots, a_n, c_{n+1}) \mid c = c_1 \xrightarrow{a_1} c_2 \xrightarrow{a_2} \cdots \xrightarrow{a_n} c_{n+1}\}$$

is compact.[5]

Proof: This proposition is proved by induction on n. For $n = 0$ the proposition is vacuously true. Let $n > 0$. Let $c \in C$. Because the metric labelled transition system is compactly branching, for all $c_n \in C$, the set $\mathcal{CB}(c_n)$ is compact. Consequently, for all $c_2, \ldots, c_n \in C$ and $a_1, a_2, \ldots, a_{n-1} \in A$, the set

$$\{(a_1, c_2, a_2, \ldots, a_n, c_{n+1}) \mid c_n \xrightarrow{a_n} c_{n+1}\}$$

is also compact. Since the metric labelled transition system is nonexpansive, the function corresponding to the above set is nonexpansive in $(a_1, c_2, a_2, \ldots, a_{n-1}, c_n)$. By induction, the set $\mathcal{CB}^{n-1}(c)$ is compact. Because the nonexpansive image of a compact set is compact,

$$\{\{(a_1, c_2, a_2, \ldots, a_n, c_{n+1}) \mid c_n \xrightarrow{a_n} c_{n+1}\} \mid (a_1, c_2, a_2, \ldots, a_{n-1}, c_n) \in \mathcal{CB}^{n-1}(c)\}$$

is a compact set of compact sets. It follows from Michael's theorem [19, Theorem 2.5][6] that the set

$$\mathcal{CB}^n(c) =$$

$$\bigcup\{\{(a_1, c_2, a_2, \ldots, a_n, c_{n+1}) \mid c_n \xrightarrow{a_n} c_{n+1}\} \mid (a_1, c_2, a_2, \ldots, a_{n-1}, c_n) \in \mathcal{CB}^{n-1}(c)\}$$

is compact. ❑

Now we are ready to prove the main result of this paper.

THEOREM 2.9: The operational semantics induced by a compactly branching and nonexpansive metric labelled transition system is compact.

Proof: Let $(C, A, \rightarrow)$ be a compactly branching and nonexpansive metric labelled transition system. We prove that the induced operational semantics $\mathcal{O}$ is compact, *i.e.*, for all $c \in C$, the set $\mathcal{O}(c)$ is compact.

Let $c \in C$. Let $(\sigma_n)_n$ be a sequence in $\mathcal{O}(c)$. We show that there exists a subsequence $(\sigma_{s(n)})_n$ of $(\sigma_n)_n$ converging to some $\sigma \in \mathcal{O}(c)$.

The subsequence $(\sigma_{s(n)})_n$ will be constructed from a collection of subsequences $(\sigma_{s_m(n)})_n$ satisfying

$$(\forall m \in \mathbb{N} : Q(m)) \vee (\exists k \in \mathbb{N} : \forall 1 \leq m < k : Q(m) \wedge R(k)) \qquad (1)$$

where

$$Q(m) \Leftrightarrow$$

$$\forall n \in \mathbb{N} : \sigma_{s_m(n)} = a_{1,s_m(n)}\, a_{2,s_m(n)} \cdots a_{m,s_m(n)} \sigma_{m,s_m(n)} \wedge$$

[5]To be precise, $\mathcal{CB}^n(c)$ is a compact subset of $A \times \frac{1}{2} \cdot (C \times A \times \frac{1}{2} \cdot (\cdots A \times \frac{1}{2} \cdot C))$. We leave it to the reader to fill in these details in the proof.

[6]Let X be a 1-bounded metric space. If $\mathcal{A} \in \mathcal{P}_{nc}(\mathcal{P}_{nc}(X))$ then $\bigcup\mathcal{A} \in \mathcal{P}_{nc}(X)$.

$$c = c_{1,s_m(n)} \xrightarrow{a_{1,s_m(n)}} c_{2,s_m(n)} \xrightarrow{a_{2,s_m(n)}} \cdots \xrightarrow{a_{m,s_m(n)}} c_{m+1,s_m(n)} \rightarrow \wedge$$

$$\sigma_{m,s_m(n)} \in \mathcal{O}(c_{m+1,s_m(n)}) \wedge$$

$$\forall 1 \leq j \leq m\colon \lim_h a_{j,s_m(h)} = a_j \wedge$$

$$\forall 1 \leq j \leq m+1\colon \lim_h c_{j,s_m(h)} = c_j \wedge$$

$$c = c_1 \xrightarrow{a_1} c_2 \xrightarrow{a_2} \cdots \xrightarrow{a_m} c_{m+1} \rightarrow$$

and

$R(m) \Leftrightarrow$

$$\forall n \in \mathbb{N}\colon \sigma_{s_m(n)} = a_{1,s_m(n)}\, a_{2,s_m(n)} \cdots a_{m,s_m(n)} \wedge$$

$$c = c_{1,s_m(n)} \xrightarrow{a_{1,s_m(n)}} c_{2,s_m(n)} \xrightarrow{a_{2,s_m(n)}} \cdots \xrightarrow{a_{m,s_m(n)}} c_{m+1,s_m(n)} \not\rightarrow \wedge$$

$$\forall 1 \leq j \leq m\colon \lim_h a_{j,s_m(h)} = a_j \wedge$$

$$\forall 1 \leq j \leq m+1\colon \lim_h c_{j,s_m(h)} = c_j \wedge$$

$$c = c_1 \xrightarrow{a_1} c_2 \xrightarrow{a_2} \cdots \xrightarrow{a_m} c_{m+1} \not\rightarrow.$$

The existence of the subsequences $(\sigma_{s_m(n)})_n$ is verified by proving, for all $i \in \mathbb{N}$,

$$P(i) \Leftrightarrow (\forall 1 \leq m \leq i\colon Q(m)) \vee (\exists 1 \leq k \leq i\colon \forall 1 \leq m < k\colon Q(m) \wedge R(k))$$

by induction on i.

To prove $P(0)$ it suffices to show $Q(0) \vee R(0)$. Obviously, the sequence $(\sigma_n)_n$ satisfies $Q(0) \vee R(0)$.

Let $i > 0$. To prove $P(i-1) \Rightarrow P(i)$ it suffices to show $Q(i-1) \Rightarrow Q(i) \vee R(i)$. If $Q(i-1)$, then

$$\forall n \in \mathbb{N}\colon ((\sigma_{s_{i-1}(n)} = a_{1,s_{i-1}(n)}\, a_{2,s_{i-1}(n)} \cdots a_{i,s_{i-1}(n)}\, \sigma_{i,s_{i-1}(n)} \wedge$$

$$c = c_{1,s_{i-1}(n)} \xrightarrow{a_{1,s_{i-1}(n)}} c_{2,s_{i-1}(n)} \xrightarrow{a_{2,s_{i-1}(n)}} \cdots \xrightarrow{a_{i,s_{i-1}(n)}} c_{i+1,s_{i-1}(n)} \rightarrow \wedge$$

$$\sigma_{i,s_{i-1}(n)} \in \mathcal{O}(c_{i+1,s_{i-1}(n)})) \vee$$

$$(\sigma_{s_{i-1}(n)} = a_{1,s_{i-1}(n)}\, a_{2,s_{i-1}(n)} \cdots a_{i,s_{i-1}(n)} \wedge$$

$$c = c_{1,s_{i-1}(n)} \xrightarrow{a_{1,s_{i-1}(n)}} c_{2,s_{i-1}(n)} \xrightarrow{a_{2,s_{i-1}(n)}} \cdots \xrightarrow{a_{i,s_{i-1}(n)}} c_{i+1,s_{i-1}(n)} \not\rightarrow)) \wedge$$

$$\forall 1 \leq j \leq i-1\colon \lim_h a_{j,s_{i-1}(h)} = a_j \wedge$$

$$\forall 1 \le j \le i: \lim_h c_{j,s_{i-1}(h)} = c_j \wedge$$

$$c = c_1 \xrightarrow{a_1} c_2 \xrightarrow{a_2} \cdots \xrightarrow{a_{i-1}} c_i \rightarrow.$$

Since the sequence

$$(a_{1,s_{i-1}(n)}, c_{2,s_{i-1}(n)}, a_{2,s_{i-1}(n)}, \ldots, a_{i,s_{i-1}(n)}, c_{i+1,\, s_{i-1}(n)})_n$$

is a sequence in $\mathcal{CB}^i(c)$ and by Proposition 2.8 the set $\mathcal{CB}^i(c)$ is compact, the sequence has a subsequence

$$(a_{1,s'_{i-1}(n)}, c_{2,s'_{i-1}(n)}, a_{2,s'_{i-1}(n)}, \ldots, a_{i,s'_{i-1}(n)}, c_{i+1,s'_{i-1}(n)})_n$$

which converges to $(a_1, c_2, a_2, \ldots, a_i, c_{i+1})$ in $\mathcal{CB}^i(c)$ for some $a_i \in A$ and $c_{i+1} \in C$, i.e.,

$$c = c_1 \xrightarrow{a_1} c_2 \xrightarrow{a_2} \cdots \xrightarrow{a_i} c_{i+1}.$$

If $c_{i+1} \rightarrow$ (or $c_{i+1} \not\rightarrow$), then there exists a subsequence

$$(a_{1,s_i(n)}, c_{2,s_i(n)}, a_{2,s_i(n)}, \ldots, a_{i,s_i(n)}, c_{i+1,s_i(n)})_n$$

of the sequence

$$(a_{1,s'_{i-1}(n)}, c_{2,s'_{i-1}(n)}, a_{2,s'_{i-1}(n)}, \ldots, a_{i,s'_{i-1}(n)}, c_{i+1,\, s'_{i-1}(n)})_n$$

satisfying $c_{i+1,\, s_i(n)} \rightarrow$ (or $c_{i,\, s_i(n)} \not\rightarrow$), since the nonterminal and terminal configurations are distance 1 apart according to Proposition 2.7. Consequently, $Q(i)$ (or $R(i)$).

From the subsequences $(\sigma_{s_m(n)})_n$ satisfying (1) we construct the subsequence $(\sigma_{s(n)})_n$ distinguishing the following two cases.

1. If $\forall m \in \mathbb{N}: Q(m)$, then we define $s(n) = s_n(n)$. In this case, the sequence $(\sigma_{s(n)})_n$ converges to $\sigma = a_1 a_2 \cdots$ in $\mathcal{O}(c)$.
2. If $\exists k \in \mathbb{N}: \forall 1 \le m < k: Q\ (m) \wedge R(k)$, then we define $s = s_k$. The sequence $(\sigma_{s(n)})_n$ converges to $\sigma = a_1 a_2 \cdots\ a_k$ in $\mathcal{O}(c)$. ❑

Given a finitely branching labelled transition system $(C, A, \rightarrow)$, we endow the action set A with the discrete metric (consequently, the metric on A^∞ is the usual Baire-like metric) and the configuration set C with an arbitrary 1-bounded metric. We obtain a compactly branching and nonexpansive metric labelled transition system. According to the above theorem the corresponding operational semantics is compact. Hence, the folklore result that a finitely branching labelled transition system induces a compact operational semantics is a consequence of the above theorem.

The operational semantics induced by a compactly branching and nonexpansive metric labelled transition system has another property besides being compact: it is nonexpansive. The nonexpansiveness of a compact operational semantics is exploited when we want to apply the unique fixed point proof principle (the details will be supplied in Section 3).

THEOREM 2.10: The compact operational semantics induced by a compactly branching and nonexpansive metric labelled transition system is nonexpansive.

Proof: Let $(C, A, \rightarrow)$ be a compactly branching and nonexpansive metric labelled transition system. Let $\mathcal{O}$ be the induced compact operational semantics. To prove the nonexpansiveness of $\mathcal{O}$, a sequence $(\mathcal{O}_n)_n$ of nonexpansive functions converging to $\mathcal{O}$ is introduced. Because the set of nonexpansive functions $C \rightarrow^1 \mathcal{P}_{nc}(A^\infty)$ is closed (a consequence of the completeness of A and Lemma 3 of Kuratowski's [18]), we can conclude that $\mathcal{O}$ is nonexpansive. The function $\mathcal{O}_n\colon C \rightarrow \mathcal{P}_n(A^\infty)$ is defined by

$$\mathcal{O}_n(c) = \{a_1a_2\cdots a_{k-1} \mid c = c_1 \xrightarrow{a_1} c_2 \xrightarrow{a_2} \cdots \xrightarrow{a_{k-1}} c_k \not\rightarrow \wedge\ k \leq n+1\} \cup$$

$$\{a_1a_2\cdots a_{n-1} \mid c = c_1 \xrightarrow{a_1} c_2 \xrightarrow{a_2} \cdots \xrightarrow{a_{n-1}} c_n \rightarrow\}.$$

We have left to prove that, for all n, $\mathcal{O}_n \in C \rightarrow^1 \mathcal{P}_{nc}(A^\infty)$. We prove this by induction on n. Obviously, $\mathcal{O}_0 \in C \rightarrow^1 \mathcal{P}_{nc}(A^\infty)$. Assume $n > 0$. Let $c \in C$. By definition,

$$\mathcal{O}_n(c) = \begin{cases} \{\varepsilon\} & \text{if } c \not\rightarrow \\ \{a\sigma \mid c \xrightarrow{a} c' \wedge \sigma \in \mathcal{O}_{n-1}(c')\} & \text{otherwise.} \end{cases}$$

Clearly, the set $\mathcal{O}_n(c)$ is nonempty.

Next, we show that the set $\mathcal{O}_n(c)$ is compact. By induction, $\mathcal{O}_{n-1}$ delivers compact sets. One can easily verify that, for all $a \in A$ and $c' \in C$, the set

$$\{a\sigma \mid \sigma \in \mathcal{O}_{n-1}(c')\}$$

is compact. By induction, $\mathcal{O}_{n-1}$ is nonexpansive. As a consequence, the function corresponding to the above set is nonexpansive in a and c'. Because the metric labelled transition system is compactly branching and the nonexpansive image of a compact set is compact,

$$\{\{a\,\sigma \mid \sigma \in \mathcal{O}_{n-1}(c')\} \mid c \xrightarrow{a} c'\}$$

is a compact set of compact sets. According to Michael's theorem, the set

$$\mathcal{O}_n(c) = \bigcup\{\{a\,\sigma \mid \sigma \in \mathcal{O}_{n-1}(c')\} \mid c \xrightarrow{a} c'\}$$

is compact. Also $\{\varepsilon\}$ is a compact set. Hence, the set $\mathcal{O}_n(c)$ is compact.

Finally, the nonexpansiveness of $\mathcal{O}_n$ is shown. We have to show that, for all $c_1, c_2 \in C$,

$$d(\mathcal{O}_n(c_1), \mathcal{O}_n(c_2)) \leq d(c_1, c_2).$$

If both c_1 and c_2 are terminal configurations then the above is vacuously true. Because the nonterminal and terminal configurations are distance 1 apart (Proposition 2.7), the above is also true if one of the configurations is a nonterminal configuration and the other one is a terminal configuration. That leaves us only the case that both c_1 and c_2 are nonterminal configurations. In that case,

$$d(\mathcal{O}_n(c_1), \mathcal{O}_n(c_2))$$

$$= d(\{a_1\sigma_1 \mid c_1 \xrightarrow{a_1} c_1' \wedge \sigma_1 \in \mathcal{O}_{n-1}(c_1')\}, \{a_2\sigma_2 \mid c_2 \xrightarrow{a_2} c_2' \wedge \sigma_2 \in \mathcal{O}_{n-1}(c_2')\})$$

$$= d(\bigcup\{\{a_1\sigma_1 \mid \sigma_1 \in \mathcal{O}_{n-1}(c_1')\} \mid c_1 \xrightarrow{a_1} c_1'\}, \bigcup\{\{a_2\sigma_2 \mid \sigma_2 \in \mathcal{O}_{n-1}(c_2')\} \mid c_2 \xrightarrow{a_2} c_2'\})$$

$$\leq d(\{\{a_1\sigma_1 \mid \sigma_1 \in \mathcal{O}_{n-1}(c_1')\} \mid c_1 \xrightarrow{a_1} c_1'\}, \{\{a_2\sigma_2 \mid \sigma_2 \in \mathcal{O}_{n-1}(c_2')\} \mid c_2 \xrightarrow{a_2} c_2'\})$$

[$\bigcup$ is nonexpansive]

$$\leq d(\{(a_1, \mathcal{O}_{n-1}(c_1')) \mid c_1 \xrightarrow{a_1} c_1'\}, \{(a_2, \mathcal{O}_{n-1}(c_2')) \mid c_2 \xrightarrow{a_2} c_2'\})$$

$$\leq d(\{(a_1, c_1') \mid c_1 \xrightarrow{a_1} c_1'\}, \{(a_2, c_2') \mid c_2 \xrightarrow{a_2} c_2'\})$$

[by induction, $\mathcal{O}_{n-1}$ is nonexpansive]

$$\leq d(c_1, c_2)$$

[the metric labelled transition system is nonexpansive] ❑

3. SEMANTICS TRANSFORMATIONS

In the previous section we have shown that the operational semantics induced by a compactly branching and nonexpansive metric labelled transition system is an element of a metric space. To apply the unique fixed point proof principle we have to introduce a contractive function from the metric space to itself with the operational semantics as fixed point. We call this function a semantics transformation: a function transforming a semantics into another semantics. Like an operational semantics, a semantics transformation is induced by a metric labelled transition system.

DEFINITION 3.1: A *semantics transformation* induced by a metric labelled transition system $(C, A, \rightarrow)$ is a function

$$\mathfrak{T}: (C \rightarrow \mathcal{P}_n(A^\infty)) \rightarrow (C \rightarrow \mathcal{P}_n(A^\infty))$$

defined by

$$\mathfrak{T}(S)(c) = \begin{cases} \{\varepsilon\} & \text{if } c \not\rightarrow \\ \{a\sigma \mid c \xrightarrow{a} c' \wedge \sigma \in S(c')\} & \text{otherwise.} \end{cases}$$

The semantics transformation $\mathfrak{I}$ transforms the semantics S into the semantics $\mathfrak{I}(S)$. This semantics $\mathfrak{I}(S)$ assigns to a terminal configuration the singleton set consisting of the empty sequence ε. To a nonterminal configuration c, the semantics $\mathfrak{I}(S)$ assigns the set of sequences $a\sigma$ obtained from the label a of a transition (of the metric labelled transition system inducing the semantics transformation) from the nonterminal configuration c to some configuration c', and a sequence σ of $S(c')$.

PROPOSITION 3.2: The operational semantics $\mathcal{O}$ induced by a metric labelled transition system is a fixed point of the semantics transformation $\mathfrak{I}$ induced by the metric labelled transition system, i.e.,

$$\mathcal{O} = \mathfrak{I}(\mathcal{O}).$$

Proof: Let $\mathcal{O}$ and $\mathfrak{I}$ be the operational semantics and the semantics transformation induced by the metric labelled transition system $(C, A, \rightarrow)$. Let $c \in C$. Obviously $\mathfrak{I}(\mathcal{O})(c) = \mathcal{O}(c)$ if $c \not\rightarrow$. Otherwise, for all $\sigma \in \mathcal{P}_n(A^\infty)$,

$$\begin{aligned} &\sigma \in \mathfrak{I}(\mathcal{O})(c) \\ &\Leftrightarrow \exists a \in A : \exists \sigma' \in \mathcal{P}_n(A^\infty) : \exists c' \in C : \sigma = a\sigma' \wedge c \xrightarrow{a} c' \wedge \sigma' \in \mathcal{O}(c') \\ &\Leftrightarrow \sigma \in \mathcal{O}(c). \quad \square \end{aligned}$$

According to the above proposition a semantics transformation has a fixed point. This fixed point is not necessarily unique.

EXAMPLE 3.3: Consider the semantics transformation $\mathfrak{I}$ induced by the metric labelled transition system of Example 2.6. According to Proposition 3.2, the corresponding operational semantics $\mathcal{O}$ given by

$$\begin{aligned} \mathcal{O}(0) &= \{0^m\, 1/n \mid m, n \in \mathbb{N}\} \cup \{0^\omega\} \\ \mathcal{O}(1/n) &= \{\varepsilon\} \qquad \text{for } n \in \mathbb{N} \end{aligned}$$

is a fixed point of $\mathfrak{I}$. Also, the semantics S defined by

$$\begin{aligned} S(0) &= \{0^m\, 1/n \mid m, n \in \mathbb{N}\} \\ S(1/n) &= \{\varepsilon\} \qquad \text{for } n \in \mathbb{N} \end{aligned}$$

is a fixed point of $\mathfrak{I}$.

We restrict ourselves to semantic transformations transforming compact and nonexpansive semantics into compact and nonexpansive semantics.

DEFINITION 3.4: A semantics transformation

$$\mathfrak{I} : (C \rightarrow \mathcal{P}_n(A^\infty)) \rightarrow (C \rightarrow \mathcal{P}_n(A^\infty))$$

is called *compactness and nonexpansiveness preserving* if

$$\Im \in (C \rightarrow^1 \mathcal{P}_{nc}(A^\infty)) \rightarrow (C \rightarrow^1 \mathcal{P}_{nc}(A^\infty)).$$

Not every metric labelled transition system induces a compactness and nonexpansiveness preserving semantics transformation.

EXAMPLE 3.5: Consider the metric labelled transition system

$$0 \xrightarrow{1/n} 1/n \text{ for } n \in \mathbb{N}$$

depicted by

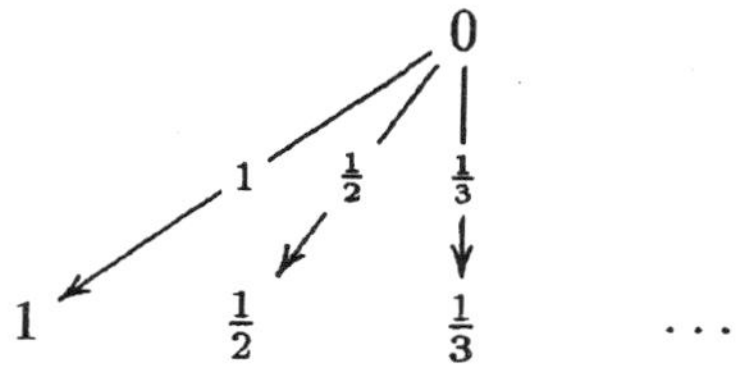

with the set of configurations and the set of actions endowed with the Euclidean metric. Although the semantics S, defined by, for all c,

$$S(c) = \{\varepsilon\}$$

is compact, the semantics $\Im(S)$ is not compact.

Not even a compactly branching metric labelled transition system necessarily induces a compactness and nonexpansiveness preserving semantics transformation.

EXAMPLE 3.6: The metric labelled transition system of Example 2.6 is compactly branching. The semantics S, defined by, for all c,

$$S(c) = \{\varepsilon\}$$

is compact and nonexpansive. The semantics $\Im(S)$ is compact but not nonexpansive.

But a compactly branching and nonexpansive metric labelled transition system gives rise to a compactness and nonexpansiveness preserving semantics transformation.

THEOREM 3.7: The semantics transformation induced by a compactly branching and nonexpansive metric labelled transition system is compactness and nonexpansiveness preserving.

Proof: Similar to the induction step of the proof of Theorem 2.10. ❑

A compactness and nonexpansiveness preserving semantics transformation is a function from a metric space to itself. According to Proposition 3.2 and Theorems 2.9 and 2.10, the corresponding operational semantics is a fixed point of the semantics transformation. To be able to apply the unique fixed point proof principle we have left to prove that the semantics transformation is contractive.

PROPOSITION 3.8: A compactness and nonexpansiveness preserving semantics transformation is contractive.

Proof: Let $\mathfrak{I} : (C \to^1 \mathcal{P}_{nc}(A^\infty)) \to (C \to^1 \mathcal{P}_{nc}(A^\infty))$ be a compactness and nonexpansiveness preserving semantics transformation. Let $S_1, S_2 \in C \to^1 \mathcal{P}_{nc}(A^\infty)$ and $c \in C$. We show that

$$d(\mathfrak{I}(S_1)(c), \mathfrak{I}(S_2)(c)) \leq \tfrac{1}{2} \cdot d(S_1, S_2).$$

We distinguish two cases.

1. If $c \not\rightarrow$, then

$$\begin{aligned} d(\mathfrak{I}(S_1)(c), \mathfrak{I}(S_2)(c)) &= d(\{\varepsilon\}, \{\varepsilon\}) \\ &\leq \tfrac{1}{2} \cdot d(S_1, S_2). \end{aligned}$$

2. If $c \rightarrow$, then

$$\begin{aligned} & d(\mathfrak{I}(S_1)(c), \mathfrak{I}(S_2)(c)) \\ &= d(\bigcup\{\{a\sigma_1 \mid \sigma_1 \in S_1(c')\} \mid c \xrightarrow{a} c'\}, \bigcup\{\{a\sigma_2 \mid \sigma_2 \in S_2(c')\} \mid c \xrightarrow{a} c'\}) \\ &\leq d(\{\{a\sigma_1 \mid \sigma_1 \in S_1(c')\} \mid c \xrightarrow{a} c'\}, \{\{a\sigma_2 \mid \sigma_2 \in S_2(c')\} \mid c \xrightarrow{a} c'\}) \\ &\quad [\textstyle\bigcup \text{ is nonexpansive}] \\ &\leq \sup\{d(\{a\sigma_1 \mid \sigma_1 \in S_1(c')\}, \{a\sigma_2 \mid \sigma_2 \in S_2(c')\}) \mid c \xrightarrow{a} c'\} \\ &= \sup\{\tfrac{1}{2} \cdot d(S_1(c'), S_2(c')) \mid c \xrightarrow{a} c'\} \\ &\leq \tfrac{1}{2} \cdot d(S_1, S_2). \quad \square \end{aligned}$$

Combining the above results we arrive at:

THEOREM 3.9: The operational semantics $\mathcal{O}$ induced by a compactly branching and nonexpansive metric labelled transition system is the unique fixed point of the semantics transformation $\mathfrak{I}$ induced by the metric labelled transition system, i.e.,

$$\mathcal{O} = \mathit{fix}(\mathfrak{I}).$$

The above theorem generalizes the result of Kok and Rutten that the operational semantics induced by a finitely branching labelled transition system is the unique fixed point of the semantics transformation induced by the labelled transition system.

We conclude this section with an example motivating our restriction to (compact and) nonexpansive semantics.

EXAMPLE 3.10: The metric labelled transition system

$$c \xrightarrow{a} c' \quad \text{for } c, c' \in [0, 1] \text{ and } a \in [0, 1],$$

with the configurations and the actions endowed with the Euclidean metric, is compactly branching and nonexpansive. Given the compact semantics S defined by

$$S(c) = \begin{cases} \{1^n\} \text{ if } c = 1/n & \text{for some } n \in \mathbb{N} \\ \{\varepsilon\} \text{ otherwise,} & \end{cases}$$

the semantics $\mathfrak{I}(S)$ is not compact.

CONCLUSION

Already in the early sixties the problem what structure to add to an abstract machine — like a labelled transition system — to obtain a topological machine was formulated by Ginsburg [14]. Shreider [23] introduced a particular topological machine — a compact automaton — to study dynamic programming. A general and detailed study of topological machines was developed by Brauer [7]. Our metric labelled transition systems are a special instance of his topological machines. Another instance are Kent's metrical transition systems [16]. A metrical transition system is a labelled transition system with the configurations endowed with a (generalized) ultrametric (the labels are not provided with any additional structure). Neither Brauer nor Kent uses the topological machines to give (operational) semantics as we have done in this paper.

This paper describes only part of a theory of metric labelled transition systems developed in the author's thesis [10]. We have shown how to generalize from finitely branching labelled transition systems to compactly branching and nonexpansive metric labelled transition systems. Similarly *image finite* labelled transition systems can be generalized to *image compact* and *binonexpansive* metric labelled transition systems. The semantic models we have considered in this paper are *linear*: they assign to each configuration a set of sequences. By means of (metric) labelled transition systems one can also define *branching* operational semantic models. These semantic models assign to each configuration a tree-like object. The details are provided in [10]. For an overview we refer the reader to [9].

If we replace nonexpansiveness by continuity (in Definition 1.6) Theorem 2.9 still holds. In that case the induced operational semantic models are continuous, but not necessarily nonexpansive (cf. Theorem 2.10). In several applications of the theory the nonexpansiveness of the operational semantics is crucial (e.g., in application 2 below). The results of Section 3 can also be adapted to the continuous setting.

The presented theory has been applied in semantics. We mention three applications.

1. An operational and a denotational semantics for a fragment of the real-time language ACPrρ have been proved equal by uniqueness of fixed point exploiting the theory in [8].
2. The theory has also been used to relate an operational and a denotational semantics for a higher-order language in [5].
3. Metric labelled transition systems have turned out to be very convenient to define abstraction operators as has been shown in [11].

We are interested to see whether a similar theory can be developed if we replace the metric spaces of a metric labelled transition system by *algebraic complete generalized* metric spaces. To develop a theory of generalized metric labelled transition systems we have to restrict ourselves to nonexpansive and continuous semantics (for generalized metric spaces nonexpansiveness does not imply continuity as has been shown in, e.g., [22]) rather than nonexpansive semantics, and instead of using the hyperspace of nonempty and compact sets endowed with the Hausdorff metric we have to employ the generalized convex powerdomain as defined in [6].

ACKNOWLEDGMENTS

The author thanks Jaco de Bakker, Jan Rutten, and Erik de Vink for their comments on a preliminary version of this paper and the referees for their constructive reports.

REFERENCES

1. AMERICA, P. & J.J.M.M. RUTTEN. 1989. Solving Reflexive Domain Equations in a Category of Complete Metric Spaces. Journal of Computer and System Sciences. **39:** (3) 343–375.
2. BAIRE, R. 1909. Sur la Représentation des Fonctions Discontinues. Acta Mathematica. **32:** (1) 97–176.
3. BANACH, S. 1922. Sur les Opérations dans les Ensembles Abstraits et leurs Applications aux Equations Intégrales. Fundamenta Mathematicae. **3:** 133–181.
4. BAETEN, J.C.M. & J.A. BERGSTRA. 1991. Real Time Process Algebra. Formal Aspects of Computing. **3:** (2) 142–188.
5. DE BAKKER, J.W. & F. VAN BREUGEL. 1993. Topological Models for Higher Order Control Flow. *In:* Proceedings of the 9th International Conference on Mathematical Foundations of Programming Semantics, New Orleans (S. Brookes, M. Main, A. Melton, M. Mislove, & D. Schmidt, Eds.). Lecture Notes in Computer Science. **802:** 122–142. Springer-Verlag. New York.

6. BONSANGUE, M.M., F. VAN BREUGEL, & J.J.M.M. RUTTEN. 1995. Generalized Ultrametric Spaces: completion, topology, and powerdomains via the Yoneda embedding. Report CS-R9560, CWI, Amsterdam. Available at **ftp.cwi.nl** as **pub/CWIreports/AP/CS-R9560.ps.Z**.
7. BRAUER, W. 1970. Zu den Grundlagen einer Theorie Topologischer Sequentieller Systeme und Automaten. Berichte der Gesellschaft für Mathematik und Datenverarbeitung, vol. 31. Gesellschaft für Mathematik und Datenverarbeitung, Bonn.
8. VAN BREUGEL, F. 1991. Comparative Semantics for a Real-Time Programming Language with Integration. *In:* Proceedings of the 4th International Conference on Theory and Practice of Software Development, Brighton (S. Abramsky & T.S.E. Maibaum, Eds.). Lecture Notes in Computer Science. **493:** 397–411. Springer-Verlag. New York.
9. VAN BREUGEL, F. 1994. Generalizing Finiteness Conditions of Labelled Transition Systems. *In:* Proceedings of the 21th International Colloquium on Automata, Languages, and Programming, Jerusalem (S. Abiteboul & E. Shamir, Eds.). Lecture Notes in Computer Science. **820:** 376–387. Springer-Verlag. New York.
10. VAN BREUGEL, F. 1994.Topological Models in Comparative Semantics. PhD thesis. Vrije Universiteit, Amsterdam.
11. VAN BREUGEL, F. 1995. From Branching to Linear Metric Domains (and back). Report RS-95-30, BRICS, Aarhus. Available at **ftp.brics.aau.dk** as **pub/BRICS/RS/95/30/BRICS-RS-95-30.ps.gz**.
12. DE BAKKER, J.W. & E.P. DE VINK. 1996. Control Flow Semantics. Foundations of Computing Series. The MIT Press. Cambridge, Massachusetts.
13. DE BAKKER, J.W. & J.I. Zucker. 1982. Processes and the Denotational Semantics of Concurrency. Information and Control. **54:** (1/2) 70–120.
14. GINSBURG, S. 1962. An Introduction to Mathematical Machine Theory. Series in Computer Science and Information Processing. Addison-Wesley. Reading, Massachusetts.
15. HAUSDORFF, F. 1914. Grundzüge der Mengenlehre. Leipzig.
16. KENT, R.E. 1987. The Metric Closure Powerspace Construction. *In:* Proceedings of the 3rd Workshop on Mathematical Foundations of Programming Language Semantics, New Orleans (M. Main, A. Melton, M. Mislove, & D. Schmidt, Eds.). Lecture Notes in Computer Science. **298:** 173–199. Springer-Verlag. New York.
17. KOK, J.N. & J.J.M.M. RUTTEN. 1990. Contractions in Comparing Concurrency Semantics. Theoretical Computer Science. **76:** (2/3) 179–222.
18. KURATOWSKI, K. 1956. Sur une Méthode de Métrisation Complète des Certains Espaces d'Ensembles Compacts. Fundamenta Mathematicae. **43:** (1) 114–138.
19. MICHAEL, E. 1951. Topologies on Spaces of Subsets. Trans. Amer. Math. Soc. **71:** (1) 152–182.
20. NIVAT, M. 1979. Infinite Words, Infinite Trees, Infinite Computations. *In:* Foundations of Computer Science III. Part 2: Languages, Logic, Semantics (J.W. de Bakker & J. van Leeuwen, Eds.). Mathematical Centre Tracts. **109:** 3–52. Mathematical Centre, Amsterdam.
21. Plotkin, G.D. 1981. A Structural Approach to Operational Semantics. Report DAIMI FN-19, Aarhus University, Aarhus.
22. Rutten, J.J.M.M. 1995. Elements of Generalized Ultrametric Domain Theory. Report CS-R9507, CWI, Amsterdam. Theoretical Computer Science. To appear. Available at **ftp.cwi.nl** as **pub/CWIreports/AP/CS-R9507.ps.Z**.
23. Shreider, Yu.A. 1964. Automata and the Problem of Dynamic Programming. Problems of Cybernetics. **5:** (1) 33–58.

Shape Theory for Arbitrary C^*-algebras

ZVONKO ČERIN[a]
Kopernikova 7
10020 Zagreb, Croatia

ABSTRACT: In this paper, we describe the $\star$-shape category $Sh_\star$ with objects $C^\star$-algebras and with morphisms $\star$-homotopy classes of fundamental $\star$-sequences. In other words, we shall define shape theory for arbitrary $C^\star$-algebras. All previous descriptions of shape theories for $C^\star$-algebras only work with separable $C^\star$-algebras. Our method is formally similar to the original Borsuk's description of the shape category of compact metric spaces and uses the $\star^\varepsilon$-homomorphisms which are the nonexpansive functions that satisfy conditions of the $\star$-homomorphism only approximately.

1. INTRODUCTION

Borsuk's shape theory has been improved and adapted to various situations in numerous ways. An interesting improvement of shape theory was proposed by Effros and Kaminker [11]. Their idea was to view shape theory via Gelfand's isomorphism, as generalized $\star$-homotopy theory for commutative $C^\star$-algebras and extend it to arbitrary $C^\star$-algebras. In this way they launched the study of noncommutative shape theory.

The method of [11] follows the inverse limit approach to shape theory [14]. In order to compare shapes of $C^\star$-algebras A and B, we assume that A and B are represented as limits of direct sequences $\{A_i, p_j^i\}$ and $\{B_i, q_j^i\}$ of $C^\star$-algebras with extremely nice properties. That A and B have the same shape means the existence of $\star$-homomorphism back and forth between $C^\star$-algebras A_i and B_j which make certain diagrams commute. This mimics the commutative case of [14] where spaces are expanded into inverse systems of absolute neighborhood retracts (or ANRs shortly).

In [11], the $C^\star$-algebras A_i and B_j above are assumed projective, while in [1] this was improved with the assumption that they are semiprojective and even further, that bonding $\star$-homomorphism p_j^i and q_j^i are semiprojective. Since a commutative $C^\star$-algebra is semiprojective if and only if it is $\star$-isomorphic to the $C^\star$-algebra $C_0(X)$ of all continuous functions which vanish at infinity from an ANR X into the $C^\star$-algebra $\mathbb{C}$ of complex numbers [1], the analogy with shape theory of [14] is obvious.

Mathematics Subject Clasification: Primary 46L85, 54C56, 55P55.
Key words and phrases: $C^\star$-algebra, shape theory, $\star$-shape theory, $\star$-shape category.
[a]E-mail address: zcerin@X400.srce.hr.

Both [11] and [1] can deal only with separable $C^\star$-algebras. Moreover, they do not define analogs of shape morphisms and thus do not consider shape category of $C^\star$-algebras. The same is true for the extensive paper [10], where the connection between K-theory of $C^\star$-algebras and their shape theory was thoroughly explored. Everything started with observations in [11] and [1] that $C^\star$-algebras of same shape have isomorphic K-groups.

We describe shape theory for arbitrary $C^\star$-algebras following the original Borsuk's method based on the notion of a fundamental sequence [3]. More precisely, we describe the $\star$-shape category with objects $C^\star$-algebras and with morphisms $\star$-homotopy classes of fundamental $\star$-sequences. Our prime objective is to improve $\star$-homotopy theory of $C^\star$-algebras by relaxing the requirement that we must use $\star$-homomorphisms. Instead, we use so-called $\star^\varepsilon$-homomorphisms, i.e., nonexpansive functions which satisfy conditions for $\star$-homomorphisms only approximately.

A fundamental $\star$-sequence $\varphi: A \to B$ is a collection $\varphi = \{f_i\}_{i=1}^\infty$ of nonexpansive functions $f_i: A \to B$ with the property that for every $\varepsilon > 0$ there is an $i \in \mathbb{N}$ such that f_j is joined with f_i for every $j \geq i$ by a $\star^\varepsilon$-homomorphism called $\star^\varepsilon$-homotopy h from A into the $C^\star$-algebra $C(I; B)$ of all continuous functions of the unit closed segment I into B. Fundamental $\star$-sequences $\varphi, \psi: A \to B$ are $\star$-homotopic whenever for every $\varepsilon > 0$ for almost every index i the functions f_i and g_i are $\star^\varepsilon$-homotopic.

An analogous approach to strong shape theory of separable $C^\star$-algebras based on the notion of an asymptotic homomorphism was considered earlier by A. Connes and N. Higson [9].

Briefly, the organization of this paper is as follows. Section 2 explains our notation and recalls some standard conventions in our exposition. In Section 3 we define $\star^\varepsilon$-homomorphisms and the relation of $\star^\varepsilon$-homotopy and prove four useful technical lemmas about them. Section 4 introduces fundamental $\star$-sequences and the notion of $\star$-homotopy for them, and we establish two propositions that will be used later over and over again. Section 5 shows how to define composition of $\star$-homotopy classes of fundamental $\star$-sequences. Our method is very similar to the approach taken in [7]. The tedious work in this section is summarized in Section 6 which completes the task of setting up the $\star$-shape category $Sh_\star$. We close with Section 7, where we consider some special fundamental $\star$-sequences with desire to identify $C^\star$-algebras on which $\star$-homotopy theory and $\star$-shape theory coincide. The answer is provided with the help of so called internally movable, internally calm, and calm $C^\star$-algebras. The more detailed study of these classes of $C^\star$-algebras and others which are suggested by some important classes of spaces from shape theory will be the subject of another paper. This paper gives foundations for shape theory of arbitrary $C^\star$-algebras. Many natural questions have not been discussed. The author hopes to consider some of those in future works. The most obvious question is that of relationship of our approach to those of Effros and Kaminker [11] and Blackadar [1]. Are separable $C^\star$-algebras $\star$-shape equivalent if and only if they have the same shape in the sense of [1]? Another important problem is to see whether our $\star$-shape theory is a noncommutative shape theory. In spite of all the effort the author was unable to prove or disprove the existence of a contravariant equivalence of categories

$$\pi_S\colon Sh_p^{lc} \to Sh_\star^{co}$$

between the proper shape category for locally compact Hausdorff spaces [7] and the $\star$-shape category of commutative $C^\star$-algebras which would make the following diagram commutative.

$$\begin{array}{ccc}
S_p^{lc} & \xrightarrow{\ \mu\ } & S_\star^{co} \\
\downarrow & & \downarrow \\
\mathcal{H}_p^{lc} & \xrightarrow{\ \pi_H\ } & \mathcal{H}_\star^{co} \\
\downarrow & & \downarrow \\
Sh_p^{lc} & \xrightarrow{\ \pi_S\ } & Sh_\star^{co}
\end{array}$$

The upper rectangle of this diagram has the categories S_p^{lc} and $\mathcal{H}_p^{lc}$ of locally compact Hausdorff spaces and proper maps and proper homotopy classes of proper maps and $S_\star^{co}$ and $\mathcal{H}_\star^{co}$ of commutative $C^\star$-algebras and $\star$-homomorphisms and $\star$-homotopy classes of $\star$-homomorphisms as vertices. Its horizontal arrows are contravariant equivalences of categories provided by Gelfand's theorem, while all vertical arrows are self-evident natural functors.

Let us also observe that Corollary 4.13 in [1] states that for locally compact metrizable spaces X and Y the $C^\star$-algebras $C_0(X)$ and $C_0(Y)$ have the same shape if and only if spaces X and Y have the same shape. By the above remarks, one direction of this statement is obviously not true because shape of algebras should be related to proper shape and not to shape of spaces.

2. PRELIMINARIES AND NOTATION

In this paper by a *$C^\star$-algebra* we mean a complete normed algebra A over the field $\mathbb{C}$ of complex numbers with an involution $\star$ such that:

(1) $x^{\star\star} = x$,
(2) $(\lambda x + \mu y)^\star = \bar{\lambda} x^\star + \bar{\mu} y^\star$,
(3) $(xy)^\star = y^\star x^\star$,
(4) $(\|x\|_A)^2 = \|x^\star x\|_A$,

for all $x, y \in A$ and all $\lambda, \mu \in \mathbb{C}$, where $\bar{\lambda}$ is complex conjugate of λ and $\| \ \|_A$ denotes the norm on A. Any algebraic $\star$-homomorphism (i.e., respecting the involution) between two $C^\star$-algebras is norm-decreasing, thus uniformly continuous and every $\star$-isomorphism between two $C^\star$-algebras is isometric. When speaking of homomorphisms between $C^\star$-algebras, we always assume that they are $\star$-homomorphisms. We recommend the books [5], [12], [15], and [16], as general references for the theory of $C^\star$-algebras.

The symbol 0_A denotes the zero element of the $C^\star$-algebra A. For a $C^\star$-algebra B and a compact topological space X, let $C(X; B)$ denote the $C^\star$-algebra of all continuous functions from X into B. The norm $\| \ \|_{C(X;B)}$ on $C(X; B)$ is given by

$$\|f\|_{C(X;B)} = \sup\{\|f(t)\|_B | t \in X\}.$$

Let I denote the unit closed segment $[0, 1]$ of real numbers. For every t in I, there is a natural evaluation $\star$-homomorphism $e_t^B: C(I; B) \to B$ defined by $e_t^B(f) = f(t)$ for every f in $C(I; B)$.

Our theory will be an improvement of the $\star$-homotopy theory for $C^\star$-algebras, which studies the equivalence relation of $\star$-homotopy on $\star$-homomorphisms. Recall that $\star$-homomorphisms f and g between $C^\star$-algebras A and B are *$\star$-homotopic* and we write $f \simeq_\star g$ provided there is a $\star$-homomorphism $h: A \to C(I; B)$ such that $h_0 = e_0^B \circ h = f$ and $h_1 = e_1^B \circ h = g$. The $\star$-homomorphism h is said to be a *$\star$-homotopy* that joins f and g or which realizes the relation $f \simeq_\star g$. For an efficient introduction to some aspects of $\star$-homotopy theory the reader should consult P. Kohn's thesis [13], J. Rosenberg's excellent expository article [17], and reference [8].

3. THE $\star^\varepsilon$-HOMOMORPHISMS

Let A and B be $C^\star$-algebras. Let ε be a positive real number. A function $f: A \to B$ is an *α^ε-homomorphism* provided:

($0(f)$) f takes the zero element 0_A of A into the zero element 0_B of B,

($ne(f)$) f is nonexpansive, i.e., the relation $\|f(x) - f(y)\|_B \leq \|x - y\|_A$ holds for all $x, y \in A$, and

($\alpha^\varepsilon(f)$) $\|f(x + y) - f(x) - f(y)\|_B < \varepsilon(\|x\|_A + \|y\|_A)$ for all $x, y \in A$.

By replacing the condition $\alpha^\varepsilon(f)$ with conditions $\lambda^\varepsilon(f)$, $\mu^\varepsilon(f)$, and $\kappa^\varepsilon(f)$ below we shall get notions of *λ^ε-homomorphism*, *μ^ε-homomorphism*, and *κ^ε-homomorphism*, respectively.

($\lambda^\varepsilon(f)$) $\|f(bx) - bf(x)\|_B < \varepsilon\|x\|_A$ for each $x \in A$ and each $b \in \mathbb{C}$.

($\mu^\varepsilon(f)$) $\|f(xy) - f(x)f(y)\|_B < \varepsilon\|x\|_A \|y\|_A$ for all $x, y \in A$.

($\kappa^\varepsilon(f)$) $\|f(x^\star) - f(x)^\star\|_B < \varepsilon\|x\|_A$ for each $x \in A$.

When f satisfies all of the above conditions it is called a *$\star^{\varepsilon}$-homomorphism*. One can introduce notions of an *α^{ε}-homomorphism*, a *λ^{ε}-homomorphism*, a *μ^{ε}-homomorphism*, and a *κ^{ε}-homomorphism* analogously.

Observe that a $\star^{\varepsilon}$-homomorphism is a uniformly continuous function. Moreover, for every real number ε between 0 and 1 and every $C^{\star}$-algebra A the function f from A into itself which takes an $x \in A$ into the product of ε and x is an example of a $\star^{\varepsilon}$-homomorphism which is not a $\star$-homomorphism.

In the first version of this paper on the right side of conditions $\alpha^{\varepsilon}(f)$, $\lambda^{\varepsilon}(f)$, $\mu^{\varepsilon}(f)$, and $\kappa^{\varepsilon}(f)$ we took simply the number ε. However, M. Dadarlat has observed that a $\star^{\varepsilon}$-homomorphism by this definition is in fact a $\star$-homomorphism. We use this occasion to thank him for this lucid remark. The proposed version is not the only one possible. The litmus test for selections is the possibility to prove lemmas of the present section. For example, acceptable set is to take $\mu^{\varepsilon}(f)$ as above and others with just ε on the right. Each such choice formally will give a different $\star$-shape theory for which all of the results in this paper hold. It is not clear which of these choices is the best.

Let $\varepsilon > 0$. Nonexpansive functions f and g between $C^{\star}$-algebras A and B are *α^{ε}-homotopic* and we write $f \overset{\varepsilon}{\simeq}_{\alpha} g$ provided there is an α^{ε}-homomorphism $h: A \to C(I; B)$ with $h_0 = e_0^B \circ h = f$ and $h_1 = e_1^B \circ h = g$. We also say that h is an α^{ε}-homotopy which *joins* f and g or that it *realizes* the relation or α^{ε}-homotopy $f \overset{\varepsilon}{\simeq}_{\alpha} g$. In a similar way we can introduce the relations $\overset{\varepsilon}{\simeq}_{\lambda}$, $\overset{\varepsilon}{\simeq}_{\mu}$, $\overset{\varepsilon}{\simeq}_{\kappa}$, and $\overset{\varepsilon}{\simeq}_{\star}$ and notions of *λ^{ε}-homotopy*, *μ^{ε}-homotopy*, *κ^{ε}-homotopy*, and *$\star^{\varepsilon}$-homotopy*.

LEMMA 3.1: Let A and B be $C^{\star}$-algebras and let $\varepsilon \geq 0$. Let f, g, and h be nonexpansive functions from A into B. If $f \overset{\varepsilon}{\simeq}_{\star} g$ and $g \overset{\varepsilon}{\simeq}_{\star} h$, then $f \overset{\varepsilon}{\simeq}_{\star} h$.

Proof: Let $a, b: A \to C(I; B)$ be $\star^{\varepsilon}$-homotopies joining f and g and g and h, respectively. Define $c: A \to C(I; B)$ by the rule

$$c(x)(t) = \begin{cases} a(x)(2t), & x \in A, \quad 0 \leq t \leq 1/2 \\ b(x)(2t-1), & x \in A, \quad 1/2 \leq t \leq 1. \end{cases}$$

Observe that for every $x \in A$, the function $c(x)$ is a continuous function from I into B because $a(x)$ and $b(x)$ and continuous functions and $c(x)$ is constructed from $a(x)$ and $b(x)$ by the standard pasting procedure.

Next, observe that for every $x \in A$ we have

$$e_0^B \circ c(x) = c(x)(0) = a(x)(0) = e_0^B \circ a(x) = f(x),$$

and similarly

$$e_1^B \circ c(x) = c(x)(1) = b(x)(1) = e_1^B \circ b(x) = h(x).$$

In other words, c joins f and h. Therefore, it remains to check that c is a $\star^{\varepsilon}$-homomorphism. Thus, we must verify that: (1) $c(0_A) = 0_B$, (2) the function c is nonexpansive, and (3) conditions $\alpha^{\varepsilon}(c)$, $\lambda^{\varepsilon}(c)$, $\mu^{\varepsilon}(c)$, and $\kappa^{\varepsilon}(c)$ hold.

ADD (1). This condition is clearly satisfied because both a and b take the zero element of A into the zero element of B. The requirement $0(f)$ from the definition of an α^{ε}-homomorphism is always easy to verify so that in the sequel the verification that it holds will be left out.

ADD (2). We must show that for all $x, y \in A$ the relation $\|c(x) - c(y)\|_B \leq \|x - y\|_A$ holds. Consider elements x and y of A. Then

$$\|c(x) - c(y)\|_{C(I;B)} = \sup\{\|c(x)(t) - c(y)(t)\|_B \,|\, t \in I\} =$$
$$\max\{\sup\{\|[a(x) - a(y)](2t)\|_B \,|\, t \in [0,1/2]\},$$
$$\sup\{\|[b(x) - b(y)](2t - 1)\|_B \,|\, t \in [1/2, 1]\}\} =$$
$$\max\{\|a(x) - a(y)\|_B, \|b(x) - b(y)\|_B\} \leq \|x - y\|_A.$$

ADD $\alpha^{\varepsilon}(c)$. For any $x, y \in A$, we obtain

$$D = \|c(x + y) - c(x) - c(y)\|_{C(I;B)} = \sup\{\|[c(x + y) - c(x) - c(y)](t)\|_B \,|\, t \in I\}.$$

But,

$$[c(x + y) - c(x) - c(y)](t) = \begin{cases} [a(x + y) - a(x) - a(y)](2t), & 0 \leq t \leq 1/2 \\ [b(x + y) - b(x) - b(y)](2t - 1), & 1/2 \leq t \leq 1. \end{cases}$$

Since conditions $\alpha^{\varepsilon}(a)$ and $\alpha^{\varepsilon}(b)$ are true, the numbers

$$\|a(x + y) - a(x) - a(y)\|_{C(I;B)} = \sup\{\|[a(x + y) - a(x) - a(y)](t)\|_B \,|\, t \in I\}$$

and

$$\|b(x + y) - b(x) - b(y)\|_{C(I;B)} = \sup\{\|[b(x + y) - b(x) - b(y)](t)\|_B \,|\, t \in I\}$$

are both less than $\varepsilon(\|x\|_A + \|y\|_A)$. It follows that $D < \varepsilon(\|x\|_A + \|y\|_A)$. The conditions $\lambda^{\varepsilon}(c)$, $\mu^{\varepsilon}(c)$, and $\kappa^{\varepsilon}(c)$ are verified similarly. ❑

LEMMA 3.2: Let ε be a positive real number. Let $h: A \to C(I; B)$ be a $\star^{\varepsilon}$-homotopy joining functions f and g between $C^{\star}$-algebras A and B. Then f and g are $\star^{\varepsilon}$-homomorphisms.

Proof: It suffices to prove that f is a $\star^{\varepsilon}$-homomorphism because the proof for g is analogous. In other words, we must verify conditions $ne(f)$, $\alpha^{\varepsilon}(f)$, $\lambda^{\varepsilon}(f)$, $\mu^{\varepsilon}(f)$, and $\kappa^{\varepsilon}(f)$.

ADD $ne(f)$. The condition $ne(h)$ gives that for all elements x and y in A the relation

$$\|h(x) - h(y)\|_{C(I;B)} = \sup\{\|h(x)(t) - h(y)(t)\|_B \,|\, t \in I\} \leq \|x - y\|_A$$

holds. In particular, for $t = 0$, we get $\|f(x) - f(y)\|_B \le \|x - y\|_A$.

ADD $\alpha^\varepsilon(f)$. The condition $\alpha^\varepsilon(h)$ implies that

$$\|h(x+y) - h(x) - h(y)\|_{C(I;B)} = \sup\{\|[h(x+y) - h(x) - h(y)](t)\|_B | t \in I\} < \varepsilon(\|x\|_A + \|y\|_A)$$

for all elements x and y from A. In particular, for $t = 0$, we get

$$\|f(x+y) - f(x) - f(y)\|_B < \varepsilon(\|x\|_A + \|y\|_A).$$

The other conditions follow in a similar fashion. ❑

LEMMA 3.3: Let ε be a positive real number. A $\star^\varepsilon$-homomorphism $f: A \to B$ between $C^\star$-algebras induces a $\star^{2\varepsilon}$-homomorphism $\hat{f}: C(I; A) \to C(I; B)$ given by $\hat{f}(\pi) = f \circ \pi$ for every $\pi \in C(I; A)$.

Proof: We must show that the conditions $ne(\hat{f})$, $\alpha^{2\varepsilon}(\hat{f})$, $\lambda^{2\varepsilon}(\hat{f})$, $\mu^{2\varepsilon}(\hat{f})$, and $\kappa^{2\varepsilon}(\hat{f})$ hold.

ADD $ne(f)$. It suffices to prove that for all $a, b \in C(I;A)$ the relation

$$\|\hat{f}(a) - \hat{f}(b)\|_{C(I;B)} \le \|a - b\|_{C(I;A)}$$

holds. But, since f is nonexpansive, we obtain

$$\|\hat{f}(a) - \hat{f}(b)\|_{C(I;B)} = \sup\{\|\hat{f}(a)(t) - \hat{f}(b)(t)\|_B | \; t \in I\} = \sup\{\|f(a(t)) - f(b((t))\|_B | \; t \in I\} \le \sup\{\|a(t) - b(t)\|_A | \; t \in I\} = \|a - b\|_{C(I;A)}.$$

ADD $\alpha^{2\varepsilon}(\hat{f})$. Let $a, b \in C(I;A)$. Since the condition $\alpha^\varepsilon(f)$ is true, we have

$$\|f(a(t) + b(t)) - f(a(t)) - f(b(t))\|_B < \varepsilon(\|a(t)\|_A + \|b(t)\|_A)$$

for every $t \in I$. It follows that

$$\|\hat{f}(a+b) - \hat{f}(a) - \hat{f}(b)\|_{C(I;B)} = \sup\{\|[\hat{f}(a+b) - \hat{f}(a) - \hat{f}(b)](t)\|_B | t \in \mathrm{I}\} \le \varepsilon \sup\{\|a(t)_A + \|b(t)\|_A | \; t \in I\} < 2\varepsilon(\|a\|_{C(I;A)} + \|b\|_{C(I;A)}).$$

LEMMA 3.4: Let ε be a positive real number and let A, B, and C be $C^\star$-algebras. If $f: A \to B$ and $g: B \to C$ are $\star^\varepsilon$-homomorphisms, then the composition $g \circ f$ is a $\star^{2\varepsilon}$-homomorphism from A into C.

Proof: As the composition of two nonexpansive functions $g \circ f$ is a nonexpansive function. Thus, it remains to check conditions $\alpha^{2\varepsilon}(g \circ f)$, $\lambda^{2\varepsilon}(g \circ f)$, $\mu^{2\varepsilon}(g \circ f)$, and $\kappa^{2\varepsilon}(g \circ f)$.

ADD $\alpha^{2\varepsilon}(g \circ f)$. Let $x, y \in A$. Let $u = g(f(x)) + g(f(y))$, $v = g(f(x) + f(y))$, and $w = g(f(x+y))$. Since $\alpha^\varepsilon(f)$ holds, we have $\|f(x+y) - f(x) - f(y)\|_B < \varepsilon(\|x\|_A + \|y\|_A)$. It follows that $\|w - v\|_C < \varepsilon(\|x\|_A + \|y\|_A)$ because g is nonexpansive. On the other hand, $\|v - u\|_C < \varepsilon$ because $\alpha^\varepsilon(g)$ is true. The triangle inequality now implies the required estimate $\|w - u\|_C < 2\varepsilon(\|x\|_A + \|y\|_A)$.

The conditions $\lambda^{2\varepsilon}(g \circ f)$, $\mu^{2\varepsilon}(g \circ f)$, and $\kappa^{2\varepsilon}(g \circ f)$ are verified analogously. ❑

4. FUNDAMENTAL ⋆-SEQUENCES

After the preparations in Section 3, we now can introduce fundamental ⋆-sequences and define the relation of ⋆-homotopy for them. These definitions correspond to Borsuk's [3] definitions of fundamental sequence and homotopy for fundamental sequences.

Let A and B be $C^\star$-algebras. A family $\varphi = \{f_i\}_{i=1}^\infty$ of nonexpansive functions $f_i: A \to B$ is a *fundamental* α*-sequence* from A into B provided for every $\varepsilon \geq 0$ there is an $i \in \mathbb{N}$ such that $f_j \overset{\varepsilon}{\simeq}_\alpha f_i$ for every $j \geq i$. By replacing the letter α with λ, μ, κ, and ⋆, we shall get four other kinds of fundamental sequences. We use functional notation $\varphi: A \to B$ to indicate that φ is a fundamental α-sequence from A into B. Let $F_\alpha(A,B)$ denote all fundamental α-sequences $\varphi: X \to Y$. By replacing the letter α with λ, μ, κ, and ⋆, we get notation for other kinds of fundamental sequences between $C^\star$-algebras.

Two families $\varphi = \{f_i\}_{i=1}^\infty$ and $\psi = \{g_i\}_{i=1}^\infty$ of nonexpansive functions f_i, g_i: $A \to B$ are α-*homotopic* and we write $\varphi \simeq_\alpha \psi$, provided for every $\varepsilon \geq 0$ there is an $i \in \mathbb{N}$ such that $f_j \overset{\varepsilon}{\simeq}_\alpha g_j$ for every $j \geq i$. The relations $\simeq_\psi$, $\simeq_\mu$, $\simeq_\kappa$, and $\simeq_\star$ are defined analogously.

It follows from Lemma 3.1 that the relation of ⋆-homotopy is an equivalence relation on the set $F_\star(A,B)$. The ⋆-homotopy class of a fundamental ⋆-sequence φ is denoted by $[\varphi]_\star$ and the set of all ⋆-homotopy classes by $Sh_\star(A,B)$.

Let us agree that an *increasing* function f: $P \to P$ of a partially ordered set $(P, <)$ into itself is a function which satisfies $x < f(x)$ for every $x \in P$ and $x < y$ in P implies $f(x) < f(y)$. In the case when the domain and the codomain of a function f are different, the first requirement is dropped. When defining increasing functions that connect indexing sets, we repeatedly use the following simple lemma.

LEMMA 4.1: Let $\{f_1, \ldots, f_n\}$ be functions from a cofinite directed set $(M, <)$ into a directed set $(L, <)$. Then there is an increasing function $g: M \to L$ such that $g(x) \geq f_1(x), \ldots, f_n(x)$ for every $x \in M$.

The following two results provide useful tools in dealing with fundamental ⋆-sequences and their ⋆-homotopy classes. The first provides a simple method of building fundamental ⋆-sequences ⋆-homotopic to a given fundamental ⋆-sequence through "subsequences" analogous to [4, p. 49]. The second gives a characterization of the relation of $\star$-homotopy for fundamental $\star$-sequences similar to Theorems 8.1 and 2.1 in [4, pp. 57 & 206].

PROPOSITION 4.2: Let $\varphi = \{f_i\}_{i \in \mathbb{N}}$ be a fundamental $\star$-sequence between $C^\star$-algebras A and B and let $\lambda\colon \mathbb{N} \to \mathbb{N}$ be an increasing function. Then $\psi = \{g_i\}_{i \in \mathbb{N}}$, where $g_i = f_{\lambda(i)}$ for every $i \in \tilde{\mathbb{N}}$, is a fundamental $\star$-sequence from A into B $\star$-homotopic to φ.

Proof: In order to see that ψ is a fundamental $\star$-sequence, let an $\varepsilon > 0$ be given. Select an $i \in \mathbb{N}$ such that $f_v \overset{\varepsilon}{\simeq}_\star f_w$ for all $v, w \geq i$. Since $\lambda(v), \lambda(w) \geq \lambda(i) \geq i$, we get $g_v \overset{\varepsilon}{\simeq}_\star g_w$ for all $v, w \geq i$. The fundamental $\star$-sequences φ and ψ are $\star$-homotopic because for a given $\varepsilon > 0$, the index i selected above will have the property that $j \geq i$ implies $\lambda(j) \geq j \geq i$ and $g_j = f_{\lambda(j)} \overset{\varepsilon}{\simeq}_\star f_j$. ❑

For a positive real number ε and a natural number n, let

$$\varepsilon^{\langle n \rangle} = \{i \in \mathbb{N},\ i > n/\varepsilon\}.$$

The sets $\varepsilon^{\langle 1 \rangle}$ and $\varepsilon^{\langle 2 \rangle}$ are denoted by ε^* and ε^{**}, respectively.

PROPOSITION 4.3: Let $\varphi = \{f_i\}_{i \in \mathbb{N}}$ and $\psi = \{g_i\}_{i \in \mathbb{N}}$ be fundamental $\star$-sequences from a $C^\star$-algebra A into a $C^\star$-algebra B. Then φ and ψ are $\star$-homotopic if and only if there is a fundamental $\star$-sequence $\chi = \{h^i\}_{i \in \mathbb{N}}\colon A \to C(I; B)$ and an increasing function $\lambda\colon \mathbb{N} \to \mathbb{N}$ such that $h^i_0 = f_{\lambda(i)}$ and $h^i_1 = g_{\lambda(i)}$ for every $i \in \mathbb{N}$.

Proof: ($\Rightarrow$) For every $i \in \mathbb{N}$, choose $\mu(i), \nu(i) \in \mathbb{N}$ such that $u, v \geq \mu(i)$ implies that f_u and f_v are joined by a $\star^{1/i}$-homotopy k^u_v, while $u \geq \nu(i)$ implies that f_u and g_u are joined by a $\star^{1/i}$-homotopy L^u. Pick an increasing function $\lambda\colon \mathbb{N} \to \mathbb{N}$ such that $\lambda(i) \geq \mu(i), \nu(i)$ for every $i \in \mathbb{N}$. Define $h^i = L^{\lambda(i)}$ for every $i \in \mathbb{N}$.

In order to check that $\chi = \{h^i\}_{i \in \mathbb{N}}$ is a fundamental $\star$-sequence, let an $\varepsilon > 0$ be given. Pick an $i \in \varepsilon^*$. Let $u, v \geq i$. Observe that $1/i$, $1/u$, and $1/v$ are smaller than ε. We shall define a $\star^{1/i}$-homotopy $m\colon A \to C(I; C(I; B))$ joining h^u and h^v. The function m will be the composition $m = q \circ \tilde{r} \circ n$, where $n\colon A \to C(U; B)$ is a $\star^{1/i}$-homomorphism described below, U is a subset $I \times \{0\} \cup \{0, 1\} \times I$ of the square $I^2 = I \times I$, the function $\tilde{r}\colon C(U; B) \to C(I^2; B)$ is a $\star$-homomorphism induced by a retraction $r\colon I^2 \to U$, and $q\colon C(I^2; B) \to C(I; C(I; B))$ is a standard $\star$-isomorphism.

For every $x \in A$, we let $n(x)$ to be a continuous function from U into B defined by the rule

$$n(x)(s, t) = \begin{cases} k^{\lambda(u)}_{\lambda(v)}(s), & (s, t) \in I \times \{0\}, \\ L^{\lambda(u)}(t), & (s, t) \in \{0\} \times I, \\ L^{\lambda(v)}(t), & (s, t) \in \{1\} \times I. \end{cases}$$

(⇐) Let an $\varepsilon > 0$ be given. Since χ is a fundamental $\star$-sequence, there is an $i \in \mathbb{N}$ such that h^j is a $\star^\varepsilon$-homomorphism for every $j \geq i$. It follows that for such indices j, the functions $f_{\lambda(j)}$ and $g_{\lambda(j)}$ are joined by a $\star^\varepsilon$-homotopy. Hence, fundamental $\star$-sequences $\{f_{\lambda(i)}\}$ and $\{g_{\lambda(i)}\}$ are $\star$-homotopic. We now can use Proposition 4.2 and transitivity of the relation of $\star$-homotopy to conclude that φ and ψ are $\star$-homotopic. ❑

5. COMPOSITION OF $\star$-HOMOTOPY CLASSES

In order to organize $C^\star$-algebras and $\star$-homotopy classes of fundamental $\star$-sequences into a $\star$-shape category $Sh_\star$, we must define a composition for $\star$-homotopy classes of fundamental $\star$-sequences and establish it's associativity. This section is devoted solely to this task. Our definition of the composition will be with the help of two increasing functions associated with every fundamental $\star$-sequence, and will be done using representatives of $\star$-homotopy classes. We begin with the description of the first function. In Claim 5.1 we prove the existence of the second function, Claim 5.2 verifies that the sequence of functions that we associate to representatives of classes is a fundamental $\star$-sequence, Claim 5.3 shows that this definition is independent of the choice of representatives, and Theorem 5.4 at last proves that the composition is associative.

All these steps and checking are entirely elementary but somewhat tedious. However, they appear in every description of every shape theory in one form or the other. The reader is well advised to study them carefully because they contain basic techniques of working with morphisms of the $\star$-shape category.

The definition of the composition is the only tricky part in setting up the category $Sh_\star$. Our idea is to associate to every fundamental $\star$-sequence φ: $A \to B$ two increasing functions φ: $\mathbb{N} \to \mathbb{N}$ and f: $\mathbb{N} \to \mathbb{N}$. The first function associates to an index $i \in \mathbb{N}$ of the sequence $\varphi = \{f_i\}$ a much larger index $\varphi(i)$ in $\mathbb{N}$ such that f_j and f_k are joined by a $\star^{1/i}$-homotopy whenever $j, k \geq \varphi(i)$. The second function associates to an $i \in \mathbb{N}$ an element $f(i)$ of $\mathbb{N}$ such that the reciprocal value $1/f(i)$ of $f(i)$ is sufficiently small. This is a rough description of these functions and now we proceed with the details.

Let $\varphi = \{f_i\}: A \to B$ be a fundamental $\star$-sequence between $C^\star$-algebras. Let φ: $\mathbb{N} \to \mathbb{N}$ be an increasing function such that for every $i \in \mathbb{N}$ the relation $j, k \geq \varphi(i)$ implies the relation $f_j \overset{1/i}{\simeq}_\star f_k$. The multiple use of notation here can not lead to confusion provided one keeps in mind that fundamental $\star$-sequences act only between $C^\star$-algebras and that they cannot be evaluated in an index (which is a natural number).

Let $\mathcal{L}_\varphi = \{(i, j, k) \mid i, j, k \in \mathbb{N},\ \ j, k \geq \varphi(i)\}$. Then $\mathcal{L}_\varphi$ is a subset of $\mathbb{N} \times \mathbb{N} \times \mathbb{N}$ that becomes a cofinite directed set when we define that $(i, j, k) \geq (m, n, p)$

if and only if $i \geq m$, $j \geq n$, and $k \geq p$. We shall use the same notation φ for an increasing function φ: $\mathcal{L}_\varphi \to \mathbb{N}$ such that $\varphi(i, j, k) \geq \varphi(i)$ whenever $(i, j, k) \in \mathcal{L}_\varphi$.

CLAIM 5.1: There is an increasing function f: $\mathbb{N} \to \mathbb{N}$ such that:

(1) $f(i) \geq \varphi(i, \varphi(i), \varphi(i))$ for every $i \in \mathbb{N}$, and

(2) f is cofinal in φ, i.e., for every $(i, j, k) \geq \mathcal{L}_\varphi$ there is an $m \in \mathbb{N}$ with $f(m) \geq \varphi(i, j, k)$.

Proof: Let $\mathcal{D}_\varphi = \{\varphi(i, j, k) \mid (i, j, k) \in \mathcal{L}_\varphi\}$. The set $\mathcal{D}_\varphi$ is countable so that there is a surjection g: $\mathbb{N} \to \mathcal{D}_\varphi$. Let f: $\mathbb{N} \to \mathbb{N}$ be an increasing function such that $f(i) \geq g(i), \varphi(i, \varphi(i), \varphi(i))$ for every $i \in \mathbb{N}$. ❑

The above discussion shows that every fundamental $\star$-sequence φ: $A \to B$ determines two increasing functions φ: $\mathbb{N} \to \mathbb{N}$ and f: $\mathbb{N} \to \mathbb{N}$. With the help of these functions we shall define the composition of $\star$-homotopy classes of fundamental $\star$-sequences as follows.

Let $\varphi = \{f_i\}_{i \in \mathbb{N}}$: $A \to B$ and $\psi = \{g_i\}_{i \in \mathbb{N}}$: $B \to C$ be fundamental $\star$-sequences. Let $\psi \circ \varphi$ denote the collection $\chi = \{h_i\}_{i \in \mathbb{N}}$, where we define $h_i = g_{\psi(i)} \circ f_{\varphi(g(i))}$ for every $i \in \mathbb{N}$. Observe that each h_i is a nonexpansive function because it is the composition of two nonexpansive functions.

CLAIM 5.2: The collection $\psi \circ \varphi$ is a fundamental $\star$-sequence from A into C.

Proof: Let a $\sigma > 0$ be given. We must find a $u \in \mathbb{N}$ such that

$$h_v \overset{\sigma}{\simeq}_\star h_u \quad \text{for every } v \geq u. \tag{3}$$

Let $u \in \sigma^{\langle 4 \rangle}$. Consider an index $v \geq u$. In order to prove (3), we shall find an index $c \in \tilde{\mathbb{N}}$ so that

$$h_v \overset{\sigma}{\simeq}_\star g_x \circ f_c, \tag{4}$$

$$g_x \circ f_c \overset{\sigma}{\simeq}_\star g_y \circ f_c, \tag{5}$$

and

$$g_y \circ f_c \overset{\sigma}{\simeq}_\star h_u, \tag{6}$$

where $x = \psi(v)$ and $y = \psi(u)$. Repeated use of Lemma 3.1 will give (3) from the relations (4)–(6).

ADD (4) AND (6). Since $x = \psi(v)$, the triple (v, x, x) belongs to $\mathcal{L}_\psi$, so that the function g_x is joined to itself by a $\star^{1/v}$-homotopy. It follows from Lemma 3.2 that g_x is a $\star^{1/v}$-homomorphism. By the Lemma 3.3, the function g_x induces a $\star^{2/v}$-homomorphism $\hat{g_x}$: $C(I; B) \to C(I; C)$.

On the other hand, for $c \geq \varphi(g(v))$ the functions $f_{\varphi(g(v))}$ and f_c are joined by a $\star^{1/g(v)}$-homotopy K: $A \to C(I; B)$. The property (1) of the Claim 5.1 implies that $g(v) \geq \psi(v, x, x) \geq \psi(v) \geq v$. Therefore, K is a $\star^{2/v}$-homotopy too, so that by the Lemma 3.4 the composition $\widehat{g_x} \circ K$ is a $\star^{4/v}$-homotopy which realizes the relation (4). In an analogous fashion, one can show that (6) is also true.

ADD (5). Since $(u, x, y) \in \mathcal{L}_\psi$, it follows that the functions g_x and g_y are joined by a $\star^{1/u}$-homotopy $K: B \to C(I; C)$. Invoking the property (2) of Claim 5.1, choose a $t \geq v$ so that $g(t) \geq \psi(u, x, y)$. Notice that the last inequality implies $g(t) \geq u$ because $\psi(u, x, y) \geq \psi(u)$ and $\psi(u) \geq u$. Let $c \geq \varphi(g(t))$ Then f_c is a $\star^{1/g(t)}$-homomorphism so that by Lemma 3.4 the composition $K \circ f_c$ is a $\star^{2/u}$-homotopy which realizes the relation (5). ❑

We now define the composition of $\star$-homotopy classes of fundamental $\star$-sequences by the rule $[\psi]_\star \circ [\varphi]_\star = [\psi \circ \varphi]_\star$.

CLAIM 5.3: The composition of $\star$-homotopy classes of fundamental $\star$-sequences is well-defined.

Proof: Let $\kappa = \{k_i\}$ and $\lambda = \{\ell_i\}$ be fundamental $\star$-sequences $\star$-homotopic to φ and ψ, respectively, and let $\mu = \{m_i\}$, where $m_i = \ell_{\lambda(i)} \circ k_{\kappa(\ell(i))}$ for every $i \in \mathbb{N}$. We must show that fundamental $\star$-sequences χ and μ are $\star$-homotopic, where $\chi = \psi \circ \varphi$. In other words, that for every $\sigma > 0$ there is an $s \in \mathbb{N}$ such that:

$$H_t \overset{\sigma}{\simeq}_\star M_t \quad \text{for every } t \geq s. \tag{7}$$

Let a $\sigma > 0$ be given. Let $s \in \sigma^{\langle 4 \rangle}$. In order to prove (7), we shall argue that for every $t \geq s$ we can find indices $c \in \mathbb{N}$ and $u \in \mathbb{N}$ such that:

$$h_t \overset{\sigma}{\simeq}_\star g_x \circ f_c, \tag{8}$$

$$g_x \circ f_c \overset{\sigma}{\simeq}_\star g_u \circ f_c, \tag{9}$$

$$g_u \circ f_c \overset{\sigma}{\simeq}_\star \ell_u \circ f_c, \tag{10}$$

$$\ell_u \circ f_c \overset{\sigma}{\simeq}_\star \ell_u \circ k_c, \tag{11}$$

$$\ell_u \circ k_c \overset{\sigma}{\simeq}_\star \ell_y \circ k_c, \tag{12}$$

$$\ell_y \circ k_c \overset{\sigma}{\simeq}_\star m_t, \tag{13}$$

where we put $x = \psi(t)$ and $y = \lambda(t)$. From the relations (8)–(13) with the help of Lemma (3.1) we get (7).

We now describe how big c and u must be chosen for relations (8), (9), (10), and (11) to hold separately. Relations (12) and (13) are analogous to relations (9) and (8), respectively. We leave to the reader the task of making a cumulative choice for c and u which accomplishes our goal. It is important to notice that u is selected first while c is selected only once u is already known. Moreover, observe that except for relations (10) and (11), all other relations are analogous to either (4) or (5). Hence, we must treat only (10) and (11).

ADD (10). Since fundamental $\star$-sequences ψ and λ are $\star$-homotopic, there is a $u \in \mathbb{N}$ and a $\star^{\sigma/2}$-homotopy $S\colon B \to C(I; B)$ joining g_u and ℓ_u. Let $i \in (\sigma/2)^*$ and $c \geq \varphi(i)$. The function f_c is a $\star^{1/i}$-homomorphism from A into B, so that the composition $S \circ f_c$ is a $\star^{\sigma}$-homotopy joining compositions which appear in the relation (10).

ADD (11). Let $u \geq y = \lambda(t)$. Then ℓ_u is a $\star^{1/t}$-homomorphism. It follows from Lemma 3.3 that the function ℓ_t induces a $\star^{2/t}$-homomorphism $\widehat{\ell_t}\colon C(I; B) \to C(I; C)$

Since φ and κ are $\star$-homotopic, there is an index $c \in \mathbb{N}$ as large as we please so that f_c and k_c are joined by a $\star^{1/t}$-homotopy $w\colon A \to C(I; B)$. The composition $\widehat{\ell_t} \circ w$ realizes the relation (11). ❑

The following theorem is the main result in this section and a key step in the description of the $\star$-shape category $Sh_\star$.

THEOREM 5.4: The above composition of $\star$-homotopy classes of fundamental $\star$-sequences is associative.

Proof: Let A, B, C, and D be $C^\star$-algebras. Let $\varphi = \{f_i\}$, $\psi = \{g_i\}$, and $\chi = \{h_i\}$ be fundamental $\star$-sequences from A into B, from B into C, and from C into D, respectively. Let $\mu = \{m_i\}$, $\nu = \{n_i\}$, $\kappa = \{k_i\}$, and $\lambda = \{\ell_i\}$, where $m_i = g_{\psi(i)} \circ f_{\varphi(g(i))}$, $n_i = h_{\chi(i)} \circ g_{\psi(h(i))}$, $k_i = h_{\chi(i)} \circ m_{\mu(h(i))}$, and $\ell_i = n_{\nu(i)} \circ f_{\varphi(n(i))}$ for every $i \in \mathbb{N}$. We must show that fundamental $\star$-sequences κ and λ are $\star$-homotopic, i.e., that for every $\pi > 0$ there is a $p \in \widetilde{\mathbb{N}}$ such that

$$k_q \overset{\pi}{\simeq}_\star \ell_q \quad \text{for every } q \geq p. \tag{14}$$

Let a $\pi > 0$ be given. Let $p \in \pi^{\langle 8 \rangle}$. In order to prove (14), we shall show that for every $q \geq p$ we can find indices $c \in \mathbb{N}$ and $s \in \mathbb{N}$ such that:

$$k_q \overset{\pi}{\simeq}_\star h_x \circ g_y \circ f_c, \tag{15}$$

$$h_x \circ g_y \circ f_c \overset{\pi}{\simeq}_\star h_x \circ g_s \circ f_c, \tag{16}$$

$$h_x \circ g_s \circ f_c \overset{\pi}{\simeq}_\star h_z \circ g_s \circ f_c, \tag{17}$$

$$h_z \circ g_s \circ f_c \overset{\pi}{\simeq}_\star n_w \circ f_c, \tag{18}$$

and

$$n_w \circ f_c \overset{\pi}{\simeq}_\star \ell_q, \tag{19}$$

where $x = \chi(q)$, $y = \psi(\mu(h(q)))$, $z = \chi(\nu(q))$, and $w = \nu(q)$. Repeated use of Lemma 3.1 will give (14) from the relations (15)–(19).

The method of proof is similar to the proof of the Claim 5.3. We shall only describe for each of the relations (15)–(19) how large the indices u and c must be in order that this $\star^\pi$-homotopy holds. An easy exercise of putting together all these selections is once again left to the reader. Since relations (18) and (19) are analogous with relations (16) and (15), respectively, it suffices to consider only relations (15)–(17).

ADD (15). Observe that the function h_x is a $\star^{1/q}$-homomorphism while the function g_y is a $\star^{1/v}$-homomorphism, where $u = h(q)$ and $v = \mu(u)$. Since $v = \mu(u) \geq u$ and $u = h(q) \geq \chi(q, x, x) \geq \chi(q) \geq q$ according to the part (1) of Claim 5.1, by Lemma 3.4 it follows that the composition $h_x \circ g_y$ is a $\star^{2/q}$-homomorphism. If $c \geq \varphi(g(v))$, then $f_{\varphi(g(v))}$ and f_c are joined by a $\star^{1/g(v)}$-homotopy $P: A \to C(I; B)$. But, $g(v)$ is greater than q by the property (1) of Claim 1. Hence, the composition of P and the lift $\widehat{h_x \circ g_y}: C(I; B) \to C(I; D)$ of the composition $h_x \circ g_y$ is a $\star^{8/q}$-homotopy between k_q and $h_x \circ g_y \circ f_c$.

ADD (16). As above, h_x is a $\star^{1/q}$-homomorphism Since $\mu(t) \geq t$ for every $t \in \mathbb{N}$, we get $y \geq \psi(u)$. Therefore, if we take $s \geq y$, then g_y and g_s are joined by a $\star^{1/u}$-homotopy $Q: B \to C(I; C)$ However, as we have seen above, $u \geq \chi(q, x, x)$ so that the composition $\widehat{h_x} \circ Q$ of $Q$and the lift $\widehat{h_x}: C(I; C) \to C(I; D)$ is a $\star^{4/q}$-homomorphism. Finally, for $c \geq \varphi(\psi(u, y, s))$ we see that the composition $\widehat{h_x} \circ Q \circ f_c$ realizes the relation (16).

ADD (17). Since $\nu(r) \geq r$ for every $r \in \widetilde{W}$, we get $z \geq x$ so that h_x and h_z are joined by a $\star^{1/q}$-homotopy $T: C \to C(I; D)$. Let $r = \chi(q, x, z)$. Let $s \geq \psi(r)$. Then g_s is a $\star^{1/q}$-homomorphism. Let $e = \psi(r, s, s)$ and $c \geq \varphi(e)$. The composition $T \circ g_s \circ f_c$ realizes the relation (17). ❑

6. THE $\star$-SHAPE CATEGORY $Sh_\star$

For a $C^\star$-algebra A, let $\iota^A = \{I_i\}: A \to A$ be the identity fundamental $\star$-sequence defined by $I_i = id_A$ for every $i \in \mathbb{N}$. It is easy to show that for every fundamental $\star$-sequence $\varphi: A \to B$, the following relations hold.

$$[\varphi]_\star \circ [\iota^A]_\star = [\varphi]_\star = [\iota^B]_\star \circ [\varphi]_\star$$

In other words, the $\star$-homotopy classes of the identity fundamental $\star$-sequences ι^A will be identity morphisms in our $\star$-shape category. Therefore, we can summarize the above with the following theorem which was our main objective.

THEOREM 6.1: The $C^\star$-algebras as objects together with the $\star$-homotopy classes of fundamental $\star$-sequences as morphisms, the $\star$-homotopy classes $[\iota^A]_\star$ as identity morphisms, and the above composition of $\star$-homotopy classes form the $\star$-shape category $Sh_\star$.

7. SPECIAL FUNDAMENTAL $\star$-SEQUENCES

In this section, we look for conditions under which a $\star$-homotopy class has a representative of a special kind. As a corollary of these results we identify a class of $C^\star$-algebras on which $\star$-shape theory and $\star$-homotopy theory of $C^\star$-algebras coincide.

A $C^\star$-algebra B is *internally movable* provided for every $\varepsilon > 0$ there is a $\delta > 0$ such that every $\star^\delta$-homomorphism $f: A \to B$ from any $C^\star$-algebra A into B is $\star^\varepsilon$-homotopic to a $\star$-homomorphism.

A $C^\star$-algebra B is *internally calm* provided there is a $\gamma > 0$ such that $\star$-homomorphism $f, g: A \to B$ from any $C^\star$-algebra A into B which are $\star^\gamma$-homotopic are $\star$-homotopic.

A $C^\star$-algebra B is *it calm* provided there is a $\gamma > 0$ such that for every $\sigma > 0$ there is a $\tau > 0$ with the property that $\star^\tau$-homomorphisms $f, g: A \to B$ from any $C^\star$-algebra A into B which are $\star^\gamma$-homotopic are also $\star^\sigma$-homotopic.

Of course, the first and the third of the above three classes of spaces are analogues of the corresponding classes of internally movable spaces (see [2] and [19]) and calm spaces (see [6]). Also, one can prove that every semi-projective [1] separable $C^\star$-algebra has all three properties.

Let A and B be $C^\star$-algebras. A fundamental $\star$-sequence $\varphi = \{f_i\}_{i \in \mathbb{N}}$ is *regular* provided each function f_i is a $\star$-homomorphism. It is called *simple* when there is a $\star$-homomorphism g such that $g = f_i$ for every $i \in \mathbb{N}$.

THEOREM 7.1: If a $C^\star$-algebra B is internally movable, then every fundamental $\star$-sequence $\varphi = \{f_i\}_{i=1}^{\infty}: A \to B$ from any $C^\star$-algebra A into B is $\star$-homotopic to a regular fundamental $\star$-sequence.

Proof: Since B is internally movable, for every $i \in \mathbb{N}$ there is an $h(i) \geq i$ in $\mathbb{N}$ such that every $\star^{1/h(i)}$-homomorphism into B is $\star^{1/i}$-homotopic to a $\star$-homomorphism. Let $\lambda: \mathbb{N} \to \mathbb{N}$ be an increasing function such that $\lambda(i) \geq \varphi(h(i))$ for every $i \in \mathbb{N}$. The function $f_{\lambda(i)}$ is $\star^{1/h(i)}$-homomorphism so that we can select a $\star$-homomorphism

$$g_i: A \to B \text{ with } g_i \overset{1/i}{\simeq}_\star f_{\lambda(i)}, \text{ for every } i \in \mathbb{N}.$$

In order to verify that the collection $\psi = \{g_i\}_{i \in \mathbb{N}}$ is a fundamental $\star$-sequence, let an $\varepsilon > 0$ be given. Let $i \in \varepsilon^*$. For every $j \geq i$ we have

$$g_j \overset{1/j}{\simeq}_\star f_{\lambda(j)} \overset{1/h(i)}{\simeq}_\star f_{\lambda(i)} \overset{1/i}{\simeq}_\star g_i.$$

Hence, $g_j \overset{\varepsilon}{\simeq}_\star g_i$ for every $j \geq i$.

It remains to check that the fundamental $\star$-sequences φ and ψ are $\star$-homotopic.

Let an $\varepsilon > 0$ be given. Choose an index $i_0 \in \mathbb{N}$ such that $f_j \overset{\varepsilon}{\simeq}_\star f_i$ for all $i, j \geq i_0$.

Let $i \geq i_0$ be such that $\varepsilon \geq 1/i$. For every $j \geq i$, we get $g_j \overset{1/j}{\simeq}_\star f_{\lambda(j)}$ by construction, while $f_{\lambda(j)} \overset{\varepsilon}{\simeq}_\star f_j$ because $\lambda(j) \geq j \geq i_0$. Hence, $g_j \overset{\varepsilon}{\simeq}_\star f_j$ for every $j \geq i$. ❑

THEOREM 7.2: If a $C^\star$-algebra B is both internally movable and calm, then every fundamental $\star$-sequence $\varphi = \{f_i\}_{i=1}^\infty: A \to B$ from any $C^\star$-algebra A into B is $\star$-homotopic to a simple fundamental $\star$-sequence.

Proof: Since B is calm, there is a $\gamma > 0$ such that for every $\varepsilon > 0$ there is a $\delta > 0$ with the property that $\star^\gamma$-homotopic $\star^\delta$-homomorphisms into B are in fact $\star^\varepsilon$-homotopic.

Since B is also internally movable, there is an $0 < \eta < \gamma$ such that every $\star^\eta$-homomorphism into B is $\star^\delta$-homotopic to a $\star$-homomorphism. Let $i \in \eta^*$ and put $j = \varphi(i)$. Then the function f_j is a $\star^\eta$-homomorphism so that it is $\star^\gamma$-homotopic to a $\star$-homomorphism g. Let ψ denote the simple fundamental $\star$-sequence determined by the $\star$-homomorphism g.

In order to check that the fundamental $\star$-sequences φ and ψ are $\star$-homotopic, let an $\varepsilon > 0$ be given. Choose a $\delta > 0$ as above. Since φ is a fundamental $\star$-sequence, there is an index $k \geq j$ such that f_ℓ is a $\star^\delta$-homomorphism for every $\ell \geq k$. Thus, for every $\ell \geq k$ we get

$$f_\ell \overset{1/i}{\simeq}_\star f_j \overset{\gamma}{\simeq}_\star g.$$

Hence,

$$f_\ell \overset{\gamma}{\simeq}_\star g \text{ so that } f_\ell \overset{\varepsilon}{\simeq}_\star g. \quad ❑$$

THEOREM 7.3: Let B be an internally calm $C^\star$-algebra and let $\varphi = \{f\}$ and $\psi = \{g\}$ be simple fundamental $\star$-sequences from a $C^\star$-algebra A into B generated by $\star$-homomorphism f and g. If φ and ψ are $\star$-homotopic, then the f and g are $\star$-homotopic.

Proof: Since B is internally calm, there is an $\varepsilon > 0$ such that $\star^\varepsilon$-homotopic $\star$-homomorphism into B are in fact $\star$-homotopic. But, the assumption that φ and ψ are $\star$-homotopic gives $f \overset{\varepsilon}{\simeq}_\star g$. Hence, f and g are $\star$-homotopic. ❑

There is an obvious covariant functor L from the category $\mathcal{H}_\star$ of $C^\star$-algebras and $\star$-homotopy classes of $\star$-homomorphism into the $\star$-shape category $Sh_\star$. On objects the functor L is the identity while on morphisms it associates to a $\star$-homotopy class of a $\star$-homomorphism $f: A \to B$ the $\star$-homotopy class of a fundamental $\star$-sequence $\underline{f} = \{f_i\}: A \to B$, where $f_i = f$ for every $i \in \mathbb{N}$. The last three theorems imply that the functor L is a covariant equivalence of categories when we restrict to $C^\star$-algebras that have the above properties.

Let $\mathcal{A}$ denote the collection of all $C^\star$-algebras that are at the same time internally movable, internally calm, and calm. We already observed that every semiprojective $C^\star$-algebra is in the class $\mathcal{A}$.

Let $\mathcal{H}_\star^{\mathcal{A}}$ be the full subcategory of $\mathcal{H}_\star$ with objects precisely the members of the collection $\mathcal{A}$. The category $Sh_\star^{\mathcal{A}}$ is defined similarly. Let $L^{\mathcal{A}}$: $\mathcal{H}_\star^{\mathcal{A}} \to Sh_\star^{\mathcal{A}}$ be the restriction of the functor L to the category $\mathcal{H}_\star^{\mathcal{A}}$.

THEOREM 7.4: The functor $L^{\mathcal{A}}$ is a covariant equivalence of categories.

Proof: We construct a functor $M^{\mathcal{A}}$: $Sh_\star^{\mathcal{A}} \to \mathcal{H}_\star^{\mathcal{A}}$ which satisfies the relations $L^{\mathcal{A}} \circ M^{\mathcal{A}} = Id$ and $M^{\mathcal{A}} \circ L^{\mathcal{A}} = Id$. The functor $M^{\mathcal{A}}$ leaves the objects unchanged, and on morphisms, is defined as follows. Let C be a $\star$-homotopy class of fundamental $\star$-sequences between two members A and B of $\mathcal{A}$. Let φ be a representative of C and let $g: A \to B$ be a $\star$-homomorphism such that the simple fundamental $\star$-sequence ψ determined by g is $\star$-homotopic to φ. The functor $M^{\mathcal{A}}$ associates to C the $\star$-homotopy class of the $\star$-homomorphism g. It follows from the above results that this definition is correct and that $M^{\mathcal{A}}$ has the required properties. ❑

COROLLARY 7.5: On $C^\star$-algebras which are at the same time internally movable, internally calm, and calm the $\star$-homotopy theory and our $\star$-shape theory coincide.

COROLLARY 7.6: On semiprojective separable $C^\star$-algebras $\star$-homotopy theory and $\star$-shape theory coincide.

REFERENCES

1. BLACKADAR, B. 1985. Shape theory for $C^\star$-algebras. Math. Scand. **56:** 249–275.
2. BOGATYI, S. 1974. Approximate and fundamental retracts. Mat. Sbornik. **93:** 90–102.
3. BORSUK, K. 1968. Concerning homotopy properties of compacta. Fund. Math. **62:** 223–254
4. BORSUK, K. 1977. Theory of Shape. *In:* Monografie Matem. 59. Polish Scientific Publishers. Warszawa.
5. BRATTELI, O. & D.W. ROBINSON. 1979. Operator Algebras and Quantum Statistical Mechanics. Springer Verlag . New York.
6. ČERIN, Z. 1978. Homotopy properties of locally compact spaces at infinity–calmness and smoothness. Pacific J. Math. **79:** 69–91
7. ČERIN, Z. 1994. Proper shape theory. Acta Sci. Math. **59:** 679–711
8. ČERIN, Z. 1993. Homotopy groups for $C^\star$-algebras. Bolyai Society Mathematical Studies. **4:** 29–45
9. CONNES, A. & N. HIGSON. 1990. Deformations, morphismes asymptotiques et K-theorie bivariante. C.R. Acad. Sci. Paris. **313:** 163–170

10. DADARLAT, M. & A. NEMETHI. 1990. Shape theory and (connective) K-theory. J. Operator Theory. **23:** 207–291
11. EFFROS, E.G. & J. KAMINKER. 1985. Homotopy continuity and shape theory for C^*-algebras. *In:* Geometric methods in operator algebras, U. S.-Japan seminar at Kyoto 1983. Pitman.
12. KADISON, R. & R. RINGROSE. 1991. Fundamentals of the theory of operator algebras. Academic Press. New York.
13. KOHN, P. 1972. A homotopy theory of C^*-algebras. Ph. D. Dissertation, Univ. of Pennsylvania, Philadelphia.
14. MORITA, K. . 1975. On shapes of topological space. Fund. Math. **86:** 251–259
15. MURPHY, G.J. 1990. C^*-algebras and operator theory. Academic Press. New York.
16. PEDERSEN, G.K. . 1979. C^*-algebras and their automorphism groups. Academic Press. London.
17. ROSENBERG, J. 1982. The role of K-theory in noncommutative algebraic topology; Operator algebras and K-theory. Contemporary Math. **10:** 155–182.
18. ROSENBERG, J. 1990. K and KK: Topology and operator algebras. Proceedings of Symposia in Pure Mathematics. **51:** 445–480.
19. WATANABE, T. 1987. Approximate Shape I–IV. Tsukuba J. Math. **11:** 17–59.

Metrisation, Topological Groups, and Compacta

P.J. COLLINS AND P.M. GARTSIDE
St. Edmund Hall
Oxford, OX1 4AR, UK

ABSTRACT: The paper discusses metrisation and generalised metric space properties of topological groups and compacta, with special reference to recent work of Cairns on reflection theorems, of Cairns, Knight, and Rudin on monotone normality, and of Gartside and Rezničenko on set-Mal'tsev operators and stratifiable topological groups.

INTRODUCTION

It can be argued that the main thrust of general topology has been to isolate the fundamental properties of topological spaces and to investigate their interrelationships. This has led to relatively few deep theorems and a rich theory of the construction of counter-examples. However, if we consider the fundamental properties in restricted classes of spaces, then we may expect the balance between theorems and examples to swing back in the direction of proof.

In particular, consider the classes of compact spaces and topological groups. Topological spaces that arise naturally in other areas of mathematics tend either to carry an external structure (often algebraic) which interacts with the topology; or to have very strong topological properties. Thus, the theories of compacta and topological groups promise to be both rich and to have applications to many other parts of mathematics.

Given the qualitative difference between metric and nonmetric topology, it is of fundamental importance to obtain tests for metrisability in compact spaces and topological groups. To set the scene, consider the following well-known results.

WE ASSUME THROUGHOUT THAT ALL SPACES ARE TYCHONOFF. The diagonal in X^n will be denoted by Δ_n.

KATĚTOV THEOREM: [25] Every compact space X for which X^3 is hereditarily normal is metrisable.

Mathematics Subject Classification: Primary 54C60, 54D15, 54D30, 54H13; Secondary 54B99, 54D05, 54D99, 54E18.
Key words and phrases: metrisable, topological group, compact, set-Mal'tsev operator, stratifiable, monotonically normal, shallow *MN* operator, resolution.

GRUENHAGE THEOREM: [18] A compact space X is metrisable if and only if $X^2 \backslash \Delta_2$ is paracompact.

DOW REFLECTION THEOREM: [8] If every subspace of size $\leq \aleph_1$ of a countably compact space X is metrisable, then X itself is metrisable.

BIRKHOFF-KAKUTANI THEOREM: [1], [24] Every first countable topological group is metrisable.

Observe that these theorems all demonstrate how much better behaved topological groups and compacta are compared with general topological spaces. Nonetheless, it is not the case that every topological group or compact space is metrisable. It would seem natural, then, to seek tests not only for metrisability but also for topological properties as close as is possible to metrisability. Such "near metric" properties are called generalised metric properties.

The central aim of this paper is to discuss recent progress of the authors and others on some aspects of metrisation and generalised metric properties in compacta and topological groups. Indeed, the first section demonstrates an unexpected link between compact topological groups and the metrisation theorems of Katětov and Gruenhage quoted above. As well as this connexion between apparently unrelated areas of topology, the theorems of Gruenhage and Katětov are generalised beyond compact spaces and shown to be merely different aspects of a unifying theorem.

Section 2 continues the search for metrisability criteria for compacta. Instead of the classical theorems above, the starting point is the elegant reflection theorem of Dow. The meaning of reflection in this context, is that nonmetrisability of a countably compact space is "reflected" in a "small" subspace. "Small" in Dow's result means "of size $\leq \aleph_1$." However, the same question may be asked for other meanings of the word "small." In particular, Cairns has studied this problem in the case when small means "closed and nowhere dense," and his findings are presented here.

Next we consider compact spaces which are not metrisable. All the preceding results make it clear that a compact space which is "nearly" metrisable very often is metrisable. Possibly the strongest generalised metric property not automatically resulting in metrisability is monotone normality. The formal definition is given in Section 3, and it suffices for the present to know that a monotonically normal space is "strongly" normal. It is known that every metrisable space and every linearly ordered space is monotonically normal. It is also known that every compact metrisable space is the continuous image of a compact metrisable linearly ordered space. Since the continuous image of a compact monotonically normal space is again monotonically normal, it seemed natural to Nikiel to ask whether every monotonically ordered compactum is the continuous image of a linearly ordered compact space. On the other hand, linearly ordered compacta are very well understood, and their continuous images have many properties which *a priori* seem unlikely to be possessed by a compact monotonically normal space. In Section 3, appear deep and highly complex results of Rudin and Knight. Also mentioned are results of Nyikos and Purisch which, with hindsight, may be seen to be a start in this direction. These show that, in many cases, compact monotonically normal spaces are indeed continuous images of compact linearly ordered spaces. Nonetheless, a counter-example to Nikiel's question still seems plausible. In this context, Cairns' attempts at constructing compact monotonically normal spaces by resolutions are described in Section 4.

The final section considers topological groups. The Birkhoff-Kakutani theorem quoted above provides a simple test for metrisability of a topological group. Interesting results have been obtained concerning weakening the first countability condition, or the algebraic structure, but we do not deal with them here (but see [6]). Instead, we move straight on to the question of generalised metric properties of topological groups. As many generalised metric properties imply first countability, this may seem to be somewhat fruitless. Fortunately, perhaps the most powerful of all generalised metric properties, stratifiability, does not imply first countability. The definition of stratifiability is given in Section 5. For the present it is convenient to think of stratifiability as being "monotonically perfectly normal." Thus, it bears the same relation to perfect normality as monotone normality does to normality. As explained in Section 5, stratifiable spaces have powerful extension properties. These work especially well in topological groups. Section 5 contains a necessary and sufficient condition for the free Abelian group of a space to be stratifiable (due to Sipachëva) and also sufficient conditions for the space of continuous real-valued functions on a space to be stratifiable in the compact-open topology (due to Gartside and Reznichenko).

Any concepts not defined below may be found in Engelking's book [11].

1. SET-MAL'TSEV OPERATORS AND SPACES

The results in this section provide an unexpected connection between Katětov's Theorem, Gruenhage's Theorem (see the Introduction), and compact topological groups.

Recall that a space X is *dyadic* if it is the continuous image of 2^κ, some cardinal κ. We follow Comfort [7] in using Haydon's characterisation [20] as our preferred definition of Dugundji compact: a compact space X is *Dugundji compact* if every continuous mapping $f: A \to X$, where A is a closed subspace of 2^κ, can be extended to a continuous mapping $\bar{f}: 2^\kappa \to X$. From this definition, it is easy to see that Dugundji compact spaces are dyadic. It is important to know that both compact topological groups and compact metric spaces are Dugundji compact (see [22], [39], [27], and [32]); and that every nonmetrisable dyadic space contains a compact subspace which is not dyadic (see [11]).

First we indicate how closely Dugundji compactness and topological groups are linked. To do this, a structure weaker than the group structure is introduced.

DEFINITION 1.1: [14] An upper semicontinuous set-valued mapping M: $X^3 \to X$ is a *set-Mal'tsev operator* if

$$M(x, y, y) = M(y, y, x) = \{x\}$$

and is a 2-set-Mal'tsev operator if, in addition,

$$M(x, y, z) \subseteq \{x, z\}.$$

A space X is *(2-)set-Mal'tsev* if it has a (2-)set-Mal'tsev operator.

Observe that if G is a topological group, then M: $G^3 \to G$, defined by $M(x, y, z) = \{xy^{-1}z\}$, is a set-Mal'tsev operator for G. Spaces with a single-valued Mal'tsev operator have been extensively studied (see [6], [16], [33], [37], and [38]) because they provide a convenient setting for generalising many theorems about topological groups into a useful wider setting. The following result is strictly nontrivial. It is of interest from a number of viewpoints: it provides an internal characterisation of Dugundji compactness; it generalises the theorems of Kuz'minov and Pasynkov, telling us that a compact topological group is Dugundji compact; and it demonstrates that every Dugundji compact space has a weak kind of algebraic structure. Uspenskii has pointed out that the sphere S^2 is a Dugundji compact space which does not admit a single-valued Mal'tsev operator.

THEOREM 1.2: [17] A compact space is Dugundji compact if and only if it is set-Mal'tsev.

The property of being 2-set-Mal'tsev (though not that of being set-Mal'tsev) is hereditary. Thus every closed subspace of a compact 2-set-Mal'tsev space is 2-set-Mal'tsev, and hence Dugundji. As Dugundji compact spaces are dyadic, by our observation above that a hereditarily dyadic compact space is metrisable, the following metrisation theorem follows.

THEOREM 1.3: [14] Every 2-set-Mal'tsev compactum is metrisable.

Now we make the connection between the apparently rather mysterious 2-set-Mal'tsev spaces, and the theorems of Katětov and Gruenhage. Some of the ideas in the genesis of this result are due to Knight. Let us use the notation:

$$\Pi_1 = \{(x, y, y): x, y \in X\}$$

$$\Pi_3 = \{(y, y, x): x, y \in X\}$$

for subsets Π_1 and Π_3 of X^3.

THEOREM 1.4: [14] X is 2-set-Mal'tsev if and only if there exist disjoint open U,V in X^3 such that

$$\Pi_1 \setminus \Pi_3 \subseteq U \text{ and } \Pi_3 \setminus \Pi_1 \subseteq V.$$

The following corollary is then immediate.

COROLLARY 1.5: If either $X^3 \setminus \Delta_3$ or $X \times (X^2 \setminus \Delta_2)$ is normal, then X is 2-set-Mal'tsev.

We are now in a position to state one of the main results in this paper.

THEOREM 1.6: [14] The following statements about a compact space X are equivalent:

(i) X is metrisable,
(ii) X is 2-set-Mal'tsev,
(iii) X^3 is hereditarily normal,
(iv) $X^3 \setminus \Delta_3$ is normal,
(v) $X \times (X^2 \setminus \Delta_2)$ is normal,
(vi) $X^2 \setminus \Delta_2$ is paracompact.

NOTE: (iii) $\Leftrightarrow$ (i) is Katětov's Theorem and (i) $\Leftrightarrow$ (vi) is Gruenhage's Theorem.

Proof: Clearly (iii) implies both (iv) and (v), and (vi) implies (v) because the product of a compact space with a paracompact space is paracompact. Corollary 1.5 demonstrates that (ii) follows from either (iv) or (v), and (i) clearly implies all the other statements, in particular, (iii) and (vi). The circle is completed by applying Theorem 1.3 to show that (ii) implies (i). ❑

Theorem 1.6 is the promised unification of the Gruenhage and Katětov metrisation theorems. Its proof is interesting in that it provides a link, via set-Mal'tsev spaces and Dugundji compacta, to compact topological groups. However, the proofs of the classical metrisation theorems are direct and consequently much simpler. Aware of this, Gartside and Reznichenko went on to provide direct proofs. These direct proofs allowed them to prove similar results for much broader classes of spaces. For completeness, the definitions of Σ-spaces and σ-spaces are now included. But, from the point of view of metrisation, the more significant definition is that of an M-space. Observe that every countably compact space is an M-space, as is every metrisable space.

(a) X is a *Σ-space* if there exists a σ-locally finite closed collection $\mathcal{F}$ in X and a cover C of closed countably compact sets such that

$$C \subseteq U,\ C \in \mathcal{C},\quad U \text{ open} \Rightarrow C \subseteq F \subseteq U, \quad \text{for some } F \in \mathcal{F}.$$

(b) X is a *σ-space* if X has a σ-discrete network.

(c) X is an M-space if it is the quasiperfect pre-image of a metric space (a map being *quasi*-perfect if it is closed and has *countably* compact fibres).

THEOREM 1.7: [14] Suppose X is a 2-set-Mal'tsev space. Then
(i) if X is a Σ-space, then X is a σ-space;
(ii) if X is an M-space, then X is metrisable.

Part (ii) of Theorem 1.7 follows from Part (i) because every M-space is a Σ-space and every M-space which is also a σ-space is metrisable. Theorem 1.3 is then immediate (as compact spaces are clearly M-spaces).

2. CAIRNS' REFLECTION THEOREMS

As mentioned in the introduction, here reflection is meant to indicate that a property held by all "small" subspaces is "reflected" in the fact that it is also held by the whole space. The theorem of Dow, cited in the introduction, uses reflection on subspaces of size $\leq\aleph_1$. In his thesis, Cairns also investigates reflection. However, this time, small subspaces are not determined by their cardinality but by the property of being closed nowhere dense. Unlike Dow's result, Cairns discovered that his results are independent of the axioms of set theory. More precisely they depend on the existence, or otherwise, of certain types of Lusin spaces.

DEFINITION 2.1: [3] A space X has *boundary-$\mathcal{P}$* if every closed nowhere dense subspace has property $\mathcal{P}$.

Recall that a space is *Lusin* if it is uncountable, has at most countably many isolated points and all its nowhere dense subspaces are countable, and that a *Lusin set* is a Lusin subspace of the reals. Lusin sets have interesting category properties, while Lusin spaces may be used to construct *L-spaces*, that is to say, hereditarily Lindelöf nonseparable spaces.

THEOREM 2.2: [3] (a) If there are no Lusin spaces, every compact boundary-metrisable space is metrisable.

(b) If there exists a Lusin set, then there also exists a compact boundary-metrisable nonmetrisable space.

Though unrelated to metrisation and the main thrust of this paper, the following results provide an interesting comparison.

THEOREM 2.3: [3] If there exist no L-spaces, then every boundary-separable space is separable.

THEOREM 2.4: [3] Every boundary-hereditarily separable space is separable if and only if there are no nonseparable Lusin spaces.

We are grateful to Murat Tuncali for drawing the following recent result to our attention. Recall that a space is *rim-$\mathcal{P}$* if it has a base of open sets with boundaries having property $\mathcal{P}$. Recall, also, that a space is *scattered* if every subspace contains an isolated point.

THEOREM 2.3: [9] There exists a rim-scattered, locally connected continuum which is not rim-countable.

3. MONOTONICALLY NORMAL COMPACTA

Monotone normality, although at first glance quite innocuous, has turned out to be one of the most subtle generalised metric properties. The current paper is not intended to cover all recent advances concerning monotonically normal spaces, we focus our attention on compact monotonically normal spaces. The interested reader is warmly encouraged to investigate the survey articles [5] and [19].

DEFINITION 3.1: [21] A space X is *monotonically normal*, or *MN*, if to each x in X and each open subset U of X, containing x, there is an open set $V(x, U)$ such that

(i) $x \in V(x, U) \subseteq U$;

(ii) $x \in U \subseteq U'$, U and U' open $\Rightarrow V(x, U) \subseteq V(x, U')$;

(iii) $V(x, X\setminus\{y\}) \cap V(y, X\setminus\{x\}) = \varnothing$ whenever $x \neq y$;

In these circumstances $V(.,\ .)$ is called an *MN operator*.

To see that every monotonically normal space is normal, let A and B be disjoint closed sets in a space X, and suppose $V(.,\ .)$ is as above in the definition of monotone normality. Then it is easily checked that

$$U = \bigcup_{a \in A} V(a, X\setminus B) \quad \text{and} \quad V = \bigcup_{b \in B} V(b, X\setminus A)$$

are disjoint open sets containing A and B, respectively. Of course every compact space is normal. However, rather few compact spaces are monotonically normal. In trying to pin down precisely how compact monotonically spaces arise, we are led to Nikiel's conjecture. For notational convenience, we write *CICLOTS* for "continuous image of a compact linearly ordered space."

PROBLEM 3.2: [29] Is every compact *MN* space a *CICLOTS*?

The reader is reminded that a linearly ordered topological space is *MN*, and that the continuous image of a compact monotonically normal space is again monotonically normal. Examples of nonmetrisable compact *MN* spaces include the double arrow space, compact uncountable ordinals, the Alexandrov duplicate of the unit interval, and the one-point compactification of an uncountable discrete space. At first (and even second and third glances!), Nikiel's conjecture appears to be highly optimistic. The authors of the present paper consider the proofs of recent results of Rudin and Knight to be among the most intricate in general topology and the theorems themselves to be highly significant. The first step was due to Rudin.

THEOREM 3.3: [34] Every separable zero-dimensional compact *MN* space is a CICLOTS.

Merging earlier work he had done on the possible structure of a counter-example to Nikiel's problem and Rudin's theorem, Knight has established a more general theorem which also includes some results of Nyikos and Purisch (see [30]). To state the theorem, we require a technical definition.

Suppose $V(.,.)$ is an *MN* operator. We use the notation

$$V^1(x, U) = V(x, U)$$

$$V^{n+1}(x, U) = V(x, V^n(x, U))$$

Let ∂A denote the boundary of a set A.

DEFINITION 3.3: [26] An *MN* operator $V(.,.)$ is *shallow* if, given sequences (x_n) of distinct points, (k_n) of positive integers, and (U_n) of open sets such that $x_n \in U_n$ for each n and $x_{n+1} \in V^{k_n}(x_n, U_n)$, then

$$\left| \partial \bigcap_{n<\omega} V^{k_n}(x_n, U_n) \right| \leq 1.$$

An intuitive description of shallow *MN* operators is elusive: one can regard *MN* spaces which have no shallow operator as being "uncountably locally complicated." Knight has observed to the authors that there are compact linearly ordered spaces with no shallow *MN* operators. However, one easily observes that all scattered spaces and all metric spaces have shallow *MN* operators.

THEOREM 3.4: [26] Suppose that the compact space X has a shallow *MN* operator. Then X is a CICLOTS.

Knight's result also encompasses an earlier result of Nyikos and Purisch [30] to the effect that compact scattered *MN* spaces are CICLOTS. More recently, Purisch strengthened the result by showing that any compact *MN* space, in which each component is a CICLOTS and the union of the boundaries of the components is scattered, is a CICLOTS. This strongly suggests that some sort of recursive approach to proving that a space is a CICLOTS ought to be possible.

The authors are grateful to Knight for the following outline of the principle concepts behind his proof. The key idea is to construct an inverse system of compact pre-images of a compact space X, which has a shallow MN operator. These pre-images are intended to become closer and closer to being linearly ordered. At the first stage of the recursion, one cuts X up into (not necessarily disjoint) compact subsets which are arranged into a line. At the second stage, one applies this process to each of the pieces, and so on.

The first ω stages of such a process could, of course, be applied to any compact space. The problems (and interest!) start when one attempts the first limit stage, which is the inverse limit of the preceding stages. In general, the order which has been recursively constructed hitherto, breaks down. We require a shallow MN operator to avoid this occuring.

Two cases need to be examined. The first is when the space is hereditarily paracompact. In this case, Knight applies Rudin's result to show that the space is a continuous image of a compact monotonically normal space, "most" of which is zero-dimensional and separable. If the space is not hereditarily paracompact, then it contains a homeomorphic copy of an uncountable cardinal, which can be detached from the space and ordered. Repeating this operation, one is left with a remainder which is hereditarily paracompact, and we are back in the first case.

4. ON THE EXISTENCE OF MONOTONICALLY NORMAL RESOLUTIONS

The technique of resolutions was introduced by Fedorčuk.

DEFINITION 4.1: [12] (see [40]) Suppose that X is a space, that $\{Y_x: x \in X\}$ is a family of spaces and that $f_x: X\backslash\{x\} \to Y_x$ is a continuous mapping, for every x in X. For $x \in U$, U open in X, and V open in Y_x, define the triple $\langle x, U, V\rangle$ by

$$\langle x, U, V\rangle = (\{x\} \times V) \cup \bigcup (\{x'\} \times Y_{x'}: x' \in U \cap f_x^{-1}(V)).$$

The collection of all such $\langle x, U, V\rangle$ is a base for a topology $\mathcal{T}$ on

$$Z = \bigcup_{x\in X} (\{x\} \times Y_x).$$

The space $(Z, \mathcal{T})$ is called "the *resolution* of X at each point $x \in X$ into Y_x by the mapping f_x."

For the rest of this section, we use the notation of Definition 4.1. The fundamental theorem on resolutions ([12], or see [40]) is that, if X and each Y_x are compact, then the resolution Z is also compact. As Watson has convincingly demonstrated, resolutions provide a powerful framework in which the theory of the construction of topological spaces may be studied. In the light of Nikiel's conjecture, it is of some interest to determine if a compact MN space which is not a CICLOTS may be constructed using resolutions. We note that double arrow spaces and Alexandroff duplicates of intervals are compact *MN* spaces obtained by resolving points of the unit interval into the 2-point space. The occurrence of the number "2" is not an accident, as we shall see. Nikiel and Treybig have investigated circumstances where the resolution is not monotonically normal. Let ran f_x denote the range of the function f_x.

THEOREM 4.2: [31] If $|\text{ran } f_x| > 2$ for uncountably many x, then Z, if separable, cannot be monotonically normal.

In contrast, consider the following result of Cairns where, under natural conditions, a compact *MN* resolution does exist.

THEOREM 4.3: [3] Suppose X is a locally connected *MN* continuum and that each Y_x is *MN* and such that

(a) if x is a non-cut-point of X, then Y_x is the singleton $\{x\}$

(b) if x is a cut-point of X, then $|\text{ran } f_x| \leq 2$ and each component of $X \setminus \{x\}$ is mapped to a singleton by f_x.

Then Z is *MN*.

Unfortunately for any hopes of constructing a compact *MN* space which is not a CICLOTS via Theorem 4.3, Cairns has also shown the space Z in 4.3 is a CICLOTS whenever X and all the Y_x's are CICLOTS.

5. STRATIFIABLE TOPOLOGICAL GROUPS

We now move to consider some important classes of spaces which are not metrisable. Because many generalised metric spaces are first countable (for example, Moore spaces) and first countable topological groups are metrisable (see the introduction), rather few generalised metric space properties are of interest in the context of topological groups. One property that is, is stratifiability.

DEFINITION 5.1: [4] Let X be a space. Then X is *stratifiable* provided that for each closed subset A of X and integer $n \in \omega$, there is assigned an open set $G(n, A)$, such that

(i) $\bigcap_{n \in \omega} G(n, A) = A = \bigcap_{n \in \omega} \overline{G(n, A)}$, and

(ii) $G(n, A) \subseteq G(m, B)$ whenever $A \subseteq B$ and $n \geq m$.

Observe that a space is perfectly normal if and only if it has $G(n, A)$'s satisfying condition (i). Condition (ii) is a monotonicity requirement, and it is in this sense that stratifiable spaces may be considered to be "monotonically perfectly normal." Alternative characterisations of stratifiability stress the connexion with metrisable spaces via the Nagata-Smirnov metrisation theorem.

Certainly, stratifiable spaces share many of the desirable features of metrisable spaces. For example, they possess good extension properties. So that, every closed subspace of a strongly zero-dimensional stratifiable space is a retract. These extension properties are especially effective in locally convex topological vector spaces.

Write $C(X,Y)$ for the set of all continuous maps from a space X into another space Y. We may consider this set as a topological space when endowed either with the pointwise topology, or the compact-open topology.

DUGUNDJI–MICHAEL–BORGES THEOREM: ([10], [28], [2]) Let X be stratifiable, A a closed subspace of X and let L be a locally convex topological vector space. Then, there is a map $e: C(A, L) \to C(X, L)$ such that

(1) $e(f)|A = f$

(2) $e(f)(X)$ is contained in the closed convex hull of $f(A)$

(3) e is linear

(4) e is continuous when the function spaces are both given either the pointwise topology or the compact-open topology.

Thus, for example, every closed convex subspace of a stratifiable locally convex topological vector space is a retract. This indicates the desirability of determining when topological groups in general, and locally convex topological vector spaces in particular, are stratifiable.

A large step forward as regards topological groups is due to Sipachëva. We recall that, given a space X, the free Abelian topological group of X, denoted $A(X)$, is algebraically the free Abelian group over the set X, and has the coarsest compatible group topology such that, whenever f is a continuous map of X into an Abelian topological group G, the natural extension of f, denoted Af, to a group homomorphism from $A(X)$ to G is continuous. As is clear from the definition, free Abelian topological groups play as fundamental a role in the theory of Abelian topological groups as free Abelian groups play in the theory of Abelian groups. It is also clear that the description of the free Abelian topological group topology is not easy to construct. For this reason, any nontrivial result about the free Abelian topological group of a space is rarely straightforward to prove. Sipachëva's result is no exception.

THEOREM 5.2: [36] The free Abelian topological group of a stratifiable space is stratifiable.

Thus, every stratifiable space can be embedded in an Abelian stratifiable topological group. We can also deduce that every strongly zero-dimensional stratifiable space is the retract of an (Abelian) topological group. Every such space can easily be seen to be Mal'tsev (see Section 1). Heath has informed the authors that, on studying Sipachëva's proof, he can show that the free topological group over a stratifiable space is stratifiable. (The free topological group of a topological space is defined analogously to the definition given above for the free Abelian topological group.)

Now we discuss some recent results of Gartside and Rezničenko on certain topological vector spaces. To set the scene, we recall that previously Gartside [13] and, independently, Shkarin [35] had shown that $C_p(X)$, the set of continuous real-valued functions on a space X, with the topology of pointwise convergence, contains a dense stratifiable subspace if and only if X is countable, in which case $C_p(X)$ is in fact metrisable. Similarly, a Banach space in its weak topology is stratifiable only if it is metrisable.

The situation for function spaces with the compact-open topology is significantly more complex. Let $C_k(X)$ denote the set of continuous real-valued functions on X, with the compact-open topology.

THEOREM 5.3: [15] If X is a σ-compact, completely metrisable space, then $C_k(X)$ is stratifiable.

Among metrisable spaces X, $C_k(X)$ is metrisable if and only if X is locally compact. A typical example of a subspace of the plane which is σ-compact and completely metrisable, but not locally compact, is the open unit disc along with one point on its boundary. Denoting such a space Y, $C_k(Y)$ is stratifiable but not metrisable.

An alternative description of σ-compact completely metrisable spaces is that they are the separable metrisable spaces with the property that every closed subspace contains a point of local compactness. Comparing this with the characterisation of metrisability of $C_k(X)$ mentioned above, it is reasonable to conjecture that this result is close to best possible.

CONJECTURE 5.4: [15] If X is separable metrisable and $C_k(X)$ is stratifiable, then X is σ-compact and completely metrisable.

Although stratifiability of the entirety of $C_k(X)$ appears to impose very strong conditions on X, in more general circumstances many subspaces of $C_k(X)$ are always stratifiable.

THEOREM 5.5: [15] If X is separable and metrisable, then every countable subset of $C_k(X)$ is stratifiable.

It is interesting here to compare pointwise and compact convergence. If a *dense* subspace of $C_p(X)$ is stratifiable, then X is countable and so $C_p(X)$ is metrisable (see [13]). However, if X is any separable metrisable space which is not locally compact, then $C_k(X)$ is not metrisable but does contain a dense stratifiable subspace.

The following conjecture now seems rather natural. However, the method of proof of Theorem 5.5 relies on the fact that, for any countable family of continuous (bounded) real-valued functions on a separable metrisable space X, say, there is a metrisable compactification of X to which all the members of the countable family may be extended. The analogous result for σ-compact families is false.

CONJECTURE 5.6: [15] If X is separable and metrisable, then every σ-compact subset of $C_k(X)$ is stratifiable.

All of these results depend on a simple characterisation of stratifiability in separable topological groups. This characterisation is in a similar vein to the Birkhoff-Kakutani metrisation theorem for topological groups. The interested reader is referred for details to [15].

CONCLUSIONS AND OPEN PROBLEMS

We consider that the results outlined in this paper in many ways speak for themselves and, collectively, support the authors' contention that the study of metrisability and near metric properties in compact spaces and topological groups is vibrant, rich and deep. However, in each area touched on above, there remain a number of unresolved questions. We mention a few which we consider promising.

Metrisation of Compacta

QUESTION A: Is every pseudocompact 2-set-Mal'tsev space metrisable?

Gartside and Rezničenko show that every countably compact 2-set-Mal'tsev space is metrisable. But the gap between countably compact and pseudocompact can be hard to bridge.

QUESTION B: For what other types of subspaces will a reflection theorem (such as Cairns' Theorem 2.2) hold?

Question B is rather vague. A more specific version might be: suppose X is a compact space with a "nice" measure, and every measure zero subspace of X is metrisable, then is it true that X is metrisable?

Monotonically Normal Compact Spaces

Concerning Nikiel's Conjecture 3.2, the authors tentatively suggest that a careful examination of Knight's proof of Theorem 3.4 will lead to a better understanding of compact spaces with a shallow *MN* operator, thence to the construction of compact *MN* spaces without shallow operators, and eventually to a counter-example to the Nikiel conjecture.

Although the results of Section 4 place strong restrictions on a possible counter-example to the Nikiel conjecture, extending the definition of resolutions to multivalued functions, over sets, may also provide some hope of progress.

Stratifiable Groups

Besides Conjectures 5.4 and 5.6, perhaps the most interesting problem in this area is whether the free locally convex topological vector space of a stratifiable space is stratifiable. (The free locally convex topological vector space of a space is defined analogously to the free Abelian topological group, with the space forming a Hamel basis.) A positive answer would have the following interesting corollary: if X is metrisable and $C_p(X)$ is linearly homeomorphic to $C_p(Y)$, then Y is stratifiable. Note that there are two countable spaces, one of which is stratifiable, the other not, with linearly homeomorphic function spaces.

REFERENCES

1. BIRKHOFF, G. 1936. A note on topological groups. Compositio Math. **3:** 427–430.
2. BORGES, C.J.R. 1966. On stratifiable spaces. Pacific J. Math. **17:** 1–16.
3. CAIRNS, P.A. 1995. Boundary properties and construction techniques in general topology. Doctoral thesis. Oxford.
4. CEDER, J.G. 1961. Some generalizations of metric spaces. **11:** 105–125.
5. COLLINS, P.J. 1997. Monotone normality. Topol. Appl. To appear.
6. COLLINS, P.J. & P.M. GARTSIDE. 1997. Mal'tsev spaces, retral spaces and rectifiable diagonals. To appear.
7. COMFORT, W.W., K.-H. HOFMANN, & D. REMUS. 1992. Topological groups and semigroups. *In:* Recent Progress in General Topology (M. Husek & J. van Mill, Eds.). North Holland. Amsterdam. 239–274.
8. DOW, A. 1988. An empty class of nonmetric spaces. Proc. Amer. Math. Soc. **104:** 999–1001.
9. DROZDOVSKII, A. & V.V. FILIPPOV. 1994. An example of a rim-rarefied locally connected continuum that is not rim-countable. Mat. Sbornik. **185:** (10) 27–38 (in Russian).
10. DUGUNDJI, J. 1951. An extension of Tietze's theorem. Pacific J. Math. **1:** 353–367.
11. ENGELKING, R. 1989. General Topology. Heldermann Verlag. p. 232, part (i).
12. FEDORČUK, V.V. 1968. Bicompacta with noncoinciding dimensionalities. Soviet Math. Dokl. **9:** (5) 1148–1150.
13. GARTSIDE, P.M. 1993. Monotonicity in analytic topology. Doctoral thesis. Oxford,.
14. GARTSIDE, P.M. & E.A. REZNIČENKO. 1995. Around Katetov's theorem. Preprint.
15. GARTSIDE, P.M. & E.A. REZNIČENKO. 1995. Stratifiability of $C_K(X)$. Preprint.
16. GARTSIDE, P.M., E.A. REZNIČENKO, & O. SIPACHËVA. Mal'tsev and retral spaces. Top. Appl. To appear.
17. GARTSIDE, P.M., E.A. REZNIČENKO, & V.V. USPENSKII. 1995. Topological consequences of algebraic structure. Preprint.
18. GRUENHAGE, G. 1984. Covering properties of $X^2\backslash\Delta$, *W*-sets and compact subsets of *S*-products. Topology Appl. **17:** 287–304.
19. GRUENHAGE, G. 1992. Generalized metric spaces and metrization. *In:* Recent Progress in General Topology (M. Husek & J. van Mill, Eds.). North Holland. Amsterdam. 239–274.
20. HAYDON, R.G. 1974. On a problem of Pelczyński: Milutin spaces, Dugundji spaces and *AE* (*O*-dim). Studia Math. **52:** 23–31.

21. HEATH, R.W., D.J. LUTZER, & P.L. ZENOR. 1973. Monotonically normal spaces. Trans. Amer. Math. Soc. **178:** 481–493.
22. IVANOVSKII, L.N. 1958. On a hypothesis of P.S. Alexandroff. Doklady Akad. Nauk SSS. **123:** 785–786 (in Russian).
23. JASCHENKO, I. 1994. Private communication.
24. KAKUTANI, S. 1936. Über die Metrisation der Topologischen Gruppen. Proc. Imperial Acad. Tokyo. **12:** 82–84.
25. KATĚTOV, M. 1948. Complete normality of Cartesian products. Fund. Math. **36:** 271–274.
26. KNIGHT, R.W. 1996. Compact MN spaces. Preprint.
27. KUZ'MINOV, V. 1959. On a hypothesis of P.S. Alexandroff in the theory of topological groups. Doklady Akad. Nauk SSSR N.S. **125:** 727–729.
28. MICHAEL, E.A. 1953. Some extension theorems for continuous functions. Pacific J. Math. **3:** 789–806.
29. NIKIEL, J. 1986/87. Some problems on continuous images of compact ordered spaces. Q and A in Gen. Topology. **4:** 117–128.
30. NYIKOS, P.J. & S. PURISCH. 1989. Monotone normality and paracompactness in scattered spaces. Annals of the New York Acad. Sci. **552:** 124–137.
31. NIKIEL, J. & L.B. TREYBIG. 1995. Null families of subsets of monotonically normal compacta. Preprint.
32. PASYNKOV, B.A. 1976. On spaces with a compact group of transformations. Soviet Math. Dokl. **17:** 1522–1526.
33. REZNIČENKO, E.A. & V.V. USPENSKII. 1995. Private communication.
34. RUDIN, M.E. 1995. Private communication.
35. SHKARIN, S.A. 1995. Private communication.
36. SIPACHËVA, O. 1993. On the stratifiability of free topological groups. Top. Proc. **18:** 271–311.
37. USPENSKII, V.V. 1982. The topological group generated by a Lindelof *S*-space has the Souslin property. Soviet Math. Dokl. **26:** 166–169.
38. USPENSKII, V.V. 1990. Topological groups and Dugundji compacta. Math. USSR Sbornik. **67:** 555–580.
39. VILENKIN, N.Y. 1958. On the dyadicity of the group space of bicompact commutative groups. Uspehi Mat. Nauk. **13:** 79–80 (in Russian).
40. WATSON, W.S. 1992. The construction of topological spaces: planks and resolutions. *In:* Recent Progress in General Topology (M. Husek & J. van Mill, Eds.). North Holland. Amsterdam. 683–757.

Intervals of Totally Bounded Group Topologies

W.W. COMFORT[a] AND DIETER REMUS[b]

[a]*Department of Mathematics*
Wesleyan University
Middletown, Connecticut 06459
and
[b]*Institut für Mathematik*
Universität Hannover
Welfengarten 1, D-30167
Hannover, Germany

ABSTRACT: Let G be a group and let $\mathcal{T}_i$ $(i = 0, 1)$ be totally bounded Hausdorff group topologies on G such that $\mathcal{T}_0 \subseteq \mathcal{T}_1$ and $w(G, \mathcal{T}_i) = \alpha_i$ with $\omega \le \alpha_0 < \alpha_1$. Let $[\mathcal{T}_0, \mathcal{T}_1]$ denote the (partially ordered) "interval" of group topologies $\mathcal{T}$ for G such that $\mathcal{T}_0 \subseteq \mathcal{T} \subseteq \mathcal{T}_1$. Then:

(a) every cardinal γ such that $\alpha_0 \le \gamma < \alpha_1$ satisfies $|\{\mathcal{T} \in [\mathcal{T}_0, \mathcal{T}_1]: w(G, \mathcal{T}) = \gamma\}| \ge \gamma^+$; and

(b) $[\mathcal{T}_0, \mathcal{T}_1]$ contains a well-ordered set which is order-isomorphic to the cardinal number α_1.

The present proof, based in part on the Peter-Weyl-van Kampen theorem, rests on the fact that every compact Hausdorff group K with $w(K) = \alpha \ge \omega$ contains homeomorphically the space $\{0, 1\}^\alpha$; thus the present emphasis is topological. In contrast, the second-listed author had earlier achieved the same result using the theory of unitary representations (Habilitationsschrift, Fall, 1995).

1. NOTATION, DEFINITIONS, AND PRELIMINARIES

The symbols α, β, and γ (sometimes with subscripts) denote infinite cardinals, and ζ, η, and ξ denote ordinals. The cardinal successor of a cardinal α is denoted α^+, and the ordinal successor of an ordinal ζ is denoted $\zeta + 1$. Of course for each ordinal ζ we have $|\zeta| \le \zeta < \zeta + 1 < |\zeta|^+$.

The notation $A \subset B$ means that A is a proper subset of B. The notation $A \subseteq B$ means $A \subset B$ or $A = B$.

It is well known (cf. [10, Section 8]) that every Hausdorff topological group is a completely regular space, i.e., is a Tychonoff space. In this paper, except for a brief remark in 2.7(c), we consider only Hausdorff topological groups.

The identity of a group G is denoted 1 or 1_G. When G is a topological group, the set of open neighborhoods of 1_G is denoted by $\mathcal{N}_G(1)$, or by $\mathcal{N}(1)$ if confusion is unlikely.

Mathematics Subject Classification: Primary 22A05, 54H11, 54A10.
Key words and phrases: topological group, group topology; totally bounded group topology, precompact group topology; weight, local weight; Weil completion.

We say that a topological group G is *totally bounded* if for every $U \in \mathcal{N}(1)$ there is a finite subset F of G such that $G = FU$.

NOTATION: For a topological group G, the weight and the local weight (at any of its points) of G are denoted $w(G)$ and $\chi(G)$, respectively.

NOTATION: Let G be a group. Then $\mathcal{B}(G)$ denotes the class of totally bounded group topologies on G. For a cardinal number α, we write

$$\mathcal{B}_\alpha(G) = \{\mathcal{T} \in \mathcal{B}(G)\colon w(G, \mathcal{T}) = \alpha\}.$$

NOTATION: Given a group G and $\mathcal{T} \in \mathcal{B}(G)$, the *Weil completion* [18] of the totally bounded group $\langle G, \mathcal{T}\rangle$, denoted $\overline{\langle G, \mathcal{T}\rangle}$, is the completion of G in (say) its left uniformity. This is in a natural sense the only compact Hausdorff topological group in which $\langle G, \mathcal{T}\rangle$ is a dense topological subgroup.

LEMMA 1.1: Let G be an infinite totally bounded group with Weil completion $\overline{G}$. Then the cardinal numbers $\chi(G)$, $\chi(\overline{G})$, $w(G)$ and $w(\overline{G})$ are equal.

Proof: This is well known. For $\chi(G) = \chi(\overline{G})$ see [3, (2.7)], and for

$$\chi(G) \le w(G) \le w(\overline{G}) = \chi(\overline{G})$$

see [3, (3.5(ii)]). ❑

LEMMA 1.2: Let G be a group and let $\mathcal{T}_i \in \mathcal{B}(G)$ $(i = 0, 1)$ satisfy $\mathcal{T}_0 \subseteq \mathcal{T}_1$. Then $w(G, \mathcal{T}_0) \le w(G, \mathcal{T}_1)$.

Proof: The (continuous) identity function id: $(G, \mathcal{T}_1) \twoheadrightarrow (G, \mathcal{T}_0)$ extends continuously to $\overline{\text{id}}\colon \overline{(G, \mathcal{T}_1)} \twoheadrightarrow \overline{(G, \mathcal{T}_0)}$, and since a continuous surjection between compact Hausdorff spaces cannot raise weight (cf. [9, (3.1.22)]) we have from Lemma 1.1 that

$$w(G, \mathcal{T}_1) = w\,\overline{(G, \mathcal{T}_1)} \ge w\,\overline{(G, \mathcal{T}_0)} = w(G, \mathcal{T}_0). \quad ❑$$

NOTATION 1.3: Let X be a set and let $\{Y_i\colon i \in I\}$ be a set of topological spaces. If for each $i \in I$ a function $f_i\colon X \to Y_i$ is given, then with $\mathcal{F} = \{f_i\colon i \in I\}$ we denote by $e_\mathcal{F}$ the *evaluation map* from X into $\Pi_{i \in I} Y_i$ defined as usual: $e_\mathcal{F}(x)_i = f_i(x) \in Y_i$ for $x \in X$, $i \in I$, $f_i \in \mathcal{F}$. In this context we reserve the notation $\mathcal{T}_\mathcal{F}$ to denote the smallest topology on X making $e_\mathcal{F}$ continuous. If the spaces Y_i are Hausdorff spaces, then $\mathcal{T}_\mathcal{F}$ is a Hausdorff topology if and only if the family $\mathcal{F}$ separates points of X.

The following version of Kelley's evaluation lemma [12, (4.5)] is taken from Willard [19, (8.12)].

LEMMA 1.4: Let $\langle X, \mathcal{T}\rangle$ be a Hausdorff space, for $i \in I$ let f_i be a continuous function from X to a Hausdorff space Y_i, and set $\mathcal{F} = \{f_i\colon i \in I\}$. Then

$\mathcal{T} = \mathcal{T}_{\mathcal{F}}$ if and only if the evaluation map $e_{\mathcal{F}}: X \to \Pi_{i \in I} Y_i$ is a homeomorphism from X onto $e_{\mathcal{F}}[X] \subseteq \Pi_{i \in I} Y_i$. ❑

2. THE INTERVALS $[\mathcal{T}_0, \mathcal{T}_1]$

We denote by $\mathfrak{U}_n$ the group of $n \times n$ complex unitary matrices, and we write $\mathfrak{U} = \Pi_{n < \omega} \mathfrak{U}_n$. For a topological group $G = \langle G, \mathcal{T} \rangle$ we denote by $\mathbb{H}(G)$ (or by $\mathbb{H}(G, \mathcal{T})$ if specificity is required) the set of continuous homomorphisms from G into $\mathfrak{U}$. It is known that every infinite compact group G satisfies $|\mathbb{H}(G)| \geq \omega$. When G is Abelian this follows from Pontrjagin-van Kampen duality; when G is non-Abelian there is an irreducible continuous representation $h: G \to \mathfrak{U}_n$ for some $n \geq 2$, and an elementary application of Schur's Lemma (as exposed for example in [11, (27.9) & (27.10)]) then shows that in fact $|\mathbb{H}(G)| \geq \mathfrak{c}$.

LEMMA 2.1: Let G be a totally bounded group with $w(G) = \alpha \geq \omega$. Then there is $\mathcal{H} \subseteq \mathbb{H}(G)$ such that $|\mathcal{H}| = \alpha$ and $e_{\mathcal{H}}: G \to \mathfrak{U}^{\mathcal{H}}$ is a homeomorphism from G into $\mathfrak{U}^{\mathcal{H}}$.

Proof: *Case 1.* G is compact. We claim first that there is $I \subseteq \mathbb{H}(G)$ such that $|I| \geq \alpha$ and $e_I: G \to \mathfrak{U}^I$ is a homeomorphism from G into $\mathfrak{U}^I$.

According to the Peter-Weyl theorem as generalized by van Kampen to arbitrary compact groups (see, e.g, [10, (22.14)]), for $1 \neq x \in G$ there is a $h_x \in \mathbb{H}(G)$ such that $h_x(x) \neq 1_{\mathfrak{U}}$. Now define $I = \{h_x: 1 \neq x \in G\}$; then the evaluation map $e_I: G \to \mathfrak{U}^I$ separates points and is therefore a homeomorphism. If $\alpha > \omega$ we have $|I| = w(\mathfrak{U}^I) \geq w(G) = \alpha$, as required; if $\alpha = \omega$ and I is finite, we augment I (using the fact that $|\mathbb{H}(G)| \geq \omega$) to achieve $|I| = \alpha = \omega$. The claim is proved. Now let $\mathcal{B}$ be a countable base for $\mathfrak{U}$. The set S of all sets of the form $N(F, \mathcal{F}) = \bigcap_{k \leq n} h_k^{-1}(B_k)$ with $F = \{h_k: k \leq n\} \subseteq I$ and $\mathcal{F} = \{B_k: k \leq n\} \subseteq \mathcal{B}$ is a base for G. Since (for every space X) every base for the topology of X contains a base of cardinality $w(X)$ (cf. [9, (1.1.15)]), there is $\mathcal{M} \subseteq S$ such that $\mathcal{M}$ is a base for G and $|\mathcal{M}| = \alpha$. Then

$$\mathcal{H} := \bigcup\{F: N(F, \mathcal{F}) \in \mathcal{M}\} \subseteq \mathbb{H}(G)$$

satisfies $|\mathcal{H}| = \alpha$ and separates points of G, so $e_{\mathcal{H}}$ is a homeomorphism as required.

Case 2. G is totally bounded and not compact. By Case 1 applied to the Weil completion $\overline{G}$ of G there is $\overline{\mathcal{H}} \subseteq \mathbb{H}(\overline{G})$ such that $|\overline{\mathcal{H}}| = \alpha$ and the evaluation map $e_{\overline{\mathcal{H}}}: \overline{G} \to \mathfrak{U}^{\overline{\mathcal{H}}}$ is a homeomorphism. Define $\mathcal{H} = \{\overline{h}|G: \overline{h} \in \overline{\mathcal{H}}\}$. Then $\mathcal{H} \subseteq \mathbb{H}(G)$, and since G is dense in $\overline{G}$ the map $\overline{\mathcal{H}} \twoheadrightarrow \mathcal{H}$ given by $\overline{h} \to h$ is one-to-one. Then from Lemma 1.1 we have $|\mathcal{H}| = |\overline{\mathcal{H}}| = w(\overline{G}) = w(G)$, and since $e_{\overline{\mathcal{H}}}$ is a homeomorphism on $\overline{G}$ its restriction to G, which is $e_{\mathcal{H}}$, is a homeomorphism on G. ❑

REMARK 2.2: It was the exposition in [11, Section 27] of the so-called *dual object* $\Sigma = \Sigma(G)$ (of a locally compact group G) which prompted us to consider the group $\mathfrak{U}$ and the set $\mathbb{H}(G)$ used in Lemma 2.1. For a proof that $|\Sigma(G)| = w(G)$ for infinite compact groups G, see [11, (28.2)].

THEOREM 2.3: Let G be a group and let $\mathcal{T}_i \in \mathcal{B}_{\alpha_i}(G)$ $(i = 0, 1)$ with $\alpha_0 < \alpha_1$ and $\mathcal{T}_0 \subset \mathcal{T}_1$. Then for $\alpha_0 \leq \gamma < \alpha_1$ there is $\mathcal{T} \in \mathcal{B}_\gamma(G)$ such that $\mathcal{T}_0 \subset \mathcal{T} \subset \mathcal{T}_1$.

Proof: First, use Lemma 2.1 to find $\mathcal{H}_i \subseteq \mathbb{H}(G,\mathcal{T}_i)$ such that $|\mathcal{H}_i| = \alpha_i$ and $e^{\mathcal{H}_i}\colon \langle G,\mathcal{T}_i\rangle \to \mathfrak{U}^{\mathcal{H}_i}$ is a homeomorphism; replacing $\mathcal{H}_1$ by $\mathcal{H}_0 \cup \mathcal{H}_1$ if necessary, we assume without loss of generality that $\mathcal{H}_0 \subseteq \mathcal{H}_1$. Now we consider two cases.

Case 1. $\gamma = \alpha_0$. Some $h \in \mathcal{H}_1$ is not $\mathcal{T}_0$-continuous, since otherwise (in the notation of Lemma 2.1) we have $\mathcal{T}_1 = \mathcal{T}_{\mathcal{H}_1} \subseteq \mathcal{T}_{\mathcal{H}_0} = \mathcal{T}_0$. Choosing such h we set $\mathcal{H} = \mathcal{H}_0 \cup \{h\}$ and we define $\mathcal{T} = \mathcal{T}_{\mathcal{H}}$. It is clear that $\mathcal{T} \subseteq \mathcal{T}_1$ and that $\mathcal{T}_0 \subset \mathcal{T}$ (since h is $\mathcal{T}$-continuous). Since $e_{\mathcal{H}}\colon (G,\mathcal{T}_{\mathcal{H}}) \to \mathfrak{U}^{\mathcal{H}}$ is a homeomorphism we have

$$w(G, \mathcal{T}) = w(G, \mathcal{T}_{\mathcal{H}}) \leq |\mathcal{H}| = |\mathcal{H}_0| = \alpha_0 < \alpha_1 = w(G, \mathcal{T}_1)$$

and hence $\mathcal{T} \neq \mathcal{T}_1$; the inequality $w(G, \mathcal{T}) \geq \alpha_0$ follows from Lemma 1.2.

Case 2. $\alpha_0 < \gamma < \alpha_1$. Let $K_i = \overline{\langle G, \mathcal{T}_i\rangle}$ $(i = 0, 1)$, and denote by F_i the closure in $\mathfrak{U}^{\mathcal{H}_i}$ of the set $e^{\mathcal{H}_i}[G]$. It follows from the uniqueness aspect of Weil's theorem [18] that F_i is homeomorphic to K_i. Each $h \in \mathcal{H}_i$ extends uniquely to $\overline{h} \in \mathbb{H}(K_i)$ and with $\overline{\mathcal{H}}_i = \{\overline{h}\colon h \in \mathcal{H}_i\}$ the map $e_{\overline{\mathcal{H}}_i}\colon K_i \twoheadrightarrow F_i \subseteq \mathfrak{U}^{\overline{\mathcal{H}}_i}$ is a homeomorphism.

It is well known that the compact, Hausdorff group F_1 contains a homeomorph of the space $\{0, 1\}^{\alpha_1}$. (To our best knowledge this result is due to Pełczyński [13, (8.10)]; see Shakhmatov [17] for a direct proof using modern techniques, and for references to relevant papers of Cleary and Morris, of Efimov, of Gerlits, of Hagler, of Šapirovskiĭ and of Uspenskiĭ.) Let A denote the one-point compactification of the discrete space of cardinality γ. Since A is a compact zero-dimensional Hausdorff space with $w(A) = \gamma$, the space $\{0,1\}^{\alpha_1}$ contains (a homeomorph of) A. We write $A \subseteq \{0,1\}^{\alpha_1} \subseteq F_1 \subseteq \mathfrak{U}^{\overline{\mathcal{H}}_1}$ with $w(A) = |A| = \gamma$, and we choose $\overline{\mathcal{I}} \subseteq \overline{\mathcal{H}}_1$ such that $|\overline{\mathcal{I}}| = \gamma$ and the projection function $\pi_{\overline{\mathcal{I}}}\colon \mathfrak{U}^{\overline{\mathcal{H}}_1} \twoheadrightarrow \mathfrak{U}^{\overline{\mathcal{I}}}$ is one-to-one on the set A (so, $\pi_{\overline{\mathcal{I}}}|A$ is a homeomorphism); we assume without loss of generality that each $h \in \mathcal{H}_0$ satisfies $\overline{h} \in \overline{\mathcal{I}}$. Writing $F = \pi_{\overline{\mathcal{I}}}[F_1]$ we have from $\pi_{\overline{\mathcal{I}}}[A] \subseteq F \subseteq \mathfrak{U}^{\overline{\mathcal{I}}}$ that

$$\gamma = w(A) = w(\pi_{\overline{\mathcal{I}}}[A]) \leq w(F) \leq |\overline{\mathcal{I}}| = \gamma,$$

so the space

$$F = \pi_{\overline{\mathcal{I}}}[F_1] = \pi_{\overline{\mathcal{I}}} \circ e_{\overline{\mathcal{H}}_1}[K_1] = e_{\overline{\mathcal{I}}}[K_1]$$

satisfies $w(F) = \gamma$. We note that since $\overline{\mathcal{H}}_0 \subseteq \overline{I}$ the function $e_{\overline{I}}$ is one-to-one on G.

Writing $I = \{h \in \mathcal{H}_1 : \overline{h} \in \overline{I}\}$ and denoting as usual by $\mathcal{T} = \mathcal{T}_I$ the topology induced on G by I, we have from $\mathcal{H}_0 \subseteq I \subseteq \mathcal{H}_1$ and Lemma 1.4 that

$$\mathcal{T}_0 = \mathcal{T}_{\mathcal{H}_0} \subseteq \mathcal{T} \subseteq \mathcal{T}_{\mathcal{H}_1} = \mathcal{T}_1.$$

According to the same Lemma the function $e_I : \langle G, \mathcal{T}\rangle \to \mathfrak{U}^I$ is a homeomorphism. Since e_I and $e_{\overline{I}}|\langle G, \mathcal{T}\rangle$ agree on $\langle G, \mathcal{T}\rangle$, with $e_{\overline{I}}|\langle G,\mathcal{T}\rangle$ a homeomorphism of $\langle G, \mathcal{T}\rangle$ onto a dense subset of F, we have from the uniqueness aspect of Weil's theorem that F is the Weil completion of $e_I(G,\mathcal{T})$. It then follows from Lemma 1.1 that $w(G, \mathcal{T}) = w(e_I(G, \mathcal{T})) = w(F) = \gamma$, as required. ❑

REMARK 2.4: We note concerning Theorem 2.3, that if a fixed $k \in \mathbb{H}(G,\mathcal{T}_1)$ is given in advance, then the topology $\mathcal{T}$ may be chosen so that $k \in \mathbb{H}(G,\mathcal{T})$. To see this, it is enough to treat the case $\gamma = \alpha_0$. If $k \in \mathbb{H}(G, \mathcal{T}_0)$ there is nothing to prove, and if $k \notin \mathbb{H}(G, \mathcal{T}_0)$ one may take $h = k$ in the proof of Theorem 2.3.

The theorem just proved has as a consequence the assertions made in our abstract. We give these now in a more comprehensive form.

THEOREM 2.5: Let G be a group and let $\mathcal{T}_i \in \mathcal{B}_{\alpha_i}(G)$ $(i = 0, 1)$ with $\alpha_0 < \alpha_1$ and $\mathcal{T}_0 \subset \mathcal{T}_1$. Then there is a set $\mathcal{W} = \{\mathcal{U}_\xi : \alpha_0 \leq \xi \leq \alpha_1\} \subseteq [\mathcal{T}_0,\mathcal{T}_1]$ such that

(i) $\mathcal{U}_{\alpha_0} = \mathcal{T}_0$ and $\mathcal{U}_{\alpha_1} = \mathcal{T}_1$;
(ii) the map $\xi \to \mathcal{U}_\xi$ is an order-isomorphism from the well-ordered set $[\alpha_0,\alpha_1]$ of ordinals onto $\mathcal{W} \subseteq [\mathcal{T}_0, \mathcal{T}_1]$; and
(iii) $w(G, \mathcal{U}_\xi) = |\xi|$ for $\xi \in [\alpha_0, \alpha_1]$

Proof: [Remark. Condition (ii) implies *inter alia* that every limit ordinal ξ such that $\alpha_0 < \xi \leq \alpha_1$ satisfies $\mathcal{U}_\xi = \sup\{\mathcal{U}_\eta : \alpha_0 \leq \eta < \xi\} = \bigcup_{\alpha_0 \leq \eta < \xi} \mathcal{U}_\eta$. In particular every cardinal γ with $\alpha_0 < \gamma \leq \alpha_1$ has $\mathcal{U}_\gamma \in \mathcal{B}_\gamma(G)$ realized as the supremum (the union) of γ-many smaller topologies of weight less than γ, and $\mathcal{T}_1$ itself is $\sup(\mathcal{W}\backslash\{\mathcal{T}_1\})$.]

First for cardinals γ such that $\alpha_0 \leq \gamma \leq \alpha_1$ we use Theorem 2.3 to choose $\mathcal{V}_\gamma \in \mathcal{B}_\gamma(G) \cap [\mathcal{T}_0,\mathcal{T}_1]$, with $\mathcal{V}_{\alpha_0} = \mathcal{T}_0$, $\mathcal{V}_{\alpha_1} = \mathcal{T}_1$. For such γ let $\mathcal{U}'_\gamma$ be the smallest group topology on G containing $\bigcup\{\mathcal{V}_\beta : \alpha_0 \leq \beta \leq \gamma, \beta \text{ a cardinal}\}$. Then $w(G, \mathcal{U}'_\gamma) \leq \Sigma_{\alpha_0 \leq \beta \leq \gamma}\, w(G, \mathcal{V}_\beta) = \gamma$, and $w(G, \mathcal{U}'_\gamma) \geq w(G, \mathcal{V}_\gamma) = \gamma$ by Lemma 1.2. Thus $w(G, \mathcal{U}'_\gamma) = \gamma$ for $\alpha_0 \leq \gamma \leq \alpha_1$. Clearly, for $\alpha_0 \leq \beta \leq \gamma \leq \alpha_1$ we have

$$\mathcal{T}_0 = \mathcal{U}'_{\alpha_0} \subseteq \mathcal{U}'_\beta \subseteq \mathcal{U}'_\gamma \subseteq \mathcal{U}'_{\alpha_1} = \mathcal{T}_1.$$

For $\alpha_0 \leq \gamma \leq \alpha_1$ there is by Lemma 2.1 a set $\mathcal{H}_\gamma \subseteq \mathbb{H}(G,\mathcal{U}'_\gamma)$ such that $|\mathcal{H}_\gamma| = \gamma$ and $\mathcal{U}'_\gamma = \mathcal{T}_{\mathcal{H}_\gamma}$. We assume without loss of generality, replacing if necessary $\mathcal{H}_\gamma$ by $\cup_{\alpha_0 \le \beta \le \gamma} \mathcal{H}_\beta$, that $\mathcal{H}_{\alpha_0} \subseteq \mathcal{H}_\beta \subseteq \mathcal{H}_\gamma \subseteq \mathcal{H}_{\alpha_1}$ when $\alpha_0 \le \beta \le \gamma \le \alpha_1$; the relation $\mathcal{U}'_\gamma = \mathcal{T}_{\mathcal{H}_\gamma}$ survives throughout this replacement.

We will define the family $\mathcal{W}$ by recursion, arranging *en route* that $\mathcal{U}_\gamma = \mathcal{U}'_\gamma$ for cardinals γ with $\alpha \le \gamma \le \alpha_1$. Noting for $\alpha_0 \le \gamma < \alpha_1$ that $|\mathcal{H}_{\gamma^+} \backslash \mathcal{H}_\gamma| = \gamma^+$ and that $\mathcal{H}_{\alpha_1} = \cup_{0 \le \gamma < \alpha_1}(\mathcal{H}_{\gamma^+} \backslash \mathcal{H}_\gamma)$, we write $\mathcal{H}_{\alpha_1} = \{k_\xi: \alpha_0 \le \xi < \alpha_1\}$ with $\mathcal{H}_{\gamma^+} \backslash \mathcal{H}_\gamma = \{k_\xi: \gamma \le \xi < \gamma^+\}$ for $\alpha_0 \le \gamma < \alpha$.

Define $\mathcal{U}_{\alpha_0} = \mathcal{U}'_{\alpha_0} = \mathcal{T}_0$. Let $\alpha_0 \le \xi < \alpha_1$ and suppose for $\alpha_0 \le \eta < \xi$ that $\mathcal{U}_\eta$ has been defined so that:

(i) $\{\mathcal{U}_\eta: \alpha \le \eta < \xi\}$ is order-isomorphic to the set $[\alpha_o, \xi)$ of ordinals;

(ii) $w(G, \mathcal{U}_\eta) = |\eta|$ for $\alpha_0 \le \eta < \xi$;

(iii) $k_\eta \in \mathbb{H}(G, \mathcal{T}_{\eta+1})$ for $\alpha_0 \le \eta < \eta + 1 < \xi$;

(iv) $\mathcal{U}_\eta \subseteq \mathcal{U}'_{|\eta|^+}$ for $\alpha_0 \le \eta < \xi$; and

(v) $\mathcal{U}_\gamma = \mathcal{U}'_\gamma$ for cardinals γ such that $\alpha_0 \le \gamma < \xi$.

To define $\mathcal{U}_\xi$ we consider two cases.

Case 1. $\xi = \zeta + 1$. Since $\mathcal{U}_\zeta \subseteq \mathcal{U}'_{|\xi|^+}$ and $k_\zeta \in \mathbb{H}(G, \mathcal{U}'_{|\xi|^+})$, it follows from Theorem 2.3 and Remark 2.4 that there is $\mathcal{U}_\xi = \mathcal{U}_{\zeta+1} \in [\mathcal{U}_\zeta, \mathcal{U}'_{|\xi|^+}]$ such that $w(G, \mathcal{U}_\xi) = w(G, \mathcal{U}_\zeta)$ and $\mathcal{U}_\xi \neq \mathcal{U}_\zeta$ and $k_\zeta \in \mathbb{H}(G, \mathcal{U}_\xi)$.

Case 2. ξ is a limit ordinal. We define $\mathcal{U}_\xi = \cup_{\alpha_0 \le \eta < \xi} \mathcal{U}_\eta$.

The definition of $\mathcal{U}_\xi$ for $\alpha_0 \le \xi \le \alpha_1$ is complete. It is clear that the function $\xi \to \mathcal{U}_\xi$ is an order-isomorphism from the ordinal interval $[\alpha_0, \alpha_1]$ onto $\mathcal{W} = \{\mathcal{U}_\xi: \alpha_0 \le \xi \le \alpha_1\}$, so to complete the proof it suffices to show these two statements.

(A) $\mathcal{U}_\gamma = \mathcal{U}'_\gamma$ for γ a cardinal, $\alpha_0 \le \gamma \le \alpha_1$.

(B) $w(G, \mathcal{U}_\xi) = |\xi|$ for $\xi \in [\alpha_0, \alpha_1]$.

(A) This is clear for $\gamma = \alpha_0$. When $\gamma > \alpha_0$ we have from (iv) that $\mathcal{U}_\eta \subseteq \mathcal{U}'_{|\eta|^+} \subseteq \mathcal{U}'_\gamma$ for $\eta < \gamma$, so $\mathcal{U}_\gamma = \cup_{\alpha_0 \le \eta < \gamma} \mathcal{U}_\eta \subseteq \mathcal{U}'_\gamma$. Each $k_\eta \in \mathcal{H}_\gamma$ satisfies $k_\eta \in \mathcal{H}_{\alpha^+} \backslash \mathcal{H}_\alpha$ for some cardinal $\alpha < \gamma$, so $k_\eta \in \mathbb{H}(G, \mathcal{U}_{\eta+1}) \subseteq \mathbb{H}(G, \mathcal{U}_\gamma)$; from $\mathcal{U}'_\gamma = \mathcal{T}_{\mathcal{H}_\gamma}$ then follows $\mathcal{U}'_\gamma \subseteq \mathcal{U}_\gamma$.

(B) Since $w(G, \mathcal{U}_{\zeta+1}) = w(G, \mathcal{U}_\zeta)$ for all ζ, the first failure of (B) (if any) must occur at a limit ordinal ξ. Clearly then $w(G, \mathcal{U}_\xi) \le \Sigma_{\alpha_0 \le \eta < \xi} w(G, \mathcal{U}_\eta) = \Sigma_{\alpha_0 \le \eta < \xi} |\eta| = |\xi|$. If ξ is not a cardinal there is $\eta < \xi$ such that $|\eta| = |\xi|$, and from $\mathcal{U}_\eta \subseteq \mathcal{U}_\xi$ we have $w(G, \mathcal{U}_\xi) \ge w(G, \mathcal{U}_\eta) = |\eta| = |\xi|$ from Lemma 1.2. If ξ is the cardinal γ we have from (A) that $w(G, \mathcal{U}_\gamma) = w(G, \mathcal{U}'_\gamma) = \gamma$, as required. ❑

COROLLARY 2.6: Let G be a group and let $\mathcal{T}_i \in \mathcal{B}_{\alpha_i}(G)$ $(i = 0, 1)$ with $\alpha_0 < \alpha_1$ and $\mathcal{T}_0 \subset \mathcal{T}_1$. Then

(a) $|[\mathcal{T}_0, \mathcal{T}_1]| \ge \alpha_1$; and

(b) every cardinal γ such that $\alpha_0 \le \gamma < \alpha_1$ satisfies $|\mathcal{B}_\gamma(G) \cap [\mathcal{T}_0, \mathcal{T}_1]| \ge \gamma^+$. ❑

REMARK 2.7: We say as usual that a (discrete) group G is *maximally almost periodic* in the sense of von Neumann if for $1_G \neq x \in G$ there is a homomorphism h_x from G into some compact group (equivalently: into $\mathfrak{U}$) such that $h_x(x) \neq 1$. For such groups G, two facts are evident:

(a) There is a largest totally bounded group topology (here denoted $\mathcal{T}^*$) on G; and

(b) there is a totally bounded group topology $\mathcal{T}$ on G such that $w(G, \mathcal{T}) \leq |G|$.

(To prove these statements let $\mathrm{Hom}(G,\mathfrak{U})$ be the set of homomorphisms from G into $\mathfrak{U}$, and let $\mathcal{H}$ be a point-separating subfamily of $\mathrm{Hom}(G,\mathfrak{U})$ with $|\mathcal{H}| \leq |G|$. Although $\mathrm{Hom}(G, \mathfrak{U})$ and $\mathcal{H}$ are in general not groups, the evaluation mappings $e_{\mathrm{Hom}(G,\mathfrak{U})}: G \to \mathfrak{U}^{\mathrm{Hom}(G,\mathfrak{U})}$ and $e_{\mathcal{H}}: G \to \mathfrak{U}^{\mathcal{H}}$ are isomorphisms; the induced topologies $\mathcal{T}^* = \mathcal{T}_{\mathrm{Hom}(G,\mathfrak{U})}$ and $\mathcal{T} = \mathcal{T}_{\mathcal{H}}$ are as required in (a) and (b), respectively.

For comments about the class of maximally almost periodic groups and related classes, the reader may consult [3, p. 1229ff] and [4, Sections 3.1 & 3.9]. Now we note the relevance of Theorem 2.5 to that class of groups.

COROLLARY 2.8: Let G be an infinite (discrete) maximally almost periodic group and let $\mathcal{T}^*$ be the largest totally bounded group topology on G. Then

(a) $|\mathcal{B}(G)| \geq w(G, \mathcal{T}^*)$; and

(b) every cardinal γ such that $|G| \leq \gamma < w(G, \mathcal{T}^*)$ satisfies $|\mathcal{B}_\gamma(G)| \geq \gamma^+$.

Proof: Let $\mathcal{T}$ be a totally bounded group topology $\mathcal{T}$ on G such that $w(G, \mathcal{T}) \leq |G|$, as given in 2.7(b); then apply Corollary 2.6 with $\mathcal{T}_0 = \mathcal{T}$ and $\mathcal{T}_1 = \mathcal{T}^*$. ❑

REMARK 2.9(a): We have shown in [6] and [16] for G, $\mathcal{T}_0$ and $\mathcal{T}_1$ as in Theorem 2.5 that the estimate $|[\mathcal{T}_0,\mathcal{T}_1]| \geq \alpha_1$ of Corollary 2.6 can be improved to $|[\mathcal{T}_0,\mathcal{T}_1]| = 2^{\alpha_1}$ in case the compact group $\overline{\langle G, \mathcal{T}_1\rangle}$ is connected, but we do not know whether this latter equality (or even $|[\mathcal{T}_0,\mathcal{T}_1]| > \alpha_1$) holds in general.

REMARK 2.9(b): For certain groups G, other order-theoretic properties of $\mathcal{B}(G)$ and of some of the sets $\mathcal{B}_\gamma(G)$ have appeared already in the literature. (1) In [2], for example, for every Abelian group G with $|G| = \alpha$ every cardinal number $|\mathcal{B}_\gamma(G)|$ is computed explicitly as a function of α and γ, together with the depth, width and height of the partially ordered set $\mathcal{B}(G)$. (2) Similar computations are given for a wide class of non-Abelian groups in [7], [8], [1], [15], [16], [5], and [6]. (3) In the proof of [1, Lemma 7.3] an argument may be found for deriving our Theorem 2.3 for $\alpha_0 < \gamma$. The main tools there are taken from [15], where representation theory is used.

REMARK 2.9(c): Let us say that a (not necessarily Hausdorff) topological group G is *precompact* if for every $U \in \mathcal{N}(1)$ there is a finite subset F of G such that $G = FU$. Using the fact (from[14]) that the complete lattice of precompact group topologies on a group G is anti-isomorphic to the lattice of closed, normal subgroups of the Bohr compactification of the discrete group G and facts from the theory of unitary representations, statements (a) and (b) of our abstract were proved in [16, Satz 3.25] for precompact group topologies $\mathcal{T}_0$, $\mathcal{T}_1$ with $\mathcal{T}_0 \subseteq \mathcal{T}_1$, $w(G, \mathcal{T}_i) = \alpha_i$, and $\omega \leq \alpha_0 < \alpha_1$. These results can be achieved by minor modification of the methods of the present paper. (For the possibly non-Hausdorff analogue of Theorem 2.3, one applies Lemma 2.1 to the Hausdorff group $\langle G/N_i, \mathcal{T}_i/N_i \rangle$, where N_i denotes the $\mathcal{T}_i$-closure of $\{1_G\}$ and $\mathcal{T}_i/N_i$ is the quotient topology. The proof of the required modification of Theorem 2.5 is then routine.)

ACKNOWLEDGMENT

We thank the referee for several helpful expository suggestions.

REFERENCES

1. BERARDUCCI, A., D. DIKRANJAN, M. FORTI, & S. WATSON. 1996. Cardinal invariants and independence results in the poset of precompact group topologies. J. Pure and Applied Algebra. To appear.
2. BERHANU, S., W.W. COMFORT, & J.D. REID. 1985. Counting subgroups and topological group topologies. Pacific J. Math. **116:** 217–241.
3. COMFORT, W.W. 1984. Topological groups. *In:* Handbook of Set-theoretic Topology (Kenneth Kunen &Jerry E. Vaughan, Eds.). North-Holland, Amsterdam. pp.1143–1263.
4. COMFORT, W.W, K.-H. HOFMANN, & D.REMUS. 1992. Topological groups and semigroups. *In:* Recent Progress in General Topology (M.Hušek & J.van Mill, Eds.). Elsevier Science Publishers B.V. Amsterdam. 57–144.
5. COMFORT, W.W, & D. REMUS. 1991. Long chains of Hausdorff topological group topologies. J. Pure & Appl. Algebra. **70:** 53–72.
6. COMFORT, W.W, & D. REMUS. 1996. Long chains of group topologies: A continuation. 1995. Topology and Its Applications. To appear.
7. DIKRANJAN, D. 1993. On the poset of precompact group topologies. *In:* Mathematica Societas János Bolyai Studies 4 (Topology with Applications). Szekszárd. Hungary. 135–149.
8. DIKRANJAN, D. 1995. The lattice of compact representations of an infinite group. *In:* Groups '93, Proc. Galway/St.Andrews Conference (Cambridge) (C.M. Campbell *et al.*, Eds.). Cambridge University Press. Cambridge. 138–155.
9. ENGELKING, R. 1989. General Topology. Heldermann Verlag. Berlin.
10. HEWITT, E. & K. A. ROSS. 1963. Abstract Harmonic Analysis, Vol. I. Die Grundlehren der mathematischen Wissenschaften in Einzeldarstellungen, Vol. 115. Springer-Verlag, Berlin-*G*öttingen-Heidelberg.
11. HEWITT, E. & K. A. ROSS. 1970. Abstract Harmonic Analysis, Vol. II. Die Grundlehren der mathematischen Wissenschaften in Einzeldarstellungen, Vol.152. Springer-Verlag, Berlin-Heidelberg-New York.
12. KELLEY, J.L. 1955. General Topology. D. Van Nostrand Co. New York.

13. PEŁCZYŃSKI, A. 1958. Linear extensions, linear averagings, and their applications to linear topological classification of spaces of continuous functions. Dissertationes Math. **58:** 89 pages. Rozprawy Mat. Polish Scientific Publishers, Warszawa.
14. REMUS, D. 1983. Zur Struktur des Verbandes der Gruppentopologien. Doctoral Dissertation, Universität Hannover, Hannover, Germany. [English summary: 1983. Resultate Math. **6:** 151–152].
15. REMUS, D. 1991. Minimal and precompact group topologies on free groups. J. Pure & Appl. Algebra.**70:** 147–157.
16. REMUS, D. 1995. Anzahlbestimmungen von gewissen präkompakten bzw. nichtpräkompakten hausdorffschen Gruppentopologien. Habilitationsschrift. Universität Hanover, Hanover, Germany.
17. SHAKHMATOV, D.B. 1994. A direct proof that every infinite compact group G contains $\{0, 1\}^{w(G)}$. *In:* Papers on General Topology and Applications, "Annals of the New York Academy of Sciences." (Susan Andima, Gerald Itzkowitz, T. Yung Kong, R. Kopperman, P.R. Misra, L. Narici, & A. Todd, Eds.). **728:** pp. 276–283. Proc. June, 1992 Queens College Summer Conference on General Topology and Applications. New York.
18. WEIL, A. 1938. Sur les Espaces à Structure Uniforme et Sur la Topologie Générale. Publ. Math. Univ. Strasbourg. Hermann & Cie. Paris. .
19. WILLARD, S. 1970. General Topology. Addison-Wesley Publishing Co. Reading, Massachusetts.

Weak P-points and Cancellation in βS

MAHMOUD FILALI

[a]*Department of Mathematical Sciences*
University of Oulu
SF 90570 Finland

ABSTRACT: Let S be a cancellative semigroup and let βS be the Stone-Čech compactification of S. Then βS is a semigroup with an operation which extends that of S and which is continuous only in one variable. The points s of S are easily shown to be right (left) cancellative in βS, i.e., ys and zs (sy and sz) are different elements of βS whenever y and z are. It is known that such a property is not valid for all the elements of $\beta S \backslash S$. However, we will see that the set of points in $\beta S \backslash S$ which are cancellative in βS is dense in $\beta S \backslash S$. In particular, we will see that the (weak) p-points of $\beta S \backslash S$, which are contained in the closure of some countable subset of S, are (right) cancellative in βS.

1. INTRODUCTION AND NOTATIONS

Throughout the paper, S is an infinite discrete semigroup written multiplicatively. Let βS be the Stone-Čech compactification of S. Several constructions of βS are known. We cite, for example, that βS can be described as the maximal ideal space of the space $\ell^\infty(S)$ of bounded complex-valued functions on S or as the space of ultrafilters of S (see [12]). The closure in βS of a subset A of βS will be denoted by $\overline{A}$. If A is a subset of S, then A^* will denote $\overline{A} \backslash A$. In particular, $S^* = \beta S \backslash S$.

The compact space βS can be made into a semigroup with an operation which extends that of S and which is continuous *only* on one side. This can be obtained using the characteristic property of βS, which asserts that every mapping of S into a compact space extends uniquely to βS [12, Theorem 1.46]. Given $s \in S$, the mapping $t \mapsto st$ of S into βS extends to a continuous mapping $y \mapsto sy$ of βS into βS. Then, for each $y \in \beta S$, we extend the mapping $s \mapsto sy$ from S into βS to a continuous mapping $x \mapsto xy$ from βS to βS; and so we obtain a binary operation on βS. It is easy to verify that this operation is associative, and so βS is a semigroup. It follows then that βS is a compact right topological semigroup in the sense that the mapping $x \mapsto xy$ of βS into βS is continuous. The semigroup βS has a rich algebraic structure; see [1] or the survey in [9]. In this note, we look at the cancellation property in this semigroup. We say that an element x of βS is *right* (respectively, *left*) *cancellative* in βS if the identity $yx = zx$ (respectively, $xy = xz$) holds if and only

Mathematics Subject Classification: 22A15, 54D35.
Key words and phrases: cancellative, sparse, Stone-Čech compactification, weak p-points.

if $y = z$. When S is a cancellative semigroup, the elements of S are all left and right cancellative in βS (Lemma 4). Such a property is not, however, true for all elements of βS. For instance, let $s \in S$ and let e be an idempotent in S^* (which exists by [3]). Then $e(es) = es$ and $(se)e = se$ whereas $es \neq s$ and $se \neq s$ in general. In fact, if S is cancellative then se and es are in S^* since S^* is a two-sided ideal of βS as we will see below in Lemma 3. Therefore, e is neither left nor right cancellative in βS.

Various authors have addressed the problem of finding points in S^* that are cancellative in βS. In [2], Butcher has proved, when S is a countable group, that the p-points of S^* are right cancellative in βS. So, under the continuum hypothesis, the set of points in S^* which are right cancellative in βS is a dense subset of S^*. In [5] and [8] (see also [6]), the set of right cancellative points in βS is shown to be dense in S^* when S is a countable cancellative semigroup. The proof makes no use of p-points, and it is true without the assumption of the continuum hypothesis. Recently, we have managed to show in [4] that the hypothesis of S being countable is not necessary for the validity of these results. This is achieved by showing that the weak p-points of S^*, which are contained in the closure of some countable subset of S, are right cancellative in βS.

We start this paper by collecting some basic results in the following section. In Section 3, we follow the methods applied in [5] and [8], and prove directly in Theorem 1 that the set of points in S^* that are right cancellative in βS is dense in S^* when S is either a commutative cancellative semigroup or a group. In Theorem 2, we see that if x is a weak p-point of S^*, then

$$(x S^*) \cap (x S) = \varnothing \quad \text{and} \quad (S^* x) \cap (S x) = \varnothing.$$

We then deduce from $(S^* x) \cap (Sx) = \varnothing$ that x is right cancellative in βS, (a result which was proved in [4]). Finally, Theorem 3 shows that the p-points which are in the closure of a countable subset of S are also left cancellative in βS.

2. PRELIMINARIES

In a topological space X, a point x is said to be *a p-point* if every countable intersection of neighborhoods of x is also a neighborhood of x. A point x in X is said to be *a weak p-point* if x is not a limit point of any countable subset of $X \setminus \{x\}$. Under the assumption of the continuum hypothesis, Rudin has proved in [10] that p-points exist in $\mathbb{N}^*$ and concluded that $\mathbb{N}^*$ is not homogeneous. In [7], Kunen introduced the weak p-points and proved in ZFC that such points exist in $\mathbb{N}^*$. This fact implies also that $\mathbb{N}^*$ is not homogeneous.

LEMMA 1: Let X be a compact, Hausdorff, extremally disconnected space (βS is a space of this type). Let A and B be countable subsets of X. Then $\overline{A} \cap \overline{B} = \varnothing$ if and only if $\overline{A} \cap B = \varnothing$ and $A \cap \overline{B} = \varnothing$.

Proof: This can be found, for example, in Lemma 1.1 of [6] or Lemma 1 of [11]. ❑

LEMMA 2: If S is cancellative, then $xs \neq xt$ and $sx \neq tx$ for all $x \in \beta S$ and s, t distinct elements of S.

Proof: This follows from (a more general) Theorem 2.5 of [6]. ❑

LEMMA 3: S^* is a closed two-sided ideal of βS if and only if for each s and t in S, each of the equations $su = t$ and $us = t$ has only a finite number of solutions in S. In particular, S^* is a two-ideal of βS when S is cancellative.

Proof: This lemma is known and it is in fact true for any semigroup compactification X (following [1, Definition 3.1.1]) with $S \subset X$, see [4, Lemma 1]. ❑

LEMMA 4: Right (left) cancellation in βS holds at each element s of S which is right (left) cancellative in S.

Proof: Let $s \in S$ be right cancellative in S. Let x and y be distinct elements of βS, and let A and B be disjoint subsets of S with $x \in \overline{A}$ and $y \in \overline{B}$. Then As and Bs are also disjoint in S, and so are their closures in βS. Thus right cancellation holds in βS at the point s.

The proof is as straightforward on the other side. ❑

LEMMA 5: Let S be an infinite, discrete, cancellative semigroup, $x \in S^*$, and $s \in S$. Then the following statements are equivalent:

(1) xs is a weak p-point in S^*,
(2) sx is a weak p-point in S^*,
(3) x is a weak p-point in S^*.

Proof: We prove that (1) and (3) are equivalent. The necessity is deduced as follows. If x is not a weak p-point then $x \in \overline{C}$ for some countable subset C of S^* which does not contain x. Thus $xs \in \overline{Cs}$ where Cs is clearly countable, and it is a subset of S^* since S^* is an ideal of βS by Lemma 3. Furthermore, $xs \notin Cs$ by Lemma 4. So xs is not a weak p-point of S^*.

For the sufficiency, suppose that xs is not a weak p-point in S^*, and let C be a countable subset of S^* such that $xs \in \overline{C} \setminus C$. Then the set

$$D = \{w \in S^*: ws \in C\}$$

does not contain x by assumption. Since $\overline{A}s = \overline{As}$ for every subset A of S, it is easy to verify that $y \mapsto ys$ is an open mapping from βS into βS. It follows that the set D has x as a limit point. Finally, by Lemma 4, for each $y \in C$

there is at most one element $w \in S^*$ for which $ws = y$; so the set D is at most countable. So x is not a weak p-point in S^*.

With the fact that, for each $s \in S$, the mapping $y \mapsto sy$ is also continuous and open from βS into βS, we show in the same way that the claim is true with sx in place of xs. ❑

3. THE RESULTS

We say that a subset V of S is *sparse* if $sV \cap tV$ is finite whenever s and t are distinct elements of S. When S is a cancellative countable semigroup, Hindman in [5] and Parsons in [8] arrived to the conclusion that the set of points in S^* which are right cancellative in βS is dense in S^* after they had proved two facts:

(1) the points belonging to V^* for some sparse subset V of S form a dense subset of S^*, and

(2) the points in the closure of sparse sets are right cancellative in βS.

In Theorem 1, below, we show first that (2) is true for any cancellative semigroup, and then we show that (1) is also true when S is either a commutative cancellative semigroup or a group. It follows then that the set of points in S^* which are right cancellative in βS is dense in S^*. This result will also be deduced later on (Corollary 2), but with a less elementary proof. In Theorem 2, we see that the weak p-points x, which are in $\overline{V}$ for some countable subset V of S, have the properties that

$$(x\,S^*) \cap (xS) = \varnothing \quad \text{and} \quad (S^*x) \cap (Sx) = \varnothing.$$

The latter property is equivalent to the right cancellativity of x in βS as already seen in [4]. Although the former property seems to be close to the claim that the weak p-points are left cancellative in βS, we are still not able to deduce this result. However, we end with Theorem 3 which shows that the p-points contained in V with V a countable subset of S are also left cancellative in βS.

From this point on, our semigroup S will be cancellative.

THEOREM 1: Let S be an infinite, discrete, cancellative semigroup. Let $\mathcal{V}$ be the set of points which belong to the closure of some countable sparse subset of S. Then

(1) each point of $\mathcal{V}$ is right cancellative in βS;

(2) when S is either commutative or a group, then $\mathcal{V} \cap S^*$ is dense in S^*; in particular, the right cancellative elements of βS form a dense subset of S^*.

Proof: Let $x \in \overline{V}$, where V is a countable sparse subset of S. That each element of V is (right) cancellative in βS follows immediately from Lemma 4. So we may assume that $x \in V^*$, and let y and z be two distinct elements of

βS. We partition S into two disjoint subsets A and B with $y \in \overline{A}$ and $z \in \overline{B}$. We shall find a $\{0, 1\}$-valued mapping f on S whose continuous extension $\bar{f}$ to βS has values 0 at zx and 1 at yx.

We start by defining an equivalence relation in S as follows: two elements s and t of S are congruent modulo V if and only if there exist finite subsets $\{t_1, t_2, \ldots, t_m\}$ of S, $\{u_1, u_2, \ldots, u_{m-1}\}$ and $\{v_1, v_2, \ldots, v_{m-1}\}$ of V with $t_1 = s$ and $t_m = t$ such that

$$t_i u_i = t_{i+1} v_i \text{ for each } i = 1, 2, \ldots, m-1.$$

For each class, choose a representative $s \in S$, and denote the class by $\tilde{s}$. Note that each equivalence class $\tilde{s}$ is countable since S is cancellative, and so for each element u and v in the countable set V, there is at most one element t in S satisfying the identity $su = tv$. Define now f on sV by

$$f(sv) = \begin{cases} 1 & \text{if } s \in A \\ 0 & \text{if } s \in B. \end{cases}$$

Observe that the mapping f is well defined since the sets sV and $s'V$ are disjoint whenever s and s' represent different classes. Having defined f on sV, we now define f on tV for t in the rest of the class $\tilde{s}$. Put $\tilde{s} = \{s_1, s_2, \ldots\}$ with repetition if necessary and with $s_1 = s$. With the fact that the sets s_1V and s_2V have finite intersection in mind, we define f on s_2V by

$$f(s_2v) = \begin{cases} 1 & \text{if } s_2 \in A \\ 0 & \text{if } s_2 \in B \end{cases}$$

except for those finitely many elements v of V which satisfy $s_2v = s_1u$ for some $u \in V$. In the latter case, we let

$$f(s_2v) = f(s_1u).$$

Inductively, the mapping f is defined on tV for all elements t in $\tilde{s}$. If f is defined on s_mV for all $m \in \{1, 2, \ldots, n\}$, then

$$f(s_{n+1}v) = \begin{cases} f(s_mu) & \text{if } s_{n+1}v = s_mu \text{ for some } m \in \{1, 2, \ldots, n\} \text{ and } u \in V, \\ 1 & \text{if } s_{n+1}v \notin \bigcup_{m=1}^{n} s_mV \text{ and } s_{n+1} \in A, \\ 0 & \text{if } s_{n+1}v \notin \bigcup_{m=1}^{n} s_mV \text{ and } s_{n+1} \in B. \end{cases}$$

Extend now f to $\bar{f}$ on βS, and let (x_α), (y_β), and (z_γ) be nets in V, A, and B converging to x, y, and z, respectively. Then

$$\bar{f}(yx) = \lim_\beta \lim_\alpha f(y_\beta x_\alpha)$$

and

$$f(zx) = \lim_{\gamma} \lim_{\alpha} f(z_\gamma x_\alpha).$$

We prove that $\tilde{f}(zx) = 0$. Let γ be fixed. Then $z_\gamma \in \tilde{s} = \{s_1, s_2, \ldots\}$ for some $s = s_1 \in S$. Let $z_\gamma = s_n$. Then since $z_\gamma V \cap s_k V$ is finite for each $k = 0, 1, \ldots, n-1$, we can choose α_γ such that

$$z_\gamma x_\alpha \notin \cup_{k=0}^{n-1} s_k V \quad \text{for all } \alpha \geq \alpha_\gamma.$$

Accordingly, we have $f(z_\gamma x_\alpha) = 0$ for all $\alpha \geq \alpha_\gamma$ (remember that $z_\gamma \in B$), and so

$$\tilde{f}(z_\gamma x) = \lim_{\alpha} f(z_\gamma x_\alpha) = 0.$$

Thus $\tilde{f}(zx) = 0$. In a similar fashion, we show that $\tilde{f}(yx) = 1$.

We turn now to the second part of the claim, and show that $\overline{\mathcal{V} \cap S^*} = S^*$. Let x be in S^*, and let U be an open neighborhood of x. We need to verify that $U \cap (\mathcal{V} \cap S^*) \neq \varnothing$. Let A be a countable subset of S such that $A^* \subset U$, and let $T = \{s_1, s_2, \ldots\}$ be a countable subsemigroup of S containing A. Then, as shown in the proof of Theorem 4.3, page 305 of [5], or the proof of Proposition 9.3 of [8], A contains an infinite sparse subset V of T. Here, the proof is simplified by the fact that S is cancellative, and is included for completeness. For each integer n, we put $T_n = \{s_1, s_2, \ldots, s_n\}$, and we construct inductively the subset $V = \{v_1, v_2, \ldots\}$ of A such that

$$(T_{n+1} v_{n+1}) \cap (T_n\{v_1, v_2, \ldots, v_n\}) = \varnothing \text{ for each } n = 1, 2, \ldots.$$

We start with an arbitrary v_1 in A. Then take $v_2 \in A$ different than v_1 such that $(T_2 v_2) \cap (T_1 v_1) = \varnothing$. If $v_1, v_2, \ldots, v_n$ are elements of A satisfying

$$(T_{i+1} v_{i+1}) \cap (T_i\{v_1, v_2, \ldots, v_i\}) = \varnothing \quad \text{for each } i = 1, 2, \ldots, n-1.$$

Then the element v_{n+1} is simply chosen in A such that

$$(T_{n+1} v_{n+1}) \cap (T_n\{v_1, v_2, \ldots, v_n\}) = \varnothing.$$

Let s_n and s_m be elements of T with $n > m$. Then

$$\{s_n v_i \colon i \geq n\} \cap \{s_m v_i \colon i \geq n\} = \varnothing.$$

In fact, $s_n v_i \neq s_m v_i$ since S is cancellative, and if $i > j \geq n$ then

$$s_n v_j \in T_n v_j \subset T_{i-1} v_j \text{ and } s_m v_i \in T_m v_i \subset T_i v_i,$$

and so $s_n v_j \neq s_m v_i$ since $T_i v_i \cap T_{i-1} v_j = \varnothing$. Hence $s_n V \cap s_m V$ is finite. In other words, V is a sparse subset of T.

Suppose now that S is a commutative semigroup, and that for some distinct elements s and t of S, $sV \cap tV$ is infinite. Then there are infinitely many pairs (u_n, v_n) in $V \times V$ such that $u_n \neq v_n$ and $su_n = tv_n$. Then

$$tv_1 u_n = su_1 u_n = su_n u_1 = tv_n u_1,$$

and so $v_1 u_n = v_n u_1$. Thus $u_1 V \cap v_1 V$ is infinite with u_1 and v_1 elements of V, whence a contradiction. Therefore V is a sparse subset of S.

When S is a group, we let T be the subgroup generated by A. Then T is countable, and as above A contains a sparse subset V of T. It follows that V is also a sparse subset of S since $sV \cap tV$ is finite if $t^{-1}s$ is in T, and empty otherwise. Accordingly $V^* \subset A^* \subset U$. In other words, $U \cap (\mathcal{V} \cap S^*) \neq \emptyset$, as required. ❑

THEOREM 2: Let S be an infinite, discrete, cancellative semigroup. Let x be a weak p-point of S^* and suppose that it is contained in the closure of a countable subset V of S. Then for all $s \in S$ and $w \in \beta S$ with $w \neq s$, we have $sx \neq wx$ and $xs \neq xw$.

Proof: Note that if s and w are distinct elements in S, then this theorem follows immediately from Lemma 2. So we assume without loss of generality that $w \in S^*$, and suppose that $sx = wx$ for some $s \in S$. Then the point w must be in the closure of some countable subset of S. In fact, we consider again the equivalence relation in S as defined in the proof of Theorem 1. Let W be the class defined by V to which s belongs. As already explained, W is countable since V is countable. We show that $w \in \overline{W}$. Let $\overline{A}$ be a neighborhood of w (here $A \subset S$). Then $\overline{Ax} \cap \overline{sV} \neq \emptyset$. Since $\overline{sV}$ is an open set in βS, this implies that $Ax \cap \overline{sV} \neq \emptyset$. Therefore $ax \in \overline{sV}$ for some $a \in A$, and accordingly, $\overline{aV} \cap \overline{sV} \neq \emptyset$. Thus $(aV) \cap (sV) \neq \emptyset$, which means that $a \in W$. This shows that for every neighborhood $\overline{A}$ of w, the sets A and W meet. In other words, $w \in \overline{W}$. So $sx = wx \in \overline{W}x = \overline{Wx}$. This is not possible since Wx is a countable subset of S^* and sx is a weak p-point of S^* by Lemma 5.

If $xs = xw$ then $xs \in \overline{V}w = \overline{Vw}$. Since Vw is a countable subset of S^* and xs is a weak p-point, there must exist $v \in V$ such that $xs = vw$. This implies that vw is also a weak p-point in S^*, which implies in turn that w is a weak p-point in S^*. Therefore $xw \neq vw = xs$, as required. ❑

COROLLARY 1: [4] Let $x \in S^*$ be a weak p-point contained in the closure of a countable subset of S. Then x is right cancellative in βS.

Proof: Note first that since $sx \neq wx$ for all $s \in S$ and all $w \neq s$ in βS, it follows that Sx is a discrete subspace of βS. Let y and z be distinct elements of βS, and let Y and $Z = S \backslash Y$ be subsets of S such that $y \in \overline{Y}$ and $z \in \overline{Z}$. By Lemma 2, $Yx \cap Zx = \emptyset$, and so

$$cl_{\beta(Sx)}(Yx) \cap cl_{\beta(Sx)}(Zx) = \emptyset.$$

By Theorem 1 of [4], we know that Sx is C^*-embedded in βS, (this result was obtained with a technique similar to the one used in the proof of Theorem 1). Therefore,

$$\overline{Y}x \cap \overline{Z}x = \overline{Yx} \cap \overline{Zx} = cl_{\beta(Sx)}(Yx) \cap cl_{\beta(Sx)}(Yx) = \varnothing.$$

This implies that yx and zx are different elements of βS, and completes the proof. ❑

COROLLARY 2: [4] The set of points in S^* at which right cancellation holds in βS is dense in S^*.

Proof: For each open subset U of S^*, let C be a countable subset of S such that $C^* \subset U$. Since C^* is open in S^*, it follows that weak p-points of C^* are also weak p-points of S^*, and so U contains weak p-points, as required. ❑

REMARK 1: By Corollary 1, the p-points are right cancellative. When S is commutative or a group, this follows also from the fact that each p-point belongs to the closure of some sparse set, and then apply Theorem 1. This was first noticed by Butcher [2, p.42] for countable groups, then by the author in Theorem 2 of [4]. One can also proceed as follows. Let x be a p-point contained in the closure of some countable subset V of S. As in the proof of the second part of Theorem 1, we can find a sparse set V_1 contained in V. If $V \setminus V_1$ is finite, then x is in $\overline{V_1}$ and so we are done. Otherwise, let V_2 be a sparse subset of $V \setminus V_1$, which contains the first element of $V \setminus V_1$, and consider again whether $x \in \overline{V_2}$. We carry on this process, and we make sure at each time to include in V_n the first element of $V \setminus \bigcup_{k=1}^{n-1} V_k$. We obtain a countable family $V_1, V_2, \ldots$, of sparse subsets of V such that $V \setminus \bigcup_{n=1}^{\infty} V_n$ is finite. If x is not in the closure of any of the sets V_n, then there exists for each positive integer n a neighborhood U_n of x which does not meet V_n. Thus, the intersection of the neighborhoods U_n (which is a neighborhood of x since x is a p-point) meets none of the sets V_n, and so cannot meet their union. This is not possible since x is a member of the closure of V.

REMARK 2: The p-points are also left cancellative in βS when S is the additive semigroup of the integers $\mathbb{N}$. This is done by Hindman and Strauss in [6]. Below, we show with a simpler proof that this is true for any discrete cancellative semigroup.

THEOREM 3: Let S be an infinite, discrete, cancellative semigroup. Then the p-points of S^* which are in the closure of countable subsets of S are left cancellative in βS.

Proof: Let x be a p-point and let V be a countable subset of S with $x \in \overline{V}$. Let y and z be distinct points in βS. We show that xy and xz are different. Now there is at most one element s of S with $xy = sz$. For, otherwise, $sz = s'z$ for different elements s and s' of S which is not possible by Lemma 2. Similarly, $xz = ty$ for, at most, one element t of S. If such cases occur, we consider $V \setminus \{s, t\}$ instead of V, and so without loss of generality we may assume that xy is not a member of Vz and that xz is not a member of Vy.

For each $s \in V$, let

$$A_s = \{w \in S^* : wz = sy\} \text{ and } B_s = \{w \in S^* : wy = sz\}.$$

Then A_s and B_s are closed subsets of βS since the operation in βS is continuous in the left variable, i.e., the mapping $x \mapsto xy$ is continuous on βS. Moreover, the point x belongs to neither A_s nor B_s by assumption. Accordingly, for each $s \in S$, the sets $\beta S \backslash A_s$ and $\beta S \backslash B_s$ are open neighborhoods of x, and so is $U_s = (\beta S \backslash A_s) \cap (\beta S \backslash B_s)$. Since V is countable and x is a p-point, it follows that $U = \cap_{s \in V} U_s$ is an open neighborhood of x. Now choose W a subset of V such that $x \in \overline{W} \subset U$. Then

$$\overline{W}y \cap Wz = \varnothing \quad \text{and} \quad Wy \cap \overline{W}z = \varnothing,$$

which implies, by Lemma 1, that $\overline{W}y \cap \overline{W}z = \varnothing$ since Wy and Wz are countable subsets of βS. Thus $xy \neq xz$, and so the proof is complete. ❑

COROLLARY 3: Under the assumption of the continuum hypothesis, the set of points in S^* at which cancellation holds in βS is dense in S^*. ❑

REMARK: Right (left) cancellative elements in $\beta\mathbb{N}$ which are not left (right) cancellative in $\beta\mathbb{N}$ are produced in Theorems 5.3 and 5.4 of [6].

ACKNOWLEDGMENTS

The author thanks Professor J. W. Baker and Professor N. Hindman for their helpful comments during the preparation of this paper. Thanks also to the referee for the careful reading of the paper and for several suggestions which have made the paper more readable.

REFERENCES

1. BERGLUND, J.F., H.D. JUNGHENN, & P. MILNES. 1989. Analysis on Semigroups: Function Spaces, Compactifications, Representations. Wiley. New York.
2. BUTCHER, R.J. 1975. The Stone-Čech compactification of a topological semigroup and its algebra of measures. Ph.D. Thesis. The University of Sheffield.
3. ELLIS, R. 1969. Lectures on Topological Dynamics. Benjamin. New York.
4. FILALI, M. 1996. Right cancellation in βS and UG. Semigroup Forum. **52:** 381–388.
5. HINDMAN, N. 1982. Minimal ideals and cancellation in $\beta\mathbb{N}$. Semigroup Forum. **125:** 291–310
6. HINDMAN, N. & D. STRAUSS. 1994. Cancellation in the Stone-Čech compactification of a discrete semigroup. Proc. Edinburgh Math. Soc. **37:** 379–397
7. KUNEN, K. 1978. Weak P-points in $\mathbb{N}^*$. Coll. Math. Soc. János Bolyai. **23** Topology. Budapest. Hungary. 741–749
8. PARSONS, D.J. 1984. The centre of the second dual of a commutative semigroup algebra. Math. Proc. Cambridge Philos. Soc. **95:** 71–92

9. PYM, J.S. 1990. Compact semigroups with one-sided continuity. *In:* The Analytical and Topological Theory of Semigroups — Trends and Developments (K.H. Hofmann, J. Lawson, & J.S. Pym, Eds.). de Gruyter. Berlin. 197–217
10. RUDIN, W. 1956. Homogeneity Problems in the theory of Čech compactifications \jour Duke Math. J. **23:** 409–419
11. STRAUSS, D. 1992. Semigroup structures on $\beta\mathbb{N}$. Semigroup Forum. **44:** 238–244
12. WALKER, R.C. 1974. The Stone-Čech Compactification. Springer-Verlag. Berlin-New York.

Axiomatic Characterizations of Hyperuniverses and Applications[a]

MARCO FORTI,[b,e] FURIO HONSELL,[c,f] AND MARINA LENISA[d,g]

[b]*Dipartimento di Matematica Applicata "U. Dini"*
Università di Pisa, Italy

[c]*Dipartimento di Matematica e Informatica*
Università di Udine, Italy
and
[d]*Dipartimento di Informatica*
Università di Pisa, Italy

ABSTRACT: *Hyperuniverses* are topological structures exhibiting strong closure properties under formation of subsets. They have been used both in computer science, for giving denotational semantics *à la* Scott-de Bakker, and in mathematical logic, in order to show the consistency of set theories that do not abidc by the "limitation-of-size" principle. We present correspondences between set-theoretic properties and topological properties of hyperuniverses. We give existence theorems and discuss applications and generalizations to the *non-κ-compact* case.

INTRODUCTION

Natural frameworks for dicussing self-reference and other circular phenomena are extremely useful in areas such as the *foundations of mathematics and computer science* (see, e.g., [2], [8], [11], [20], [24]), or the *semantics of natural and programming languages* (see, e.g., [1], [3], [4], [15], [19], [21], [22]).

Hyperuniverses can play the role of such frameworks. From the foundational point of view, hyperuniverses are transitive models of nonstandard set theories which do not abide by the "limitation-of-size" principle and, hence, they provide comprehensive but intuitive "universes of sets" (see [8], [11], [12], [13]). Many classes having the size of the universe, such as those necessary for an adequate self-description, are sets of the hyperuniverse; more-

Mathematics Subject Classification: 03E65, 54B20.
Key words and phrases: non-well-founded sets, metric semantics for programming languages, hyperuniverses, exponential κ-uniformities.
[a]Work partially supported by 40% and 60% MURST grants, CNR grants, and EEC Science MASK, and BRA Types 6453 contracts.
[e]Member of GNSAGA of CNR. E-mail: forti@dm.unipi.it.
[f]E-mail: honsell@dimi.uniud.it.
[g]E-mail: lenisa@di.unipi.it.

over, it is closed under many basic set theoretic operations. From the semantical point of view, hyperuniverses provide a uniform setting for solving many recursive domain equations such as those which arise in giving denotational semantics to programming languages *à la* Scott-de Bakker (see [14], [15]).

In this paper we discuss various classes of generalized hyperuniverses which exhibit different topological properties. We investigate their closure properties under set theoretic operations and we provide axiomatic characterizations of many of them. In view of the applications, we consider hyperuniverses with *atoms* (*Urelements*). Atoms are needed to model nonreductionist foundational theories, where many important mathematical concepts (operations, relations, properties, etc.) are taken as primitive and are not represented by ingenious, but artificial, set theoretic codings (see [11]). In the semantics of programming languages, atoms are convenient in modeling directly primitive data types (such as natural numbers with the discrete topology or real numbers with the Euclidean metric) or simply actions whose internal structure is not analyzed further at the level of abstraction under consideration. Such structures could be embedded in atomless hyperuniverses only assuming specific axioms which postulate particular non-well-founded sets. In fact, the uniform structure of the closure of the well-founded sets is uniquely determined (see [12]).

In view of the above remarks, we work in a weakly extensional Zermelo-Frænkel-like set theory. Moreover, since hyperuniverses are necessarily non-well-founded, we give up the Axiom of Foundation in the ground theory, and we assume instead a suitable *Free Construction Principle* as in [11], [12]. Thus, the axioms of our ground theory are:

(ZF_0^-C) *Pairing, Union, Powerset, Infinity, Replacement, and Choice* as in Zermelo-Frænkel Set Theory ZFC;

(ExtA) *Weak Extensionality with respect to a (possibly empty) set A of atoms*:

$$\forall x z(x \in A \rightarrow z \notin x) \wedge \exists x \notin A .\, \forall y .\, y \notin x \wedge$$

$$\forall x y(x \notin A \wedge y \notin A \wedge \forall z(z \in x \leftrightarrow z \in y) \rightarrow x = y).$$

(FCA) *Unique Free Construction with respect to A*:
Given a function $f\colon X \rightarrow \mathcal{P}(X \cup A)$, where $X \cap A = \varnothing$, there exists a unique function $g\colon X \rightarrow V$ such that for all $x \in X$

$$g(x) = (f(x) \cap A) \cup \{g(y) \mid y \in f(x) \cap X\}.$$

The Axiom (FCA), used in [15], generalizes to a set theory with atoms the Axiom (X_1) of [9], which is equivalent to the Axiom (AFA) of [2].

We call *sets* objects which do not belong to A. The way Axiom (ExtA) is phrased allows us to introduce the *empty set* $\varnothing$ as the unique $x \notin A$ such that

$\forall y.\, y \notin x$. We take the *subset* relation $\subseteq$ to be defined only between sets. Therefore the *powerset* operator $\mathcal{P}(\ )$ applied to atoms yields the empty set. We refer to [17] for standard definitions and results in *set theory*.

Hyperuniverses have been used to show the consistency of the Comprehension Principle (GPC) of [8]. Axiom schema (GPC) asserts the existence of the *set* $\{x \mid \phi\}$ for any *generalized positive* formula ϕ. Since we allow for the existence of a set of atoms, we extend, in this paper, the language used to define the class *GPF* of generalized positive formulae, to include a constant denoting the set of atoms A. Hence, the class *GPFA* is defined as the least class of formulae which includes the *atomic* formulae, is closed under *conjunction*, *disjunction*, *existential* and *universal quantification* and also under the following rules of *bounded quantification* (which, strictly speaking, are nonpositive):

if ϕ is *GPFA*, then both $\forall x(x \in y \rightarrow \phi)$ and $\forall x(\theta \rightarrow \phi)$ are *GPFA*,

where θ is any formula with only x free.

The *Generalized Positive Comprehension Schema with Atoms* becomes:

(GPCA) $\exists y\, \forall x\, (x \in y \leftrightarrow \phi)$ where ϕ is *GPFA* and y is not free in ϕ.

In Section 1 we give the basic axiomatic characterizations of hyperuniverses. In particular we present and discuss various correspondences between set-theoretic properties and topological properties of hyperuniverses. In Section 2 we introduce and discuss the weaker notion of quasihyperuniverse. In Section 3 we give a construction for generating quasihyperuniverses, and we isolate conditions for obtaining hyperuniverses. In Section 4 we discuss briefly applications of hyperuniverses to semantics of programming languages, presenting general fixed point and final coalgebra theorems. Final remarks appear in Section 5. In the appendix we list definitions and basic facts from *set theory* and *general topology* used in the paper.

1. CHARACTERIZATION THEOREMS

In this Section we define *topological hyperuniverses* and we investigate their set-theoretic properties. For topological notions we refer to the appendix and [5], [23]. Throughout the paper we assume that κ is an *infinite*, *regular* cardinal.

DEFINITION 1.1: (*topological hyperuniverse*) Let A be a set of atoms. A set $N \subseteq \mathcal{P}(N) \cup A$ is a κ-*topological hyperuniverse* if the elements of $N \cap \mathcal{P}(N)$ are the closed subsets of a κ-additive, Hausdorff topology τ on N, such that A is a clopen subset of N and τ induces on $N \setminus A$ the Vietoris κ-topology.

Notice that every κ-topological hyperuniverse is *κ-compact* (see Proposition 1 of the appendix). The "formidable comprehensiveness" of topological hyperuniverses is illustrated by the following theorem, which generalizes Theorem 2 of [14] to structures with atoms.

THEOREM 1.1: Let N be a κ-topological hyperuniverse. Then we have the following.

(i) Any subset of N of size less than κ belongs to N. The Cartesian product of less than κ elements of $N \cap \mathcal{P}(N)$ belongs to N. If $X \in N \cap \mathcal{P}(N)$, then $\bigcup X$, $N \cap \mathcal{P}(X) = \Box(X)$, $\{Y \in N \mid X \cap Y \neq \varnothing\} = \Diamond(X)$, $dom(X)$ and $cod(X)$ are elements of N.

(ii) N is a transitive model of (GPCA). In particular $N \cap \mathcal{P}(N)$ is closed under all Gödel operations, excluding complement. Moreover, both the relations of membership and inclusion between members of N are members of N, as well as the identity, the singleton map, and all projections, permutations and regroupings of n-tuples.

(iii) A function f belongs to N if and only if it is continuous with closed domain, and then f is a closed uniformly continuous map. More generally, if $X \subseteq N$, a function $g: X \longrightarrow N$ is extendable to a function $f \in N$ if and only if g is uniformly continuous on X (the domain of the closure of g being then the closure of X).

(iv) For any $X \in N \cap \mathcal{P}(N)$ and for any $Y \subseteq N$, the function space $Y^X \cap N$ is the set $C(X, Y)$ of all uniformly continuous functions from X into Y, endowed with the Vietoris κ-topology on their graphs. Moreover, a set of functions $F \subseteq N^X$ is an element of N if and only if F is equicontinuous and X is closed. In particular, if $|Y| > 1$, then $C(X, Y)$ belongs to N if and only if X is discrete (i.e., if $|X| < \kappa$), and then $C(X, Y) = Y^X$.

(v) Internal and external cardinalities agree below κ. Moreover, κ is the set of all ordinals of N and its closure is the set $\bar{\kappa} = \kappa \cup \{\bar{\kappa}\} \in N$. Finally $\kappa \rightarrow (\kappa)^2_2$ and the cardinality of N is greater than or equal to 2^κ.

(vi) A point $X \in N \cap \mathcal{P}(N)$ is isolated if and only if it is made up of less than κ isolated points. Then $T = TC(X)$ is isolated and all elements of $\Pi_\kappa(T)$ are isolated. Moreover, if $T \in N$ is transitive and clopen, then $\Pi(T) \cap N \subseteq \Pi_\kappa(T)$, and equality holds if and only if $|T| < \kappa$.

Proof: The proofs of (i)–(v) follow closely the corresponding ones of Theorem 2 in [14], taking into account the presence of atoms and the fact that we have not assumed that the topology is 0-dimensional. Hence, in case $\kappa = \omega$, the topology of N is induced by a uniformity $\mathcal{U}$, which cannot have in general a basis made out of equivalences (see the appendix). To complete the proof of (v), we only need to prove that $\kappa \rightarrow (\kappa)^2_2$. Clearly, κ is regular by assumption. Moreover, κ is strong limit, since, if $X \subseteq N$ and $|X| = \xi < \kappa$, then X is closed and discrete; hence $\mathcal{P}(X)$ is closed and discrete, whence $2^{\xi} < \kappa$. Finally κ has the *tree property* since the hyperuniverse N includes κ and induces on κ the same topology as the κ-hyperuniverse N_κ of [11], [12]; if κ does not have the tree property, then the same counterexamples of [11, Section 2] yield that N is not κ-compact.

The first assertion of (vi) follows immediately from the fact that a set $x \in N \cap \mathcal{P}(N)$ is a cluster point of N if and only if it contains a cluster point of N. This can be proved as in [14]. Assume that x has no cluster point. Then $|x| < \kappa$ and all elements of x are isolated in N. Therefore, there is an entourage $U \in \mathcal{U}$ such that, for all $y \in x$, $U(y) = \{y\}$. Hence, x itself is isolated, being the only element of $U^+(x)$. On the other hand, if a cluster point of N, say y, belongs to x, then for any open entourage $U \in \mathcal{U}$ we have that $U(y) \neq \{y\}$. Hence, we can pick an element $y_U \in U(y)\setminus\{y\}$, and put $x_U = (x\setminus U(y)) \cup \{y_U\}$. Clearly, $x_U \in U^+(x)$, hence, x is not isolated. By inaccessibility of κ, any element of $\Pi_\kappa(T)$ has size less than κ if $|T| < \kappa$, and hence it is isolated if T is. What remains to be proved is that, if $x \in \Pi(T)\setminus\Pi_\kappa(T)$ and T is clopen, then x is not closed. One can easily see by transfinite induction that, if there is such a set x in N, then there exists also an $x \in N \cap (\Pi_{\kappa+1}(T)\setminus\Pi_\kappa(T))$. Clearly, $x \cap \Pi_\alpha(T) \neq x$ for all $\alpha < \kappa$. For all $\alpha < \kappa$, pick $y_\alpha \in x\setminus\Pi_\alpha(T)$: the set $Y = \{y_\alpha \mid \alpha < \kappa\}$ cannot have cluster points in $\Pi_\kappa(T) \cap N$, since $\Pi_\alpha(T) \cap N$ is open for all $\alpha < \kappa$. Hence, either Y has size less than κ, whence $x \not\subseteq \Pi_\kappa(T)$, or x is not closed. ❑

The remarkable fact that few set-theoretic properties can capture such a rich variety of subsets suggests the following definition.

DEFINITION 1.2: (*hyperuniverse*) Let A be a set of atoms. A set $N \subseteq \mathcal{P}(N) \cup A$ is a *hyperuniverse* if the following conditions are satisfied:

(H0) $A, N\setminus A \in N$ (*atoms*);
(H1) if $X \in N$, then $\{X\} \in N$ (*singletons*);
(H2) if $X, Y \in N \cap \mathcal{P}(N)$, then $X \cup Y \in N$ (*binary union*);
(H3) if $X \in \mathcal{P}(N\setminus A)$, then $\bigcap X \in N$ (*intersection*);
(H4.1) if $X \in N \cap \mathcal{P}(N)$, then $N \cap \mathcal{P}(X) = \Box(X) \in N$ (*powerset*);
(H4.2) if $X \in N \cap \mathcal{P}(N)$, then $\{Y \in N \mid X \cap Y \neq \emptyset\} = \Diamond(X) \in N$ (*dual powerset*);
(H5) if $X \in N \cap \mathcal{P}(N)$, then $\bigcup X \in N$ (*internal union*);
(H6) if $X, Y \in N \cap \mathcal{P}(N)$ and $X \cap Y = \emptyset$, then $\exists Z, W \in N \cap \mathcal{P}(N)$ such that $Z \cup W = N$, $X \cap Z = \emptyset$ and $Y \cap W = \emptyset$ (*separation*).

Actually, the conditions (H1)–(H5) in the above definion are quite natural and *prima facie* rather weak. Everybody should agree that a set-theoretic structure sufficiently self-descriptive and comprehensive to deserve the name "hyperuniverse" would satisfy all of them and, moreover, some kind of "weak complementation," such as the properties (H6.2) or (QC) below, both of which imply (H6).

DEFINITION 1.3: (*additivity*) The cardinal $\kappa = \min\{|X| \mid X \subseteq \mathcal{P}(N), \bigcup X \notin N\}$ is called the *additivity* of the hyperuniverse N.

Clearly, the additivity of a hyperuniverse is at least ω because of (H2). Moreover, by property (H5), the definition of additivity is equivalent to that of *fullness*, namely $\kappa = \min\{|X| \mid X \subseteq N, X \notin N\}$.

THEOREM 1.2: N is a κ-topological hyperuniverse if and only if N is a hyperuniverse of additivity κ.

Proof: First we show that a κ-topological hyperuniverse N satisfies the properties (H0)–(H6). Properties (H1), (H2), (H4.1), (H4.2), (H5) are instances of the generalized positive comprehension schema (GPCA), hence they follow from (ii) of Theorem 1.1. For the sake of completeness, we give a direct proof here. Property (H0) holds by definition. Properties (H2) and (H3) trivially hold for the closed sets of any topology. Property (H6) merely rephrases in terms of closed sets the separation axiom T_4, which follows by κ-compactness. Then Property (H1) also follows at once. Properties (H4.1)–(H4.2) hold since the sets $\Box(x)$ and $\Diamond(x)$ are closed in the Vietoris κ-Topology if x is closed. Moreover, since $\in$ is closed whenever the space is uniformizable (see, e.g., [18, Chap. 1, Sect. 17-IV, Theorem 1]) and the domain of a closed set is closed, by κ-compactness, $\bigcup x = dom\ (\in \cap (N \times x))$ is in N for all $x \in N \cap \mathcal{P}(N)$. Finally the additivity of N is obviously κ.

Next we show that a hyperuniverse N according to Definition 1.2 satisfies the properties of Definition 1.1, where κ is taken to be the additivity of N according to Definition 1.3.

Clearly, the complements, relative to N, of the elements of $N \cap \mathcal{P}(N)$ are the open sets of a κ-additive topology τ. This topology is T_4 by Property (H6) and it induces on $N \cap \mathcal{P}(N)$ a topology finer than the Vietoris κ-topology by Properties (H4.1)–(H4.2). Hence it suffices to prove that the Vietoris κ-topology is κ-compact, since then the topology τ is also κ-compact. As a topological space, N is in fact homeomorphic to the closed subspace of $N \cap \mathcal{P}(N)$ consisting of the singletons of the elements of N. Now, the Vietoris κ-topology is Hausdorff, since τ is T_4, and, therefore, it cannot be strictly coarser than τ on $N \cap \mathcal{P}(N)$.

Assume that the Vietoris κ-topology is not κ-compact: then there exists a regular cardinal $\mu \geq \kappa$, and a μ-sequence of closed sets g_α, $\alpha < \mu$, which we can assume strictly decreasing and continuous at limits, such that $g_0 = N$ and $g_\mu = \bigcap_{\alpha < \mu} g_\alpha = \emptyset$.

Put $G_\beta = \{g_\alpha \mid \alpha < \beta\}$, $\mathcal{G} = \overline{G_\beta}\ \{\mid \beta < \mu\}$, $\mathcal{F} = \{(g_\beta, \overline{G_\beta}) \mid \beta < \mu\}$.

Now we show that $\mathcal{F}$ is closed while $\mathcal{G}$ is not, and since we have that $\mathcal{G} \subset \bigcup\bigcup\mathcal{F}$ and G_μ is a cluster point of $\mathcal{G}$ not belonging to $\bigcup\bigcup\mathcal{F}$, we obtain a contradiction. We break up the proof in several steps.

STEP 1: $\forall\alpha < \kappa\ \forall\gamma < \alpha$ there is a neighborhood of g_α disjoint from $\{g_\delta \mid$

$\delta \le \gamma$ or $\delta > \alpha\}$; hence, only g_α's with limit α and $cof\,\alpha \ge \kappa$ can be cluster points.

Proof: Pick $z \in g_\gamma \backslash g_\alpha$ and put $W = \Box(N\backslash\{z\}) \cap \Diamond(N\backslash g_{\alpha+1})$. Clearly, W is an open neighborhood of g_α disjoint from the set $\{g_\delta \mid \delta \le \gamma$ or $\delta > \alpha\}$.

STEP 2: G_β is closed if and only if g_β is not a cluster point; otherwise $\overline{G_\beta} = G_{\beta+1}$.

Proof: Only nonempty sets can be limit points of g_β's. Now if $x \notin G_\mu$ is a nonempty set, then there exists δ such that $x \not\subseteq g_\alpha$ if and only if $\alpha \ge \delta$. Then $\delta = \beta + 1$, since the μ-sequence $\{g_\alpha\}_{\alpha<\mu}$ is continuous at limits. Thus take $U = (N\backslash\Box(g_\delta)) \cap \Box(N\backslash\{z\})$, where $z \in g_\beta\backslash x$. Clearly, U is an open neighborhood of x disjoint from G_μ. Hence, G_μ is closed. By Step 1, if $\alpha > \beta$ then g_α has a neighborhood disjoint from G_β. On the other hand, if g_β is a cluster point, then, by Step 1, β is a limit ordinal and $cof(\beta) \ge \kappa$, hence, $g_\beta = \bigcap_{\alpha<\beta} g_\alpha$. But this amounts to the fact that in the Vietoris κ-topology g_β is a limit point of G_β.

STEP 3: If X is a cluster point of $\mathcal{G}$, then there is a limit λ such that $X = \overline{G_\lambda}$.

Proof: Since G_μ is closed, $\Box(G_\mu)$ is a closed set including $\mathcal{G}$, hence $X \subseteq G_\mu$. First we show that X is an "initial segment" of G_μ. Assume that there exist $\alpha < \beta < \mu$ such that $g_\alpha \notin X$ and $g_\beta \in X$; then put $U = \Box(N\backslash\{g_\alpha\}) \cap \Diamond(W)$, where W is an open neighborhood of g_β disjoint from $\{g_\delta \mid \delta \le \alpha\}$, which exists by Step 1. Clearly, U is an open neighborhood of X, which is disjoint from $\mathcal{G}$. If X has no largest γ such that $g_\gamma \in X$, then $X = G_\lambda$ for limit λ and being closed it is also equal to $\overline{G_\lambda}$. On the other hand, suppose that there exists a largest γ such that $g_\gamma \in X$. By Step 1 and Step 2, we need only to show that g_γ is a cluster point of G_μ. Assuming the contrary, let W be an open neighborhood of g_γ such that $G_\mu \cap W = \{g_\gamma\}$ and let U be an open neighborhood of g_γ including X and disjoint from $\{g_\delta \mid \delta > \gamma\}$, which exists by Step 1. Clearly, $\Diamond(W) \cap \Box(U)$ is a neighborhood of X disjoint from $\mathcal{G}$.

STEP 4: $G_\mu = \overline{G_\mu}$ is a cluster point of $\mathcal{G}$ not belonging to $\mathcal{G}$.

Proof: The fact that G_μ is closed was proved in Step 2, and clearly, G_μ does not belong to $\mathcal{G}$. In order to show that it is a cluster point of $\mathcal{G}$, pick a neighborhood of G_μ of the form $W = \bigcap_{\alpha<\eta<\kappa} \Box(U_\alpha) \cap \bigcap_{\beta<\lambda<\kappa} \Diamond(V_\beta)$, where the U_α's are open sets including G_μ and the V_β's are open sets intersecting G_μ. For each $\beta < \lambda$, pick $\gamma_\beta < \mu$ such that $g_{\gamma_\beta} \in V_\beta \cap G_\mu$ and put $\gamma = (\sup_{\beta<\lambda} \gamma_\beta) + 1$. Then $\gamma < \mu$, since μ is regular, and clearly $G_\gamma \in W$.

STEP 5: $\mathcal{F}$ is closed.

Proof: A cluster point of $\mathcal{F}$ has to be a pair (x, X), where x is a cluster point of G_μ and X is a cluster point of $\mathcal{G}$. By Step 1 and Step 3 there are limit ordinals α and λ such that $x = g_\alpha$ and $X = \overline{G_\lambda}$. If $\alpha = \lambda$, then $(x, X) \in \mathcal{F}$. Otherwise, assuming $\alpha < \lambda$, let U be a neighborhood of g_α disjoint from $\{g_\delta \mid$

$\delta > \alpha\}$, existing by Step 1, and let V be a neighborhood of $\overline{G_\lambda}$ disjoint from $\overline{G_\beta}\ \{| \beta \leq \alpha\}$, say $V = \Diamond(W)$, where W is an open neighborhood of $g_{\alpha+1}$ disjoint from $\{g_\delta \mid \delta \leq \alpha\}$. Clearly, $U \times V$ is an open neighborhood of (x, X) disjoint from $\mathcal{F}$, contradiction. The case $\alpha > \lambda$ is dealt with similarly. ❑

We point out the following facts:

REMARK 1.1: Notice that the notion of hyperuniverse is not "first order," but rather *topological* or *set-theoretical*, since condition (H3) refers to the intersection of *arbitrary* families of subsets. A "first order approximation" of property (H3) can be obtained by introducing the schema

(H3.θ) $\forall x_1 \ldots \forall x_n\, \exists y\, \forall z\, (z \in y \leftrightarrow \forall x\, (\theta(x, x_1, \ldots, x_n) \rightarrow z \in x))$

where θ is *any* formula whose free variables are among $x, x_1, \ldots, x_n$ (see [6]).

REMARK 1.2: One can easily see from the proof above that condition (H5), which is crucial in obtaining κ-compactness, can be replaced equivalently by any of the following:

(H5.1) if $G \in N \cap \mathcal{P}(N^2)$, then $dom(G) \in N$ (*domain*);

(H5.2) if $G \in N \cap \mathcal{P}(N^2)$, then $cod(G) \in N$ (*codomain*);

(H5.3) if $F, G \in N \cap \mathcal{P}(N^2)$, then $F \circ G \in N$ (*composition*);

(H5.4) if $f \in N$ is a function and $X \in N$, then $\{ f(x) \mid \in X\} \in N$ (*replacement*);

and also by the apparently weaker

(H5.5) if $G \in N \cap \mathcal{P}(N^2)$ is one-to-one, then $G \circ G^{-1} \in N$ (*identity's restrictions*).

REMARK 1.3: Clearly, the axiom of complementation cannot be assumed consistently. Weak forms of "complementation" can be achieved via "separation" axioms. The property (H6) amounts to the Axiom of Separation T_4, which implies that the Vietoris topology is regular, hence Hausdorff. Equivalently one could have postulated the weaker property corresponding to T_3:

(H6.1) if $x \in N$, $Y \in N \cap \mathcal{P}(N)$ and $x \notin Y$, then there exist $Z, W \in N \cap \mathcal{P}(N)$ such that $Z \cup W = N$, $x \notin Z$ and $Y \cap W = \varnothing$ (*regularity*).

Moreover, if $\kappa > \omega$, then N is 0-dimensional, hence it satisfies the property:

(H6.2) if $X, Y \in N \cap \mathcal{P}(N)$ and $X \cap Y = \varnothing$, then $\exists Z \in N \cap \mathcal{P}(N)$ such that $N \setminus Z \in N$, $X \cap Z = \varnothing$ and $Y \subseteq Z$ (*0-dimensionality*).

The property (H6.2) has the flavor of an *approximation property*, and it amounts to asserting that the complementable sets of the Universe are enough to distinguish every set, namely:

(QC) Every set is the intersection of a family of complementable sets (*quasi-complementation*).

A surprisingly strong property can be obtained from the Axiom (QC) by postulating that families of size κ of complementable sets are sufficient to approximate sets:

THEOREM 1.3: N is a κ-ultrametrizable hyperuniverse if and only if N satisfies the properties (H0.5) and

($QC^{\leq \kappa}$) every element of $\mathcal{P}(N) \cap N$ is the intersection of at most κ complementable elements of N.

Proof: Clearly, ($QC^{\leq \kappa}$) implies (H6). By κ-compactness, any nested family of κ clopen subsets of N^2, whose intersection is Δ_N, is a uniformity basis for N, hence N is κ-metrizable. ❑

Clearly, the clopen subsets of N include all clopen sets of atoms and are stable with respect to complementation, $\Box$'s, $\Diamond$'s, $\bigcup_{<\kappa}$, and $\bigcap_{<\kappa}$. A strong "Minimality Principle" for hyperuniverses can be obtained by adding to the clopen approximability property (QC) the following *Axiom of Constructibility*:

(CD) the clopen subsets of N are the least subset of $\mathcal{P}(N)$ including a clopen subbasis of A, N, and stable with respect to $\Box$'s, $\Diamond$'s, $\bigcup_{<\kappa}$ and $\bigcap_{<\kappa}$ (*clopen describability*).

REMARK 1.4: In hyperuniverses, Choice and Ordering Principles are strictly connected to metrizability, as shown in [14]. In fact, we have the following theorem (whose proof is a mere rephrasing) to the case with atoms, of that of Theorems 3 and 4 of [14].

THEOREM 1.4: A topological κ-hyperuniverse N is κ-metrizable if and only if it satisfies any of the following properties:

(H7.1) there exists $R \in N \cap \mathcal{P}(N^2)$ which is a total ordering of N (*ordering principle*);

(H7.2) there exists $F \in N$, $F\colon N^2 \to N$ such that $F(x, y) = F(y, x) \in \{x, y\}$ for all $x, y \in N$ (*weak selection*);

(H7.3) there exists $F \in N$, $F\colon \Diamond(N) \to N$ such that $F(x) \in x$ for all $x \in \Diamond(N)$ (*universal choice*).

REMARK 1.5: Given a binary relation $R \subseteq \mathcal{P}((N\backslash A)^2)$, define the relation:

$$R^+ = \{(x, y) \mid x \cap A = y \cap A \wedge$$
$$\forall t \in (x\backslash A).\ \exists s \in (y).\ tRs \wedge \forall s \in (y\backslash A).\ \exists t \in (x\backslash A).\ tRs\}.$$

Compare the definition of R^+ with the definition of exponential uniformity of Definition 3(v). A *bisimulation* on N is a relation R such $R \subseteq R^+$. One can introduce the axiom of *strong extensionality up to atoms*, as follows:

(H8) two sets are equal if they belong to some bisimulation.

Then a standard κ-compactness argument yields

THEOREM 1.5: A hyperuniverse N satisfies the property (H8) if and only if no element of the uniformity of N is a bisimulation.

2. QUASIHYPERUNIVERSES

The κ-compactness property of κ-hyperuniverses is crucial in showing strong comprehension properties, but it is clearly very demanding. On one hand, if $\kappa = \omega$, it excludes important spaces of atoms such as the natural numbers or the reals. On the other hand, if $\kappa > \omega$, κ-compactness necessitates "large cardinal" assumptions. In this section we generalize the notion of hyperuniverse to noncompact spaces.

DEFINITION 2.1: (κ-*topological quasihyperuniverse*) Let A be a set of atoms. A set $N \subseteq \mathcal{P}(N) \cup A$ is a κ-*topological quasihyperuniverse* if the elements of $N \cap \mathcal{P}(N)$ are the closed subsets of a κ-additive T_1 topology τ on N, such that A is a clopen subset of N.

The interest of quasihyperuniverses appears once suitable topological properties are assumed. Clearly the κ-topological hyperuniverses are precisely those κ-topological quasihyperuniverses whose topology is κ-compact.

DEFINITION 2.2: (*special quasihyperuniverses*) Let N be a κ-topological quasihyperuniverse, τ the natural topology on N, υ the corresponding Vietoris κ-topology on $N \backslash A$, and τ_+ the topology induced by τ on $N \backslash A$:

(i) N is κ-*Vietoris* if τ_+ is finer than υ;

(ii) N is κ-*Bolzano* if every subset of N of size κ has a cluster point;

(iii) N is κ-*Tychonoff* if τ is induced by a κ-uniformity $\mathcal{U}$ such that τ_+ is induced by the exponential uniformity $\mathcal{U}^+$;

(iv) N is κ-*Hausdorff* if τ is induced by a bounded κ-metric ρ such that τ_+ is induced by the Hausdorff κ-distance ρ_H.

Various classes of κ-topological quasihyperuniverses can be characterized by purely set-theoretic properties, as was the case for κ-hyperuniverses.

THEOREM 2.1: N is a κ-topological quasihyperuniverse if and only if it has additivity $\geq \kappa$ and satisfies the properties (H0)–(H3). Moreover:

(i) N is κ-Vietoris if and only if it satisfies (H0)–(H4);

(ii) N is both κ-Bolzano and κ-Vietoris if and only if it has additivity κ and it satisfies the properties (H0)–(H4) and

(H5$^{\leq\kappa}$) if $X \in N \cap \mathcal{P}(N)$ and $|X|\,|\bigcup X| \leq \kappa$, then $\bigcup X \in N$;

(iii) if N is κ-Tychonoff, then N satisfies (H4.1) and (H6.1) ((H6.2) if $\kappa > \omega$).

Proof: (i) Immediate.

(ii) We call κ-Bolzano a space which satisfies the property that every subset of size at least κ has a cluster point. The validity of (H0)–(H4) in κ-Bolzano quasihyperuniverses is proved as in Theorem 1.2. The validity of (H5$^{\leq\kappa}$) can be shown following the lines of Theorem 1.2. Simply take into account that a closed subset of a κ-Bolzano space is κ-Bolzano, that κ-Bolzano sets of size less than or equal to κ are closed, and that images under continuous functions of κ-Bolzano subsets are κ-Bolzano.

In order to prove the reverse implication, we recall that the κ-Bolzano Property is equivalent to the nonexistence of a strictly decreasing κ-sequence of closed subsets with empty intersection. Therefore, we can proceed as in the proof of Theorem 1.2 and take $\mu = \kappa$, thus obtaining that $\mathcal{F}$, $\mathcal{G}$, and all G_β's have size not exceeding κ.

(iii) Property (H6.1) is immediate (and if $\kappa > \omega$, also (H6.2), since then any κ-uniformity has a basis of equivalences). To show (H4.1) we prove that $\diamond(G)$ is open for G open. Let $X \cap G \neq \varnothing$ and pick $x \in X \cap G$. Let $U \in \mathcal{U}$ be such that $U(x) \subseteq G$. Then $U^+(X) \subseteq \diamond(G)$. In fact, if $Y \in U^+(X)$, then there is $y \in Y$ such that $(y, x) \in U$ and hence, $y \in G \cap Y$. ❑

THEOREM 2.2: Let N be a κ-topological quasihyperuniverse:

(i) N is κ-Hausdorff if and only if N is κ-Tychonoff with respect to a κ-metrizable uniformity;

(ii) if N is paracompact (in particular, if N is κ-Hausdorff or 0-dimensional) and κ-Tychonoff with respect to the universal uniformity, then N is κ-Vietoris;

(iii) if N is κ-Bolzano, κ-Vietoris, and κ-metrizable, then N is a κ-hyperuniverse, hence also κ-Hausdorff;

(iv) there exist κ-Hausdorff quasihyperuniverses which are not κ-Vietoris.

Proof: (i) Use the fact that if a bounded κ-metric ρ induces the uniformity $\mathcal{U}$, then the Hausdorff κ-metric ρ_H induces the exponential uniformity $\mathcal{U}^+$.

(ii) We only need to show property (H4.2). Since N is paracompact, the universal uniformity $\mathcal{U}$ includes all open entourages of the diagonal [5, Theorem 5.1.12]. We show that $\square(G)$ is open for G open. Let $X \in N$ be such that $X \subseteq G$. For each $x \in X$, pick $U_x \in \mathcal{U}$ such that $U_x(x) \subseteq G$. Put

$$U = (G\setminus X)^2 \cup \bigcup_{x\in X}(U_x(x))^2,$$

then $U^+(X) \subseteq \Box(G)$. In fact, if $(Y, X) \in U^+$, then $\forall t \in Y.\ \exists s \in X\ (t, s) \in U$, hence there exists $x \in X$ such that both $t, s \in U_x(x)$; therefore, $t \in G$.

(iii) Since N is κ-metrizable, the κ-Bolzano property implies κ-compactness of the topology τ. Now, the Vietoris κ-topology on N is Hausdorff and coarser than τ; hence, it coincides with τ on $N \cap \mathcal{P}(N)$. Finally, if ρ is any bounded κ-metric inducing τ on N, then ρ_H induces the Vietoris κ-topology on $N \cap \mathcal{P}(N)$.

(iv) Let $^*\mathbf{R}$ be a (nonstandard) model of the *reals* of cofinality κ, let A be isometric to $^*\mathbf{R}$ with the Euclidean κ-metric ρ, and let $\mathcal{U}$ be the κ-uniformity induced by ρ. In the quasihyperuniverse $\mathcal{N}_\kappa(\mathcal{U})$ of Theorem 3.1 below the set $\Diamond(\Delta_A)$ is not closed. For example, pick an increasing unbounded κ-sequence $\langle x_\alpha\rangle_{\alpha<\kappa}$ in $^*\mathbf{R}^+$ and put $X = \{(x, x + \frac{1}{x}) \mid x \in {}^*\mathbf{R}^+\}$. Clearly, the κ-sequence $\langle X \cup \{(x_\alpha, x_\alpha)\}\rangle_{\alpha<\kappa}$ converges to X in the Hausdorff κ-metric. ❑

Although non-κ-compact quasihyperuniverses lack many comprehension properties, we still have the following theorem, whose proof we omit since it is similar to that of Theorem 1.1:

THEOREM 2.3: Let N be a κ-Tychonoff quasihyperuniverse. Then:

(i) Any subset of N of size less than κ belongs to N. The Cartesian product and the union of less than κ elements of $N \cap \mathcal{P}(N)$ belong to N.

(ii) $N \cap \mathcal{P}(N)$ is closed under arbitrary intersections and □'s (and also ◇'s if N is κ-Vietoris). Moreover, both the relations of membership and inclusion between members of N are members of N, as well as the identity, the singleton map, and all projections, permutations and regroupings of n-tuples.

(iii) All continuous functions with closed domain belong to N.[1]

(iv) Internal and external cardinalities agree below κ. Moreover, κ is the set of all ordinals of N and its closure is the set $\bar{\kappa} = \kappa \cup \{\bar{\kappa}\} \in N$. Finally, the cardinality of N is greater than or equal to 2^κ.

3. EXISTENCE THEOREMS

In this section we give existence theorems for various kinds of quasihyperuniverses. More precisely, we describe a construction for obtaining quasihyperuniverses starting from any uniform structure on the set A of atoms, which generalizes the one for κ-ultrametric spaces given in [11]. Under suitable assumptions, which are specified in the theorems below, this con-

[1] *Caveat*: also continuous functions with nonclosed domain, and even noncontinuous functions, can have closed graphs and hence can be in N.

struction produces quasihyperuniverses satisfying several of the axioms of Section 1.

Let $\mathcal{U}$ be a uniformity on the set of atoms A. For any cardinal μ, let $e_0(\mu) = \mu + 1$, $e_{\alpha+1}(\mu) = \mu + 2^{e_\alpha(\mu)}$, and $e_\lambda(\mu) = \Pi_{\alpha<\lambda}\, e_\alpha(\mu)$ for limit λ. Put $\nu = e_{\kappa+1}(|A|)$ and $S = H(\nu^+)$, where $H(\nu^+) = \{x \mid |TC(x)| \leq \nu\}$. Since we are assuming (FCA), S is a set (see, e.g., [10, I]). We endow the set $S \cup A$ with a "pseudo-uniform" topology. To this end, define inductively, for each $U \in \mathcal{U}$, the κ-sequence of entourages $\langle U_\alpha \rangle_{\alpha<\kappa}$ as follows: $U_0 = U \cup (S \times S)$, $U_{\alpha+1} = U \cup (U_\alpha{}^+ \cap (S \times S))$ and $U_\lambda = \bigcap_{\alpha<\lambda} U_\alpha$ for limit λ.[2] Similarly to [11], we define on $S \cup A$ the relations $\approx_{\mathcal{U}}$ $\mathcal{U}$-*equivalence*) and $\in_{\mathcal{U}}$ ($\mathcal{U}$-*membership*) as follows:

$$x \approx_{\mathcal{U}} y \text{ if and only if } \forall U \in \mathcal{U}\ \forall \alpha < \kappa\ (x, y) \in U_\alpha;$$

$$x \in_{\mathcal{U}} y \text{ if and only if } \begin{cases} \exists x' \approx_{\mathcal{U}} x\ \exists y' \approx_{\mathcal{U}} y.\ x' \in y' \\ \text{or equivalently} \\ \forall U \in \mathcal{U} \quad \forall \alpha < \kappa\ \exists x' \in y.\ (x, x') \in U_\alpha . \end{cases}$$

The relation $\approx_{\mathcal{U}}$ is indeed an equivalence and it induces the identity on atoms. Clearly, it is reflexive and symmetric. Moreover, since $\mathcal{U}$ is a uniformity, for all $U \in \mathcal{U}$ there exists $W \in \mathcal{U}$ such that $W \circ W \subseteq U$ and, by induction, the same inclusion holds at level α, using the fact that $W^+ \circ W^+ \subseteq (W \circ W)^+$. Hence $\approx_{\mathcal{U}}$ is transitive.

The family $\{U_\alpha \mid \alpha < \kappa,\ U \in \mathcal{B}\}$, where $\mathcal{B}$ is any basis of the uniformity $\mathcal{U}$, is a *pseudo-κ-uniformity basis* on $S \cup A$, which allows for nice topological characterizations of both $\mathcal{U}$-membership and $\mathcal{U}$-equivalence. It is easily seen that $x, y \in S$ are $\mathcal{U}$-equivalent if and only if they have the same closure. Moreover, the $\mathcal{U}$-members of any $x \in S$ are exactly the *ordinary* members of its closure (compare to [11, Lemma 1.1]).

Define the function $f\colon S \to \mathcal{P}(S \cup A)$ by $f(x) = \{y \in S \cup A \mid y \in_{\mathcal{U}} x\}$ and let g be the unique function given by the Axiom (FCA). The transitive set $\mathcal{N}_\kappa(\mathcal{U}) = A \cup cod(g)$ provides a *transitive collapse* of the *quotient structure* $(S \cup A, \in_{\mathcal{U}}) \approx_{\mathcal{U}}$. Put $\sigma = g \cup \Delta_A$: a straightforward generalization of [11] yields the following.

THEOREM 3.1: The function $\sigma\colon S \cup A \to \mathcal{N}_\kappa(\mathcal{U})$ is the unique function satisfying the following properties:
(i) for all $x \in S \cup A$, $x \approx_{\mathcal{U}} \sigma(x)$;
(ii) for all $x, y \in S \cup A$, $\sigma(y) = \sigma(x)$ if and only if $x \approx_{\mathcal{U}} y$;
(iii) for all $x, y \in S \cup A$, $\sigma(y) \in \sigma(x)$ if and only if $x \in_{\mathcal{U}} y$.
In particular $\in_{\mathcal{U}}$-membership on $\mathcal{N}_\kappa(\mathcal{U})$ agrees with ordinary membership. The entourages $\{U_\alpha \cap (\mathcal{N}_\kappa(\mathcal{U}) \times \mathcal{N}_\kappa(\mathcal{U})) \mid U \in \mathcal{U},\ \alpha < \kappa\}$ are the basis of

[2]See Definition 3(v) of the appendix for the notation U^+.

a κ-uniformity $\mathcal{W}$ on $\mathcal{N}_\kappa(\mathcal{U})$ inducing on A the original uniformity. For all $x \subseteq \mathcal{N}_\kappa(\mathcal{U})$ the closure of x is $\sigma(x)$. Moreover, $\mathcal{W}^+$ is the uniformity induced by $\mathcal{W}$ on $\mathcal{N}_\kappa(\mathcal{U})\setminus A$. Hence, $\mathcal{N}_\kappa(\mathcal{U})$ is a κ-Tychonoff quasihyperuniverse, including as a clopen subset the space $N_\kappa = \{x \in \mathcal{N}_\kappa(\mathcal{U}) \mid TC(x) \cap A = \varnothing\}$.[3]

Proof: First of all we show that $|\mathcal{N}_\kappa(\mathcal{U})| \leq \nu$ and hence $\mathcal{P}(\mathcal{N}_\kappa(\mathcal{U})) \subseteq S$, in particular $\sigma(x) \in S$ for all $x \in S$. To this end we introduce the decreasing κ-sequence of equivalence relations $\langle E_\alpha \rangle_{\alpha \leq \kappa}$ as follows: $E_0 = \Delta_A \cup (S \times S)$, $E_{\alpha+1} = \Delta_A \cup (E_\alpha{}^+ \cap (S \times S))$, and $E_\lambda = \bigcap_{\alpha<\lambda} E_\alpha$ for limit λ. Clearly, $E_\alpha \subseteq U_\alpha$ for all $\alpha < \kappa$ and all $U \in \mathcal{U}$, hence $\approx_{\mathcal{U}}$ is coarser than E_κ. Call ν_α the number of equivalence classes *modulo* E_α. An easy computation yields: $\nu_0 = |A| + 1$, $\nu_{\alpha+1} = |A| + 2^{\nu_\alpha}$, and $\nu_\lambda \leq \Pi_{\alpha<\lambda} \nu_\alpha$ for limit λ; hence, $\nu_\kappa \leq \nu$.

To prove item (i) we show, by induction on α, that for all $U \in \mathcal{U}$ and all $x \in S \cup A$, $(x, \sigma(x)) \in U_\alpha$. The initial and the limit steps are true by definition. In order to show the successor step, given $U \in \mathcal{U}$, pick $W \in \mathcal{U}$ such that $W \circ W \subseteq U$. Now, for all $s \in x$, we have that $\sigma(s) \in \sigma(x)$ and $(s, \sigma(s)) \in U_\alpha$ by induction hypothesis. On the other hand, for any $t \in \sigma(x)$, there is $y \in_{\mathcal{U}} x$ such that $t = \sigma(y)$. Pick $s \in x$ such that $(s, y) \in W_\alpha$: then $(s, t) \in W_\alpha \circ W_\alpha$, therefore, $(x, \sigma(x)) \in U_\alpha{}^+$.

The remaining assertions are straightforward, taking into account that $\sigma(a) = a$ if a is an atom and $\sigma(x) = \{\sigma(y) | y \in_{\mathcal{U}} x\}$ for $x \in S$. ❑

Taking into account the characterizations of Theorem 2.2, we obtain the following embedding theorems in special quasihyperuniverses:

COROLLARY 3.1: Any κ-additive Tychonoff space X is homeomorphic to the space of atoms of a κ-Tychonoff quasihyperuniverse N. If, moreover, X is paracompact, then N can be taken κ-Vietoris. Finally, N can be taken κ-Hausdorff if and only if X is κ-metrizable.

If $\kappa = \omega$ and the topology on the space X is compact, the previous construction yields a compact hyperuniverse. Actually, we have:

THEOREM 3.2: A topological space X is homeomorphic to the space of atoms of a ω-hyperuniverse N if and only if X is compact. Moreover:

(i) N can be taken (ultra)metrizable if and only if X is (ultra)metrizable;
(ii) N can be taken 0-dimensional if and only if X is 0-dimensional;
(iii) in all cases N can be chosen so as to satisfy the axioms (H8) and (FCA);
(iv) if X is 0-dimensional, then N can be chosen so as to satisfy the axioms (QC) and (CD).

Proof: Assertions (i) and (ii) follow from the fact that the construction outlined in this section preserves compactness, (ultra)metrizability, and 0-

[3]The space N_κ was introduced in [8].

dimensionality. In order to prove (iii), one can adapt the proof of Theorem 5 in [12], so as to take care of the presence of atoms. To show (iv), notice that X, being compact and 0-dimensional, has a uniformity basis $\mathcal{B}$ made up by equivalences having a finite number of clopen classes. The proof then runs exactly as in the case of uncountable κ (see below). ❑

If $\kappa > \omega$, even if the topolgy on the space X is κ-compact, the previous construction does not provide in general a κ-compact hyperuniverse. By Theorem 1.1(v), if $\kappa \not\to (\kappa)^2_2$, then there are no κ-hyperuniverses. However, we have:

THEOREM 3.3: Let κ be an uncountable regular cardinal. Every κ-additive Tychonoff space X is homeomorphic to the space of atoms of a κ-Tychonoff, κ-Vietoris, 0-dimensional quasihyperuniverse N. Moreover:

(i) if κ is strongly compact, then N can be taken to be a κ-hyperuniverse if and only if X is κ-compact;
(ii) N can be taken to be a κ-Hausdorff hyperuniverse if and only if $\kappa \to (\kappa)^2_2$, X is κ-metrizable and κ-compact;
(iii) in all cases the hyperuniverse N can be chosen so as to satisfy the axioms (H8), (FCA), (QC), and (CD).

Proof: Since $\kappa > \omega$, the universal uniformity $\mathcal{U}$ on any κ-Tychonoff space has a basis $\mathcal{B}$ made up of equivalences. Then for all $U \in \mathcal{B}$ and $\alpha < \kappa$ the U_α's are equivalences, hence the quasihyperuniverse $N = N_\kappa(\mathcal{U})$ is κ-Tychonoff, κ-Vietoris, and 0-dimensional. Moreover, if the atom space is κ-compact, then every U_α has less than κ clopen classes.

In order to achieve (i), we need only to prove that N is κ-compact. Let C be a family of closed subsets of N such that the intersection of less than κ members of C is nonempty: we have to show that C has nonempty intersection. Let $\mathcal{D}$ be the least κ-complete filter including C, and let $\mathcal{F}$ be a κ-complete ultrafilter refining $\mathcal{D}$, which exists since κ is strongly compact. For each $U \in \mathcal{B}$ and $\alpha < \kappa$ there is exactly one U_α-class in $\mathcal{F}$, say F^U_α. Moreover, either all F^U_α's are included in A (and we call $\mathcal{F}$ *atomlike*), or all of them are disjoint from A (and we call $\mathcal{F}$ *setlike*). By κ-compactness of A, if $\mathcal{F}$ is atomlike, then $\bigcap F^U_\alpha \neq \emptyset$. We now show that the same holds for setlike ultrafilters. Let S be the set of all setlike κ-complete ultrafilters on N, and define the function $f\colon S \to \mathcal{P}(S \cup A)$ by

$$f(x) = \{\mathcal{G} \in S \mid \forall U \in \mathcal{B}\ \forall \alpha < \kappa\ \exists s \in G^U_\alpha\ \exists x \in F^U_{\alpha+1}.\ s \in x\} \cup \{a \in A \mid \forall U \in \mathcal{B}\ \forall \alpha < \kappa\ .\ a \in F^U_\alpha\}.$$

Let g be the unique function given by the axiom (FCA): an easy cardinality argument proves that the range of g is a transitive set of size not exceeding κ, hence, $g\colon S \to S$. The same inductive argument used in the proof of

Theorem 3.1 yields that $(\sigma\,(g(\mathcal{F})), F^U_\alpha) \in U_\alpha$ for all $U \in \mathcal{B}$ and all $\alpha < \kappa$. Therefore, $\sigma\,(g(\mathcal{F})) \in \bigcap F^U_\alpha \subseteq \bigcap C$, and (i) is proved.

The proof of (ii) is given in [11], where the construction of $N_\kappa(\mathcal{U})$ is performed using the canonical basis of the uniformity $\mathcal{U}$ induced by a κ-ultrametric on A.

In order to prove (iii), one can adapt the proof of Theorem 5 in [12] so as to take care of atoms, as for the case of $\kappa = \omega$; one then obtains that (FCA) holds in N, hence, (H8) also follows.

Let $\mathcal{A}$ be the least family of subsets of N which includes all equivalence classes modulo U for all $U \in \mathcal{B}$, and is closed under $\Box$'s, $\Diamond$'s, $\bigcup_{<\kappa}$ and $\bigcap_{<\kappa}$. Denoting by $[x]^U_\alpha$ the equivalence class of x modulo U_α, we have by definition

$$[x]^U_{\alpha+1} = \bigcap_{s \in x} \Diamond([s]^U_\alpha) \cap \Box(\bigcup_{s \in x} [s]^U_\alpha)$$

and, for limit λ, $[x]^U_\lambda = \bigcap_{\alpha<\lambda} [x]^U_\alpha$. Since all U_α's have less than κ classes, it follows that all classes belong to $\mathcal{A}$. Now, by κ-compactness, every clopen set is the union of less than κ classes, and the property (CD) follows. Finally, (QC) holds in any 0-dimensional space. ❑

4. APPLICATIONS

4.1. Fixed Point Theorems in Hyperuniverses

In this subsection we generalize the fixed point theorems of [15] to arbitrary hyperuniverses. Thereby, also general hyperuniverses can be used as frameworks for providing semantics for concurrent languages, in that they allow to solve naturally the recursive (domain) equations used for defining processes.

The first theorem can be seen as a generalization of Tarski's Fixed Point Theorem; in general it does not yield uniqueness of fixed points.

THEOREM 4.1: Let N be a κ-hyperuniverse, let $f \in N$ be a function and let $\leq\, \in N$ be a partial order such that f is monotone with respect to $\leq$. Put $D = \{x \in dom(f) \mid x \leq f(x)\}$ and assume that any λ-chain in D with $\lambda < \kappa$ has an upper bound in D. Then, for all $x \in D$, there is $z \geq x$ such that $z = f(z)$.

Proof: Given $x \in D$, pick a κ-sequence $\langle x_\alpha \rangle_{\alpha<\kappa}$ in D such that $x_0 = x$, $x_{\alpha+1} = f(x_\alpha)$, and, for limit $\lambda < \kappa$, $x_\lambda \geq x_\gamma\ \forall \gamma < \lambda$.

If the κ-sequence is not eventually constant, by κ-compactness it has a cluster point z, say. Since f is uniformly continuous, $(z, f(z))$ is a cluster point of $\{(x_\alpha, x_{\alpha+1})\}_{\alpha<\kappa}$. Similarly $(f(z), z)$ is a cluster point of $\{(x_{\alpha+1}, x_\beta)\}_{\alpha+1<\beta<\kappa}$. Hence, both $(z, f(z))$ and $(f(z), z)$ are in $\leq$. ❑

It is interesting to notice that Theorem 4.1 holds also for κ-*Bolzano quasi-*

hyperuniverses, since only κ-sequences are involved in its proof.

Using the above theorem, a plethora of reflexive domain equations, involving many different set-theroretic operators, are immediatly solvable in the setting of κ-hyperuniverses. Theorem 4.1 applies in particular whenever $(D, \leq)$ is a κ-complete upper semilattice (i.e., every subset of size $< \kappa$ has a least upper bound), and it provides a κ-complete upper semilattice of fixed points. In case $\leq$ is $\subseteq$ ($\supseteq$), it amounts to a special case of Tarski's Fixed Point Theorem. It yields immediately fixed points to e.g., the following functions:

$$F_1(X) = (X_0 \cup X) \times \mathcal{P}(X) \times \Diamond(X), \quad F_2(X) = X_0 \cup cod(X) \cup X^{-1}.$$

It is interesting to consider also Banach-type fixed point theorems in the context of general hyperuniverses. To this end we need to introduce notions corresponding to the "k-Lipschitz" condition.

DEFINITION 4.1: Let N be a κ-Tychonoff quasihyperuniverse, let $\mathcal{B}$ be a basis for the κ-uniformity $\mathcal{U}$ of N and let $\nu: \mathcal{B} \rightarrow \mathcal{U}$ be a monotone map, such that $\{\nu(U) \mid U \in \mathcal{B}\}$ is again a uniformity basis.

A function $f \in N$ is *ν-controlled* if $\forall U \in \mathcal{B}. (x, y) \in U \Rightarrow (f(x), f(y)) \in \nu(U)$. The function $f \in N$ is *$\mathcal{B}$-nonexpansive* if f is $\Delta_{\mathcal{B}}$-controlled. A function $f \in N$ is *$\mathcal{B}$-contractive* if f is ν-controlled for some ν satisfying $\bigcap_{\alpha < \kappa} U_\alpha = \Delta_N$, where $U_0 = N^2$, $U_{\alpha+1} = \nu(U_\alpha)$ and $U_\lambda = \bigcap_{\alpha < \lambda} U_\alpha$.

In every κ-Tychonoff quasihyperuniverse N, the set $X \rightarrow_\nu Y$ of all ν-controlled functions from to X to Y is an element of N for any ν. Hence, in particular, the set of all $\mathcal{B}$-nonexpansive functions belongs to N. Theorem 4.1 allows to solve domain equations involving function spaces such as $X = \{f: N \rightarrow_\nu N \mid f(x) \in A \text{ for all } x \in X\}$.

A map ν, as in the definition of $\mathcal{B}$-contractive functions, can exist in a κ-hyperuniverse N only if N is κ-metrizable. Hence, Banach's Fixed Point Theorem can be stated as follows:

THEOREM 4.2: Let N be a hyperuniverse and let $\mathcal{B}$ be a uniformity basis for N. Any $\mathcal{B}$-contractive $f \in N$ such that $cod(f) \subseteq dom(f)$ has a unique fixed point.

The above theorem is surprisingly powerful: it allows to solve domain equations which involve "mixed variance" constructors (e.g., $X = A \cup (X \rightarrow_{\Delta_{\mathcal{B}}} X)$). Notice that Theorem 4.2 holds for any *complete* ω-Hausdorff quasihyperuniverse: it would be interesting to compare its strength to that of [3].

4.2. Hyperuniversal Categories

A categorical viewpoint is often very useful for giving and investigating

[4]If (X, δ) is a complete metric space and $f: X \rightarrow X$ is k-Lipschitz for $k < 1$ (i.e., $\delta(f(x), f(y)) \leq k\delta(x, y)\ \forall x, y \in X$), then f has a unique fixed point.

the semantics of programs. In fact, in providing semantics to programming languages, one has to introduce simultaneously both *domains* of denotations as well as *interpretation functions*. In this subsection we associate to hyperuniverses suitable categories and we attempt to recast, in the setting of hyperuniverses, the *final semantics* approach of [2], [22], [16]. Final coalgebra theorems are crucial tools in this approach. They allow to prove uniformly the existence of interpretation functions and they lead to coinductive characterizations of significant equivalences on programs.

Using the strong closure properties of hyperuniverses, we can immediately view them as categories as follows:

DEFINITION 4.2: (*hyperuniversal categories*) Let N be a κ-Tychonoff quasihyperuniverse and let $\mathcal{B}$ be a uniformity basis for N.

(i) The category $\mathbf{C}_N$ having the elements of N as objects and the functions of N as arrows is called the *external category* associated to N.
(ii) The category $\mathbf{C}_N^{\mathcal{B}}$ having the elements of N as objects and the $\mathcal{B}$-nonexpansive functions of N as arrows is called the *internal category* associated to N, $\mathcal{B}$.

Using the hyperuniversal categories defined above, and the fixed points theorems of Subsection 4.1, we recover two *final coalgebra theorems* in the style of [2], [22] for suitable endofunctors. First of all we give a *special final coalgebra theorem*:

THEOREM 4.3: Let N be a κ-topological quasihyperuniverse satisfying (FCA). Let $F = (F_{obj}, F_{arr})$ be an endofunctor on $\mathbf{C}_N$, which is inclusion preserving and uniform on maps.[5] Then F_{obj} has a greatest fixed point X, such that (X, Δ_X) is the final F-coalgebra.

Proof: The function F_{obj} has a greatest fixed point X by Theorem 4.1. Using Axiom (FCA), one can adapt the proof of Theorem 4.22 of [22] and show that (X, Δ_X) is the final F-coalgebra. ❑

Actually, in proving the above theorem one simply uses Tarski's Fixed Point Theorem. Using Theorem 4.1 in its full generality, one can get many "weakly" final coalgebras theorems. Using instead Theorem 4.2, we can state a *metric final coalgebra theorem*:

THEOREM 4.4: Let N be a hyperuniverse and let $\mathcal{B}$ be a uniformity basis for N. Let $F = (F_{obj}, F_{arr})$ be an endofunctor on $\mathbf{C}_N^{\mathcal{B}}$, such that F_{obj} and all F_{arr}'s are $\mathcal{B}$-contractive. Then F_{obj} has a unique fixed point X, such that (X, Δ_X) is the final F-coalgebra.

[5] That is, if $j: A \to B$ is an inclusion map, then $F_{arr}(A, B)(j): F_{obj}(A) \to F_{obj}(B)$ is also an inclusion map. The technical notion of "uniform on maps functor" is given in [2].

Proof: The function F_{obj} has a unique fixed point X by Theorem 4.2. The F-coalgebra (X, Δ_X) can be seen to be final by modifying the proof of Theorem 5.5 of [22]. ❑

REMARK 4.1: The category $\mathbf{C}_N$ of Definition 4.2(i) is indeed *external*, in the sense that the hom-sets $C_N(A, B)$ in general *are not* elements of N. On the other hand, the hyperuniversal category $\mathbf{C}_N^{\mathcal{B}}$ of Definition 4.2(ii) is *internal* in the sense that all *hom-sets* are *objects* of $\mathbf{C}_N^{\mathcal{B}}$ and, moreover, the operations of *domain* and *codomain* of arrows, the operation which assigns to every object the *identity* arrow, and all *composition operations* $Comp_{ABC}$: $\mathbf{C}_N^{\mathcal{B}}(B, C) \times \mathbf{C}_N^{\mathcal{B}}(A, B) \rightarrow \mathbf{C}_N^{\mathcal{B}}(A, C)$, are themselves *arrows* of $\mathbf{C}_N^{\mathcal{B}}$. It would be interesting to pursue further the possibility of adequately internalizing categorical notions in hyperuniverses.

5. FINAL REMARKS

REMARK 1. Interesting game theoretic accounts of extensionality in hyperuniverses can be given by reformulating the game originally introduced in [7]. For $x,y \in V$, we define the following two-players game $G_{x,y}$. If x,y are both nonempty sets, then Player I picks a set E_1 such that $dom(E_1) = x$ and $cod(E_1) = y$. Presented with E_1, Player II picks a pair $(x_1, y_1) \in E_1$. The game is over if Player I is presented with a pair not consisting of two nonempty sets. In this case, Player I wins if presented with $(x, y) \in \Delta_{A \cup \{\varnothing\}}$, otherwise Player II wins. If the game goes on indefinitely, then Player I wins. One can easily see that the Axiom of strong extensionality (H8) can be rephrased as follows:

(H8.1) Two sets x and y are equal if and only if Player I has a winning strategy in the game $G_{x,y}$.

Using the notation of Section 3 one can see that, whenever two sets x,y are in the relation E_{n+1} of the proof of Theorem 3.1, then Player I can postpone defeat in $G_{x,y}$ for n moves. Hence, we can characterize the equivalence $\approx_{\mathcal{U}}$ on S, in the case $\kappa = \omega$, as follows:

THEOREM 5.1: Let $\mathcal{U}$ be a uniformity on A having a countable basis. For all $x,y \in S$, $x \approx_{\mathcal{U}} y$ if and only if for all $n \in \omega$ Player I can postpone defeat for n moves in $G_{x,y}$.

In particular the following theorem holds in $\mathcal{N}_{\omega}(\mathcal{U})$:

THEOREM 5.2: If Player I can postpone defeat arbitrarily, then Player I has a winning strategy in $G_{x,y}$. In particular $\mathcal{N}_{\omega}(\mathcal{U})$ satisfies the axiom:

(H8.2) Two sets x and y are equal if and only if for all $n \in \omega$ Player I can postpone defeat for n moves in $G_{x,y}$.

REMARK 2. In this paper we consider at the outset theories with atoms, because they seem more natural in view of the applications. However, topological spaces could be embedded faithfully also in atomless hyperuniverses by suitably modifying the theorems in [13]. This amounts to use some kind of non-well-founded sets in place of *atoms*, e.g., *self-singletons*. Hence, one is forced to give up the axiom (FCA) and to assume instead in the ground universe more involved Anti-foundation Axioms. In general the resulting hyperuniverses do not satisfy precious structural principles such as the principle of clopen describability (CD) and the strong extensionality axiom (H8).

REMARK 3. The κ-Bolzano condition follows from the "Maximal Union Principle" of [24], which asserts that the cardinality of the hyperuniverse is *minimal* with respect to its additivity. Namely, we have:

THEOREM 5.3: Any κ-Vietoris quasihyperuniverse of cardinality 2^κ is κ-Bolzano. In particular N is a κ-metrizable κ-hyperuniverse if and only if N has cardinality 2^κ and satisfies the conditions (H0)–(H4) together with any of (H7.1)–(H7.3).

Proof: We proceed by contradiction. Assume that $I \subseteq N$ is a set of κ isolated points. Then $\mathcal{P}(I) \subseteq N$ consists of clopen sets. Then every filter over I is closed, and hence is in N. To see this, suppose that J is a cluster point of a filter $\mathcal{F}$. Then $J \subseteq I$ since $\square(I) = \mathcal{P}(I)$ is closed and hence, $\square(J)$ is a clopen neighborhood of J. Therefore, there exists $F \in \mathcal{F}$ such that $F \subseteq J$. Since there are 2^{2^κ} filters over I, the hyperuniverse N cannot have cardinality 2^κ. ❑

REMARK 4. It would be interesting to investigate further *approximation properties* in quasihyperuniverses, such as those which have been dealt with in [12] for the special case of the hyperuniverses N_κ:

(H9.1) Every set is approximated by sets wellfounded over A (*weak foundation*).

(H9.2) Every set is approximated by isolated sets (*pseudo-foundation*).

(H9.3) Every set is approximated by sets of hereditary cardinal < κ (*small approximation*).

For instance, under the hypothesis on X and κ of Theorems 3.2 and 3.3, one can replace (FCA) by (H9.1) in item (iii) of each of the two theorems: simply take the closure in $\mathcal{N}_\kappa(\mathcal{U})$ of the set $A \cup WF$, where WF consists of the wellfounded sets over A.

APPENDIX

Here, we recall some basic definitions and facts from *set theory and general topology*. For more details and complete proofs we refer to [5], [17].

Throughout this appendix let X be a set, let τ be a topology on X, and let κ be an infinite regular cardinal.

DEFINITION 1: Let κ be an infinite regular cardinal.

(i) κ has the *partition property* $\kappa \to (\kappa)^2_2$ if for any function p: $[\kappa]^2 \to \{0, 1\}$ there is a subset $H \subseteq \kappa$ of size κ such that p is constant on $[H]^2$; or equivalently if every tree of size κ, in which each chain has less than κ immediate successors, has a κ-branch.

(ii) κ is *strongly compact* if any κ-complete filter on a κ-complete field of sets is included in a κ-complete ultrafilter.

Notice that ω is strongly compact by the Boolean Prime Ideal Theorem. Notice also that if κ is strongly compact, then $\kappa \to (\kappa)^2_2$, and hence, κ is strongly inaccessible.

DEFINITION 2:

(i) A *filter* on X is a nonempty set $\mathcal{F} \subset \mathcal{P}(X)$ such that:

  (a) $A \in \mathcal{F}$ and $B \supseteq A \quad \Rightarrow \quad B \in \mathcal{F}$;

  (b) $A, B \in \mathcal{F} \quad \Rightarrow \quad A \cap B \in \mathcal{F}$.

(The filter generated by a set S is the least filter containing S).

(ii) A filter is κ-*complete* if it is closed under intersections of length $< \kappa$.

(iii) A *uniformity* on X is a filter $\mathcal{U}$ of *entourages* (i.e., symmetric and reflexive binary relations on X) satisfying the following conditions:

  (a) $\bigcap \mathcal{U} = \Delta_X \equiv \{(x, y) \in X \times X \mid x = y\}$ (separation);

  (b) $\forall U \in \mathcal{U}\ \exists V \in \mathcal{U}.\ V \circ V \subseteq U$ (triangular inequality), (where $\circ$ is the composition of relations.

(iv) The *topology* $\tau_{\mathcal{U}}$ induced by the uniformity $\mathcal{U}$ has as open sets $\{A \subseteq X \mid \forall x \in A\ \exists U \in \mathcal{U}\ U(x) \subseteq A\}$, where $U(x) = \{y \mid (x, y) \in U\}$.

(v) A κ-*uniformity* is a uniformity which is *a* κ-*complete filter*. A topology is κ-*additive* (shortly a κ-topology) if the intersection of less than κ open sets is open. If $\mathcal{U}$ is a κ-uniformity then $\tau_{\mathcal{U}}$ is a κ-topology.

(vi) A (κ-)*uniformity basis* $\mathcal{B}$ is a set of entourages such that the filter generated by $\mathcal{B}$ is a (κ-)uniformity.

(vii) A κ-metric (respectively, κ-ultrametric) on X is a symmetric function ρ: $X^2 \to {}^*\mathbf{R}$ such that ${}^*\mathbf{R}$ has cofinality κ, satisfying $\rho(x, y) = 0 \Rightarrow x = y$ and the triangular inequality (respectively, $\rho(x, z) \le \max\{\rho(x, y), \rho(y, z)\}$) inequality.

(viii) The (κ)-uniformity *generated* by the (κ-)metric ρ has a canonical basis $\mathcal{B} = \{U_\alpha \mid \alpha < \kappa\}$, where $U_\alpha = \{(x, y) \in X \times X \mid \delta(x, y) < \varepsilon_\alpha\}$ and $\langle \varepsilon_\alpha \rangle_{\alpha < \kappa}$ is a strictly decreasing κ-sequence with infimum 0.

DEFINITION 3:

(i) A topology is *0-dimensional* if it has a basis of clopen (i.e., simultaneously closed and open) sets.

(ii) A κ-topology is κ-compact if every open cover of X has a subcover of cardinality less than κ.

(iii) The *exponential space* $\mathcal{P}_{cl}(X)$ of X is the space of all closed subsets of X.

(iv) The *Vietoris* (κ-)*topology* on the exponential space $\mathcal{P}_{cl}(X)$ is the coarsest (κ-)topology such that $\Box(A) = \mathcal{P}_{cl}(X) \cap \mathcal{P}(A)$ is open for every open set A of X and $\Box(F)$ is closed for every closed set F of X. A natural subbasis for this topology is given by the sets $\Box(A)$ and $\Diamond(A) = \{C \in \mathcal{P}_{cl}(X) \mid A \cap C \neq \varnothing\}$, for A open.

(v) The *exponential uniformity* $\mathcal{U}^+$ on the exponential space $\mathcal{P}_{cl}(X)$ corresponding to the κ-uniformity $\mathcal{U}$ on X has a basis consisting of the sets

$$U^+ = \{(A, B) \in \mathcal{P}_{cl}(X) \times \mathcal{P}_{cl}(X) \mid (\forall x \in A\ \exists y \in B.\ (x, y) \in U) \wedge (\forall y \in B\ \exists x \in A.\ (y, x) \in U)\}, \text{ where } U \in \mathcal{U}.$$

$\mathcal{U}^+$ is a κ-uniformity whenever $\mathcal{U}$ is.

(vi) The *Hausdorff* κ-*metric* ρ_H on $\mathcal{P}_{cl}(X)$ corresponding to the bounded κ-metric ρ on X is defined as

$$\rho_H(A, B) = \max\{\sup_{x \in A} \inf_{y \in B} \rho(x, y), \sup_{y \in B} \inf_{x \in A} \rho(y, x)\}.$$

PROPOSITION 1: (See [5, 2.7.20 & 3.12.26].) Let τ be a κ-topology on X and let υ be the Vietoris κ-topology on $\mathcal{P}_{cl}(X)$. Then:

(i) υ is T_1 if and only if τ is T_1;

(ii) υ is Hausdorff if and only if τ is T_3;

(iii) υ is T_3 if and only if τ is T_4;

(iv) υ is T_4 if and only if τ is κ-compact;

(v) υ is κ-compact if and only if τ is κ-compact.

PROPOSITION 2: Let τ be a (κ-)topology on X.

(i) If τ is κ-compact, and σ is a κ-topology on X strictly coarser than τ, then σ is not Hausdorff.

(ii) There can be *more than one* or possibly *no* uniformity which induces τ. However, if (X, τ) is κ-compact, then there is *exactly one* κ-uniformity inducing τ.

(iii) If $\mathcal{U}$ induces τ on X, then the topology induced on $\mathcal{P}_{cl}(X)$ by the exponential uniformity $\mathcal{U}^+$ can be different from the Vietoris topology. If (X, τ) is κ-compact then the Vietoris κ-topology is induced by the exponential κ-uniformity.

(iv) If the topology τ is induced by a uniformity $\mathcal{U}$, then τ is 0-dimensional if and only if $\mathcal{U}$ has a basis of equivalences. This is always the case if $\mathcal{U}$ is a κ-uniformity for some κ > ω. If $\mathcal{U}$ is generated by a bounded κ-metric ρ, then $\mathcal{U}^+$ is generated by the corresponding Hausdorff κ-metric ρ_H.

(v) The topology τ is κ-metrizable (respectively, κ-ultrametrizable) if and only if τ is induced by a κ-uniformity $\mathcal{U}$ having a nested basis of size κ (respectively, consisting of equivalences). If κ is uncountable, then any κ-metrizable space is κ-ultrametrizable.

ACKNOWLEDGMENTS

The authors are grateful to Franck van Breugel, Jan Pelant, and Jan Rutten for helpful discussions, and to the referee for useful remarks and suggestions.

REFERENCES

1. ABRAMSKY, S. 1988. A Cook's tour of finitary non-wellfounded sets. Unpublished manuscript.
2. ACZEL, P. 1988. Non-well-founded sets. CSLI Lecture Notes. **14**. Stanford 1988
3. AMERICA, P. & J.J.M.M. RUTTEN. 1989. Solving domain equations in a category of complete metric spaces. J. Comp. Sys. Sci. **39:** 343–375.
4. DE BAKKER, J.W. & J.I.ZUCKER. 1982. Processes and denotational semantics of concurrency. Info. Contr. **54:** 70–120.
5. ENGELKING, R. 1977. General Topology. Polish Scientific Publishers. Warszawa.
6. ESSER, O. 1955. An interpretation of the Zermelo-Frænkel set theory and the Kelley-Morse set theory in a positive theory. Manuscript.
7. FORSTER, T. 1983. Axiomatizing set Theory with a Universal Set. Manuscript.
8. FORTI, M., & R.HINNION. 1989. The consistency problem for positive comprehension principles. J. Symb. Logic. **54:** 1401–1418.
9. FORTI, M. & F. HONSELL. 1983. Set theory with free construction principles. Ann. Scuola Norm. Sup. Pisa, Cl. Sci. **10:** (4) 493–522.
10. FORTI, M. & F. HONSELL. 1984/85/87. Axioms of choice and free construction principles I. Bull. Soc. Math. Belg. **36:** (B) 69–79; II, Ibid. **37:** (B) fasc.2, 1–12; III, Ibid. **39:** (B) , 259–76.
11. FORTI, M. & F. HONSELL. 1989. Models of Selfdescriptive Set Theories. *In:* Partial Differential Equations and the Calculus of Variations, Essays in Honor of Ennio De Giorgi, I (F. Colombini *et al.* Eds.). Birkhäuser. Boston, Massachusetts. 473–518.
12. FORTI, M. & F. HONSELL. 1992. Weak Foundation and Antifoundation Properties of Positively Comprehensive Hyperuniverses. *In:* L'Anti-fondation en Logique et en Théorie des Ensembles (R.Hinnion, Ed.). Cahiers du Centre de Logique. Louvain-la-Neuve. **7:** 31–43.
13. FORTI, M. & F. HONSELL. 1996. A general construction of Hyperuniverses. Theoretical Computer Science. **156:** 203–215.
14. FORTI, M. & F. HONSELL. 1996. Choice Principles in Hyperuniverses. Annals of Pure & Applied Logic. **77:** 35–52.

15. FORTI, M., F. HONSELL, & M.LENISA. 1994. Processes and Hyperuniverses. *In:* Mathematical Foundations of Computer Science. (I.Prívara *et al.*, Eds.). Springer Lecture Notes in Computer Science. Berlin. **841:** 352–361.
16. HONSELL, F. & M.LENISA. 1995. Final Semantics for Untyped Lambda Calculus. *In:* TLCA '95 (M. Dezani & G. Plotkin, Eds.). Springer Lecture Notes in Computer Science. Berlin. **902:** 249–265.
17. JECH, T.J. 1978. Set theory. Academic Press. New York. 1978.
18. KURATOWSKI, K. 1966. *Topology.* Academic Press. New York.
19. LENISA, M. 1966. Final Semantics for a Higher Order Concurrent Language. *In:* CAAP '96 (H. Kirchner, Ed.). Springer Lecture Notes in Computer Science. Berlin. **1059:** 202–218.
20. MALITZ, R.J. 1976. Set theory in which the axiom of foundation fails. Ph.D. Thesis, UCLA, Los Angeles. Unpublished.
21. MISLOVE, M.W., L.S. MOSS, & F.J.OLES. 1991. Non-wellfounded Sets Modeled as Ideal Fixed Points. Information and Computation. **93:** 16–54.
22. RUTTEN, J.J.M.M. & D.TURI. 1993. On the Foundations of Final Semantics: Non-Standard Sets, Metric Spaces, Partial Orders(REX Conference Proceedings) (J.de Bakker *et al.,* Eds.). Springer Lecture Notes in Computer Science. **666:** 477–530.
23. STEVENSON, F.W. & W.J.THRON. 1969. Results on ω_μ-metric spaces. Fundam. Math. **65:** 317–324.
24. WEYDERT, E. 1989. How to approximate the naive comprehension scheme inside of classical logic. Bonner mathematische Schriften. **194**. Bonn.

On the Completion of a MAP Group

JORGE GALINDO[a,c] AND SALVADOR HERNÁNDEZ[b,d]

Departamento de Matemáticas
Campus de Penyeta Roja
Universidad Jaume I
12071-Castellón, Spain

ABSTRACT: The following question is adressed in this paper: Given a maximally almost periodic (MAP) group, is the bilateral Raĭkov completion of G also a MAP group?

It is proved by means of an example that the answer is negative in general. Then a characterization of the MAP groups having a MAP completion is obtained so that some sufficient conditions follow from it.

A topological group G is said to be maximally almost periodic (MAP) in the sense of von Neumann when for every $e_G \neq x \in G$ there is a continuous homomorphism h_x into some compact Hausdorff group such that $h_x(x) \neq e_k$. The Raĭkov (or bilateral) completion of a topological group G is the topological group, $\overline{G}$, which results of completing the group with its Raĭkov uniformity. When G is Abelian, the Raĭkov completion coincides with the completion of G with its left uniformity and is a complete group (for questions related to completeness and completions of topological groups see [5] or [2]).

We shall begin by presenting an example of an (Abelian) MAP group the completion of which is not MAP. To this end we begin with an example given by Hooper in [4].

Let G be the topological vector space c_0 formed by all sequences of real numbers converging to 0 endowed with the supremum norm, and consider the following two subgroups of G:

$$K := \langle \{e_n : n < \omega\} \rangle$$

$$K_1 := \{x \in K : x = \textstyle\sum_{i=1}^{p} m_i e_{n_i} \text{ and the integer } \sum_{i=1}^{p} m_i \text{ is a multiple of } 2\}$$

Mathematics Subject Clasification: 22A05, 54H11, 54D35.

Key words and phrases: Raĭkov completion, maximally almost periodic group, Bohr topology.

[a]Research supported by a Generalitat Valenciana F.P.I. grant.

[b]Research partially supported by Generalitat Valenciana grant number GV-2223/94, and by Fundació Caixa Castelló grant number P1B95-18.

[c]E-mail: jgalindo@nuvol.uji.es.

[d]E-mail: hernande@nuvol.uji.es.

where, if A is a set, $\langle A\rangle$ is the smallest group containing A (that is, the group generated by A) and where the element e_n is the sequence with every term equal to zero but its n-th, which is equal to one.

The following facts have been proved by Hooper in [4].

FACT 1: The group G/K is maximally almost periodic.

FACT 2: The group G/K_1 is not maximally almost periodic. In fact, the set of all elements in G/K which cannot be separated from the identity element by means of continuous homomorphisms into compact groups is precisely K/K_1.

EXAMPLE 1: Let A be a Hamel basis for the real numbers over the rationals such that $1 \in A$ and set $B := A\setminus\{1\}$. The vector group $H := Lin_{\mathbb{Q}}B$ (i.e., $Lin_{\mathbb{Q}}B$ is the linear space over the rationals generated by B) is clearly a proper dense subgroup of the real numbers with $H \cap \mathbb{Q} = \{0\}$. Indeed, if $q \in \mathbb{Q} \cap H$ then $q = \sum_{i=1}^{n} q_i b_i$ with $q_i \in \mathbb{Q}$, and $b_i \in B$, $1 \le i \le n$. Thus $1 \in Lin_{\mathbb{Q}}\{b_1, \ldots, b_n\}$ which is a contradiction.

If we define $H_0 := \langle\{he_n : h \in H, n < \omega\}\rangle$, then we conclude that the group $(H_0 + K_1)/K_1$ is a MAP group but its completion is not.

Indeed, let $\pi: G \longrightarrow G/K_1$ be the canonical quotient map and take an arbitrary element $\bar{x} \neq \bar{0}$, where $\bar{x} \in (H_0 + K_1)/K_1$. From the definition of H_0 it is clear that $H_0 \cap K = \{0\}$. Therefore, if $h \in H_0$ and $k_1 \in K_1$ are such that $\pi(h + k_1) = \bar{x}$, then $h + k_1 \notin K$, that is, $\bar{x} \notin K/K_1$. This and Fact 2 prove that $(H_0 + K_1)/K_1$ is a MAP group.

Thus, again by Fact 2, we are done if we prove that the completion of $(H_0 + K_1)/K_1$ contains G/K_1 or, equivalently, if we prove that $(H_0 + K_1)/K_1$ is dense in G/K_1.

Let $\{x_n\}_{n<\omega}$ be an element of G. Since $\lim x_n = 0$, given any $\epsilon > 0$ there exists $n_0 < \omega$ such that $|x_n| < \epsilon$ for all $n > n_0$.

As H is dense in $\mathbb{R}$, for any $n \le n_0$ we can find $h_n \in H$ with $|h_n - x_n| < \epsilon$. Let $\{y_n\}_{n<\omega}$ be the sequence in H_0 defined by $y_n = h_n$ for $1 \le n \le n_0$ and $y_n = 0$ for $0 < n < \omega$. It is clear that $\|(x_n) - (y_n)\|_\infty < \epsilon$ and, consequently, that H_0 is dense in G. Hence, $(H_0 + K_1)/K_1$ is dense G/K_1, and this completes the proof. ❑

Hereafter, $(G, \mathcal{T})$ will denote a MAP group and $(\overline{G}, \overline{\mathcal{T}})$ its Raĭkov completion. We recall that the Bohr topology of a MAP group is the topology that it inherits from its Bohr compactification. (See [2], [3] to find the basic features of these concepts.)

LEMMA 1: For $(G, \mathcal{T})$ as above, the following statements are equivalent:

(a) $(\overline{G}, \overline{\mathcal{T}})$ is MAP.

(b) A Cauchy net contained in G is convergent if, and only if, it is also convergent in the Bohr topology.

(c) A Cauchy filter contained in G is convergent if, and only if, it is also convergent in the Bohr topology.

Proof: Since the Bohr topology of $(G, \mathcal{T})$ is weaker than $\mathcal{T}$ and the equivalence between (b) and (c) is obvious, all we need to prove is that (b) implies (a).

Suppose that $(\overline{G}, \overline{\mathcal{T}})$ is not MAP. Then there will be some $x \in \overline{G} \backslash G$ such that for every continuous homomorphism ϕ of $\overline{G}$ into a compact group K, $\phi(x) = \phi(e_G) = e_K$. Let $\{x_\delta\}_{\delta \in \Delta}$ be a net in G converging to x. Then $\lim_\delta \phi(x_\delta) = \phi(x) = \phi(e_G)$ and it follows that $\{x_\delta\}_{\delta \in \Delta}$ converges to e_G in the Bohr topology. Since $\{x_\delta\}_{\delta \in \Delta}$ cannot converge within G, this is a contradiction with (b), and the proof is complete. ❑

Next, we are going to give a characterization of when the group $\overline{G}$ is MAP provided that G has this property. In order to do it, the following concept will be useful.

DEFINITION 1: Let $\mathcal{A}$ be a family of sets in a topological group $(G, \mathcal{T})$. We say that $\mathcal{A}$ is divergent when no Cauchy filter in $(G, \mathcal{T})$ contains $\mathcal{A}$.

In the sequel we shall denote the Bohr topology of a topological group $(G, \mathcal{T})$ by $(G, \mathcal{T}^+)$ and the closure of a subset $A \subseteq (G, \mathcal{T}^+)$, by $\overline{A}^+$.

THEOREM 1: For a MAP group, $(G, \mathcal{T})$, the following assertions are equivalent:

(a) $(\overline{G}, \mathcal{T})$ is MAP;

(b) There is a neighborhood base $\mathcal{U}$ of e_G such that the family

$$\mathcal{A} := \{\overline{U}^+ \backslash V : U \in \mathcal{U}\}$$

is divergent for every neighborhood of the identity V.

Proof: Suppose first that $(\overline{G}, \mathcal{T})$ is MAP and consider $\mathcal{U}$ a neighborhood base of e_G. If V is a neighborhood of the identity in G such that the set $\mathcal{A} = \{\overline{U}^+ \backslash V : U \in \mathcal{U}\}$ is not divergent then there must exist a Cauchy filter $\mathcal{F}$ in $(G, \mathcal{T})$ containing $\mathcal{A}$. It is then clear that $\mathcal{F}$ must also contain the family $\{\overline{U}^+ : U \in \mathcal{U}\}$. Thus, $\mathcal{F}$ converges to e_G in the Bohr topology of G which as a consequence of Lemma 1, above, means that $\mathcal{F}$ converges to e_G in $(G, \mathcal{T})$. And this is a contradiction since $G \backslash V$ also belongs to $\mathcal{F}$.

Conversely, assume that there is a neighborhood base of the identity $\mathcal{U}$ in $(G, \mathcal{T})$ such that the family $\mathcal{A} = \{\overline{U}^+ \backslash V : U \in \mathcal{U}\}$ is divergent for every neighborhood, V, of the identity in $(G, \mathcal{T})$. In order to apply Lemma 1, assume that $\mathcal{F}$ is a Cauchy filter in $(G, \mathcal{T})$ which converges to e_G in the Bohr topology of $(G, \mathcal{T})$ and take $U \in \mathcal{U}$. Since $\mathcal{F}$ is a Cauchy filter there is $F \in \mathcal{F}$ with $F^{-1}F \subseteq U$. Also, from the definition of $\mathcal{F}$ we obtain that $e_G \in \overline{F}^+$, for every $F \in \mathcal{F}$. Thus, we have that

$$F \subseteq \overline{F^{-1}F^{+}} \subseteq \overline{U}^{+}.$$

Now take V any neighborhood of the identity in $(G, \mathcal{T})$ and suppose that $F \cap (\overline{U}^{+} \backslash V) \neq \varnothing$ for all $F \in \mathcal{F}$. Then the filter generated by the family

$$\{F \cap (\overline{U}^{+} \backslash V): F \in \mathcal{F}, \quad U \in \mathcal{U}\}$$

will be a Cauchy filter containing $\{\overline{U}^{+} \backslash V: U \in \mathcal{U}\}$, which is impossible, since the latter is a divergent family by hypothesis. Hence, there must exist some $F \in \mathcal{F}$ with $F \subseteq V$. As V was arbitrarily chosen this means that $\mathcal{F}$ converges to e_G. From Lemma 1 the result now follows. ❑

COROLLARY 1: Let $(G, \mathcal{T})$ be a MAP group which admits a neighborhood base of the identity consisting of Bohr closed sets. Then the group $(\overline{G}, \mathcal{T})$ is also MAP.

NOTE: In connection with Corollary 1 above, we notice that in [1], W. Banaszczyk and E. Martín-Peinador prove that the completion of a locally quasiconvex group is also locally quasiconvex. A locally quasiconvex group is a Hausdorf topological group which has a base of the identity formed by quasiconvex sets. A quasiconvex set is a set U, such that for every $x \in G \backslash U$ there exists some character $\chi \in \widehat{G}$ such that $|\chi(U)| \leq 1/4$ and $|\chi(x)| > 1/4$. Obviously, locally quasiconvex groups are maximally almost periodic. Thus Banaszczyk and Marín-Peinador proved a special case of our result about completions of MAP groups. In fact, we should note that quasiconvex sets are always Bohr closed. Finally, our Example 1 gives an example of a MAP nonlocally quasiconvex group with MAP completion. In fact, the group $(H_0 + K)/K$, using the notation introduced before Example 1, is not locally quasiconvex, since it is topologically isomorphic to $(H_0 + K)/K_1$, a subgroup of which is the nonlocally quasiconvex group $(H_0 + K_1)/K_1$.

There is still another sufficient condition for $(\overline{G}, \mathcal{T})$ to be MAP which can be derived fromTheorem 1.

COROLLARY : Let $(G, \mathcal{T})$ be a MAP metrizable group such that $\mathcal{T}$ and $\mathcal{T}^{+}$ have the same convergent sequences. Then $(\overline{G}, \overline{\mathcal{T}})$ is MAP.

Proof: Let $\mathcal{U} = \{U_n: n < \omega\}$ be any countable base of neighborhoods of the identity in $(G, \mathcal{T})$ and suppose that there is some neighborhood of the identity V such that the family $\mathcal{A} = \{\overline{U_n}^{+} \backslash V: n < \omega\}$ is not divergent, that is, there exists a Cauchy filter $\mathcal{F}$ containing $\mathcal{A}$.

By hypothesis $\mathcal{T}$ is metrizable, consequently we can find a Cauchy sequence $\{x_n\}_{n<\omega}$ such that $x_n \in \overline{U_n}^{+} \backslash V$ for every $n < \omega$. This implies that $\{x_n\}_{n<\omega}$ converges to e_G in the Bohr topology and, therefore, also in $\mathcal{T}$. This is a contradiction since $\{x_n\}_{n<\omega}$ is contained in the complement of a neighborhood of e_G, and the proof is complete. ❑

REFERENCES

1. BANASZCZYK, W. & E. MARTÍN-PEINADOR. 1994. Weakly pseudocompact subsets of nuclear groups. Preprint.
2. COMFORT, W.W., K .H. HOFMANN, & D. REMUS. 1992. Topological groups and semigroups. *In:* Recent progress in General Topology (M. Husek and J. van Mill, Eds.). Elsevier Science Publishers B.V. Amsterdam. 59–144.
3. HEWITT, E. & K.A. ROSS. 1963. Abstract Harmonic Analysis, Vol. I. Die Grundlehren der mathematischen Wissenschaften in Einzeldarstellungen, Vol. 115. Springer-Verlag, Berlin-Göttingen-Heidelberg.
4. HOOPER, R.C. 1968. Topological groups and integer valued norms. J. Funct. Anal. **2:** 243-257.
5. ROELKE, W. & S. DIERLOF. 1981. Uniform Structures on Topological Groups and Their Quotients. McGraw-Hill International Book Company. New York-Toronto.
6. TRIGOS-ARRIETA, F.J. 1991. Pseudocompactness on groups. Ph.D. Thesis. Wesleyan University. Middletown, Connecticut.

On Volterra Spaces II

DAVID GAULD,[a] SINA GREENWOOD,[a] AND ZBIGNIEW PIOTROWSKI[b]

[a]*Department of Mathematics*
University of Auckland
Private Bag 92019
Auckland, New Zealand
and
[b]*Department of Mathematics*
Youngstown State University
Youngstown, Ohio 44555

ABSTRACT: We say that a topological space X is Volterra if for each pair $f, g: X \to \mathbb{R}$ for which the sets of points at which f, respectively g, are continuous are dense, there is a common point of continuity; and X is strongly Volterra if in the same circumstances the set of common points of continuity is dense in X. For both of these concepts equivalent conditions are given and the situation involving more than two functions is explored.

For any function $f: X \to Y$ we denote by $C(f)$ the set of points of X at which f is continuous. Motivated by the proof in [3] and observations in [1] we make the following definitions.

DEFINITION 1: Let X be a topological space. Say that X is *Volterra* if $C(f) \cap C(g) \neq \varnothing$ whenever $f, g: X \to \mathbb{R}$ are two functions for which $C(f)$ and $C(g)$ are dense in X.

DEFINITION 2: Let X be a topological space. Say that X is *strongly Volterra* if $C(f) \cap C(g)$ is dense in X whenever $f, g : X \to \mathbb{R}$ are two functions for which $C(f)$ and $C(g)$ are dense in X.

It should be noted that these definitions conflict with those given in [2] where what we call strongly Volterra is called Volterra. However, in Theorem 2 below we show that the two concepts, Volterra and strongly Volterra, defined in [2] are in fact equivalent. The definitions given here conform more closely to Volterra's original proof.

LEMMA: (Cf. Proposition 18 of [2].) Let X be a topological space and A a dense G_δ subset of X. Then there is a function $f: X \to \mathbb{R}$ for which $C(f) = A$.

Proof: As A is G_δ we may write $A = \bigcap_{i=1}^{\infty} A_i$ where each A_i is open,

Mathematics Subject Classification: Primary 54E99; Secondary 26A15, 54C05, 54C30, 54E52.

Key words and phrases: Volterra spaces, strongly Volterra spaces, Baire spaces, developable spaces, sets of continuity points.

$A_{n+1} \subset A_n$ for each n and $A_1 = X$. Define $f : X \to \mathbb{R}$ by

$$f(x) = \begin{cases} 0 & \text{if } x \in A, \\ 1/n & \text{if } x \in A_n - A_{n+1}. \end{cases}$$

We have $C(f) \subset A$ for if $x \notin A$ then $x \in A_n - A_{n+1}$ for some n. Then $(0, \infty)$ is a neighborhood of $f(x)$ but $f^{-1}(0, \infty)$ cannot be a neighborhood of x as it contains no point of the dense set A. Thus $x \notin C(f)$. Conversely, it is clear that $A \subset C(f)$. ❑

THEOREM 1: For any topological space X, the following three conditions are equivalent:

(a) X is Volterra;
(b) for each pair A, B of dense G_δ subsets of X the set $A \cap B$ is nonempty;
(c) for each pair Y, Z of developable spaces and each pair $f : X \to Y$ and $g: X \to Z$ of functions for which $C(f)$ and $C(g)$ are dense in X, the set $C(f) \cap C(g)$ is nonempty.

Proof: (a) $\Rightarrow$ (b). Suppose A, B are dense G_δ subsets of X. By the lemma there are functions $f, g: X \to \mathbb{R}$ with $C(f) = A$ and $C(g) = B$. The result is immediate from the definition.

(b) $\Rightarrow$ (c). Suppose Y and Z are developable spaces and $f: X \to Y$ and $g: X \to Z$ are such that $C(f)$ and $C(g)$ are dense. By Remark 6 of [2], both $C(f)$ and $C(g)$ are G_δ sets so by (b) $C(f) \cap C(g) \neq \varnothing$.

(c) $\Rightarrow$ (a). This follows from the developability of $\mathbb{R}$. ❑

THEOREM 2: For any topological space X, the following three conditions are equivalent:

(a) X is strongly Volterra.
(b) for each pair A, B of dense G_δ subsets of X the set $A \cap B$ is dense;
(c) for each pair Y, Z of developable spaces and each pair $f: X \to Y$ and $g: X \to Z$ of functions for which $C(f)$ and $C(g)$ are dense in X the set $C(f) \cap C(g)$ is dense.

Proof: Similar to the proof of Theorem 1. ❑

THEOREM 3: Let X be a nonempty topological space.

(a) If X is Baire, then X is strongly Volterra.
(b) If X is strongly Volterra, then X is Volterra.
(c) If X is of second category, then X is Volterra.

Proof:

(a) This is the content of Theorem 8 of [2].
(b) Obvious.
(c) This follows from (a) and Theorem 1, since any space of second cat-

egory contains a nonempty open Baire subspace. ❑

Theorem 3 may be summarized by the following diagram where the arrows represent implications. (It is known that a Baire space is of second category.)

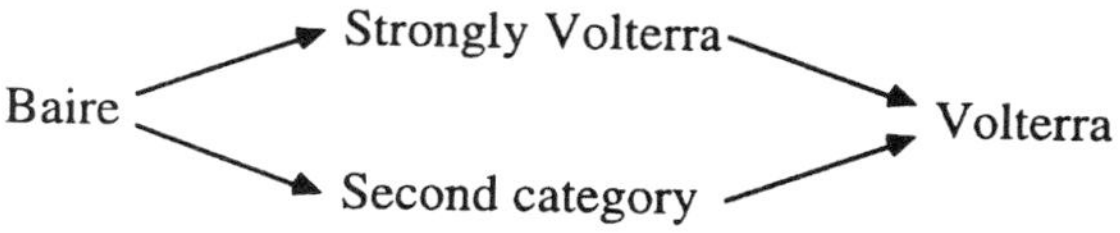

The following examples show that the converse implications in Theorem 3 fail, even when X is a T_1-space, and that there is no relation between strongly Volterra spaces and spaces of second category.

EXAMPLE 1: Let $X = [0, \infty)$ with the following topology:

$$\{\phi\} \cup \{[a, \infty) - F/a \in X \text{ and } F \text{ is a finite subset of } X\}.$$

X is not of second category (hence, not a Baire space) because X is the union of countably many closed, nowhere dense sets $\{[0, n)/n \in \mathbb{N}\}$. However, each dense G_δ set is of the form $[a, \infty) - C$ where $a \in X$ and C is a countable subset of X and the intersection of any pair of such sets is of the same form, hence dense. Thus X is strongly Volterra (and, therefore, also Volterra). Note that singletons are closed so X is T_1.

EXAMPLE 2: Let $X = \mathbb{R}_- \cup \mathbb{Q}_+$ topologized as a subspace of the reals, where $\mathbb{R}_- = \{x \in \mathbb{R}/x \le 0\}$ and $\mathbb{Q}_+ = \{x \in \mathbb{Q}/x \ge 0\}$. Then X is not strongly Volterra because if we let $\mathbb{Q}_{+o}$ and $\mathbb{Q}_{+e}$ denote, respectively, the members of $\mathbb{Q}_+$ with odd and even denominators when in lowest terms then $\mathbb{R}_- \cup \mathbb{Q}_{+o}$ and $\mathbb{R}_- \cup \mathbb{Q}_{+e}$ are dense G_δ subsets of X whose intersection is not dense. On the other hand X is Volterra because if $f, g: X \to \mathbb{R}$ are such that $C(f)$ and $C(g)$ are dense in X then $C(f|\mathbb{R}_-)$ and $C(g|\mathbb{R}_-)$ are dense in the strongly Volterra space $\mathbb{R}_-$, hence have nonempty intersection. X is also of second category.

Note that in Example 2, X is metrizable. Is there a metrizable example of a strongly Volterra, non-Baire space, or better still, a non-Baire subspace of the reals? As a possible contender, illustrating a possible way of constructing such a space, we offer the following. Let C be the Cantor set and C_n the Cantor set translated by n ($n \in \mathbb{Z}$). Let $U_1 = \mathbb{R} - \bigcup_{n=-\infty}^{\infty} C_n$. Define U_n ($n \in \mathbb{N}$) inductively by $U_n = \{x \in \mathbb{R}/3x \in U_{n-1}\}$. Then each U_n is open and dense in $\mathbb{R}$ so $\bigcap_{n=1}^{\infty} U_n$ is a dense G_δ. Let $X = \mathbb{R} - \bigcap_{n=1}^{\infty} U_n$. Then X is not Baire because if $X_n = X \cap U_n$, then X_n is open in X. X_n is dense in $\mathbb{R}$, being the intersection of two dense sets one of which is open. Thus X_n is a dense open subset of X. However, $\bigcap_{n=1}^{\infty} X_n = \varnothing$. Thus, $\{X_n / n \in \mathbb{N}\}$ is a countable collection of dense open subsets of X having empty, hence nondense, intersection. Therefore, X is not Baire.

As a corollary of Theorem 2, we have the following result.

THEOREM 4: Let X be a strongly Volterra space, Y_i ($i \in \mathbb{N}$) developable spaces and $f_i : X \to Y_i$ ($i \in \mathbb{N}$) functions such that for each i the set $C(f_i)$ is dense in X. Then $\bigcap_{i=1}^{n} C(f_i)$ is dense in X.

Proof: Each set $C(f_i)$ is a dense G_δ subset of X so by Theorem 2(b) and induction on n we have $\bigcap_{i=1}^{n} C(f_i)$ is dense. ❑

Note that if X is a Baire space then the conclusion of Theorem 4 holds even for a countably infinite family of functions, but Example 1 shows that this is not the case if X is merely strongly Volterra.

We cannot relax "strongly Volterra" in this theorem to "Volterra" and then merely require that $\bigcap_{i=1}^{n} C(f_i)$ is nonempty when $n > 2$, even when X is a T_1 space, as the following example shows.

EXAMPLE 3: Let $A = \{(x, y) \in \mathbb{R}^2 / y \geq 0\}$ and for each $r \geq 0$ let $A_r = \{(x, y) \in \mathbb{R}^2 / y + r > 0\}$. Define B and B_r to be the sets obtained by rotating A and A_r 120° about (0, 0) in the anticlockwise direction and C and C_r by a similar rotation in the clockwise direction. Define the sets D and E by

$$D = (A_0 \cap B_0) \cup (B_0 \cap C_0) \cup (C_0 \cap A_0)$$

and

$$E = (A_0 - B - C) \cup (B_0 - C - A) \cup (C_0 - A - B).$$

Let

$$\mathcal{B}_1 = \{A_r \cap B_s \cap C_t - F / r, s, t > 0 \quad \text{and} \quad F \subset \mathbb{R}^2 \quad \text{is finite}\}$$

$$\mathcal{B}_2 = \{A_r \cap B_s \cap C_t \cap D - F / r, s, t > 0 \quad \text{and} \quad F \subset \mathbb{R}^2 \quad \text{is finite}\}$$

$$\mathcal{B}_3 = \{A_r \cap B_s \cap C_t \cap E - F / r, s, t > 0 \quad \text{and} \quad F \subset \mathbb{R}^2 \quad \text{is finite}\}.$$

Setting $\mathcal{B} = \mathcal{B}_1 \cup \mathcal{B}_2 \cup \mathcal{B}_3$ it may be verified that $\mathcal{B}$ is closed under finite intersections and covers $\mathbb{R}^2$ so forms a basis for a topology, $\mathcal{T}$, on $\mathbb{R}^2$.

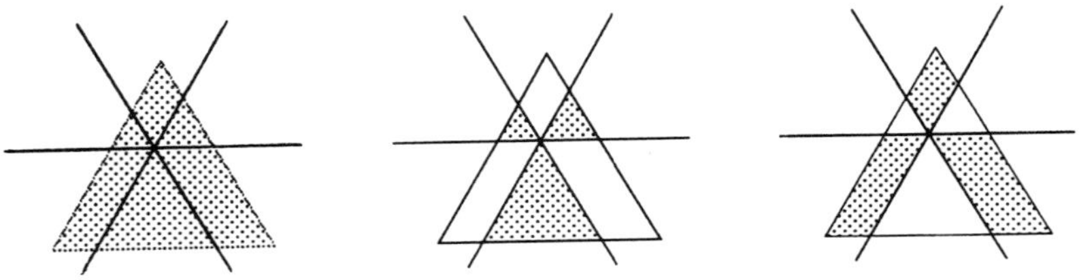

FIGURE

The figure shows the three types of basic open subsets of $(\mathbb{R}^2, \mathcal{T})$. Clearly, $\mathcal{T}$ is T_1 as it contains all cofinite sets. The sets A, B, and C are dense G_δ sets. So also are the sets $A \cap B_r \cap C_s - F$ for $r, s > 0$ and F countable, as well as those obtained by permuting A, B, and C. However, $A \cap B$ is not dense as $A \cap B \cap E = \varnothing$. Thus, the sets $A \cap B_r \cap C_s - F$ and their (A, B, C)-permutations are the "smallest" dense G_δ sets, where $r, s > 0$ and F is countable. Note that the intersection of any two of these sets is of the form $A \cap B \cap C_r - F$, or an (A, B, C)-permutation, where $r > 0$ and F is countable, so $(\mathbb{R}^2, \mathcal{T})$ is a Volterra space.

Consider the three subsets $A - \{(0, 0)\}$, B and C. These are dense G_δ subsets so by the lemma there are functions $f, g, h : \mathbb{R}^2 \to \mathbb{R}$ for which $C(f) = A - \{(0, 0)\}$, $C(g) = B$ and $C(h) = C$. As $C(f) \cap C(g) \cap C(h) = \varnothing$, it is seen that these functions violate the Volterra analogue of Theorem 4.

REFERENCES

1. GAULD, D.B. 1993. Did the young Volterra know about Cantor? Math. Magazine. **66:** 246–247.
2. GAULD, D.B., & Z. PIOTROWSKI. 1993. On Volterra spaces. Far East J. Math. Sci. **1:** 209–214.
3. VOLTERRA, V. 1881. Alcune osservasioni sulle funzioni punteggiate discontinue. G.di Mat. **19:** 76–86; 1954 Opera Matematiche di Vito Volterra. Accad. Naz. dei Lincei, Roma. **1:** 7–15.

Spaces of Valuations

REINHOLD HECKMANN[a]

FB 14 — Informatik
Universität des Saarlandes
Postfach 151150
D-66041 Saarbrücken, Germany

ABSTRACT: Valuations are measurelike functions mapping the open sets of a topological space X into positive real numbers. They can be classified into finite, point continuous, and Scott continuous valuations. We define corresponding spaces of valuations $\mathsf{V}_{\mathrm{f}}X \subseteq \mathsf{V}_{\mathrm{p}}X \subseteq VX$. The main results of the paper are that $\mathsf{V}_{\mathrm{p}}X$ is the soberification of $\mathsf{V}_{\mathrm{f}}X$, and that $\mathsf{V}_{\mathrm{p}}X$ is the free sober locally convex topological cone over X. From this universal property, the notion of the integral of a real-valued function over a Scott continuous valuation can be easily derived. The integral is used to characterize the spaces $\mathsf{V}_{\mathrm{p}}X$ and VX as dual spaces of certain spaces of real-valued functions on X.

1. INTRODUCTION

For a topological space X, a *valuation* on X is a function ν which maps the open sets of X to real numbers in the range from zero to infinity (inclusively) with the following properties [11], [8], [9], [10], [13]:

(1) The empty set is mapped to zero: $\nu\varnothing = 0$ (strictness).

(2) The values assigned to binary union and intersection are related by the following equation:

$$\nu(U \cup V) + \nu(U \cap V) = \nu U + \nu V \quad \text{for all opens } U \text{ and } V \text{ (modularity).}$$

(3) Bigger sets are mapped to bigger numbers: if $U \subseteq V$, then $\nu U \leq \nu V$ (monotonicity).

Most often, we consider *Scott continuous valuations* which enjoy the additional property $\nu(\bigcup_{i \in I} V_i) = \bigsqcup_{i \in I} \nu V_i$ for every directed family $(V_i)_{i \in I}$ of opens V_i of X.

Some authors [8], [9] write *evaluations* instead of valuations, and some authors immediately require Scott continuity.

The concept of valuations has some similarity with the concept of *measures*. Borel measures are defined for all Borel sets of a space, and every open set is a Borel set. Hence, every measure can be restricted to a valuation.

Mathematics Subject Classification: 54B99, 28A25, 46A99, 46E99, 54D35.
Key words and phrases: topological spaces of valuations, topological cones, integration, soberification.

[a]E-mail: heckmann@cs.uni-sb.de.

This valuation is not Scott continuous in general, since measures have to satisfy a weaker property where only countable directed families $(V_i)_{i \in I}$ are considered. On the other hand, a valuation cannot always be extended to a measure.

Previously, valuations were already used in the following contexts:

(1) The *probabilistic power domain* of Jones and Plotkin [8], [9] over some dcpo X is the dcpo of Scott continuous valuations on X which are bounded by 1. It is used to model the semantics of probabilistic programs. If the semantics of a program p is ν, this means that for every open set U, the number νU is the probability that the result of running p is in U. The difference $1 - \nu(X)$ is the probability that running p yields no result at all, i.e., does not terminate.

(2) In [6], the author presented a *lower bag domain* as an analogue to the lower power domain, but without idempotence of addition. It can be used to specify a bag semantics for nondeterministic programs which takes multiplicities of results into account. One possible description of the lower bag domain consists of Scott continuous integer-valued valuations.

(3) In [3], Edalat used Scott continuous valuations to obtain a domain-theoretic treatment of measures and Riemann-like integrals. In [1], he connected dynamical systems and fractals with domain theory. The probabilistic power domain, i.e., the collection of Scott continuous valuations bounded by 1, plays a major role in this connection which leads to better algorithms for fractal image generation [2].

Because of these applications, we feel that the concept of valuations deserves further interest.

In this paper, we investigate valuations in a topological setting: the set $\mathsf{V}X$ of Scott continuous valuations on a space X is made into a topological space, whose structure and properties are studied. Thus, we present some background information about dcpo's and topological spaces in Section 2.

In Section 3, valuations are defined formally. In addition to Scott continuity, we introduce *point continuity* as a possible property of valuations which is stronger than Scott continuity. Some special valuations are defined, e.g., the *point valuations* $\hat{x}$ which map open sets containing x to 1 and all other open sets to 0. We also consider some operations on valuations such as addition, multiplication by a constant from $\overline{\mathbf{R}}_+$, and restriction to an open set.

In Section 4, we study the relationship between point continuous and Scott continuous valuations. In particular, we prove that for continuous dcpo's, all Scott continuous valuations are point continuous. We also introduce *finite valuations* which are finite linear combinations of point valuations. Their representation as linear combination is unique if X is a $\mathcal{T}_0$-space.

In Section 5, we define the space $\mathsf{V}X$ of Scott continuous valuations on X, and the subspaces of point continuous ($\mathsf{V}_{\mathrm{p}}X$) and finite valuations ($\mathsf{V}_{\mathrm{f}}X$).

We also prove one of the main results of this paper: the space $\mathsf{V}_{\mathrm{p}}X$ is the soberification of $\mathsf{V}_{\mathrm{f}}X$.

The other main results are shown in Section 6. For a $\mathcal{T}_0$-space X, the space $\mathsf{V}_{\mathrm{f}}X$ of finite valuations is the free locally convex $\mathcal{T}_0$-cone over X, and the space $\mathsf{V}_{\mathrm{p}}X$ of point continuous valuations is the free locally convex sober cone over X. Here, a $\mathcal{T}_0$-cone is an $\mathbf{R}_+$-module with a $\mathcal{T}_0$-topology such that addition and multiplication are continuous, and a sober cone is a $\mathcal{T}_0$-cone with sober topology. We have not found a universal property for V.

Nevertheless, the universal property of V_{p} suffices to find a novel definition of integration $\int_X$: $[X \to \overline{\mathbf{R}}_+] \times \mathsf{V}X \to \overline{\mathbf{R}}_+$ of a real-valued function with respect to a Scott continuous valuation (see Section 7). This definition allows for elegant proofs of the properties of integration.

In Section 8, integration is used to prove that the spaces of valuations are isomorphic to the dual spaces of certain function spaces, namely

$$\mathsf{V}X \cong [[X \to \overline{\mathbf{R}}_+]_{\mathrm{i}} \xrightarrow{lin} \overline{\mathbf{R}}_+]_{\mathrm{p}} \quad \text{and} \quad \mathsf{V}_{\mathrm{p}}X \cong [[X \to \overline{\mathbf{R}}_+]_{\mathrm{p}} \xrightarrow{lin} \overline{\mathbf{R}}_+]_{\mathrm{p}},$$

where the index "$[\ ]_{\mathrm{p}}$" means a function space with pointwise topology, and "$[\ ]_{\mathrm{i}}$" a function space with Isbell topology.

2. TOPOLOGICAL BACKGROUND

The main purpose of this section is to introduce some notion and notation which may not be familiar to everybody since it does not occur in usual topological textbooks. In particular, we introduce dcpo's, d-spaces, sober spaces, spaces of opens, and function spaces. Then we present a characterization of the soberification of a space which seems to be novel. Finally, we briefly look at the space $\overline{\mathbf{R}}_+$ of positive real numbers.

2.1. Some Notation

Some Set-Theoretic Notation. We only mention some slightly nonstandard notation. If A is a subset of a fixed set X, then the complement of A in X is written $\neg A$.

If $f: X \to Y$ is a function, then we denote the image of a set $A \subseteq X$ by $f^+A = \{fx \mid x \in A\}$, and the inverse image of a set $B \subseteq Y$ by $f^-B = \{x \in X \mid fx \in B\}$.

We use λ-*notation* for the definition of functions, e.g., $\lambda x.\ x + y$ is the function which maps the argument x into the result $x + y$, where y is a fixed constant.

Topological Spaces. We write the set of opens of a topological space X as ΩX, and denote the topological closure of a subset A of X by $\mathrm{cl}\,A$. For a subset A of a space X, let $\mathcal{O}(A) = \{O \in \Omega X \mid A \subseteq O\}$. We abbreviate $\mathcal{O}(\{x\})$ to $\mathcal{O}(x)$.

Preorders and Posets. A *preorder* $(X, \sqsubseteq)$ is a set X together with a reflexive and transitive relation "$\sqsubseteq$." For a subset A of a preorder X, we define $\downarrow A = \{x \in X \mid \exists a \in A: x \sqsubseteq a\}$ and $\uparrow A = \{x \in X \mid \exists a \in A: a \sqsubseteq x\}$. We shall often abbreviate $\downarrow\{a\}$ by $\downarrow a$ and $\uparrow\{a\}$ by $\uparrow a$. The set A is *lower* if $A = \downarrow A$, and *upper* if $A = \uparrow A$.

A *poset* is a preorder with anti-symmetric relation "$\sqsubseteq$." In a poset, the *least upper bound* or *join* of a subset A is denoted by $\sqcup A$ (if it exists).

Directed Sets and Dcpo's. A subset D of a poset is *directed* if it is nonempty, and for all x,y in D, there is z in D with $x, y \sqsubseteq z$. A *dcpo* is a poset where every directed set has a least upper bound. A function f: $\mathbf{X} \to \mathbf{Y}$ between dcpo's $\mathbf{X}$ and $\mathbf{Y}$ is *Scott continuous* if for all directed subsets D of $\mathbf{X}$, $f(\sqcup D) = \sqcup f^+ D$ holds.

A subset O of a dcpo $\mathbf{X}$ is *Scott open* if it is upper, and for all directed sets D, $\sqcup D \in O$ implies $D \cap O \neq \varnothing$. With this definition, every dcpo becomes a topological space. A function f: $\mathbf{X} \to \mathbf{Y}$ between dcpo's is Scott continuous in the sense defined above iff it is topologically continuous [7, II.1.10]. *Continuous dcpo's* can be topologically characterized as follows: if x is in an open set U, then there are an open V and a point y such that $x \in V \subseteq \uparrow y \subseteq U$.

The Specialization Preorder. Every topological space can be preordered by defining $x \sqsubseteq x'$ iff every open set which contains x also contains x'. This is called the *specialization preorder* of the space. Every continuous function between two spaces is monotonic in their specialization preorders.

A space is a $\mathcal{T}_0$-space iff its specialization preorder is a partial order, i.e., antisymmetric. It is a $\mathcal{T}_1$-*space* iff the preorder coincides with equality. In this paper, most spaces will be $\mathcal{T}_0$, but not $\mathcal{T}_1$. The specialization preorder of a dcpo with its Scott topology is the original order of the dcpo. Hence, every dcpo is a $\mathcal{T}_0$-space.

If we use order notions such as lower, upper, $\downarrow A$, and $\uparrow A$ in a topological space, this always refers to the specialization preorder. All open sets are upper sets, and all closed sets are lower sets. For finite F, $\mathrm{cl}\, F = \downarrow F$ holds.

Joint Continuity versus Separate Continuity. For topological spaces X, Y, and Z, a function f: $X \times Y \to Z$ is *jointly continuous* if it is continuous in the product topology of $X \times Y$. It is *separately continuous* if all the functions f_x: $Y \to Z$ with $f_x y = f(x, y)$ and f^y: $X \to Z$ with $f^y x = f(x, y)$ are continuous. Every jointly continuous function is separately continuous; the opposite implication does not hold in general. We consider joint continuity as the default case, i.e., when we say "f: $X \times Y \to Z$ is continuous," we mean jointly continuous.

2.2. Spaces of Opens and Function Spaces

Spaces of Open Sets. The set ΩX of open sets of a space X can be topologized in several different ways.

First, $(\Omega X, \subseteq)$ is a dcpo, whence it can be endowed with the *Scott topology*. We call the resulting space $\Omega_s X$.

Second, ΩX can be given the *point topology* with subbase $\{\mathcal{O}(x) \mid x \in X\}$. A set $\mathcal{O}$ of opens is open in the point topology (*point open*) iff for every O in $\mathcal{O}$ there is a finite set F such that $O \in \mathcal{O}(F) \subseteq \mathcal{O}$. We call the resulting space $\Omega_p X$.

Since every set $\mathcal{O}(x)$ is Scott open, the topology of $\Omega_p X$ is contained in that of $\Omega_s X$. The two spaces $\Omega_s X$ and $\Omega_p X$ have the same specialization preorder, namely subset inclusion.

The following two properties are easily proved.

PROPOSITION 2.1: If $f: X \to Y$ is continuous, then $f^-: \Omega Y \to \Omega X$ is continuous from $\Omega_s Y$ to $\Omega_s X$ as well as from $\Omega_p Y$ to $\Omega_p X$.

Proof: Scott continuity of f^- is obvious. For every x in X and V in ΩY, $f^- V \in \mathcal{O}(x)$ iff $fx \in V$ iff $V \in \mathcal{O}(fx)$. ❑

PROPOSITION 2.2: Binary intersection $\cap: \Omega X \times \Omega X \to \Omega X$ is separately continuous in $\Omega_s X$, and jointly continuous in $\Omega_p X$.

Proof: For fixed V in ΩX, the function $\lambda U.\ U \cap V: \Omega X \to \Omega X$ is Scott continuous. For every x in X, $U \cap V$ is in $\mathcal{O}(x)$ iff (U, V) is in $\mathcal{O}(x) \times \mathcal{O}(x)$. ❑

The Pointwise Function Space. For two spaces X and Y, the *pointwise function space* $[X \to Y]_p$ consists of all continuous functions $f: X \to Y$ with subbase $\{\langle x \to V \rangle \mid x \in X, V \in \Omega Y\}$ where $\langle x \to V \rangle = \{f: X \to Y \mid fx \in V\}$. It is a subspace of the product $\prod_{x \in X} Y$ of copies of Y. The function $\Omega_p: [X \to Y]_p \to [\Omega_p Y \to \Omega_p X]_p$ with $\Omega_p f = f^-$ is continuous.

The Isbell Function Space. For two spaces X and Y, the *Isbell function space* [12] $[X \to Y]_i$ consists of all continuous functions $f: X \to Y$ with subbase $\{\langle \mathcal{U} \leftarrow V \rangle \mid \mathcal{U} \in \Omega(\Omega_s X), V \in \Omega Y\}$ where $\langle \mathcal{U} \leftarrow V \rangle = \{f: X \to Y \mid \mathcal{U} \ni f^- V\}$. Since $\langle x \to V \rangle = \langle \mathcal{O}(x) \leftarrow V \rangle$, the topology of $[X \to Y]_i$ includes that of $[X \to Y]_p$. Both function spaces have the same specialization preorder, namely $f \sqsubseteq g$ iff $fx \sqsubseteq gx$ for all x in X.

If X and Y are $\mathcal{T}_0$-spaces, the function $\Omega_s: [X \to Y] \to [\Omega_s Y \to \Omega_s X]$ with $\Omega_s f = f^-$ is injective, no matter which topologies are chosen for the two function spaces. If $[\Omega_s Y \to \Omega_s X]$ is equipped with the pointwise topology, the Isbell topology on $[X \to Y]$ is characterized as the topology which makes Ω_s into an embedding. For, $\Omega_s f \in \langle V \to \mathcal{U} \rangle$ iff $f^- V \in \mathcal{U}$ iff $f \in \langle \mathcal{U} \leftarrow V \rangle$. Hence, we obtain a continuous embedding $\Omega_s: [X \to Y]_i \hookrightarrow [\Omega_s Y \to \Omega_s X]_p$.

2.3. *D*-Spaces and Sober Spaces

A space X is a *d-space* if its specialization preorder is a dcpo and all open sets of X are Scott open with respect to this dcpo. (These spaces are called *monotone convergence spaces* in [5, Definition II.3.9]). A continuous function $f: X \to Y$ between two d-spaces is Scott continuous. Every dcpo with its Scott topology is a d-space, and every $\mathcal{T}_1$-space is a d-space. All d-spaces are $\mathcal{T}_0$-spaces.

A closed subset C of a space X is *irreducible* if, whenever $C \subseteq \bigcup_{i \in I} C_i$ for some finite family $(C_i)_{i \in I}$ of closed sets, then $C \subseteq C_i$ for some i in I. A space X is *sober* if all irreducible closed subsets of X are the closure cl $\{x\}$ of a unique point x of X [5, p. 79]. Every Hausdorff space is sober. Every sober space is a d-space [5, II.3.17]. (The proof relies on the fact that closures of directed sets are irreducible.)

The properties sober and d-space are preserved by products and $\mathcal{T}_0$-equalizer subspaces. A $\mathcal{T}_0$-equalizer subspace of a space X is a subspace $\{x \in X \mid fx = gx\}$ where $f, g: X \to Y$ are continuous and Y is $\mathcal{T}_0$. If Y is sober or a d-space, then so is $[X \to Y]_{\mathrm{p}}$ (no matter what X is). If Y is a d-space, then so is $[X \to Y]_{\mathrm{i}}$. The proofs of the statements about sobriety are contained in Chapter 5 of [13], following ideas of the author. The proofs of the d-space properties can be performed in a similar fashion.

2.4. Soberification

Every space X has a *soberification* $\overline{X}$ which is characterized by the following property:

THEOREM 2.3: X can be embedded into $\overline{X}$, and every continuous $f: X \to Z$ to a sober space Z has a unique continuous extension $\overline{f}: \overline{X} \to Z$.

There are several known ways to construct the soberification, which we do not consider here [5, V.4.7, V.4.9]. Instead, we are interested in a criterion which tells us whether a given sober space Y is (homeomorphic to) the soberification of a subspace X. Since this criterion seems to be novel, we include its derivation.

If X is a subspace of a space Y, the function $\lambda V. X \cap V$ is a surjective frame homomorphism from ΩY to ΩX. A sober space Y is the soberification of X iff this function is a frame isomorphism. This is the starting point for the following criterion:

THEOREM 2.4: For a subspace X of a sober space Y, the following are equivalent:

(1) Y is the soberification of X.

(2) For all opens U and V of Y, $X \cap U \subseteq V$ implies $U \subseteq V$.

(3) For every y in Y and O in ΩY with $y \in O$, there is some x in X with $x \in O$ and $x \sqsubseteq y$.

Proof: By the discussion above, (1) holds if $\lambda O. X \cap O$ is a frame isomorphism from ΩY to ΩX. Since it is a surjective homomorphism anyway, this is equivalent to $\lambda O.\ X \cap O$ being an order embedding, i.e., $X \cap U \subseteq X \cap V \Rightarrow U \subseteq V$. This implication is equivalent to (2).

Assume (3) and $X \cap U \subseteq V$. For u in U, there is x in $X \cap U \subseteq V$ with $x \sqsubseteq u$. Since V is upper, u in V follows.

Assume (2) and let $y \in O$. If $x \not\sqsubseteq y$ held for all x in $X \cap O$, then $X \cap O \subseteq \neg\downarrow y$, whence $O \subseteq \neg\downarrow y$ by (2). Then $y \in O$ would imply $y \in \neg\downarrow y$ — a contradiction. Thus there is x in $X \cap O$ with $x \sqsubseteq y$. ❑

In the sequel, we present some consequences of the theorem.

PROPOSITION 2.5: Let Y be the soberification of its subspace X, and let Z be a $\mathcal{T}_0$-space. Two continuous functions f, g: $Y \to Z$ with $fx = gx$ for all x in X are equal.

Proof: Let y in Y. We prove $fy \sqsubseteq gy$. If $fy \in V$ open, then $y \in f^{-}V$. By Theorem 2.4(3), there is x in X such that $x \sqsubseteq y$ and $x \in f^{-}V$. Then $gy \sqsupseteq gx = fx \in V$. Similarly, $gy \sqsubseteq fy$ is shown, whence $fy = gy$ since Z is $\mathcal{T}_0$. ❑

Soberification commutes with binary products:

PROPOSITION 2.6: If X is the soberification of $A \subseteq X$ and Y is the soberification of $B \subseteq Y$, then $X \times Y$ is the soberification of $A \times B$.

Proof: Let (x, y) be in an open set W of $X \times Y$. Then there are open sets U of X and V of Y such that $(x, y) \in U \times V \subseteq W$. By hypothesis, there is a in A with $a \sqsubseteq x$ and $a \in U$, and b in B with $b \sqsubseteq y$ and $b \in V$. Thus, $(a, b) \sqsubseteq (x, y)$ and $(a, b) \in U \times V \subseteq W$. ❑

2.5. Numbers

Let $\mathbf{R}_+$ be the set of positive real numbers including 0, but without ∞, and let $\overline{\mathbf{R}}_+$ be $\mathbf{R}_+$ together with ∞. Arithmetic is extended to $\overline{\mathbf{R}}_+$ by $x + \infty = \infty + x = \infty$ for all x, $x \cdot \infty = \infty \cdot x = \infty$ for all $x \neq 0$, and $0 \cdot \infty = \infty \cdot 0 = 0$. Subtraction $x - y$ is only defined if $x \geq y$ and $x \neq \infty$.

The set $\overline{\mathbf{R}}_+$ is ordered in the standard way, which yields a dcpo. It is given the Scott topology. Hence, the open sets of $\overline{\mathbf{R}}_+$ are $\emptyset$, $\overline{\mathbf{R}}_+$ itself, and all the sets $\{x \in \overline{\mathbf{R}}_+ \mid x > r\}$ for fixed numbers $r < \infty$. This space is sober. Addition and multiplication as defined above are continuous.

Sometimes, we shall need the split lemma for real numbers.

LEMMA 2.7: (Split lemma). Let $(r_i)_{i \in I}$ and $(s_j)_{j \in J}$ be two families of members of $\mathbf{R}_+$, where the index sets I and J are finite, and let $R \subseteq I \times J$ be a relation. For $T \subseteq I$, we write $R^+(T)$ for $\{j \in J \mid \exists i \in T\colon (i, j) \in R\}$.

If for all $T \subseteq I$, $\sum_{i \in T} r_i \leq \sum_{j \in R^+(T)} s_j$ holds, then there are numbers t_{ij} in $\mathbf{R}_+$ for i in I and j in J with

(1) $\sum_{j \in J} t_{ij} = r_i$ for all i in I,

(2) $\sum_{i \in I} t_{ij} \leq s_j$ for all j in J,

(3) if $t_{ij} > 0$, then $(i, j) \in R$.

Proof: This is essentially the proof of the Splitting Lemma 4.10 of [8] or Lemma 9.2 of [9]. The Max-Flow Min-Cut Theorem 5.1 of [4] is applied to a graph with nodes $\perp$ (source), i in I, j in J, and $\top$ (sink); the index sets I and J are assumed to be disjoint. There are edges from $\perp$ to i with capacities r_i, from i to j with "large" capacity C if $(i, j) \in R$ and 0 otherwise, and from j to $\top$ with capacities s_j, where C is a constant which is bigger than the sums of all occurring numbers. The remainder of the proof is in analogy to [8], [9], and thus omitted. ❑

In contrast to addition, subtraction is not continuous since it is not even monotonic. Nevertheless, the following lemma holds.

LEMMA 2.8: Let Y be a topological space with a separately continuous binary meet operation $\sqcap: Y \times Y \to Y$ (with respect to the specialization preorder), and let $f, g: Y \to \mathbf{R}_+$ be two functions with $f \sqsupseteq g$. If f is continuous, and both g and $f - g$ are monotonic, then $f - g$ is continuous.

Proof: For r in $\mathbf{R}_+$, we have to show that $W = \{y \in Y \mid (f-g)(y) > r\}$ is open. Let y be in W, i.e., $fy > r + gy$. By continuity of f, $U = \{a \in Y \mid fa > r + gy\}$ is open. By hypothesis, $h: Y \to Y$ with $hb = b \sqcap y$ is continuous. Hence, $V = h^{-}U$ is open. Since $hy = y \in U$, y is in V. We show $V \subseteq W$. Let b be in V.

$$\begin{aligned}(f-g)(b) &\geq (f-g)(b \sqcap y) && (f-g \text{ is monotonic})\\ &= f(b \sqcap y) - g(b \sqcap y) &&\\ &> r + gy - g(b \sqcap y) && (b \sqcap y = hb \text{ is in } U)\\ &\geq r && (g \text{ is monotonic}).\end{aligned}$$ ❑

This result is a bit surprising since g need not even be continuous. Note that the lemma applies in particular to the spaces $Y = \Omega_S X$ and $Y = \Omega_P X$ because of Proposition 2.2.

3. VALUATIONS

In this section, we define valuations and their potential continuity properties. Then, some operations on valuations are introduced, e.g., addition of two valuations, multiplication by a real number, restriction and corestriction to an open set.

3.1. Definition and Continuity Properties

A *valuation* on a topological space X is a function $\nu: \Omega X \to \overline{\mathbf{R}}_+$ with the following properties:

(a) $\nu\varnothing = 0$ (strictness).
(b) $\nu(U \cup V) + \nu(U \cap V) = \nu U + \nu V$ for all opens U and V (modularity).
(c) $\nu U \leq \nu V$ for all opens U and V with $U \subseteq V$ (monotonicity).

Valuations are partially ordered by defining $\nu \sqsubseteq \nu'$ iff $\nu O \leq \nu' O$ for all O in ΩX. A valuation ν is *bounded* if $\nu(X) < \infty$.

Mostly, we shall consider valuations with an additional continuity condition. There are several such conditions according to which topology is chosen for ΩX.

(1) A valuation ν is *Scott continuous* iff ν: $\Omega_s X \to \overline{\mathbf{R}}_+$ is continuous. Equivalently, for every directed family $(V_i)_{i \in I}$ of opens, $\nu(\bigcup_{i \in I} V_i) = \bigsqcup_{i \in I} \nu V_i$ holds.

Since every Scott continuous function is monotonic, the condition of monotonicity in the definition of valuations becomes redundant once we consider Scott continuous valuations.

(2) A valuation ν is *point continuous* iff ν: $\Omega_p X \to \overline{\mathbf{R}}_+$ is continuous. Equivalently, $\nu^-\{s \in \overline{\mathbf{R}}_+ \mid s > r\}$ is point open for every r in $\mathbf{R}_+$, or: for every open O and number r in $\mathbf{R}_+$ with $\nu O > r$, there is some finite $F \subseteq O$ such that $F \subseteq O'$ implies $\nu O' > r$.

Since the topology of $\Omega_p X$ is a subset of that of $\Omega_s X$, we obtain:

PROPOSITION 3.1: Every point continuous valuation is Scott continuous.

While many authors [8], [10], [13] have considered Scott continuous valuations, the notion of point continuity seems to be new.

3.2. Operations on Valuations

In this subsection, we present some basic ways to obtain valuations from other valuations or from scratch.

(1) The *zero function* $\lambda U. 0$, which maps every open set to 0, is a bounded valuation. As a constant function, it has all continuity properties you like; thus, it is point continuous.

(2) For every point x of X, there is a bounded valuation $\hat{x}$, where $\hat{x}(U)$ is 1 if $x \in U$, and 0 otherwise. Valuations of the form $\hat{x}$ are called *point valuations*. Since $x \in U$ and $x \sqsubseteq y$ implies $y \in U$, $x \sqsubseteq y$ implies $\hat{x} \sqsubseteq \hat{y}$.

Every point valuation is point continuous. For, if $\hat{x}(U) > r$, then $r < 1$ and $x \in U$. Choose $\{x\}$ as the finite set F in the characterization of point continuity.

(3) If r is a constant from $\overline{\mathbf{R}}_+$ and ν is a valuation, then $r \cdot \nu$ is a valuation, where $(r \cdot \nu)(U) = r \cdot \nu U$. If $r < \infty$ and ν is bounded, then $r \cdot \nu$ is bounded.

Since $\lambda s.\, r \cdot s$: $\overline{\mathbf{R}}_+ \to \overline{\mathbf{R}}_+$ is continuous and compositions of continuous functions are continuous, we obtain:

If ν is Scott continuous/point continuous, then so is $r \cdot \nu$.

(4) If ν_1 and ν_2 are two valuations on X, then so is $\nu_1 + \nu_2$, where $(\nu_1 + \nu_2)(U) = \nu_1 U + \nu_2 U$. If both ν_1 and ν_2 are bounded, then so is $\nu_1 + \nu_2$.

Since $+$: $\overline{\mathbf{R}}_+ \times \overline{\mathbf{R}}_+ \to \overline{\mathbf{R}}_+$ is continuous and $\nu_1 + \nu_2 = (+) \circ (\nu_1 \times \nu_2)$, we obtain:

If ν_1 and ν_2 are Scott continuous/point continuous, then so is $\nu_1 + \nu_2$.

(5) Every directed family $(\nu_i)_{i \in I}$ of valuations has a least upper bound, namely the valuation $\bigsqcup_{i \in I}\nu_i$, which is defined by $(\bigsqcup_{i \in I}\nu_i)(U) = \bigsqcup_{i \in I}\nu_i\, U$. Even if all ν_i are bounded, $\bigsqcup_{i \in I}\nu_i$ may be unbounded.

For all open sets W of $\overline{\mathbf{R}}_+$, $(\bigsqcup_{i \in I}\nu_i)(U)$ is in W iff $\nu_i(U)$ in W for some i in I. Hence, $(\bigsqcup_{i \in I}\nu_i)^- W = \bigcup_{i \in I}\nu_i^- W$. Thus we obtain:

If all ν_i are Scott continuous/point continuous, then so is $\bigsqcup_{i \in I}\nu_i$.

(6) If $f: X \to Y$ is continuous, then every valuation ν on X induces a valuation $\nu \circ f^-$ on Y. This operation maps point valuations to point valuations since $\hat{x} \circ f^- = \widehat{fx}$. The valuation $\nu \circ f^-$ is bounded iff ν is bounded. Using Proposition 2.1, we obtain:

If ν is Scott continuous/point continuous, then so is $\nu \circ f^-$.

(7) Let ν_1 and ν_2 be valuations on X, where ν_1 is bounded and $\nu_1 \sqsupseteq \nu_2$ holds. Then $\nu_1 - \nu_2$ with $(\nu_1 - \nu_2)(U) = \nu_1 U - \nu_2 U$ is a bounded strict modular function from ΩX to $\overline{\mathbf{R}}_+$. We require that ν_1 is bounded to avoid differences involving ∞. The condition $\nu_1 \sqsupseteq \nu_2$ is needed to ensure that $\nu_1 - \nu_2$ yields values in $\overline{\mathbf{R}}_+$.

Even if ν_1 and ν_2 are monotonic, the difference $\nu_1 - \nu_2$ may not be monotonic. On the other hand, monotonicity of the difference is sufficient to derive stronger continuity results.

PROPOSITION 3.2: Let ν_1 and ν_2 be valuations on X, where ν_1 is bounded, $\nu_1 \sqsupseteq \nu_2$ holds, and $\nu_1 - \nu_2$ is monotonic. Then $\nu_1 - \nu_2$ is a valuation. If ν_1 is Scott continuous/point continuous, then so is $\nu_1 - \nu_2$.

Proof: The continuity properties follow from Lemma 2.8 with Proposition 2.2. ❑

Interestingly, this holds without requiring the corresponding kind of continuity for ν_2.

3.3. Restriction and Corestriction of Valuations

Let ν be a valuation on X, and let W be an open set of X. The *restriction* $\nu|_W$ of ν to W is defined by $\nu|_W(U) = \nu(W \cap U)$ for every U in ΩX. This is again a valuation on X. It is bounded iff $\nu(W) < \infty$. This holds in particular if ν is bounded.

By Proposition 2.2, the function $\lambda U.\, W \cap U: \Omega X \to \Omega X$ is continuous in $\Omega_s X$ as well as in $\Omega_p X$. Thus, we obtain:

If ν is Scott continuous/point continuous, then so is $\nu|_W$.

Let ν be a *bounded* valuation on X, and W an open set of X. The *corestriction* $\nu|^W$ of ν to W is defined by $\nu|^W = \nu - \nu|_W$. Hence, $\nu|^W(U) = \nu U - \nu(U \cap W)$. By modularity, this is equal to $\nu(U \cup W) - \nu W$.

We require ν to be bounded in order to avoid problems with undefined differences. Monotonicity of ν guarantees $\nu \sqsupseteq \nu|_W$. Because of $\nu|^W(U) = \nu(U \cup W) - \nu W$, monotonicity of $\nu|^W$ follows from monotonicity of ν. Using Proposition 2.8, we obtain:

If ν is Scott continuous/point continuous, then so is $\nu|^W$.

It is not difficult to prove some more properties of restriction and corestriction:

PROPOSITION 3.3:

$W \cap U = W \cap V \Rightarrow \nu\|_W U = \nu\|_W V$	$\neg W \cap U = \neg W \cap V \Rightarrow \nu\|^W U = \nu\|^W V$
$(\nu\|_U)\|_V = (\nu\|_V)\|_U = \nu\|_{U \cap V}$	$(\nu\|^U)\|^V = (\nu\|^V)\|^U = \nu\|^{U \cup V}$
$(\nu\|_U)\|^V = (\nu\|^V)\|_U$	$\nu = \nu\|_W + \nu\|^W$

We shall use the abbreviation $\nu|_U^V$ for $\nu|_U|^V = \nu|^V|_U$.

4. CLASSES OF VALUATIONS

In this section, we study the relationship among some classes of valuations. We already know the classes of Scott continuous, point continuous, and bounded valuations ($\nu X < \infty$).

4.1. Valuations on Locally Finitary Spaces

There is a class of spaces where every Scott continuous valuation is point continuous.

A subset F of a space X is *finitary* iff $F = \uparrow E$ for some finite set E. The space X is *locally finitary* iff for every point x in X and open U of X with $x \in U$, there are a finitary set F and an open V such that $x \in V \subseteq F \subseteq U$.

Since finitary sets are compact, every locally finitary space is locally compact. Every continuous dcpo (with its Scott topology) is locally finitary. A locally finitary $\mathcal{T}_1$-space is discrete.

THEOREM 4.1: Every Scott continuous valuation on a locally finitary space is point continuous.

Proof: Let ν be a Scott continuous valuation and assume $\nu U > r$ for some open U and $r < \infty$. Let $\mathcal{V}$ be the set of all open sets V such that there is a finitary set F with $V \subseteq F \subseteq U$. Since the union of two open/finitary sets is again so, $\mathcal{V}$ is directed. Because of local finitariness, the union of $\mathcal{V}$ is U.

Since ν is Scott continuous, there is some V in $\mathcal{V}$ such that $\nu V > r$. Let E be a finite set with $V \subseteq \uparrow E \subseteq U$. If $E \subseteq O'$, then $V \subseteq O'$, whence $\nu O' > r$. ❑

In general, the notions of point continuity and Scott continuity differ. For instance, the length or Lebesgue measure on the unit interval of the reals with the standard Hausdorff topology induces a bounded Scott continuous valuation which is not point continuous.

4.2. Approximation by Bounded Valuations

THEOREM 4.2: Every Scott continuous/point continuous valuation can be obtained as a directed join of bounded Scott continuous/point continuous valuations.

Proof: Let ν be a Scott continuous valuation. Let $\mathbf{\Phi}(\nu)$ be the set of all opens O with $\nu O < \infty$, and let C be $\neg \bigcup \mathbf{\Phi}(\nu)$. For V in $\mathbf{\Phi}(\nu)$, $F \subseteq_{fin} C$, and $n \in \mathbf{N}_0$, let $\nu_{V,F,n} = \nu|_V + \sum_{x \in F} n \cdot \hat{x}$. This is a bounded valuation. If ν is point continuous, then so is $\nu_{V,F,n}$. The set $D = \{\nu_{V,F,n} \mid V \in \mathbf{\Phi}(\nu),\ F \subseteq_{fin} C,\ n \in \mathbf{N}_0\}$ is directed. Let $\nu' = \sqcup D$. We claim $\nu' = \nu$.

If there is an x in $O \cap C$, then $\nu' O \geq (n \cdot \hat{x})(O) = n$ for all n in $\mathbf{N}_0$, whence $\nu' O = \infty$. On the other hand, if νO were finite, then $O \subseteq \bigcup \mathbf{\Phi}(\nu) = \neg$C in contradiction to $O \cap C \neq \emptyset$.

If $O \subseteq \bigcup \mathbf{\Phi}(\nu)$, then $O = \bigcup\{O \cap V \mid V \in \mathbf{\Phi}(\nu)\}$. By modularity, $\mathbf{\Phi}(\nu)$ is directed.

$$
\begin{aligned}
\nu O &= \sqcup_{V \in \Phi(\nu)} \nu(O \cap V) && (\nu \text{ is Scott continuous}) \\
&= \sqcup_{V \in \Phi(\nu)} \nu|_V (O) \\
&= \sqcup_{\nu^* \in D} \nu^*(O) && (\nu_{V,F,n}(O) = \nu|_V(O)) \\
&= \nu' O \quad \square
\end{aligned}
$$

4.3. Finite Valuations

A valuation φ on a space X is *finite* if it is a finite linear combination of point valuations, i.e., if $\varphi = \sum_{x \in F} r_x \cdot \hat{x}$ where F is a finite subset of X and $0 < r_x < \infty$ for all x in F. This representation is unique provided that the underlying space X is $\mathcal{T}_0$.

THEOREM 4.3: (Unique Representation Theorem). Let X be a $\mathcal{T}_0$-space. If $\sum_{x \in F} r_x \cdot \hat{x} = \sum_{x \in G} s_x \cdot \hat{x}$ where F and G are finite subsets of X, and $0 < r_x < \infty$ for all x in F and $0 < s_x < \infty$ for all x in G, then $F = G$ and $r_x = s_x$ for all x in F.

Proof: This is Lemma 2.1 in[10]. We sketch Kirch's proof. Let a be maximal in $F \cup G$, and let U be an open set with $U \cap (F \cup G) = \{a\}$. If a is in G, then $\sum_{x \in F} r_x \cdot \hat{x}(U) = s_a$, whence a is in F as well, and $r_a = s_a$ holds. With induction, the claim follows. ❑

The $\mathcal{T}_0$ property is really needed. Consider the space $X = \{a, b\}$ where $\varnothing$ and X are the only open sets. The finite valuations $1 \cdot \hat{a}$ and $1 \cdot \hat{b}$ both map $\varnothing$ to 0 and X to 1. Hence, they are identical despite their different representation.

The theorem also relies on the fact that finite valuations have finite coefficients. Consider for instance the Sierpinski space $X = \{\bot, \top\}$ where $\varnothing$, $\{\top\}$, and X are open. The valuations $\infty \cdot \hat{\top}$, $1 \cdot \hat{\bot} + \infty \cdot \hat{\top}$, and $\infty \cdot \hat{\bot} + \infty \cdot \hat{\top}$ all map $\varnothing$ to 0, and $\{\top\}$ and X to ∞.

5. SPACES OF VALUATIONS

In this section, we define various topological spaces of valuations and study their relationship.

Let $\mathsf{V}X$ be the set of all *Scott continuous valuations* on X, $\mathsf{V}_{\mathrm{p}}X$ the set of all *point continuous* valuations, and $\mathsf{V}_{\mathrm{f}}X$ the set of *finite* valuations. We topologize these sets as subspaces of the pointwise function space $[\Omega_{\mathrm{s}}X \to \overline{\mathbf{R}}_{+}]_{\mathrm{p}}$. Thus, the topology of $\mathsf{V}X$ is generated by the subbasic opens $\langle U > r\rangle = \{\nu \in \mathsf{V}X \mid \nu U > r\}$ where U ranges over the opens of X and r ranges over $\overline{\mathbf{R}}_{+}$ with $0 < r < \infty$. The order defined by this topology is $\nu \sqsubseteq \nu'$ iff $\nu O \leq \nu' O$ for all opens O.

In general, continuous operations on a space Y can be lifted to continuous operations on pointwise function spaces $[X \to Y]_{\mathrm{p}}$. Hence, addition $+ : \mathsf{V}X \times \mathsf{V}X \to \mathsf{V}X$ and multiplication $\cdot : \overline{\mathbf{R}}_{+} \times \mathsf{V}X \to \mathsf{V}X$ are continuous.

The function $\mathbf{s}: X \to \mathsf{V}X$ with $\mathbf{s}x = \hat{x}$ is continuous. For, $\mathbf{s}^{-}\langle U > r\rangle = U$ if $r < 1$, and $= \varnothing$ otherwise. This also shows that $\mathbf{s}^{-}$ is surjective, whence $\mathbf{s}$ is a topological embedding for $\mathcal{T}_0$-spaces X.

Every continuous function $f: X \to Y$ induces a function $\mathsf{V}f: \mathsf{V}X \to \mathsf{V}Y$ where $\mathsf{V}f(\nu) = \nu \circ f^{-}$. The function $\mathsf{V}f$ is linear with respect to addition and scalar multiplication of $\mathsf{V}X$. It is continuous since $\mathsf{V}\, f(\nu) \in \langle V > r\rangle$ iff $\nu \in \langle f^{-}V > r\rangle$. Thus, $(\mathsf{V}\, f)^{-}\langle V > \mathrm{r}\rangle = \langle f^{-}V > r\rangle$. From this equation, we see that $(\mathsf{V}f)^{-}$ is surjective if f^{-} is surjective. Hence, $\mathsf{V}f$ is a topological embedding if f is an embedding. The operation V is functorial, i.e., $\mathsf{V}\,\mathrm{id} = \mathrm{id}$ and $\mathsf{V}(g \circ f) = \mathsf{V}g \circ \mathsf{V}f$, and $\mathsf{V}f \circ \mathbf{s} = \mathbf{s} \circ f$ holds.

Let us now consider the topological properties of the various spaces of valuations.

PROPOSITION 5.1: For every space X, the spaces $\mathsf{V}X$ and $\mathsf{V}_{\mathrm{p}}X$ are sober, and $\mathsf{V}_{\mathrm{f}}X$ is a $\mathcal{T}_0$-space.

Proof: The space $\mathsf{V}X$ is a subspace of the pointwise function space $F = [\Omega_{\mathrm{s}}X \to \overline{\mathbf{R}}_{+}]_{\mathrm{p}}$ which is sober as $\overline{\mathbf{R}}_{+}$ is sober. Since the elements of $\mathsf{V}X$ are characterized by continuous equations with values in $\overline{\mathbf{R}}_{+}$ (strictness and modularity), $\mathsf{V}X$ is a $\mathcal{T}_0$-equalizer subspace of F and thus sober. For $\mathsf{V}_\mathrm{p}X$, we start with $F' = [\Omega_\mathrm{p}X \to \overline{\mathbf{R}}_+]_\mathrm{p}$. The space $\mathsf{V}_\mathrm{f}X$ is $\mathcal{T}_0$ as a subspace of $\mathsf{V}_\mathrm{p}X$. ❑

Now, we come to one of the main results of this paper: For every space X, $\mathsf{V}_\mathrm{p}X$ is the soberification of $\mathsf{V}_\mathrm{f}X$. By Theorem 2.4(3), we have to prove: For every point continuous valuation ν and open set $\mathcal{O}$ of $\mathsf{V}_\mathrm{p}X$ with ν in $\mathcal{O}$, there is a finite valuation $\varphi \sqsubseteq \nu$ with φ in $\mathcal{O}$. The proof of this statement is structured into several parts. The results of these parts are presented as auxiliary lemmas.

The first lemma contains the step from arbitrary to bounded valuations.

LEMMA 5.2: For every point continuous valuation ν and open set $\mathcal{O}$ of $\mathsf{V}_\mathrm{p}X$ with ν in $\mathcal{O}$, there is a bounded point continuous valuation $\nu' \sqsubseteq \nu$ with ν' in $\mathcal{O}$.

Proof: By Theorem 4.2, ν is a directed join of bounded point continuous valuations. In a sober space such as $\mathsf{V}_\mathrm{p}X$, every open set is Scott open. Hence, there is some bounded point continuous valuation $\nu' \sqsubseteq \nu$ with ν' in $\mathcal{O}$. ❑

The next lemma deals with the step from bounded to finite valuations in a special case.

LEMMA 5.3: Let ν be a bounded point continuous valuation with $\nu W > r$ for some open set W and real number r. Then there is a finite valuation $\varphi \sqsubseteq \nu$ with $\varphi W > r$.

Proof: Choose a real number r' such that $\nu W > r' > r$. Since ν is point continuous, there is a finite set $F = \{x_1, \ldots, x_n\} \subseteq W$ such that $F \subseteq O$ implies $\nu O > r'$ for all open sets O. Since $\nu\varnothing = 0$, n cannot be 0. Let $\varepsilon = (r' - r)/n$. For every point x_i, choose an open set U_i such that $x_i \in U_i$ and $\nu U_i < \varepsilon + \sqcap_{O \ni x_i} \nu O$. We also need the unions $V_i = \bigcup_{j=1}^{i} U_i$ for $0 \le i \le n$; in particular, $V_0 = \varnothing$.

Using restriction and corestriction, for every i with $1 \le i \le n$,

$$\nu|^{V_{i-1}} = \nu|^{V_{i-1}}|_{U_i} + \nu|^{V_{i-1}}|^{U_i} = \nu|_{U_i}^{V_{i-1}} + \nu|^{V_i}.$$

Let $\nu_i = \nu|_{U_i}^{V_{i-1}}$. Starting from $\nu = \nu|^{V_0}$, we obtain by iteration $\nu = (\sum_{i=1}^{n}\nu_i) + \nu|^{V_n}$. On the other hand, $\nu = \nu|_{V_n} + \nu|^{V_n}$ holds, whence $\sum_{i=1}^{n}\nu_i = \nu|_{V_n}$.

Let $a_i = \nu_i(X)$ and $b_i = 0 \sqcup (a_i - \varepsilon)$. With these numbers, let $\varphi = \sum_{i=1}^{n} b_i \cdot \hat{x}_i$.

To prove $\varphi W > r$, we first compute $\sum_{i=1}^{n} a_i = \sum_{i=1}^{n}\nu_i(X) = \nu|_{V_n}(X) = \nu(V_n) > r'$ since $F \subseteq V_n$. Since all x_i are in W, we obtain $\varphi W = \sum_{i=1}^{n} b_i \ge \sum_{i=1}^{n}(a_i - \varepsilon) = \sum_{i=1}^{n} a_i - n \cdot \varepsilon > r' - (r' - r) = r$.

To prove $\varphi \sqsubseteq \nu$, note that $\nu O \ge \sum_{i=1}^{n}\nu_i\, O \ge \sum_{i: x_i \in O}\nu_i\, O$ and $\varphi O = \sum_{i: x_i \in O} b_i$. Hence, it suffices to show $\nu_i O \ge b_i$ for all i with $x_i \in O$. By definition, $\nu_i\, O = \nu(O \cap U_i) - \nu(O \cap U_i \cap V_{i-1})$ holds. Since x_i in $O \cap U_i$, we obtain $\nu(O \cap U_i)$

$> \nu U_i - \varepsilon$ by the choice of U_i. Thus, $\nu_i O > \nu U_i - \varepsilon - \nu(U_i \cap V_{i-1}) = \nu_i(X) - \varepsilon = a_i - \varepsilon$. Since $\nu_i O \geq 0$ also holds, $\nu_i O \geq b_i$ follows. ❑

The next lemma generalizes Lemma 5.3 from one open set W to any finite number of open sets O_i.

LEMMA 5.4: Let ν be a bounded point continuous valuation with $\nu O_i > r_i$ for some open sets $O_1, \ldots, O_n$ and real numbers $r_1, \ldots, r_n$. Then there is a finite valuation $\varphi \sqsubseteq \nu$ with $\varphi O_i > r_i$ for all i with $1 \leq i \leq n$.

Proof: The valuation ν can be partitioned along O_1 into $\nu = \nu|_{O_1} + \nu|^{O_1}$. Both parts can be partitioned along O_2 into $\nu|_{O_1} = \nu|_{O_1 \cap O_2} + \nu|^{O_2}_{O_1}$ and $\nu|^{O_1} = \nu|^{O_1}_{O_2} + \nu|^{O_1 \cup O_2}$. Iterating this process, we obtain $\nu = \sum_T \nu_T$ where T ranges over the subsets of $I = \{1, \ldots, n\}$ and $\nu_T = \nu|^{V_T}_{U_T}$ with $U_T = \bigcap_{i \in T} O_i$ and $V_T = \bigcup_{i \in I \setminus T} O_i$. By construction, $\nu_T U_T = \nu_T X$ holds.

If i in T, then $U_T \subseteq O_i$, whence $\nu_T O_i = \nu_T X$. If i not in T, then $O_i \subseteq V_T$, whence $\nu_T O_i = 0$. Thus, $r_i < \nu O_i = \sum_T \nu_T O_i = \sum_{T \ni i} \nu_T X$. Choose some real number ρ with $0 < \rho < 1$ such that $r_i < \rho \cdot (\sum_{T \ni i} \nu_T X)$ still holds for all i. Let $q_T = \rho \cdot \nu_T X$. Then $\sum_{T \ni i} q_T > r_i$ for all i.

If $\nu_T X \neq 0$, then $\nu_T U_T = \nu_T X > q_T$. By Lemma 5.3, there is a finite valuation $\varphi_T \sqsubseteq \nu_T$ such that $\varphi_T U_T > q_T$. If $\nu_T X = 0$, then $q_T = 0$, and we set $\varphi_T = 0$. In both cases, we obtain a finite $\varphi_T \sqsubseteq \nu_T$ such that $\varphi_T U_T \geq q_T$.

Let $\varphi = \sum_T \varphi_T$. This is a finite valuation below ν. For all i, $\varphi O_i = \sum_T \varphi_T O_i \geq \sum_{T \ni i} \varphi_T U_T \geq \sum_{T \ni i} q_T > r_i$ holds. ❑

With these lemmas, we can now prove:

THEOREM 5.5: For every space X, $\mathsf{V}_p X$ is the soberification of $\mathsf{V}_f X$.

Proof: Let ν be in $\mathsf{V}_p X$ and $\mathcal{O}$ in $\Omega(\mathsf{V}_p X)$ with $\nu \in \mathcal{O}$. By Lemma 5.2, there is a bounded $\nu' \sqsubseteq \nu$ with $\nu' \in \mathcal{O}$.

Using the subbase of $\mathsf{V}_p X$, we have $\nu' \in \bigcap_{i \in I} \langle O_i > r_i \rangle \subseteq \mathcal{O}$ for some finite I, open sets O_i of X, and r_i in $\mathbf{R}_+$ with $0 < r_i < \infty$. From Lemma 5.4, we obtain a finite $\varphi \sqsubseteq \nu'$ with $\varphi \in \bigcap_{i \in I} \langle O_i > r_i \rangle \subseteq \mathcal{O}$. ❑

6. UNIVERSAL PROPERTIES

In the sequel, we look for universal properties of the valuation spaces. We shall prove that $\mathsf{V}_f X$ is the free locally convex $\mathcal{T}_0$-cone, and $\mathsf{V}_p X$ is the free locally convex sober cone. We have not found a universal property for $\mathsf{V} X$.

6.1. Cones

A *cone* or $\mathbf{R}_+$-module is an algebraic structure $(M, +, 0, \cdot)$ where $+ : M \times M \to M$ is a commutative associative operation with neutral element $0 \in M$, and $\cdot : \mathbf{R}_+ \times M \to M$ is an operation satisfying the module (or vector space) axioms:

$$\begin{array}{ll} r \cdot 0 = 0 & r \cdot (m_1 + m_2) = r \cdot m_1 + r \cdot m_2 \\ 0 \cdot m = 0 & (r + s) \cdot m = r \cdot m + s \cdot m \\ 1 \cdot m = m & (r \cdot s) \cdot m = r \cdot (s \cdot m) \end{array}$$

A *topological cone* is a cone with a topology such that "+" and "·" are continuous if $\mathbf{R}_+$ is given the Scott topology. Often, we shall omit the word "topological" if there is already a topological notion around such as "sober."

Homomorphisms between cones are linear functions, where linearity means $f(m + m') = fm + fm'$ and $f(r \cdot m) = r \cdot fm$ as usual. Homomorphisms between topological cones are continuous linear functions.

If $(M)_{i \in I}$ is a family of (topological) cones, then the product $\prod_{i \in I} M_i$ is a (topological) cone with $(m_i)_{i \in I} + (m_i')_{i \in I} = (m_i + m_i')_{i \in I}$, $0 = (0_i)_{i \in I}$, and $r \cdot (m_i)_{i \in I} = (r \cdot m_i)_{i \in I}$.

In every topological cone, 0 is the least element since continuous functions are monotonic, whence $0 = 0 \cdot m \sqsubseteq 1 \cdot m = m$ holds for all m in M. Thus, nontrivial topological cones cannot be $\mathcal{T}_1$-spaces.

Standard examples of topological cones are given by powers of $\mathbf{R}_+$ or $\overline{\mathbf{R}}_+$, and linear subspaces thereof. On the other hand, there are quite strange cones which have nothing to do with real numbers. Let $(L, \vee, \wedge)$ be a distributive lattice with least element F and greatest element T. Define $a + b = a \vee b$ for a and b in L, $0 = \mathsf{F}$, and for r in $\mathbf{R}_+$ and a in L, define $r \cdot a = \mathsf{F}$ if $r = 0$, and $= a$ otherwise. With these operations, L becomes a cone. If L is endowed with a topology which makes "$\vee$" and "·" continuous, then L is a topological cone. A suitable topology is the Scott topology if L is a continuous lattice.

6.2. Uniqueness Properties

The notions introduced so far are sufficient to state the uniqueness parts of the universal properties of V_f and V_p.

THEOREM 6.1: Let X be a topological space.

(1) Every linear function from $\mathsf{V}_\mathrm{f}X$ to some cone is uniquely determined by its values on point valuations.

(2) Every continuous linear function from $\mathsf{V}_\mathrm{p}X$ to a $\mathcal{T}_0$-cone is uniquely determined by its values on point valuations.

Proof: Part (1) is obvious since every finite valuation is a finite linear combination of point valuations. Part (2) follows from part (1) and Proposition 2.5, using the fact that $\mathsf{V}_\mathrm{p}X$ is the soberification of $\mathsf{V}_\mathrm{f}X$ (Theorem 5.5). ❑

Unfortunately, we do not know whether similar properties hold for $\mathsf{V}X$, the space of all Scott continuous valuations on X. We do not even know the answer for the special case that the target cone is $\overline{\mathbf{R}}_+$.

PROBLEM 1: Are continuous linear functions from $\mathsf{V}X$ to $\overline{\mathbf{R}}_+$ uniquely determined by their values on point valuations?

If the answer to this problem is yes, then also the continuous linear functions from $\mathsf{V}X$ to $\mathsf{V}Y$ are uniquely determined by their values on point valuations. For, the functions $\lambda\nu.\ \nu V$: $\mathsf{V}Y \to \overline{\mathbf{R}}_+$ are continuous and linear for every open set V of Y.

For continuous dcpo's, the question in Problem 1 can be answered affirmatively. For, continuous dcpo's X are locally finitary, hence $\mathsf{V}_\mathrm{p}X$ and $\mathsf{V}X$ coincide by Theorem 4.1, and thus Theorem 6.1(2) applies.

6.3. Convexity and Local Convexity

In order to formulate the universal properties for V_f and V_p, we need some more notions connected with cones.

A *convex combination* is a linear combination $r_1 \cdot x_1 + \cdots + r_n \cdot x_n$ whose coefficients sum up to 1. A subset S of a cone is *convex* if all convex combinations of points of S are back in S again. Intersections of convex sets are convex, hence every subset S of a cone has a least convex superset, the *convex hull* $\mathrm{con}\, S$. It is routine to show the following properties.

PROPOSITION 6.2: Let f: $M \to M'$ be a linear map between two cones.
(1) If S is convex in M', then $f^{-}S$ is convex in M.
(2) $\mathrm{con}\, f^{-}S \subseteq f^{-}(\mathrm{con}\, S)$ holds for all $S \subseteq M$.

A topological cone is *locally convex* if whenever a point x is in an open set U, there is an open set V such that $x \in V \subseteq \mathrm{con}\, V \subseteq U$. It is easily verified that it suffices to consider open sets U from a subbase S.

As long as only continuous dcpo's are considered, local convexity is not an issue.

PROPOSITION 6.3: A topological cone whose underlying space is a continuous dcpo is locally convex.

Proof: Let x be in an open set U. By continuity, there are an open set V and a point y such that $x \in V \subseteq \uparrow y \subseteq U$. By monotonicity of addition and multiplication, $\uparrow y$ is convex. Hence, $x \in V \subseteq \mathrm{con}\, V \subseteq \uparrow y \subseteq U$. ❑

Thus, the topological cones $\mathbf{R}_+$ and $\overline{\mathbf{R}}_+$ are locally convex. Also, distributive lattices with the Scott topology are locally convex since convex combinations are finite nonempty joins, and open sets are upper sets. For the moment, we do not have any examples of cones which are not locally convex.

In the sequel, we present three properties of local convexity.

PROPOSITION 6.4: (Products). If all topological cones M_i are locally convex, then so is $\prod_{i \in I} M_i$.

Proof: The product $M = \prod_{i \in I} M_i$ has a subbase $\{\pi_i^- O \mid i \in I, O \in \Omega M_i\}$. If x in $\pi_i^- O$, then $\pi_i x \in O$. Since M_i is locally convex, there is an open V in M_i such that $\pi_i x \in V \subseteq \mathrm{con}\, V \subseteq O$, whence $x \in \pi_i^- V \subseteq \pi_i^-(\mathrm{con}\, V) \subseteq \pi_i^- O$. The projection π_i is linear, whence $\mathrm{con}\, \pi_i^- V \subseteq \pi_i^-(\mathrm{con}\, V)$ by Proposition 6.2. ❑

PROPOSITION 6.5: (Linear subspaces). If M and M' are topological cones, $e: M \hookrightarrow M'$ is a linear topological embedding, and M' is locally convex, then so is M.

Proof: Let x in O for some open O of M. Since e is an embedding, $O = e^- U$ holds for some $U \in \Omega M'$. Thus ex is in U. Since M' is locally convex, there is an open V in M' such that $ex \in V \subseteq \mathrm{con}\, V \subseteq U$, whence $x \in e^- V \subseteq e^-(\mathrm{con}\, V) \subseteq e^- U$. Since e is linear, $\mathrm{con}\, e^- V \subseteq e^-(\mathrm{con}\, V)$ holds by Proposition 6.2. ❑

From the two propositions above, we may conclude that the spaces $\mathsf{V}X$, $\mathsf{V}_{\mathrm{p}}X$, and $\mathsf{V}_{\mathrm{f}}X$ are locally convex as linear subspaces of products of $\overline{\mathbf{R}}_+$.

PROPOSITION 6.6: (Linear retracts). If M and M' are topological cones, $e: M \to M'$ is continuous, and $r: M' \to M$ is linear and continuous with $r \circ e = \mathrm{id}$, then local convexity of M' implies local convexity of M.

Proof: If $x = r\,(ex)$ in U where U in ΩM, then $ex \in r^- U$. Because M' is locally convex, there is an open V in $\Omega M'$ such that $ex \in V \subseteq \mathrm{con}\, V \subseteq r^- U$. Hence, $x \in e^- V$. We claim $\mathrm{con}(e^- V) \subseteq U$.

We show that every convex combination $\sum t_i \cdot x_i$ of points x_i from $e^- V$ is in U. Since ex_i is in V, $\sum t_i \cdot ex_i$ is in $\mathrm{con}\, V \subseteq r^- U$. Thus, $r\,(\sum t_i \cdot ex_i) = \sum t_i \cdot x_i$ is in U. Here, linearity of r is used. ❑

6.4. A Universal Property for the Space of Finite Valuations

In this subsection, we present a universal property for the space $\mathsf{V}_{\mathrm{f}}X$ of finite valuations.

THEOREM 6.7: $\mathsf{V}_{\mathrm{f}}X$ is the free locally convex cone over X in $\mathcal{T}_0$, the category of $\mathcal{T}_0$-spaces and continuous maps.

This means: for a $\mathcal{T}_0$-space X, $\mathsf{V}_{\mathrm{f}}X$ is itself a locally convex $\mathcal{T}_0$-cone, and for every continuous function $f: X \to M$ from X to a locally convex $\mathcal{T}_0$-cone M, there is a unique continuous linear function $\overline{f}: \mathsf{V}_{\mathrm{f}}X \to M$ with $\overline{f} \circ \mathbf{s} = f$, i.e., $\overline{f}(\hat{x}) = fx$ for all x in X.

Proof: We already know that $\mathsf{V}_{\mathrm{f}}X$ is a locally convex $\mathcal{T}_0$-cone. The uniqueness statement is given by Theorem 6.1. We still have to show existence of $\overline{f}$.

Every finite valuation φ can be written as $\sum_{x \in F} r_x \cdot \hat{x}$ for some finite set F and some numbers r_x with $0 < r_x < \infty$. By Theorem 4.3, this representation is unique. Hence, $\bar{f}(\varphi) = \sum_{x \in F} r_x \cdot fx$ is a well-defined element of M. The function $\bar{f}$: $\mathsf{V}_{\mathrm{f}}X \to M$ defined in this manner is obviously linear and satisfies $\bar{f} \circ \mathbf{s} = f$. The only remaining task is to prove continuity of $\bar{f}$. This turns out to be quite complex.

Let U be an open set of M, and assume $\bar{f}(\varphi) \in U$. Then $\sum_{x \in F} r_x \cdot fx$ is in U. Since addition and multiplication are continuous in M, there are open sets R_x of $\mathbf{R}_+$ and V_x of M such that $r_x \in R_x$, $fx \in V_x$, and whenever $s_x \in R_x$ and $m_x \in V_x$, then $\sum_{x \in F} s_x \cdot m_x \in U$.

Choose numbers $r_x' > 0$ such that $r_x > r_x' \in R_x$, and applying local convexity of M, choose open sets W_x of M such that $fx \in W_x \subseteq \operatorname{con} W_x \subseteq V_x$, and let $O_x = f^- W_x$. By continuity of f, the sets O_x are open sets of X with x in O_x for all x in F. For every nonempty $T \subseteq F$, $\varphi(\bigcup_{x \in T} O_x) \geq \sum_{x \in T} r_x > \sum_{x \in T} r_x'$ holds. Hence, φ is in $\mathcal{O} = \bigcap_{\varnothing \neq T \subseteq F} \langle \bigcup_{x \in T} O_x > \sum_{x \in T} r_x' \rangle$, which is an open set of $\mathsf{V}_{\mathrm{f}}X$. We have to show that for every ψ in $\mathcal{O}$, $\bar{f}(\psi)$ is in U.

Let $\psi = \sum_{y \in G} s_y \cdot \hat{y}$ be in $\mathcal{O}$. Let $R \subseteq F \times G$ be the relation given by $(x, y) \in R$ iff $O_x \ni y$, whence $R^+(T) = \{y \in G \mid \exists x \in T\colon y \in O_x\}$ for subsets T of I. Since ψ is in $\mathcal{O}$,

$$\sum_{y \in R^+(T)} s_j = \psi\left(\bigcup_{x \in T} O_x\right) > \sum_{x \in T} r_x'$$

holds for all nonempty subsets T of I. For $T = \varnothing$, both sides are zero. Thus, "$\geq$" instead of "$>$" holds for all subsets T of I. Applying the Split Lemma 2.7, we obtain numbers $t_{xy} \in \mathbf{R}_+$ for x in F and y in G such that

(1) $\sum_{y \in G} t_{xy} = r_x'$ for all x in F,
(2) $\sum_{x \in F} t_{xy} \leq s_y$ for all y in G,
(3) if $t_{xy} > 0$, then $y \in O_x$.

Let $\psi' = \sum_{x \in F} \sum_{y \in G} t_{xy} \cdot \hat{y}$. Then

$$\bar{f}(\psi') = \sum_{y \in G} \left(\sum_{x \in F} t_{xy}\right) \cdot fy \sqsubseteq \sum_{y \in G} s_y \cdot fy = \bar{f}(\psi)$$

using monotonicity of addition and multiplication in M. The valuation ψ' may alternatively be written as

$$\psi' = \sum_{x \in F} r_x' \cdot \psi_x \quad \text{where } \psi_x = \sum_{y \in G \cap O_x} (t_{xy}/r_x') \cdot \hat{y}$$

The coefficients of ψ_x sum up to 1. Thus, $\bar{f}(\psi_x)$ is a convex combination of the points fy where y in $G \cap O_x$. All these points are in W_x, whence $\bar{f}(\psi_x)$ in V_x by choice of W_x.

Since $\bar{f}(\psi') = \sum_{x \in F} r_x' \cdot \bar{f}(\psi_x)$ where r_x' in R_x and $\bar{f}(\psi_x)$ in V_x, it is in U. Since $\bar{f}(\psi)$ is above $\bar{f}(\psi')$, it is in U as well. ❑

In Theorem 6.7, local convexity cannot be dispensed with: if M is a topological cone with the property that identity id: $M \to M$ has a continuous linear extension $\overline{\mathsf{id}}$: $\mathsf{V}_{\mathrm{f}}M \to M$, then M is a linear retract of $\mathsf{V}_{\mathrm{f}}M$, whence locally convex by Proposition 6.6. As a subspace of $\mathsf{V}_{\mathrm{f}}M$, it is also $\mathcal{T}_0$.

6.5. A Universal Property for the Space of Point Continuous Valuations

Here, we present a universal property for the space $\mathsf{V}_{\mathrm{p}}X$ of point continuous valuations.

THEOREM 6.8: $\mathsf{V}_{\mathrm{p}}X$ is the free locally convex sober cone over X in $\mathcal{T}_0$.

Proof: We already know that $\mathsf{V}_{\mathrm{p}}X$ is a locally convex sober cone. Let M be an arbitrary locally convex sober cone, and let $f\colon X \to M$ be continuous. By Theorem 6.7, there is a unique continuous linear function $f^*\colon \mathsf{V}_{\mathrm{f}}X \to M$ with $f^* \circ \mathbf{s} = f$. By Theorem 5.5, $\mathsf{V}_{\mathrm{p}}X$ is the soberification of $\mathsf{V}_{\mathrm{f}}X$. Hence, by Theorem 2.3, the continuous function $f^*\colon \mathsf{V}_{\mathrm{f}}X \to M$ has a unique continuous extension $\bar{f}\colon \mathsf{V}_{\mathrm{p}}X \to M$. Since $\bar{f}$ extends f^*, $\bar{f} \circ \mathbf{s} = f$ follows. The only thing which remains to be proved is linearity of $\bar{f}$.

Let r be a fixed element of $\overline{\mathbf{R}}_+$. Consider the two functions F, $G\colon \mathsf{V}_{\mathrm{p}}X \to \overline{\mathbf{R}}_+$ with $F(\nu) = \bar{f}(r \cdot \nu)$ and $G(\nu) = r \cdot \bar{f}(\nu)$. They are continuous and coincide on $\mathsf{V}_{\mathrm{f}}X$ because of linearity of f^*. By Proposition 2.5, $F = G$ follows.

For addition, consider the two functions F, $G\colon \mathsf{V}_{\mathrm{p}}X \times \mathsf{V}_{\mathrm{p}}X \to \overline{\mathbf{R}}_+$ with $F(\nu, \nu') = \bar{f}(\nu + \nu')$ and $G(\nu, \nu') = \bar{f}(\nu) + \bar{f}(\nu')$. They are continuous and coincide on $\mathsf{V}_{\mathrm{f}}X \times \mathsf{V}_{\mathrm{f}}X$ because of linearity of f^*. By Proposition 2.6, $\mathsf{V}_{\mathrm{p}}X \times \mathsf{V}_{\mathrm{p}} X$ is the soberification of $\mathsf{V}_{\mathrm{f}}X \times \mathsf{V}_{\mathrm{f}}X$. By Proposition 2.5, $F = G$ follows. ❑

6.6. The Continuous Case

If X is a continuous dcpo, then $\mathsf{V}X$ equals $\mathsf{V}_{\mathrm{p}}X$ by Theorem 4.1, and is sober by Proposition 5.1. Thus, its specialization preorder forms a dcpo, and every open set of $\mathsf{V}X$ is Scott open. It is, however, not immediately obvious that the topology of $\mathsf{V}X$ actually is identical with the Scott topology, and thus $\mathsf{V}X$ is the continuous dcpo of valuations considered by Jones [8].

Our universal property can be used to obtain a simple proof of the coincidence. For the purpose of this proof, let $\mathsf{V}_{\mathrm{S}}X$ be the dcpo of Scott continuous valuations on X equipped with the Scott topology. As shown by Jones, continuity of X implies continuity of $\mathsf{V}_{\mathrm{S}}X$, and $\mathbf{s}_{\mathrm{S}}\colon X \to \mathsf{V}_{\mathrm{S}}X$ with $\mathbf{s}_{\mathrm{S}}x = \hat{x}$ is Scott continuous. For continuous dcpo's, the Scott topology of the product is the product of the Scott topologies, whence $\mathsf{V}_{\mathrm{S}}X$ is a topological cone in our sense. By Proposition 6.3, it is locally convex. Thus, Theorem 6.8 applies, and we obtain a continuous linear function $F\colon \mathsf{V}X \to \mathsf{V}_{\mathrm{S}}X$ with $F \circ \mathbf{s} = \mathbf{s}_{\mathrm{S}}$. Let $G\colon \mathsf{V}_{\mathrm{S}}X \to \mathsf{V}X$ be the function which operates as identity on elements. It is obviously linear, and continuous since every open of $\mathsf{V}X$ is Scott open. Thus, $G \circ F\colon \mathsf{V}X \to \mathsf{V}X$ is continuous and linear. By Theorem 6.1(2), it equals the identity function of $\mathsf{V}X$. Thus, $F\colon \mathsf{V}X \to \mathsf{V}_{\mathrm{S}}X$ must also operate as identity on elements. Its continuity implies that every open set of $\mathsf{V}_{\mathrm{S}}X$ is open in $\mathsf{V}X$. Therefore, the topologies of $\mathsf{V}X$ and $\mathsf{V}_{\mathrm{S}}X$ coincide.

Summarizing, we obtain:

THEOREM 6.9: If X is a continuous dcpo with its Scott topology, then so is $\mathsf{V}X = \mathsf{V}_{\mathrm{p}}X$.

7. INTEGRATION

Several authors [8], [9], [10], [13] already defined integration of real-valued functions with respect to a valuation. Since they defined integration from scratch, the proofs of its properties are quite involved. Here, we present a novel definition of integration which is so simple that most proofs become trivial. (The complexity has not disappeared, though; it is now in the proofs of Theorems 2.3 and 6.7, which are needed to prove Theorem 6.8).

For every space X, integration will be a separately continuous function $\int_X: [X \to \overline{\mathbf{R}}_+]_i \times \mathsf{V}X \to \overline{\mathbf{R}}_+$. Note that the function space $[X \to \overline{\mathbf{R}}_+]_i$ is not topologized by the pointwise topology, but by the *Isbell topology* which has more open sets (see Subsection 2.2).

The function $\int_X$ is built from the following pieces:

(1) The function Ω_s: $[X \to \overline{\mathbf{R}}_+]_i \to [\Omega_s \overline{\mathbf{R}}_+ \to \Omega_s X]_p$ with $\Omega_s f = f^-$ is continuous. In fact, the Isbell topology was chosen to guarantee this.

(2) Using Ω_s, we map from

$$[X \to \overline{\mathbf{R}}_+]_i \times \mathsf{V}X \text{ to } [\Omega_s \overline{\mathbf{R}}_+ \overset{\cup,\cap}{\to} \Omega_s X]_p \times [\Omega_s X \overset{\text{mod}}{\to} \overline{\mathbf{R}}_+]_p,$$

where the labels at the arrows indicate the properties of the resulting functions. Now, we can use function composition to reach $[\Omega_s \overline{\mathbf{R}}_+ \overset{\text{mod}}{\to} \overline{\mathbf{R}}_+]_p = \mathsf{V}_S \overline{\mathbf{R}}_+$. Function composition

$$\circ: [X \to Y]_p \times [Y \to Z]_p \to [X \to Z]_p$$

is separately continuous.

(3) Since $\overline{\mathbf{R}}_+$ is a continuous dcpo, it is locally finitary, whence $\mathsf{V}\overline{\mathbf{R}}_+ = \mathsf{V}_p \overline{\mathbf{R}}_+$ by Theorem 4.1.

(4) $\overline{\mathbf{R}}_+$ is a locally convex sober cone. By Theorem 6.8, identity $\overline{\mathsf{id}}$: $\overline{\mathbf{R}}_+ \to \overline{\mathbf{R}}_+$ can be extended to a continuous linear function $\overline{\mathsf{id}}$: $\mathsf{V}_p \overline{\mathbf{R}}_+ \to \overline{\mathbf{R}}_+$ with the property $\overline{\mathsf{id}}(\hat{r}) = r$ for all r in $\overline{\mathbf{R}}_+$.

Putting all pieces together, we yield a separately continuous function

$$\int_X: [X \to \overline{\mathbf{R}}_+]_i \times \mathsf{V}X \to \overline{\mathbf{R}}_+ \text{ with } \int_X(f, \nu) = \overline{\mathsf{id}}(\nu \circ f^-).$$

Of course, this function can be restricted to a "pointwise" function $[X \to \overline{\mathbf{R}}_+]_i \times \mathsf{V}_p X \to \overline{\mathbf{R}}_+$ which is also separately continuous. This continuity is not destroyed if the Isbell topology on the real-valued functions is replaced by the smaller pointwise topology. For, the function Ω_p: $[X \to \overline{\mathbf{R}}_+]_p \to [\Omega_p \overline{\mathbf{R}}_+ \to \Omega_p X]_p$ with $\Omega_p f = f^-$ is continuous. Using Ω_p, we map from

$$[X \to \overline{\mathbf{R}}_+]_p \times \mathsf{V}_p X \text{ to } [\Omega_p \overline{\mathbf{R}}_+ \overset{\cup,\cap}{\to} \Omega_p X]_p \times [\Omega_p X \overset{\text{mod}}{\to} \overline{\mathbf{R}}_+]_p,$$

and composition can be used to reach $[\Omega_p \overline{\mathbf{R}}_+ \overset{\text{mod}}{\to} \overline{\mathbf{R}}_+]_p = \mathsf{V}_p \overline{\mathbf{R}}_+$.

Thus, we obtain two variants of integration with the same definition $\int_X(f, \nu) = \overline{\mathrm{id}}\,(\nu \circ f^-)$, but different continuity properties; the Isbell variant $\int_X$: $[X \to \overline{\mathbf{R}}_+]_i \times \mathsf{V}X \to \overline{\mathbf{R}}_+$, and the pointwise variant $\int_X$: $[X \to \overline{\mathbf{R}}_+]_p \times \mathsf{V}_pX \to \overline{\mathbf{R}}_+$.

In the sequel, we derive the essential properties of integration. They hold for both variants because the defining equations are the same. They are collected in Theorem 7.1 at the end of this section.

From the construction of the two variants of $\int_X$, we know that they are separately continuous. Since $[X \to \overline{\mathbf{R}}_+]_i$, $[X \to \overline{\mathbf{R}}_+]_p$, $\mathsf{V}X$, V_pX, and $\overline{\mathbf{R}}_+$ are d-spaces, they are also Scott continuous in both arguments. Integration is linear in the valuation argument, since $\int_X(f, \nu) = \overline{\mathrm{id}}(\nu \circ f^-)$, and $\overline{\mathrm{id}}$ is linear. The effect of integration on point valuations is as follows:

$$\int_X (f, \hat{x}) = \overline{\mathrm{id}}\,(\hat{x} \circ f^-) = \overline{\mathrm{id}}(\widehat{fx}) = fx.$$

A kind of "substitution theorem" is easily proved for continuous functions h: $X \to Y$, f: $Y \to \overline{\mathbf{R}}_+$, and valuations ν in $\mathsf{V}X$:

$$\int_X(f \circ h, \nu) = \overline{\mathrm{id}}\,(\nu \circ (f \circ h)^-) = \overline{\mathrm{id}}\,(\nu \circ h^- \circ f^-) = \int_Y(f, \nu \circ h^-).$$

Our final goal is to show that integration is also linear in its functional argument. If we only considered point continuous valuations, the proof would be quite easy: Both sides of the equation $\int_X(f + g, \nu) = \int_X(f, \nu) + \int_X(g, \nu)$ are continuous and linear in $\nu \in \mathsf{V}_pX$. By Theorem 6.1, it suffices to consider the special case $\nu = \hat{x}$. In this case, both sides are $fx + gx$. The equation $\int_X(r \cdot f, \nu) = r \cdot \int_X(f, \nu)$ would be handled similarly.

Unfortunately, this elegant proof is not possible in the general case since we do not have an analogous property for continuous linear functions defined on $\mathsf{V}X$ (cf. Problem 1). Fortunately, there is a way around the problem. Before we can present it, we have to consider the continuous functions f: $X \to \overline{\mathbf{R}}_+$ a bit closer.

A continuous function f: $X \to \overline{\mathbf{R}}_+$ is *simple* if its image f^+X is finite and does not contain ∞. An arbitrary continuous function is the directed join of all the simple functions below it. Every finite linear combination $r_1 \cdot \widehat{U_1} + \cdots + r_n \cdot \widehat{U_n}$ of characteristic functions $\widehat{U_i}$ of open sets U_i is simple. Conversely, every simple function can be written as such a linear combination.

We need some auxiliary statements for our proof of the linearity of integration in the functional argument.

(1) For r in $\overline{\mathbf{R}}_+$: $\int_X (r \cdot f, \nu) = r \cdot \int_X (f, \nu)$.

Proof: Let $(r\,\cdot)$: $\overline{\mathbf{R}}_+ \to \overline{\mathbf{R}}_+$ be the function defined by $(r\,\cdot)(s) = r \cdot s$. Then $r \cdot f = (r\,\cdot) \circ f$. By the substitution property, we obtain $\int_X(r \cdot f, \nu) = \int_{\overline{\mathbf{R}}_+}((r\,\cdot), \nu \circ f^-)$, and by the definition of integration, $r \cdot \int_X (f, \nu) = r \cdot \overline{\mathrm{id}}\,(\nu \circ f^-)$ holds. We claim $\int_{\overline{\mathbf{R}}_+}((r\,\cdot), \sigma) = r \cdot \overline{\mathrm{id}}(\sigma)$ for all valuations σ in $\mathsf{V}\overline{\mathbf{R}}_+ = \mathsf{V}_p\overline{\mathbf{R}}_+$. Since both sides of the equation are linear continuous functions in σ, it suffices by Theorem 6.1 to prove the equation for the special case of point valuations σ

$= \hat{s}$ where s in $\overline{\mathbf{R}}_+$. In this case, the left-hand side is $\int_{\overline{\mathbf{R}}_+}((r\,\cdot), \hat{s}) = (r\,\cdot)(s) = r \cdot s$, and the right hand side is $r \cdot \overline{\mathrm{id}}\,(\hat{s}) = r \cdot s$. ❑

(2) $\int_X (\underline{0}, \nu) = 0$ where $\underline{0}\,(x) = 0$ for all x in X.

Proof: By (1), using $\underline{0} = 0 \cdot \underline{0}$. ❑

(3) For r in $\overline{\mathbf{R}}_+$: $\int_X(r + f, \nu) = r \cdot \nu(X) + \int_X (f, \nu)$.

Proof: Apply the same idea as in the proof of (1). Since $X = f^{-}(\overline{\mathbf{R}}_+)$, Eq. (3) is equivalent to $\int_{\overline{\mathbf{R}}_+}((r\,+), \sigma) = r \cdot \sigma(\overline{\mathbf{R}}_+) + \overline{\mathrm{id}}(\sigma)$ where $\sigma = \nu \circ f^{-}$ in $\mathsf{V}\overline{\mathbf{R}}_+$. This equation holds for all σ in $\mathsf{V}\overline{\mathbf{R}}_+$, since both sides are continuous and linear in σ, and $\int_{\overline{\mathbf{R}}_+}((r\,+), \hat{s}) = (r\,+)(s) = r + s$, and $r \cdot \hat{s}\,(\overline{\mathbf{R}}_+) + \overline{\mathrm{id}}\,(\hat{s}) = r \cdot 1 + s$. ❑

(4) If $fx = gx$ for all x in the open set W, then $\int_X(f, \nu|_W) = \int_X(g, \nu|_W)$.

Proof: By hypothesis, $W \cap f^{-}V = W \cap g^{-}V$ holds for all V in $\Omega\overline{\mathbf{R}}_+$. By Proposition 3.3, $\nu|_W(f^{-}V) = \nu|_W(g^{-}V)$ follows, whence $\nu|_W \circ f^{-} = \nu|_W \circ g^{-}$. ❑

(5) If $fx = gx$ for all x in $X \backslash W$ where W is open, and ν is bounded, then $\int_X(f, \nu|^W) = \int_X(g, \nu|^W)$.

Proof: The valuation ν must be bounded so that the corestriction $\nu|^W$ is well defined. By hypothesis, $\neg W \cap f^{-}V = \neg W \cap g^{-}V$ holds for all V in $\Omega\overline{\mathbf{R}}_+$. By Proposition 3.3, $\nu|^W \circ f^{-} = \nu|^W \circ g^{-}$ follows. ❑

(6) For r in $\overline{\mathbf{R}}_+$ and W in ΩX, $\int_X(r \cdot \widehat{W} + f, \nu) = r \cdot \nu(W) + \int_X(f, \nu)$.

Proof: First assume that ν is bounded. Then it can be partitioned along W into $\nu = \nu|_W + \nu|^W$. Since integration is linear in the valuation argument, $\int_X(r \cdot \widehat{W} + f, \nu) = \int_X(r \cdot \widehat{W} + f, \nu|_W) + \int_X(r \cdot \widehat{W} + f, \nu|^W)$ holds. Since $\widehat{W}\,(x) = 0$ for x in $X \backslash W$, the second summand equals $\int_X(f, \nu|^W)$ by (5). Since $\widehat{W}\,(x) = 1$ for x in W, the first summand equals $\int_X(r + f, \nu|_W)$ by (4). By (3), this is $r \cdot \nu|_W\,(X) + \int_X(f, \nu|_W)$. Hence, we obtain $r \cdot \nu(W) + \int_X(f, \nu|_W) + \int_X(f, \nu|^W) = r \cdot \nu(W) + \int_X(f, \nu)$ for the sum.

The equation holds for every valuation ν in $\mathsf{V}X$, since integration is Scott continuous in the valuation argument, and ν is a directed join of bounded members of $\mathsf{V}X$ by Theorem 4.2. ❑

(7) $\int_X(r_1 \cdot \widehat{U_1} + \cdots + r_n \cdot \widehat{U_n}, \nu) = r_1 \cdot \nu(U_1) + \cdots + r_n \cdot \nu(U_n)$.

In particular, $\int_X(\widehat{U}, \nu) = \nu(U)$.

Proof: Apply (6) n times, then (2). ❑

(8) $\int_X(f + g, \nu) = \int_X(f, \nu) + \int_X(g, \nu)$

Proof: First assume that f and g are simple. Since every simple function is a finite linear combination of characteristic functions of opens, the statement follows from (7).

The equation for general f and g follows, since every continuous function is a directed join of simple functions, and integration is Scott continuous in the functional argument. ❑

Summarizing our results, we obtain:

THEOREM 7.1:
(1) Integration $\int_X: [X \to \overline{\mathbf{R}}_+]_i \times \mathsf{V}X \to \overline{\mathbf{R}}_+$ is separately continuous.
(2) The variant $\int_X: [X \to \overline{\mathbf{R}}_+]_p \times \mathsf{V}_p X \to \overline{\mathbf{R}}_+$ is also separately continuous.
(3) Integration (in both variants) is Scott continuous in both arguments.
(4) Integration is linear in both arguments.
(5) For $f: X \to \overline{\mathbf{R}}_+$ and x in X, $\int_X(f, \hat{x}) = fx$ holds.
(6) $\int_X (\widehat{U}, \nu) = \nu(U)$ where $\widehat{U}$ is the characteristic function of an open U of X.
(7) Let $h: X \to Y$ and $f: Y \to \overline{\mathbf{R}}_+$ be continuous, and let ν be in $\mathsf{V}X$. Then $\int_X(f \circ h, \nu) = \int_Y(f, \nu \circ h^-)$.

Comparison with other Notions of Integral. Jones [8, Section 3.9], Kirch [10, Chapter 4], and Tix [13, Chapter 4] define integration on a space X with respect to a Scott continuous valuation. While Jones only deals with bounded continuous functions and bounded valuations, Kirch and Tix work with the extended nonnegative reals $\overline{\mathbf{R}}_+$ as we do. Jones and Kirch define their integrals first for simple functions, using each their own standard representation. The integral notions are extended to all (bounded) continuous functions using the fact that these are directed joins of simple functions. The definition of Tix bears some similarity with ours: $\int_X(f, \nu)$ is defined as the Riemann integral $\int_0^\infty f_\nu(r)\,dr$ where $f_\nu(r) = \nu(f^-\{s \in \overline{\mathbf{R}}_+ \mid s > r\})$.

Jones's integral is Scott continuous and linear [8, Theorem 3.13, 3.14] in its functional argument, and $\int_X(\widehat{U}, \nu) = \nu U$ follows from its definition. Tix's integral has the same properties [13, Theorem 4.4]. Hence, these integrals coincide with ours since continuous functions from X to $\overline{\mathbf{R}}_+$ are directed joins of simple functions, which in turn are finite linear combinations of characteristic functions $\widehat{U}$ of opens.

Kirch's integral also coincides with ours because his definitions are instances of parts of our Theorem 7.1. He proves continuity and linearity of his integral for core-compact spaces only so that an argument as above is not possible.

Edalat [3] defines a Riemann-like integral on compact metric spaces X which differs from the integral notions mentioned so far in that the functions to be integrated are not necessarily continuous, and the integration is done with respect to a measure, not a valuation. There are some connections, though, as he uses the fact that every Borel measure on X can be approximated by finite valuations on a power space of X.

8. ISOMORPHIC DESCRIPTIONS

Integration may be used to derive isomorphic descriptions of $\mathsf{V}X$ and V_pX. Kirch [10, Section 8.1] and Tix [13, Section 4.3] prove similar descriptions; Kirch for dcpo's and Tix for the weak topology. In contrast, we do not start with the weak topology, but obtain it a posteriori (Theorem 8.3).

THEOREM 8.1: For every space X, the topological cone of Scott continuous valuations on X, i.e., strict modular Scott continuous functions from ΩX to $\overline{\mathbf{R}}_+$, is isomorphic to the topological cone of linear continuous functions from $[X \to \overline{\mathbf{R}}_+]_i$ to $\overline{\mathbf{R}}_+$ with the pointwise topology.

$$\mathsf{V}X = [\Omega_s X \overset{mod}{\to} \overline{\mathbf{R}}_+]_p \cong [\, [X \to \overline{\mathbf{R}}_+]_i \overset{lin}{\to} \overline{\mathbf{R}}_+]_p$$

Proof: For the proof, let $\mathcal{F} = [X \to \overline{\mathbf{R}}_+]_i$.

One isomorphism is constructed from integration: Define α: $\mathsf{V}\,X \to [\mathcal{F} \overset{lin}{\to} \overline{\mathbf{R}}_+]_p$ by $\alpha(\nu) = \lambda f.\ \int_X(f, \nu)$. This function has the claimed type since integration is continuous and linear in its functional argument. The function α itself is linear since integration is linear in its valuation argument. It is continuous since integration is separately continuous.

The inverse isomorphism is defined using characteristic functions: For F in $[\mathcal{F} \overset{lin}{\to} \overline{\mathbf{R}}_+]_p$, let $\beta(F) = F \circ \chi$ where χ: $\Omega X \to \mathcal{F}$ with $\chi U = \hat{U}$, the characteristic function of U. Strictness and modularity of $\beta(F)$ are easily verified. $\beta(F)$ is Scott continuous since both F and χ are Scott continuous. F is Scott continuous since it is continuous and both $\mathcal{F}$ and $\overline{\mathbf{R}}_+$ are d-spaces.

The function β itself is obviously linear. Since $[\mathcal{F} \overset{lin}{\to} \overline{\mathbf{R}}_+]_p$ is a subspace of $\prod_{f \in \mathcal{F}} \overline{\mathbf{R}}_+$, the function β_U: $[\mathcal{F} \overset{lin}{\to} \overline{\mathbf{R}}_+]_p \to \overline{\mathbf{R}}_+$ with $\beta_U(F) = F(\hat{U})$ is continuous for every fixed U in ΩX. Since $[\Omega_s X \overset{mod}{\to} \overline{\mathbf{R}}_+]_p$ is a subspace of $\prod_{U \in \Omega X} \overline{\mathbf{R}}_+$, continuity of the function β: $[\mathcal{F} \overset{lin}{\to} \overline{\mathbf{R}}_+]_p \to [\Omega_s X \overset{mod}{\to} \overline{\mathbf{R}}_+]_p$ follows.

At this point, we know that both α and β are continuous linear functions. We still have to show that they are inverse to each other.

For ν in $\mathsf{V}\,X$ and U in ΩX, $\beta(\alpha\nu)(U) = (\alpha\nu)(\hat{U}) = \int_X(\hat{U}, \nu) = \nu(U)$ holds using Theorem 7.1(6), whence $\beta \circ \alpha = \mathrm{id}$.

For F in $[\mathcal{F} \overset{lin}{\to} \overline{\mathbf{R}}_+]_p$, $\alpha(\beta F)(g) = \int_X(g, \beta F)$ holds for all g in $\mathcal{F}$. Hence, we have to show $\int_X(g, \beta F) = F(g)$ for all g in $\mathcal{F}$. Since both integration and the function F are Scott continuous, it suffices to show the equation for all simple functions g. Since both integration and the function F are linear, and every simple function is a finite linear combination of characteristic functions, it even suffices to show the equation for all functions $\hat{U}$ with U in ΩX. By Proposition 7.1(6), $\int_X(\hat{U}, \beta F) = \beta F U = F\hat{U}$ holds as required. ❑

An analogous theorem can be formulated for point continuous valuations.

THEOREM 8.2: For every space X, the topological cone of point continuous valuations on X, i.e., strict modular continuous functions from $\Omega_p X$ to

$\overline{\mathbf{R}}_+$, is isomorphic to the topological cone of linear continuous functions from $[X \to \overline{\mathbf{R}}_+]_p$ to $\overline{\mathbf{R}}_+$ with the pointwise topology.

$$\mathsf{V}_p X = [\Omega_p X \overset{mod}{\to} \overline{\mathbf{R}}_+]_p \cong [\,[X \to \overline{\mathbf{R}}_+]_p \overset{lin}{\to} \overline{\mathbf{R}}_+]_p.$$

Proof: Algebraically, the isomorphisms are the same as in the proof of Theorem 8.1. The difference is that the "pointwise" version of integration is used. Hence, α defined by $\alpha(\nu) = \lambda f.\ \int_X(f, \nu)$ has type $\mathsf{V}_p X \to [[X \to \overline{\mathbf{R}}_+]_p \overset{lin}{\to} \overline{\mathbf{R}}_+]_p$ in this case.

For F in $[[X \to \overline{\mathbf{R}}_+]_p \overset{lin}{\to} \overline{\mathbf{R}}_+]_p$, we have to show that $\beta(F) = F \circ \chi$ is point continuous, i.e., continuous with type $\Omega_p X \to \overline{\mathbf{R}}_+$. We prove continuity of χ: $\Omega_p X \to [X \to \overline{\mathbf{R}}_+]_p$. For x in X and V in $\Omega\overline{\mathbf{R}}_+$, $\chi^-\langle x \to V\rangle$ is $\Omega_p X$ if $0 \in V$, $\mathcal{O}(x)$ if $0 \notin V$ and $1 \in V$, and $\varnothing$ if $1 \notin V$.

The remainder of the proof of Theorem 8.1 can be taken over unchanged. ❑

In general, the topology of a pointwise function space $[Y \to Z]_p$ is the least such that for every y in Y, the function $\lambda f.\ fy$: $[Y \to Z]_p \to Z$ is continuous. Applying this to $[[X \to \overline{\mathbf{R}}_+]_i \overset{lin}{\to} \overline{\mathbf{R}}_+]_p$, we see that its topology is the least such that for every continuous f: $X \to \overline{\mathbf{R}}_+$, the function $\lambda F.\ F(f)$ is continuous. Using the isomorphism α of the proof of Theorem 8.1, we conclude:

THEOREM 8.3: The topology on $\mathsf{V}X$ is the least such that for every continuous function f: $X \to \overline{\mathbf{R}}_+$, the function f^*: $\mathsf{V}X \to \overline{\mathbf{R}}_+$ with $f^*(\nu) = \int_X (f, \nu)$ is continuous.

This characterizes the topology of $\mathsf{V}X$ as a "weak topology." Notice that in contrast to classical results which look similar, the space $\overline{\mathbf{R}}_+$ is equipped with the Scott topology instead of the usual Hausdorff topology.

REFERENCES

1. EDALAT, A. 1993. Dynamical systems, measures and fractals via domain theory. *In:* Proceedings of the First Imperial College, Department of Computing, Workshop on Theory and Formal Methods, Workshops in Computing (G.L. Burn, S.J. Gay, & M.D. Ryan, Eds.). Springer-Verlag, 1993.
2. EDALAT, A. 1993. Power domain algorithms for fractal image decompressing. Technical Report Doc 93/44, Imperial College, 1993.
3. EDALAT, A. 1995. Domain theory and integration. Theoretical Computer Science. **151:** 163–193.
4. FORD, L.R. & D.R. FULKERSON. 1962. Flows in Networks. Princeton Univ. Press. Princeton, New Jersey.
5. GIERZ, G., K.H. HOFMANN, K. KEIMEL, J. D. LAWSON, M. MISLOVE, & D.S. SCOTT. 1980. A Compendium of Continuous Lattices. Springer-Verlag.
6. HECKMANN, R. 1995. Lower bag domains. Fundamenta Informaticae. **24:** (3)259–281.
7. JOHNSTONE, P. 1982. Stone Spaces. Cambridge University Press. New York.
8. JONES, C.J. 1990. Probabilistic Non-Determinism. PhD thesis, Univ. of Edinburgh.

9. JONES, C.J. & G.D. PLOTKIN. 1989. A probabilistic powerdomain of evaluations. *In*: LICS 1989. IEEE Computer Society Press. 186–195.
10. KIRCH, O. 1993. Bereiche und Bewertungen. Master's thesis. Technische Hochschule Darmstadt.
11. LAWSON, J.D. 1982. Valuations on continuous lattices. *In:* Continuous Lattices and Related Topics (R.E. Hoffmann, Ed.). Mathematik Arbeitspapiere, Vol. 27. Universität Bremen.
12. LAWSON, J.D. 1988. The versatile continuous order. *In:* Mathematical Foundations of Programming Language Semantics (MFPLS 1987) (M.G. Main, A. Melton, M. Mislove, & D. Schmidt, Eds.). Lecture Notes in Computer Science, 298. Springer-Verlag. 565–622.
13. TIX, R. 1995. Stetige Bewertungen auf topologischen Räumen. Master's thesis. Technische Hochschule Darmstadt.

An Effective Construction of a Free z-ultrafilter

HORST HERRLICH

University of Bremen
Fb. 3, P.f. 330440
28334 Bremen, Germany

ABSTRACT: It is known that free ultrafilters cannot be constructed effectively (i.e., in ZF set theory without any additional choice principle). In this paper a nonconvergent, thus free, z-ultrafilter is constructed effectively in a suitable zero-dimensional topological space X. Thus the implication

$$C\text{-compact} \Rightarrow B\text{-compact}$$

(which under AC is an equivalence for completely regular spaces) fails to be an equivalence in ZF even for zero-dimensional spaces, where B-compact (respectively, C-compact) means that every ultrafilter (respectively, z-ultrafilter) converges.

0. INTRODUCTION

Free ultrafilters cannot be constructed effectively. In fact, Blass [2] has constructed a model of ZF in which no free ultrafilters exist. Since ultrafilters are precisely the maximal clopen filters, respectively, the z-ultrafilters on discrete topological spaces, Blass's result raises the following questions in ZF:

(A1) Does there exist a free maximal clopen filter on some zero-dimensional topological space?

(A2) Does there exist a free z-ultrafilter on some completely regular topological space?

Since fixed maximal clopen filters in a zero-dimensional space, respectively fixed z-ultrafilters in a completely regular space, are automatically convergent, the above questions may be reformulated into the following more topological-oriented ones in ZF:

(B1) Does there exist a nonconvergent maximal clopen filter on some zero-dimensional topological space?

(B2) Does there exist a nonconvergent z-ultrafilter on some completely regular topological space?

In this paper both questions will be answered positively. This result throws some light on the relations between various concepts of *compactness* in ZF (that are equivalent under AC). In particular, if a space X is called

Mathematics Subject Classification: 03E25, 04A25, 54D30.

Key words and phrases: axiom of choice, z-ultrafilter, maximal clopen filter, compact space.

(a) *Boolean*, provided that X is zero-dimensional and in X every maximal clopen filter converges (see [1]),
(b) *C-compact* provided that X is completely regular and in X every z-ultrafilter converges (see [3], [6], [5], [1]),
(c) *B-compact* provided that in X every ultrafilter converges (see [5]),

then the following implications hold:
(I1) Boolean $\Rightarrow$ C-compact and zero-dimensional,
(I2) C-compact $\Rightarrow$ B-compact and completely regular.

The positive answer to question (B2) above implies that the implication (I2) fails to be an equivalence. The perhaps even more interesting question whether the implication (I1) is an equivalence remains open.

1. PRELIMINARIES

Assume ZF.

DEFINITION 1.1: For any set M, $\wp_f M$ denotes the set of all finite subsets of M.

LEMMA 1.2: If a set M is well-orderable, then so is $\wp_f M$.

Proof: Denote the symmetric difference of A and B by $A \Delta B$. Let the relation $\leq$ well-order M. Then the relation $\sqsubseteq$, defined by

$$A \sqsubseteq B \Leftrightarrow \begin{cases} A = B \text{ or} \\ \text{Card}\, A < \text{Card}\, B \text{ or} \\ A \neq B \text{ and Card}\, A = \text{Card}\, B \text{ and } \text{Min}_{\leq}(A \,\Delta\, B) \in A, \end{cases}$$

well-orders $\wp_f M$. Only transitivity of $\sqsubseteq$ is nonobvious. To show this, assume that there exist A, B, and C in $\wp_f M$ with $A \sqsubset B$, $B \sqsubset C$, and $C \sqsubseteq A$. Then $\text{Card}\, A = \text{Card}\, B = \text{Card}\, C$, $a = \text{Min}_{\leq}(A \,\Delta\, B) \in A$, $b = \text{Min}_{\leq}(B \,\Delta\, C) \in B$ and $c = \text{Min}_{\leq}(A \,\Delta\, C) \in C$.

If $a \in C$, then $a \in C \setminus B$, thus $b < a$.
If $a \notin C$, then $a \in A \setminus C$, thus $c < a$.

Thus ($b < a$ or $c < a$). Similarly one obtains ($a < b$ or $c < b$) and ($a < c$ or $b < c$). However these three assertions are incompatible. ❑

REMARK 1.3: Whereas finite unions of countable sets are again countable, countable unions of finite sets may fail to be so. Thus, in the construction below we need to keep track of the "counting functions." The following terminology helps to do this.

DEFINITION 1.4: (1) A set A is called (*well-*)*countable* provided that there is some (distinguished) surjective map $f: \mathbb{N} \longrightarrow A$.

(2) A set is called *uncountable* provided that it is neither empty nor countable.

LEMMA 1.5: (1) Well-ordered nonempty finite sets are well-countable.

(2) Well-countable unions of well-countable sets are well-countable.

(3) The union of a sequence of well-ordered nonempty finite sets is well-countable.

(4) The set $\bigcup_{n \in \mathbb{N}} \mathbb{N}^n$ is well-countable.

(5) If a set M is well-countable, so is $\wp_f M$.

(6) There exist uncountable well-ordered sets.

Proof: (1), (2), and (4) are obvious.

(3) follows from (1) and (2).

(5) follows from (4).

(6) Hartogs [4] has constructed a well-ordered set $(M, \leq)$ such that $\mathrm{Card}\, M \leq \aleph_0$ fails. Such M cannot be countable, since if $f: \mathbb{N} \longrightarrow M$ would be a surjection, then the map $g: M \longrightarrow \mathbb{N}$, defined by $g(m) = \mathrm{Min}\, f^{-1}(m)$, would be an injection, implying $\mathrm{Card}\, M \leq \aleph_0$. ❑

2. RESULTS

THEOREM 2.1: There exists a zero-dimensional topological space X with a nonconvergent z-ultrafilter.

Proof:

(1) The construction of X:

Let $(M, \leq)$ be a well-ordered uncountable set. (By Lemma 1.5 (6) such a set exists.) Let $(\wp_f M, \sqsubseteq)$ be well-ordered. (This is possible by Lemma 1.2.) Let D be the discrete space with underlying set $\{0, 1\}$. Let $Y = D^M$ be the product with projections:

$$\pi_m^Y: Y \longrightarrow D \quad \text{for } m \in M \text{ and}$$

$$\pi_K^Y: Y \longrightarrow D^K \quad \text{for } K \subset M.$$

Let p (respectively, q) be the point of Y with $\pi_m^Y(p) = 1$ (respectively, $\pi_m^Y(q) = 0$) for each $m \in M$.

Let X be the subspace obtained from Y by removal of the point p.

Let π_m respectively, π_K denote the restriction of π_m^Y respectively, π_K^Y to X. For $x \in X$ and $A \subset M$ denote by x_A the point in X with $\pi_A(x_A) = \pi_A(x)$ and $\pi_{M \setminus A}(x_A) = \pi_{M \setminus A}(q)$, and by $U(x, A)$ the set of all $y \in X$ with $\pi_A(y) = \pi_A(x)$.

Let $\mathfrak{z}$ be the collection of all zero-sets Z of X with $p \in cl_Y Z$.

(2) For every continuous map $f: X \longrightarrow \mathbb{R}$ there exists a well-countable

subset K of M such that the implication

$$\pi_K(x) = \pi_K(y) \Rightarrow f(x) = f(y)$$

holds.

Proof: For each point $x \in X$ and each $n \in \mathbb{N}^+$ the set $V(x, n) = \{y \in X \,|\, |f(y) - f(x)| < 1/n\}$ is open in X. Thus the set $\mathcal{K}(x, n) = \{A \in \wp_f M \,|\, A \neq \emptyset$ and $U(x, A) \subset V(x, n)\}$ is not empty. Let $K(x, n)$ be the smallest member of $\mathcal{K}(x, n)$ in the well-ordered set $(\wp_f M, \sqsubseteq)$. ❑

By Lemma 1.5(3), the set $K(x) = \bigcup\{K(x, n) \,|\, n \in \mathbb{N}^+\}$ is well-countable. By construction, any point y of X with $y_{K(x)} = x_{K(x)}$ satisfies $f(y) = f(x)$.

Next an increasing sequence (K_n) of well-countable subsets of M will be constructed by induction:

(a) $K_0 = K(q)$.

(b) Let K_n be constructed. Then the set $K_{n+1} = K_n \cup \bigcup\{K(p_B) \,|\, B \in \wp_f K_n\}$ is well-countable by Lemma 1.5 (2) and (5).

Thus, by Lemma 1.5 (2) again, the set $K = \bigcup\{K_n | n \in \mathbb{N}\}$ is a well-countable subset of M.

It remains to be shown that $\pi_K(x) = \pi_K(y)$ implies $f(x) = f(y)$. Assume on the contrary that x and y are points in X with $\pi_K(x) = \pi_K(y)$ and $f(x) \neq f(y)$. Then $f(x)$ and $f(y)$ have disjoint open neighborhoods U and V in $\mathbb{R}$. Thus there exists a finite subset A of M with $U(x, A) \subset f^{-1}[U]$ and $U(y, A) \subset f^{-1}[V]$. Thus, $f(x_A) \in U$ and $f(y_A) \in V$, hence $f(x_A) \neq f(y_A)$. Consider the point $z = x_{A \cap K}$ of X. By construction, any point v of X with $v_K = z_K$ satisfies $f(v) = f(z)$. Since in particular x_A and y_A are such v's, we conclude $f(x_A) = f(z) = f(y_A)$. This contradicts $f(x_A) \neq f(y_A)$.

(3) X is C-embedded in Y.

Proof: Let $f: X \longrightarrow \mathbb{R}$ be a continuous map. By (2) there exists a well-countable subset K of M such that the implication $\pi_K(x) = \pi_K(y) \Rightarrow f(x) = f(y)$ holds. Since M is uncountable, K is a proper subset of M. Thus the point p_K belongs to X. Consequently, the map $g: Y \longrightarrow \mathbb{R}$, defined by

$$g(y) = \begin{cases} f(y), & \text{if } y \in X \\ f(p_K), & \text{if } y = p. \end{cases}$$

is a continuous extension of f. ❑

(4) If Z_1 and Z_2 belong to $\mathfrak{z}$, then $Z_1 \cap Z_2 \neq \emptyset$.

Proof: Assume that $Z_1 \cap Z_2 = \emptyset$. Then there exists a continuous map $f: X \longrightarrow \mathbb{R}$ with $Z_1 = f^{-1}(0)$ and $Z_2 = f^{-1}(1)$. By (3), f has a continuous extension $g: Y \longrightarrow \mathbb{R}$. Thus, $p \in (cl_Y Z_1 \cap cl_Y Z_2)$ implies $g(p) = 0$ and $g(p) = 1$, which is not possible. ❑

(5) If Z_1 and Z_2 belong to $\mathfrak{z}$, then so does $Z_1 \cap Z_2$.

Proof: If $Z_1 \cap Z_2$ would not belong to $Z_1 \cap Z_2$, then $p \notin cl_Y (Z_1 \cap Z_2)$. Thus there would exist a zero-set neighborhood U of p in Y with $U \cap (Z_1 \cap Z_2) = \varnothing$. Thus $U \cap Z_1$ and $U \cap Z_2$ would be disjoint elements of $\mathfrak{z}$ — contradicting (4). ❑

(6) For any zero-set Z of X that does not belong to $\mathfrak{z}$, there exists a member $\overline{Z}$ of $\mathfrak{z}$ with $Z \cap \overline{Z} = \varnothing$.

Proof: If $Z \notin \mathfrak{z}$, then $p \notin cl_Y Z$. Thus, there exists a continuous map f: $Y \longrightarrow \mathbb{R}$ with $f(p) = 0$ and $f[Z] \subset \{1\}$. Consequently, p has a zero-set neighborhood U that fails to meet Z. Hence, $\overline{Z} = U \cap X$ has the desired properties. ❑

(7) $\mathfrak{z}$ generates a nonconvergent z-ultrafilter on X.

Proof: Immediate from (4)–(6). ❑

COROLLARY 2.2: Zero-dimensional B-compact spaces may fail to be C-compact.

Proof: The space X, constructed in Theorem 2.1, is zero-dimensional and fails to be C-compact in any model of ZF. However, in any model without free ultrafilters, X is B-compact. ❑

THEOREM 2.3: There exists a zero-dimensional topological space with a nonconvergent maximal clopen filter.

Proof: If $\mathfrak{U}$ is a nonconvergent z-filter in a zero-dimensional space X, then the filter $\mathfrak{W}$, generated by all clopen members of $\mathfrak{U}$, is a nonconvergent maximal clopen filter in X. Thus the result follows immediately from Theorem 2.1. In fact, it can be seen easily that in this case $\mathfrak{W}$ is the trace on X of the neighborhood-filter of p in Y. ❑

COROLLARY 2.4: Zero-dimensional B-compact spaces may fail to be Boolean.

REFERENCES

1. BENTLEY, H.L. & H. HERRLICH. Compactness and rings of continuous functions — without the axiom of choice. To appear.
2. BLASS, A. 1977. A model without ultrafilters. Bull. Acad. Sci. Polon., Sér. Sci. Math. Astr. Astr. Phys. **25:** 325–331.
3. COMFORT, W.W. 1968. A theorem of Stone-Čech type, and a theorem of Tychonoff type, without the axiom of choice; and their realcompact analogues. Fund. Math. **63:** 97–110.
4. HARTOGS, F. 1915. Über das Problem der Wohlordnung. Math. Annalen. **76:** 138–143.

5. HERRLICH, H. 1996. Compactness and the axiom of choice. Appl. Categ. Structures. **4:** 1–14.
6. SALBANY, S. 1974. On compact* spaces and compactifications. Proc. Amer. Math. Soc. **45:** 274–280.

Topological Symmetry and the Theory of Metrization

H.H. HUNG

Department of Mathematics
Concordia University
Montréal
Québec, Canada H4B 1R6

Respectfully dedicated to Professor Jun-iti Nagata on his 70th birthday.

ABSTRACT: We reaffirm the importance of the role of *topological symmetry* in the Theory of Metrization and obtain, as a consequence, a number of elegant and interesting metamorphoses of the Double Sequence Theorem of Nagata, all of which improve significantly upon some classical theorems in ways very obvious.

In 1977, 20 years after Nagata's Double Sequence Theorem [8], I gave a unification of the classical metrization theorems, in terms of disjoint subsets [4, Theorem 2.1], [11, Corollary 1.13], a more conventional expression of it being Corollary 2.3 of the same paper. It was noticed then that if a property, to be called *topological symmetry* here, is isolated and if it is taken to be part of the condition, Corollary 2.3 splits into two possibilities, each implying the other and either suffices for metrizability, resulting in Corollary 2.4, a result since rediscovered by Junnila [7, Theorem 4.19] and, in part, by Nagata [10, Theorem 1].[1]

Almost twenty years later, it occurred to me to split Nagata's Double Sequence condition instead, in the same manner (3.1). Nagata's Double Sequence Condition being a stronger condition than Corollary 2.3 of [4], some very good results follow. Although one of the two branches (3.2) is quite sterile, the other is fructuous, and we could strengthen it so that *topological symmetry* is automatically taken care of and disappears, leaving behind a very simple metrizability condition (3.3). Condition (3.3) can in turn be weakened by shedding the neighborhood requirements (3.4) and (3.5). Al-

Mathematics Subject Classification: 54E35.
Key words and phrases: topological symmetry, metrization.

[1]Part of it also appeared as Theorem 3 in Collins and Roscoe, Coll. Math. Soc. János Bolyai, 41, Topology and Applications, Eger (Hungary) 1983, 177–181, amid some equivalent formulations. The condition of Theorem 6 in the same paper clearly includes, of the two conditions, (i) and (ii)(b), in Corollary 2.4 of [4], (i) for an increasing subsequence in ℕ for each x and (ii)(b) in its entirety.

ternative formulations are possible ((3.6) and (3.7)) if we let the *duals* take over the requirement of being *basic*, as in the case of semistratifiable spaces. (3.8) is yet another metamorphosis of the Double Sequence Theorem into the realm of *annihilators* [11, 1.5]. All our results here strengthen some classical results in obvious ways.

The drama of the metamorphoses of the Double Sequence Theorem are recounted in the Remarks on and the Proofs of successive theorems in Section 3. Proofs, when deemed necessary, confine themselves to pointing out where to look for, at each stage, the apparently vanished topological symmetry and neighborhood requirements and are, therefore, very brief and (apparently) very simple.

1. NOTATIONS AND TERMINOLOGY

(1.1) In the following, X denotes the T_0-space under discussion; x, y, ξ, and η are points on X; n, ν are nonnegative integers; $U_n(x)$, $V_n(x)$, $W_n(x)$ are *neighborhoods* of x. On the other hand, $Y_n(x)$ and $Y_n'(x)$ are *subsets* containing x and do *not* have to be neighborhoods of x.

(1.2) Given a family $\{U_n(x): x \in X\}$. If $U_n^*(x) \equiv \{y \in X: x \in U_n(y)\}$ is also a neighborhood of x, whatever x, the given family is said to be *topologically symmetric* (cf. Junnila's *cushioned neighbornet* [7] and Nagata's (a) in [12]). It is of course necessary and sufficient, for $\{U_n(x): x \in X\}$ to be *topologically symmetric*, that the family $\{\{x\}: x \in X\}$ be *cushioned* in it, although, we can also speak of $\{\{x\}: x \in X\}$ and indeed of $\{Y_n'(x): x \in X\}$ being *cushioned* in $\{Y_n(x): x \in X\}$ and write $\{x\} \ll Y_n(x)$ and $Y_n'(x) \ll Y_n(x)$, respectively.

(1.3) The countable collection of families of neighborhoods $\{\{U_n(x): x \in X\}: n \in \omega\}$ is said to be a *millefeuille* if $\forall n$, $U_{n+1}(x) \ll U_n(x)$ (see (1.2) above for notation).

(1.4) The sequence $\langle Y_n(x)\rangle$ at x is said to be *basic* if for every neighborhood U of x, there contained in U could be found an $Y_\nu(x)$.

2. THE METRIZABILITY CONDITION OF THE DOUBLE SEQUENCE THEOREM OF NAGATA

For every n, there are $\{U_n(x): x \in X\}$, $\{V_n(x): x \in X\}$, and $\{W_n(x): x \in X\}$ such that

I. $\forall x$, $\langle U_n(x)\rangle$ is *basic*,

IIA. $\forall n, x$, $\xi \notin U_n(x) \Rightarrow W_n(x) \cap V_n(\xi) = \varnothing$, and

IIB. $\forall n, x, \quad \eta \in W_n(x) \Rightarrow V_n(\eta) \subset U_n(x)$.

REMARK 1: We have here the original version of Nagata's Theorem [8, Theorem 1] which involves *three* sequences rather than two. It is probably because we can always let the intersection $V_n(x) \cap W_n(x)$ play the role of both $V_n(x)$ and $W_n(x)$ that the more common version [12, Theorem 1] or [2, Section 3]) is preferred.

REMARK 2: Hodel, coming upon the symmetry between IIA and IIB, is particularly fond of the factorization of metrizability into Nagata spaces (I + IIA) and the Nagata first countable (I + IIB) [2, Section 3].

REMARK 3: The symmetry spoken of in (2) above allows either IIA or IIB to be dropped if the $V_n(x)$'s are *symmetric*, i.e., if $x \in V_n(y) \Leftrightarrow y \in V_n(x)$. Symmetric neighborhoods have been explicitly used in Theorem 7 of [8] and in [3]. In particular, if the $V_n(x)$'s are *stellar*, i.e., if, for each n, there is an open cover $\mathcal{U}_n$ of X such that $V_n(x) = \mathrm{St}(x, \mathcal{U}_n)$ for every x, Theorem of Nagata above becomes that of Moore [1, Section 1.4]. It is in this manner that the former improves upon the latter.

3. METAMORPHOSES OF THE DOUBLE SEQUENCE METRIZABILITY CONDITION

(3.1) For every n, there are $\{U_n(x): x \in X\}$, $\{V_n(x): x \in X\}$, and $\{W_n(x): x \in X\}$ such that

I. $\forall x, \quad \langle U_n(x) \rangle$ is *basic*,

II.A. $\forall n, x, \quad \xi \notin U_n(x) \Rightarrow W_n(x) \cap V_n(\xi) = \varnothing$, or

B. $\forall n, x, \quad \eta \in W_n(x) \Rightarrow V_n(\eta) \subset U_n(x)$,

III. $\forall n$, $\{V_n(x): x \in X\}$ is *topologically symmetric.*

REMARK 1: Note that, by III, $[x]_n \equiv V_n(x) \cap V_n{}^*(x)$ is a symmetric *neighborhood* and can be used in place of $V_n(x)$ in IIA and IIB. Note further that IIA and IIB thus modified imply each other, and either *alone* would suffice for metrizability according to Remark 3 on Section 2.

REMARK 2: (3.1) should be compared with Corollary 2.4 of [4] mentioned in the Introduction.

REMARK 3: With III, the $V_n{}^*(x)$'s are neighborhoods and we might as well assume that $W_n(x) \subset V_n{}^*(x)$ and let III read:

III′. $\forall n, \quad \eta \in W_n(x) \Rightarrow x \in V_n(\eta)$.

Because IIB and III' share a common hypothesis (that $\eta \in W_n(x)$) they give the impression of being one condition when they collapse into one line ((3.2) below). That, of course, is an illusion.

REMARK 4: More interesting is the other alternative, viz., IIA + III, as (3.3) below testifies. Note the fact that IIA implies that $V_n(\xi) \subset U_n{}^*(\xi)$ and therefore the *topological symmetry* of $\{U_n(x): x \in X\}$.

(3.2) For every n, there are $\{U_n(x): x \in X\}$, $\{V_n(x): x \in X\}$, and $\{W_n(x): x \in X\}$ such that

I. $\forall x, \quad \langle U_n(x)\rangle$ is *basic*,

II. $\forall n, x, \quad \eta \in W_n(x) \Rightarrow x \in V_n(\eta) \subset U_n(x)$.

REMARK 1: II here is of course IIB + III in (3.1).

REMARK 2: Note that $\langle V_n(\eta)\rangle$ does not have to be *decreasing* (cf. the second of the two theorems referred to in footnote 1 to the introductory paragraphs).

(3.3) At every x, there is a *basic* sequence $\langle U_n(x)\rangle$ such that

$$\forall n, x, \quad \xi \notin U_n(x) \Rightarrow W_n(x) \cap U_{n+1}(\xi) = \varnothing \text{ for some } W_n(x), \qquad (\dagger)$$

that is,

$$\forall n, x, \quad x \notin Cl \bigcup\{U_{n+1}(\xi): \xi \notin U_n(x)\}. \qquad (*)$$

Proof: Noting that IIA of (3.1) implies the *topological symmetry* of $\{U_n(x): x \in X\}$ (Remark 4 on (3.1) above) and that in (†) above $U_{n+1}(x)$ plays the role that $V_n(x)$ plays in IIA of (3.1), we quickly see that III of (3.1) is taken care of in this new circumstance. ❑

REMARK 1: Mindful of the symmetry between IIA and IIB in Section 2, the very sight of (†) reminds us of the *missing*

$$\forall n, x, \quad \eta \in W_n(x) \Rightarrow U_{n+1}(\eta) \subset U_n(x). \qquad (@)$$

We are of course aware that a slightly stronger version

$$\forall n, x, \quad \eta \in U_{n+1}(x) \Rightarrow U_{n+1}(\eta) \subset U_n(x) \qquad (@@)$$

(together with I) characterizes quasimetrizability [1, 10.2] and I + @ is enough to ensure a γ-space. The question is then: What is I + @? (A topological characterization of quasimetrizability has been found [6].)

REMARK 2: The statement in 23H of Willard [13] the validity of which is to be shown by counterexamples is shown here to be *wrong*. The metrizability condition in 23.5 of [13] is therefore unnecessarily strong. (See [5].)

(3.4) At every x, there is a *basic* sequence $\langle Y_n(x)\rangle$ (see (1.1)) such that
(i) $\forall n, x, \quad \xi \notin Y_n(x) \Rightarrow Y_\nu(x) \cap Y_{n+1}(\xi) = \varnothing$, for some ν dependent on n and x, and
(ii) $\forall n, x, \quad \{x\} \ll Y_n(x)$.

Proof: By (ii), the $Y_n{}^*(x)$'s are neighborhoods of x (although the $Y_n(x)$'s may not be). Noting that $Y_{n+1}{}^*(x) \subset Y_n(x)$, we see that the $Y_n(x)$'s are neighborhoods of x after all. ❑

(3.5) At every x, there is a *basic* sequence $\langle Y_n(x)\rangle$ (see (1.1)) such that

$$\forall n, x, \quad Y_{n+1}(x) \ll Y_n{}^*(x). \qquad (\sharp)$$

Proof: Noting that there is a neighborhood $W_n(x) \subset Y_n(x) \quad \forall n, x$, in such a manner that $W_n(x) \cap Y_{n+1}(\xi) = \varnothing$ if $\xi \notin Y_n(x)$, we see that $Y_n(x)$ is a neighborhood of x, $\forall x$ and $\forall n$. Metrizability follows from (3.3). ❑

REMARK 1: (3.5) above is probably formally the weakest. Condition (3.3) can be seen to satisfy it. Alexandroff-Urysohn, when held beside it, is seen to have redundancies everywhere: The stars need not be stars, or even neighborhoods — all that is necessary is the cushioning and *some* symmetry. In Nagami [4, Corollary 2.5], we have $\mathrm{St}(x, \mathcal{F}_n) \ll \mathrm{St}(x, \mathcal{F}_n)$. By comparison, (3.5) is mild.

(3.6) For every n, there are $\{U_n(x): x \in X\}$ and $\{V_n(x): x \in X\}$ such that
(i) $\forall x, \quad \langle U_n{}^*(x)\rangle$ is *basic*, and
(ii) $\forall n, x, \quad \{x\} \ll V_n(x) \ll U_n(x)$.

Proof: It follows from (ii) that $\{V_n(x): x \in X\}$ is *topologically symmetric* (1.2) and there is a neighborhood $W_n(x) \subset U_n{}^*(x)$ $\forall n, x$, in such a manner that $W_n(x) \cap V_n(\xi) = \varnothing$ if $\xi \notin U_n{}^*(x)$. Upon identifying this $U_n{}^*(x)$ with the $U_n(x)$ of (3.1), one sees immediately that we have a metrizability condition in the above. ❑

REMARK 1: Theorems 7 and 11 of [8], corollaries of the Double Sequence Theorem of Nagata in his original paper and Morita (Corollary iii to Theorem VI.2 of [9]) are obvious special cases of (3.6). In the light of (3.6), the *symmetry* on the $U_n(x)$'s that is required in Theorem 7 of [8] is *largely* superfluous. Theorem 11 of [8] is of course an extension of the classical Alexandroff-Urysohn.

REMARK 2: Borges' Theorem [11, 1.3] is a special case of (3.6) above with $V_n(x)$ closed and equal to $U_n(x)$ and $\{V_n(x): x \in X\}$ closure-preserving.

(3.7) At every x, there is a sequence $\langle U_n(x)\rangle$ such that
(i) $\{\{U_n(x): x \in X\}: n \in \omega\}$ is a *millefeuille* (1.3) and
(ii) $\forall x, \quad \langle U_n{}^*(x)\rangle$ is *basic*.

REMARK 1: In some sense, (3.7) complements (3.3) above and is particularly interesting if one has in mind (*ss*) of [12].

REMARK 2: One wonders what kind of a space we have if a *basic* $\langle U_n{}^*(x)\rangle$ is replaced by a *basic* $\langle U_n(x)\rangle$.

REMARK 3: Alexandroff-Urysohn [13, 23.4] is of course a special case of (3.7). Our theorem is a very *unusual* generalization of Alexandroff-Urysohn.

(3.8) X has a *monotone, lower semi-continuous annihilator* φ (1.5 of [11] ; here we do *not* require that $\varphi(x, F) = 0 \Rightarrow x \in F$ or that F be closed) on the collection of all singletons so that

(i) for every $x \in X$ and a neighborhood U, there is such an n that

$$[x, n] \equiv \{y \in X\colon \varphi(y, \{x\}) \geq \frac{1}{2^n}\}$$

contains $\backslash U$, and

(ii) φ can be extended to the collection $\{[x, n]\colon x \in X,\ n \in \omega\}$ in such a manner that

$$\bar{\varphi}(x, [x, n]) > \frac{1}{2} \cdot \frac{1}{2^n}.$$

Proof: $\forall n,x$, let $Y_n(x) = \backslash[x, n]$. There is such a neighborhood W of x that

$$\bar{\varphi}(y, [x, n]) > \frac{1}{2} \cdot \frac{1}{2^n}, \quad \forall y \in W$$

(by lower semicontinuity of $\bar{\varphi}$). We have, therefore (by monotonicity of $\bar{\varphi}$),

$$\varphi(y, \{\xi\}) > \frac{1}{2} \cdot \frac{1}{2^n}, \quad \forall y \in W,\ \xi \in [x, n],$$

i.e., $W \cap Y_{n+1}(\xi) = \varnothing$, $\forall \xi \notin Y_n(x)$, i.e., $Y_{n+1}(\xi) \ll Y_n{}^*(\xi)$ and metrizability of X by (3.5). ❑

REFERENCES

1. GRUENHAGE, G. 1984. Generalized Metric Spaces. *In:* Handbook of Set-Theoretic Topology (Kunen & Vaughan, Eds.). Elsevier Science Publishers B.V. Amsterdam. Chap. 10.
2. HODEL, R.E. 1974. Some Results in Metrization Theory, 1950–1972. Proc. Top. Conf. VPI and State Univ. 1973. Springer Lect. Notes **375:** 120–136.
3. HUNG, H.H. 1976. One More Metrization Theorem. Proc. Amer. Math. Soc. **57:** 351–353.

4. HUNG, H.H. 1977. A Contribution to the Theory of Metrization. Canad. J. Math. **29:** 1145–1151.
5. HUNG, H.H. 1995. A Companion to the Double Sequence Theorem of Nagata. Q & A, Gen. Top. **13:** 223.
6. HUNG, H.H. 1996. Quasi-metrizability. Abstracts Amer. Math. Soc. **17:** To appear.
7. JUNNILA, H.J.K. 1978. Neighbornets. Pacif. J. Math. **76:** 83–108.
8. NAGATA, J. 1957. A Contribution to the Theory of Metrization. J. Inst. Poly., Osaka City U. **8:** 185–192.
9. NAGATA, J. 1985. Modern General Topology. 2nd rev'd. Edit. North-Holland. Amsterdam.
10. NAGATA, J. 1986/87. Characterizations of Metrizable and Lašnev Spaces in Terms of *g*-functions. Q & A, Gen. Top.**4:** 129–139.
11. NAGATA, J. 1989. Metrization I. *In:* Topics in General Topology (K. Morita & J. Nagata, Eds.). North-Holland, Amsterdam. Chapter 7.
12. NAGATA, J. 1992. A Survey of Metrization Theory II. Q & A, Gen. Top. **10:** 15–30.
13. WILLARD, S. 1970. General Topology. Addison-Wesley. Reading, Massachusetts.

On the Duality of Compact vs. Open

ACHIM JUNG[a,c] AND PHILIPP SÜNDERHAUF[b,d,e]

[a]*School of Computer Science*
University of Birmingham
Edgbaston
Birmingham, B15 2TT, UK
and
[b]*Department of Computing*
Imperial College
180 Queens Gate
London SW7 2BZ, UK

ABSTRACT: It is a pleasant fact that Stone-duality may be described very smoothly when restricted to the category of compact spectral spaces: The Stone-duals of these spaces, arithmetic algebraic lattices, may be replaced by their sublattices of compact elements thus discarding infinitary operations.

We present a similar approach to describe the Stone-duals of coherent spaces, thus dropping the requirement of having a base of compact-opens (or, alternatively, replacing algebraicity of the lattices by continuity). The construction via *strong proximity lattices* is resembling the classical case, just replacing the order by an *order of approximation*. Our development enlightens the fact that "open" and "compact" are dual concepts which merely happen to coincide in the classical case.

INTRODUCTION

Pointless topology relies on the duality between the category of sober spaces with continuous functions and the category of spatial frames with frame-homomorphisms. The origin of this subject, however, dates back to the classical Representation Theorem by Marshall Stone [15], which establishes a duality between compact zero-dimensional Hausdorff spaces and Boolean algebras. So the original theory deals just with *finitary* operations on the algebraic side, whereas one has to pay a price for the broader generality of frames: *infinitary* operations are involved.

The key property of Stone spaces which makes it possible to restrict to finitary operations is the fact that their lattices of open sets are *algebraic*. More precisely, they are exactly the lattices arising as ideal completions of Boolean algebras. Having realized this, one can immediately generalize

Mathematics Subject Classification: Primary 54H99; Secondary 06E15, 06B35.
Key words and phrases: coherent space, (strong) proximity lattice, Stone duality.
[c]E-mail: A.Jung@cs.bham.ac.uk.
[d]E-mail: P.Sunderhauf@doc.ic.ac.uk.
[e]Partly supported by the Deutsche Forschungsgemeinschaft.

Stone's Theorem to distributive lattices, yielding spectral spaces as their duals (see, e.g., [6, II-3]). But this is the end of the road: Ideal completions of posets are always algebraic domains, and hence the spaces which are described in this fashion will always have a base of compact-open sets.

The way out of this cul-de-sac is via the theory of *R-structures* or *abstract bases* as introduced in [12] and developed in [1]. These sets are "ordered" by an *order of approximation* which need not be reflexive. Ideal completion of these structures leads to continuous domains, and, in our setting, to continuous lattices.

Note the shift of emphasis as we pass to the continuous setting: While Stone was concerned with a representation of Boolean algebras in topological spaces, we are more interested in the spaces and seek to *describe* them via a finitary algebraic structure. It is our goal, then, to make this description as faithful as possible.

Similar attempts already exist in the literature; we mention [4] and [14] where one can find pointers to yet earlier work. Our approach differs from theirs in that we require stronger axioms; so strong in fact, that the lattice of open sets no longer qualifies. It is a particularly striking result of our work that there is nevertheless a representation theorem for all coherent spaces. The representation takes both open and compact subsets into account. We feel that this sheds fresh light on the foundations of pointless topology.

Quite a number of equivalences between certain topological spaces and certain complete lattices have been established and it is therefore no surprise that there is no commonly accepted naming convention for the concepts involved. In what follows, we will essentially use the terminology of [1]. It is also our main reference for background information.

1. COHERENT SPACES

We begin with the following corollary to the Hofmann-Mislove Theorem (see [5], [8]):

THEOREM 1: Let X be a sober space.

(1) If K is compact and if $(O_i)_{i \in I}$ is a directed family of open sets such that $K \subseteq \bigcup_{i \in I} O_i$ then K is contained in some O_i already.

(2) If O is open and if $(K_i)_{i \in I}$ is a filtered family of compact saturated sets such that $\bigcap_{i \in I} K_i \subseteq O$ then some K_i is contained in O already.

Except for the necessity to restrict to *saturated* compact subsets, this theorem highlights a fundamental duality between open and compact sets. (Recall that a subset of a topological space is called *saturated*, if it coincides with the intersection of all its neighborhoods.) A more familiar formulation would employ arbitrary families rather than directed ones. But this does not work as the (even finite) intersection of compact saturated sets is not necessarily compact again. If we require this property then we indeed have a complete lattice of compact saturated sets and the following formulation of Theorem 1 is true:

THEOREM 2: Let X be a sober space for which finite intersections of compact saturated subsets are compact.

(1) Every open cover of a compact set contains a finite subcover.
(2) Whenever the intersection of compact saturated sets is contained in an open set then the same is true for an intersection of finitely many of them.

More connections between opens and compacts appear if we also require local compactness. We arrive at what we choose to call *coherent spaces*, our principal objects of interest.

DEFINITION 3: A topological space X is called *coherent* if it is sober, locally compact and if intersections of finite families of compact saturated sets are compact.

We denote the set of open sets of X by $\mathcal{T}_X$ and order it by inclusion. The set of compact saturated sets is denoted by $\mathcal{K}_X$ and is ordered by reversed inclusion.

THEOREM 4: Let X be a coherent space.

(1) $\mathcal{T}_X$ and $\mathcal{K}_X$ are arithmetic lattices.
(2) In $\mathcal{T}_X$ we have $O \ll O'$ if and only if there is $K \in \mathcal{K}_X$ with $O \subseteq K \subseteq O'$.
(3) In $\mathcal{K}_X$ we have $K \ll K'$ if and only if there is $O \in \mathcal{T}_X$ with $K' \subseteq O \subseteq K$.

The terminology *arithmetic lattice* is adopted from [1] (it was used in [3] for the algebraic version). It comprises completeness, distributivity, continuity, and the following multiplicativity property:

$$x \ll y, z \Rightarrow x \ll y \wedge z.$$

We will also assume $1 \ll 1$ for convenience. We cite the following converse to Theorem 4 from [3]:

THEOREM 5: Arithmetic lattices are spatial and their spaces of points are coherent.

Our theory of *strong proximity lattices* will connect up with coherent spaces via arithmetic lattices. In particular, we will not need to use the Axiom of Choice for the correspondence. It is present, of course, through the duality of coherent spaces and arithmetic lattices.

The basic topological concepts on the two sides of Stone duality take the following forms:

(spatial) frame	(sober) space
completely prime filter	element
Scott-open filter	compact saturated set
element	open set

We may point out that on the side of frames the duality between opens and compacts is rather opaque. Proximity lattices, as we will see below, behave far better in this respect.

2. PROXIMITY LATTICES

DEFINITION 6: A *proximity lattice* is given by a distributive bounded lattice $(B; \vee, \wedge, 0, 1)$ together with a binary relation $\prec$ on B satisfying $\prec^2 = \prec$. We call $(B; \vee, \wedge, 0, 1)$ the *algebraic structure* of the proximity lattice and $(B; \prec)$ the *approximation structure*. The two structures are connected through the following two axioms:

$(\vee\text{-}\prec)$ $\forall a \in B \, \forall M \subseteq_{\text{fin}} B.\ M \prec a \Leftrightarrow \bigvee M \prec a$;

$(\prec\text{-}\wedge)$ $\forall a \in B \, \forall M \subseteq_{\text{fin}} B.\ a \prec M \Leftrightarrow a \prec \bigwedge M$.

Here we write $a \prec M$ for $\forall m \in M.\ a \prec m$ and $M \prec a$ for $\forall m \in M.\ m \prec a$. If $x \prec y$, we also say that *x approximates y*.

There are various sets of axioms in the literature featuring under the name of proximity lattices. The particular choice we made is very close to the ones found in [4], [13], [14]. Unlike these previous accounts, we do not require the order of approximation $\prec$ to be contained in the order $\leq$ derived from the lattice structure. In fact, we rather think of $\prec$ *replacing* $\leq$. We do not employ the lattice order in our arguments. This is reflected by the fact that the notations $\uparrow$ and $\downarrow$ refer to $\prec$ rather than to $\leq$: For subsets A of B, we define $\uparrow A = \{x \in X \mid \exists a \in A.\ a \prec x\}$ and $\downarrow A = \{x \in X \mid \exists a \in A.\ x \prec a\}$. Moreover, $\uparrow x$ and $\downarrow x$ stand for $\uparrow\{x\}$ and $\downarrow\{x\}$, respectively.

Note that our definition is self-dual, i.e., $B^{op} = (B; \wedge, \vee, 1, 0; \succ)$ is a proximity lattice if B is. Let us start with stating some simple consequences of the axioms enlightening the interplay of algebraic and approximation structure.

LEMMA 7: If $(B; \vee, \wedge, 0, 1; \prec)$ is a proximity lattice and $a, b, a', b' \in B$, then:

(1) $0 \prec a \prec 1$,

(2) $a \prec b \Rightarrow a \prec b \vee b'$,

(3) $a \prec b \Rightarrow a \wedge a' \prec b$,

(4) $a \prec a'$ and $b \prec b' \Rightarrow a \vee b \prec a' \vee b'$,
(5) $a \prec a'$ and $b \prec b' \Rightarrow a \wedge b \prec a' \wedge b'$.

Proof: Put $M = \emptyset$ in ($\vee$-$\prec$) and ($\prec$-$\wedge$) to get (1). The second assertion holds by ($\prec$-$\wedge$) and the fact that $b = b \wedge (b \vee b')$. If we assume $a \prec a'$ and $b \prec b'$, then (2) gives us $\{a, b\} \prec a' \vee b'$. Now ($\vee$-$\prec$) yields (4). Finally, (3) and (5) follow by symmetry. ❑

3. IDEALS AND OPEN SETS

DEFINITION 8: Suppose $(B; \vee, \wedge, 0, 1; \prec)$ is a proximity lattice. We define the set of all *ideals* on B by

$$\mathrm{Idl}(B) = \{I \subseteq B \mid I = {\downarrow}I, M \subseteq_{\text{fin}} I \Rightarrow \bigvee M \in I\}.$$

Observe that the condition $I = {\downarrow}I$ implies that for each $x \in I$ there is $y \in I$ with $x \prec y$. Since $\prec$ is not necessarily reflexive this is a nontrivial condition. Some authors emphasize this by using the term "round ideal."

Lemma 9: Let B be a proximity lattice and $I \in \mathrm{Idl}(B)$. Then for all $a, b \in B$:
(1) $a \vee b \in I \Leftrightarrow (a \in I$ and $b \in I)$,
(2) $a \in I \Rightarrow a \wedge b \in I$.

Proof: To prove (1), assume $a \vee b \in I$. Then there is $x \in I$ with $a \vee b \prec x$ since $I \subseteq {\downarrow}I$. Now ($\vee$-$\prec$) implies $a \prec x$ and $b \prec x$. Thus $\{a, b\} \subseteq I$ since $x \in I$ and ${\downarrow}I \subseteq I$. The reverse implication holds by definition. By $a = a \vee (a \wedge b)$, (2) is a trivial consequence of (1). ❑

Let us now have a closer look at the set of ideals.

LEMMA 10: Suppose $(B; \vee, \wedge, 0, 1; \prec)$ is a proximity lattice. Then ${\downarrow}x$ is an ideal for each $x \in B$. Moreover $I = \bigcup_{x \in I} {\downarrow}x$ for every $I \in \mathrm{Idl}(B)$, and this union is directed.

Proof: The first claim is immediate from the axioms. Clearly, ${\downarrow}x \subseteq {\downarrow}I \subseteq I$ for each $x \in I$, hence $\bigcup_{x \in I} {\downarrow}x \subseteq I$. In order to see that this union is directed suppose $x, y \in I$. Then $x \vee y \in I$. Since $I \subseteq {\downarrow}I$, this implies the existence of $z \in I$ with $x \vee y \prec z$, hence $x \prec z$ and $y \prec z$. Therefore, we have ${\downarrow}x \cup {\downarrow}y \subseteq {\downarrow}z$, i.e., directedness. Finally, choosing $x = y$ in this proof gives for each $x \in I$ a $z \in I$ with $x \in {\downarrow}z$, thus $I \subseteq \bigcup_{x \in I} {\downarrow}x$. ❑

THEOREM 11: Suppose that $(B; \vee, \wedge, 0, 1; \prec)$ is a proximity lattice. Then $(\mathrm{Idl}(B), \subseteq)$ is an arithmetic lattice. Finite infima are intersections, general infima are given by

$$\bigwedge_{j \in J} I_j = \downarrow \bigcap_{j \in J} I_j .$$

Directed suprema are unions, general suprema are calculated by

$$\bigvee_{j \in J} I_j = \downarrow \{ \bigvee M \mid M \subseteq_{\text{fin}} \bigcup_{j \in J} I_j \} .$$

The order of approximation is given by $I \ll J \Leftrightarrow \exists x \in B.\, I \subseteq \downarrow x \subseteq J$.

Proof: By Lemma 7(4), $\downarrow C$ is closed under suprema if $C \subseteq B$ is. Hence, $\downarrow \bigcap_{j \in J} I_j$ is an ideal, clearly being the greatest lower bound of $\{I_j \mid j \in J\}$. It is a standard observation that the directed union of a family of ideals is an ideal, hence directed suprema are indeed just unions. To see that finite infima are intersections, it suffices to prove $A \cap A' = \downarrow(A \cap A')$. One always has $\downarrow(A \cap A') \subseteq \downarrow A \cap \downarrow A' = A \cap A'$. For the other inclusion, observe that if $x \in A \cap A'$ then there exist $a \in A$, $a' \in A'$ with $x \prec a, a'$, so $x \prec a \wedge a' \in A \cap A'$.

By Lemma 10, we have $I = \bigvee^{\uparrow} \{ \downarrow x \mid x \in I \}$. This implies that $I \ll J$ iff there is some $x \in B$ with $I \subseteq \downarrow x \subseteq J$ and moreover that the lattice is continuous. Multiplicativity follows from the characterization of the approximation order and axiom ($\prec$-$\wedge$).

Next we verify the correctness of the second formula. To this end let $(I_j)_{j \in J}$ be a collection of ideals and denote $\downarrow \{ \bigvee M \mid M \subseteq_{\text{fin}} \bigcup_{j \in J} I_j \}$ by A. We show that this defines an ideal. It is clear that $\downarrow A = A$ because $\prec^2 = \prec$. For the closure under suprema let $N \subseteq_{\text{fin}} A$. By definition, each $n \in N$ approximates the supremum of some $M_n \subseteq_{\text{fin}} \bigcup_{j \in J} I_j$. By Lemma 7(4), $\bigvee N \prec \bigvee \bigcup_{n \in N} M_n$ and so $\bigvee N \in A$.

Each I_j is contained in $\bigcup_{j \in J} I_j$ and as $\downarrow$ is monotone we get $I_j = \downarrow I_j \subseteq \downarrow \bigcup_{j \in J} I_j$. The last set is contained in A because the definition includes suprema of singleton sets. So A is an upper bound for the I_j.

If, on the other hand, A' is an ideal which contains all I_j then it must also contain suprema of finite subsets of $\bigcup_{j \in J} I_j$. Hence it will contain A.

Finally, distributivity. We have to show $A \wedge (C \vee C') \subseteq (A \wedge C) \vee (A \wedge C')$. So assume that $x \in B$ belongs to $A \wedge (C \vee C') = A \cap (C \vee C')$. Then $x \in A$ and $x \prec \bigvee M$ for some $M \subseteq_{\text{fin}} C \cup C'$. Because $A \subseteq \downarrow A$ there exists $a \in A$ with $x \prec a$. Now we can use ($\prec$-$\wedge$) and we get $x \prec a \wedge \bigvee M = \bigvee_{m \in M} (a \wedge m)$. All elements $a \wedge m$ belong to either $A \cap C = A \wedge C$ or $A \cap C' = A \wedge C'$. Hence x is in $(A \wedge C) \vee (A \wedge C')$. ❑

From Theorem 5 it follows that proximity lattices are indeed a finitary description of coherent spaces:

COROLLARY 12: Suppose that $(B; \vee, \wedge, 0, 1; \prec)$ is a proximity lattice. Then $\mathrm{pt}(\mathrm{Idl}(B))$ is a coherent space and $\mathrm{Idl}(B)$ is isomorphic to its lattice of open subsets.

4. FILTERS AND COMPACT SATURATED SUBSETS

Filters in a proximity lattice are defined dually to ideals. They are denoted by filt(B). Because our definition of proximity lattice is self-dual the results of the preceding section hold also for filters and the filter completion.

We want to show that filters correspond to compact saturated subsets of the coherent space described. In preparation for this we look at a more general correspondence.

DEFINITION 13: Let $(B; \vee, \wedge, 0, 1; \prec)$ be a proximity lattice. We call a subset U of B *upper* if $U = \uparrow U$ holds. The collection of all upper sets is denoted by upper(B).

LEMMA 14: Let $(B; \vee, \wedge, 0, 1; \prec)$ be a proximity lattice. Then $\sigma_{\mathrm{Idl}(B)}$ (the Scott-topology on Idl(B)) is isomorphic to the set of upper sets of B. The isomorphisms are

$$\phi: \sigma_{\mathrm{Idl}(B)} \to \mathrm{upper}(B), \quad \mathcal{U} \mapsto \{x \in B \mid \downarrow x \in \mathcal{U}\}$$

and

$$\psi: \mathrm{upper}(B) \to \sigma_{\mathrm{Idl}(B)}, \quad U \mapsto \{A \in \mathrm{Idl}(B) \mid A \cap U \neq \varnothing\}.$$

Proof: We first show that ϕ is well-defined. Clearly, $\phi(\mathcal{U})$ is upwards closed with respect to $\prec$. It equals $\uparrow\phi(\mathcal{U})$ because $\downarrow x \in \mathcal{U}$ and $\downarrow x = \bigvee^{\uparrow}_{y \prec x} \downarrow y$ imply $\downarrow y \in \mathcal{U}$ for some $y \prec x$ as $\mathcal{U}$ is Scott-open.

Next, let us check the well-definedness of ψ. Again, it is clear that $\psi(U)$ is an upwards closed subset of Idl(B). We show that $\psi(U)$ is Scott-open. Assume $\bigvee^{\uparrow}_{j \in J} A_j = \bigcup_{j \in J} A_j \in \psi(U)$. This means $\bigcup_{j \in J} A_j \cap U \neq \varnothing$ and so there is some A_j for which the intersection with U is nonempty. This ideal then belongs to $\psi(U)$.

It is clear that both ϕ and ψ are monotone.

The two functions compose to identities. We first check this for $\psi \circ \phi$:

$$\begin{aligned} \psi \circ \phi(\mathcal{U}) &= \{A \mid A \cap \phi(\mathcal{U}) \neq \varnothing\} \\ &= \{A \mid A \cap \{x \mid \downarrow x \in \mathcal{U}\} \neq \varnothing\} \\ &= \{A \mid \exists x \in A.\ \downarrow x \in \mathcal{U}\}. \end{aligned}$$

The last set is equal to $\mathcal{U}$. One inclusion is true because $\mathcal{U}$ is upwards closed, the other because $A = \bigvee^{\uparrow}_{x \in A} \downarrow x$ and $\mathcal{U}$ is Scott-open.

The calculation for $\phi \circ \psi$ reads

$$\begin{aligned} \phi \circ \psi(U) &= \{x \mid \downarrow x \in \psi(U)\} \\ &= \{x \mid \downarrow x \in \{A \mid A \cap U \neq \varnothing\}\} \\ &= \{x \mid \downarrow x \cap U \neq \varnothing\}. \end{aligned}$$

The last set is equal to U. For one inclusion use that $\uparrow U \subseteq U$, for the other that $U \subseteq \uparrow U$. This completes the proof. ❑

Let us now turn to compact saturated sets. By the Hofmann-Mislove Theorem (cf. [1, Theorem 7.2.9], and also [8], [5]) we know that $\mathcal{K}_X$ is isomorphic to the set of Scott-open filters on the lattice of opens. We relate these to filters on proximity lattices as follows:

LEMMA 15: Let $(B; \vee, \wedge, 0, 1; \prec)$ be a proximity lattice. The isomorphisms in Lemma 14 cut down to isomorphisms between the set of Scott-open filters on $\mathrm{Idl}(B)$ and $\mathrm{filt}(B)$.

Proof: All we need to show is the well-definedness of ϕ and ψ. First of all, both functions preserve the property that the set they are applied to is nonempty, simply because they are isomorphisms and map the empty set onto the empty set. In the case of ϕ we use the fact that $\downarrow(x \wedge y) = \downarrow x \cap \downarrow y$ by ($\prec$-$\wedge$). Hence $x \wedge y$ belongs to $\phi(\mathcal{F})$ if x and y do. Finally, let us show that $\psi(F)$ is a filter. Assume $A, A' \in \psi(F)$. This is by definition equivalent to $A \cap F \neq \varnothing$ and $A' \cap F \neq \varnothing$. Let a and a' be elements from the intersections. By Lemma 9(2), it follows that $a \wedge a'$ belongs to $(A \cap A') \cap F$. ❑

THEOREM 16: If $(B; \vee, \wedge, 0, 1; \prec)$ is a proximity lattice then $(\mathcal{K}_{\mathrm{pt}(\mathrm{Idl}(B))}, \supseteq) \cong (\mathrm{filt}(B), \subseteq)$.

5. STRONG PROXIMITY LATTICES

The results achieved so far are quite pleasing. The duality between compact and open sets is apparent. There are two issues, however, in which we fall short of a good finitary description of coherent spaces. The first is that we haven't yet said how to recover the points themselves from a proximity lattice. Indeed, a rather complicated definition is needed as can be seen from [14]. The second shortcoming is more subtle. We ask what the tokens of a proximity lattice stand for. One would think that an element $a \in B$ represents the open set $\downarrow a$ and, dually, also the compact saturated set $\uparrow a$. This, however, conflicts with the algebraic structure of the proximity lattice, unless we add further axioms.

PROPOSITION 17: Suppose $(B; \vee, \wedge, 0, 1; \prec)$ is a proximity lattice. Then the map $\downarrow : B \to \mathrm{Idl}(B)$ preserves finite infima and 0. It preserves binary suprema (and hence is a lattice homomorphism) iff B satisfies

$$(\prec\text{-}\vee)\ \forall a, x, y \in B.\ a \prec x \vee y \Rightarrow \exists x', y' \in B.\ x' \prec x,\ \ y' \prec y \text{ and } a \prec x' \vee y'.$$

Proof: Preservation of 0 and 1 is immediate. Moreover, $\downarrow(a \wedge b) = \downarrow a \cap \downarrow b$ by ($\prec$-$\wedge$). By Theorem 11, the latter equals $\downarrow a \wedge \downarrow b$.

By the formula for the supremum in Theorem 11, we have $x \in {\downarrow}a \vee {\downarrow}b$ iff there are $a' \prec a$, $b' \prec b$ with $a \prec a' \vee b'$. Requiring $\downarrow$ to preserve $\vee$ means requiring that this holds iff $x \in {\downarrow}(a \vee b)$, i.e., iff $x \prec a \vee b$. This condition is exactly ($\prec$-$\vee$). ❑

It appears that we should add ($\prec$-$\vee$) to the list of our axioms. But an analogous result holds for filters and the map $\uparrow: B \to \text{filt}(B)$. So we should also include

$$(\wedge\text{-}\prec)\quad \forall a, x, y \in B.\, x \wedge y \prec a \Rightarrow \exists x', y' \in B.\, x \prec x',\quad y \prec y' \text{ and } x' \wedge y' \prec a.$$

We arrive at our main definition:

DEFINITION 18: A *strong proximity lattice* is a proximity lattice additionally satisfying ($\prec$-$\vee$) and its dual ($\wedge$-$\prec$).

From Proposition 17 we see that the tokens of a strong proximity lattice are very concrete in the sense that each of them stands for a particular open set and also for a particular compact saturated set. The algebraic operations in the lattice translate into actual union and intersection under this reading. Furthermore, the order of approximation corresponds to the way-below relation on the respective lattices. As a by-product, we can now also recover the points of the coherent space in a simple fashion:

DEFINITION 19: Suppose $(B; \vee, \wedge, 0, 1; \prec)$ is a strong proximity lattice. Let the *spectrum* of B comprise all *prime filters* of B:

$$\text{spec}(B) = \{F \in \text{filt}(B) \mid (M \subseteq_{\text{fin}} B \text{ and } \bigvee M \in F) \Rightarrow M \cap F \neq \varnothing\}.$$

For $x \in B$ define the *basic open set*

$$\mathcal{O}_x = \{F \in \text{spec}(B) \mid x \in F\}.$$

Finally, let $\mathcal{T}_B$ denote the topology on spec(B) generated by the sets $\mathcal{O}_x$, $x \in B$. We refer to it as the *canonical topology*.

LEMMA 20: Suppose $(B; \vee, \wedge, 0, 1; \prec)$ is a strong proximity lattice and $a, b \in B$. Then $\mathcal{O}_a \cap \mathcal{O}_b = \mathcal{O}_{a \wedge b}$ and $\mathcal{O}_a \cup \mathcal{O}_b = \mathcal{O}_{a \vee b}$. Hence the $\mathcal{O}_a$ form indeed a base for the canonical topology on spec(B) (rather than merely a subbase).

Proof: The first assertion holds by the dual of Lemma 9(1), the second by primeness of elements of spec(B). ❑

THEOREM 21: Let $(B; \vee, \wedge, 0, 1; \prec)$ be a strong proximity lattice. The isomorphism in Lemma 14 cuts down to a homeomorphism between pt(Idl(B)) and spec(B).

Proof: For bijectivity it remains to show that ϕ and ψ preserve finitary primeness of filters. To this end let $M \subseteq_{\text{fin}} B$ and $\bigvee M \in \phi(\mathcal{F})$. We use the fact that $\downarrow$ preserves finite suprema and get $\bigvee_{m \in M} \downarrow m = \downarrow \bigvee M \in \mathcal{F}$. Since $\mathcal{F}$ is prime, some $\downarrow m_0$ belongs to it and its generator m_0 belongs to $\phi(\mathcal{F})$.

Assuming that F is a prime filter in B, we show that $\psi(F)$ is completely prime. Assume $\bigvee_{j \in J} A_j \in \psi(F)$. This means $\bigvee_{j \in J} A_j \cap F \neq \emptyset$ and so there is $M \subseteq_{\text{fin}} \bigcup_{j \in J} A_j$ and $a \in F$ with $a \prec \bigvee M$. Since F is upwards closed it also contains $\bigvee M$ and since it is prime it contains some $m_0 \in M$. This m_0 came from some A_{j_0} which therefore belongs to $\psi(F)$.

Recall that the topology on $\mathrm{pt}(\mathrm{Idl}(B))$ is given by the collection of all $\mathcal{O}_A = \{\mathcal{F} \in \mathrm{pt}(\mathrm{Idl}(B)) \mid A \in \mathcal{F}\}$, $A \in \mathrm{Idl}(B)$. We check that it translates into the canonical topology on $\mathrm{spec}(B)$ under ψ^{-1}:

$$\begin{aligned}\psi^{-1}(\mathcal{O}_A) &= \psi^{-1}(\{\mathcal{F} \in \mathrm{pt}(\mathrm{Idl}(B)) \mid A \in \mathcal{F}\}) \\ &= \{F \in \mathrm{spec}(B) \mid A \in \psi(F)\} \\ &= \{F \in \mathrm{spec}(B) \mid A \cap F \neq \emptyset\} \\ &= \bigcup_{a \in A} \{F \in \mathrm{spec}(B) \mid a \in F\} = \bigcup_{a \in A} \mathcal{O}_a .\end{aligned}$$

The translation for ϕ^{-1} reads:

$$\begin{aligned}\phi^{-1}(\mathcal{O}_a) &= \phi^{-1}(\{F \in \mathrm{spec}(B) \mid a \in F\}) \\ &= \{\mathcal{F} \in \mathrm{pt}(\mathrm{Idl}(B)) \mid \downarrow a \in \mathcal{F}\} = \mathcal{O}_{\downarrow a}. \quad \square\end{aligned}$$

We summarize the situation in the following table which should be contrasted to the one at the end of Section 1.

strong proximity lattice	coherent space
prime filter	element
ideal	open set
filter	compact saturated set

6. THE REPRESENTATION THEOREM

So far, we did only one half of the work due: all strong proximity lattices are descriptions of certain coherent spaces. What is missing is a construction which assigns to an arbitrary coherent space X a strong proximity lattice B with $\mathrm{spec}(B) \cong X$. In the classical case of spectral spaces one takes the lattice of all compact-open subsets of X. There is no base of compact-opens in our situation, so this does not work. The naive approach of taking for B (a base of) the topology $\mathcal{T}_X$ of X together with the lattice operations on $\mathcal{T}_X$ and $\ll$ as order of approximation (i.e., $O \prec O'$ iff there is some $K \in \mathcal{K}_X$ with $O \subseteq K \subseteq O'$) does not work either; Axiom ($\wedge$-$\prec$) is violated.

EXAMPLE 22: The unit interval with the standard Hausdorff topology is a coherent space. One has $[0, \frac{1}{2}) \cap (\frac{1}{2}, 1] = \emptyset \ll \emptyset$, but for no open sets O, O' with $[0, \frac{1}{2}) \ll O$ and $(\frac{1}{2}, 1] \ll O'$, the intersection $O \cap O'$ is empty.

This failure is not surprising; the classical construction relies on *both* properties of compact-open sets, openness and compactness. If we want to capture both, we have to consider pairs (O, K) consisting of an open set O and a compact set K. What ought to be the order of approximation, replacing subset inclusion in the classical case? Now $O \subseteq O'$ for compact-opens means in fact that O' is a compact neighborhood of O. These thoughts lead to the idea that one should define (O, K) to approximate (O', K') if $K \subseteq O'$. These definitions do indeed work. We define

(1) $B := \{(O, K) \in \mathcal{T}_X \times \mathcal{K}_X \mid O \subseteq K\}$
(2) $(O, K) \vee (O', K') := (O \cup O', K \cup K')$
(3) $(O, K) \wedge (O', K') := (O \cap O', K \cap K')$
(4) $0 := (\emptyset, \emptyset)$; $1 := (X, X)$
(5) $(O, K) \prec (O', K') :\Leftrightarrow K \subseteq O'$

THEOREM 23: If X is a coherent space, then the above defined structure is a strong proximity lattice with $X \cong \text{spec}(B)$.

Proof: It is clear that $(B; \vee, \wedge, 0, 1)$ is a distributive lattice and that $\prec \circ \prec \subseteq \prec$. On the other hand, local compactness of X implies that whenever $K \subseteq O$ for compact K and open O, there are $O' \in \mathcal{T}_X$ and $K' \in \mathcal{K}_X$ with $K \subseteq O' \subseteq K' \subseteq O$, hence $\prec$ is interpolating.

Now for the axioms: ($\vee$-$\prec$) and ($\prec$-$\wedge$) hold trivially since all operations and relations involved are set-theoretic.

For ($\prec$-$\vee$) assume that $K \subseteq O_1 \cup O_2$. Each $x \in K$ belongs to either O_1 or O_2. For $i = 1, 2$, we have O_x and K_x with $x \in O_x \subseteq K_x \subseteq O_i$ by local compactness. Since K is compact, it is covered by finitely many O_x which may be grouped as belonging to O_1 or O_2. This gives the interpolating neighborhoods.

For ($\wedge$-$\prec$) assume $K_1 \cap K_2 \subseteq O$ for an open set O and compact saturated sets K_1, K_2. Since $\mathcal{K}_X$ is arithmetic, we have $K_i = \bigcap\{K_i' \mid K_i' \ll K_i\}$, $i = 1, 2$, and hence $K_1 \cap K_2 = \bigcap\{K'_1 \cap K_2' \mid K_1' \ll K_1, K_2' \ll K_2\}$. This is a filtered intersection of compact saturated sets and Theorem 1 tells us that some $K_1' \cap K_2'$ is contained in O already.

Finally, we have to verify $X \cong \text{spec}(B)$. As both spaces are sober, it suffices to prove their topologies isomorphic. To this end, define $\Psi: (\mathcal{T}_X, \subseteq) \to \text{Idl}(B, \subseteq)$ and $\Phi: \text{Idl}(B, \subseteq) \to (\mathcal{T}_X, \subseteq)$ by

$$\Psi(O) = \downarrow(O, X) \quad \text{and} \quad \Phi(I) = \bigcup\{O \mid \exists K \in \mathcal{K}_X.\ (O, K) \in I\}.$$

It is clear that these maps are well-defined and monotone. It remains to verify them being inverses: $\Phi(\Psi(O)) = \bigcup\{O' \mid \exists K \in \mathcal{K}_X.\ O' \subseteq K \subseteq O\} = O$ holds by local compactness, hence $\Phi \circ \Psi = \mathrm{id}_{\mathcal{T}_X}$. Moreover, we have

$$\begin{aligned}(O, K) \in \Psi(\Phi(I)) &\Leftrightarrow (O, K) \prec (\Phi(I), X)\\ &\Leftrightarrow K \subseteq \bigcup\{O' \in \mathcal{T}_X \mid \exists K' \in \mathcal{K}_X.\ (O', K') \in I\}\\ &\Leftrightarrow K \subseteq O' \quad \text{for some } O' \text{ with } (O', K') \in I\}\\ &\Leftrightarrow (O, K) \in I,\end{aligned}$$

where the last equivalence holds since $I = \downarrow I$ and the second last by compactness of K and I's being an ideal. Thus $\Psi \circ \Phi = \mathrm{id}_{\mathrm{Idl}(B)}$. ❑

THEOREM 24: The spectra of strong proximity lattices are precisely the coherent spaces.

In case the order of approximation on the strong proximity lattice is reflexive, we find ourselves in the world of algebraic lattices and totally disconnected spaces. The Stone duality of this situation has been described in [6, II-3]. Note the strong resemblance between our constructions and the classical situation. Indeed, if we take the lattice-order as order of approximation, then all axioms are satisfied automatically and the situation reduces to the classical one.

7. MORPHISMS

We have already emphasized that it is not our purpose to represent proximity lattices in the category of topological spaces but rather the other way round, we developed strong proximity lattices to get a finitary and faithful representation of coherent spaces. It is therefore our task to describe arbitrary continuous functions between coherent spaces. It is well-known that this can not be done with functions between proximity lattices but that one has to resort to certain relations.

DEFINITION 25: Let $(A; \vee, \wedge, 0, 1; \prec_A)$ and $(B; \vee, \wedge, 0, 1; \prec_B)$ be strong proximity lattices. A binary relation $G \subseteq A \times B$ is called *approximable* if the following conditions are satisfied:

(G-$\prec$) $G \circ \prec_B = G$;
($\prec$-G) $\prec_A \circ\, G = G$;
($\vee$-G) $\forall M \subseteq_{\mathrm{fin}} A\ \forall b \in B.\ M\ G\ b \Leftrightarrow \bigvee M\ G\ b$;
(G-$\wedge$) $\forall a \in A\ \forall M \subseteq_{\mathrm{fin}} B.\ a\ G\ M \Leftrightarrow a\ G \bigwedge M$;
(G-$\vee$) $\forall a \in A\ \forall M \subseteq_{\mathrm{fin}} B.\ a\ G \bigvee M \Rightarrow \exists N \subseteq_{\mathrm{fin}} A.\ (a \prec_A \bigvee N$ and $\forall n \in N\ \exists m \in M.\ n\ G\ m)$.

We write $G\colon A \to B$ for an approximable relation from A to B. Composition of approximable relations is via the relational product $\circ$.

Note the strong resemblance between these axioms and the definition of strong proximity lattices themselves. (We could have also included the empty supremum and the empty infimum in the axioms ($\prec$-$\vee$) and ($\wedge$-$\prec$), respectively: These cases trivially hold.) Indeed, the order of approximation on a strong proximity lattice gives rise to the identity approximable relation. We also observe that the definition does not include an analogue of ($\wedge$-$\prec$), the dual of (G-$\vee$). This is due to the fact that the concept of a continuous function is inherently nonsymmetric.

THEOREM 26: The category of strong proximity lattices and approximable relations is equivalent to the category of coherent spaces and continuous functions.

Proof: One first has to check that approximable relations indeed give rise to a category. As we have said already, the orders of approximation themselves play the role of identity morphisms. The laws for a category are then straightforward to verify.

We show the one-to-one correspondence between approximable relations and continuous functions by once again making use of the already established duality with the category of arithmetic lattices and frame homomorphisms.

If $G \subseteq A \times B$ is an approximable relation then we let

$$h_G: \mathrm{Idl}(B) \to \mathrm{Idl}(A), \quad h_G(I) := \{a \in A \mid \exists b \in I.\ a\ G\ b\}$$

be the corresponding frame homomorphism. The result of applying h_G gives an ideal because of ($\prec$-G) and ($\vee$-G). It is clear that h_G is monotone. Simple calculations show that it preserves finite infima and arbitrary suprema.

For the converse, assume that h: $\mathrm{Idl}(B) \to \mathrm{Idl}(A)$ is a frame homomorphism. We define a relation $G_h \subseteq A \times B$ as follows:

$$a\ G_h\ b :\Leftrightarrow a \in h(\downarrow b).$$

Again, it is straightforward that this relation satisfies the axioms for an approximable relation.

The two translations are inverses of each other:

$$\begin{aligned} h_{G_h}(I) &= \{a \in A \mid \exists b \in I.\ a\ G_h\ b\} \\ &= \{a \in A \mid \exists b \in I.\ a \in h(\downarrow b)\} \\ &= \bigcup_{b \in I} h(\downarrow b) = \bigvee_{b \in I} h(\downarrow b) \\ &= h(\bigvee_{b \in I} \downarrow b) = h(I) \end{aligned}$$

and

$$\begin{aligned} a\ G_{h_G}\ b &\Leftrightarrow a \in h_G(\downarrow b) = \{x \in A \mid \exists b' \prec_B b.\ x\ G\ b'\} \\ &\Leftrightarrow \exists b' \prec_B b.\ a\ G\ b' \end{aligned}$$

$$\Leftrightarrow a\ G\ b$$

where the last equivalence holds because of (G-$\prec$). ❑

8. THE COCOMPACT TOPOLOGY

Coherent spaces have been studied in many different guises and there are at least three different names for them in the literature. The first appearance is in [11] where they are called *compact ordered spaces*. Connections with Stone duality are contained in [3, Sections V.5 and VII.3]. Finally, they have been characterized as *supersober spaces* in [3, Section VII.1] and [9]. In more recent work, coherent spaces are investigated as the adequate substitute for compact Hausdorff spaces in nonsymmetric topology [10], [7]. From all these sources, we collect the following facts:

FACTS: If $(X, \mathcal{T})$ is a coherent space, then $\mathcal{T}_c$, the collection of all complements of compact saturated sets forms the *cocompact topology* on X. The *patch topology* $\mathcal{T}_p = \mathcal{T} \vee \mathcal{T}_c$ is a compact Hausdorff topology on X such that $(X, \mathcal{T}_p, \leq_{\mathcal{T}})$ is a compact ordered space. Starting with a compact ordered space $(X, \mathcal{T}, \leq)$, the open upper sets form a topology $\mathcal{T}_\#$ such that $(X, \mathcal{T}_\#)$ is coherent. The cocompact topology for $\mathcal{T}_\#$ consist of all $\mathcal{T}$-open lower subsets and $(\mathcal{T}_\#)_p$ coincides with $\mathcal{T}$. ❑

In this section, we want to make the relationship between a strong proximity lattice B, its spectrum spec(B), its dual B^{op}, and the cocompact topology on spec(B) explicit. We know from Theorem 16 that for any proximity lattice B, the set $\mathcal{K}_{\mathrm{pt}(\mathrm{Idl}(B))}$ is isomorphic to filt(B). So the cocompact topology on pt(Idl(B)) equals filt(B) which in turn is the topology on pt(Idl(B^{op}). For strong proximity lattices, this isomorphism is manifested on the point-level, too. There it turns out to be essentially complementation of prime filters/ideals. But let us first turn our attention to the compact saturated subsets.

THEOREM 27: Let $(B; \vee, \wedge, 0, 1; \prec)$ be a strong proximity lattice. Then the map comp : $(\mathrm{filt}(B), \subseteq) \to (\mathcal{K}_{\mathrm{spec}(B)}, \supseteq)$ with

$$\mathrm{comp}(K) = \{F \in \mathrm{spec}(B) \mid K \subseteq F\}$$

is an isomorphism. Its inverse is given by $\mathfrak{k} \mapsto \bigcap \mathfrak{k}: \mathcal{K}_{\mathrm{spec}(B)} \to \mathrm{filt}(B)$.

Proof: The isomorphism of the Hofmann-Mislove-Theorem maps a compact saturated set $\mathfrak{k}$ to its filter of open neighborhoods; if the elements of $\mathfrak{k}$ are given as completely prime filters, then this is their intersection. Thus we get for $\mathfrak{k} \in \mathcal{K}_{\mathrm{spec}(B)}$:

$$
\begin{aligned}
\mathfrak{k} &\longmapsto \{\psi(F) \mid F \in \mathfrak{k}\} && \text{via Theorem 21} \\
&\longmapsto \bigcap\{\psi(F) \mid F \in \mathfrak{k}\} && \text{via the HM-Theorem} \\
&\longmapsto \phi(\bigcap\{\psi(F) \mid F \in \mathfrak{k}\}) && \text{via Lemma 15} \\
&= \bigcap\{\phi(\psi(F)) \mid F \in \mathfrak{k}\} && \text{since } \phi \text{ preserves } \bigcap \\
&= \bigcap \mathfrak{k} && \text{since } \phi = \psi^{-1}.
\end{aligned}
$$

Going the other way, the Hofmann-Mislove-Isomorphism maps a Scott-open filter G of opens to its intersection which consists of all points with neighborhood filter containing G. Therefore, we calculate for $K \in \mathrm{filt}(B)$:

$$
\begin{aligned}
K &\longmapsto \psi(K) && \text{via Lemma 15} \\
&\longmapsto \{\mathcal{F} \in \mathrm{pt}(\mathrm{Idl}(B)) \mid \psi(K) \subseteq \mathcal{F}\} && \text{via the HM-Theorem} \\
&= \{\mathcal{F} \in \mathrm{pt}(\mathrm{Idl}(B)) \mid K \subseteq \phi(\mathcal{F})\} && \text{since } \phi = \psi^{-1} \\
&\longmapsto \{\phi(\mathcal{F}) \mid \mathcal{F} \in \mathrm{pt}(\mathrm{Idl}(B)); K \subseteq \phi(\mathcal{F})\} && \text{via Theorem 21.}
\end{aligned}
$$

As ϕ: $\mathrm{pt}(\mathrm{Idl}(B)) \to \mathrm{spec}(B)$ is in particular surjective, the latter set equals $\mathrm{comp}(K)$. ❑

THEOREM 28: Let $(B; \vee, \wedge, 0, 1; \prec)$ be a strong proximity lattice. Then the spaces $(\mathrm{spec}(B), (\mathcal{T}_{\mathrm{spec}(B)})_c)$ and $(\mathrm{spec}(B^{op}), \mathcal{T}_{\mathrm{spec}(B^{op})})$ are homeomorphic via

$$F \mapsto \downarrow(B \backslash F).$$

Moreover, the frame isomorphism between $\mathrm{filt}(B) \cong (\mathcal{T}_{\mathrm{spec}(B)})_c$ and $\mathrm{Idl}(B^{op}) \cong \mathcal{T}_{\mathrm{spec}(B^{op})}$ arising from this homeomorphism is the identity.

Proof: We first have to check that $I := \downarrow(B \backslash F)$ is a prime ideal for $F \in \mathrm{spec}(B)$. Clearly $I = \downarrow I$. Moreover, $B \backslash F$ is closed under suprema since F is prime. Hence I is indeed an ideal by Lemma 7(4). To see primeness, we have to employ $(\wedge\text{-}\prec)$: Suppose $a \wedge b \in I$. Then there is $x \in B \backslash F$ with $a \wedge b \prec x$. Axiom $(\wedge\text{-}\prec)$ gives us $a', b' \in B$ with $a \prec a'$, $b \prec b'$ and $a' \wedge b' \prec x$. The last relation implies $a' \wedge b' \notin F$ since otherwise we had $x \in F$. Hence one of a', b' is not an element of F because this is a filter. Thus $a \in I$ or $b \in I$.

By symmetry, $\uparrow(B \backslash I)$ is a prime filter if I is a prime ideal. For bijectivity, it remains to check $\uparrow(B \backslash \downarrow(B \backslash F)) = F$ which is routine.

By Theorem 21, every open set on $\mathrm{spec}(B^{op})$ is of the form

$$\mathcal{O}^{op} = \{I \in \mathrm{spec}(B^{op}) \mid I \cap K \neq \varnothing\}$$

for some $K \in \mathrm{Idl}(B^{op}) = \mathrm{filt}(B)$. We calculate for $F \in \mathrm{spec}(B)$ and $K \in \mathrm{filt}(B)$:

$$\begin{aligned} \downarrow(B\backslash F) \in \mathcal{O}_K^{op} &\Leftrightarrow \downarrow(B\backslash F) \cap K \neq \varnothing \\ &\Leftrightarrow (B\backslash F) \cap K \neq \varnothing \qquad \text{since } K = \uparrow K \\ &\Leftrightarrow K \nsubseteq F \\ &\Leftrightarrow F \in \mathrm{spec}(B)\backslash \mathrm{comp}(K). \end{aligned}$$

So the map is indeed a homeomorphism and gives rise to the identity as the corresponding frame-isomorphism. ❑

Hence the table of Section 5 should be augmented by a third column: the one at the end of Section 1.

strong proximity lattice	spectrum	dual spectrum
prime filter or prime ideal	element	element
ideal	open set	compact saturated set
filter	compact saturated set	open set

Again, restricting to the reflexive case leads to familiar territory. *Priestley duality* combines the correspondence between spectral spaces and distributive lattices with that between spectral spaces and their patches, totally order-disconnected compact ordered spaces. An account of this theory may be found in Chapter 10 of [2].

REFERENCES

1. ABRAMSKY, S. & A. JUNG. 1994. Domain theory. *In:* Handbook of Logic in Computer Science (S. Abramsky, D. M. Gabbay, & T.S.E. Maibaum, Eds.). **3:** 1–168. Clarendon Press.
2. DAVEY, B.A. & H. A. PRIESTLEY. 1990. Introduction to Lattices and Order. Cambridge University Press, Cambridge.
3. GIERZ, G., K.H. HOFMANN, K. KEIMEL, J.D. LAWSON, M. MISLOVE, & D.S. SCOTT. 1980. A Compendium of Continuous Lattices. Springer Verlag, New York.
4. GIERZ, G. & K. KEIMEL. 1981. Continuous ideal completions and compactifications. *In:* Continuous Lattices, Proceedings Bremen 1979 (B. Banaschewski and R.-E. Hoffmann, Eds.). Lecture Notes in Mathematics. **871:** 97–124. Springer Verlag. New York.
5. HOFMANN, K.H. & M. MISLOVE. 1981. Local compactness and continuous lattices. *In:* Continuous Lattices, Proceedings Bremen 1979 (B. Banaschewski and R.-E. Hoffmann, Eds.). Lecture Notes in Mathematics. **871:** 209–248. Springer Verlag. New York.
6. JOHNSTONE, P.T. 1982. Stone Spaces. Cambridge Studies in Advanced Mathematics, Vol. 3. Cambridge University Press. Cambridge.
7. KOPPERMAN, R. 1994. The skew-fields of topology. Manuscript.
8. KEIMEL, K. & J. PASEKA. A direct proof of the Hofmann-Mislove theorem. Proceedings of the Amer. Math. Soc. **120:**301–303.

9. LAWSON, J.D. 1988. The versatile continuous order. *In:* Mathematical Foundations of Programming Language Semantics (M. Main, A. Melton, M. Mislove, & D. Schmidt, Eds.). Lecture Notes in Computer Science. **298:** 134–160. Springer Verlag. New York.
10. LAWSON, J.D. 1991. Order and strongly sober compactification. *In:* Topology and Category Theory in Computer Science (G.M. Reed, A.W. Roscoe, & R.F. Wachter, Eds.). 179–205. Clarendon Press. Oxford.
11. NACHBIN, L. 1965. Topology and Order. Van Nostrand. Princeton, N.J. (Reprinted by Robert E. Kreiger Publ. Co. Huntington, NY. 1967.)
12. SMYTH, M.B. 1977. Effectively given domains. Theoretical Computer Science. **5:** 257–274.
13. SMYTH, M.B. 1986. Finite approximation of spaces. *In:* Category Theory and Computer Programming (D. Pitt, S. Abramsky, A. Poigné, & D. Rydeheard, Eds.). Lecture Notes in Computer Science. **240:** 225–241. Springer Verlag.
14. SMYTH, M.B. 1992. Stable compactification I. J. London Math. Soc. **45:** 321–340.
15. STONE, M.H. 1936. The theory of representations for Boolean algebras. Trans. American Math. Soc. **40:** 37–111.

Completeness of the Quasi-uniformity of Quasi-uniform Convergence

HANS-PETER A. KÜNZI[a, c,e] AND SALVADOR ROMAGUERA[b,d, e]

[a]*Department of Mathematics*
University of Berne
Sidlerstrasse 5, CH-3012, Berne, Switzerland
and
[b]*Escuela de Caminos*
Departamento de Matemática Aplicada
Universidad Politécnica de Valencia
46071 Valencia, Spain

ABSTRACT: We discuss the completeness of the quasi-uniformity of quasi-uniform convergence on function spaces. In particular we observe that the notions of quasi-uniform completeness based on the convergence of several types of stable filters provide satisfactory results in this context.

1. INTRODUCTION

In the last years several authors have attacked well-known problems concerning topological properties of function spaces with the help of quasi-uniform structures (see [10], [11], [12], [8], [16], [13], [1], [9], etc.) Of course, a basic problem in this direction consists of obtaining an appropriate quasi-uniform generalization of the classical result that if X is a topological space and $(Y, \mathcal{U})$ is a complete uniform space, then the uniformity of uniform convergence is complete. In our paper we study this problem and observe that the notions of quasi-uniform completeness based on the convergence of several types of stable filters provide satisfactory results in this context. We also show that other known notions of quasi-uniform completeness appear more intractable in this setting.

Mathematics Subject Classification: 54E15, 54C35.
Key words and phrases: quasi-uniformity of quasi-uniform convergence, completeness.
[c]E-mail: kunzi@math-stat.unibe.ch.
[d]E-mail: sromague@mat.upv.es.
[e]The first author thanks the support of the Swiss National Science Foundation, under grant 2000-041745.94/1. The second author is supported by the Conselleria de Educacio i Ciencia, Generalitat Valenciana, under grant *GV*-2223/94.

2. CONCEPTS OF COMPLETENESS

We refer the reader to [6] for basic information about quasi-uniform and quasi-pseudometric spaces.

Let us recall that if $\mathcal{U}$ is a quasi-uniformity on a set X, then $\mathcal{U}^{-1} = \{U^{-1}: U \in \mathcal{U}\}$ is also a quasi-uniformity on X. As usual $\mathcal{U}^*$ will denote the coarsest uniformity finer than $\mathcal{U}$.

Let $(X, \mathcal{U})$ be a quasi-uniform space and let $\mathcal{F}$ be a filter on X. Then $\mathcal{F}$ is called:

(i) a *$\mathcal{U}$-stable filter* if for each $U \in \mathcal{U}$, $\cap\{U(F) : F \in \mathcal{F}\} \in \mathcal{F}$ [2].

(ii) a *Cauchy filter* if for each $U \in \mathcal{U}$ there is $x \in X$ such that $U(x) \in \mathcal{F}$ [6].

(iii) a *left K-Cauchy filter* if for each $U \in \mathcal{U}$ there is $F \in \mathcal{F}$ such that $U(x) \in \mathcal{F}$ for all $x \in F$ [14].

(iv) a *right K-Cauchy filter* if it is left K-Cauchy on $(X, \mathcal{U}^{-1})$ [14].

(v) a *$\mathcal{U}^*$-Cauchy filter* if it is a Cauchy filter on the uniform space $(X, \mathcal{U}^*)$ [6].

Now let $(\mathcal{G}, \mathcal{F})$ be an ordered pair of filters on X. Then the pair $(\mathcal{G}, \mathcal{F})$ is called:

(vi) a *Cauchy pair of filters* if $(\mathcal{G}, \mathcal{F}) \to 0$, where $(\mathcal{G}, \mathcal{F}) \to 0$ provided that for each $U \in \mathcal{U}$ there exist $F \in \mathcal{F}$ and $G \in \mathcal{G}$ such that $G \times F \subseteq U$ [3], [4], [5].

(vii) a *stable pair of filters* if $\mathcal{F}$ is $\mathcal{U}$-stable and $\mathcal{G}$ is $\mathcal{U}^{-1}$-stable [3].

A quasi-uniformity $\mathcal{U}$ on a set X is convergence complete [6] provided that each Cauchy filter is $T(\mathcal{U})$-convergent; $\mathcal{U}$ is left (right) K-complete [14] provided that each left (right) K-Cauchy filter is $T(\mathcal{U})$-convergent; $\mathcal{U}$ is half complete [3] provided that each $\mathcal{U}^*$-Cauchy filter is $T(\mathcal{U})$-convergent; $\mathcal{U}$ is bicomplete [6] provided that the uniformity $\mathcal{U}^*$ is complete; $\mathcal{U}$ is D-complete [4], [5], provided that if $(\mathcal{G}, \mathcal{F}) \to 0$ then the filter $\mathcal{F}$ is $T(\mathcal{U})$-convergent; $\mathcal{U}$ is strongly D-complete [7] provided that if $(\mathcal{G}, \mathcal{F}) \to 0$ then the filter $\mathcal{G}$ has a $T(\mathcal{U})$-cluster point; $\mathcal{U}$ is S-complete [3] provided that each stable Cauchy pair of filters $(\mathcal{G}, \mathcal{F})$ converges to a point $x \in X$ (i.e., $\mathcal{F}$ is $T(\mathcal{U})$-convergent to x and $\mathcal{G}$ is $T(\mathcal{U}^{-1})$-convergent to x) and $\mathcal{U}$ is U-complete [3] provided that each stable Cauchy pair of ultrafilters is convergent to a point $x \in X$.

3. COMPLETENESS OF FUNCTION SPACES

Let (X, T) be a topological space and $(Y, \mathcal{U})$ a quasi-uniform space. We shall denote by Y^X the set of all (not necessarily continuous) functions from X into Y and by $C(X, Y)$ the set of all continuous functions from (X, T) into $(Y, T(\mathcal{U}))$. If $\mathcal{A}$ is a family of subsets of X we denote by $\mathcal{U}_{\mathcal{A}}$ the quasi-uniformity on Y^X which has as a subbase the family of sets of the form $(A, U) = \{(f, g) \in Y^X \times Y^X : (f(x), g(x)) \in U$ for all $x \in A\}$ whenever $A \in \mathcal{A}$ and $U \in \mathcal{U}$. The quasi-uniformity $\mathcal{U}_{\mathcal{A}}$ is called *the quasi-uniformity of quasi-uniform convergence of* $\mathcal{A}$. If $\mathcal{A} = \{X\}$, $\mathcal{U}_{\mathcal{A}}$ is called, simply, *the quasi-uniformity of quasi-uniform convergence* and is denoted by $\mathcal{U}_X$. If $\mathcal{A} = \{K \subseteq X$: K is a compact subset of $X\}$, $\mathcal{U}_A$ is called *the quasi-uniformity of compact convergence*.

Given two sets X and Y we shall use, in the rest of the paper, the following notation:

(a) If $\mathcal{F}$ is a filter on Y^X, $F \in \mathcal{F}$ and $x \in X$ we denote by F_x the set $\{f(x): f \in F\}$ and by $\mathcal{F}_x$ the filter on Y generated by $\{F_x : F \in \mathcal{F}\}$.

(b) If $\mathcal{F}$ is a filter on Y, $F \in \mathcal{F}$ and $y \in F$ we denote by f_y the function from X into Y such that $f_y(x) = y$ for all $x \in X$, and by $B(F)$ the set $\{f_y: y \in F\}$. Thus, $B(\mathcal{F}) = \{B(F): F \in \mathcal{F}\}$ is a filterbase on Y^X.

THEOREM 3.1: Let (X, T) be a topological space, $(Y, \mathcal{U})$ a quasi-uniform space, and $\mathcal{A}$ a family of subsets of X which covers X. Then:

(1) The quasi-uniformity of quasi-uniform convergence of $\mathcal{A}$ is half complete if and only if $\mathcal{U}$ is half complete.

(2) The quasi-uniformity of quasi-uniform convergence of $\mathcal{A}$ is bicomplete if and only if $\mathcal{U}$ is bicomplete.

(3) The quasi-uniformity of quasi-uniform convergence of $\mathcal{A}$ is right K-complete if and only if $\mathcal{U}$ is right K-complete.

(4) The quasi-uniformity of quasi-uniform convergence of $\mathcal{A}$ is S-complete if and only if $\mathcal{U}$ is S-complete.

(5) The quasi-uniformity of quasi-uniform convergence of $\mathcal{A}$ is U-complete if and only if $\mathcal{U}$ is U-complete.

Proof: We first prove (1). Suppose that $(Y, \mathcal{U})$ is half complete. Let $\mathcal{F}$ be a $(\mathcal{U}_{\mathcal{A}})^*$-Cauchy filter on Y^X. Then for each $U \in \mathcal{U}$ and each $A \in \mathcal{A}$ there is $F \in \mathcal{F}$ such that $F \times F \subseteq (A, U)$. Fix $x \in X$. Then x is in some $A_x \in \mathcal{A}$. Hence, $\mathcal{F}_x$ is a $\mathcal{U}^*$-Cauchy filter on Y so that it is $T(\mathcal{U})$-convergent to a point $f_o(x) \in Y$.Thus, we have defined a function f_o from X into Y, and we shall show that $\mathcal{F}$ converges to f_o with respect to the topology of quasi-uniform convergence of $\mathcal{A}$. In fact, given $U \in \mathcal{U}$ and $A \in \mathcal{A}$ choose a $V \in \mathcal{U}$ such that $V^2 \subseteq U$. There is $F \in \mathcal{F}$ with $F \times F \subseteq (A, V)$. Now let $x \in A$; then there is $G \in \mathcal{F}$ with $G_x \subseteq V(f_o(x))$. For each $f \in F$ and each $g \in G \cap F$ we obtain $(g(x), f(x)) \in V$. Since $(f_o(x), g(x)) \in V$ we conclude that $(f_o, f) \in (A, U)$. Conversely, suppose that $\mathcal{F}$ is a $\mathcal{U}^*$-Cauchy filter on Y. Then for each $U \in \mathcal{U}$ there is $F \in \mathcal{F}$ with $F \times F \subseteq U$. Thus, $B(F) \times B(F) \subseteq (X, U) \subseteq (A, U)$ for each $A \in \mathcal{A}$. Hence, $B(\mathcal{F})$ is a $(\mathcal{U}_{\mathcal{A}})^*$-Cauchy filterbase on Y^X so that it is convergent with respect to the topology of quasi-uniform convergence of $\mathcal{A}$ to a function $f_o \in Y^X$. Fix $x_o \in X$ and let $y_o = f(x_o)$. It easily follows that $\mathcal{F}$ is $T(\mathcal{U})$-convergent to y_o.

The sufficiency in (2) and (3) is proved in [9], and the necessity follows similarly to (1).

Finally, we prove (4) and (5) in parallel: Suppose that $(Y, \mathcal{U})$ is (U-complete) S-complete. Let $(\mathcal{G}, \mathcal{F})$ be a stable Cauchy pair of (ultra)filters on $(Y^X, \mathcal{U}_{\mathcal{A}})$. Fix $x \in X$. Then x is in some $A_x \in \mathcal{A}$. Therefore $(\mathcal{G}_x, \mathcal{F}_x)$ is a stable Cauchy pair of (ultra)filters on $(Y, \mathcal{U})$. Hence, $(\mathcal{G}_x, \mathcal{F}_x)$ converges to some point $f_o(x) \in Y$. Thus, we have defined a function f_o from X into Y and we shall show that the (ultra)filter $\mathcal{F}$ converges to f_o with respect to the topology of quasi-uniform convergence of $\mathcal{A}$. In fact, given $U \in \mathcal{U}$ and $A \in \mathcal{A}$ choose a $V \in \mathcal{U}$ such that $V^2 \subseteq U$. There is $H \in \mathcal{F}$ such that $H \subseteq (A, V)(F)$ for all $F \in \mathcal{F}$. Now let $x \in A$; then there is $F' \in \mathcal{F}$ such that $F'_x \subseteq V(f_o(x))$. Since for each $h \in H$ there is $f \in F'$ with $(f, h) \in (A, V)$ and $(f_o(x), f(x)) \in V$, it follows that $(f_o(x), h(x)) \in V^2 \subseteq U$. We conclude that $H \subseteq (A, U)(f_o)$, so that $\mathcal{F}$ converges to f_o with respect to $T(\mathcal{U}_{\mathcal{A}})$. Similarly, we show that the (ultra)filter $\mathcal{G}$ is $T((\mathcal{U}_{\mathcal{A}})^{-1})$-convergent to f_o. Conversely, suppose that $\mathcal{U}_{\mathcal{A}}$ is (U-complete) S-complete and let $(\mathcal{G}, \mathcal{F})$ be a stable Cauchy pair of (ultra)filters on $(Y, \mathcal{U})$. Then for each $U \in \mathcal{U}$ there are $G \in \mathcal{G}$ and $F \in \mathcal{F}$ with $G \times F \subseteq U$. Thus $B(G) \times B(F) \subseteq (X, U) \subseteq (A, U)$ for each $A \in \mathcal{A}$. Denote by $\mathcal{F}_1$ and $\mathcal{G}_1$ the two (ultra)filters generated on Y^X by $B(\mathcal{F})$ and $B(\mathcal{G})$, respectively. Then $(\mathcal{G}_1, \mathcal{F}_1)$ is a Cauchy pair of (ultra)filters on $(Y^X, \mathcal{U}_{\mathcal{A}})$. Furthermore, $\mathcal{F}_1$ is $\mathcal{U}_{\mathcal{A}}$-stable and $\mathcal{G}_1$ is $(\mathcal{U}_{\mathcal{A}})^{-1}$-stable. Hence $(\mathcal{G}_1, \mathcal{F}_1)$ converges to a function $f_o \in Y^X$. Fix $x_o \in X$. Then it is easy to see that the pair of (ultra)filters $(\mathcal{G}, \mathcal{F})$ converges to $f_o(x_o)$. ❑

In [11] B.K. Papadopoulos showed that if (X, T) is a topological space and $(Y, \mathcal{U})$ is a quiet D-complete quasi-uniform space, then the quasi-uniformity of quasi-uniform convergence is D-complete. (See [4] for the notion of a quiet quasi-uniform space.)The following example shows that *quietness* cannot be omitted in this result, not even on $C(X, Y)$. Therefore, the property of D-completeness cannot be added to the list of completeness properties given in the preceding theorem.

EXAMPLE 3.1: Let X be the set of positive integers and let T be the discrete topology on X. Put $Y = X$ and define a quasi-metric d on Y by $d(x, y) = 1$ if $x < y$, $d(x, y) = (1/2)((1/x) + (1/y))$ if $x > y$ and $y \neq 1$, $d(x, 1) = 1/x$ if $x \neq 1$, and $d(x, x) = 0$ for all $x \in Y$. It is easy to check that $T(d)$ is the discrete topology on Y. Moreover, the quasi-uniformity $\mathcal{U}(d)$ generated by d (see [6, p.3]) is D-complete. In fact, suppose $(\mathcal{G}, \mathcal{F}) \to 0$. In case the filter $\mathcal{G}$ is generated by a singleton, $\mathcal{F}$ is $T(d)$-convergent. So we consider the other case. Since $G_1 \times F_1 \subseteq \{(x, y) \in Y \times Y: d(x, y) < 1\}$ for some $G_1 \in \mathcal{G}$ and some $F_1 \in \mathcal{F}$, the filter $\mathcal{F}$ contains necessarily a finite set. Let E be such a set of minimal cardinality. If its cardinal is greater than 1, then it contains a point x different from 1. This contradicts that $(\mathcal{G}, \mathcal{F}) \to 0$ because the filter $\mathcal{G}$ does not contain the singleton $\{x\}$. So E is a singleton and the filter $\mathcal{F}$ is $T(d)$-convergent.

Now define two sequences $\{g_n\}_{n\in\omega}$ and $\{f_n\}_{n\in\omega}$ of functions from X into Y as follows: $g_0(x) = g_1(x) = f_0(x) = f_1(x) = 1$ for all $x \in X$, $g_n(x) = n$ if $x < n$, $g_n(x) = n + 1$ if $x = n$, $g_n(x) = x + 1$ if $x > n$, $f_n(x) = 1$ if $x < n - 1$ and $f_n(x) = n$ if $x \geq n-1$, whenever $n = 2, 3, \ldots$.

An easy computation of the different cases shows that $d(g_n(x), f_m(x)) \leq 1/k$ for $n, m > k$ and for all $x \in X$. Thus the filter $\mathcal{F}$ generated on $Y^X(= C(X, Y))$ by the sequence $\{f_n\}_{n\in\omega}$ is D-Cauchy. Suppose that $\mathcal{F}$ converges to some function $f_0 \in C(X, Y)$ quasi-uniformly. Then $f_0(x) = 1$ for any $x \in X$. But the convergence is clearly not quasi-uniformly to the constant function 1. We conclude that the quasi-uniformity of quasi-uniform convergence is not D-complete.

The next example shows that the properties of convergence completeness, left K-completeness and strong D-completeness cannot be added to the list of completeness properties given in Theorem 3.1.

EXAMPLE 3.2: Let $X = Y = \{0\} \cup \{1/n : n \in \mathbf{N}\}$, let T be the restriction of the usual topology on X and let d be the quasi-metric defined on Y by $d(0, 1/n) = 1/n$ and $d(1/n, 0) = 1$ for all $n \in \mathbf{N}$, $d(1/n, 1/m) = 1$ whenever $n \neq m$, and $d(x, x) = 0$ for all $x \in Y$. It is readily seen that $\mathcal{U}(d)$ is a quiet locally symmetric convergence complete quasi-uniformity. For each $n \in \mathbf{N}$ define $g_n(x) = x$ if $x \geq 1/n$ and $g_n(x) = 0$ if $x < 1/n$. Let $\mathcal{F}$ be the filter generated by the sets $F_n = \{g_k : k > n\}$. In Example 3.8 of [13], H. Render observes that for each $n \in \mathbf{N}$, $F_n \subseteq (X, U_n)(g_n)$ where $U_n = \{(y, z) \in Y \times Y : d(y, z) < 1/n\}$. He also shows that $\mathcal{F}$ is not convergent in Y^X with respect to the topology $T(\mathcal{U}(d)_X)$ of quasi-uniform convergence but it converges to the identity function with respect to $T((\mathcal{U}(d)_X)^{-1})$. Since (X, T) is compact we deduce that the quasi-uniformity of compact convergence is not point-symmetric on $C(X, Y)$. It is easy to check that $\mathcal{F}$ is a left K-Cauchy filter on $C(X, Y)$ so that the quasi-uniformity of quasi-uniform convergence is not left K-complete, hence not convergence complete.

Note that the quasi-uniformity $\mathcal{U}(d)$ of the above example is strongly D-complete while the quasi-uniformity of quasi-uniform convergence is not strongly D-complete. However, we have the following positive result.

PROPOSITION 3.1: Let (X, T) be a topological space, $(Y, \mathcal{U})$ a strongly D-complete quasi-uniform space and $\mathcal{A}$ be a family of subsets of X which covers X. Then the quasi-uniformity of quasi-uniform convergence of $\mathcal{A}$ is D-complete.

Proof: Let $(\mathcal{G}, \mathcal{F})$ be a Cauchy pair of filters on $(Y^X, \mathcal{U}_{\mathcal{A}})$. Fix $x \in X$. Then x is in some $A_x \in \mathcal{A}$. Therefore, $(\mathcal{G}_x, \mathcal{F}_x)$ is a Cauchy pair of filters on $(Y, \mathcal{U})$, so that $\mathcal{G}_x$ has a $T(\mathcal{U})$-cluster point $f_o(x) \in Y$. Thus we have defined a function f_o from X into Y and we shall show that the filter $\mathcal{F}$ is $T(\mathcal{U}_{\mathcal{A}})$-convergent to f_o. In fact, given $U \in \mathcal{U}$ and $A \in \mathcal{A}$ choose a $V \in \mathcal{U}$ such that $V^2 \subseteq U$. There are $G' \in \mathcal{G}$ and $F' \in \mathcal{F}$ such that $G' \times F' \subseteq (A, V)$. We shall show that $F' \subseteq (A, U)(f_o)$. Let $f \in F'$ and $x \in A$. Since $f_o(x)$ is a $T(\mathcal{U})$-cluster point of $\mathcal{G}_x$ there exists $g \in G'$ such that $g(x) \in V(f_o(x))$. Since $(g(x), f(x)) \in V$, we obtain that $(f_o(x)\ , f(x)) \in V^2 \subseteq U$. We conclude that $f \in (A, U)(f_o)$. Hence, $\mathcal{U}_{\mathcal{A}}$ is D-complete. ❑

In Corollary 1.3 of [12] B.K. Papadopoulos states that if (X, T) is a topological space and $(Y, \mathcal{U})$ is a quiet locally symmetric D-complete quasi-uniform space, then the quasi-uniformity of quasi-uniform convergence is D-complete on $C(X, Y)$. However, *quietness* of $(Y, \mathcal{U})$ can be omitted as the following result shows.

COROLLARY 3.1: Let (X, T) be a topological space and $(Y, \mathcal{U})$ a locally symmetric D-complete quasi-uniform space. Then the quasi-uniformity of quasi-uniform convergence is D-complete on $C(X, Y)$.

Proof: Since every locally symmetric D-complete quasi-uniform space is strongly D-complete [5, Proposition 3.1], it follows from Proposition 3.1 that the quasi-uniformity of quasi-uniform convergence is D-complete on Y^X. By Corollary 1.2 of [12], $C(X, Y)$ is $T(\mathcal{U}_X)$-closed in Y^X. We conclude that $\mathcal{U}_X$ is D-complete on $C(X, Y)$. ❑

REFERENCES

1. CAO, J. 1994. Quasi-uniform convergence for multifunctions. J.Math.Research and Exposition. **14:** 327–331.
2. CSÁSZÁR, A. 1981. Extensions of quasi-uniformities. Acta Math. Hungar. **37:** 121–145.
3. DEÁK, J. 1995. A bitopological view of quasi-uniform completeness I, II, III. Studia Sci. Math. Hungar. **30:** (1) 389–409; **30:** (2) 411–431; (3), to appear.
4. DOITCHINOV, D. 1988. On completeness of quasi-uniform spaces. C.R. Acad. Bulg. Sci. **41:** 5–8.
5. FLETCHER, P. & W. HUNSAKER. 1992. Completeness using pairs of filters. Topol. Appl. **44:** 149–155.
6. FLETCHER, P. & W.F. LINDGREN. 1982. Quasi-Uniform Spaces. Lecture Notes in Pure and Appl. Math. **77**. Marcel Dekker, New York, 1982.
7. KOPPERMAN, R.D. 1990. Total boundedness and compactness for filter pairs. Ann. Univ. Sci. Budapest. **33:** 25–30.
8. KÜNZI, H.K.P.A. 1995. On quasi-uniform convergence and quiet spaces,. Q & A in General Topology. **13:** 87–92.
9. KÜNZI, H.K.P.A. & S. Romaguera. 199?. Spaces of continuous functions and quasi-uniform convergence. Preprint.
10. PAPADOPOULOS, B.K. 1993. Ascoli's theorem in a quiet quasi-uniform space. Panamer. Math. J. **3:** 19–22.
11. PAPADOPOULOS, B.K. 1994. Quasi-uniform convergence on function spaces. Q & A in General Topology. **12:** 121–131.

12. PAPADOPOULOS, B.K. 1995. A note on the paper "Quasi-uniform convergence on function spaces." Q & A in General Topology. **13:** 55–56.
13. RENDER, H. Generalized uniform spaces and applications to function spaces. Preprint.
14. ROMAGUERA, S. 1995. An example on quasi-uniform convergence and quiet spaces. Q & A in General Topology **13:** 169–171.
15. ROMAGUERA, S. On hereditary precompactness and completeness in quasi-uniform spaces. Acta Math. Hungar. To appear.
16. ROMAGUERA, S. & M. Ruiz-Gómez. 1995. Bitopologies and quasi-uniformities on spaces of continuous functions. Publ. Math. Debrecen. **47:** 81–93.

On Transition Systems and Non-well-founded Sets

R.S. LAZIĆ AND A.W. ROSCOE

Oxford University Computing Laboratory
Wolfson Building, Parks Road
Oxford OX1 3QD, UK

ABSTRACT: (Labelled) transition systems are relatively common in theoretical computer science, chiefly as vehicles for operational semantics. The first part of this paper constructs a hierarchy of canonical transition systems and associated maps, aiming to give a strongly extensional theory of transition systems, where any two points with equivalent behaviors are identified. The cornerstone of the development is a notion of convergence in arbitrary transition systems, generalizing the idea of finite (n-step) approximations to a given point. In particular, our canonical transition systems are also uniform spaces.

The resulting hierarchy has very rich combinatorial (and topological) structure, and a lot of the first part of the paper is devoted to its study. We also discuss fixed points in this framework.

In the second part of the paper, we show how to obtain a model of set theory with Aczel's Anti-Foundation Axiom (AFA) from canonical transition systems constructed earlier. We study further the structure of the model thus obtained, and also give a few more abstract results, concerning consistency and independence in the presence of AFA.

1. INTRODUCTION

(Labelled) transition systems are relatively common in theoretical computer science, chiefly as vehicles for operational semantics. The first part of this paper, which grew out of some work done in [13, 215–230] and some more recent work by the same author on unbounded nondeterminism in CSP, constructs a hierarchy of canonical transition systems and associated maps, aiming to give a strongly extensional theory of transition systems, where any two points with equivalent behaviors are identified. The cornerstone of the development is a notion of convergence in arbitrary transition systems, generalizing the idea of finite (n-step) approximations to a given point. In particular, our canonical transition systems are also uniform spaces.

The resulting hierarchy has very rich combinatorial (and topological) structure, and a lot of the first part of the paper is devoted to its study. We also discuss fixed points in this framework.

Mathematics Subject Classification: 68Q90 (also, 04A99, 54E15).

Key words and phrases: transition systems, bisimulation, fixed points, non-well-founded set theory, uniform spaces.

[a]The authors gratefully acknowledge that the work reported in this paper was supported by a grant from Hajrija & Boris Vukobrat and Copechim France S.A. (to R.S. Lazić) and ONR grant N00014-87-G-0242 (to A.W. Roscoe).

Aczel (among others) observed that this kind of study of transition systems is very closely connected to non-well-founded set theory. Based on Milner's work on operational semantics of Synchronous CCS, in [1] he gave a quotient construction of a model of set theory with the Axiom of Foundation replaced by an Anti-Foundation Axiom (*AFA*).

In the second part of this paper, we show how to obtain a model of set theory with *AFA* from canonical transition systems constructed earlier. This gives the model a rich structure, which we then study further, building on the work in the first part. We also give a few more abstract results, concerning consistency and independence in the presence of *AFA*.

Non-well-founded set theory has been worked on long before [1]. In particular, *AFA* was probably first introduced as X_1 by Forti and Honsell in [5], which investigates a number of axioms derived from a Free Construction Principle. In subsequent papers (especially in [7] and [8]), they study structures which correspond to our canonical transition systems at regular cardinal heights (and for a singleton alphabet). They regard them primarily as quotients of a universe satisfying *AFA*.

Working in set theory with *AFA*, Aczel and Mendler obtain a Terminal Coalgebra Theorem (see Chapter 7 of [1], and [2]), which can be used to obtain spaces for semantics of process algebras based on the idea that $[\![P]\!] = \{\langle\delta, [\![P']\!]\rangle \mid P \xrightarrow{\delta} P'\}$ (see Chapter 8 of [1], and, e.g., [14]). By generalizing the structures from [7] and [8], Forti, Honsell, and Lenisa arrive at hyperuniverses, which are models of a strong Comprehension Scheme with topological structure, and can be similarly used for denotational semantics (see [9]).

2. TRANSITION SYSTEMS, MORPHISMS, AND BISIMULATIONS

We work within ZFC^-, i.e., Zermelo–Fraenkel set theory with the Axiom of Choice, but without the Axiom of Foundation. We drop the Axiom of Foundation, because we will sometimes want to adopt an axiom which contradicts it as one of our basic axioms. We follow the usual use of proper classes, i.e., classes which are not sets.[1]

We fix a set $\Sigma^+ = \Sigma \cup \{\tau\}$ of events, where $\tau \notin \Sigma$.[2] This will be an implicit parameter of almost everything we do from now on.

DEFINITION 1: A *transition system* is a class S together with a family of binary relations $\xrightarrow{\delta}$ on S indexed by Σ^+, such that $a_{S,\delta} = \{b \in S \mid a \xrightarrow{\delta} b\}$ is a set for all $a \in S$, $\delta \in \Sigma^+$. A transition system is *small* iff its underlying class is a set.

[1]The reader is referred to Chapter 1 of [10].

[2]In the usual terminology of process algebras, τ is an internal (invisible) event. However, its purpose in this paper is merely to ensure that Σ^+ is nonempty.

If S is a transition system and $S' \subseteq S$, then S' is a *subsystem* of S iff $a \in S' \wedge b \in S \wedge a \xrightarrow{\delta} b \Rightarrow b \in S'$, i.e., iff $a_{S',\delta} = a_{S,\delta}$ for all $a \in S'$, $\delta \in \Sigma^+$.

An *accessible pointed system* (*aps*) is a transition system S with a designated *point* $a \in S$ such that, given any $b \in S$, there is a finite sequence of transitions $a \xrightarrow{\delta_0} \ldots \xrightarrow{\delta_{n-1}} b$. Given a transition system S and any $a \in S$, let S a be the aps whose point is a and which consists of all $b \in S$ reachable from a in a finite number of transitions.

It is easily seen that any aps must be a small transition system.

DEFINITION 2: Given any transition systems S, S', a map $\mathcal{F}$: $S \rightarrow S'$ is a *morphism* iff:

$$a \xrightarrow{\delta} b \Rightarrow \mathcal{F}(a) \xrightarrow{\delta} \mathcal{F}(b),$$

and

$$\mathcal{F}(a) \xrightarrow{\delta} b' \Rightarrow \exists b \,.\, a \xrightarrow{\delta} b \wedge \mathcal{F}(b) = b'.$$

Alternatively, that is equivalent to saying that $\mathcal{F}(a_{S,\delta}) = (\mathcal{F}(a))_{S',\delta}$ for all $a \in S$, $\delta \in \Sigma^+$. The idea is that a point and its image under a morphism cannot be told apart by an experimentor who can only observe the passing of events (both internal and external). A closely linked notion is that of a binary relation which relates pairs of points with the same behaviors (in the same sense).

DEFINITION 3: A binary relation R on a transition system S is a *bisimulation* iff:

$$aRb \wedge a \xrightarrow{\delta} a' \Rightarrow \exists b' \,.\, b \xrightarrow{\delta} b' \wedge a'Rb',$$

and

$$aRb \wedge b \xrightarrow{\delta} b' \Rightarrow \exists a' \,.\, a \xrightarrow{\delta} a' \wedge a'Rb'.$$[3]

A maximum bisimulation exists on any transition system S. It is given by the union of all small bisimulations[4] on S, and it is an equivalence relation. If S is small, the maximum bisimulation on S is also given by the set of all pairs $\langle a, b\rangle$ such that there is a morphism $\mathcal{F}$ with domain S with $\mathcal{F}(a) = \mathcal{F}(b)$.

DEFINITION 4: A transition system S is *strongly extensional* iff the maximum bisimulation on S is the identity relation (i.e., the diagonal) on S.[5]

[3]These have sometimes been called "strong bisimulations," because they treat internal and external events in the same way.

[4]A bisimulation is small iff it is a set.

[5]A transition system S is sometimes said to be "weakly extensional" iff $a = b$ whenever $a_{S,\delta} = b_{S,\delta}$ for all $\delta \in \Sigma^+$.

LEMMA 1:
(a) Given a transition system S, there exists a strongly extensional transition system $\widetilde{S}$ (its quotient) and a unique surjective morphism $\mathcal{F}$: $S \rightarrow \widetilde{S}$.
(b) If $\mathcal{G}$: $S \rightarrow S'$ is any surjective morphism, then there is a unique surjective morphism $\mathcal{H}$: $S' \rightarrow \widetilde{S}$ such that $\mathcal{H} \circ \mathcal{G} = \mathcal{F}$.

In (a), if S is a proper class, the formalization becomes nontrivial if we want to avoid using a stronger choice principle, since we do not assume the Axiom of Foundation. For any equivalence class C, we need to consider the class of all aps's on well-founded sets which are isomorphic to Sa for a point $a \in C$, and then take its subset consisting of all of its elements which have minimal rank as the representation for C (see the proof of Lemma 2.17 in [1]).

LEMMA 2: For a transition system S, the following are equivalent:
(a) S is strongly extensional.
(b) For any small transition system S', there is at most one morphism from S' into S.
(c) For any transition system S', there is at most one morphism from S' into S.
(d) Any morphism with domain S is injective.

For more results and examples about morphisms, the reader can look at [13, pp. 215–230]. It uses "." instead of "τ," and deals only with small transition systems, which it calls "P,Q-spaces." Lemma 1 for small transition systems is proved there. Milner [11], as well as Aczel [1] (which proves Lemma 2 for $\Sigma = \varnothing$ — see Theorem 2.19 there), have more about bisimulations.

3. THE SPACES OF CANONICAL APPROXIMATIONS

From now on, α, β, γ, ζ, η, θ, ν, and ξ (and their variations) will always denote ordinals. γ will always be a limit ordinal. κ and λ will always be cardinals.

We also abbreviate "transition system" simply to "system" in the future. In this section and the next, we assume that all systems are small (i.e., "system" will mean "small system").

Given a point a in a system S, we can think of all the sequences of transitions of length at most n that a can perform as determining an n-step approximation to a. We generalize this idea as follows.

DEFINITION 5: Given a system S, we define the following maps on S by transfinite recursion starting from 1:

$$\mathcal{H}_1^S(a) = \varnothing.$$

$$\mathcal{H}^S_{\alpha+1}(a) = \{\langle \delta, \mathcal{H}^S_\alpha(b)\rangle \mid a \xrightarrow{\delta} b\}.$$

$$\mathcal{H}^S_\gamma(a) = \langle \mathcal{H}^S_\alpha(a) \mid 0 < \alpha < \gamma\rangle.$$

LEMMA 3: If $\mathcal{F}\colon S \rightarrow S'$ is a morphism, then $\mathcal{H}^{S'}_\alpha \circ \mathcal{F} = \mathcal{H}^S_\alpha$ for all $\alpha > 0$.

Proof: The proof is by transfinite induction on α. The base case and the limit case are trivial. For the successor case, suppose $\mathcal{H}^{S'}_\alpha \circ \mathcal{F} = \mathcal{H}^S_\alpha$ for some $\alpha > 0$. Then, for any $a \in S$, we have:

$$\begin{aligned}\mathcal{H}^{S'}_{\alpha+1}(\mathcal{F}(a)) &= \bigcup_{\delta \in \Sigma^+} \{\langle \delta, \mathcal{H}^{S'}_\alpha(b')\rangle \mid \mathcal{F}(a) \xrightarrow{\delta} b'\} \\ &= \bigcup_{\delta \in \Sigma^+} \{\langle \delta, \mathcal{H}^{S'}_\alpha(\mathcal{F}(b))\rangle \mid a \xrightarrow{\delta} b\} \\ &= \bigcup_{\delta \in \Sigma^+} \{\langle \delta, \mathcal{H}^{S}_\alpha(b)\rangle \mid a \xrightarrow{\delta} b\} = \mathcal{H}^S_{\alpha+1}(a). \quad \square\end{aligned}$$

Since a subset S' of a system S is a subsystem of S iff the identity map from S' into S is a morphism, an immediate corollary of Lemma 3 is that, if S' is a subsystem of S, then $\mathcal{H}^S_\alpha \upharpoonright S' = \mathcal{H}^{S'}_\alpha$ for all $\alpha > 0$.

We can now define the spaces of canonical approximations as the ranges of the $\mathcal{H}^S_\alpha$ maps.

DEFINITION 6: For any $\alpha > 0$, let

$$\mathcal{T}_\alpha = \{\mathcal{H}^S_\alpha(a) \mid a \text{ is a point in a system } S\}.^6$$

We make each $\mathcal{T}_\alpha$ into a system as follows:

$$\varnothing \in \mathcal{T}_1 \text{ has no transitions.}$$

$$a, b \in \mathcal{T}_{\alpha+1} \Rightarrow (a \xrightarrow{\delta} b \Leftrightarrow \exists b'.\ \langle \delta, b'\rangle \in a \wedge b = \mathcal{H}^{\mathcal{T}_\alpha}_{\alpha+1}(b')).$$

$$\underline{a}, \underline{b} \in \mathcal{T}_\gamma \Rightarrow (\underline{a} \xrightarrow{\delta} \underline{b} \Leftrightarrow \forall 0 < \alpha < \gamma\ .\ \langle \delta, b_\alpha\rangle \in a_{\alpha+1}).^7$$

EXAMPLE 1: Let $S = \{a_n \mid n < \omega\} \cup \{a^*\}$ be a system whose transitions are given by $a_{n+1} \xrightarrow{\tau} a_n$ and $a^* \xrightarrow{\tau} a_n$ for all $n < \omega$. Then, for any nonzero $m < \omega$, we have:

$$\mathcal{H}^S_m(a_n) = \overbrace{\{\langle \tau, \ldots \{\langle \tau, \varnothing\rangle\} \ldots\}}^{n \cap (m-1)} \quad (n < \omega)$$

$$\mathcal{H}^S_1(a^*) = \varnothing$$

[6] Since $\mathcal{T}_1 = \{\varnothing\}$, $\mathcal{T}_{\alpha+1} \subseteq \mathcal{P}(\Sigma^+ \times \mathcal{T}_\alpha)$ and $\mathcal{T}_\gamma \subseteq (\bigcup_{0<\alpha<\gamma}\mathcal{T}_\alpha)^{\gamma\setminus\{0\}}$, it is easy to see (by transfinite induction on α) that $\mathcal{T}_\alpha$ is a set for all $\alpha > 0$.

[7] We use $\underline{a}$ (etc.) as an abbreviation for the sequence $\langle a_\alpha \mid 0 < \alpha < \gamma\rangle$.

$$\mathcal{H}^S_{m+1}(a^*) = \{\langle\tau, \mathcal{H}^S_m(a_n)\rangle \mid n < \omega\} = \{\langle\tau, \mathcal{H}^S_m(a_n)\rangle \mid n \leq m-1\}.$$

Hence, for any $n < \omega$, we have

$$\mathcal{H}^S_\omega(a_{n+1}) = \langle\varnothing\rangle^\frown\langle\mathcal{H}^S_{m+1}(a_{n+1}) \mid 0 < m < \omega\rangle =$$
$$\langle\varnothing\rangle^\frown\langle\{\langle\tau, \mathcal{H}^S_m(a_n)\rangle\} \mid 0 < m < \omega\rangle,$$

so that

$$(\mathcal{H}^S_\omega(a_{n+1}))\mathcal{T}_{\omega},\tau = \{\mathcal{H}^S_\omega(a_n)\} \quad (\text{and } (\mathcal{H}^S_\omega(a_0))\mathcal{T}_{\omega},\tau = \varnothing).$$

Also, we have

$$\begin{aligned}\mathcal{H}^S_\omega(a^*) &= \langle\varnothing\rangle^\frown\langle\mathcal{H}^S_{m+1}(a^*) \mid 0 < m < \omega\rangle \\ &= \langle\varnothing\rangle^\frown\langle\{\langle\tau, \mathcal{H}^S_m(a_n)\rangle \mid n \leq m-1\} \mid 0 < m < \omega\rangle.\end{aligned}$$

Now, if $0 < m < \omega$, $n \leq m-1$ and $n' \leq m$, then $\mathcal{H}^S_m(a_n) = \mathcal{H}^{\mathcal{T}_{m+1}}_m(\mathcal{H}^S_{m+1}(a_{n'}))$ iff either $n = n'$ or $n = m-1$ and $n' = m$. Hence, it follows that $(\mathcal{H}^S_\omega(a^*))\mathcal{T}_{\omega},\tau$ $= \{\mathcal{H}^S_\omega(a_n) \mid n < \omega\} \cup \{\underline{b}\}$, where $\underline{b} = \langle\mathcal{H}^S_m(a_{m-1}) \mid 0 < m < \omega\rangle$ is such that $\underline{b}\mathcal{T}_{\omega},\tau = \{\underline{b}\}$.

Since $\mathcal{H}^S_\omega$ is a morphism on the subsystem $\{a_n \mid n < \omega\}$ of S, it is not difficult to see using Lemma 3 that $\mathcal{H}^S_{\omega+1}$ is a morphism (on S).

LEMMA 4:

(a) For any $\alpha > 0$, we have $\mathcal{H}^{\mathcal{T}_\alpha}_\alpha(a) = a$ for all $a \in \mathcal{T}_\alpha$.

(b) If $0 < \beta \leq \alpha$, then $\mathcal{H}^{\mathcal{T}_\alpha}_\beta \circ \mathcal{H}^S_\alpha = \mathcal{H}^S_\beta$ for any system S.

Proof: We prove (a) and (b) simultaneously by transfinite induction on α.

Base case: Trivial.

Successor case: Suppose (a) and (b) hold for some $\alpha > 0$. For (b), we argue by transfinite induction on nonzero $\beta \leq \alpha + 1$. The base case and the limit case are trivial, so suppose $\mathcal{H}^{\mathcal{T}_{\alpha+1}}_\beta \circ \mathcal{H}^S_{\alpha+1} = \mathcal{H}^S_\beta$ for some $0 < \beta < \alpha + 1$ and all systems S. Then, for any a in a system S, we have:

$$\begin{aligned}\mathcal{H}^{\mathcal{T}_{\alpha+1}}_{\beta+1}(\mathcal{H}^S_{\alpha+1}(a)) &= \{\langle\delta, \mathcal{H}^{\mathcal{T}_{\alpha+1}}_\beta(b)\rangle \mid \mathcal{H}^S_{\alpha+1}(a) \xrightarrow{\delta} b\} \\ &= \{\langle\delta, \mathcal{H}^{\mathcal{T}_{\alpha+1}}_\beta(\mathcal{H}^{\mathcal{T}_\alpha}_{\alpha+1}(\mathcal{H}^S_\alpha(b')))\rangle \mid a \xrightarrow{\delta} b'\} \\ &= \{\langle\delta, \mathcal{H}^{\mathcal{T}_\alpha}_\beta(\mathcal{H}^S_\alpha(b'))\rangle \mid a \xrightarrow{\delta} b'\} = \mathcal{H}^S_{\beta+1}(a).\end{aligned}$$

For (a), if $a \in \mathcal{T}_{\alpha+1}$, we have:

$$\begin{aligned}\mathcal{H}^{\mathcal{T}_{\alpha+1}}_{\alpha+1}(a) &= \{\langle\delta, \mathcal{H}^{\mathcal{T}_{\alpha+1}}_\alpha(b)\rangle \mid a \xrightarrow{\delta} b\} \\ &= \{\langle\delta, \mathcal{H}^{\mathcal{T}_{\alpha+1}}_\alpha(\mathcal{H}^{\mathcal{T}_\alpha}_{\alpha+1}(b'))\rangle \mid \langle\delta, b'\rangle \in a\} \\ &= \{\langle\delta, \mathcal{H}^{\mathcal{T}_\alpha}_\alpha(b')\rangle \mid \langle\delta, b'\rangle \in a\} \\ &= a.\end{aligned}$$

Limit case: Suppose (a) and (b) hold for all nonzero $\alpha < \gamma$. We prove that $\mathcal{H}_\alpha^{\mathcal{T}_\gamma}(\underline{a}) = a_\alpha$ for all $\underline{a} \in \mathcal{T}_\gamma$, $0 < \alpha < \gamma$ by transfinite induction on α.

(i) The base case is trivial.

(ii) Suppose the claim holds for some nonzero $\alpha < \gamma$, and consider some $\underline{a} \in \mathcal{T}_\gamma$. Then:

$$\mathcal{H}_{\alpha+1}^{\mathcal{T}_\gamma}(\underline{a}) = \{\langle\delta, \mathcal{H}_\alpha^{\mathcal{T}_\gamma}(\underline{b})\rangle \,|\, \underline{a} \xrightarrow{\delta} \underline{b}\} = \{\langle\delta, b_\alpha\rangle \,|\, \underline{a} \xrightarrow{\delta} \underline{b}\}.$$

Now $\langle\delta, b_\alpha\rangle \in a_{\alpha+1}$ whenever $\underline{a} \xrightarrow{\delta} \underline{b}$, so it suffices to show that, if $\langle\delta, b^*\rangle \in a_{\alpha+1}$, then $b^* = b_\alpha$ for some $\underline{b} \in \underline{a}_{\mathcal{T}_\gamma, \delta}$. But we know that $\underline{a} = \mathcal{H}_\gamma^S(a')$ for some point a' in a system S, so if $\langle\delta, b^*\rangle \in a_{\alpha+1} = \mathcal{H}_{\alpha+1}^S(a')$, then $b^* = \mathcal{H}_\alpha^S(b')$ for some $b' \in a'_{S,\delta}$, which gives us what we want since $\underline{a} = \mathcal{H}_\gamma^S(a') \xrightarrow{\delta} \mathcal{H}_\gamma^S(b')$ (just observe that $\forall 0 < \beta < \gamma \,.\, \langle\delta, \mathcal{H}_\beta^S(b')\rangle \in \mathcal{H}_{\beta+1}^S(a')$).

(iii) Suppose the claim holds for all nonzero $\alpha < \gamma'$, where $\gamma' < \gamma$. If $\underline{a} \in \mathcal{T}_\gamma$, then:

$$\mathcal{H}_{\gamma'}^{\mathcal{T}_\gamma}(\underline{a}) = \langle \mathcal{H}_\alpha^{\mathcal{T}_\gamma}(\underline{a}) \mid 0 < \alpha < \gamma'\rangle = \langle a_\alpha \,|\, 0 < \alpha < \gamma'\rangle = a_{\gamma'}.$$

Now that the claim is established, (a) and (b) for γ in place of α follow at once. ❑

If A is any subset of $\Sigma^+ \times \mathcal{T}_\alpha$, consider a system $S = \{a\} \cup \mathcal{T}_\alpha$ (where $a \notin \mathcal{T}_\alpha$) which inherits the transitions on $\mathcal{T}_\alpha$ from $\mathcal{T}_\alpha$ and such that $a_{S,\delta} = \{b \,|\, \langle\delta, b\rangle \in A\}$. Then $\mathcal{H}_{\alpha+1}^S(a) = \{\langle\delta, \mathcal{H}_\alpha^S(b)\rangle \,|\, \langle\delta, b\rangle \in A\} = A$ by the remark after Lemma 3 and by (a) of Lemma 4. Therefore, for any $\alpha > 0$, we have

$$\mathcal{T}_{\alpha+1} = \mathcal{P}(\Sigma^+ \times \mathcal{T}_\alpha).$$

Whenever $\underline{a} \in \mathcal{T}_\gamma$, (b) of Lemma 4 gives us

$$0 < \beta \leq \alpha < \gamma \Rightarrow a_\beta = \mathcal{H}_\beta^{\mathcal{T}_\alpha}(a_\alpha),$$

so that any $\mathcal{T}_\gamma$ is a subset of the inverse limit $\mathcal{T}_\gamma^0$ of the sets $\mathcal{T}_\alpha$ $(0 < \alpha < \gamma)$ with respect to the maps $\mathcal{H}_\beta^{\mathcal{T}_\alpha}\colon \mathcal{T}_\alpha \to \mathcal{T}_\beta (0 < \beta \leq \alpha < \gamma)$. Transitions on $\mathcal{T}_\gamma^0$ (and any $S \subseteq \mathcal{T}_\gamma^0$) are defined as in Definition 6.

If $\mathcal{F}\colon \mathcal{T}_\alpha \to S$ is any morphism, then $\mathcal{H}_\alpha^S \circ \mathcal{F} = \mathcal{H}_\alpha^{\mathcal{T}_\alpha}$ is the identity map on $\mathcal{T}_\alpha$, and so $\mathcal{F}$ is injective. The following theorem now follows from Lemma 2.

THEOREM 5: For any $\alpha > 0$, $\mathcal{T}_\alpha$ is strongly extensional.

Given any $S \subseteq \mathcal{T}_\gamma^0$, let

$$\Phi(S) = \{\underline{a} \in S \,|\, 0 < \alpha < \gamma \wedge \langle\delta, b^*\rangle \in a_{\alpha+1} \Rightarrow \exists \underline{b} \in \underline{a}_{S,\delta} \,.\, b^* = b_\alpha\}.$$

Then Φ is a monotonic map on the complete lattice $\langle\mathcal{P}(\mathcal{T}_\gamma^0), \subseteq\rangle$, and we have the following result.[8]

THEOREM 6: $\mathcal{T}_\gamma$ is the greatest fixed point of Φ.

Proof: We have already seen in (ii) in the proof of Lemma 4 that $\mathcal{T}_\gamma$ is a fixed point of Φ.

Suppose $S \subseteq \mathcal{T}_\gamma^0$ is a fixed point of Φ. Then it follows as in the limit case of the same proof that $\mathcal{H}_\gamma^S(\underline{a}) = \underline{a}$ for all $\underline{a} \in S$, and hence, $S \subseteq \mathcal{T}_\gamma$. ❑

We now know by Knaster–Tarski Theorem that

$$\mathcal{T}_\gamma = \bigcup\{S \subseteq \mathcal{T}_\gamma^0 \mid S \subseteq \Phi(S)\}.$$

Alternatively, any complete lattice with its order reversed is a complete partial order. Hence, if we define by transfinite recursion:

$$\mathcal{T}_\gamma^{\alpha+1} = \Phi(\mathcal{T}_\gamma^\alpha),$$

$$\mathcal{T}_\gamma^{\gamma'} = \bigcap_{\alpha<\gamma'} \mathcal{T}_\gamma^\alpha,$$

then the $\mathcal{T}_\gamma^\alpha$ form a decreasing chain of subsets of $\mathcal{T}_\gamma^0$ which becomes constant at some $\alpha^* < |\mathcal{T}_\gamma^0|^+$,[9] and we have that $\mathcal{T}_\gamma = \mathcal{T}_\gamma^{\alpha^*}$.

The definition of $\mathcal{T}_\gamma$ we gave can in principle be replaced by either of these, which are in a sense "more direct."

We continue this kind of study of $\mathcal{T}_\gamma^0$ after we establish a few results of a different kind.

From now on, we will usually omit the superscript in $\mathcal{H}_\alpha^S$, provided that does not introduce ambiguity which is not covered by Lemma 3.

Given a system S, the decreasing chain of sets

$$U_\alpha = \{\langle a, b\rangle \in S \times S \mid \mathcal{H}_\alpha(a) = \mathcal{H}_\alpha(b)\}$$

for $0 < \alpha < \gamma$ forms a fundamental system of entourages of a uniformity $\mathcal{U}_\gamma^S$ on S.[10] In this way, for any γ, the maps $\mathcal{H}_\alpha$ $(0 < \alpha < \gamma)$ give rise to a notion of convergence of points in an arbitrary system S.

Let $\mathcal{V}_\gamma$ be the uniformity on $\mathcal{T}_\gamma^0$, the inverse limit of the discrete uniformities on $\mathcal{T}_\alpha$ $(0 < \alpha < \gamma)$. Then $\mathcal{U}_\gamma^{\mathcal{T}_\gamma}$ is the subspace uniformity on $\mathcal{T}_\gamma$ induced by $\mathcal{V}_\gamma$.

Whenever $\mathcal{F}: S \to S'$ is a morphism, $\mathcal{U}_\gamma^S$ is by Lemma 3 the inverse image under $\mathcal{F}$ of $\mathcal{U}_\gamma^{\mathcal{F}(S)}$ (which is the subspace uniformity on $\mathcal{F}(S)$ induced by $\mathcal{U}_\gamma^{S'}$, since $\mathcal{F}(S)$ is a subsystem of S'). In particular, given any system S, the

[8]For partial orders, we refer the reader to [4], where Chapter 4 concentrates on fixed points.

[9]Since $|\mathcal{T}_\gamma^0| \geq \omega$ (see, e.g., Corollary 8), $|\mathcal{T}_\gamma^0|^+$ is an infinite regular cardinal. (Suppose that κ is infinite and regular, and that X_α for $\alpha < \kappa$ form a decreasing sequence of sets such that $\forall\alpha < \kappa \,.\, |X_\alpha| < \kappa$. For any $x \in X_0' = X_0 \backslash \bigcap_{\alpha<\kappa} X_\alpha$, let $\alpha^x < \kappa$ be minimal such that $x \notin X_{\alpha^x}$. Then, letting $\alpha^* = \bigcup_{x \in X_0'} \alpha^x$, we have that the X_α for $\alpha^* \leq \alpha < \kappa$ are all identical.)

[10]For both uniform spaces and inverse limits, we refer the reader to [3]

uniformity $\mathcal{U}_\gamma^S$ on it is an inverse image of the uniformity $\mathcal{U}_\gamma^{\tilde{S}}$ on its strongly extensional quotient. ❑

We now turn to the question of when are the maps $\mathcal{H}_\alpha^S\colon S \to \mathcal{T}_\alpha$ morphisms.

THEOREM 7: If $\mathcal{H}_\alpha$ is a morphism on a system S and $\beta \geq \alpha$, then $\mathcal{H}_\beta$ is also a morphism on S.

Proof: The proof is by transfinite induction on $\beta \geq \alpha$. ❑

COROLLARY 8: If $0 < \alpha \leq \beta$, then $\mathcal{H}_\beta$ is a morphism on $\mathcal{T}_\alpha$.

Proof: $\mathcal{H}_\alpha$ is a morphism on $\mathcal{T}_\alpha$ by Lemma 4(a). ❑

DEFINITION 7: Given a system S, let $i(S)$, the *index of nondeterminism* of S, be the smallest infinite regular cardinal which is strictly greater than $|a_{S,\delta}|$ for all $a \in S$, $\delta \in \Sigma^+$.

THEOREM 9: Suppose S is any system.
(a) $\mathcal{H}_{i(S)}$ is a morphism on S.
(b) If S is countable, then $\mathcal{H}_\alpha$ is a morphism on S for some countable α.

Proof: (a) Observe first that $i(S)$ is a limit ordinal, and hence,

$$a \xrightarrow{\delta} b \Rightarrow \mathcal{H}_{i(S)}(a) \xrightarrow{\delta} \mathcal{H}_{i(S)}(b).$$

So suppose $\mathcal{H}_{i(S)}(a) \xrightarrow{\delta} \underline{c}$. For any $0 < \alpha < i(S)$, let

$$X_\alpha = \{b \in a_{S,\delta} \mid \mathcal{H}_\alpha(b) = c_\alpha\}.$$

Now $0 < \alpha < i(S) \Rightarrow \langle\delta, c_\alpha\rangle \in \mathcal{H}_{\alpha+1}(a)$, so each X_α is nonempty. Also, if $\alpha \leq \beta$ and $b \in X_\beta$, then $\mathcal{H}_\alpha(b) = \mathcal{H}_\alpha(\mathcal{H}_\beta(b)) = \mathcal{H}_\alpha(c_\beta) = c_\alpha$, so that $b \in X_\alpha$. Hence, the X_α form a decreasing chain and so, since $|X_\alpha| \leq |a_{S,\delta}| < i(S)$ for all α, we can pick a $b \in \bigcap_{0<\alpha<i(S)} X_\alpha$, for which we will have $\underline{c} = \mathcal{H}_{i(S)}(b)$.
(b) Proved similarly, by considering the sets

$$Y_\alpha^a = \{b \in S \mid \mathcal{H}_\alpha(b) = \mathcal{H}_\alpha(a)\}$$

for $a \in S$, $0 < \alpha < \omega_1$. ❑

If $\mathcal{H}_\gamma$ is a morphism on a strongly extensional system S, we know by Lemma 2 that $\mathcal{H}_\gamma$ is injective on S, and so S is isomorphic to $\mathcal{H}_\gamma(S)$ which is a subsystem of $\mathcal{T}_\gamma$. Also, the uniformity $\mathcal{U}_\gamma^S$ is isomorphic to the subspace uniformity on $\mathcal{H}_\gamma(S)$ induced by $\mathcal{U}_\gamma^{\mathcal{T}_\gamma}$. In this sense, the $\mathcal{T}_\alpha$ are canonical systems, and the $\mathcal{U}_\gamma^{\mathcal{T}_\gamma}$ are canonical uniformities.

From now on, unless stated otherwise, we assume that, for any γ, the uniformity on $\mathcal{T}_\gamma$ is $\mathcal{U}_\gamma^{\mathcal{T}_\gamma}$, and the uniformity on $\mathcal{T}_\gamma^0$ is $\mathcal{V}_\gamma$.

If $\underline{a} \in \mathcal{T}^0_\gamma$ and $\delta \in \Sigma^+$, then $\underline{a}_{\mathcal{T}^0_\gamma, \delta}$ consists of all sequences in the inverse limit of the sets $\{b \mid \langle \delta, b \rangle \in a_{\alpha+1}\}$ (with respect to the maps $\mathcal{H}^{\mathcal{T}_\beta}_\alpha$, for $0 < \alpha \le \beta < \gamma$). If $\underline{a} \in \mathcal{T}_\gamma$, then $\underline{a}_{\mathcal{T}_\gamma, \delta}$ is the set of all sequences from the same inverse limit which are also in $\mathcal{T}_\gamma$.

LEMMA 10: Suppose a is a point in a system S and $\delta \in \Sigma^+$.

(a) $(\mathcal{H}_\gamma(a))_{\mathcal{T}^0_\gamma, \delta} = \overline{\mathcal{H}_\gamma(a_{S,\delta})}^{\mathcal{T}^0_\gamma}$.

(b) $(\mathcal{H}_\gamma(a))_{\mathcal{T}_\gamma, \delta} = \overline{\mathcal{H}_\gamma(a_{S,\delta})}^{\mathcal{T}_\gamma}$.

Proof: By the remarks above, it suffices to observe that

$$\{b_\alpha \mid \underline{b} \in \mathcal{H}_\gamma(a_{S,\delta})\} = \mathcal{H}_\alpha(a_{S,\delta}) = \{b' \mid \langle \delta, b' \rangle \in a_{\alpha+1}\}$$

for all $0 < \alpha < \gamma$.[11] ❑

THEOREM 11:

(a) If $X_\delta \subseteq \mathcal{T}^0_\gamma$ for each $\delta \in \Sigma^+$, then there exists an $\underline{a} \in \mathcal{T}^0_\gamma$ such that $\forall \delta \in \Sigma^+ . \, a_{\mathcal{T}^0_\gamma, \delta} = X_\delta$ iff each X_δ is closed.

(b) If $X_\delta \subseteq \mathcal{T}_\gamma$ for each $\delta \in \Sigma^+$, then there exists an $\underline{a} \in \mathcal{T}_\gamma$ such that $\forall \delta \in \Sigma^+ . \, a_{\mathcal{T}_\gamma, \delta} = X_\delta$ iff each X_δ is closed in $\mathcal{T}_\gamma$.

The strong extensionality of $\mathcal{T}_\gamma$ gives us that the correspondence between points in $\mathcal{T}_\gamma$ and Σ^+-tuples of closed subsets of $\mathcal{T}_\gamma$ in (*b*) of Theorem 11 is 1–1.

In their setting, Forti and Honsell take these ideas further, resulting in their theory of comprehension properties of hyperuniverses (see [7]).

Suppose $\kappa < cf(\gamma)$ and Y_ξ for $\xi < \kappa$ are closed subsets of $\mathcal{T}^0_\gamma$. If $\underline{a} \notin \bigcup_{\xi<\kappa} Y_\xi$, then, for each $\xi < \kappa$, there exists a nonzero $\alpha_\xi < \gamma$ such that $\{\underline{b} \in \mathcal{T}^0_\gamma \mid b_{\alpha_\xi} = a_{\alpha_\xi}\} \cap Y_\xi = \varnothing$. Letting $\alpha^* = \bigcup_{\xi<\kappa} \alpha_\xi$, we have that $\{\underline{b} \in \mathcal{T}^0_\gamma \mid b_{\alpha^*} = a_{\alpha^*}\} \cap \bigcup_{\xi<\kappa} Y_\xi = \varnothing$. Hence, $\bigcup_{\xi<\kappa} Y_\xi$ is closed. We conclude that, in $\mathcal{T}^0_\gamma$ (and hence, in $\mathcal{T}_\gamma$), unions of strictly less than $cf(\gamma)$ many closed sets are closed.

In particular, the X_δ in Theorem 11 are closed whenever $|X_\delta| < cf(\gamma)$ for all $\delta \in \Sigma^+$.

DEFINITION 8: For any $0 < \alpha$, let $\mathcal{T}'_\alpha = \mathcal{T}_\alpha \setminus \bigcup_{0<\beta<\alpha} \mathcal{H}_\alpha(\mathcal{T}_\beta)$. For any γ, let $\mathcal{T}^c_\gamma = \{\mathcal{H}^{\mathcal{T}_\alpha}_\gamma(a) \mid 0 < \alpha < \gamma \wedge a \in \mathcal{T}_\alpha\}$.

It seems very plausible to conjecture that $0 < \alpha < \beta \wedge a \in \mathcal{T}'_\beta \Rightarrow \mathcal{H}_\alpha(a) \in \mathcal{T}'_\alpha$. However, that is false, essentially because the statement $a \in \mathcal{T}'_{\alpha+1} \Leftrightarrow \exists \delta \in \Sigma^+ . \, \exists b \in \mathcal{T}'_\alpha . \, \langle \delta, b \rangle \in a$ fails whenever α is a limit ordinal, which is easily seen by Theorem 11, once we observe that $\mathcal{T}'_\alpha = \mathcal{T}_\alpha \setminus \mathcal{T}^c_\alpha$ for such α. In particular, if we fix γ and pick $a \in \mathcal{T}'_{\gamma+1}$ (recall that $|\mathcal{T}_{\gamma+1}| = |\mathcal{P}(\Sigma^+ \times \mathcal{T}_\gamma)| \ge 2^{|\mathcal{T}_\gamma|} > |\mathcal{T}_\gamma|$), it is not difficult to see that, for any $n < \omega$, we have

[11] See Corollary to Proposition 9 of Chapter I, Section 4.4 in [3].

$$0 < m \leq n \Rightarrow \mathcal{H}_{\gamma+m}(\overbrace{\tau.\ldots.\tau}^{n}.a) = \mathcal{H}_{\gamma+m}(\overbrace{\tau.\ldots.\tau}^{n}.\mathcal{H}_{\gamma}(a)) \notin \mathcal{T}'_{\gamma+m},^{12}$$

in spite of the fact that

$$\mathcal{H}_{\gamma+n+1}(\overbrace{\tau.\ldots.\tau}^{n}.a) \in \mathcal{T}'_{\gamma+n+1}.$$

To construct a counter-example to the statement "if $\underline{a'} \in \mathcal{T}'_{\gamma'}$, then there is a nonzero $\alpha < \gamma'$ such that $\alpha \leq \beta < \gamma' \Rightarrow a_{\beta}' \in \mathcal{T}_{\beta}'$," first let $b_0 = a$ and $b_{n+1} = a + \tau.\tau.b_n$ for each $n < \omega$. Then we have

$$0 < m < \omega \Rightarrow (\mathcal{H}_{\gamma+m}(b_n) \in \mathcal{T}'_{\gamma+m} \Leftrightarrow (2 \nmid m \wedge m \leq 2n+1)).$$

Hence, it follows that

$$0 < m < \omega \Rightarrow (\mathcal{H}_{\gamma+m+1}\Big(\sum_{n<\omega} \tau.b_n\Big) \in \mathcal{T}'_{\gamma+m+1} \Leftrightarrow 2 \nmid m)^{13}$$

A counter-example to "if $\underline{a'} \in \mathcal{T}^c_{\gamma'}$, then $\underline{a'}$ is an isolated point of $\mathcal{T}'_{\gamma'}$ can be constructed as follows. For any $n < \omega$, let

$$c_n = (\overbrace{\tau.\ldots.\tau}^{n+1}.a) + \Big(\sum_{m \in \omega \setminus \{n\}} \overbrace{\tau.\ldots.\tau}^{m+1}.\mathcal{H}_{\gamma}(a)\Big),$$

and let

$$d = \sum_{m<\omega} \overbrace{\tau.\ldots.\tau}^{m+1}.\mathcal{H}_{\gamma}(a).$$

Then $\mathcal{H}_{\gamma+n+1}(c_n) = \mathcal{H}_{\gamma+n+1}(d)$, but $\mathcal{H}_{\gamma+n+2}(c_n) \in \mathcal{T}'_{\gamma+n+2}$, so that $\mathcal{H}_{\gamma+n+2}(c_n) \neq \mathcal{H}_{\gamma+n+2}(d)$ (since $\mathcal{H}_{\gamma+n+2}(d) = \mathcal{H}_{\gamma+n+2}(\mathcal{H}_{\gamma+1}(d))$). Hence, $\mathcal{H}_{\gamma+\omega}(d)$ is not an isolated point of $\mathcal{T}_{\gamma+\omega}$ (although $\mathcal{H}_{\gamma+\omega}(d) \in \mathcal{T}^c_{\gamma+\omega}$).

THEOREM 12: Given $\underline{a} \in \mathcal{T}^c_{\gamma^*}$, let α^* be the smallest ordinal such that $0 < \alpha^* < \gamma^*$ and $\alpha^* \leq \alpha < \gamma^* \Rightarrow a_{\alpha} \notin \mathcal{T}'_{\alpha}$.

(a) $\alpha^* = \alpha' + 1$ for some $\alpha' > 0$, and we have $\alpha' \leq \alpha < \gamma^* \Rightarrow a_{\alpha} = \mathcal{H}_{\alpha}(a_{\alpha'})$.

(b) $0 < n < \omega \wedge n < \alpha^* \Rightarrow a_n \in \mathcal{T}'_n$.

(c) $\gamma < \alpha^* \Rightarrow a_{\gamma} \in \mathcal{T}'_{\gamma}$.

(d) If $\gamma + \omega < \alpha^*$, then $a_{\gamma+n} \in \mathcal{T}'_{\gamma+n}$ for infinitely many $n < \omega$.

[12]Here, e.g.,

$$\overbrace{\tau.\ldots.\tau}^{n}.a$$

is a "new" point which can only perform n consecutive $\xrightarrow{\tau}$ transitions and then behave like $a \in \mathcal{T}_{\gamma+1}$.

[13]Similarly, the transitions of $\sum_{n<\omega} \tau.b_n$ are given by $\sum_{n<\omega} \tau.b_n \xrightarrow{\tau} b_n$ for all $n < \omega$.

Proof: We claim:

(i) If $0 < n < m < \omega$ and $a_m \in \mathcal{T}'_m$, then $a_n \in \mathcal{T}_n'$.

(ii) If $\beta < \beta' < \gamma^*$ are such that $\beta < \xi \leq \beta' \Rightarrow a_\xi \notin \mathcal{T}'_\xi$, then $a_{\beta'} = \mathcal{H}_{\beta'}(a_\beta)$.

(iii) If $\gamma < \gamma^*$, then $a_\gamma \in \mathcal{T}'_\gamma$ iff $\{\beta \mid 0 < \beta < \gamma \wedge a_\beta \in \mathcal{T}'_\beta\}$ is cofinal in γ.

(iv) If $n < \omega$, $\gamma < \gamma^*$ and $a_{\gamma+n} \in \mathcal{T}'_{\gamma+n}$, then $a_\gamma \in \mathcal{T}'_\gamma$.

It is easy to see that (a)–(d) follow from (i)–(iv), so it remains to prove (i)–(iv).

For (i), observe that, if $0 < k < \omega$ and $b \in \mathcal{T}_k$, then $b \in \mathcal{T}'_k$ iff b can perform $k - 1$ consecutive transitions. (ii) follows by proving that $\beta \leq \xi \leq \beta' \Rightarrow a_\xi = \mathcal{H}_\xi(a_\beta)$ by transfinite induction on ξ. (iii) is then trivial.

To prove (iv), consider first $a_{\gamma+n} \in \mathcal{T}'_{\gamma+n}$ with $n \geq 2$. Then

$$a_{\gamma+n} \xrightarrow{\delta_{n-1}} \cdots \xrightarrow{\delta_1} \mathcal{H}_{\gamma+n}(b)$$

for some $b \in \mathcal{T}'_{\gamma+1}$ (since $a' \in \mathcal{T}'_{\gamma+m+1} \Leftrightarrow \exists\delta \in \Sigma^+ . \exists b' \in \mathcal{T}'_{\gamma+m} . \langle\delta, b'\rangle \in a'$ whenever $m \geq 1$), and so

$$a_\gamma = \mathcal{H}_\gamma(a_{\gamma+n}) \xrightarrow{\delta_{n-1}} \cdots \xrightarrow{\delta_1} \mathcal{H}_\gamma(\mathcal{H}_{\gamma+n}(b)) = \mathcal{H}_\gamma(b).$$

Now $\mathcal{T}^c_\gamma$ is a subsystem of $\mathcal{T}_\gamma$, and hence it suffices to show that $a_{\gamma+1} \in \mathcal{T}'_{\gamma+1} \Rightarrow a_\gamma \in \mathcal{T}'_\gamma$. This will in turn follow once we establish that $\mathcal{H}_\gamma$ is a morphism on $\mathcal{X} = (\mathcal{H}_\gamma^{\mathcal{T}_{\gamma+1}})^{-1}(\mathcal{T}^c_\gamma)$ (which is a subsystem of $\mathcal{T}_{\gamma+1}$), because $\mathcal{H}_\gamma$ must then be injective on $\mathcal{X}$, so that $\mathcal{X} = \mathcal{H}_{\gamma+1}(\mathcal{T}^c_\gamma)$, and hence, $a_\gamma \notin \mathcal{T}'_\gamma \Rightarrow a_{\gamma+1} \in \mathcal{X} \subseteq \mathcal{H}_{\gamma+1}(\mathcal{T}_\gamma)$.

We claim that, for any $0 < \beta < \gamma$, $\mathcal{H}_\gamma(\mathcal{T}_\beta)$ is a closed subset of $\mathcal{T}_\gamma$. Suppose $\underline{b} \in \mathcal{T}_\gamma \backslash \mathcal{H}_\gamma(\mathcal{T}_\beta)$. Then $b_{\beta'} \neq \mathcal{H}_{\beta'}(b_\beta)$ for some $\beta < \beta' < \gamma$, which gives us that $\{\underline{b'} \in \mathcal{T}_\gamma \mid b'_{\beta'} = b_{\beta'}\} \cap \mathcal{H}_\gamma(\mathcal{T}_\beta) = \emptyset$, and the claim is established. Now, if $c \in \mathcal{X}$, then $\mathcal{H}_\gamma(c) \in \mathcal{H}_\gamma(\mathcal{T}_\beta)$ for some nonzero $\beta < \gamma$, and hence

$$(\mathcal{H}_\gamma(c))_{\mathcal{T}_\gamma, \delta} = \overline{\mathcal{H}_\gamma(c_{\mathcal{X},\delta})}^{\mathcal{T}_\gamma} = \overline{\mathcal{H}_\gamma(c_{\mathcal{X},\delta})}^{\mathcal{H}_\gamma(\mathcal{T}_\beta)} = \mathcal{H}_\gamma(c_{\mathcal{X},\delta})$$

for each $\delta \in \Sigma^+$ by Lemma 10(b) (observe that $\mathcal{H}_\gamma(\mathcal{T}_\beta)$ is a discrete subspace of $\mathcal{T}_\gamma$). ❑

If $\underline{a}$ is an isolated point in $\mathcal{T}_\gamma$, then $\{\underline{a'} \in \mathcal{T}_\gamma \mid a'_\alpha = a_\alpha\} = \{\underline{a}\}$ for some nonzero $\alpha < \gamma$, so that $\underline{a} = \mathcal{H}_\gamma(a_\alpha) \in \mathcal{T}^c_\gamma$. In the other direction, if $\underline{a} \in \mathcal{T}^c_\gamma$ and γ is not of the form $\gamma' + \omega$, then it is an easy consequence of Theorem 12 that $\underline{a}$ is isolated.

Let t_α for ordinals α be points such that the transitions of any t_α are given by $t_\alpha \xrightarrow{\tau} t_\beta$ for all $\beta < \alpha$. Then $\mathcal{O}_\alpha = \{t_\beta \mid \beta < \alpha\}$ is a system for any α.

THEOREM 13: For any $\alpha > 0$, $\mathcal{H}_\alpha$ is an injective morphism on $\mathcal{O}_\alpha$, but not a morphism on $\mathcal{O}_{\alpha+1}$.[14]

[14]A version of this result was known to Forti and Honsell (see [7, Remark 1.5]).

Proof: It easily follows by transfinite inductions that every $\mathcal{O}_\alpha$ is strongly extensional, and that, for any $\alpha > 0$, $\mathcal{H}_\alpha$ is a morphism on $\mathcal{O}_\alpha$. We claim that:

(i) $(\mathcal{H}_{\alpha+1}(t_{\alpha+1}))_{\mathcal{T}_{\alpha+1},\tau} = \mathcal{H}_{\alpha+1}(\{t_\beta \mid \beta < \alpha\} \cup \{\mathcal{H}_\alpha(t_\alpha)\})$ for any $\alpha > 0$.

(ii) $(\mathcal{H}_\gamma(t_\gamma))_{\mathcal{T}_\gamma,\tau} = \{\mathcal{H}_\gamma(t_\beta) \mid \beta < \gamma\} \cup \{\mathcal{H}_\gamma(t_\gamma)\}$ for any γ.

We then will have that $\mathcal{H}_\alpha(t_\alpha)$ can perform an infinite sequence of $\xrightarrow{\tau}$ transitions whenever $\alpha \geq \omega$. Also, for any nonzero $n < \omega$, t_n can perform n consecutive transitions, whereas any $a \in \mathcal{T}_n$ can perform at most $n - 1$. Hence, we will have that no $\mathcal{H}_\alpha$ for $\alpha > 0$ is a morphism on $\mathcal{O}_{\alpha+1}$ (observe that no t_α can perform infinitely many consecutive transitions). Therefore, it suffices to prove (i) and (ii). (i) is straightforward, and (ii) follows by first establishing

$$0 < \alpha \leq \beta \Rightarrow \mathcal{H}_\alpha(t_\beta) = \mathcal{H}_\alpha(t_\alpha)$$

by transfinite induction on α. ❑

If $0 < n < \omega$, we have that $\mathcal{H}_n(t_n) \in \mathcal{T}'_n$. Consider any γ. Then $\mathcal{H}_\gamma(t_\gamma) \in \mathcal{T}'_\gamma$, but $\mathcal{H}_{\gamma+1}(t_{\gamma+1}) = \mathcal{H}_{\gamma+1}(\mathcal{H}_\gamma(t_\gamma)) \notin \mathcal{T}'_{\gamma+1}$. Whenever $2 \leq n < \omega$, we again have $\mathcal{H}_{\gamma+n}(t_{\gamma+n}) \in \mathcal{T}'_{\gamma+n}$ (since $\langle \tau, \mathcal{H}_{\gamma+n-1}(t_{\gamma+n-2})\rangle \in \mathcal{H}_{\gamma+n}(t_{\gamma+n})$ and $\mathcal{H}_{\gamma+n-1}(t_{\gamma+n-2}) \in \mathcal{T}'_{\gamma+n-1}$).

For any γ, we know that

$$|\mathcal{T}_\gamma| \leq |\mathcal{T}^0_\gamma| \leq |\mathcal{T}^c_\gamma|^{|\gamma|} \leq 2^{|\mathcal{T}^c_\gamma| \times |\gamma|} = 2^{|\mathcal{T}^c_\gamma|}$$

($|\mathcal{T}^c_\gamma| \geq |\gamma|$ by Theorem 13). If $\mathcal{X}, \mathcal{X}' \subseteq \mathcal{T}^c_\gamma$ are distinct, it is not difficult to see that

$$\mathcal{H}_\gamma\Big(\sum_{0<\alpha<\gamma \wedge a \in \mathcal{X} \cap \mathcal{T}'_\alpha} \prec t_\alpha, a \succ\Big) \neq \mathcal{H}_\gamma\Big(\sum_{0<\alpha<\gamma \wedge a' \in \mathcal{X}' \cap \mathcal{T}'_\alpha} \prec t_\alpha, a' \succ\Big),$$

where $\prec d, e \succ$ is an abbreviation for $(\tau.\tau.d) + \tau.(\tau.d + \tau.e)$.[15] Hence, in fact $|\mathcal{T}_\gamma| = 2^{|\mathcal{T}^c_\gamma|}$.

DEFINITION 9: We say that a tree $\langle W, \leq\rangle$[16] is a γ-*special tree* iff:

[15]Note how this expression corresponds to $\{\{d\}, \{d, e\}\}$, which is the standard Kuratowski's set-theoretic definition of an ordered pair $\langle d, e\rangle$.

[16]A tree is a partial order $\langle W, \leq\rangle$ such that $\{y \in W \mid y < x\}$ is well-ordered by $<$ for each $x \in W$. For any $x \in W$, we write $ht(x, W)$ for $type(\langle\{y \in W \mid y < x\}, <\rangle)$. For any α, $Lev_\alpha(W) = \{x \in W \mid ht(x, W) = \alpha\}$, and we take $ht(W)$ to be the smallest α with $Lev_\alpha(W) = \varnothing$. A subtree of $\langle W, \leq\rangle$ is a downwards-closed $W' \subseteq W$ with the order induced by $\leq$.

For any maximal chain $C \subseteq W$, let $h(C) \leq ht(W)$ be such that C contains exactly one element of $Lev_\alpha(W)$ for $\alpha < h(C)$, and no elements of $Lev_\alpha(W)$ for $h(C) \leq \alpha < ht(W)$. A path through W is a maximal chain $C \subseteq W$ with $h(C) = ht(W)$.

It will sometimes be convenient to "relabel" the indices so that $W = \bigcup_{0<\alpha<ht(W)} Lev_\alpha(W)$ and $\forall x \in Lev_\alpha(W)\ .\ ht(x, W) = \alpha$ whenever $0 < \alpha < ht(W)$.

If κ is regular, a κ-Aronszajn tree is a tree $\langle W, \leq\rangle$ such that $ht(W) = \kappa$, $\forall\alpha < \kappa\ .\ |Lev_\alpha(W)| < \kappa$, and there are no paths through W.

For an introductory account of trees, see, e.g., Chapter 2 of [10].

(a) $ht(W) = \gamma$,
(b) $(x \in W \wedge ht(x, W) < \alpha < \gamma) \Rightarrow \exists y \in Lev_\alpha(W) \,.\, x \leq y$,
(c) There exists a strictly increasing sequence $\alpha \mapsto \eta_\alpha : \gamma \to \gamma \backslash \{0\}$ such that $\forall \alpha < \gamma \,.\, |Lev_\alpha(W)| \leq |\mathcal{T}_{\eta_\alpha}|$, and
(d) There are no paths through W.

We say that γ is *ω-like* iff either $cf(\gamma) = \omega$ or γ is a weakly compact cardinal.[17]

In Section 2 of [7], Forti and Honsell essentially establish the following. (It is obvious that if $\gamma' > \gamma$, $cf(\gamma') = cf(\gamma)$ and a γ-special tree exists, then a γ'-special tree exists.)

THEOREM 14: There are no γ-special trees iff γ is ω-like.

Given any γ,

$$\underline{a} \Leftrightarrow \mathcal{K}_{\underline{a}} = \{X \subseteq \mathcal{T}_\gamma \mid \exists \alpha \,.\, 0 < \alpha < \gamma \wedge \{\underline{b} \in \mathcal{T}_\gamma \mid b_\alpha = a_\alpha\} \subseteq X\}$$

gives a 1–1 correspondence between points of $\mathcal{T}^0_\gamma$ and minimal Cauchy filters on $\mathcal{T}_\gamma$, such that $\mathcal{K}_{\underline{a}}$ converges iff $\underline{a} \in \mathcal{T}_\gamma$, and in that case the limit point of $\mathcal{K}_{\underline{a}}$ is $\underline{a}$. Hence, $\mathcal{T}^0_\gamma$ with the identity mapping from $\mathcal{T}_\gamma$ into $\mathcal{T}^0_\gamma$ is a completion of $\mathcal{T}_\gamma$. In particular, $\mathcal{T}_\gamma$ is complete iff $\mathcal{T}_\gamma = \mathcal{T}^0_\gamma$.

THEOREM 15: $\mathcal{T}_\gamma$ is complete iff there are no γ-special trees.[18]

Proof: We give a proof only of the "only if" part. Suppose that $\langle W, \leq \rangle$ is a γ-special tree. Suppose also that $\underline{u} \in \mathcal{T}^0_\gamma$ and $n^* < \omega$ are such that $\gamma' < \gamma \wedge n \geq n^* \Rightarrow u_{\gamma'+n} \in \mathcal{T}'_{\gamma'+n}$. (We can take $\underline{u} = \mathcal{H}_\gamma(t_\gamma)$ and $n^* = 2$.) We fix a mapping $\alpha \mapsto \eta_\alpha : \gamma \to \gamma \backslash \{0\}$ associated to W such that $\forall \alpha < \gamma \,.\, \neg(\exists \gamma' < \gamma \,.\, \exists n \,.\, 0 \leq n < n^* \wedge \eta_\alpha = \gamma' + n)$,[19] and then we fix an injective mapping F with domain W such that $F(Lev_\alpha(W)) \subseteq \mathcal{T}'_{\eta_\alpha+1}$ for all $\alpha < \gamma$.

For any $x \in W$ and $c \in \mathcal{X} = \bigcup_{0<\alpha<\gamma} \mathcal{T}'_{\eta_\alpha+1}$, we define an ordinal $\zeta_{x,c}$ as follows. If $c = F(x')$ for some $x' \leq x$, then let $\zeta_{x,c} = \eta_{ht(x',W)}$. Otherwise, let $\zeta_{x,c} = \eta_{ht(x,W)} + 1$. Now, for any $x \in W$, let

$$G_{\underline{u}}(x) = \mathcal{H}_{\eta_{ht(x,W)}+4}\Big(\sum_{c<\mathcal{X}} \tau.\prec c, u_{\zeta_{x,c}} \succ\Big).$$

It is not difficult to see the following:

(i) $(ht(x, W) = ht(y, W) \wedge x \neq y) \Rightarrow G_{\underline{u}}(x) \neq G_{\underline{u}}(y)$.

[17] κ is weakly inaccessible iff κ is a regular limit cardinal.
κ is strongly inaccessible iff κ > ω, κ is a regular cardinal, and $2^\lambda < \kappa$ whenever $\lambda < \kappa$. (In particular, κ is then weakly inaccessible.)
κ is weakly compact iff κ is strongly inaccessible and there are no κ-Aronszajn trees.

[18] This result is essentially established in [7]. We give a part of an alternative proof which provides us with some additional information to be used later in the paper.

[19] It is trivial to obtain such a mapping from any mapping associated with W.

(ii) $x \leq y \Rightarrow \mathcal{H}_{\eta_{ht(x,W)}+4}(G_{\underline{u}}(y)) = G_{\underline{u}}(x)$.

Let W' be the tree of height γ given by

$$Lev_\alpha(W') = \{\langle \alpha, G_{\underline{u}}(x)\rangle \mid x \in Lev_\alpha(W)\}$$

for all $\alpha < \gamma$, with the order induced by the maps $\mathcal{H}_{\eta_\alpha+4}^{\mathcal{T}_{\eta_{\alpha'}+4}} (\alpha < \alpha' < \gamma)$.[20] (i) and (ii) give us that

$$ht(x, W) \leq ht(y, W) \wedge \mathcal{H}_{\eta_{ht(x,W)}+4}(G_{\underline{u}}(y)) = G_{\underline{u}}(x) \Rightarrow x \leq y,$$

and so we have that $x \mapsto \langle ht(x, W), G_{\underline{u}}(x)\rangle$ is an isomorphism between W and W'.

Now, for any $\alpha < \gamma$, let

$$a_{\eta_\alpha+5} = \{\langle \tau, G_{\underline{u}}(x)\rangle \mid x \in Lev_\alpha(W)\}.$$

Then $\alpha < \alpha' \Rightarrow \mathcal{H}_{\eta_\alpha+5}(a_{\eta_{\alpha'}+5}) = a_{\eta_\alpha+5}$ (by (i) and Definition 9(b)), and hence, the $a_{\eta_\alpha+5}$ can be uniquely extended to an $\underline{a} \in \mathcal{T}_\gamma^0$. To see that $\underline{a} \notin \mathcal{T}_\gamma$, pick an $x \in Lev_0(W)$. Then $\langle \tau, G_{\underline{u}}(x)\rangle \in a_{\eta_0+5}$, but any $\underline{b} \in \underline{a}_{\mathcal{T}_\gamma^0,\tau}$ would clearly yield a path $\{\langle \alpha, b_{\eta_\alpha+4}\rangle \mid \alpha < \gamma\}$ through W', which is a contradiction. Hence, $\underline{a} \notin \mathcal{T}_\gamma^1 \supseteq \mathcal{T}_\gamma$. ❑

In particular, as long as $cf(\gamma) = \omega$, all the $\mathcal{T}_\gamma$ are complete. In fact, in [13], $\mathcal{T}_\gamma = \mathcal{T}_\gamma^0$ served as the definition of $\mathcal{T}_\gamma$ for such γ.

For any γ, let:

$$\mathcal{Q}_\gamma = \{\underline{a}^0 \in \mathcal{T}_\gamma^0 \mid n < \omega \wedge \underline{a}^0 \xrightarrow{\delta_0} \cdots \xrightarrow{\delta_{n-1}} \underline{a}^n \wedge \langle \delta_n, b^*\rangle \in a^n_{\alpha+1} \Rightarrow \exists \underline{b} \in \underline{a}^n_{\mathcal{T}_\gamma^0,\delta_n} . b^* = b_\alpha\},$$

$$\mathcal{Q}'_\gamma = \{\underline{a} \in \mathcal{T}_\gamma \mid \mathcal{H}_\gamma^{\mathcal{T}_\gamma^0}(\underline{a}) = \underline{a}\},$$

$$\mathcal{Q}''_\gamma = \mathcal{H}_\gamma^{\mathcal{T}_\gamma^0}(\mathcal{T}_\gamma^0).$$

THEOREM 16:

(a) $\mathcal{Q}_\gamma$ is the largest subset of $\mathcal{T}_\gamma$ which is a subsystem of $\mathcal{T}_\gamma^0$.

(b) $\mathcal{Q}'_\gamma$ is the largest subset of $\mathcal{T}_\gamma^0$ containing $\mathcal{T}_\gamma^c$ on which $\mathcal{V}_\gamma$ and $\mathcal{U}_\gamma^{\mathcal{T}_\gamma^0}$ induce the same topology.

(c) If γ is ω-like, then $\mathcal{T}_\gamma^c \subset \mathcal{Q}_\gamma = \mathcal{Q}'_\gamma = \mathcal{Q}''_\gamma = \mathcal{T}_\gamma = \mathcal{T}_\gamma^0$.

(d) If γ is not ω-like, then $\mathcal{T}_\gamma^c \subset \mathcal{Q}_\gamma \subset \mathcal{Q}'_\gamma \subset \mathcal{Q}''_\gamma \subseteq \mathcal{T}_\gamma \subset \mathcal{T}_\gamma^0$.

Proof: For (*a*), observe first that $\underline{a}^0 \in \mathcal{Q}_\gamma \wedge \underline{a}^0 \xrightarrow{\delta_0} \underline{a}^1 \Rightarrow \underline{a}^1 \in \mathcal{Q}_\gamma$, so that $\mathcal{Q}_\gamma$ is a subsystem of $\mathcal{T}_\gamma^0$. Then $\mathcal{Q}_\gamma$ is a fixed point of Φ, and hence $\mathcal{Q}_\gamma \subseteq \mathcal{T}_\gamma$ by Theorem 6. If $S \subseteq \mathcal{T}_\gamma$ is a subsystem of $\mathcal{T}_\gamma^0$, then S is a fixed point of

[20] The pairing with α ensures explicitly that $\alpha \neq \alpha' \Rightarrow Lev_\alpha(W') \cap Lev_{\alpha'}(W') = \emptyset$.

Φ since $\mathcal{T}_\gamma$ is, and so $S \subseteq \mathcal{Q}_\gamma$ by the definition of $\mathcal{Q}_\gamma$.

Now, we claim that $\mathcal{T}^c_\gamma \subset \mathcal{Q}_\gamma \subseteq \mathcal{Q}'_\gamma \subseteq \mathcal{Q}''_\gamma \subseteq \mathcal{T}_\gamma \subseteq \mathcal{T}^0_\gamma$ for any γ. Since $\mathcal{T}^c_\gamma$ is a subsystem of $\mathcal{T}^0_\gamma$, we have $\mathcal{T}^c_\gamma \subseteq \mathcal{Q}_\gamma$, and, e.g., $\mathcal{H}_\gamma(t_\gamma) \in \mathcal{Q}_\gamma \backslash \mathcal{T}^c_\gamma$. Also, $\mathcal{H}^{\mathcal{T}^0_\gamma}_\gamma \upharpoonright \mathcal{Q}_\gamma = \mathcal{H}^{\mathcal{T}_\gamma}_\gamma \upharpoonright \mathcal{Q}_\gamma$, and so $\mathcal{Q}_\gamma \subseteq \mathcal{Q}'_\gamma$. The remaining inclusions are trivial.

Since $a_\alpha = \mathcal{H}^{\mathcal{T}^0_\gamma}_\alpha(\underline{a})$ for all $\underline{a} \in \mathcal{Q}'_\gamma$, $0 < \alpha < \gamma$, the uniformities (and hence, the topologies) on $\mathcal{Q}'_\gamma$ induced by $\mathcal{V}_\gamma$ and $\mathcal{U}^{\mathcal{T}^0_\gamma}_\gamma$ are the same. So suppose that $\mathcal{T}^c_\gamma \subseteq S \subseteq \mathcal{T}^0_\gamma$ and that $\mathcal{V}_\gamma$ and $\mathcal{U}^{\mathcal{T}^0_\gamma}_\gamma$ induce the same topology on S. Consider any $\underline{a} \in S$, and let

$$\mathcal{X} = \{\mathcal{H}_\gamma(\mathcal{H}^{\mathcal{T}^0_\gamma}_\alpha(\underline{a})) \mid 0 < \alpha < \gamma\}.$$

Then any $\mathcal{O} \subseteq S$ which is open with respect to $\mathcal{U}^{\mathcal{T}^0_\gamma}_\gamma$ and such that $\underline{a} \in \mathcal{O}$ intersects $\mathcal{X}$. Hence, any $\mathcal{O} \subseteq S$ open with respect to $\mathcal{V}_\gamma$ such that $\underline{a} \in \mathcal{O}$ intersects $\mathcal{X}$, and so $\underline{a} = \mathcal{H}^{\mathcal{T}^0_\gamma}_\gamma(\underline{a})$. Therefore, $S \subseteq \mathcal{Q}'_\gamma$, which establishes (b).

If γ is ω-like, we know that $\mathcal{T}_\gamma = \mathcal{T}^0_\gamma$, and so (c) follows at once from (a).

For (d), suppose γ is not ω-like. Then we can pick $\underline{a} \in \mathcal{T}^0_\gamma \backslash \mathcal{T}_\gamma$ such that $\mathcal{H}^{\mathcal{T}^0_\gamma}_\gamma(\underline{a}) = \mathcal{H}_\gamma(\varnothing)$ (such an $\underline{a}$ is given by the proof of Theorem 15), and let $\underline{b} = \mathcal{H}_\gamma(\sum_{0<\alpha<\gamma} \tau \,.\, a_\alpha)$. By Lemma 10(a), we have

$$\underline{b}_{\mathcal{T}^0_\gamma, \tau} = \overline{\{\mathcal{H}_\gamma(a_\alpha) \mid (0 < \alpha < \gamma)\}}^{\mathcal{T}^0_\gamma} = \{\mathcal{H}_\gamma(a_\alpha) \mid 0 < \alpha < \gamma\} \cup \{\underline{a}\},$$

and so $\underline{b} \notin \mathcal{Q}_\gamma$. Also, since $\mathcal{H}^{\mathcal{T}^0_\gamma}_\gamma(\underline{a}) = \mathcal{H}_\gamma(\varnothing) = \mathcal{H}^{\mathcal{T}^0_\gamma}_\gamma(\mathcal{H}_\gamma(a_1))$, it follows that $\mathcal{H}^{\mathcal{T}^0_\gamma}_\gamma(\underline{b}) = \underline{b}$, and hence, $\underline{b} \in \mathcal{Q}'_\gamma \backslash \mathcal{Q}_\gamma$.

It remains to prove that $\mathcal{Q}''_\gamma \backslash \mathcal{Q}'_\gamma \neq \varnothing$. Let $\langle W, \leq\rangle$ be a γ-special tree. We take $n^* = 4$, and fix $\langle \eta_\alpha \mid \alpha < \gamma\rangle$ as in the proof of Theorem 15. Recalling Definition 9, there exist maximal chains $C_\xi \subseteq W$ for $\xi < cf(\gamma)$ such that $\langle h(C_\xi) \mid \xi < cf(\gamma)\rangle$ is a strictly increasing sequence of limit ordinals cofinal in γ.

For any $\xi < cf(\gamma)$, we add a new point x_ξ to $Lev_{h(C_\xi)}(W)$ such that $\forall y \in C_\xi \,.\, y \leq x_\xi$. This gives us a tree $\langle W', \leq\rangle$, which clearly still satisfies $\forall \alpha < \gamma \,.\, |Lev_\alpha(W')| \leq |\mathcal{T}_{\eta_\alpha}|$. In the same way that $G_{\underline{u}}$ was defined in the proof of Theorem 15, we can find $a_\xi \in \mathcal{T}_{\eta_{h(C_\xi)}+4}$ for $\xi < cf(\gamma)$ such that the tree Z given by

$$Lev_\alpha(Z) = \{\mathcal{H}_{\eta_\alpha+4}(a_\xi) \mid \xi < cf(\gamma) \wedge h(C_\xi) \geq \alpha\}$$

(for all $\alpha < \gamma$) with the order induced by the maps

$$\mathcal{H}^{\mathcal{T}_{\eta_{\alpha'}+4}}_{\eta_\alpha+4} \quad (\text{for } \alpha < \alpha' < \gamma)$$

is isomorphic to the subtree $\bigcup_{\xi < cf(\gamma)} C_\xi \cup \{x_\xi\}$ of W' (the isomorphism being given by mapping any $y \in C_\xi \cup \{x_\xi\}$ to $\mathcal{H}_{\eta_{ht(y, W')}+4}(a_\xi)$).

We now fix a mapping F with domain W as in the proof of Theorem 15. For any $\xi < cf(\gamma)$, let $\underline{b}^\xi \in \mathcal{T}^0_\gamma \backslash \mathcal{T}_\gamma$ be constructed as in that proof, with $\underline{u}$ replaced by $\underline{u}^\xi = \mathcal{H}_\gamma(\prec t_\gamma, a_\xi \succ)$. Then, given any ξ, we have $\langle \tau, G_{\underline{u}^\xi}(x)\rangle \in b^\xi_{\eta_{h(C_\xi)+5}+5}$ for some $x \in Lev_{h(C_\xi)+5}(W)$ such that

$$\langle \tau, \mathcal{H}_{\eta_{h(C_\xi)+5}+3}(\prec c, \mathcal{H}_{\eta_{h(C_\xi)+5}+1}(\prec t_\gamma, a_\xi \succ)\succ)\rangle \in G_{\underline{u}^\xi}(x)$$

for some c. Observing that $\eta_{h(C_\xi)+5} + 1 \geq \eta_{h(C_\xi)} + 6$, it follows that

$$b^\xi_{\eta_{h(C_\xi)+5}+5} \neq b^{\xi'}_{\eta_{h(C_\xi)+5}+5} \text{ for all } \xi' \neq \xi.$$

Also, $\{\underline{b}^\xi \mid \xi < cf(\gamma)\}$ is closed in $\mathcal{T}^0_\gamma$.

Let $\underline{c} \in \mathcal{T}^0_\gamma \backslash \mathcal{T}_\gamma$ be constructed with $\underline{u}$ replaced by $\underline{u*} = \mathcal{H}_\gamma(\prec\varnothing, t_\gamma\succ)$ (F is still fixed). For any $\xi < cf(\gamma)$, let $\eta_\xi' = \eta_{h(C_\xi)+5} + 5$ and let $\underline{d}^\xi \in \mathcal{T}^0_\gamma$ be given by

$$d^\xi_{\beta+1} = b^\xi_{\beta+1} \cup \mathcal{H}_{\beta+1}(c_{\eta_\xi'})$$

for all nonzero $\beta < \gamma$. The way we chose $\underline{u}^\xi$ and $\underline{u*}$ ensures that

$$b^\xi_{\eta_\alpha+5} \text{ and } c_{\eta_\alpha+5}$$

are disjoint whenever $\eta_\alpha + 1 \geq 4$ (for all $\xi < cf(\gamma)$). It is then not difficult to see that $\{\underline{d}^\xi \mid \xi < cf(\gamma)\}$ is closed in $\mathcal{T}^0_\gamma$. Also, we have that $\mathcal{H}^{\mathcal{T}^0_\gamma}_\gamma(\underline{d}^\xi) = \mathcal{H}_\gamma(c_{\eta_\xi'})$ for all ξ.

By Theorem 11(a), let $\underline{e} \in \mathcal{T}^0_\gamma$ be such that $\underline{e}_{\mathcal{T}^0_\gamma,\tau} = \{\underline{d}^\xi \mid \xi < cf(\gamma)\}$ and $\underline{e}_{\mathcal{T}^0_\gamma,\delta} = \varnothing$ for all $\delta \in \Sigma$. Then Lemma 10(a) gives us that

$$(\mathcal{H}^{\mathcal{T}^0_\gamma}_\gamma(\underline{e}))_{\mathcal{T}^0_\gamma,\tau} = \overline{\{\mathcal{H}^{\mathcal{T}^0_\gamma}_\gamma(\underline{d}^\xi) \mid \xi < cf(\gamma)\}}^{\mathcal{T}^0_\gamma} = \overline{\{\mathcal{H}_\gamma(c_{\eta_\xi'}) \mid \xi < cf(\gamma)\}}^{\mathcal{T}^0_\gamma}$$
$$= \{\mathcal{H}_\gamma(c_{\eta_\xi'}) \mid \xi < cf(\gamma)\} \cup \{\underline{c}\}.$$

Hence, we have $\mathcal{H}_\gamma(\varnothing) = \mathcal{H}^{\mathcal{T}^0_\gamma}_\gamma(\underline{c}) \in (\mathcal{H}^{\mathcal{T}^0_\gamma}_\gamma(\mathcal{H}^{\mathcal{T}^0_\gamma}_\gamma(\underline{e})))_{\mathcal{T}^0_\gamma,\tau}$. But $\mathcal{H}_\gamma(\varnothing) \notin (\mathcal{H}^{\mathcal{T}^0_\gamma}_\gamma(\underline{e}))_{\mathcal{T}^0_\gamma,\tau}$, and so

$$\mathcal{H}^{\mathcal{T}^0_\gamma}_\gamma(\mathcal{H}^{\mathcal{T}^0_\gamma}_\gamma(\underline{e})) \neq \mathcal{H}^{\mathcal{T}^0_\gamma}_\gamma(\underline{e}), \text{ so that } \mathcal{H}^{\mathcal{T}^0_\gamma}_\gamma(\underline{e}) \in \mathcal{Q}''_\gamma \backslash \mathcal{Q}'_\gamma. \quad \square$$

If γ is not ω-like, Theorem 16 leaves open the question of whether and when is the inclusion $\mathcal{Q}''_\gamma \subseteq \mathcal{T}_\gamma$ proper. We conjecture that the method used in the proof of $\mathcal{Q}'_\gamma \subset \mathcal{Q}''_\gamma$ above can be used to establish that in fact $\mathcal{Q}''_\gamma = \mathcal{T}_\gamma$.

4. FIXED POINTS

In a variety of mathematical treatments of process algebras, the notions of *nondestructiveness* and *constructiveness* have been important in the study of fixed points (which are used to model recursion). If we have a family of restriction maps $\downarrow n$ for $n < \omega$ on a set X so that: (i) $x \downarrow 0 = y \downarrow 0$, (ii) $(x \downarrow n) \downarrow m = x \downarrow (n \cap m)$, and (iii) $(\forall n < \omega \,.\, x \downarrow n = y \downarrow n) \Rightarrow x = y$, then a function is said to be nondestructive iff $(f(x)) \downarrow n = (f(x \downarrow n)) \downarrow n$, and constructive iff $(f(x)) \downarrow n + 1 = (f(x \downarrow n)) \downarrow n + 1$ (for all $x \in X$, $n < \omega$). If we define a metric d on X by $d(x, y) = inf\{2^{-n} \mid x \downarrow n = y \downarrow n\}$ (which is usually complete), then the nondestructive maps are precisely the nonexpanding maps, and the constructive maps are just the contraction maps (to which the Banach Fixed Point Theorem applies when the space is complete).

We can generalize these ideas as follows.

DEFINITION 10:
For any $\underline{a} \in \mathcal{T}_\gamma$, $0 < \alpha < \gamma$, let $\underline{a} \downarrow \alpha = \mathcal{H}_\gamma(a_\alpha)$.

A function f: $\mathcal{T}_\gamma \to \mathcal{T}_\gamma$ is *nondestructive* iff $(f(\underline{a})) \downarrow \alpha = (f(\underline{a} \downarrow \alpha)) \downarrow \alpha$ for all $\underline{a} \in \mathcal{T}_\gamma$, $0 < \alpha < \gamma$.

A function f: $\mathcal{T}_\gamma \to \mathcal{T}_\gamma$ is *constructive* iff $(f(\underline{a})) \downarrow \alpha + 1 = (f(\underline{a} \downarrow \alpha)) \downarrow \alpha + 1$ for all $\underline{a} \in \mathcal{T}_\gamma$, $0 < \alpha < \gamma$.

THEOREM 17: Suppose f: $\mathcal{T}_\gamma \to \mathcal{T}_\gamma$ is constructive.
(a) f has at most one fixed point in $\mathcal{T}_\gamma$.
(b) If γ is countable, then f has a unique fixed point in $\mathcal{T}_\gamma$.

Proof: The proof is by constructing $\underline{a} \downarrow \alpha$ by transfinite recursion on nonzero $\alpha < \gamma$, where $\underline{a}$ is a hypothetical fixed point. ❑

DEFINITION 11: Say a predicate R on $\mathcal{T}_\gamma$ is *continuous* iff, whenever $\neg R(\underline{a})$, there is a maximal nonzero $\alpha < \gamma$ such that $\exists \underline{b} \in \mathcal{T}_\gamma . \underline{b} \downarrow \alpha = \underline{a} \downarrow \alpha \wedge R(\underline{b})$.

If R is a predicate on $\mathcal{T}_\gamma$, it is easy to see that R is continuous iff $\mathcal{X} = \{\underline{a} \in \mathcal{T}_\gamma \mid R(\underline{a})\}$ is nonempty and $\mathcal{H}_{\gamma'}(\mathcal{X})$ is closed in $\mathcal{T}_{\gamma'}$ for all $\gamma' \leq \gamma$. Nonempty finite disjunctions preserve continuity, but arbitrary conjunctions (even consistent pairwise conjunctions) do not seem to. This is an interesting topic for future research.

THEOREM 18: If f: $\mathcal{T}_\gamma \to \mathcal{T}_\gamma$ is constructive with fixed point $\underline{a}$ and R is a continuous predicate such that $\forall \underline{b} \in \mathcal{T}_\gamma . R(\underline{b}) \Rightarrow R(f(\underline{b}))$, then $R(\underline{a})$ holds.

Forti, Honsell, and Lenisa have studied fixed points in the context of hyperuniverses (see [9, Section 4]).

5. TRANSITION TO NON-WELL-FOUNDED SETS

From now on, we assume that $\Sigma = \emptyset$ (so that $\Sigma^+ = \{\tau\}$), and we drop the assumption that all systems are small. We simplify some of the notation as follows:

A system is now a class S with a binary relation $\to$ on S such that $a_S = \{b \in S \mid a \to b\}$ is a set for all $a \in S$.

For any $\alpha > 0$, we now have $\mathcal{H}^S_{\alpha+1}(a) = \{\mathcal{H}^S_\alpha(b) \mid a \to b\}$ for all $a \in S$, and so $\mathcal{T}_{\alpha+1} = \mathcal{P}(\mathcal{T}_\alpha)$. (This gives rise to the obvious changes in the definitions of transitions on the $\mathcal{T}_\alpha$, and elsewhere — essentially, any $\langle \delta, b \rangle$ with $\delta \in \Sigma^+$ is replaced by just b.)

For any system S, we can regard $\langle S, \leftarrow \rangle$[21] as a model for the language of set theory (the first-order language with equality having only the binary predicate symbol $\in$). This observation provides the link between the study of transition systems and set theory.

DEFINITION 12: A system S is *universal* iff, for any small system S', there exists a unique morphism $S' \to S$.

In particular, any universal system is strongly extensional. Also, a system S is universal iff a unique morphism $S' \to S$ exists for *any* system S'.

ANTI-FOUNDATION AXIOM (AFA): $\langle V, \ni \rangle$[22] is a universal system.

DEFINITION 13: Let $\mathcal{T} = \bigcup_{\alpha > 0} \mathcal{T}'_\alpha$ and, for any $a \in \mathcal{T}$, let θ_a be the unique $\theta > 0$ such that $a \in \mathcal{T}'_\theta$. If $a, b \in \mathcal{T}$, let $a \to b$ iff $\mathcal{H}_{\theta_a \cup \theta_b}(a) \to \mathcal{H}_{\theta_a \cup \theta_b}(b)$.

THEOREM 19: $\mathcal{T}$ is universal.

Proof: Suppose S is a small system. By Theorem 9, $\mathcal{H}_{i(S)}$ is a morphism on S. For any $a \in \mathcal{T}_{i(S)}$, let $\alpha_a > 0$ be minimal such that $a \in \mathcal{H}_{i(S)}(\mathcal{T}_{\alpha_a})$. Then $\forall a \in \mathcal{T}_{i(S)} \,.\, \mathcal{H}_{\alpha_a}(a) \in \mathcal{T}'_{\alpha_a}$, so it follows that $\mathcal{F}: a \mapsto \mathcal{H}_{\alpha_a}(a)$ is an isomorphism

$$\mathcal{T}_{i(S)} \to \bigcup_{0 < \alpha \leq i(S)} \mathcal{T}'_\alpha \subseteq \mathcal{T}.$$

Hence, $\mathcal{F} \circ \mathcal{H}_{i(S)} : S \to \mathcal{T}$ is a morphism.

By Lemma 2, it remains to prove that any morphism $\mathcal{G}$ of $\mathcal{T}$ is injective. Suppose not, and let $a, b \in \mathcal{T}$ be such that $a \neq b$ and $\mathcal{G}(a) = \mathcal{G}(b)$. Let $\theta = \theta_a \cup \theta_b$. Then $\mathcal{G} \restriction \mathcal{X}$ $(= \bigcup_{0 < \alpha \leq \theta} \mathcal{T}_\alpha')$ is a morphism (observe that $\mathcal{X}$ is certainly a subsystem of $\mathcal{T}$) of $\mathcal{X}$ which is not injective. But, as above, $\mathcal{X}$ is isomorphic to $\mathcal{T}_\theta$, contradicting Theorem 5. ❑

[Hence, any system "can be found" inside $\mathcal{T}$ uniquely. If we consider the systems $\mathcal{T}_\alpha$ as having been defined recursively, this means that the construction (which built the $\mathcal{T}_\alpha$'s from scratch) "reaches" every system without introducing any unwanted garbage on the way.]

DEFINITION 14: A system S is *full* iff, for any set $X \subseteq S$, there exists a unique $a \in S$ such that $a_S = X$.

[21] When S is a proper class, this is a metatheoretic abuse of notation.

[22] Here $V = \{x \mid x = x\}$ is the universal class.

LEMMA 20: Any universal system is full.

The following is Rieger's Theorem — for a proof, see either [12] or [1, Appendix B].

THEOREM 21: Suppose $\langle S, \rightarrow\rangle$ is a full system. Then $\langle S, \leftarrow\rangle$ is a model of ZFC^-.

Suppose we have a point in a universal system S which is a small system "encoded" within S (e.g., in the way corresponding to how small systems are usually encoded as sets, that is as ordered pairs consisting of a set and a binary relation on it). That point then gives us a small system (encoded within our universe), whose underlying set is a subset of S, and then we have a unique morphism from this small system into S (since S is universal). Finally, we can encode that morphism back into S, so that it is represented by a point in S. It follows that any universal system is a model of AFA. (See the proof of Theorem 3.8 in [1].)

THEOREM 22: $\langle \mathcal{T}, \leftarrow\rangle$ is a model of $ZFC^- + AFA$.[23]

6. STRUCTURAL RESULTS

Often, it will be both more convenient and more intuitive to use AFA as if it is a basic axiom of the theory within which we are working (so far, this has been pure ZFC^-), in the sense to become apparent in the following definition. (It will be clear when exactly are we resorting to this technique.)

DEFINITION 15: ($ZFC^- + AFA$) For any aps Sa, let $\widehat{Sa}$ be the image of a under the unique morphism $\langle Sa, \rightarrow\rangle \rightarrow \langle V, \ni\rangle$. For any system S, let $\widehat{\widehat{S}} = \{\widehat{Sa} \mid a \in S\}$.
In particular, let $T_0 = \varnothing$ and $T_\alpha = \widehat{\widehat{\mathcal{T}}}_\alpha$ for each $\alpha > 0$.

The following are easy to prove:

(a) For any aps Sa, $\widehat{Sa} = \{\widehat{Sb} \mid a \rightarrow b\}$.

(b) $T_{\alpha+1} = \mathcal{P}(T_\alpha)$ for all α.

(c) $\bigcup_{\alpha<\gamma} T_\alpha \subseteq T_\gamma$ for all γ.

(d) T_α is transitive for all α.

(e) $\alpha \leq \beta \Rightarrow T_\alpha \subseteq T_\beta$.

[23]Observe that, by Gödel's 2nd Incompleteness Theorem, this result is not formalizable as a theorem of ZFC^-. Instead, in the same way as Rieger's Theorem, it is a collection of assertions in the metatheory that, for any axiom of $ZFC^- + AFA$, we can prove that it holds in $\langle \mathcal{T}, \leftarrow\rangle$, i.e., that its relativization to $\langle \mathcal{T}, \leftarrow\rangle$ is a theorem of ZFC^-. For further discussion of these and related points, see Chapter 4 of [10].

(f) $\widehat{\widehat{\mathcal{T}}} = \bigcup_{\alpha \geq 0} T_\alpha$.

Given a cardinal κ, let $exp_0(\kappa) = \kappa$, $exp_{\alpha+1}(\kappa) = 2^{exp_\alpha(\kappa)}$, and $exp_\gamma(\kappa) = \bigcup_{\alpha<\gamma} exp_\alpha(\kappa)$. Then it follows that $|T_n| = exp_n(0)$ for all $n < \omega$, and that $|T_\alpha| = exp_{\alpha+1}(0)$ for all $\alpha \geq \omega$ (recall the remarks after Theorem 13).

If the R_α form the von Neumann hierarchy (so that $R_0 = \varnothing$, $R_{\alpha+1} = \mathcal{P}(R_\alpha)$, and $R_\gamma = \bigcup_{\alpha<\gamma} R_\alpha$), then it is immediate that $R_\alpha \subseteq T_\alpha$ for all α. But also $|R_\alpha| = exp_\alpha(0)$ for all α, so that in fact $R_n = T_n$ if $n < \omega$, and $R_\alpha \subset T_\alpha$ if $\alpha \geq \omega$.

Letting

$$\nu_\alpha = \bigcup_{x \in T_\alpha \cap WF} (\text{rank}(x) + 1),^{24}$$

we know that $\nu_\alpha \geq \alpha$ for all α. Theorem 13 gives us that $\forall\alpha \,.\, \alpha \in T_{\alpha+1} \backslash T_\alpha$, and so it seems plausible to conjecture that $\forall\alpha \,.\, \nu_\alpha = \alpha$. In fact, Forti and Honsell show that this is the case whenever either $\alpha = \omega$ or α is weakly compact (see [8]). Otherwise, we have the following.

THEOREM 23: $(ZFC^- + AFA)$

(a) $\nu_0 = 0$.

(b) $\nu_{\alpha+1} = \nu_\alpha + 1$ for all α.

(c) If $\gamma \neq \omega$ and γ is not weakly compact, then $exp_\gamma(0)^+ \leq \nu_\gamma \leq exp_{\gamma+1}(0)^+$.

Proof: (a) and (b) are obvious (observe that $T_{\alpha+1} \cap WF = \mathcal{P}(T_\alpha \cap WF)$). If we had $\nu_\gamma > exp_{\gamma+1}(0)^+ = |T_\gamma|^+$, we would have an $x \in T_\gamma$ such that rank$(x) = |T_\gamma|^+$, so that $|x| = |T_\gamma|^+$, which would contradict the transitivity of T_γ. Hence, it remains to prove the first inequality in (c).

First, we claim that:

(i) If $\gamma \neq \omega$ and γ is not strongly inaccessible, then $exp_\gamma(0)$ is singular.

(ii) If $\gamma = \omega$ or γ is strongly inaccessible, then $exp_\gamma(0) = \gamma$.

For (i), suppose $exp_\gamma(0)$ is regular. Then $\gamma \leq exp_\gamma(0) = cf(exp_\gamma(0)) = cf(\gamma) \leq \gamma$, so γ is regular and $2^\lambda \leq exp_{\lambda+1}(0) < \gamma$ whenever $\lambda < \gamma$, so that either $\gamma = \omega$ or γ is strongly inaccessible. (ii) is easy to see.

Now, we construct an $x_\alpha \in T_\gamma \cap WF$ such that rank$(x_\alpha) = \alpha$ by transfinite recursion on $\alpha < exp_\gamma(0)^+$. The base case and the successor case are obvious (for the latter, recall the remarks after Theorem 11). For the limit case, suppose we have constructed such x_α for all $\alpha < \gamma'$, where $\gamma' < exp_\gamma(0)^+$, and let $\langle \alpha_\xi \mid \xi < cf(\gamma') \rangle$ be cofinal in γ'.

If $\gamma \neq \omega$ and γ is not strongly inaccessible, then $cf(\gamma') < exp_\gamma(0)$ by (i), and so $|R_\beta| \geq cf(\gamma')$ for some $\beta < \gamma$. Pick an injective mapping $\xi \mapsto b_\xi$: $cf(\gamma') \to R_\beta$. Then Lemma 10(b) gives us that

$$x_{\gamma'} = \{\langle b_\xi, x_{\alpha_\xi} \rangle \mid \xi < cf(\gamma')\} \in T_\gamma \cap WF$$

[24]Here $WF = \bigcup_{\alpha \geq 0} R_\alpha = \{x \mid x \textit{ is a well-founded set}\}$. For any $x \in WF$, rank(x) is the smallest α such that $x \subseteq R_\alpha$.

(recall that $R_\beta \subseteq T_\beta$, and see the last paragraph in the proof of Theorem 12), and clearly, $\mathrm{rank}(x_{\gamma'}) = \gamma'$.

For the remaining case, suppose γ is strongly inaccessible, but not weakly compact. Then $exp_\gamma(0) = \gamma$ by (ii). If $\{\mathcal{H}_\gamma(x_{\alpha_\xi}) \mid \xi < cf(\gamma')\}$[25] is closed in $\mathcal{T}_\gamma$ (which is always the case if $cf(\gamma') < \gamma$, by the remarks after Theorem 11), then we can take $x_{\gamma'} = \{x_{\alpha_\xi} \mid \xi < cf(\gamma')\}$ by Lemma 10(b).

Otherwise, $cf(\gamma') = \gamma$, and we have a

$$\underline{b} \in \overline{\{\mathcal{H}_\gamma(x_{\alpha_\xi}) \mid \xi < \gamma\}}^{\mathcal{T}_\gamma} \setminus \{\mathcal{H}_\gamma(x_{\alpha_\xi}) \mid \xi < \gamma\}.$$

For any nonzero $\beta < \gamma$, let $\xi_\beta < \gamma$ be such that

$$b_\beta = \mathcal{H}_\beta(x_{\alpha_{\xi_\beta}}),$$

and let $x_\beta' = x_{\alpha_{\xi_\beta}}$. Then $\beta \mapsto \xi_\beta$ is cofinal in γ, so that $\beta \mapsto \alpha_{\xi_\beta} = \mathrm{rank}(x_\beta')$ is cofinal in γ'.

Let $\langle W, \leq\rangle$ be a γ-Aronszajn tree. Since γ is strongly inaccessible, we can construct an injective mapping $y \mapsto \beta_y : W \to \gamma\setminus\{0\}$ such that $ht(y, W) < ht(y', W) \Rightarrow \beta_y < \beta_{y'}$. For any $\zeta < \gamma$, let $\beta^*_\zeta = \bigcap_{y \in Lev_\zeta(W)} \beta_y$.

Also, there exist maximal chains $C_\xi \subseteq W$ for $\xi < \gamma$ such that $\langle h(C_\xi) \mid \xi < \gamma\rangle$ is strictly increasing and cofinal in γ. Given any $\xi < \gamma$, let

$$X_\xi = \{\langle \beta, x'_{\beta^*_{h(C_\xi)}}\rangle \mid \beta \in \{\beta_y \mid y \in C_\xi\} \cup \{\beta^*_{h(C_\xi)}\}\}.$$

Then it is not difficult to see that the tree Z given by

$$Lev_\zeta(Z) = \{\mathcal{H}_{\beta^*_{\zeta+1}+3}\Big(\sum_{x'' \in X_\xi} \tau.x''\Big) \mid \xi < \gamma \wedge h(C_\xi) > \zeta\}$$

$$= \{\mathcal{H}_{\beta^*_{\zeta+1}+3}\Big(\sum_{\beta \in \{\beta_y \mid y \in C_\xi\} \cup \{\beta^*_{h(C_\xi)}\}} \tau.\prec\mathcal{H}_{\beta^*_{\zeta+1}}(t_\beta), b_{\beta^*_{\zeta+1}}\succ\Big)$$

$$\mid \xi < \gamma \wedge h(C_\xi) > \zeta\}$$

(for $\zeta < \gamma$) with the order induced by the maps

$$\mathcal{H}^{\mathcal{T}_{\beta^*_{\zeta'+1}+3}}_{\beta^*_{\zeta+1}+3} \quad (\text{for } \zeta < \zeta' < \gamma)$$

is isomorphic to the subtree $\bigcup_{\xi<\gamma} C_\xi$ of W. Hence, it follows that

$$\{\mathcal{H}_\gamma(\textstyle\sum_{x'' \in X_\xi} \tau.x'') \mid \xi < \gamma\}$$

is closed in $\mathcal{T}_\gamma$, so that $x_{\gamma'} = \{X_\xi \mid \xi < \gamma\} \in T_\gamma \cap WF$. Finally, since both $\langle h(C_\xi) \mid \xi < \gamma\rangle$ and $\langle \beta^*_\zeta \mid \zeta < \gamma\rangle$ are cofinal in γ, so is $\langle \beta^*_{h(C_\xi)} \mid \xi < \gamma\rangle$, and hence, $\mathrm{rank}(x_{\gamma'}) = \gamma'$. ❑

[25]If x is a set and $\alpha > 0$, we write $\mathcal{H}_\alpha(x)$ for $\mathcal{H}^{tc(\{x\})}_\alpha(x)$, where $tc(\{x\})$ is the transitive closure of $\{x\}$, so that x is a point in the system $\langle tc(\{x\}), \ni\rangle$.

7. ABSTRACT RESULTS

Working in ZFC, let $\mathcal{F}(x) \in \mathcal{T}$ be the image of x under the unique morphism from $\langle tc(\{x\}), \ni\rangle$ into $\langle \mathcal{T}, \rightarrow\rangle$, for any $x \in V$. Then it is not difficult to see that $\mathcal{F}$: $\langle V, \ni\rangle \rightarrow \langle WF^{\mathcal{T}, \leftarrow}, \rightarrow\rangle$ is an isomorphism.

Conversely, within $ZFC^- + AFA$, we first have that WF is a model of ZFC and that the formulae $a \in \mathcal{T}$ and $a, b \in \mathcal{T} \wedge a \rightarrow b$ are absolute for WF[26]. Theorem 3.10 in [1], which states that any two full[27] models of $ZFC^- + AFA$ are isomorphic, then gives us that $\langle V, \ni\rangle$ and $\langle \mathcal{T}, \rightarrow\rangle$ are isomorphic. (An isomorphism is obtained as in the previous paragraph.)[28]

Hence, if ϕ is a sentence which is absolute for WF and consistent with (respectively, independent of) ZFC^-, then ϕ is consistent with (independent of) $ZFC^- + AFA$. (In particular, observe that MA, $\diamond$, CH and GCH are absolute for WF, and $V = L$ is not.)

Working within a model of $ZF^- + \neg AC$,[29] we can obtain a version of Theorem 22 which states that $\langle \mathcal{T}, \leftarrow\rangle$ is a model of $ZF^- + AFA$. (Rieger's Theorem uses AC only in order to establish that AC holds in a given full system.) It then follows that $\langle \mathcal{T}, \leftarrow\rangle$ so constructed is in fact a model of $ZF^- + \neg AC + AFA$. Hence, AC is independent of the rest of $ZFC^- + AFA$ (provided ZF^- is consistent).[30]

DEFINITION 16: For any infinite κ, let $H_\kappa = \{x \mid |tc(\{x\})| < \kappa\}$,[31] and let $B_\kappa = \{x \mid \forall y \in tc(\{x\}) \,.\, |y| < \kappa\}$.

LEMMA 24: ($ZFC^- + AFA$) Suppose κ is infinite. Then:
(a) $H_\kappa \subseteq B_\kappa$.
(b) $H_\kappa = B_\kappa$ iff $\kappa > \omega$ and κ is regular.
(c) $B_\kappa \subseteq T_\kappa$ iff κ is regular.
(d) $H_\kappa \subseteq T_\kappa$.
(e) $H_\kappa = \bigcup_{\alpha<\kappa} T_\alpha$ whenever κ is strongly inaccessible.

THEOREM 25: ($ZFC^- + AFA$) If κ is strongly inaccessible, then $\bigcup_{\alpha<\kappa} T_\alpha$ is a model of $ZFC^- + AFA$.

Proof: Since κ is strongly inaccessible, Lemma 24(e) gives us that $\bigcup_{\alpha<\kappa} T_\alpha = H_\kappa$, and so it follows as when working in ZFC that $\bigcup_{\alpha<\kappa} T_\alpha$ is a model of $ZFC^- - P$.[32]

[26]For an account of relativization and absoluteness, see Chapter 4 of [10].
[27]See Definition 14.
[28]It is easy to see from this that (in ZFC^-) if $\langle S, \leftarrow\rangle$ and $\langle S', \leftarrow\rangle$ are models of $ZFC^- + AFA$ such that $\langle WF^{S, \leftarrow}, \rightarrow\rangle$ and $\langle WF^{S', \leftarrow}, \rightarrow\rangle$ are isomorphic, then $\langle S, \rightarrow\rangle$ and $\langle S', \rightarrow\rangle$ are isomorphic — this is essentially the content of Theorem 3 in [6, Part I].
[29]Here $ZF^- = ZFC^- - AC$.
[30]Reference [6] is a study of the relationships between various axioms (including AFA) contradicting the Axiom of Foundation and various choice principles.
[31]No confusion with $\mathcal{H}_\kappa$ should occur.
[32]P stands for the Power Set Axiom.

If $x \in T_\alpha$ for some $\alpha < \kappa$, then $\mathcal{P}(x) \in T_{\alpha+2}$, and hence, P holds in $\bigcup_{\alpha<\kappa} T_\alpha$.

Suppose $\langle S, \rightarrow\rangle \in H_\kappa$ is a small system, and let $\mathcal{F}$ be the unique morphism $\langle S, \rightarrow\rangle \rightarrow \langle V, \ni\rangle$. Then $\mathcal{F}$ is in fact a morphism $\langle S, \rightarrow\rangle \rightarrow \langle H_\kappa, \ni\rangle$ and $\mathcal{F} = \{\langle a, \mathcal{F}(a)\rangle \mid a \in S\} \in H_\kappa$. Hence, AFA holds in $\bigcup_{\alpha<\kappa} T_\alpha$.

For an alternative proof, observe first that $\bigcup_{\alpha<\kappa} R_\alpha = H_\kappa^{WF}$ is a model of ZFC, and so

$$\langle \mathcal{T}, \leftarrow\rangle^{\bigcup_{\alpha<\kappa} R_\alpha} \text{ is a model of } ZFC^- + AFA.$$

It is not difficult to see that

$$\langle \mathcal{T}, \rightarrow\rangle^{\bigcup_{\alpha<\kappa} R_\alpha} \text{ is in fact isomorphic to } \langle \bigcup_{\alpha<\kappa} T_\alpha, \ni\rangle. \quad \square$$

LEMMA 26: ($ZFC^- + AFA$) $\{\widehat{Sa} \mid$ *a is a point in a small system* $\langle S, \rightarrow\rangle \in \bigcup_{\alpha<\gamma} T_\alpha\} = H_{exp_\gamma(0)}$.

THEOREM 27: Each of *Inf*, *P*, and *Repl*[33] are independent of the rest of $ZFC^- + AFA$ (provided ZF^- is consistent).

Proof: We work in $ZFC^- + AFA$.[34] Letting $Th_1 = ZFC^- - Inf + \neg Inf + AFA$, $Th_2 = ZFC^- - P + \neg P + AFA$, and $Th_3 = ZFC^- - Repl + \neg Repl + AFA$, it will suffice to establish the following (which show a bit more):

(i) If κ is infinite, then H_κ is a model of Th_1 iff $\kappa = \omega$.

(ii) If κ is infinite and regular, then H_κ is a model of Th_2 iff $\exists\alpha \,.\, exp_\alpha(\omega) < \kappa \leq exp_{\alpha+1}(\omega)$.

(iii) Suppose $\langle \alpha_\beta \mid \beta < \gamma\rangle$ is strictly increasing, and let $\alpha^* = \bigcup_{\beta<\gamma} \alpha_\beta$. Suppose further that $\gamma < exp_{\alpha^*}(\omega)$ and that $\langle \alpha_\beta \mid \beta < \gamma\rangle$ is absolute for $H_{exp_{\alpha^*}}(\omega)$.[35] Then $H_{exp_{\alpha^*}}(\omega)$ is a model of Th_3.

(iv) Suppose $\langle \alpha_\beta \mid \beta < \gamma\rangle$ is strictly increasing, $\gamma < \alpha^* = exp_{\alpha^*}(0)$, and $\langle \alpha_\beta \mid \beta < \gamma\rangle$ is absolute for $\bigcup_{\beta<\gamma} T_{\alpha_\beta}$ (where α^* is as in (iii)). Then $\bigcup_{\beta<\gamma} T_{\alpha_\beta}$ is a model of Th_3.

(To obtain a sequence as in (iv), we can start from some $\langle \alpha_{0,\beta} \mid \beta < \gamma\rangle$ with $\gamma < \alpha_0^* = \bigcup_{\beta<\gamma} \alpha_{0,\beta}$ which is absolute for $\bigcup_{\beta<\alpha'} T_\beta$ for all $\alpha' \geq \alpha_0^*$. If we let $\alpha_{n+1,\beta} = exp_{\alpha_{n,\beta}}(0)$ for all $\beta < \gamma$, $n < \omega$, then the same properties are satisfied by all the sequences $\langle \alpha_{n,\beta} \mid \beta < \gamma\rangle$ for $n < \omega$. Now, if $\alpha_n^* = \bigcup_{\beta<\gamma} \alpha_{n,\beta}$ for all $n < \omega$, then $\forall n < \omega \,.\, \alpha_n^* \leq exp_{\alpha_n^*}(0) = \alpha_{n+1}^*$. Hence, letting $\alpha^* = \bigcup_{n<\omega} \alpha_n^*$, we have $exp_{\alpha^*}(0) = \bigcup_{n<\omega} exp_{\alpha_n^*}(0) = \bigcup_{n<\omega} \alpha_{n+1}^* = \alpha^*$, so that $\langle \bigcup_{n<\omega} \alpha_{n,\beta} \mid \beta < \gamma\rangle$ will have the required properties. Observe also that we can take $\langle \omega + n \mid n < \omega\rangle$ for $\langle \alpha_{0,\beta} \mid \beta < \gamma\rangle$.)

(i) and (ii) are relatively easy to see.

[33] *Inf* stands for the Axiom of Infinity, and *Repl* for the Replacement Scheme.

[34] If ZF^- is consistent, then so is ZFC^-, and hence also $ZFC^- + AFA$.

[35] What we mean is that the function $\beta \mapsto \alpha_\beta$ is given by a formula $\phi(x, y)$ ($\Leftrightarrow (x < \gamma \wedge y = \alpha_x)$) which is absolute for $H_{exp_{\alpha^*}}(\omega)$.

Suppose $\langle \alpha_\beta \mid \beta < \gamma \rangle$ and α^* are as in (iii). Then $\kappa = exp_{\alpha^*}(\omega) > \omega$ and $\lambda < \kappa \Rightarrow 2^\lambda < \kappa$, so it follows as before that H_κ is a model of $ZFC^- - Repl + AFA$. Now $\gamma \in H_\kappa$ and $exp_{\alpha_\beta}(\omega) \in H_\kappa$ for all $\beta < \gamma$. Also, $\langle exp_{\alpha_\beta}(\omega) \mid \beta < \gamma \rangle$ is absolute for H_κ, and $\{exp_{\alpha_\beta}(\omega) \mid \beta < \gamma\} \notin H_\kappa$. Hence *Repl* fails in H_κ, so that H_κ is a model of Th_3.

Suppose $\langle \alpha_\beta \mid \beta < \gamma \rangle$ and α^* are as in (iv). Then $\bigcup_{\beta < \gamma} T_{\alpha_\beta} = \bigcup_{\beta < \alpha^*} T_\beta$ is transitive and $\alpha^* > \omega$ is a limit ordinal, so it follows that $\bigcup_{\beta < \gamma} T_{\alpha_\beta}$ is a model of $ZFC^- - Repl$. To show that *AFA* holds in $\bigcup_{\beta < \gamma} T_{\alpha_\beta}$, it suffices by Lemma 26 to show that $H_{\alpha^*} = H_{exp_{\alpha^*}(0)} \subseteq \bigcup_{\beta < \gamma} T_{\alpha_\beta}$, so suppose $x \in H_{\alpha^*}$. Then, since α^* is a limit cardinal, we in fact have that $x \in H_\kappa$ for some infinite $\kappa < \alpha^*$, and hence, $x \in T_\kappa \subseteq \bigcup_{\beta < \gamma} T_{\alpha_\beta}$ by Lemma 24(d). It now remains, by considering the sequence $\langle \alpha_\beta \mid \beta < \gamma \rangle$, to observe as in the proof of (iii) that *Repl* fails in $\bigcup_{\beta < \gamma} T_{\alpha_\beta}$. ❑

[A lot of the results above suggest that, in $ZFC^- + AFA$, the smallest α such that $x \subseteq T_\alpha$ should be regarded as the rank of a set x (which is possibly non-well-founded). However, as Theorem 23 shows, this does not always coincide with $rank(x)$ for $x \in WF$.[36]]

APPENDIX

The following are some thoughts about forcing in the presence of *AFA*. Suppose that, working within $ZFC^- + AFA$, we have a countable transitive model M of *ZFC*.[37] Also, suppose $\langle P, \leq, \top \rangle \in M$ is a partial order, where $\leq$ is reflexive and transitive, but not necessarily antisymmetric, and $\top$ is a top element. We call a filter $G \subseteq P$ *generic* iff G intersects every $D \in M$ which is a dense subset of P (typically, we have that $G \notin M$). Let a set τ be a *P-name* iff τ represents an aps with arcs labelled by elements of P, in the sense that τ is a set of ordered pairs such that $\forall \langle \sigma, p \rangle \in \tau \,.\, \sigma$ *is a P-name* $\wedge\ p \in P$.

For any generic G, we define $M[G]$ to be the set of all sets which are obtained by restricting a P-name $\tau \in M$ to G and then removing the labels, i.e., we have that $M[G] = \{\tau_G \mid \tau \in M \textit{ is a P-name}\}$, where $\tau_G = \{\sigma_G \mid \exists p \in G \,.\, \langle \sigma, p \rangle \in \tau\}$ for any P-name $\tau \in M$. (Since M satisfies *ZFC*, observe that any P-name $\tau \in M$ represents a well-founded labelled aps.) Given a $p \in P$, a formula $\phi(x_1, \ldots, x_n)$ and P-names $\tau_1, \ldots, \tau_n \in M$, we write $p \Vdash \phi(\tau_1, \ldots, \tau_n)$ iff,

[36]In Section 2 of [6, Part I], Forti and Honsell define V_α to be the union of all (in our notation) $\widehat{S}$, where S is a small system such that $S \subseteq R_\alpha$ and $a_S = a$ whenever Sa is well-founded, and then use the resulting hierarchy $\langle V_\alpha \mid \alpha \geq 0 \rangle$ to define, in the obvious way, a rank function which extends the von Neumann one.

[37]For an account of the metamathematical difficulties involved here (which are again related to Gödel's 2nd Incompleteness Theorem), and of the ways of overcoming them, see Chapter 7 of [10].

for every generic G with $p \in G$, we have that $\phi(\tau_{1_G}, ..., \tau_{n_G})$ holds in $M[G]$. It turns out that, for any generic G, $M[G]$ is a model of ZFC. The crucial step in establishing this fact consists of defining a relation $\Vdash^*$ and proving that, given a formula $\phi(x_1, ..., x_n)$ and P-names $\tau_1, ..., \tau_n \in M$ as above, we have:

$$\forall p \in P \,.\, p \Vdash \phi(\tau_1, ..., \tau_n) \Leftrightarrow (p \Vdash^* \phi(\tau_1, ..., \tau_n))^M,$$

and

$$\phi(\tau_{1_G}, ..., \tau_{n_G})^{M[G]} \Leftrightarrow \exists p \in G \,.\, p \Vdash \phi(\tau_1, ..., \tau_n)$$

for every generic G.

Abusing the notation, let T^M be the model of $ZFC^- + AFA$ such that $M = WF^{T^M}$, so that we can think of T^M as being obtained by constructing $\langle \mathcal{T}, \rightarrow \rangle$ starting from M. Then T^M is countable and transitive. Hence, for any generic G,[38] we can define $T^M[G]$ by restricting every P-name in T^M (which now doesn't necessarily represent a well-founded labelled aps) to G, removing the labels, and then taking the image of the point of the resulting unlabelled aps under the unique morphism into $\langle V, \ni \rangle$. (Since any countable transitive model of $ZFC^- + AFA$ is of the form T^M for some countable transitive model M of ZFC, this effectively defines the forcing construction starting from an arbitrary countable transitive model of $ZFC^- + AFA$.) We would expect that any such $T^M[G]$ satisfies $ZFC^- + AFA$. Furthermore, we would hope to establish the commutativity of the rectangular diagram which leads to the conclusion that $T^M[G] = \mathrm{T}^{M[G]}$ for any generic G.

In order to achieve this, we seem to require a definition of a relation $p \Vdash^* \phi(\tau_1, ..., \tau_n)$, where $p \in P$, $\phi(x_1, ..., x_n)$ is a formula and $\tau_1, ..., \tau_n \in T^M$ are P-names, which satisfies the appropriate analogues of the properties above. In the well-founded case (i.e., when we restrict our attention to P-names $\tau_1, ..., \tau_n \in M$), the key part of this definition, when $\phi(x_1, ..., x_n)$ is $x_1 = x_2$, proceeds by defining recursively, for any P-names $\tau_1, \tau_2 \in M$ (which then represent well-founded labelled aps's), the set of all $p \in P$ such that $p \Vdash^* \tau_1 = \tau_2$. Given arbitrary P-names $\tau_1, \tau_2 \in T^M$, we might attempt to take the maximum assignment (under the pointwise inclusion order) of a subset of P to every pair of P-names $\pi_1, \pi_2 \in T^M$ such that π_i represents a labelled sub-aps of τ_i for $i = 1, 2$, which satisfies the "only if" part of the recursive definition mentioned above. Alternatively, it would be very pleasing if we could, instead of recursion on the structure of P-names $\tau_1, \tau_2 \in M$, use recursion on the rank of P-names $\tau_1, \tau_2 \in T^M$ suggested above.

[38]Observe that a filter $G \subseteq P$ is generic with respect to T^M iff it is generic with respect to M.

ACKNOWLEDGMENTS

We are very grateful to Stephen Blamey, Robin Knight, Angus Macintyre, and an anonymous referee for making many useful comments on earlier drafts of the paper.

REFERENCES

1. ACZEL, P. 1988. Non-well-founded sets. CSLI Lecture Notes **14**.
2. ACZEL, P. & N. MENDLER. 1989. A final coalgebra theorem. *In:* Category theory and computer science (D.H. Pitt, *et al.*, Eds.). Springer LNCS 389. 357–365.
3. BOURBAKI, N. 1966. General topology, Part 1. Hermann. (English edition published by Addison-Wesley. Reading, Massachusetts.)
4. DAVEY, B.A. & H.A. PRIESTLEY. 1990. Introduction to lattices and order. Cambridge University Press.
5. FORTI, M. & F. HONSELL. 1983. Set theory with free construction principles. Annali Scuola Normale Superiore, Pisa. **X:** 493–522.
6. FORTI, M. & F. HONSELL. 1984/85/87. Axioms of choice and free construction principles. Bull. Soc. Math. Belg., Part I: **36B:** 69–79; Part II: **37B:** 1–12; Part III: **39B:** 259–276.
7. FORTI, M. & F. HONSELL. 1989. Models of self-descriptive set theories. *In:* Partial differential equations and the calculus of variations, Vol. 1 — Essays in honor of Ennio De Giorgi (F. Colombini *et al.*, Eds.). Birkhäuser. Boston, Massachusets. 473–518.
8. FORTI, M. & F. HONSELL. 1992. Weak foundation and antifoundation properties of positively comprehensive hyperuniverses. *In:* L'antifondation en logique et en thèorie des ensembles, Cahiers du Centre de Logique 7, 31–43, Louvain-la-Neuve.
9. FORTI, M., F. HONSELL, & M. LENISA. 1994. Processes and hyperuniverses. *In:* MFCS 1994. (I. Prívara, *et al.*, Eds.) . Springer LNCS 841. 352–361.
10. KUNEN, K. 1980. Set theory — An introduction to independence proofs. North-Holland. Amsterdam.
11. MILNER, A.J.R.G. 1989. Communication and concurrency. Prentice Hall. Englewood Cliffs. New Jersey.
12. RIEGER, L. 1957. 1957. A contribution to Gödel's axiomatic set theory, I. Czechoslovak Math. J. **7:** 323–357.
13. ROSCOE, A.W. 1982. A mathematical theory of communicating processes. D.Phil. Thesis. Oxford University.
14. RUTTEN, J.J.M.M. & D. TURI. 1993. On the foundations of final semantics: nonstandard sets, metric spaces, partial orders. *In:* REX Conference Proceedings (J. de Bakker *et al.*, Eds.) Springer LNCS 666. 477–530.

Connectedness and Disconnectedness in Bitopology

HARRIET LORD[a]

Department of Mathematics
California State Polytechnic University at Pomona
Pomona, California 91768

ABSTRACT: In this paper we begin the systematic classification of connectednesses and disconnectednesses in **BiTop**, the category of bitopological spaces and bicontinuous functions. We then investigate the relationship between $\mathcal{K}$-regular morphisms and disconnectednesses in **BiTop**.

1. INTRODUCTION

The concepts of *connectedness* and *disconnectedness* were introduced by Preuß in [11]. (If $\mathcal{A}$ is a class of topological spaces, then a space X is called $\mathcal{A}$-connected if every continuous map from X into a space in $\mathcal{A}$ must be constant. A space Y is called $\mathcal{B}$-disconnected if every continuous map from a space in $\mathcal{B}$ into Y must be constant.)

If C is the class of $\mathcal{A}$-connected spaces for some class $\mathcal{A}$ of topological spaces, then C is called a *connectedness*. If $\mathcal{D}$ is the class of $\mathcal{B}$-disconnected spaces for some class $\mathcal{B}$ of spaces, then $\mathcal{B}$ is called a *disconnectedness*. These concepts have been studied by many mathematicians, not only for topological spaces, but for topological categories in general. For a more complete discussion of the early history of *connectedness* and *disconnectedness* see [9].

One of the problems in the area of *connectedness* and *disconnectedness* is the characterization of those classes of spaces that are connectednesses and of those that are disconnectednesses.

In [6], Giuli and Salbany showed that the class of bitopological spaces $(X, \mathcal{T}_1, \mathcal{T}_2)$ such that $(X, \mathcal{T}_1 \vee \mathcal{T}_2)$ is a T_0 topological space is a disconnectedness in the category **BiTop**, the category of bitopological spaces and bicontinuous maps.

Some preliminary results were obtained in [7] in the study of the relationship between the connectednesses and disconnectednesses in **Top**, the category of topological spaces and continuous functions, and those of **BiTop**.

Mathematics Subject Classification: 54B30, 54D99, 54E55, 18A40.
Key words and phrases: connectedness, disconnectedness, bitopological space, regular morphism.

[a]E-mail: HLord@CSUPomona.edu.

In this paper, we begin the systematic classification of the connectednesses and disconnectednesses in **BiTop**.

In Section 2, we present preliminary results for topological spaces. In this section we see that we can classify disconnectednesses $\mathcal{A}$ in **Top** according to the two-element spaces in $\mathcal{A}$. The classification of connectednesses and disconnectednesses for bitopological spaces appears in Section 3, while the relationship between $\mathcal{K}$-regular morphisms and disconnectednesses is in Section 4. (We recall that if $\mathcal{A}$ is a quotient-reflective subcategory of **Top**, then $\mathcal{A}$ is a disconnectedness if and only if the $\mathcal{A}$-regular morphisms in **Top** are closed under composition.) We end this paper with a discussion of those problems that are still open in the classification of connectednesses and disconnectednesses in **BiTop**.

The categorical terminology in this paper is that of [1].

2. PRELIMINARIES

The material in this section is included for the sake of completeness. A more detailed discussion, with references, appears in Section 2 of [9].

DEFINITIONS 2.1: Let $\mathcal{A}$ and $\mathcal{B}$ be classes of topological spaces. Let C and $\mathcal{D}$ be defined as follows.

$C(\mathcal{A}) = \{X \mid \text{every continuous function } f\colon X \to A \text{ is constant for all } A \in \mathcal{A}\}$.

$\mathcal{D}(\mathcal{B}) = \{Y \mid \text{every continuous function } f\colon B \to Y \text{ is constant for all } B \in \mathcal{B}\}$.

If $\mathcal{B} = C(\mathcal{A})$ for some class $\mathcal{A}$, then $\mathcal{B}$ is called a *connectedness*.

If $\mathcal{A} = \mathcal{D}(\mathcal{B})$ for some class $\mathcal{B}$, then $\mathcal{A}$ is called a *disconnectedness*.

It is not difficult to show that $\mathcal{B}$ is a connectedness if and only if $\mathcal{B} = C(\mathcal{D}(\mathcal{B}))$, and that $\mathcal{A}$ is a disconnectedness if and only if $\mathcal{A} = \mathcal{D}(C(\mathcal{A}))$.

We note that $\mathcal{A} \subset \mathcal{D}(C(\mathcal{A}))$ and $\mathcal{B} \subset C(\mathcal{D}(\mathcal{B}))$.

We now look at examples of $C(\mathcal{A})$ and $\mathcal{D}(C(\mathcal{A}))$ when $\mathcal{A}$ consists of either a singleton space or a two-point space.

EXAMPLES 2.2:

1. If $\mathcal{A}$ consists of a singleton space, then $C(\mathcal{A})$ is the class of all topological spaces, and $\mathcal{D}(C(\mathcal{A}))$ is the class of all singleton spaces.
2. If $\mathcal{A}$ consists of the two-point indiscrete space, then $C(\mathcal{A})$ is the class of all singleton spaces, and $\mathcal{D}(C(\mathcal{A}))$ is the class of all topological spaces.
3. If $\mathcal{A}$ consists of the two-point discrete space, then $C(\mathcal{A})$ is the class of all connected spaces, and $\mathcal{D}(C(\mathcal{A}))$ is the class of totally disconnected spaces.
4. If $\mathcal{A}$ consists of a two-point Sierpinski space, then $C(\mathcal{A})$ is the class

of indiscrete spaces, and $\mathcal{D}(C(\mathcal{A}))$ is the class of T_0 spaces.

5. If $\mathcal{B}$ consists of a two-point Sierpinski space, then $\mathcal{D}(\mathcal{B})$ is the class of T_1 spaces.

Consequently the following classes of topological spaces are disconnectednesses.

EXAMPLES 2.3:

1. **Top**, the class of all topological spaces;
2. $\mathbf{Top}_0$, the class of all T_0 spaces;
3. $\mathbf{Top}_1$, the class of all T_1 spaces;
4. **Sing**, the class of all singletons;
5. **TotDisc**, the class of all totally disconnected spaces.

REMARK 2.4: The class of Hausdorff spaces is not a disconnectedness. (See [13] and [8].)

The following proposition follows from the above examples.

PROPOSITION 2.5: Let $\mathcal{A}$ be a disconnectedness. Let $2 = \{0, 1\}$, and let $\mathcal{I}$ denote the indiscrete topology on 2, S a Sierpinski topology, and $\mathcal{D}$ the discrete topology. Then:

(a) if $(2, \mathcal{I}) \in \mathcal{A}$, then $\mathcal{A} = \mathbf{Top}$;
(b) if $(2, \mathcal{I}) \notin \mathcal{A}$ and $(2, S) \in \mathcal{A}$, then $\mathcal{A} = \mathbf{Top}_0$;
(c) if $(2, \mathcal{I}) \notin \mathcal{A}$ and $(2, S) \notin \mathcal{A}$ and $(2, \mathcal{D}) \in \mathcal{A}$, then $\mathcal{A} \subset \mathbf{Top}_1$.

In addition, we know from the following theorem that the examples in Examples 2.3 do not constitute all disconnectednesses.

THEOREM 2.6: [12] There is a one-to-one correspondence between the class of cardinal numbers and a subclass of the family of connectednesses.

COROLLARY 2.7: The class of all disconnectednesses is not a set.

We recall that a class $\mathcal{A}$ of spaces is called *productive* if every product of spaces in $\mathcal{A}$ is again a space in $\mathcal{A}$; it is called *mono-hereditary* if whenever $f: X \to Y$ is a continuous monomorphism and $Y \in \mathcal{A}$, then $X \in \mathcal{A}$. The proof of the following theorem is straightforward.

THEOREM 2.8: If $\mathcal{A}$ is a disconnectedness, then $\mathcal{A}$ is productive and mono-hereditary.

COROLLARY 2.9: Let **Top** denote the category of topological spaces and continuous functions. Let $\mathcal{A}$ denote the full subcategory of **Top** whose objects belong to the class $\mathcal{A}$ of topological spaces. (Here we are identifying full subcategories with their respective object classes.) Thus, if $\mathcal{A}$ is a disconnectedness then $\mathcal{A}$ is a quotient-reflective subcategory of **Top**.

3. BITOP

We now look at connectednesses and disconnectednesses for bitopological spaces.

We begin by recalling the definition of bitopological space and bicontinuous function.

DEFINITIONS 3.1: A triple $(X, \mathcal{T}_1, \mathcal{T}_2)$ is called a *bitopological space* if X is a set and $\mathcal{T}_i$ is a topology on X for $i = 1, 2$.

A function $f: (X, \mathcal{T}_1, \mathcal{T}_2) \rightarrow (Y, S_1, S_2)$ is called *bicontinuous* if $f: (X, \mathcal{T}_i) \rightarrow (Y, S_i)$ is continuous for $i = 1, 2$.

BiTop denotes the category of bitopological spaces and bicontinuous functions.

We now define $C(\mathcal{K})$, $\mathcal{D}(\mathcal{L})$, connectedness, and disconnectedness for bitopological spaces.

DEFINITIONS 3.2: Let $\mathcal{K}$ and $\mathcal{L}$ be classes of bitopological spaces. We define

$C(\mathcal{K}) = \{(X, \mathcal{T}_1, \mathcal{T}_2) \mid$ every bicontinuous function $f: (X, \mathcal{T}_1, \mathcal{T}_2) \rightarrow (Y, S_1, S_2)$ is constant for all $(Y, S_1, S_2) \in \mathcal{K}\}$.

$\mathcal{D}(\mathcal{L}) = \{(Y, S_1, S_2) \mid$ every bicontinuous function $f: (X, \mathcal{T}_1, \mathcal{T}_2) \rightarrow (Y, S_1, S_2)$ is constant for all $(X, \mathcal{T}_1, \mathcal{T}_2) \in \mathcal{L}\}$.

$\mathcal{K}$ is called a *disconnectedness* (in **BiTop**) if $\mathcal{K} = \mathcal{D}(\mathcal{L})$ for some class $\mathcal{L}$ of bitopological spaces.

$\mathcal{L}$ is called a *connectedness* (in **BiTop**) if $\mathcal{L} = C(\mathcal{K})$ for some class $\mathcal{K}$ of bitopological spaces.

We note that $\mathcal{K} \subset \mathcal{D}(C(\mathcal{K}))$ and $\mathcal{L} \subset C(\mathcal{D}(\mathcal{L}))$.

As is the case for **Top**, $\mathcal{K}$ is a disconnectedness if and only if $\mathcal{K} = \mathcal{D}(C(\mathcal{K}))$; $\mathcal{L}$ is a connectedness if and only if $\mathcal{L} = C(\mathcal{D}(\mathcal{L}))$.

We do not yet have a classification of disconnectednesses of **BiTop** similar to that which was presented for **Top** in Proposition 2.5.

The following is the bitopological version of Theorem 2.8. Its proof is straightforward.

THEOREM 3.3: If $\mathcal{K}$ is a disconnectedness in **BiTop**, then $\mathcal{K}$ is productive and mono-hereditary.

We begin by studying disconnectednesses and connectednesses of bitopological spaces that are generated by disconnectednesses in topological spaces.

If $\mathcal{A}$ and $\mathcal{B}$ are classes of topological spaces, then $\{(X, \mathcal{T}_1, \mathcal{T}_2) \mid (X, \mathcal{T}_1) \in \mathcal{A}$ and $(X, \mathcal{T}_2) \in \mathcal{B}\}$ will be denoted by $\mathbf{BiTop}_{\mathcal{A},\mathcal{B}}$.

THEOREM 3.4: If $\mathcal{A}$ is a disconnectedness in **Top**, then $\mathbf{BiTop}_{\mathcal{A},\mathbf{Top}}$ and $\mathbf{BiTop}_{\mathbf{Top},\mathcal{A}}$ are disconnectednesses in **BiTop**, and $C(\mathbf{BiTop}_{\mathcal{A},\mathbf{Top}}) = \mathbf{BiTop}_{C(\mathcal{A}),\mathbf{Top}}$, $C(\mathbf{BiTop}_{\mathbf{Top},\mathcal{A}}) = \mathbf{BiTop}_{\mathbf{Top},C(\mathcal{A})}$.

Proof: Let $\mathcal{A}$ be a disconnectedness in **Top**. We first show that $C(\mathbf{BiTop}_{\mathcal{A},\mathbf{Top}}) = \mathbf{BiTop}_{C(\mathcal{A}),\mathbf{Top}}$. If $(X, \mathcal{T}) \in C(\mathcal{A})$, then $(X, \mathcal{T}, \mathcal{T}') \in C(\mathbf{BiTop}_{\mathcal{A},\mathbf{Top}})$ for all topologies $\mathcal{T}'$. Now assume that $(X, \mathcal{T}_1, \mathcal{T}_2) \in C(\mathbf{BiTop}_{\mathcal{A},\mathbf{Top}})$. If $(X, \mathcal{T}_1) \notin C(\mathcal{A})$, then there is a continuous, non-constant function $f: (X, \mathcal{T}_1) \to (Y, S)$ for some $(Y, S) \in \mathcal{A}$. Then $f: (X, \mathcal{T}_1, \mathcal{T}_2) \to (Y, S, I)$, where I is the indiscrete topology, is a bicontinuous non-constant function from $(X, \mathcal{T}_1, \mathcal{T}_2)$ into a space in $\mathbf{BiTop}_{\mathcal{A},\mathbf{Top}}$. Consequently, $(X, \mathcal{T}_1, \mathcal{T}_2) \notin C(\mathbf{BiTop}_{\mathcal{A},\mathbf{Top}})$. We have shown that $C(\mathbf{BiTop}_{\mathcal{A},\mathbf{Top}}) = \mathbf{BiTop}_{C(\mathcal{A}),\mathbf{Top}}$. That $C(\mathbf{BiTop}_{\mathbf{Top},\mathcal{A}}) = \mathbf{BiTop}_{\mathbf{Top},C(\mathcal{A})}$ is proven similarly.

We now will show that $\mathbf{BiTop}_{\mathcal{A},\mathbf{Top}} = \mathcal{D}(C(\mathbf{BiTop}_{\mathcal{A},\mathbf{Top}}))$. Suppose that $(Y, S_1, S_2) \notin \mathbf{BiTop}_{\mathcal{A},\mathbf{Top}}$. Then $(Y, S_1) \notin \mathcal{A}$, which is a disconnectedness. Thus there exists $(X, \mathcal{T}) \in C(\mathcal{A})$ such that there is a continuous non-constant function $f: (X, \mathcal{T}) \to (Y, S_1)$. Then $f: (X, \mathcal{T}, \mathcal{D}) \to (Y, S_1, S_2)$ is a bicontinuous non-constant map from a space in $\mathbf{BiTop}_{C(\mathcal{A}),\mathbf{Top}}$ into (Y, S_1, S_2), where $\mathcal{D}$ is the discrete topology on X. Thus $(Y, S_1, S_2) \notin \mathcal{D}(C(\mathbf{BiTop}_{\mathcal{A},\mathbf{Top}}))$, and so $\mathcal{D}(C(\mathbf{BiTop}_{\mathcal{A},\mathbf{Top}})) \subset \mathbf{BiTop}_{\mathcal{A},\mathbf{Top}}$. Since $\mathbf{BiTop}_{\mathcal{A},\mathbf{Top}} \subset \mathcal{D}(C(\mathbf{BiTop}_{\mathcal{A},\mathbf{Top}}))$ we have $\mathbf{BiTop}_{\mathcal{A},\mathbf{Top}} = \mathcal{D}(C(\mathbf{BiTop}_{\mathcal{A},\mathbf{Top}}))$. Consequently, $\mathbf{BiTop}_{\mathcal{A},\mathbf{Top}}$ is a disconnectedness in **BiTop**.

The proof that $\mathbf{BiTop}_{\mathbf{Top},\mathcal{A}}$ is a disconnectedness in **BiTop** is similar. ❑

THEOREM 3.5: The intersection of two disconnectednesses is a disconnectedness.

Proof: Let $\mathcal{K}_1$ and $\mathcal{K}_2$ be two disconnectednesses in **BiTop**. Suppose that $(Y, S_1, S_2) \notin \mathcal{K}_1 \cap \mathcal{K}_2$ but $(Y, S_1, S_2) \in \mathcal{D}(C(\mathcal{K}_1 \cap \mathcal{K}_2))$. Assume, without loss of generality, that $(Y, S_1, S_2) \notin \mathcal{K}_1$. Then there exists $(X, \mathcal{T}_1, \mathcal{T}_2) \in C(\mathcal{K}_1)$ such that there is a bicontinuous non-constant function $f: (X, \mathcal{T}_1, \mathcal{T}_2) \to (Y, S_1, S_2)$. Clearly $(X, \mathcal{T}_1, \mathcal{T}_2) \in C(\mathcal{K}_1 \cap \mathcal{K}_2)$. It follows that $(Y, S_1, S_2) \notin \mathcal{D}(C(\mathcal{K}_1 \cap \mathcal{K}_2))$. Consequently, $\mathcal{K}_1 \cap \mathcal{K}_2$ is a disconnectedness in **BiTop**. ❑

COROLLARY 3.6: If $\mathcal{A}_1$ and $\mathcal{A}_2$ are disconnectednesses in **Top**, then $\mathbf{BiTop}_{\mathcal{A}_1,\mathcal{A}_2}$ is a disconnectedness in **BiTop**.

REMARK 3.7: $\mathbf{BiTop}_{\mathbf{Top}_1,\mathbf{Top}_1}$ is denoted by $\mathbf{BiTop}_1$. (See [7].)

We now look at other classes of bitopological spaces generated by disconnectednesses in **Top**.

DEFINITION 3.8: $\mathbf{BiTop}_0$ is the class of all bitopological spaces $(X, \mathcal{T}_1, \mathcal{T}_2)$ with the property that $(X, \mathcal{T}_1 \vee \mathcal{T}_2)$ is a T_0 topological space. (This class is denoted $\mathbf{2Top}_0$ by Giuli and Salbany in [6], where it was shown to be the class of those bitopological spaces $(X, \mathcal{T}_1, \mathcal{T}_2)$ such that every bicontinuous function f: $(2, I, I) \rightarrow (X, \mathcal{T}_1, \mathcal{T}_2)$ must be constant.)

PROPOSITION 3.9: $\mathbf{BiTop}_0$ is a disconnectedness in **BiTop**.

The following definitions were motivated by the definition of $\mathbf{BiTop}_0$.

DEFINITION 3.10: Let $\mathcal{A}$ be a disconnectedness in **Top**.

$\mathbf{BiTop}_{\vee\mathcal{A}}$ denotes the class of all bitopological spaces $(X, \mathcal{T}_1, \mathcal{T}_2)$ such that $(X, \mathcal{T}_1 \vee \mathcal{T}_2) \in \mathcal{A}$.

$\mathbf{BiTop}_{\wedge\mathcal{A}}$ denotes the class of all bitopological spaces $(X, \mathcal{T}_1, \mathcal{T}_2)$ such that $(X, \mathcal{T}_1 \wedge \mathcal{T}_2) \in \mathcal{A}$.

The proof of the following appears in [7].

THEOREM 3.11:

1. If $\mathcal{A}$ is a disconnectedness in **Top** then $\mathbf{BiTop}_{\vee\mathcal{A}}$ is a disconnectedness in **BiTop**.
2. If $\mathcal{B}$ is a connectedness in **Top** then $\mathbf{BiTop}_{\wedge\mathcal{B}}$ is a connectedness in **BiTop**.

The following definition of "connectedness" in bitopological spaces was given by Pervin in [10].

DEFINITION 3.12: A bitopological space $(X, \mathcal{T}_1, \mathcal{T}_2)$ is called *connected* if X is not the union of two disjoint, nonempty sets O and U such that $O \in \mathcal{T}_1$ and $U \in \mathcal{T}_2$.

Pervin then showed the following:

PROPOSITION 3.13: A bitopological space $(X, \mathcal{T}_1, \mathcal{T}_2)$ is connected if and only if every bicontinuous function f: $(X, \mathcal{T}_1, \mathcal{T}_2) \rightarrow (2, S, S^*)$ must be constant, where $2 = \{0, 1\}$, $S = \{\varnothing, \{0\}, 2\}$ and $S^* = \{\varnothing, \{1\}, 2\}$.

REMARK 3.14: We note that $(X, \mathcal{T})$ is a connected topological space if and only if $(X, \mathcal{T}, \mathcal{T})$ is connected in the sense of Pervin if and only if $(X, \mathcal{T}, \mathcal{T}) \in C(2, \mathcal{D}, \mathcal{D})$.

For the remainder of this section, we attempt to classify disconnectednesses $\mathcal{K}$ and connectednesses $C(\mathcal{K})$ according to which two-element spaces belong to $\mathcal{K}$. We soon see that the situation in bitopological spaces is significantly more complicated than the corresponding situation in topological spaces. (See Proposition 2.5.)

We recall that for $2 = \{0, 1\}$, $S = \{\varnothing, \{0\}, 2\}$ and $S^* = \{\varnothing, \{1\}, 2\}$; I denotes the indiscrete topology and $\mathcal{D}$ the discrete topology.

In addition, we recall that if $\mathcal{K}$ is a disconnectedness in **BiTop**, and $(X, \mathcal{T}_1, \mathcal{T}_2) \in \mathcal{K}$, then whenever there exists a bicontinuous monomorphism

$f: (Y, S_1, S_2) \to (X, \mathcal{T}_1, \mathcal{T}_2)$ then $(Y, S_1, S_2) \in \mathcal{K}$. In this case, (Y, S_1, S_2) is called a *mono-subspace* of $(X, \mathcal{T}_1, \mathcal{T}_2)$. (See Theorem 3.3.)

We begin with the connectednesses.

PROPOSITION 3.15: Let $\mathcal{K}$ be a disconnectedness in **BiTop**.

1. If $(2, I, I) \in \mathcal{K}$, then $C(\mathcal{K})$ = **Sing**, the class of singleton bitopological spaces.
2. If the two-element spaces in $\mathcal{K}$ are precisely the two-element mono-subspaces of $(2, S, I)$, then $C(\mathcal{K}) = \{(X, \mathcal{T}_1, \mathcal{T}_2) \mid (X, \mathcal{T}_1) \text{ is indiscrete}\}$.
3. If the two-element spaces in $\mathcal{K}$ are precisely the two-element mono-subspaces of $(2, I, S)$, then $C(\mathcal{K}) = \{(X, \mathcal{T}_1, \mathcal{T}_2) \mid (X, \mathcal{T}_2) \text{ is indiscrete}\}$.
4. If the two-element spaces in $\mathcal{K}$ are precisely the two-element mono-subspaces of $(2, S, I)$ and of $(2, I, S)$, then $C(\mathcal{K})$ is the family of all bitopological spaces $(X, \mathcal{T}_1, \mathcal{T}_2)$ such that both topologies are the indiscrete topology on X.
5. If $(2, S, S) \in \mathcal{K}$, then $C(\mathcal{K}) \subset \{(X, \mathcal{T}_1, \mathcal{T}_2) \mid (X, \mathcal{T}_1 \cap \mathcal{T}_2) \text{ is an indiscrete space}\}$.
6. If $(2, S, S^*) \in \mathcal{K}$, then $C(\mathcal{K})$ is contained in the class of spaces called "connected" by Pervin.
7. If the two-element spaces in $\mathcal{K}$ are precisely the two-element mono-subspaces of $(2, \mathcal{D}, I)$, then there is a disconnectedness $\mathcal{A} \subset$ **Top** such that $C(\mathcal{K}) = \mathbf{BiTop}_{C(\mathcal{A}), \mathbf{Top}}$.
8. If the two-element spaces in $\mathcal{K}$ are precisely the two-element mono-subspaces of $(2, I, \mathcal{D})$, then there is a disconnectedness $\mathcal{A} \subset$ **Top** such that $C(\mathcal{K}) = \mathbf{BiTop}_{\mathbf{Top}, C(\mathcal{A})}$.

Proof:

1. Since every function from every bitopological space into $(2, I, I)$ is bicontinuous, $(X, \mathcal{T}_1, \mathcal{T}_2) \in C(\mathcal{K})$ if and only if $(X, \mathcal{T}_1, \mathcal{T}_2)$ is a singleton.

2. Suppose that $(X, \mathcal{T}_1)$ is indiscrete. We must show that if $g: (X, \mathcal{T}_1, \mathcal{T}_2) \to (Y, \mathcal{P}_1, \mathcal{P}_2)$ is bicontinuous and $(Y, \mathcal{P}_1, \mathcal{P}_2) \in \mathcal{K}$, then g is constant. If $(Y, \mathcal{P}_1, \mathcal{P}_2) \in \mathcal{K}$, then $\mathcal{P}_2$ is the indiscrete topology on Y, and so g is bicontinuous if and only if $g: (X, \mathcal{T}_1) \to (Y, \mathcal{P}_1)$ is continuous. Since $(Y, \mathcal{P}_1)$ is not indiscrete unless Y is a singleton, g must be constant, and so $(X, \mathcal{T}_1, \mathcal{T}_2) \in C(\mathcal{K})$ for any $\mathcal{T}_2$.

If $(X, \mathcal{T}_1)$ is not indiscrete, there exists $U \in \mathcal{T}_1$ such that $U \neq \varnothing, X$. Define $g: (X, \mathcal{T}_1) \to (2, S)$ by $g(x) = 0$ for $x \in U$, and $g(x) = 1$ for $x \notin U$. g is continuous and not constant. $g: (X, \mathcal{T}_1, \mathcal{T}_2) \to (2, S, I)$ is bicontinuous and nonconstant. Therefore, if $(X, \mathcal{T}_1)$ is not indiscrete, $(X, \mathcal{T}_1, \mathcal{T}_2) \notin C(\mathcal{K})$ for any $\mathcal{T}_2$.

3. The proof is similar to that of (2).

4. Note that if $(Y, \mathcal{P}_1, \mathcal{P}_2) \in \mathcal{K}$ none of its subspaces other than the singletons have both topologies indiscrete since every two-element subspace must be a mono-subspace of $(2, S, I)$ or $(2, I, S)$. If $(X, \mathcal{T}_1, \mathcal{T}_2)$ is a bitopological space such that both topologies are indiscrete, then clearly the only bicontinuous functions from $(X, \mathcal{T}_1, \mathcal{T}_2)$ into a bitopological space both of whose topologies are not indiscrete are the constant functions, and so $(X, \mathcal{T}_1, \mathcal{T}_2) \in C(\mathcal{K})$.

Now let $(X, \mathcal{T}_1, \mathcal{T}_2)$ be a bitopological space such that $\mathcal{T}_1$ or $\mathcal{T}_2$ is not indiscrete. Without loss of generality, assume $\mathcal{T}_1$ is not indiscrete. Let $O \in \mathcal{T}_1$, $O \neq \varnothing, X$. Define $f: X \to 2$ by $f(x) = 0$ if $x \in O$, $f(x) = 1$ otherwise. Then $f: (X, \mathcal{T}_1) \to (2, S)$ is continuous and not constant. Therefore, $f: (X, \mathcal{T}_1, \mathcal{T}_2) \to (2, S, I)$ is bicontinuous and non-constant, and so $(X, \mathcal{T}_1, \mathcal{T}_2) \notin C(\mathcal{K})$.

5. Suppose $(X, \mathcal{T}_1, \mathcal{T}_2)$ is a bitopological space such that there exists $O \in (\mathcal{T}_1 \cap \mathcal{T}_2)$, $O \neq \varnothing, X$. Define $f: (X, \mathcal{T}_1, \mathcal{T}_2) \to (2, S, S)$ by $f(x) = 0$ if $x \in O$, $f(x) = 1$ otherwise. Then f is bicontinuous and not constant, so $(X, \mathcal{T}_1, \mathcal{T}_2) \notin C(\mathcal{K})$.

6. We note that if $\mathcal{K}_1 \subset \mathcal{K}_2$, then $C(\mathcal{K}_2) \subset C(\mathcal{K}_1)$. The result now follows because the class of bitopological spaces called "connected" by Pervin is $C((2, S, S^*))$. (See Proposition 3.13.)

7. Suppose that the two-element spaces in $\mathcal{K}$ are precisely the two-element mono-subspaces of $(2, \mathcal{D}, I)$. Let $\mathcal{A} = \{(X, \mathcal{T}_1) \mid$ there exists a topology $\mathcal{T}_2$ on X such that $(X, \mathcal{T}_1, \mathcal{T}_2) \in \mathcal{K}\}$. We will show that $C(\mathcal{K}) \subset \mathbf{BiTop}_{C(\mathcal{A}), \mathbf{Top}}$, $\mathcal{A}$ is a disconnectedness, and then that $\mathbf{BiTop}_{C(\mathcal{A}), \mathbf{Top}} \subset C(\mathcal{K})$.

Suppose that $(Y, \mathcal{P}, \mathcal{P}') \in C(\mathcal{K})$. We must show that $(Y, \mathcal{P}) \in C(\mathcal{A})$. If $(Y, \mathcal{P}) \notin C(\mathcal{A})$, there exists a non-constant continuous function $f: (Y, \mathcal{P}) \to (X, \mathcal{T}_1)$ for some $(X, \mathcal{T}_1) \in \mathcal{A}$. Since $(X, \mathcal{T}_1, I) \in \mathcal{K}$, $f: (Y, \mathcal{P}, \mathcal{P}') \to (X, \mathcal{T}_1, I)$ is bicontinuous and non-constant, which implies that $(Y, \mathcal{P}, \mathcal{P}') \notin C(\mathcal{K})$. Thus $(Y, \mathcal{P}) \in C(\mathcal{A})$, and so $C(\mathcal{K}) \subset \mathbf{BiTop}_{C(\mathcal{A}), \mathbf{Top}}$.

We now show that $\mathcal{A}$ is a disconnectedness. Suppose that $\mathcal{A}$ is not a disconnectedness. Then there exists $(X, \mathcal{T}) \in \mathcal{D}(C(\mathcal{A}))$ with $(X, \mathcal{T}) \notin \mathcal{A}$, and so $(X, \mathcal{T}, I) \notin \mathcal{K}$. There exist, therefore, $(Y, \mathcal{P}_1, \mathcal{P}_2) \in C(\mathcal{K}) \subset \mathbf{BiTop}_{C(\mathcal{A}), \mathbf{Top}}$ and a bicontinuous function $f: (Y, \mathcal{P}_1, \mathcal{P}_2) \to (X, \mathcal{T}, I)$ that is not constant. Then $f: (Y, \mathcal{P}_1) \to (X, \mathcal{T})$ is not constant, which contradicts the fact that $(Y, \mathcal{P}_1) \in C(\mathcal{A})$ and $(X, \mathcal{T}) \in \mathcal{D}(C(\mathcal{A}))$. Thus $\mathcal{A}$ is a disconnectedness.

If $(Y, \mathcal{P}, \mathcal{P}') \in \mathbf{BiTop}_{C(\mathcal{A}), \mathbf{Top}}$ then clearly every bicontinuous function into a space in $\mathcal{K}$ must be constant. Therefore, $C(\mathcal{K}) = \mathbf{BiTop}_{C(\mathcal{A}), \mathbf{Top}}$.

8. The proof is similar to that of item 7. ❑

PROPOSITION 3.16: Let $\mathcal{K}$ be a disconnectedness in **BiTop**.

1. If $(2, I, I) \in \mathcal{K}$, then $\mathcal{K} = \mathbf{BiTop}$.
2. If the two-element spaces in $\mathcal{K}$ are precisely the two-element mono-subspaces of $(2, S, I)$, then $\mathcal{K} = \mathbf{BiTop}_{\mathbf{Top}_0, \mathbf{Top}}$.

3. If the two-element spaces in $\mathcal{K}$ are precisely the two-element mono-subspaces of $(2, I, S)$, then $\mathcal{K} = \mathbf{BiTop}_{\mathbf{Top}, \mathbf{Top}_0}$.
4. If the two-element spaces in $\mathcal{K}$ are precisely the two-element mono-subspaces of $(2, S, I)$ and of $(2, I, S)$, then $\mathcal{K}$ is the family of all bitopological spaces $(X, \mathcal{T}_1, \mathcal{T}_2)$ such that $\mathcal{T}_1 \vee \mathcal{T}_2$ is a T_0 topology on X. In this case, $\mathcal{K} = \mathbf{BiTop_0}$. (See Definition 3.8.)
5. If $(2, S, S) \in \mathcal{K}$ and $(2, S, S^*) \in \mathcal{K}$, then $\mathcal{K} \subset \mathbf{BiTop}_{\mathbf{Top}_0, \mathbf{Top}_0}$.
6. If $(2, S, S^*) \in \mathcal{K}$, then $\mathcal{K}$ contains the disconnectedness that corresponds to the class of spaces called connected by Pervin.
7. If the two-element spaces in $\mathcal{K}$ are precisely the two-element mono-subspaces of $(2, \mathcal{D}, I)$, then there is a disconnectedness $\mathcal{A} \subset \mathbf{Top}$ such that $\mathcal{K} = \mathbf{BiTop}_{\mathcal{A}, \mathbf{Top}}$.
8. If the two-element spaces in $\mathcal{K}$ are precisely the two-element mono-subspaces of $(2, I, \mathcal{D})$ then there is a disconnectedness $\mathcal{A} \subset \mathbf{Top}$ such that $\mathcal{K} = \mathbf{BiTop}_{\mathbf{Top}, \mathcal{A}}$.
9. If every two-element space in $\mathcal{K}$ is $(2, \mathcal{D}, \mathcal{D})$ then $\mathcal{K} \subset \mathbf{BiTop_1}$.
10. If the two-element spaces in $\mathcal{K}$ are precisely the two-element mono-subspaces of $(2, \mathcal{D}, I)$, of $(2, I, \mathcal{D})$, and of $(2, S, S^*)$, then if $(X, \mathcal{T}_1, \mathcal{T}_2) \in \mathcal{K}$ then $\mathcal{T}_1 \vee \mathcal{T}_2$ is a T_1 topology on X.

Proof:

1. Recall that $C(\mathcal{K}) = \mathbf{Sing}$. Since every function from a singleton space to a bitopological space is bicontinuous and constant, $\mathcal{K} = \mathbf{BiTop}$.

2. This follows from the fact that $C(\mathcal{K})$ is the class of all bitopological spaces whose first topology is indiscrete.

3. This follows from the fact that $C(\mathcal{K})$ is the class of all bitopological spaces whose second topology is indiscrete.

4. Recall that $C(\mathcal{K})$ is the class of all bitopological spaces both of whose topologies are indiscrete. If $(X, \mathcal{T}_1, \mathcal{T}_2)$ is a bitopological space such that $\mathcal{T}_1 \vee \mathcal{T}_2$ is a T_0 topology, then clearly every bicontinuous function from a space in $C(\mathcal{K})$ into $(X, \mathcal{T}_1, \mathcal{T}_2)$ must be constant.

If $(X, \mathcal{T}_1, \mathcal{T}_2)$ is a bitopological space such that $\mathcal{T}_1 \vee \mathcal{T}_2$ is not a T_0 topology, then $(X, \mathcal{T}_1, \mathcal{T}_2)$ contains a two-element subspace that is homeomorphic to $(2, I, I)$. Since every function from a bitopological space to $(2, I, I)$ is continuous, $(X, \mathcal{T}_1, \mathcal{T}_2) \notin \mathcal{D}(C(\mathcal{K})) = \mathcal{K}$.

Thus $(X, \mathcal{T}_1, \mathcal{T}_2) \in \mathcal{K}$ if and only if $\mathcal{T}_1 \vee \mathcal{T}_2$ is a T_0 topology.

5. The proof is immediate.

6. This follows from the fact that if $\mathcal{K}_1 \subset \mathcal{K}_2$, then $\mathcal{D}(C(\mathcal{K}_1)) \subset \mathcal{D}(C(\mathcal{K}_2))$.

7. Suppose that the two-element spaces in $\mathcal{K}$ are precisely the two-element mono-subspace of $(2, \mathcal{D}, I)$. Let $\mathcal{A} = \{(X, \mathcal{T}_1) \mid$ there exists a topology $\mathcal{T}_2$ on X such that $(X, \mathcal{T}_1, \mathcal{T}_2) \in \mathcal{K}\}$. Recall from Proposition 3.15.7 that $\mathcal{A}$ is a disconnectedness and that $C(\mathcal{K}) = \mathbf{BiTop}_{C(\mathcal{A}), \mathbf{Top}}$. It remains to show that $\mathcal{K} = \mathbf{BiTop}_{\mathcal{A}, \mathbf{Top}}$. Clearly, $\mathcal{K} \subset \mathbf{BiTop}_{\mathcal{A}, \mathbf{Top}}$. Suppose that $(X, \mathcal{T}_1, \mathcal{T}_2)$

$\in \mathbf{BiTop}_{\mathcal{A},\mathbf{Top}}$ and $(X, \mathcal{T}_1, \mathcal{T}_2) \notin \mathcal{K}$. Then there exist $(Y, \mathcal{P}_1, \mathcal{P}_2) \in C(\mathcal{K}) = \mathbf{BiTop}_{C(\mathcal{A}),\mathbf{Top}}$ and a bicontinuous non-constant function $f\colon (Y, \mathcal{P}_1, \mathcal{P}_2) \to (X, \mathcal{T}_1, \mathcal{T}_2)$. Then $f\colon (Y, \mathcal{P}_1) \to (X, \mathcal{T}_1)$ is a non-constant continuous function from a space in $C(\mathcal{A})$ into a space in $\mathcal{A}$, which contradicts the definition of $C(\mathcal{A})$.

We have shown, therefore, that $\mathcal{K} = \mathbf{BiTop}_{\mathcal{A},\mathbf{Top}}$, where $\mathcal{A} = \{(X, \mathcal{T}_1) \mid$ there exists a topology $\mathcal{T}_2$ on X such that $(X, \mathcal{T}_1, \mathcal{T}_2) \in \mathcal{K}\}$.

8. The proof is similar to the proof of 7.

9. If $(X, \mathcal{T}_1, \mathcal{T}_2) \in \mathcal{K}$, and every two-element subspace of $(X, \mathcal{T}_1, \mathcal{T}_2)$ is $(2, \mathcal{D}, \mathcal{D})$, then both $\mathcal{T}_1$ and $\mathcal{T}_2$ must be T_1 topologies on X, and so $\mathcal{K} \subset \mathbf{BiTop}_1$.

10. Let $(X, \mathcal{T}_1, \mathcal{T}_2) \in \mathcal{K}$. Since every two-element subspace of $(X, \mathcal{T}_1, \mathcal{T}_2)$ is homeomorphic to a mono-subspace of $(2, \mathcal{D}, \mathcal{I})$ or of $(2, \mathcal{I}, \mathcal{D})$ or of $(2, \mathcal{S}, \mathcal{S}^*)$, $\mathcal{T}_1 \vee \mathcal{T}_2$ must be a T_1 topology on X. ❑

4. $\mathcal{K}$-REGULAR MORPHISMS

We begin this section with a review of the theory of $\mathcal{A}$-regular morphisms in **Top**, and then study $\mathcal{K}$-regular morphisms in **BiTop**.

$\mathcal{A}$-regular morphisms in Top

DEFINITIONS 4.1: Let $\mathcal{A}$ be a class of topological spaces. Let X be a topological space, S a subspace of X, and $m\colon S \to X$ the inclusion of S into X. m is called *$\mathcal{A}$-regular* in **Top** if and only if there exist two morphisms $f, g\colon X \to A$, $A \in \mathcal{A}$, such that $S = \{x \in X \mid f(x) = g(x)\}$. (If $\mathcal{T}$ is the topology on X, then $\mathcal{T}|_S$ denotes the induced subspace topology on S.)

(In categorical terminology, a morphism is an $\mathcal{A}$-regular morphism if and only if it is an equalizer of two morphisms whose codomain is in $\mathcal{A}$.)

REMARK 4.2: Cagliari and Mantovani have shown in [3] that for those disconnectednesses $\mathcal{A}$ that consist of T_1 spaces, the $\mathcal{A}$-regular embeddings in $\mathcal{A}$ coincide with the class of all the embeddings in $\mathcal{A}$. (This means that if m is an embedding with domain and codomain in $\mathcal{A}$, then m is an $\mathcal{A}$-regular embedding.)

This result was then used by Cagliari and Cicchese in [2] to show that a quotient-reflective subcategory $\mathcal{A}$ of **Top** is a disconnectedness if and only if the $\mathcal{A}$-regular morphisms are closed under composition. (See [2] and [3].)

$\mathcal{K}$-regular morphisms in BiTop

We now look at $\mathcal{K}$-regular embeddings, where $\mathcal{K}$ is a class of bitopological spaces.

DEFINITIONS 4.3: Let $\mathcal{K}$ be a class of bitopological spaces. Let $(X, \mathcal{T}_1, \mathcal{T}_2)$ be a bitopological space, and let $m\colon S \to X$ be the inclusion of S

into X. $m: (S, \mathcal{T}_1|_S, \mathcal{T}_2|_S) \to (X, \mathcal{T}_1, \mathcal{T}_2)$ is called *$\mathcal{K}$-regular* in **BiTop** if and only if there exist two morphisms $f, g: (X, \mathcal{T}_1, \mathcal{T}_2) \to (Y, \mathcal{P}_1, \mathcal{P}_2)$, $(Y, \mathcal{P}_1, \mathcal{P}_2) \in \mathcal{K}$, such that $S = \{x \in X \mid f(x) = g(x)\}$.

In this case, $(S, \mathcal{T}_1|_S, \mathcal{T}_2|_S)$ is called a *$\mathcal{K}$-regular subspace* of $(X, \mathcal{T}_1, \mathcal{T}_2)$.

The results of Cagliari and Mantovani, and of Cagliari and Cicchese that are discussed in Remark 4.2 also hold in **BiTop**. The proofs of these results in **Top** remain valid in **BiTop**, where we use the fact that the regular morphisms in $\mathbf{BiTop}_1$ are the embeddings. (See Theorem 3.12 in [7].)

PROPOSITION 4.4:

(a) If $\mathcal{K}$ is a disconnectedness with $\mathcal{K} \subset \mathbf{BiTop}_1$, then for each space $(X, \mathcal{T}_1, \mathcal{T}_2) \in \mathcal{K}$, the $\mathcal{K}$-regular embeddings into $(X, \mathcal{T}_1, \mathcal{T}_2)$ are the embeddings into $(X, \mathcal{T}_1, \mathcal{T}_2)$.

(b) If $\mathcal{K}$ is a quotient-reflective subcategory of **BiTop** and $\mathcal{K} \subset \mathbf{BiTop}_1$, then $\mathcal{K}$ is a disconnectedness in **BiTop** if and only if the $\mathcal{K}$-regular embeddings are closed under composition.

We note that if the $\mathcal{K}$-regular embeddings are closed under composition and $\mathcal{K}$ is a quotient-reflective subcategory of **BiTop**, then $\mathcal{K}$ is a disconnectedness. This follows from Theorem 3.7 and Corollary 4.23 in [9]. Thus the next proposition provides another proof that $\mathbf{BiTop}_0$ is a disconnectedness.

PROPOSITION 4.5: The $\mathbf{BiTop}_0$-regular embeddings are closed under composition.

(Recall that $(X, \mathcal{T}_1, \mathcal{T}_2) \in \mathbf{BiTop}_0$ if and only if $(X, \mathcal{T}_1 \vee \mathcal{T}_2)$ is T_0.)

Proof: In Proposition 3.6.1 of [6], Giuli and Salbany show that the closure operator induced by the $\mathbf{BiTop}_0$-regular embeddings is hereditary. In Proposition 3.5 of [5], Dikranjan and Giuli show that this implies that the closure operator is weakly hereditary. That this is equivalent to the class of $\mathbf{BiTop}_0$-regular embeddings being closed under composition follows from Proposition 3.2 of [5], where it is shown that an idempotent closure operator C is weakly hereditary if and only if the C-closed embeddings are closed under composition. (For more on the induced closure operator, see [5] and [9].) ❑

We now look at the regular morphisms in $\mathbf{BiTop}_{\mathcal{A}_1, \mathcal{A}_2}$.

PROPOSITION 4.6: Let $\mathcal{A}_1$, $\mathcal{A}_2$ be subcategories of **Top**, and let $\mathcal{K} = \mathbf{BiTop}_{\mathcal{A}_1, \mathcal{A}_2}$.

1. If $(S, \mathcal{T}_1|_S, \mathcal{T}_2|_S)$ is a $\mathcal{K}$-regular subspace of $(X, \mathcal{T}_1, \mathcal{T}_2)$, then $(S, \mathcal{T}_i|_S)$ is an $\mathcal{A}_i$-regular subspace of $(X, \mathcal{T}_i)$ for $i = 1, 2$.
2. Let $(X, \mathcal{T}_1, \mathcal{T}_2) \in \mathcal{K}$. If $\mathcal{A}_i$ is a quotient-reflective subcategory of **Top** for $i = 1, 2$, and $(S, \mathcal{T}_i|_S)$ is an $\mathcal{A}_i$-regular subspace of $(X, \mathcal{T}_i)$, then $(S, \mathcal{T}_1|_S, \mathcal{T}_2|_S)$ is a regular subspace of $(X, \mathcal{T}_1, \mathcal{T}_2)$.

Proof: Assume that $(S, \mathcal{T}_1|_S, \mathcal{T}_2|_S)$ is a $\mathcal{K}$-regular subspace of $(X, \mathcal{T}_1, \mathcal{T}_2)$. This means that there exist $(Y, S_1, S_2) \in \mathcal{K}$ and two bicontinuous functions f, g: $(X, \mathcal{T}_1, \mathcal{T}_2) \to (Y, S_1, S_2)$ such that $S = \{x \in X \mid f(x) = g(x)\}$. Since $(Y, S_1, S_2) \in \mathcal{K} = \mathbf{BiTop}_{\mathcal{A}_1, \mathcal{A}_2}$, it follows that $(S, \mathcal{T}_i|_S)$ is an $\mathcal{A}_i$-regular subspace of $(X, \mathcal{T}_i)$ for $i = 1, 2$.

Now assume that $(X, \mathcal{T}_1, \mathcal{T}_2) \in \mathcal{K}$ and $(S, \mathcal{T}_i|_S)$ is an $\mathcal{A}_i$-regular subspace of $(X, \mathcal{T}_i)$ for $i = 1, 2$. Define $X \sqcup X$ to be the disjoint union of X with itself, and let i_1, i_2: $X \to X \sqcup X$ denote the canonical injections of X into $X \sqcup X$.

Now define an equivalence relation "$\sim$" on $X \sqcup X$ as follows: $y \sim y$ for all $y \in X \sqcup X$ and $i_1(s) \sim i_2(s)$ for all $s \in S$.

Let $(X \sqcup X)/\sim$ denote the set of the resulting equivalence classes, and let q: $(X \sqcup X) \to (X \sqcup X)/\sim$ be the resulting identification map. Let $\mathcal{T}_i \sqcup \mathcal{T}_i$ denote the topology on $X \sqcup X$ in the disjoint topological union of $(X, \mathcal{T}_i)$ with itself, and let $(\mathcal{T}_i \sqcup \mathcal{T}_i)/\sim$, denote the quotient topology on $(X \sqcup X)/\sim$ induced by the identification map q for $i = 1, 2$. Let $v_i = qi_i$, for $i = 1, 2$. Then $S = \{x \in X \mid v_1(s) = v_2(s)\}$.

v_1 and v_2 are bicontinuous functions from $(X, \mathcal{T}_1, \mathcal{T}_2) \to ((X \sqcup X)/\sim, (\mathcal{T}_1 \sqcup \mathcal{T}_1)/\sim), (\mathcal{T}_2 \sqcup \mathcal{T}_2)/\sim)$. That $((X \sqcup X)/\sim, (\mathcal{T}_1 \sqcup \mathcal{T}_1)/\sim), (\mathcal{T}_2 \sqcup \mathcal{T}_2)/\sim) \in \mathcal{K}$ follows from the fact that since each $\mathcal{A}_i$ is quotient-reflective in **Top**, $(X \sqcup X/\sim, \mathcal{T}_i \sqcup \mathcal{T}_i)/\sim) \in \mathcal{A}_i$ for i = 1,2. (This was shown in Proposition 1.12 in [4].) Thus, we have shown that $(S, \mathcal{T}_1|_S, \mathcal{T}_2|_S)$ is a regular subspace of $(X, \mathcal{T}_1, \mathcal{T}_2)$ if $(X, \mathcal{T}_1, \mathcal{T}_2) \in \mathcal{K}$. ❑

COROLLARY 4.7: The regular embeddings in $\mathbf{BiTop}_{\mathcal{A}_1, \mathcal{A}_2}$ are closed under composition whenever $\mathcal{A}_1$ and $\mathcal{A}_2$ are disconnectednesses in **Top**.

5. OPEN PROBLEMS

Connectednesses

We do not know if the converse of Theorem 3.11.2 is true.

We do not have a good characterization of $C(\mathcal{K})$ for the disconnectednesses $\mathcal{K}$ that satisfy the following properties.

The two-element spaces in $\mathcal{K}$ are precisely the two-element mono-subspaces of $(2, \mathcal{D}, \mathcal{S})$.

The two-element spaces in $\mathcal{K}$ are precisely the two-element mono-subspaces of $(2, \mathcal{S}, \mathcal{S})$. A partial result is contained in Proposition 3.15, item 5.

The two-element spaces in $\mathcal{K}$ are precisely the two-element mono-subspaces of $(2, \mathcal{S}, \mathcal{S}^*)$. A partial result is contained in Proposition 3.15, item 6.

$\mathcal{K}$ contains two two-element spaces, neither of which is a mono-subspace of the other. The only such case in which $C(\mathcal{K})$ has been characterized is for the case where the two-element subspaces in $\mathcal{K}$ are precisely the two-element mono-subspaces of (2, S, I) and of (2, I, S). See Proposition 3.15 item 4.

$\mathcal{K}$ contains three or more two-element spaces that have the property that for any two of these spaces, neither is a mono-subspace of the other.

Disconnectednesses

We do not know if the converse of Theorem 3.11.1 is true.

We have not been able to characterize the disconnectednesses $\mathcal{K}$ that satisfy the following properties.

The two-element spaces in $\mathcal{K}$ are precisely the two-element mono-subspaces of (2,$\mathcal{D}$,S).

The two-element spaces in $\mathcal{K}$ are precisely the two-element mono-subspaces of (2, S, S) and of (2, S, S^*). See Proposition 3.16 item 5 for a partial result.

The two-element spaces in $\mathcal{K}$ are precisely the two-element mono-subspaces of (2, S, S^*). See Proposition 3.16 item 6 for a partial result.

$\mathcal{K}$ contains two two-element spaces, neither of which is a mono-subspace of the other. $\mathcal{K}$ has been characterized only in the case where every two-element space in $\mathcal{K}$ is a mono-subspace of (2, S, I) or of (2, I, S). See Proposition 3.16 item 4.

$\mathcal{K}$ contains three or more two-element spaces with the property that for any two of these spaces, neither is a mono-subspace of the other. A partial result can be found in Proposition 3.16 item 10 for the case where the two-element spaces in $\mathcal{K}$ are precisely the two-element mono-subspaces of (2, $\mathcal{D}$, I), of (2, I, $\mathcal{D}$), and of (2, S, S^*).

$\mathcal{K}$-Regular Morphisms

That a quotient-reflective subcategory $\mathcal{K}$ is a disconnectedness if and only if the $\mathcal{K}$-regular morphisms are closed under composition has been shown only for $\mathcal{K} \subset \mathbf{BiTop}_1$; $\mathcal{K} = \mathbf{BiTop}_0$; and $\mathcal{K} = \mathbf{BiTop}_{\mathcal{A},\mathcal{A}}$, where $\mathcal{A}$ is a disconnectedness in **Top**. (For the latter result, see Theorem 3.12 in [7].)

ACKNOWLEDGMENT

The author is grateful to the referee, whose comments and corrections improved both the content and presentation of this paper.

REFERENCES

1. ADÁMEK, J., H. HERRLICH, & G. STRECKER. 1990. Abstract and concrete categories. John Wiley & Sons, Inc. New York.
2. CAGLIARI, F. & M. CICCHESE. 1985. Disconnectednesses and closure operators. Proceedings of the 13th. Winter School of Abstract Analysis, Section of Topology, Supplemento ai Rendiconti del Circolo Matematico di Palermo, Serie II. **11:** 15–23.
3. CAGLIARI, F. & S. MANTOVANI. 1986. On disconnectednesses in subcategories of a topological category and related topics, II Convegno Di Topologica Taormina (ME), 4–7 Aprile 1984, ATTI. Supplemento ai Rendiconti del Circolo Matematico di Palermo, Seri II. **12:** 205–212.
4. DIKRANJAN, D. & E. GIULI. 1983. Closure operators induced by topological epireflections. Colloquia Math. Soc. J. Bolyai. **41:** 233–246.
5. DIKRANJAN, D. & E. GIULI. 1987. Closure operators I. Topology and its Applications. **27:** 129–143.
6. GIULI, E. & S. SALBANY. 1988. $2T_0$ spaces and closure operators. Seminarberichte aus dem Fachbereich Mathematik und Informatik. **29:** 11–40.
7. LORD, H. Disconnectednesses:two examples. Proceedings of the Workshop in Categorical Topology in L'Aquila, 1994. To appear.
8. LORD, H. 1989. ($\mathcal{A}$-epi, $\mathcal{M}_{\mathcal{A}}$) factorization structures on Top. *In:* Categorical Topology and its Relation to Analysis, Algebra and Combinatorics, Prague 1988. World Scientific. Singapore. 95–119.
9. LORD, H. 1995. Connectednesses and disconnectednesses. *In:* Papers on General Topology and Applications: 9th Summer Conference at Slippery Rock University. Annals of the New York Academy of Science. New York. **767:** 115–139.
10. PERVIN, W. 1967. Connectedness in bitopological spaces. Indagationes Mathematicae. **29:** 369–372.
11. PREUß, G. 1967. Über den E-Zusammenhang und seine Lokalisation, Ph.D. Thesis. Freie Universität. Berlin.
12. SALICRUP, G. & R. VÁSQUEZ. 1988. Connection categories. *In:* Categorical Topology, The Complete Work of Graciela Salicrup (H. Herrlich & C. Prieto, Eds.). Sociedad Matemática Mexicana, 1988, pp. 73–98. (This is a translation of the article: 1972. Categorías de conexión, An. Inst. Mat. U.N.A.M. **12:** 47–87.
13. SALICRUP, G. & R. VÁSQUEZ. 1988. Reflectivity and connectedness in Top. *In:* Categorical Topology, The Complete Work of Graciela Salicrup (H. Herrlich & C. Prieto, Eds.). Sociedad Matemática Mexicana. 163–208. (This is a translation of the article: 1974/75. Reflexividad y coconexidad en Top. An. Inst. Mat. U.N.A.M. **14:** 159-230 and **15:** 113–115.

Largest Proper Ideals of Transformation Semigroups

K.D. MAGILL, JR.[a]
Department of Mathematics
106 Diefendorf Hall
SUNY at Buffalo
Buffalo, New York 14214-3093

1. INTRODUCTION

In this paper, ideal means two-sided ideal and identity means two-sided identity. If a semigroup S contains a proper left ideal and a right identity, then S contains a largest proper left ideal. It is simply the union of all proper left ideals of S. In [1], we characterized this left ideal for general transformation semigroups. We also obtained analogous results for the largest proper right ideal when the semigroup has a proper right ideal and a left identity and for the largest proper ideal when the semigroup has a proper ideal and an identity. The results obtained are of such a nature that they apply, for example, to the various endomorphism semigroups of algebraic structures such as semigroups of linear transformations of a vector space, as well as to various semigroups of continuous selfmaps of topological spaces.

In Section 2 of this paper, we extend the result in [1] concerning the largest proper ideal of a transformation semigroup which contains an identity to any transformation semigroup which contains a proper ideal and which has either a left identity, a right identity, or an identity.

Let $S(X)$ denote the semigroup, under composition, of all continuous selfmaps of the topological space X and let e be a nonconstant idempotent of $S(X)$. Then e is a right identity for the subsemigroup $L(e) = S(X) \circ e$, e is a left identity for the subsemigroup $R(e) = e \circ S(X)$, and e is an identity for the subsemigroup $Loc(e) = e \circ S(X) \circ e$. It is easy to check that $L(e)$ is the largest subsemigroup of $S(X)$ for which e acts as a right identity, $R(e)$ is the largest subsemigroup of $S(X)$ for which e acts as a left identity and $Loc(e)$ is the largest subsemigroup of $S(X)$ for which e acts as an identity. Of course, $Loc(e)$ is just the *local subsemigroup* of $S(X)$ *determined by the idempotent* e and $Loc(e) = L(e) \cap R(e)$. In Section 3, we apply the main result in Section 2, as well as several of the results in [1], to the semigroups $L(e)$, $R(e)$, and

Mathemmatics Subject Classification: 20M20, 54H15.
Key words and phrases: transformation semigroups, ideals, semigroups of continuous selfmaps.
[a]E-mail: mthmagil@ubvms.bitnet.

Loc(e). Finally, in Section 4, we consider some examples. We choose specific idempotents in $S(R)$ where R is the space of real numbers, and we interpret the results in Section 3 for the subsemigroups of $S(R)$ determined by these idempotents.

2. SOME GENERAL RESULTS

$T(X)$ will denote any semigroup of transformations on a set X. Define $T(X)^1 = T(X)$ if $T(X)$ has an identity, otherwise, let $T(X)^1 = T(X) \cup \{\delta\}$ where δ is the identity map on X.

DEFINITION 2.1: Let $A, B \subseteq X$. An element $f \in T(X)^1$ is said to map A *T-isomorphically* onto B if $f[A] \subseteq B$ and there exists a $g \in T(X)^1$ such that $g[B] \subseteq A$, $f \circ g|B = \delta|B$ and $g \circ f|A = \delta|A$. In this case, we say that A and B are *T-isomorphic* and that $f|A$ is a *T-isomorphism* from A onto B.

DEFINITION 2.2: A subset A of X is a *T-retract* of X if it is the range of an idempotent of $T(X)$.

If a semigroup S has a largest proper ideal, we will denote that ideal by $M(S)$. Note that if S has a proper ideal and contains either a left identity, a right identity, or an identity τ, then $M(S)$ does indeed exist. It is simply the union of all proper ideals. This union must be proper since it cannot contain τ. As usual, the symbol J will denote Green's relation on a semigroup S where $(a, b) \in J$ if and only if both a and b generate the same principal ideal. The proof of the next result is a minor modification of the proof of Lemma 2.23 of [1] so we will omit it.

LEMMA 2.3: Suppose a semigroup S contains a proper ideal and suppose τ is either a left identity, a right identity, or an identity of S. Then $a \in S \backslash M(S)$ if and only if $a J \tau$.

There is a result from [1] which we use considerably in this paper so we state it, also without proof, for our convenience. The symbol *Ran*(f) denotes the range of a function f.

THEOREM 2.4: Suppose $f \in S(X)$ maps some subset A of X T-isomorphically onto *Ran*(f). Then both A and *Ran*(f) are T-retracts of X.

THEOREM 2.5: Suppose τ is either a left identity, a right identity, or an identity of $T(X)$ and $f \in T(X)$. Then $f J \tau$ if and only if there exists a pair of T-retracts A and B of X, both T-isomorphic to *Ran*(τ), such that f maps A T-isomorphically onto B.

Proof: We need only verify this for the cases where τ is either a left identity or a right identity as the remaining case follows from either of these.

Nevertheless, if τ happens to be an identity, then the conclusion is just Lemma 2.24 of [1]. Suppose first that τ is a right identity and $f\mathcal{J}\tau$. Then

$$\tau \in \{f\} \cup (f \circ T(X)) \cup (T(X) \circ f) \cup ((T(X) \circ f \circ T(X)).$$

Then, either

$$\tau = f, \tag{2.5.1}$$

$$\tau = f \circ h, \tag{2.5.2}$$

$$\tau = g \circ f \tag{2.5.3}$$

or

$$\tau = g \circ f \circ h. \tag{2.5.4}$$

If $\tau = f$, then it is immediate that there exist T-retracts A and B, both T-isomorphic to $Ran(\tau)$, such that f maps A T-isomorphically onto B. Since τ is idempotent, one need only take $A = B = Ran(\tau)$. Suppose $\tau = f \circ h$. Evidently, $Ran(h \circ f) \subseteq Ran(h)$. Suppose $x \in Ran(h)$. Then $x = h(y)$ for some y and we get $h \circ f(x) = h \circ f \circ h(y) = h \circ \tau(y) = h(y) = x$. This means $Ran(h) \subseteq Ran(h \circ f)$ and hence, $Ran(h) = Ran(h \circ f)$. Since $h \circ f$ is idempotent, we conclude that $Ran(h)$ is a T-retract of X. We also have $h \circ f|Ran(h) = \delta|Ran(h)$ and, of course, $f \circ h|Ran(\tau) = \delta|Ran(\tau)$. Since $f[Ran(h)] \subseteq Ran(\tau)$ and $h[Ran(\tau)] \subseteq Ran(h)$, it follows that f maps the T-retract, $Ran(h)$, T-isomorphically onto $Ran(\tau)$. This takes care of Case (2.5.2)

Cases (2.5.3) and (2.5.4) can be handled simultaneously. We can suppose that $\tau = g \circ f \circ h$, for if $\tau = g \circ f$, we can take $h = \tau$ since τ is a right identity. Now, h maps $Ran(\tau)$ into $Ran(h)$ and $g \circ f$ maps $Ran(h)$ into $Ran(\tau)$. Moreover, for any $x \in Ran(\tau)$, $g \circ f \circ h(x) = \tau(x) = x$ which means $g \circ f \circ h|Ran(\tau) = \delta|Ran(\tau)$. For any $x \in Ran(h)$, we have $x = h(y)$ for some y and $h \circ g \circ f(x) = h \circ g \circ f \circ h(y) = h \circ \tau(y) = h(y) = x$. That is $h \circ g \circ f|Ran(h) = \delta|Ran(h)$ and we conclude that h maps $Ran(\tau)$ T-isomorphically onto $Ran(h)$. We appeal to Theorem 2.4 and conclude that $Ran(h)$ is a T-retract of X.

Now then, f maps $Ran(h)$ into $Ran(f \circ h)$ and $h \circ g$ maps $Ran(f \circ h)$ into $Ran(h)$. For $x \in Ran(f \circ h)$, we have $x = f \circ h(y)$ for some y and $f \circ h \circ g(x) = f \circ h \circ g \circ f \circ h(y) = f \circ h \circ \tau(y) = f \circ h(y) = x$ which means $f \circ h \circ g|Ran(f \circ h) = \delta|Ran(f \circ h)$. Since, as we showed previously, $h \circ g \circ f|Ran(h) = \delta|Ran(h)$, it follows that f maps $Ran(h)$ T-isomorphically onto $Ran(f \circ h)$. Now, we show that $Ran(f \circ h)$ is a T-retract of X. Note that $f \circ h \circ g$ is idempotent and that $Ran(f \circ h \circ g) \subseteq Ran(f \circ h)$. We showed previously that $f \circ h \circ g(x) = x$ for each $x \in Ran(f \circ h)$ which means $Ran(f \circ h) \subseteq Ran(f \circ g \circ h)$. Thus, $Ran(f \circ h)$ coincides with $Ran(f \circ g \circ h)$ and is therefore a T-retract of X. We have now verified that if $f\mathcal{J}\tau$ where τ is a right identity of $T(X)$, then there exist two T-retracts A and B of X, both T-isomorphic to $Ran(\tau)$ such that f maps A T-isomorphically onto B.

Now consider the case where τ is a left identity of $T(X)$ and $f\mathcal{J}\tau$. Again we have

$$\tau \in \{f\} \cup (f \circ T(X)) \cup (T(X) \circ f) \cup (T(X) \circ f \circ T(X))$$

and again, one of (2.5.1) to (2.5.4) inclusive must hold. As before, if (2.5.1) holds, then it is immediate that there are two T-retracts, A and B, of X, both T-isomorphic to $Ran(\tau)$ such that f maps A T-isomorphically onto B. One again takes $A = B = Ran(\tau)$. Suppose (2.5.3) holds. Then $\tau = g \circ f$. The function f maps $Ran(\tau)$ into $Ran(f \circ \tau)$ and g maps $Ran(f \circ \tau)$ into $Ran(\tau)$. For $x \in Ran(\tau)$, we have $g \circ f(x) = \tau(x) = x$ and for $x \in Ran(f \circ \tau)$, we have $x = f \circ \tau(y)$ for some y and $f \circ g(x) = f \circ g \circ f \circ \tau(y) = f \circ \tau \circ \tau(y) = f \circ \tau(y) = x$. Consequently, f maps $Ran(\tau)$ T-isomorphically onto $Ran(f \circ \tau)$. We have also shown that $Ran(f \circ \tau) \subseteq Ran(f \circ g)$. For $x \in Ran(f \circ g)$, we have $x = f \circ g(y) = f \circ \tau \circ g(y)$ which means $Ran(f \circ g) \subseteq Ran(f \circ \tau)$. Thus, $Ran(f \circ \tau)$ is a T-retract since $Ran(f \circ \tau) = Ran(f \circ g)$ and $f \circ g$ is idempotent.

Now we consider case (2.5.4) and we suppose that $\tau = g \circ f \circ h$. This also covers case (2.5.2) because we have $\tau = \tau \circ f \circ h$ whenever $\tau = f \circ h$. We first show that

$$Ran(h \circ \tau) \text{ is a } T\text{-retract of } X. \tag{2.5.5}$$

Suppose $x \in Ran(h \circ \tau)$. Then $x = h(y)$ for some $y \in Ran(\tau)$ and $x = h \circ \tau(y) = h \circ g \circ f \circ h(y) = h \circ g \circ f \circ \tau \circ h(y) = h \circ g \circ f \circ \tau(x)$ which means $x \in Ran(h \circ g \circ f \circ \tau)$. On the other hand, if $x \in Ran(h \circ g \circ f \circ \tau)$, we have $x = h \circ g \circ f \circ \tau(y) = h \circ \tau \circ g \circ f \circ \tau(y)$ which implies $x \in Ran(h \circ \tau)$. Thus, $Ran(h \circ \tau)$ is a T-retract of X since $Ran(h \circ \tau) = Ran(h \circ g \circ f \circ \tau)$ and $h \circ g \circ f \circ \tau$ is idempotent. This verifies (2.5.5). Next, we show that

$$Ran(f \circ h \circ \tau) \text{ is a } T\text{-retract of } X. \tag{2.5.6}$$

If $x \in Ran(f \circ h \circ \tau)$, then $x = f \circ h \circ \tau(y) = f \circ h \circ g \circ f \circ h(y) = f \circ h \circ g \circ \tau \circ f \circ h(y) \in Ran(f \circ h \circ g \circ \tau)$. That is, $Ran(f \circ h \circ \tau) \subseteq Ran(f \circ h \circ g \circ \tau)$. Moreover, since $f \circ h \circ g \circ \tau = f \circ h \circ \tau \circ g \circ \tau$, we have $Ran(f \circ h \circ g \circ \tau) \subseteq Ran(f \circ h \circ \tau)$. Consequently, $Ran(f \circ h \circ \tau)$ is a T-retract of X since $f \circ h \circ g \circ \tau$ is idempotent.

It is evident that f maps $Ran(h \circ \tau)$ into $Ran(f \circ h \circ \tau)$ and $h \circ g$ maps $Ran(f \circ h \circ \tau)$ into $Ran(h \circ \tau)$. For $x \in Ran(h \circ \tau)$, we have $x = h \circ \tau(y)$ and $(h \circ g) \circ f(x) = h \circ g \circ f \circ h \circ \tau(y) = h \circ \tau \circ \tau(y) = h \circ \tau(y) = x$. For $x \in Ran(f \circ h \circ \tau)$, we have $x = f \circ h \circ \tau(y)$ and we get $f \circ (h \circ g)(x) = f \circ h \circ g \circ f \circ h \circ \tau(y) = f \circ h \circ \tau \circ \tau(y) = f \circ h \circ \tau(y) = x$ and we have shown that f maps $Ran(h \circ \tau)$ T-isomorphically onto $Ran(f \circ h \circ \tau)$.

Now we show that $Ran(h \circ \tau)$ is T-isomorphic to $Ran(\tau)$. Evidently, h maps $Ran(\tau)$ into $Ran(h \circ \tau)$ and $g \circ f$ maps $Ran(h \circ \tau)$ into $Ran(\tau)$. For $x \in Ran(\tau)$, we get $(g \circ f) \circ h(x) = \tau(x) = x$. For $x \in Ran(h \circ \tau)$, we have $x = h \circ \tau(y)$ and $h \circ (g \circ f)(x) = h \circ g \circ f \circ h \circ \tau(y) = h \circ \tau \circ \tau(y) = h \circ \tau(y) = x$. Thus, h maps $Ran(\tau)$ T-isomorphically onto $Ran(h \circ \tau)$. At this point, we have shown that if τ is either a right identity, a left identity, or an identity and $fJ\tau$, then there exist two T-retracts A and B of X, both T-isomorphic to $Ran(\tau)$, such that f maps A T-isomorphically onto B.

Now we verify the converse. So suppose that τ is either a left identity, a right identity or an identity and there exist two T-retracts A and B of X, both T-isomorphic to $Ran(\tau)$, such that f maps A T-isomorphically onto B. Since A is T-isomorphic to $Ran(\tau)$, there exist two functions $g, h \in T(X)$ such that $g[Ran(\tau)] \subseteq A$, $h[A] \subseteq Ran(\tau)$ $h \circ g|Ran(\tau) = \delta|Ran(\tau)$ and $g \circ h|A = \delta|A$. Since f maps A T-isomorphically onto B, there exists a function $k \in T(X)$ such that $k[B] \subseteq A$, $k \circ f|A = \delta|A$ and $f \circ k|B = \delta|B$. Finally, since B is T-isomorphic to $Ran(\tau)$, there exist functions p and q such that $p[B] \subseteq Ran(\tau)$, $q[Ran(\tau)] \subseteq B$, $p \circ q|Ran(\tau) = \delta|Ran(\tau)$ and $q \circ p|B = \delta|B$. One then verifies that $\tau = h \circ k \circ q \circ p \circ f \circ g \circ \tau$ which means that τ belongs to the principal ideal generated by f. Since f also belongs to the principal ideal generated by τ (which is all of $T(X)$) we conclude that $fJ\tau$. This proves the theorem. ❑

THEOREM 2.6: Suppose $T(X)$ contains a proper ideal and suppose that τ is either a left identity, a right identity or an identity of $T(X)$. Then $T(X)$ has a largest proper ideal $M(T(X))$ and $M(T(X))$ consists of all functions $f \in T(X)$ such that if A and B are two T-retracts of X, both T-isomorphic to $Ran(\tau)$, then f does not map A T-isomorphically onto B.

Proof: It is immediate that $T(X)$ has a largest proper ideal $M(T(X))$. According to Lemma 2.3, $f \in M(T(X))$ if and only if f and τ are not J equivalent and in view of Theorem 2.5, the latter is true if and only if for each pair of T-retracts A and B of X, both T-isomorphic to $Ran(\tau)$, f does not map A T-isomorphically onto B. ❑

3. SEMIGROUPS OF CONTINUOUS SELFMAPS

Denote by $S(X)$ the semigroup of all continuous selfmaps of the topological space X and let e be any nonconstant idempotent of $S(X)$. Recall from the introduction that $L(e) = S(X) \circ e$, $R(e) = e \circ S(X)$ and $Loc(e) = e \circ S(X) \circ e$. As we mentioned previously, $Loc(e)$ is just the local subsemigroup of $S(X)$ determined by the idempotent e and is the largest subsemigroup of $S(X)$ for which e is an identity. $L(e)$ and $R(e)$ are the principal left and right ideals, respectively, generated by e and they are the largest subsemigroups of $S(X)$ for which e serves, respectively, as a right and left identity. For any $f \in S(X)$, we define $\pi(f) = \{f^{-1}(f(x)): x \in X\}$ and refer to it as the *decomposition of X induced by f*. We write $\pi(f) \leq \pi(g)$ whenever, for each $A \in \pi(f)$, we have $A \subseteq B$ for some $B \in \pi(g)$ and we say that $\pi(f)$ *refines* $\pi(g)$. In what follows, e will always be a nonconstant idempotent of $S(X)$. We remark that if $e = \delta$, the identity of $S(X)$, then $L(e)$, $R(e)$, and $Loc(e)$ all coincide with $S(X)$ itself.

LEMMA 3.1: The following statements about $f \in S(X)$ are equivalent.

$$f \in L(e). \quad (3.1.1)$$

$$\pi(e) \text{ refines } \pi(f). \quad (3.1.2)$$

$$f \text{ is constant on each } A \in \pi(e). \quad (3.1.3)$$

Proof: Suppose (3.1.1) holds. Then $f \circ e = f$ and if $e(x) = e(y)$, we have $f(x) = f \circ e(x) = f \circ e(y) = f(y)$ which implies $\pi(e) \leq \pi(f)$. Now suppose (3.1.2) holds and let $A \in \pi(e)$. Then $A \subseteq B$ for some $B \in \pi(f)$ so that $f(x) = f(y)$ for any pair of points $x, y \in A$. Consequently, f is constant on A. Suppose (3.1.3) holds. Since e is idempotent, we have $e(e(x)) = e(x)$ so that x, $e(x) \in A \in \pi(e)$. Thus, $f(x) = f \circ e(x)$ for all x and we conclude that $f \in L(e)$. ❑

The proof of the next result is even more straightforward than the that of the previous lemma and will be omitted.

LEMMA 3.2: Let $f \in S(X)$. Then $f \in R(e)$ if and only if $Ran(f) \subseteq Ran(e)$.

The following corollary is an immediate consequence of Lemmas (3.1) and (3.2).

COROLLARY 3.3: Let $f \in S(X)$. Then $f \in Loc(e)$ if and only if $Ran(f) \subseteq Ran(e)$ and f is constant on each $A \in \pi(e)$.

An $L(e)$-retract is the range of an idempotent element of $L(e)$ and when two subsets A and B of X are T-isomorphic with respect to the semigroup $L(e)$, we will say they are $L(e)$-isomorphic. Analogous definitions will also hold for the semigroups $R(e)$ and $Loc(e)$.

LEMMA 3.4: Two $L(e)$-retracts, $R(e)$-retracts, $Loc(e)$-retracts are $L(e)$-isomorphic, $R(e)$-isomorphic, $Loc(e)$-isomorphic, respectively, if and only if they are homeomorphic.

Proof: It is immediate that any two subspaces whatsoever which are $L(e)$-isomorphic, $R(e)$-isomorphic or $Loc(e)$-isomorphic, are homeomorphic. Suppose A and B are two $L(e)$-retracts which are homeomorphic. Then there exist two idempotents $v, w \in L(e)$ such that $Ran(v) = A$ and $Ran(w) = B$. Let h be any homeomorphism from A onto B and let $k = h^{-1}$. It is likely that $h, k \notin S(X)$. However, $h \circ v, k \circ w \in S(X)$. In fact, since $(h \circ v) \circ e = h \circ v$ and $(k \circ w) \circ e = k \circ w$, we have $h \circ v, k \circ w \in L(e)$. For any $x \in A$, we have $k \circ w \circ h \circ v(x) = x$ and for any $x \in B$, we have $h \circ v \circ k \circ w\,(x) = x$. Thus, $h \circ v$ maps A $L(e)$-isomorphically onto B.

Now suppose A and B are two homeomorphic $R(e)$-retracts of X. Then there exist idempotents $v, w \in R(e)$ such that $Ran(v) = A$ and $Ran(w) = B$. Again, let h be a homeomorphism from A onto B and let $k = h^{-1}$. Since $e \circ$

$(w \circ h \circ v) = w \circ h \circ v$ and $e \circ (v \circ k \circ w) = v \circ k \circ w$, it follows that $w \circ h \circ v, v \circ k \circ w \in R(e)$. For $x \in A$, we have $v \circ k \circ w \circ w \circ h \circ v(x) = x$ and for $x \in B$, we have $w \circ h \circ v \circ v \circ k \circ w(x) = x$. Thus, $w \circ h \circ v$ maps A $R(e)$-isomorphically onto B. The case for $Loc(e)$-retracts is similar to the case for $R(e)$-retracts except that here, one observes that $e \circ (w \circ h \circ v) \circ e = w \circ h \circ v$ and $e \circ (v \circ k \circ w) \circ e = v \circ k \circ w$ so that $w \circ h \circ v \in Loc(e)$ and $v \circ k \circ w \in Loc(e)$. ❑

The proof of the following lemma is straightforward and is omitted.

LEMMA 3.5: Let e be any idempotent of $S(X)$. Then $A \subseteq X$ is both an $R(e)$-retract and a $Loc(e)$-retract of X if and only if it is a retract (in the usual sense) of $Ran(e)$. If $A \subseteq Ran(e)$, then A is an $L(e)$-retract of X if and only if A is a retract (in the usual sense) of $Ran(e)$.

It follows immediately from the previous result that both $R(e)$-retracts and $Loc(e)$-retracts must be subspaces of $Ran(e)$. This need not be the case for $L(e)$-retracts and we will eventually see examples of this.

For any semigroup S, we denote by $M_L(S)$ the largest proper left ideal of S (if it exists) and we denote by $M_R(X)$ the largest proper right ideal of S (if it exists). In the following lemma, $T(e)$ will denote any one of the semigroups $L(e)$, $R(e)$ or $Loc(e)$.

LEMMA 3.6: Let e be an idempotent of $S(X)$, let A be any subspace of $Ran(e)$ and suppose B is a $T(e)$-retract of X. Then $f \in T(e)$ maps A $T(e)$-isomorphically onto B if and only if it maps A homeomorphically onto B.

Proof: If f maps A $T(e)$-isomorphically onto B, it is immediate that f maps A homeomorphically onto B. Suppose, conversely, that f maps A homeomorphically onto B and for $x \in B$, define $k(x) = f^{-1}(x)$. Since B is a $T(e)$-retract of X, there exists an idempotent $v \in T(e)$ such that $Ran(v) = B$. Now, $Ran(k \circ v) = A \subseteq Ran(e)$ which implies $e \circ (k \circ v) = k \circ v$. Moreover, if $T(e)$ is either $L(e)$ or $Loc(e)$, then $v \circ e = v$ so that $(k \circ v) \circ e = k \circ v$. It then follows that $k \circ v \in T(e)$ whatever the circumstances. We have $f \circ (k \circ v)(x) = x$ for all $x \in B$ and $(k \circ v) \circ f(x) = x$ for all $x \in A$. Thus, f maps A $T(e)$-isomorphically onto B. ❑

THEOREM 3.7: Let e be a nonconstant idempotent of $S(X)$. Then $M_L(L(e))$ exists and it consists of all functions $f \in L(e)$ such that either $Ran(f)$ is not an $L(e)$-retract or f does not map $Ran(e)$ homeomorphically onto $Ran(f)$.

Proof: Since e is nonconstant, the constant functions in $L(e)$ form a proper left ideal and since e is a right identity for $L(e)$, $M_L(L(e))$ exists. According to Theorem 2.20 of [1], $M_L(L(e))$ consists of all functions $f \in L(e)$ such that f does not map $Ran(e)$ $L(e)$-isomorphically onto $Ran(f)$. Suppose $f \in M_L(L(e))$. We assert that either $Ran(f)$ is not an $L(e)$-retract of X or f

does not map $Ran(e)$ homeomorphically onto $Ran(f)$. Suppose $Ran(f)$ is an $L(e)$-retract. Then f cannot map $Ran(e)$ homeomorphically onto $Ran(f)$ for if it did, it would follow from Lemma 3.6 that f mapped $Ran(e)$ $L(e)$-isomorphically onto $Ran(f)$ and this would be a contradiction. Suppose, conversely, that either $Ran(f)$ is not an $L(e)$-retract or f does not map $Ran(e)$ homeomorphically onto $Ran(f)$. If $Ran(f)$ is not an $L(e)$-retract, it follows from Theorem 2.4 that f cannot map $Ran(e)$ $L(e)$-isomorphically onto $Ran(f)$. On the other hand, if f does not map $Ran(e)$ homeomorphically onto $Ran(f)$, it cannot map $Ran(e)$ $L(e)$-isomorphically onto $Ran(f)$ either. In either event, it follows from Theorem 2.20 of [1] that $f \in M_L(L(e))$. ❑

THEOREM 3.8: Let e be a nonconstant idempotent of $S(X)$. Then $M(L(e))$ exists and it consists of all functions $f \in L(e)$ such that if A and B are any two $L(e)$-retracts of X, both homeomorphic to $Ran(e)$, then f does not map A homeomorphically onto B.

Proof: Again, the constant functions in $L(e)$ form a proper ideal of $L(e)$ since e is nonconstant and the result is now an immediate consequence of Theorem 2.6 and Lemma 3.6. ❑

THEOREM 3.9: Let e be a nonconstant idempotent of $S(X)$. Then $M_R(R(e))$ exists and it consists of all functions $f \in R(e)$ such that f maps no subspace of $Ran(e)$ homeomorphically onto $Ran(e)$.

Proof: According to Theorem 2.22 of [1], $M_R(R(e))$ consists of all functions $f \in R(e)$ such that f maps no subspace of $Ran(e)$ $R(e)$-isomorphically onto $Ran(e)$. But according to Lemma 3.6, a function $f \in R(e)$ satisfies the latter condition if and only if it maps no subspace of $Ran(e)$ homeomorphically onto $Ran(e)$. ❑

THEOREM 3.10: Let e be a nonconstant idempotent of $S(X)$. Then $M(R(e))$ exists and it consists of all functions $f \in R(e)$ such that if A and B are any two retracts of $Ran(e)$, both homeomorphic to $Ran(e)$, then f does not map A homeomorphically onto B.

Proof: This follows from Theorem 2.6 and Lemmas 3.5 and 3.6. ❑

The proofs of the next three results are identical to the proofs of Theorem 3.7, 3.9, and 3.8, respectively.

THEOREM 3.11: Let e be a nonconstant idempotent of $S(X)$. Then $M_L(Loc(e))$ exists and it consists of all functions $f \in Loc(e)$ such that either $Ran(f)$ is not a $Loc(e)$-retract or f does not map $Ran(e)$ homeomorphically onto $Ran(f)$.

THEOREM 3.12: Let e be a nonconstant idempotent of $S(X)$. Then $M_R(Loc(e))$ exists and it consists of all functions $f \in Loc(e)$ such that f maps no subspace of $Ran(e)$ homeomorphically onto $Ran(e)$.

THEOREM 3.13: Let e be a nonconstant idempotent of $S(X)$. Then $M(Loc(e))$ exists and it consists of all functions $f \in Loc(e)$ such that if A and B are any two retracts of $Ran(e)$, both homeomorphic to $Ran(e)$, then f does not map A homeomorphically onto B.

4. SOME EXAMPLES

In this section, we choose two different idempotents of $S(R)$ where R is the space of real numbers and we study, in detail, the largest proper ideals of the three different semigroups associated with each of these idempotents.

EXAMPLE A: Define

$$v(x) = \begin{cases} 0 & \text{for } x < 0 \\ x & \text{for } 0 \le x \le 1 \\ 1 & \text{for } x > 1. \end{cases}$$

We are going to investigate the largest proper ideals of the three semigroups $L(v)$, $R(v)$, and $Loc(v)$. We begin with $L(v) = S(R) \circ v$. Now $\pi(v)$ consists of $(-\infty, 0]$ and $[1, \infty)$ together with all sets of the form $\{x\}$ with $0 < x < 1$. Consequently, the following result is an immediate consequence of Lemma 3.1.

PROPOSITION 4.1: A function $f \in S(R)$ belongs to $L(v)$ if and only if it is constant on both $(-\infty, 0]$ and $[1, \infty)$.

In order to apply the results in Section 3, we next need to determine precisely what the $L(v)$-retracts are. This is the content of our next result.

PROPOSITION 4.2: The $L(v)$-retracts of R are precisely the singletons of R together with all closed subintervals of $[0, 1]$.

Proof: It is immediate that all singletons are $L(v)$-retracts of R. Suppose r is a nonconstant idempotent of $L(v)$ and let $Ran(r) = [a, b]$. Then $r(x) = x$ for $a \le x \le b$. Suppose $a < 0$. Since r is constant on $(-\infty, 0]$, we must have $r(x) = r(0)$ for all $x \le 0$. On the other hand, we can choose c such that $a < c < \min\{0, b\} \le 0$ and we have $r(a) = a$ and $r(c) = c$ which means that $r(a)$ and $r(c)$ cannot possibly both coincide with $r(0)$. This means $a \ge 0$. In the same manner, one shows that $b \le 1$ and this implies that $Ran(r) = [a, b]$ is a subinterval of $[0, 1]$. Conversely, let $[a, b]$ be any subinterval of $[0, 1]$ and define

$$s(x) = \begin{cases} a & \text{for } x < a \\ x & a \le x \le b \\ b & \text{for } x > b. \end{cases}$$

Then $s \in L(v)$ since it is constant on both $(-\infty, 0]$ and $[1, \infty)$, it is certainly idempotent and $Ran(s) = [a, b]$. ❑

The following two results are now immediate consequences of the results in Section 3 and Proposition 4.2.

PROPOSITION 4.3: $M_L(L(v))$ consists of all functions $f \in L(v)$ which satisfy one of the two following conditions.

$$f \text{ is nonconstant and } Ran(f) \not\subseteq [0, 1] \tag{4.3.1}$$

or

$$f \text{ does not map } [0, 1] \text{ homeomorphically onto } Ran(f). \tag{4.3.2}$$

PROPOSITION 4.4: $M(L(v))$ consists of all functions $f \in L(v)$ such that f maps no nondegenerate subinterval of $[0, 1]$ homeomorphically onto another subinterval of $[0, 1]$.

PROPOSITION 4.5: $L(v)$ has no largest proper right ideal.

Proof: Define $J_n = \{f \in L(v): Ran(f) \subseteq [-n, n]\}$. Then each J_n is a proper right ideal of $L(v)$ and $L(v) = \cup \{J_n\}_{n=1}^{\infty}$. It follows from this that $L(v)$ cannot have a largest proper right ideal. ❑

Now we consider the semigroup $R(v) = v \circ S(R)$. Now $f \in S(R)$ belongs to $R(v)$ if and only if $v \circ f = f$ and this immediately implies:

PROPOSITION 4.6: A function $f \in S(R)$ belongs to $R(v)$ if and only if $Ran(f) \subseteq [0, 1]$.

Our next result is also quite easily verified.

PROPOSITION 4.7: The $R(v)$-retracts of X are precisely the closed subintervals of $[0, 1]$.

The latter two results, together with the results in Section 3, immediately yield the following two results.

PROPOSITION 4.8: $M_R(R(v)))$ consists of all functions $f \in R(v)$ such that f maps no subinterval of $[0, 1]$ homeomorphically onto $[0, 1]$.

PROPOSITION 4.9: $M(R(v))$ consists of all functions $f \in R(v)$ such that f is not injective on any subarc of $[0, 1]$.

PROPOSITION 4.10: $R(v)$ has no largest proper left ideal.

Proof: Let H denote the collection of all functions $h \in S(R)$ such that h is a homeomorphism from R into $[0, 1]$. Then $H \subseteq R(v)$ and, moreover, we assert that

$$\text{Any two elements of } H \text{ are } \mathcal{L}\text{-equivalent.} \tag{4.10.1}$$

Suppose $h, k \in H$, let $Ran(h) = (a, b)$ and $Ran(k) = (c, d)$. For $x \in (c, d)$, define $g(x) = h \circ k^{-1}(x)$ and then define

$$g(c) = \lim_{x \to c^+} g(x) \quad \text{and} \quad g(d) = \lim_{x \to d^-} g(x).$$

Note that $g(c)$ is one of the two points a, b and $g(d)$ is the other. Finally define $g(x) = g(c)$ for $x \leq c$ and $g(x) = g(d)$ for $x \geq d$. Then $g \in R(v)$ and $h = g \circ k$. That is, h belongs to the principal left ideal generated by k. Similarly, k belongs to the principal left ideal generated by h and we have verified (4.10.1).

Now let $J = R(v) \backslash H$. It is immediate that J is a left ideal of $R(v)$ and because of (4.10.1), it is maximal. We also note that it is prime since H is a subsemigroup of $R(v)$. However, J is not a largest proper left ideal of $R(v)$. J certainly does not contain $R(v) \circ H$ and our assertion will be verified when we show that

$$R(v) \circ H \text{ is a proper left ideal of } R(v). \tag{4.10.2}$$

$R(v) \circ H$ is evidently a left ideal. Define $f(x) = (1/2)(1 + \sin x)$. We will show that $f \notin R(v) \circ H$. Deny this and suppose that $f = g \circ h$ where $g \in R(v)$ and $h \in H$. Then $Ran(h) = (a, b) \subseteq [0, 1]$ and it follows that $g(x) = f \circ h^{-1}(x)$ for $x \in (a, b)$. But then

$$g(a) = \lim_{x \to a^+} f \circ h^{-1}(x) = \lim_{x \to a^+} (1/2)(1 + \sin(h^{-1}(x)))$$

and since either

$$\lim_{x \to a^+} h^{-1}(x) = \infty \quad \text{or} \quad \lim_{x \to a^+} h^{-1}(x) = -\infty,$$

it follows that

$$\lim_{x \to a^+} (1/2)(1 + \sin(h^{-1}(x)))$$

doesn't exist. This verifies (4.10.2) and completes the proof of the proposition as well. ❑

Now we collect the results for $Loc(v)$. The first such result is an immediate consequence of Propositions (4.1) and (4.6) and the fact that $Loc(v) = L(v) \cap R(v)$ and the second result is also quite easily verified.

PROPOSITION 4.11: A function $f \in S(R)$ belongs to $Loc(v)$ if and only if it is constant on both $(-\infty, 0]$ and $[1, \infty)$ and $Ran(f) \subseteq [0, 1]$.

PROPOSITION 4.12: The $Loc(v)$-retracts of X are precisely the closed subintervals of $[0, 1]$.

The next three results then follow from the latter two and Theorems 3.11, 3.12, and 3.13.

PROPOSITION 4.13: $M_L(Loc(v))$ consists of all functions $f \in Loc(v)$ such that f does not map [0, 1] homeomorphically onto $Ran(f)$.

PROPOSITION 4.14: $M_R(Loc(v))$ consists of all functions $f \in Loc(v)$ such that f maps no subarc of [0, 1] homeomorphically onto [0, 1].

PROPOSITION 4.15: $M(Loc(v))$ consists of all functions $f \in Loc(v)$ such that f is not injective on any subarc of [0, 1].

EXAMPLE B: For this example, we define an idempotent w by

$$w(x) = \begin{cases} x - n & \text{for } n \leq x \leq n + 1 \text{ when } n \text{ is an even integer} \\ 1 + n - x & \text{for } n \leq x \leq n + 1 \text{ when } n \text{ is an odd integer.} \end{cases}$$

The function w is a "sawtooth" function and, as we just stated, it is idempotent. For one thing, $Ran(w) = [0, 1]$ and $w(x) = x$ for $x \in [0, 1]$. We will examine $L(w)$, $R(w)$, and $Loc(w)$. We begin with $L(w) = S(R) \circ w$ and our first result gives us some indication what the functions in $L(w)$ look like.

PROPOSITION 4.16: Let g be any continuous function from [0, 1] into R and define a continuous selfmap f of R by

$$f(x) = \begin{cases} g(x - n) & \text{for } n \leq x \leq n + 1 \text{ when } n \text{ it is an even integer} \\ g(1 + n - x) & \text{for } n \leq x \leq n + 1 \text{ when } n \text{ is an odd integer.} \end{cases}$$

Then $f \in L(w)$ and every function in $L(w)$ is obtained in this manner.

Proof: In order to show that $f \in L(w)$, we need only show that $f \circ w = f$. Suppose $n \leq x \leq n + 1$ and n is even. Then $f(x) = g(x - n)$. However, $0 \leq x - n \leq 1$ and we have $f \circ w(x) = f(x - n) = g(x - n)$. Now suppose $n \leq x \leq n + 1$ and n is odd. In this case, we have $f(x) = g(1 + n - x)$. We also have $0 \leq 1 + n - x \leq 1$ and $f \circ w(x) = f(1 + n - x) = g(1 + n - x)$. Consequently, $f = f \circ w$ and $f \in L(w)$.

Now suppose $f \in L(w)$ and let $g = f|[0, 1]$. Suppose $n \leq x \leq n + 1$ and n is even. Then $f = f \circ w$ since $f \in L(w)$ and since $0 \leq x - n \leq 1$, we have $f(x) = f \circ w(x) = f(x - n) = g(x - n)$. If $n \leq x \leq n + 1$ and n is odd, we have $0 \leq 1 + n - x \leq 1$ and $f(x) = f \circ w(x) = f(1 + n - x) = g(1 + n - x)$. This proves the proposition. ❑

The previous result tells us that when a function f belongs to $L(w)$, the graph of $f|[1, 2]$ must be the "mirror image" of the graph of $f|[0, 1]$ and the graph of $f|[0, 2]$ is simply repeated on all intervals of the form $[2n, 2n + 2]$. Our next result tells just what the $L(w)$-retracts are.

PROPOSITION 4.17: The $L(w)$-retracts of R are precisely the closed subintervals of R which contain no integers in their interior.

Proof: Let r be an arbitrary idempotent in $L(w)$. Then $r = r \circ w$ and $r(x) = x$ for each $x \in Ran(r)$. Let $Ran(r) = [a, b]$ and suppose $a < n < b$ for some integer n. Then we can choose two points c, d such that $a < c < n < d < b$ and $w(c) = w(d)$. But then, $r(c) = r \circ w(c) = r \circ w(d) = r(d)$ and this contradicts the fact that $r(c) = c$ and $r(d) = d$. We therefore conclude that $Ran(r)$ has no integers in its interior.

Suppose, conversely, that $[a, b]$ contains no integers in its interior. Then $[a, b] \subseteq [n, n + 1]$ for some integer n. We must produce an idempotent map in $L(w)$ whose range is $[a, b]$ and we give the details only in the case where n is even, the remaining case following in a somewhat similar manner. We have $0 \leq a - n \leq b - n \leq 1$. Define a function g from $[0, 1]$ into R by

$$g(x) = \begin{cases} a & \text{for } 0 \leq x \leq a - n \\ x + n & \text{for } a - n \leq x \leq b - n \\ b & b - n \leq x \leq 1. \end{cases}$$

Now g does not belong to $S(R)$ but $g \circ w$ does belong to $S(R)$. In fact, $g \circ w \in L(w)$ since $(g \circ w) \circ w = g \circ w$ and it is evident that $Ran(g \circ w) = [a, b]$. Moreover, for $a \leq x \leq b$, we have $n \leq x \leq n + 1$ and $a - n \leq x\text{-}n \leq b - n$ which imply $(g \circ w)(x) = g(x - n) = x$. That is, $g \circ w|[a, b] = \delta|[a, b]$ and $g \circ w$ is idempotent. Consequently, $[a, b]$ is an $L(w)$-retract of X. ❑

Our next result is an immediate consequence of Theorem 3.7 and Proposition 4.17.

PROPOSITION 4.18: $M_L(L(w))$ consists of all functions $f \in L(w)$ such that either $Ran(f)$ contains an integer in its interior or f does not map $[0, 1]$ homeomorphically onto $Ran(f)$.

PROPOSITION 4.19: The following statements are equivalent for $f \in L(w)$.

$$f \in M(L(w)). \tag{4.19.1}$$

$$f \text{ is not injective on any subarc of } [0, 1]. \tag{4.19.2}$$

$$f \text{ is not injective on any subarc of } R. \tag{4.19.3}$$

Proof: Suppose (4.19.1) holds. Then according to Theorem 3.8, if A and B are two $L(w)$-retracts, both homeomorphic to $[0, 1]$, then f cannot map A homeomorphically onto B. But according to Proposition 4.17, these are precisely the closed subintervals of the intervals of the form $[n, n + 1]$ where n is an integer. It readily follows from this that (4.19.2) holds and (4.19.2) implies (4.19.3) because of Proposition 4.16. Statement (4.19.3) implies (4.19.1) by Theorem 3.8. ❑

The proof of the following proposition is identical to the proof of Proposition 4.5.

PROPOSITION 4.20: $L(w)$ has no largest proper right ideal.

The next five results follow just as Propositions 4.6 to 4.10 respectively.

PROPOSITION 4.21: A function $f \in S(R)$ belongs to $R(w)$ if and only if $Ran(f) \subseteq [0, 1]$.

PROPOSITION 4.22: The $R(w)$-retracts of X are precisely the closed subintervals of $[0, 1]$.

PROPOSITION 4.23: $M_R(R(w))$ consists of all functions $f \in R(w)$ such that f maps no subinterval of $[0, 1]$ homeomorphically onto $[0, 1]$.

PROPOSITION 4.24: $M(R(w))$ consists of all functions $f \in R(w)$ such that f is not injective on any subarc of $[0, 1]$.

PROPOSITION 4.25: $R(w)$ has no largest proper left ideal.

Now we consider the semigroup $Loc(w)$.

PROPOSITION 4.26: Let g be any continuous function from $[0, 1]$ into $[0, 1]$ and define a continuous selfmap f of R by

$$f(x) = \begin{cases} g(x - n) & \text{for } n \leq x \leq n + 1 \text{ when } n \text{ is an even integer} \\ g(1 + n - x) & \text{for } n \leq x \leq n + 1 \text{ when } n \text{ is an odd integer.} \end{cases}$$

Then $f \in Loc(w)$ and every function in $Loc(w)$ is obtained in this manner.

Proof: The proof follows from Propositions 4.16, 4.21, and the fact that $Loc(w) = L(w) \cap R(w)$.

We conclude simply by stating that Propositions 4.12 to 4.15 inclusive remain valid if the idempotent v is replaced by the idempotent w and the proofs remain valid as well.

REFERENCE

1. MAGILL, K.D., Jr. 1993. Green's equivalences and related concepts for semigroups of continuous selfmaps. General Top. and App., Proc. 1991 Northeast Conf. in honor of Mary Ellen Rudin. Annals of the New York Academy of Sciences. **704:** 246–268.

A Minimal Sober Topology Is Always Scott

A.E. McCLUSKEY[a] AND S.D. McCARTAN[b]

[a]*Department of Mathematics*
University College, Galway
Galway, Ireland
and
[b]*Department of Mathematics*
Queen's University, Belfast
BT7 1NN, Northern Ireland

ABSTRACT: Given the lattice of all topologies definable for an infinite set X, we employ the well-developed contraction technique (of intersecting a given topology with a suitably chosen principal ultratopology) for solving many minimality problems. We confirm its potential in characterising and, where possible, identifying those topologies which are minimal with respect to certain separation axioms, notably that of sobriety and its conjunction with various other axioms. Finally, we offer an alternative description of each topologically established minimal structure in terms of the behavior of the naturally occurring specialization order and the intrinsic topology on the resulting partially-ordered set.

INTRODUCTION

The family of all topologies definable for an infinite set X is a complete, atomic, and complemented lattice (under inclusion) and is often denoted by $LT(X)$. Given a topological invariant property Q, a member $\mathcal{T}$ of $LT(X)$ is said to be *minimal* Q is and only if $\mathcal{T}$ possesses property Q, but no weaker members of $LT(X)$ possess property Q. In seeking to identify those members of $LT(X)$ which minimally satisfy an invariant property, we are, in a very real sense, examining the topological essence of the invariant. In the recent past, attention has concentrated on the characterization (and, if possible, identification) of those members of $LT(X)$ which are minimal with respect to certain separation axioms lying in logical strength between T_0 and T_1 (see [5], [6], [9], [10], [11], [12], [13], [14], [15], [17]).

In the same spirit, the main purpose of the present paper is to characterize those spaces which are minimally *sober*, an invariant introduced by Grothendieck *et al.* [3], which lies in logical strength between T_0 and T_2. We observe that such topologies are always *Scott* and, further, provide sufficient order-theoretic conditions for the *Scott* topology to be sober (see [7]). Sobriety, which has arisen naturally in the study of continuous lattices (see Scott

Mathematics Subject Classification: Primary 54A10, 54D10; Secondary 06A10.
Key words and phrases: minimal T-spaces, specialization order.

et al. [18]), has been investigated by Hoffmann [4] and others, and, indeed, has been observed to play a fundamental role when one takes a topological view of predicate transformers and power domains in the theory of computing science (see Smyth [19]). When combined with some of the separation axioms referred to previously, sobriety assumes several pleasant features. In particular it may become expansive and hence qualify as a separation axiom between T_0 and T_2. Accordingly, we examine, also, minimality with respect to sobriety in conjunction with each of the axioms T_D, T_{ES}, T_{EF}, and T_1.

PRELIMINARIES

Given $\mathcal{T} \in LT(X)$, a subset A of X is said to be $\mathcal{T}$-*irreducible* if and only if, given $G_1, G_2 \in \mathcal{T}$, $A \cap G_1 \neq \phi$ and $A \cap G_2 \neq \phi$ together imply $A \cap G_1 \cap G_2 \neq \phi$. It is immediate that a $\mathcal{T}$-closed subset F of X is $\mathcal{T}$-irreducible if and only if F cannot be expressed as a union of two nonempty $\mathcal{T}$-closed proper subsets of F. If $x \in X$, then (assuming there is no danger of ambiguity) $\overline{A}$ denotes the $\mathcal{T}$-closure of A, so that $\overline{\{x\}}$ is the $\mathcal{T}$-closure of $\{x\}$, and we often refer to it as a point-closure. It is evident that $\overline{\{x\}}$ is $\mathcal{T}$-irreducible. We recall that, with increasing logical strength, $(X, \mathcal{T})$ is said to be:

(1) T_0 if and only if $\overline{\{x\}} = \overline{\{y\}}$ implies $x = y$ for all $x, y \in X$
(2) T_D if and only if $\overline{\{x\}} \backslash \{x\} = \{y \in \overline{\{x\}} : y \neq x\}$ is $\mathcal{T}$-closed for all $x \in X$ (see [1], [2], [9], [17])
(3) T_{ES} if and only if each singleton subset of X is either $\mathcal{T}$-open or $\mathcal{T}$-closed (see [10], [16])
(4) T_{EF} if and only if each finite subset of X is either $\mathcal{T}$-open or $\mathcal{T}$-closed (see [10], [16])
(5) T_1 if and only if each singleton subset of X is $\mathcal{T}$-closed.

Given $\mathcal{T} \in LT(X)$, $(X, \mathcal{T})$ is said to be a *sober* space if and only if every nonempty $\mathcal{T}$-closed $\mathcal{T}$-irreducible subset of X is the $\mathcal{T}$-closure of a unique singleton. It is immediate that a sober space is T_0, while every nonempty irreducible subset of a T_2-space is a singleton, so that a T_2-space is sober.

DEFINITION 1: ([11], [12], [13], [14]) Given $x \in X$ and $Y \subseteq X$, we define the following members of $LT(X)$:

$\mathcal{D}$ = the discrete member of $LT(X)$

$\mathcal{I}(Y) = \{G \subseteq X : \subseteq G\} \cup \{\varnothing\}$

$\mathcal{E}(Y) = \mathcal{P}(X \backslash Y) \cup \{X\}$

$\mathcal{I}(x) = \{G \subseteq X : x \in G\} \cup \{\varnothing\}$,
"included point" member of $LT(X)$

$\mathcal{E}(x) = \mathcal{P}(X\backslash\{x\}) \cup \{X\}$,
"excluded point" member of $LT(X)$

$C =$ The cofinite (or minimum T_1) member of $LT(X)$

$\mathcal{M}(Y) = \{G \subseteq X\colon Y \subseteq G \text{ or } G \subseteq Y\}$

Here $\mathcal{P}(A)$ denotes the power set of A.

The cofinite topology, C, provides an example of a space which is T_1 (and hence T_{EF}, T_{ES}, T_D, and T_0) but *not* sober, for X, itself, is C-closed, C-irreducible, and is not a singleton. The included and excluded point topologies, $\mathcal{I}(x)$ and $\mathcal{E}(x)$, provide examples of non-T_1 (yet T_{EF})-spaces which are sober. Observe, also, that $C \cap \mathcal{I}(x)$ is a sober topology (the irreducible closed sets being the singleton subsets of $X\backslash\{x\}$ together with $X = \overline{\{x\}}$), thereby demonstrating that sobriety is not an expansive invariant (in contrast to the separation axioms mentioned above).

Finally we recall that a member $\mathcal{T}$ of $LT(X)$ is said to be *nested* if and only if either $G_1 \subseteq G_2$ or $G_2 \subseteq G_1$, for all $G_1, G_2 \in \mathcal{T}$.

Three results in the literature (see [4], [6], [9], [10], [11], [12], [13], [14]) are quoted without proof.

LEMMA 2: Given $\mathcal{T} \in LT(X)$, the following statements are equivalent:
(i) $\mathcal{T}$ is nested.
(ii) For all $x,y \in X$, either $\overline{\{x\}} \subseteq \overline{\{y\}}$ or $\overline{\{y\}} \subseteq \overline{\{x\}}$.
(iii) Finite unions of point-closures are point-closures.
(iv) Every $\mathcal{T}$-closed subset of X is $\mathcal{T}$-irreducible.
(v) For all $B \in \mathcal{T}$, $\mathcal{T} \subseteq \mathcal{M}(B)$ in $LT(X)$.

LEMMA 3: Let $\mathcal{T} \in LT(X), A \subseteq X, x \neq y$ in X and $\mathcal{T}^* = T \cap (\mathcal{I}(y) \cup \mathcal{E}(x))$. Then the $\mathcal{T}^*$-closure of A is described by

$$\overline{A}^* = \begin{cases} \overline{A}, & \text{if } y \notin \overline{A} \\ \overline{A} \cup \overline{\{x\}}, & \text{if } y \in \overline{A}. \end{cases}$$

LEMMA 4: Given $\mathcal{T} \in LT(X)$, $(X, \mathcal{T})$ is T_D and sober if and only if (X, S) is T_D and sober for all $S \in LT(X)$ with $\mathcal{T} \subseteq S$.

LEMMA 5: Given $\mathcal{T} \in LT(X)$ and $x \in X$, $(X, \mathcal{T} \cap \mathcal{E}(x))$ is T_D and sober if and only if $(X, \mathcal{T})$ is T_D, sober, and $\{x\}$ is $\mathcal{T}$-closed.

Proof: For convenience write $\mathcal{T}^* = \mathcal{T} \cap \mathcal{E}(x)$.

Let $(X, \mathcal{T}^*)$ be T_D and sober; then, by Lemma 4, $(X, \mathcal{T})$ is T_D and sober. Also, $\overline{\{x\}}\backslash\{x\} = \overline{\{x\}}^*\backslash\{x\}$ which, being $\mathcal{T}^*$-closed, is empty; that is, $\{x\}$ is $\mathcal{T}$-closed.

Conversely, suppose $(X, \mathcal{T})$ is T_D and sober and that $\{x\}$ is $\mathcal{T}$-closed. Note that, given any $z \in X$,

$$\overline{\{z\}}^{\,*} = \begin{cases} \overline{\{z\}}\,, & \text{if } x \in \overline{\{z\}} \\ \overline{\{z\}} \cup \{x\}, & \text{if } x \notin \overline{\{z\}}\,. \end{cases}$$

so that

$$\overline{\{z\}}^{\,*} \backslash \{z\} = \begin{cases} \overline{\{z\}} \backslash \{z\}, & \text{if } x \in \overline{\{z\}} \\ (\overline{\{z\}} \backslash \{z\}) \cup \{x\}, & \text{if } x \notin \overline{\{z\}}\,. \end{cases}$$

is $\mathcal{T}^*$-closed and hence $(X, \mathcal{T}^*)$ is T_D.

Let F be nonempty, $\mathcal{T}^*$-closed and $\mathcal{T}^*$-irreducible, so that $x \in F$ and F is $\mathcal{T}$-closed. Either F is $\mathcal{T}$-irreducible, in which case, by hypothesis there exists a unique $t \in F$ such that $F = \overline{\{t\}} = \overline{\{t\}}^{\,*}$ (since $x \in \overline{\{t\}}$). Or F is $\mathcal{T}$-reducible, in which case there exist nonempty proper $\mathcal{T}$-closed subsets F_1, F_2 of F such that $F_1 \cup F_2 = F$; it is clear that $x \notin F_1 \cap F_2$ (otherwise F_1 and $F_2\,\mathcal{T}^*$-reduce F), so there is no loss in generality in assuming that $x \in F_1 \cap (X \backslash F_2)$; then $F_2 \cup \{x\} = F$ (otherwise F_1 and $F_2 \cup \{x\}\,\mathcal{T}^*$-reduce F) and F_2 is $\mathcal{T}$-irreducible (otherwise there exist nonempty proper $\mathcal{T}$-closed subsets F_3, F_4 of F_2 such that $F_3 \cup F_4 = F_2$, which implies that $F_3 \cup \{x\}$ and $F_4 \cup \{x\}\,\mathcal{T}^*$-reduce F) so that, by hypothesis, there exists a unique $y \in F_2$ such that $F_2 = \overline{\{y\}}$; but then $F = \cup \{x\} = \overline{\{y\}}^{\,*}$. That is, $(X, \mathcal{T}^*)$ is sober. ❑

LEMMA 6: If $\mathcal{T} \in LT(X)$ and B is a nonempty proper subset of X, let $\mathcal{T}^*$ be the member of $LT(X)$ which is generated by $\mathcal{T} \cup \{X \backslash B\}$. Then $\mathcal{T} \cap \mathcal{M}(B) = \mathcal{T}^* \cap \mathcal{M}(B)$ in $LT(X)$.

Proof: Clearly, $\mathcal{T} \cap \mathcal{M}(B) \subseteq \mathcal{T}^* \cap \mathcal{M}(B)$ in $LT(X)$. On the other hand, let $G \in \mathcal{T}^* \cap \mathcal{M}(B)$; then there exist $G_1, G_2 \in \mathcal{T}$ with $G = G_1 \cup (G_2 \cap (X \backslash B))$ and either $G \subseteq B$ or $B \subseteq G$. If $G \subseteq B$, then $G = G_1$ and so $G \in \mathcal{T} \cap \mathcal{M}(B)$. If $B \subseteq G$, then $B \subseteq G_1$ so that $G = G_1 \cup G_2$, whence $G \in \mathcal{T} \cap \mathcal{M}(B)$. ❑

LEMMA 7: If $(X, \mathcal{T})$ is T_D and sober and $B \in \mathcal{T}$, then $(X, \mathcal{T} \cap \mathcal{M}(B))$ is T_D and sober.

Proof: That $(X, \mathcal{T} \cap \mathcal{M}(B))$ is T_D has been shown by Larson [9].

Let $\mathcal{T}^*$ be the member of $LT(X)$ which is generated by $\mathcal{T} \cup \{X \backslash B\}$; then, by Lemma 4, $(X, \mathcal{T}^*)$ is sober.

For convenience, write $\mathcal{T}^{\perp} = \mathcal{T}^* \cap M(B)$ and observe that, for any $z \in X$,

$$\overline{\{z\}}^{\,\perp} = \begin{cases} \overline{\{z\}}^{\,*} = \overline{\{z\}}\,, & \text{if } z \in X \backslash B; \\ \overline{\{z\}}^{\,*} \cup (X \backslash B) = (\overline{\{z\}} \cap B) \cup (X \backslash B), & \text{if } z \in B. \end{cases}$$

Now let F be a nonempty $\mathcal{T}^{\perp}$-closed $\mathcal{T}^{\perp}$-irreducible subset of X; then either $F \subseteq X \backslash B$ or $X \backslash B \subset F$. If $F \subseteq X \backslash B$, then F is $\mathcal{T}^*$-irreducible (otherwise F is $\mathcal{T}^{\perp}$-reducible) so that, by hypothesis, there exists a unique $z \in X$ such that $F = \overline{\{z\}}^{\,*} = \overline{\{z\}}^{\,\perp}$. If $X \backslash B \subset F$, then $F = (F \cap B) \cup (X \backslash B)$, where $F \cap B$ is nonempty, $\mathcal{T}^*$-closed and hence $\mathcal{T}^*$-irreducible (otherwise F is $\mathcal{T}^{\perp}$-reducible) so that, by hypothesis, there exists a unique $z \in F \cap B$ with $F \cap B = \overline{\{z\}}^{\,*}$; then $F = \overline{\{z\}}^{\,*} \cup (X \backslash B) = \overline{\{z\}}^{\,\perp}$.

That is, $(X, \mathcal{T}^{\perp})$ is sober and the result follows by Lemma 6. ❑

MAIN RESULTS

THEOREM 8: Let $\mathcal{T} \in LT(X)$, let x, y be points in X with $y \notin \overline{\{x\}}$, and let $\mathcal{T}^* = \mathcal{T} \cap (\mathcal{E}(x) \cup \mathcal{I}(y))$. If $(X, \mathcal{T})$ is:

(i) T_0, then $(X, \mathcal{T}^*)$ is T_0 [11];

(ii) T_D, then $(X, \mathcal{T}^*)$ is T_D [11];

(iii) sober, then $(X, \mathcal{T}^*)$ is sober.

Proof: We prove only (iii).

Let F be a nonempty $\mathcal{T}^*$-closed $\mathcal{T}^*$-irreducible subset of X; then F is $\mathcal{T}$-closed and either $y \notin F$ or $x \in F$. There are three cases to consider.

Case I: $x \in F$ and F is $\mathcal{T}$-reducible: then $F = F_1 \cup F_2$ where F_1, F_2 are nonempty proper $\mathcal{T}$-closed subsets of F. Clearly, $x \notin F_1 \cap F_2$ (otherwise F_1 and F_2 $\mathcal{T}^*$-reduce F) so there is no loss of generality in assuming that $x \in F_1 \backslash F_2$, whence $y \in F_2$ (otherwise, again, F_1 and F_2 $\mathcal{T}^*$-reduce F). Thus $F = F_2 \cup \overline{\{x\}}$ (otherwise F_1 and $F_2 \cup \overline{\{x\}}$ $\mathcal{T}^*$-reduce F). Now let $A = \bigcap\{B \subseteq X$: B is $\mathcal{T}$-closed and $B \cup \overline{\{x\}} = F\}$ and observe that A is $\mathcal{T}$-closed, $A \cup \overline{\{x\}} = F$ and A is $\mathcal{T}$-irreducible (otherwise we can again $\mathcal{T}^*$-reduce F). Hence, by hypothesis, there exists $z \in X$ such that $A = \overline{\{z\}}$, whence $F = \overline{\{z\}} \cup \overline{\{x\}} = \overline{\{z\}}^{\,*}$. The uniqueness of z is guaranteed by (i) above.

Case II: $x \in F$ and F is $\mathcal{T}$-irreducible: then, immediately by hypothesis, there exists $z \in X$ such that $F = \overline{\{z\}} = \overline{\{z\}}^{\,*}$ (since $x \in \overline{\{z\}}$). Again, the uniqueness of z is guaranteed by (i) above.

Case III: $x \notin F$: then $y \notin F$ so that F is $\mathcal{T}$-irreducible (otherwise F is $\mathcal{T}^*$-irreducible) whence, by hypothesis, there exists $z \in X$ such that $F = \overline{\{z\}} = \overline{\{z\}}^{\,*}$. As before, the uniqueness of z is guaranteed by (i) above.

Thus $(X, \mathcal{T}^*)$ is sober. ❑

THEOREM 9: Given $\mathcal{T} \in LT(X)$, the following statements are equivalent:

(i) $(X, \mathcal{T})$ is minimal sober.

(ii) $(X, \mathcal{T})$ is sober and nested.

(iii) $(X, \mathcal{T})$ is minimal T_0 and sober.

Proof: (i) implies (ii). If $(X, \mathcal{T})$ is sober and not nested, then, by Lemma 2, there exist $x, y \in X$ with $x \notin \overline{\{y\}}$ and $y \notin \overline{\{x\}}$. Let $\mathcal{T}^* = \mathcal{T} \cap (\mathcal{E}(x)$

$\cup\, \mathcal{I}(y))$ in $LT(X)$. By Theorem 8, $(X, \mathcal{T}^*)$ is sober and, since $\overline{\{y\}} \subset \overline{\{y\}}^*$ $= \cup\, \overline{\{x\}}$, $\mathcal{T}^*$ is strictly weaker than $\mathcal{T}$ in $LT(X)$, so that $(X, \mathcal{T})$ is not minimal (sober).

(ii) implies (iii). By hypothesis and Lemma 2, every nonempty $\mathcal{T}$-closed subset of X is $\mathcal{T}$-irreducible and is, therefore, a unique point-closure. Consequently, the family of all point-closures forms a base for the $\mathcal{T}$-closed subsets of X so that, since $(X, \mathcal{T})$ is T_0 and nested, $(X, \mathcal{T})$ is minimal T_0 (see [9]). ❑

THEOREM 10: Given $\mathcal{T} \in LT(X)$, the following statements are equivalent:

(i) $(X, \mathcal{T})$ is minimal (T_D and sober).
(ii) $(X, \mathcal{T})$ is T_D, sober, and nested.
(iii) $(X, \mathcal{T})$ is minimal T_D and sober.
(iv) $(X, \mathcal{T})$ is minimal T_D and minimal sober.
(v) $(X, \mathcal{T})$ is minimal T_0, minimal T_D, and minimal sober.
(vi) $(X, \mathcal{T})$ is T_D and minimal sober.

Proof: (i) implies (ii). This follows, again by appealing to Lemma 2 and Theorem 8 (or, alternatively, to Lemma 2 and Lemma 7).

(ii) implies (iii). With the hypothesis, Larson [5] has shown that $(X, \mathcal{T})$ is minimal T_D.

(iii) implies (iv). By the hypothesis, Lemma 2, and Theorem 8, $(X, \mathcal{T})$ is nested, and the result follows by Theorem 9. ❑

Given $\mathcal{T} \in LT(X)$, we adopt the notation of [2] by writing $N_D(\mathcal{T}) = \{z \in X\colon \overline{\{z\}} = \{z\}\}$ and $N_O(\mathcal{T}) = \{z \in X\colon \{z\} \in \mathcal{T}\}$ (see also [11], [12], [13], [14]). If $(X, \mathcal{T})$ is T_{ES}, then evidently $N_D(\mathcal{T}) \neq \phi$, $X = N_O(\mathcal{T}) \cup N_D(\mathcal{T})$, and it has been shown in [10] that $\mathcal{E}(X\backslash Y) \cup (C \cap \mathcal{I}(Y)) \subseteq \mathcal{T}$ in $LT(X)$, where $Y = X\backslash N_D(\mathcal{T})$. (Note [10] that $\mathcal{E}(X\backslash Y) \cup (C \cap \mathcal{I}(Y))$ is a minimal T_{ES} member of $LT(X)$ provided that Y is a proper subset of X.) For any $x \in N_D(\mathcal{T}) = X\backslash Y$ it is routinely verifiable that $[\mathcal{E}(X\backslash Y) \cup (C \cap \mathcal{I}(Y))] \cup (\mathcal{T} \cap \mathcal{E}(x))$ is the supremum of $\mathcal{E}(X\backslash Y) \cup (C \cap \mathcal{I}(Y))$ and $\mathcal{T} \cap \mathcal{E}(x)$ in $LT(X)$ and is, therefore, a T_{ES} member of $LT(X)$ which is stronger than $\mathcal{E}(X\backslash Y) \cup (C \cap \mathcal{I}(Y))$ and weaker than $\mathcal{T}$.

Of course, for the important special case where $(X, \mathcal{T})$ is T_1, so that $N_D(\mathcal{T}) = X$, $Y = \phi$, $\mathcal{E}(X\backslash Y)$ is trivial, and $C \cap \mathcal{I}(Y) = C$, this means that $C \cup (\mathcal{T} \cap \mathcal{E}(x))$ is a T_1 member of $LT(X)$ which is stronger than (the minimal T_{ES}) C and weaker than $\mathcal{T}$.

LEMMA 11: (i) For any proper subset A of X, $S = \mathcal{E}(X\backslash A) \cup (C \cap \mathcal{I}(A))$ is a sober member of $LT(X)$ if and only if A is cofinite.

(ii) Given $\mathcal{T} \in LT(X)$ such that $(X, \mathcal{T})$ is T_{ES} and sober, then $(X, \mathcal{T}^*)$ is T_{ES} and sober for each $x \in N_D(\mathcal{T}) = X\backslash Y$ (where $\mathcal{T}^* = [\mathcal{E}(X\backslash Y) \cup (C \cap \mathcal{I}(Y))] \cup (\mathcal{T} \cap \mathcal{E}(x))$ in $LT(X)$).

Proof: (i) Suppose A is proper and cofinite, so that $X\backslash A$ is nonempty and finite, and let F be a $\mathcal{S}$-closed, $\mathcal{S}$-irreducible subset of X. Either $F \subseteq X\backslash A$ or $X\backslash A \subset F$. In the former case, it is clear that F must be a singleton and, therefore, a (unique) point-closure; in the latter case, $F \cap A$ must be a (unique) singleton, say $F \cap A = \{u\}$ (otherwise choose $z \in F \cap A$, with $z \neq u$, and then $\{z\} \cup (X\backslash A)$ and $(F\backslash\{z\}) \cup (X\backslash A$ $\mathcal{S}$-reduce F), whence $F = (F \cap A) \cup (F \cap (X\backslash A)) = \{u\} \cup (X\backslash A) = \overline{\{u\}}$. That is, $(X, \mathcal{S})$ is sober.

Conversely, if A is not cofinite, so that $X\backslash A$ is infinite, then $X\backslash A$, itself, is $\mathcal{S}$-closed, $\mathcal{S}$-irreducible (for the only proper $\mathcal{S}$-closed subsets of $X\backslash A$ are the finite ones), and is obviously not a point-closure. That is, $(X, \mathcal{S})$ is not sober.

(ii) By hypothesis, $(X, \mathcal{T})$ is T_D and sober so that, by Lemma 5, $(X, \mathcal{T} \cap \mathcal{E}(x))$ is T_D and sober for each $x \in N_D(\mathcal{T})$, whence $(X, \mathcal{T}^*)$ is sober by Lemma 4. ❑

THEOREM 12: Given $\mathcal{T} \in LT(X)$, the following statements are equivalent:

(i) $(X, \mathcal{T})$ is minimal (T_{ES} and sober).

(ii) $\mathcal{T} = \mathcal{E}(X\backslash Y) \cup (C \cap \mathcal{I}(Y))$ for some cofinite proper subset Y of X.

(iii) $\mathcal{T} = \mathcal{M}(Y)$ for some cofinite proper subset Y of X.

(iv) $(X, \mathcal{T})$ is minimal T_{ES} and sober.

(v) $(X, \mathcal{T})$ is minimal T_{ES} and $N_D(\mathcal{T})$ is finite.

(vi) $(X, \mathcal{T})$ is minimal T_{ES} and $N_O(\mathcal{T})$ is cofinite.

[Moreover, the representation in (ii) and (iii) is canonical; Y must be $N_O(\mathcal{T}) = X\backslash N_D(\mathcal{T})$.]

Proof: (i) implies (ii). Taking $Y = X\backslash N_D(\mathcal{T})$, then Y is proper and, by Lemma 11(ii), for each $x \in X\backslash Y$, $[\mathcal{E}(X\backslash Y) \cup (C \cap \mathcal{I}(Y))] \cup (\mathcal{T} \cap \mathcal{E}(x)) = \mathcal{T}$. Thus

$$\begin{aligned}\mathcal{T} &= \bigcap\{[\mathcal{E}(X\backslash Y) \cup (C \cap \mathcal{I}(Y))] \cup (\mathcal{T} \cap \mathcal{E}(x)) : x \in X\backslash Y\}\\ &= [\mathcal{E}(X\backslash Y) \cup (C \cap \mathcal{I}(Y))] \cup \{\cap (\mathcal{T} \cap \mathcal{E}(x) : x \in X\backslash Y)\}\\ &= [\mathcal{E}(X\backslash Y) \cup (C \cap \mathcal{I}(Y)] \cup \{\mathcal{T} \cap (\cap (\mathcal{E}(x) : x \in X\backslash Y))\}\\ &= [\mathcal{E}(X\backslash Y) \cup (C \cap \mathcal{I}(Y))] \cup (\mathcal{T} \cap \mathcal{E}((X\backslash Y))\\ &= [\mathcal{E}(X\backslash Y) \cup (C \cap \mathcal{I}(Y))] \cup \mathcal{E}(X\backslash Y)\\ &= \mathcal{E}(X\backslash Y) \cup (C \cap \mathcal{I}(Y))\end{aligned}$$

whence, by Lemma 11(i), Y is cofinite.

The equivalence of (ii) and (iii) is immediate upon observing that, when Y is cofinite, $C \cap \mathcal{I}(Y) = \mathcal{I}(Y)$ in $LT(X)$.

(ii) implies (iv). This follows by Lemma 11(i). ❑

Since, in a minimal T_{ES} space $(X, \mathcal{T})$, $N_D(\mathcal{T}) \cap N_O(\mathcal{T}) = \phi$ and $\mathcal{T} = \mathcal{E}(N_D(\mathcal{T})) \cup (C \cap \mathcal{I}(N_O(\mathcal{T})))$ (see [6]), the equivalence of (iv), (v), and (vi) follows, again by Lemma 11(i).

EXAMPLE: Let Y be an infinite subset of X such that $X\backslash Y$ is also infinite, and let $x \in X\backslash Y$. Since $Y \subseteq X\backslash\{x\}$, the topology $(C \cup \mathcal{E}(x)) \cap \mathcal{M}(Y)$ is T_{ES}. Since $C \cup \mathcal{E}(x)$ is T_2, the sobriety of $(C \cup \mathcal{E}(x)) \cap \mathcal{M}(Y)$ is guaranteed by Lemma 7. However, $Y \cup \{x\} \in \mathcal{M}(Y)$ but $Y \cup \{x\} \notin C \cup \mathcal{E}(x)$, so that $(C \cup \mathcal{E}(x)) \cap \mathcal{M}(Y)$ is strictly weaker than $\mathcal{M}(Y)$ in $LT(X)$.

This example demonstrates that, in general for any given proper subset Y of X, $\mathcal{M}(Y)$, although T_{ES} and sober, is not minimally so (unless Y is cofinite).

THEOREM 13: There are no members $\mathcal{T}$ of $LT(X)$ for which $(X, \mathcal{T})$ is minimal (T_1 and sober).

Proof: Suppose there exists $\mathcal{T} \in LT(X)$ such that $(X, \mathcal{T})$ is minimal (T_1 and sober); then, for each $x \in X$, $C \cup (\mathcal{T} \cap \mathcal{E}(x)) = \mathcal{T}$ and so $\mathcal{T} = \cap \{C \cup (\mathcal{T} \cap \mathcal{E}(x)): x \in X\} = C$, which is the desired contradiction. ❑

This conclusion has also been obtained by Hoffmann [4].

THEOREM 14: Given $\mathcal{T} \in LT(X)$, $(X, \mathcal{T})$ is minimal (T_{EF} and sober) if and only if there exists $x \in X$ such that either $\mathcal{T} = \mathcal{I}(x)$ or $\mathcal{T} = \mathcal{E}(x)$.

Proof: Let $(X, \mathcal{T})$ be minimal (T_{EF} and sober); then either $C \subseteq \mathcal{T}$ in $LT(X)$ or there exists $x \in X$ such that either $\mathcal{I}(x) \subseteq \mathcal{T}$ or $\mathcal{E}(x) \subseteq \mathcal{T}$ in $LT(X)$ (see [10]). But $C \subseteq \mathcal{T}$ implies that $(X, \mathcal{T})$ is T_1 and, therefore, minimal (T_1 and sober). Hence, by Theorem 13, the desired result follows, since each of $((X, \mathcal{I}(x))$ and $(X, \mathcal{E}(x))$ is T_{EF} and sober.

The converse is immediate upon observing that, for any $x \in X$, each of $(X, \mathcal{I}(x))$ and $(X, \mathcal{E}(x))$ is minimal T_{EF} and sober (see [10] again). ❑

ORDER

In this section, we adopt the standard definitions and notation from [11]. (See also [1], [12], [13], [14].)

DEFINITION 15: Let $(X, \leq)$ be a poset with $Y \subseteq X$, and let $n \in \omega$. If C is a chain in X with $|C| = n$, then C is said to have *length* $n - 1$. If the least upper bound, l, of the lengths of all finite chains in Y exists, then we say that Y has *length* l.

Y is said to be a *semi-tree* if and only if for each $y \in Y$, $\{z \in Y: z \leq y\}$ is a chain. Y is said to be a *tree* if and only if Y is a semi-tree with minimum element. The terms *semi-root* and *root* are defined dually.

X is said to satisfy the *ascending chain condition* if and only if every (ascending) chain has a maximum member.

$x \in X$ is said to be *ultramaximal* if and only if x is maximal and for any nonmaximal element $z \in X$, $z \leq x$.

DEFINITION 16: Given a poset $(X, \leq)$, we define the following intrinsic topologies for X:

(1) The *weak* topology, $\mathcal{W}$, whose closed sets are generated by the set $\{\emptyset, X, \downarrow\{x\}: x \in X\}$. Thus, $\mathcal{W}$ is the smallest topology on X in which all sets of the form $\downarrow\{x\}$ are closed. Note further that $\overline{\{x\}} = \downarrow\{x\}$ for all $x \in X$.

(2) The topology, denoted by $\mathcal{M}$, whose closed sets are generated by the family $\{\emptyset, X, \downarrow\{x\}, \downarrow\{x\}\backslash\{x\}: x \in X\}$.

(3) The topology, $\mathcal{L}$, which has as (open) base, the family $\mathcal{M} \cup \{\{x\}: x$ is ultramaximal$\}$.

(4) The *Scott* topology, simply denoted by *Scott*, where $Scott = \{G \subseteq X$: G is increasing and, for any nonempty directed subset S of X, if $\vee S$ exists and $\vee S \in G$, then $G \cap S \neq \emptyset\}$.

(5) The *Alexandroff* topology, $\mathcal{A}$, whose open sets are generated by sets of the form $\uparrow\{x\}$. (It is easily seen that $\mathcal{A}$ is "principal" in that arbitrary intersections of open sets are open.)

Note that $\mathcal{W} \subseteq \mathcal{M} \subseteq \mathcal{L} \subseteq \mathcal{A}$, $\mathcal{W} \subseteq Scott \subseteq \mathcal{A}$ and that for each of these topologies, $\overline{\{x\}} = \downarrow\{x\}$. Given a topological space $(X, \mathcal{T})$, its specialization order is defined by $x \leq y \Leftrightarrow x \in \overline{\{y\}}$. In fact, given a preorder $\leq$ and a topology $\mathcal{T}$ for X, it is well known that $\mathcal{T}$ will have $\leq$ as its specialization order if and only if $\mathcal{W} \subseteq \mathcal{T} \subseteq \mathcal{A}$. (See [11] or [1].) That is, $\mathcal{W}$ is the smallest and $\mathcal{A}$ is the largest of the topologies with a given specialization order and all such topologies have $\overline{\{x\}} = \downarrow\{x\}$ for all $x \in X$.

LEMMA 17: If $(X, \leq)$ is a chain, then $\mathcal{W} = Scott$.

Proof: Let F be a nonempty proper *Scott*-closed subset of X; then either F has a maximum element x say, so that $F = \downarrow\{x\} = \overline{\{x\}}^{\mathcal{W}}$, whence F is $\mathcal{W}$-closed, or F has not a maximum element, in which case $F = \bigcap\{\overline{\{y\}}^{\mathcal{W}}: y \notin F\}$ (for if $z \in \bigcap\{\overline{\{y\}}^{\mathcal{W}}: y \notin F\}$ and $z \notin F$, then $z = \vee F$ which is a contradiction since F is *Scott*-closed. The converse argument is immediate since $\leq$ is linear). Thus, F is again $\mathcal{W}$-closed so that $Scott \subseteq \mathcal{W}$ in $LT(X)$. ❑

LEMMA 18: [8] If $(X, \mathcal{T})$ is sober, then the specialization order $\leq$ has directed joins.

COROLLARY 19: If $(X, \mathcal{T})$ is sober, then $\mathcal{W} \subseteq \mathcal{T} \subseteq Scott$ (with respect to the specialization order).

In the light of [7], we note the following:

LEMMA 20: If $(X, \leq)$ is a poset with directed joins and such that no infinite diverse subset of X exists, then $\mathcal{W}$ = Scott.

Proof: Let F be a nonempty *Scott*-closed subset of X, let $x \in F$ and let C be any maximal chain in $\uparrow\{x\} \cap F$. Then, by hypothesis, $\vee C$ exists and is a member of F. Thus, every element of F is under a maximal element of F. Let M denote the set of all maximal elements of F; then M is diverse and hence finite. Since F is decreasing, $F = \cup\{\downarrow\{z\}: z \in M\} = \cup\{\overline{\{z\}}^{\mathcal{W}}: z \in M\}$. Thus, F is $\mathcal{W}$-closed and so $Scott \subseteq \mathcal{W}$ in $LT(X)$. ❑

COROLLARY 21: If $(X, \leq)$ is a poset with directed joins and such that no infinite diverse subset of X exists, then Scott is sober.

ORDER-THEORETIC MINIMALITY CHARACTERIZATIONS

We now present an order-theoretic description of the previously established minimality results. For the sake of completeness, we include also the order-theoretic characterizations for minimal T_0, T_D, T_{ES} and T_{EF}. (See [11], [13].)

Let $\mathcal{T} \in LT(X)$ with induced order $\leq$.

THEOREM 22: ([1], [4], [11]) $(X, \mathcal{T})$ is *minimal* T_0 if and only if $(X, \leq)$ is a chain and $\mathcal{T} = \mathcal{W}$.

THEOREM 23: ([4], [11]) $(X, \mathcal{T})$ is *minimal* T_D if and only if $(X, \leq)$ is a chain and $\mathcal{T} = \mathcal{M} = \mathcal{A}$.

THEOREM 24: $(X, \mathcal{T})$ is *minimal sober* if and only if $(X, \leq)$ is a chain with directed joins and $\mathcal{T}$ = Scott.

THEOREM 25: $(X, \mathcal{T})$ is *minimal* (T_D and sober) if and only if $(X, \leq)$ is a chain satisfying the ascending chain condition and $\mathcal{T}$ = Scott = $\mathcal{A}$.

THEOREM 26: ([13], [14]) $(X, \mathcal{T})$ is *minimal* T_{ES} if and only if $(X, \leq)$ is a poset such that *either*:

(i) X is diverse and $\mathcal{T} = \mathcal{W}$ or

(ii) all maximal chains in X have unit length, every maximal element is ultramaximal and $\mathcal{T} = \mathcal{L}$.

THEOREM 27: $(X, \mathcal{T})$ is *minimal* (T_{ES} and sober) if and only if $(X, \leq)$ is a poset such that:

(1) all chains in X have unit length,

(2) $\{x \in X: x \text{ is minimal}\}$ is finite and nonempty,

(3) every maximal element is ultramaximal,

and $\mathcal{T} = \mathcal{L}$.

THEOREM 28: $(X, \mathcal{T})$ is *minimal* T_{EF} if and only if $(X, \leq)$ is a poset such that *either*

(i) X is diverse and $\mathcal{T} = \mathcal{W}$ or

(ii) X is a root of length 1 and $\mathcal{T} = \mathcal{A}$ or
(iii) X is a tree of length 1 and $\mathcal{T} = \mathcal{A}$.

THEOREM 29: $(X, \mathcal{T})$ is *minimal* (T_{EF} and sober) if and only if *either*:
(i) $(X, \leq)$ is a root of length 1 and $\mathcal{T} = \mathcal{A}$ or
(ii) $(X, \leq)$ is a tree of length 1 and $\mathcal{T} = \mathcal{A}$.

REFERENCES

1. ANDIMA, S.J. & W.J. THRON. 1978. Order-induced topological properties. Pacific J. Math. **75:** (2).
2. AULL, C.E. & W.J. THRON. 1962. Separation axioms between T_0 and T_1. Indag. Math. **24:** 26–37.
3. ARTIN, M., A. GROTHENDIECK, & J. VERDIER. 1972. Théorie des topos et cohomologie étale des schémas. Springer Lect. Notes in Math. **269**. Berlin-Heidelberg-New York.
4. HOFFMANN, R.E. 1977. Irreducible filters and sober spaces. Manuscripta Math. **22:** 365–380.
5. JOHNSTON, B. & S.D. MCCARTAN. 1980. Minimal T_F-spaces and minimal T_{FF}-spaces. Proc. R. Ir. Acad. **80A:** 93–96.
6. JOHNSTON, B. & S.D. MCCARTAN. 1988. Minimal T_{YS}-spaces and minimal T_{DD}-spaces. Proc. R. Ir. Acad. **88A:** 23–28.
7. JOHNSTONE, P.T. 1981. Scott is not always sober. *In:* Continuous Lattices. Springer LNM. **871:** 282–283.
8. JOHNSTONE, P.T. 1982. Stone spaces. Cambridge University Press. New York.
9. Larson, R.E. 1969. Minimal T_0-spaces and minimal T_D-spaces. Pacific J. Math. **31:** 451–458.
10. MCCARTAN, S.D. 1979. Minimal T_{ES}-spaces and minimal T_{EF}-spaces. Proc. R. Ir. Acad. **79A:** 11–13.
11. MCCLUSKEY, A.E. & S.D. MCCARTAN. 1992. The minimal structures for T_A. Ann. New York Acad. Sci. **659:** 138–155.
12. MCCLUSKEY, A.E. & S.D. MCCARTAN. 1995. Minimality with respect to Youngs' axiom. Houston J. Math. **21:** (2) 413–428.
13. MCCLUSKEY, A.E. & S.D. MCCARTAN. 1997. Minimal structures for T_{FA}. Rend. dell'Istituto di Matematica dell'Universitá di Trieste. To appear.
14. MCCLUSKEY, A.E. & S.D. MCCARTAN. 1993. Minimality with respect to T_{SA} and T_{SD}. Topology with App. (Szekszárd, Hungary). 83–97.
15. MCCLUSKEY, A.E. & S. WATSON. 1997. Minimal T_{UD}-spaces. To be published.
16. MCSHERRY, D.M.G. 1974. On separation axioms weaker than T_1. Proc. R. Ir. Acad. **74:** (16) 115–118.
17. PAHK, KI-HYUN. 1968. Note on the characterizations of minimal T_0 and T_D spaces. Kyungpook Math. J. **8:** 5–10.
18. SCOTT, D., G. GIERZ, K.H. HOFMANN, K. KIEMEL, J.D. LAWSON, & M. MISLOVE. 1980. A compendium of continuous lattices. Springer. Berlin-Heidelberg- New York.
19. SMYTH, M.B. 1982. Power domains and predicate transformers: A topological view. Springer Lect. Notes in Comp. Sci. **154**.(1982).

Partial Metrics, Valuations, and Domain Theory

S. J. O'NEILL[a]

Department of Computer Science
University of Warwick
Coventry CV4 7AL, UK

ABSTRACT: In this paper we develop some connections between the partial metrics of Matthews and the topological aspects of domain theory. We do this by introducing the valuation spaces, which are a special class of partial metric spaces. We develop the natural duality of partial metrics and propose that a natural context in which to view a partial metric space is as a bitopological space. We then see that successive conditions on a valuation can ensure that the pmetric topology is first of all order consistent (with the underlying poset), then equivalent to the Scott topology, and finally that the induced metric topology is equivalent to the patch topology.

1. INTRODUCTION

In this paper we develop some connections between the partial metrics of [5] and the topological aspects of domain theory. We do this by introducing the valuation spaces, which are a special class of partial metric spaces, built from an auxillary function (a valuation) in much the same way as the normed spaces are a special class of metric spaces.

We make one significant change to the definition of the partial metrics of [5], and that is to extend their range from $[0, \infty)$ to $\Re$. This leads us naturally to both a dual pmetric and an induced metric on the space. In particular, we consider how the induced topologies are related, and propose that a natural context in which to view a partial metric space is as a bitopological space.

The connections with domain theory begin when we see that successive conditions on a valuation can ensure that the pmetric topology is first of all order consistent (with the underlying poset), then equivalent to the Scott topology, and finally that the induced metric topology is equivalent to the patch topology.

We end the paper with a Cartesian closed category of domains, for which we have a natural method for deriving a valuation such that:

$$\text{Scott Topology} \equiv \text{Pmetric Topology},$$

Mathematics Subject Classification: 54E35, 54E55, 06B30, 06B35.
Key words and phrases: metric, partial metric, valuation, bitopology, domain.
[a]E-mail: sjo@dcs.warwick.ac.uk.

Lawson Topology ≡ Induced Metric Topology.

Throughout this paper we will use standard notation from domain theory, for which we reference [1]. The references for the topological aspects of domain theory are [4], [8]. Toward the end of the paper we need some results from metric spaces which can be found in any standard text (such as [9]).

All the results that we prove are original. Results that we state are either simple generalizations from [5] or else can be found in the literature mentioned above.

2. PARTIAL METRICS AND VALUATIONS

In this section we introduce the partial metrics and induced topologies. We begin with the valuation spaces, which we will see are a special class of partial metric spaces.

2.1. Valuation Spaces

Rather than starting with dcpos, as in domain theory, we introduce directed completeness only when we consider completeness in a more general setting. The basic structures with which we do work are sufficient for defining a valuation space.

DEFINITION 2.1: A *consistent semilattice* is a poset $(S, \sqsubseteq)$ such that

1. $\forall x,y \in S, x \sqcap y \in S$.
2. If $\{x, y\} \subseteq S$ is consistent (bounded above), then $x \sqcup y \in S$.

DEFINITION 2.2: A *valuation space* is a consistent semilattice $(S, \sqsubseteq)$ and a function $\mu: S \rightarrow \Re$, called the *valuation*, such that

1. If $x \sqsubseteq y$ and $x \neq y$, then $\mu(x) < \mu(y)$.
2. If $\{x, y\} \subseteq S$ is consistent, then $\mu(x) + \mu(y) = \mu(x \sqcap y) + \mu(x \sqcup y)$.

The valuations we use are the positive valuations of lattice theory (as in [2]). We note that valuations in domain theory are more usually defined on the lattice of open sets of some topology, for example see [3].

2.2. The Pmetric Topology

Essentially, the partial metric generalization is that the distance of a point from itself is not necessarily zero anymore. The axioms were first introduced in [5], where the range of a pmetric was restricted to $[0, \infty)$. We extend the range to $\Re$ since this is both natural (in that there is no difficulty in extending the results from [5]) and essential for a natural dual pmetric.

DEFINITION 2.3: A *partial metric space* is a nonempty set S, and a function (the *pmetric*) $p: S \times S \rightarrow \Re$, such that for any $x,y,z \in S$:

(P1) $p(x, y) \geq p(x,x)$.

(P2) $x = y$ if and only if $p(x, x) = p(x, y) = p(y, y)$.
(P3) $p(x, y) = p(y, x)$.
(P4) $p(x, z) \leq p(x, y) + p(y, z) - p(y, y)$.

The axioms are discussed in detail in [5], along with connections to other generalized metrics. We are mainly interested in the valuation spaces, which we now see are a special class of the partial metric spaces. The analogy with normed spaces seems to be important, and will certainly motivate further research.

LEMMA 2.4: Suppose $(S, \sqsubseteq, \mu)$ is a valuation space, then $p(x, y) = -\mu(x \sqcap y)$, defines a pmetric on S.

Proof: Axioms (P1) and (P3) are immediate, and (P2) follows from:

$$p(x, y) = p(x, x) \Leftrightarrow \mu(x \sqcap y) = \mu(x) \Leftrightarrow x \sqsubseteq y.$$

We prove (P4) in detail:

$$\begin{aligned} p(x, z) + p(y, y) &= -\mu(x \sqcap z) - \mu(y) \\ &\leq -\mu(x \sqcap y \sqcap z) - \mu((x \sqcap y) \sqcup (y \sqcap z)) \\ &= -\mu(x \sqcap y) - \mu(y \sqcap z) \\ &= p(x, y) + p(y, z). \quad \square \end{aligned}$$

For a partial metric space (S, p), we immediately have a natural definition (although slightly different from the one given in [5]) for the *open balls*:

$$B_\varepsilon(x; p) = \{y \in S \mid p(x, y) < p(x, x) + \varepsilon\}, \forall x \in S, \varepsilon \in \Re \ (\varepsilon > 0).$$

These are easily seen to form the basis for a T_0-topology on S, called the *pmetric topology*, which we denote by $\mathcal{T}[p]$.

EXAMPLE 2.5:

1. Suppose we define $\mu: \Re \rightarrow \Re$ by $\mu(x) = x$, then μ is a valuation on $(\Re, \leq)$, and induces the pmetric $p(x, y) = -\min\{x, y\}$. The open balls are of the form

   $$B_\varepsilon(x; p) = \{y \in \Re \mid -\min\{x, y\} < -x + \varepsilon\} = (x - \varepsilon, \infty) \subseteq \Re.$$

2. We denote the set of finite and infinite sequences over some set Σ, with the prefix ordering, by $(\Sigma^\infty, \sqsubseteq)$. If we denote the length of a sequence $x \in \Sigma^\infty$ by $|x|$, then $\mu: \Sigma^\infty \rightarrow \Re$ defined by $\mu(x) = -2^{-|x|}$ is a valuation.
3. We denote the power set of the natural numbers, with the subset ordering, by $(P\omega, \subseteq)$. Then $\mu: P\omega \rightarrow \Re$ defined by $\mu(x) = \sum_{n \in x} 2^{-n}$ is a valuation.

2.3. Duality of Pmetrics

We now consider the dual of the pmetric topology. Unlike other generalized metrics (such as the quasimetrics) this duality is not a consequence of a lack of symmetry in the axioms. Indeed it is perhaps one of the strengths of the partial metric generalization that symmetry is preserved as an axiom.

DEFINITION 2.6: Suppose (S, p) is a partial metric space, then

$$p^*(x, y) = p(x, y) - p(x, x) - p(y, y), \text{ for all } x, y \in S,$$

is the *dual pmetric* on S.

The proof that p^* is a pmetric is straightforward, and p^* is well defined since pmetrics are $\mathfrak{R}$-valued. To see in what sense p^* is dual to p, we recall that the *specialization order* induced by a T_0-topology $\mathcal{T}$, is defined by

$$x \sqsubseteq_{\mathcal{T}} y \text{ if and only if for all } U \in \mathcal{T}, x \in U \text{ implies } y \in U.$$

Then, for a partial metric space (S, p), it is not difficult to check that:

$$\begin{aligned} x \sqsubseteq_{\mathcal{T}[p]} y &\Leftrightarrow p(x, y) = p(x, x) \\ &\Leftrightarrow p^*(x, y) = p^*(y, y) \\ &\Leftrightarrow y \sqsubseteq_{\mathcal{T}[p^*]} x. \end{aligned}$$

It is also clear that $p^{**} = p$.

LEMMA 2.7: Suppose (S, p) is a partial metric space, then

$$d(x, y) = p(x, y) + p^*(x, y), \text{ for all } x, y \in S,$$

defines a metric on S, which we call the *induced metric*. If we denote the metric topology by $\mathcal{T}[d]$, then $\mathcal{T}[d] = \mathcal{T}[p] \vee \mathcal{T}[p^*]$.

Proof: It is easy to see that d is a metric. Now $p(x, y) - p(x, x) \leq d(x, y)$, so for any $\varepsilon > 0$, $B_\varepsilon(x; d) \subseteq B_\varepsilon(x; p)$. So we have $\mathcal{T}[p] \subseteq \mathcal{T}[d]$, and dually $\mathcal{T}[p^*] \subseteq \mathcal{T}[d]$. To see that $\mathcal{T}[d] \subseteq \mathcal{T}[p] \vee \mathcal{T}[p^*]$ we notice that for any $x \in S$, $\varepsilon > 0$, we have

$$B_{\varepsilon/2}(x;p) \cap B_{\varepsilon/2}(x; p^*) \subseteq B_\varepsilon(x; d). \quad \square$$

The induced metric topology was considered in [5]. However, the dual pmetric is new, since it relies on extending the range of a pmetric to $\mathfrak{R}$.

EXAMPLE 2.8: In Example 2.5.1, the dual pmetric is $p^*(x, y) = \max\{x, y\}$ and the induced metric is $d(x, y) = |x - y|$.

BITOPOLOGY. To conclude this section, we propose that the natural context in which to view a partial metric space (S, p) is as a bitopological space $(S, \mathcal{T}[p], \mathcal{T}[d])$. We shall regard a topological property as holding on (S, p) if it holds for both the topologies, and a metric property as holding on (S, p)

if it holds on (S, d). The results we present in this paper will go some way towards justifying this point of view.

3. TOPOLOGIES ON VALUATION SPACES

In this section we concentrate on the valuation spaces, and, in particular, on the induced topologies on these spaces. Before we start, we remark that a pmetric p on a valuation space $(S, \sqsubseteq, \mu)$, will always be the pmetric defined in Lemma 2.4. Similarly, a metric d on a partial metric space (S, p), will always be the metric defined in Lemma 2.7.

3.1. Continuous Valuations

Recall that a T_0-topology $\mathcal{T}$ on a poset $(S, \sqsubseteq)$ is *order consistent* if:

1. The specialization order coincides with $\sqsubseteq$ (i.e., $\sqsubseteq_{\mathcal{T}} = \sqsubseteq$).
2. If $U \in \mathcal{T}$, $A \subseteq S$ is directed and $\bigsqcup^{\uparrow} A \in U$, then $A \cap U \neq \varnothing$.

Notice that $\bigsqcup^{\uparrow} A$ denotes the supremum of the directed set A, as in [1].

For a valuation space $(S, \sqsubseteq, \mu)$, we have already seen that

$$x \sqsubseteq_{\mathcal{T}[p]} y \Leftrightarrow p(x, y) = p(x, x) \Leftrightarrow \mu(x \sqcap y) = \mu(x) \Leftrightarrow x \sqsubseteq y,$$

so $\mathcal{T}[p]$ satisfies the first condition. To see when $\mathcal{T}[p]$ is in fact order consistent we introduce the continuous valuations. These are not a new idea, although their use in the context we propose is new.

DEFINITION 3.1: A valuation $\mu: S \rightarrow \Re$ on a consistent semilattice $(S, \sqsubseteq)$ is *continuous* if, for any directed $A \subseteq S$ with $\bigsqcup^{\uparrow} A \in S$, we have

$$\mu(\bigsqcup\nolimits^{\uparrow} A) = \sup\{\mu(a) \mid a \in A\}.$$

We also say that $(S, \sqsubseteq, \mu)$ is a *continuous valuation space.*

LEMMA 3.2: A valuation μ on a consistent semilattice $(S, \sqsubseteq)$ is continuous if, and only if, the pmetric topology is order consistent.

Proof: Suppose $A \subseteq S$ is directed and $\bigsqcup^{\uparrow} A \in S$, then

$$\begin{aligned} \mu \text{ continuous } &\Leftrightarrow \forall \varepsilon > 0\ \exists a \in A \text{ such that } \mu(\textstyle\bigsqcup^{\uparrow} A) - \varepsilon < \mu(a) \\ &\Leftrightarrow \forall \varepsilon > 0\ \exists a \in A \text{ such that } p(a, \textstyle\bigsqcup^{\uparrow} A) < p(\textstyle\bigsqcup^{\uparrow} A, \textstyle\bigsqcup^{\uparrow} A) + \varepsilon \\ &\Leftrightarrow \forall \varepsilon > 0\ \exists a \in A \text{ such that } a \in B_{\varepsilon}(\textstyle\bigsqcup^{\uparrow} A; p) \\ &\Leftrightarrow \mathcal{T}[p] \text{ is order consistent.} \quad \square \end{aligned}$$

So, for a continuous valuation space $(S, \sqsubseteq, \mu)$, since the Scott topology σ_S is the finest order consistent topology on $(S, \sqsubseteq)$, then $\mathcal{T}[p] \subseteq \sigma_S$. To see when we can strengthen this to equality, we introduce completeness.

3.2. Complete Partial Metric Spaces

From our bitopological point of view we consider a partial metric space (S, p) to be *complete* when the induced metric space (S, d) is complete. Now we already have a useful pmetric condition for $x \sqsubseteq y$ (see Section 3.1). We will also need that a sequence (x_n) converges to $a \in S$ in (S, d) if, and only if,

$$\lim_{n \to \infty} p(x_n, a) = \lim_{n \to \infty} p(x_n, x_n) = p(a, a).$$

The proof is a simple generalization from [5]. We call a the *proper limit* of (x_n), to distinguish from the limits in $(S, \mathcal{T}[p])$. Finally we have that (x_n) is Cauchy in (S, d), exactly when $\lim_{n,m \to \infty} p(x_n, x_m)$ exists.

LEMMA 3.3: Suppose $(S, \sqsubseteq, \mu)$ is a valuation space such that (S, p) is a complete partial metric space. If $A \subseteq S$ is a directed set and $\alpha = \sup\{\mu(a) \mid a \in A\}$ exists, then $\bigsqcup^{\uparrow} A \in S$ and $\mu(\bigsqcup^{\uparrow} A) = \alpha$.

Proof: First of all, we can easily find an ω-chain (a_n) in A such that, for all $n \geq 1$,

$$\alpha - 1/n < \mu(a_n) \leq \alpha.$$

Clearly, $\lim_{n,m \to \infty} p(x_n, x_m) = \lim_{n,m \to \infty} - \mu(a_n \sqcap a_m) = -\alpha$, and so (a_n) is Cauchy. Since (S, p) is complete, then (a_n) has a proper limit $c \in S$, and $\mu(c) = \alpha$. We complete the proof by showing that $c = \bigsqcup^{\uparrow} A$.

Suppose $a \in A$, then for $n \geq 1$ we let $x_n \in A$ be such that $a, a_n \sqsubseteq x_n$, and we have

$$\begin{aligned} 0 \leq p(a, c) - p(a, a) &\leq \lim_{n \to \infty}[p(a, x_n) + p(x_n,c) - p(x_n,x_n) - p(a, a)] \\ &\leq \lim_{n \to \infty}[p(x_n, a_n) + p(a_n, c) - p(a_n,a_n) - p(x_n,x_n)] \\ &\leq \lim_{n \to \infty}[p(a_n, c) + \alpha] = 0. \end{aligned}$$

So c is an upper bound for A. Now if $c' \in S$ is another upper bound, then

$$\begin{aligned} 0 \leq p(c, c') - p(c, c) &\leq \lim_{n \to \infty}[p(c, a_n) + p(a_n, c') - p(a_n, a_n) - p(c, c)] \\ &= \lim_{n \to \infty}[p(a_n, a_n) + \alpha] = 0. \end{aligned}$$

So $c \sqsubseteq c'$, which implies that $c = \bigsqcup^{\uparrow} A$ as required. ❑

Recall that a filtered set is the dual to a directed set (and we use $\sqcap^{\downarrow}$ to denote infimum). The proof of the dual lemma is then similar to the above.

LEMMA 3.4: Suppose $(S, \sqsubseteq, \mu)$ is a valuation space such that (S, p) is a complete partial metric space. If $A \subseteq S$ is a filtered set and $\alpha = \inf\{\mu(a) \mid a \in A\}$ exists, then $\sqcap^{\downarrow} A \in S$ and $\mu(\sqcap^{\downarrow} A) = \alpha$.

3.3. Complete Valuation Spaces

To define a complete valuation space, we also take into account directed completeness on the underlying consistent semilattice. We follow [4], and say that a *complete semilattice* is a bounded-complete cpo, which for us is better thought of as a consistent semilattice which is also a cpo.

DEFINITION 3.5: A valuation space $(D, \sqsubseteq, \mu)$ is *complete* if $(D, \sqsubseteq)$ is a complete semilattice, and μ induces a complete partial metric space (D, p).

LEMMA 3.6: Suppose $(D, \sqsubseteq, \mu)$ is a complete valuation space, then $\mathcal{T}[p] = \sigma_D$.

Proof: For every $A \subseteq D$ directed, $\bigsqcup^{\uparrow} A \in D$ implies that $\sup\{\mu(a) \mid a \in A\}$ exists. So by Lemma 3.3 μ is continuous, by Lemma 3.2 $\mathcal{T}[p]$ is order consistent, and we have $\mathcal{T}[p] \subseteq \sigma_D$. Now suppose for a contradiction, that there exists $U \in \sigma_D$ but with $U \notin \mathcal{T}[p]$. So there must exist some $a \in U$ such that for all $n \geq 1$, we can find $x_n \in B_{1/2^n}(a; p)\backslash U$. Since $(D, \sqsubseteq)$ is a complete semilattice, we can define

$$y_n = \sqcap\{a \sqcap x_k \mid k > n\} \notin U, \text{ for each } n \geq 1,$$

and $y_\infty = \bigsqcup^{\uparrow}\{y_n \mid n \geq 1\} \sqsubseteq a$. In fact we show that $\mu(y_\infty) = \mu(a)$, so $y_\infty = a \in U$, which then implies that $y_n \in U$, for some $n \geq 1$, which is a contradiction.

We first define, for any $n, s \geq 1$,

$$y_{n,s} = a \sqcap x_{n+1} \sqcap \ldots \sqcap x_{n+s}.$$

Then for any $n, s \geq 1$, a simple induction shows that

$$\begin{aligned} \mu(a) - \mu(y_{n,s}) &\leq \textstyle\sum_{i=1}^{s}[\mu(a) - \mu(a \sqcap x_{n+i})] \\ &= \textstyle\sum_{i=1}^{s}[p(a, x_{n+i}) - p(a, a)] \\ &\leq \textstyle\sum_{i=1}^{s} 1/2^{n+i}. \end{aligned}$$

Now, for each $n \geq 1$, $y_n = \sqcap^{\downarrow}\{y_{n,s} \mid s \geq 1\}$ so we have

$$\begin{aligned} 0 \leq \mu(a) - \mu(y_\infty) &\leq \lim_{n \to \infty}[\mu(a) - \mu(y_n)] \\ &= \lim_{n \to \infty}[\sup\{\mu(a) - \mu(y_{n,s}) \mid s \geq 1\}] \text{ by Lemma 3.4} \\ &\leq \lim_{n \to \infty} 1/2^n = 0. \end{aligned}$$

So $\mu(y_\infty) = \mu(a)$, and hence $y_\infty = a$, which gives us a contradiction as required. ❑

3.4. Compactness and Dual Topologies

For a poset $(D, \sqsubseteq)$, we recall from [4], [8] that the dual topology $\sigma_D{}^*$, has the saturated (i.e., upper sets with respect to $\sqsubseteq$) compact sets (with respect to σ_D) as a basis for the closed sets. So for a complete valuation space $(D, \sqsubseteq, \mu)$, since $\sigma_D = \mathcal{T}[p]$, then we have two *dual topologies*; σ_D^* and $\mathcal{T}[p^*]$.

LEMMA 3.7: Suppose $(D, \sqsubseteq, \mu)$ is a complete valuation space, then $\sigma_D^* \subseteq \mathcal{T}[p^*]$.

Proof: Suppose $V \subseteq D$ is a basic closed set in (D, σ_D^*). If we can show that $D\backslash V \in \mathcal{T}[p^*]$ then the result follows. Suppose $y \in D\backslash V$, then for all $x \in V$ we must have $x \not\sqsubseteq y$, and so we can define

$$\varepsilon_x = [p(x, y) - p(x,x)]/2 > 0.$$

Since $\{B_{\varepsilon_x}(x; p) | x \in V\}$ is a σ_D-open cover of V, then there exists a finite subcover $\{B_{\varepsilon_{x_n}}(x_n; p) | 1 \le n \le N, x_n \in V\}$. We define $\varepsilon = \min\{\varepsilon_{x_n} | 1 \le n \le N\} > 0$, and show that $B_\varepsilon(y; p^*) \subseteq D\backslash V$, and hence $D\backslash V \in \mathcal{T}[p^*]$.

Suppose $z \in V$, then $z \in B_{\varepsilon_{x_n}}(x_n; p)$, for some $1 \le n \le N$, and we have

$$\begin{aligned}
\varepsilon \le \varepsilon_{x_n} &= p(x_n, y) - p(x_n, x_n) - \varepsilon_{x_n} \\
&\le p(x_n, z) + p(z, y) - p(z, z) - p(x_n, x_n) - \varepsilon_{x_n} \\
&< p(z, y) - p(z, z) \text{ since } z \in B_{\varepsilon_{x_n}}(x_n; p) \\
&= p^*(z, y) - p^*(y, y).
\end{aligned}$$

So $z \notin B_\varepsilon(y; p^*)$ which implies that $B_\varepsilon(y; p^*) \subseteq D\backslash V$. ❑

To make further progress we introduce compactness. From our bitopological point of view we consider a partial metric space (S, p) to be *compact* when $\mathcal{T}[d]$ is compact (which implies that $\mathcal{T}[p]$ is compact). A valuation space $(D, \sqsubseteq, \mu)$ is then compact if the induced partial metric space (D, p) is compact.

LEMMA 3.8: Every compact valuation space $(D, \sqsubseteq, \mu)$ is complete.

Proof: Every compact metric space is complete, so we need only show that $(D, \sqsubseteq)$ is a cpo. We first show that for every $A \subseteq D$ directed, $\{\mu(a) | a \in A\}$ is bounded above, so that by Lemma 3.3 $\bigsqcup^\uparrow A \in D$. It then follows by a similar argument that $\{\mu(x) | x \in D\}$ is bounded below, and since D is a filtered set, then by Lemma 3.4 $\bot \in D$ and so $(D, \sqsubseteq)$ is a cpo.

For a contradiction, suppose that we have $A \subseteq D$ directed, such that $\{\mu(a) | a \in A\}$ is unbounded. Then clearly there exists an ω-chain (a_n) in A with $\{\mu(a_n) | n \ge 1\}$ unbounded. But (D, d) is compact, so that (a_n) has a Cauchy subsequence (a_{n_k}). But then $\sup\{\mu(a_n) | n \ge 1\} = \lim_{k \to \infty} \mu(a_{n_k})$, which must exist since (a_{n_k}) is Cauchy, and so we have a contradiction. ❑

LEMMA 3.9: Suppose $(D, \sqsubseteq, \mu)$ is a compact valuation space, then $\sigma_D^* = \mathcal{T}[p^*]$.

Proof: By Lemmas 3.8 and 3.7, $\sigma_D{}^* \subseteq \mathcal{T}[p^*]$. Now suppose that $U \in \mathcal{T}[p^*]$, then $U \in \mathcal{T}[d]$ and so $D \backslash U$ is compact in $\mathcal{T}[d]$. Now any $\mathcal{T}[p]$-open cover of $D \backslash U$ is a $\mathcal{T}[d]$-open cover and so there exists a finite subcover. So $D \backslash U$ is compact in $\mathcal{T}[p] = \sigma_D$, and since $D \backslash U$ is clearly an upper set with respect to $\sqsubseteq$, then $U \in \sigma_D{}^*$ and so $\sigma_D{}^* = \mathcal{T}[p^*]$. ❑

COROLLARY 3.10: Suppose $(D, \sqsubseteq, \mu)$ is a compact valuation space, then the patch topology is equivalent to the induced metric topology: $\sigma_D \vee \sigma_D{}^* = \mathcal{T}[d]$.

4. THE CATEGORY ωp-ALG

We conclude this paper by showing that there are "domains" for which there exist compact valuations. Furthermore, for the spaces in this section we show that the induced metric topology is in fact the Lawson topology, and that we have a Cartesian closed category of such spaces.

DEFINITION 4.1: If $(D, \sqsubseteq)$ is a dcpo, then $q \in D$ is *prime* if, whenever $B \subseteq D$ $(B \neq \varnothing)$ and $q \sqsubseteq \bigsqcup B$, then $\uparrow q \cap B \neq \varnothing$. If every element is the supremum of primes, then we say that $(D, \sqsubseteq)$ is *prime algebraic.*

We denote the prime elements of D by $\mathcal{P}(D)$, and for each $x \in D$, we denote the set of prime elements below x by $\mathcal{P}_x = \{a \sqsubseteq x \mid a \text{ prime}\}$. There is some variation in the literature in the definition of a prime element (see, e.g., [6]), so we remark that our primes are sometimes called complete primes, with "primes" being reserved for when B in the definition is finite. We also insist that $B \neq \varnothing$, so for a cpo, $\bot$ is prime.

EXAMPLE 4.2: The dcpos $(P\omega, \subseteq)$ and $(\Sigma^\infty, \sqsubseteq)$ are prime algebraic, with primes the singleton sets $\{n\}$ and the finite sequences respectively (as well as the bottom element in each case).

We denote the category of countably-based prime algebraic complete semilattices (or prime algebraic Scott-domains) and Scott-continuous functions, by ωp-ALG. That this category is Cartesian closed is well known. For conciseness, we will refer to the objects of ωp-ALG as domains. We show that each domain has a natural compact valuation.

LEMMA 4.3: Suppose $(D, \sqsubseteq)$ is a domain and that $\sum_{n=1}^{\infty} \alpha_n$ is a convergent series of positive terms. We enumerate $\mathcal{P}(D) = \{q_1, q_2, \ldots\}$, and define

$$\mu(x) = \sum_{q_n \in \mathcal{P}_x} \alpha_n < \infty.$$

Then $(D, \sqsubseteq, \mu)$ is a valuation space. We say that μ is a *constructable valuation.*

Proof: Since D is prime algebraic, if $x \sqsubsetneq y$ in D, then $\mathcal{P}_x \subsetneq \mathcal{P}_y$, and

$$\mu(x) = \sum_{q_n \in \mathcal{P}_x} \alpha_n < \sum_{q_n \in \mathcal{P}_y} \alpha_n = \mu(y).$$

If $\{x, y\} \subseteq D$ is consistent, then

$$\begin{aligned} \mu(x) + \mu(y) &= \sum_{q_n \in \mathcal{P}_x} \alpha_n + \sum_{q_n \in \mathcal{P}_x} \alpha_n \\ &= \sum_{q_n \in \mathcal{P}_x \cup \mathcal{P}_y} \alpha_n + \sum_{q_n \in \mathcal{P}_x \cap \mathcal{P}_y} \alpha_n \\ &= \sum_{q_n \in \mathcal{P}_{x \sqcup y}} \alpha_n + \sum_{q_n \in \mathcal{P}_{x \sqcap y}} \alpha_n \\ &= \mu(x \sqcup y) + \mu(x \sqcap y). \quad \square \end{aligned}$$

Continuing with the notation of the lemma, we note that the induced metric is easily checked to be:

$$d(x, y) = \sum_{q_n \in \mathcal{P}_x \Delta \mathcal{P}_y} \alpha_n,$$

where $\mathcal{P}_x \Delta \mathcal{P}_y = \mathcal{P}_x \backslash \mathcal{P}_y \cup \mathcal{P}_y \backslash \mathcal{P}_x$.

LEMMA 4.4: Suppose $(D, \sqsubseteq)$ is a domain, and μ is a constructable valuation, then the induced metric space (D, d) is complete.

Proof: Suppose (x_n) is a Cauchy sequence in (D, d), and let $y_n = \sqcap\{x_k \mid k > n\}$, for all $n \geq 1$. We let $x = \bigsqcup^{\uparrow}\{y_n \mid n \geq 1\}$ and show that $x_n \to x$ in (D, d). For any $\varepsilon > 0$, there exists $K > 1$ such that $\sum_{k=K}^{\infty} \alpha_j < \varepsilon$. If we let

$$\delta = \min\{\alpha_k \mid 1 \leq k < K\} > 0,$$

then we can find $N \geq 1$ such that $d(x_n, x_m) < \delta$ for $n, m \geq N$. Clearly, if $q_k \in \mathcal{P}_{x_n} \Delta \mathcal{P}_{x_m}$ then we must have $k \geq K$. So we see that, for all $n \geq N$,

$$d(x_n, x) = \sum_{q_k \in \mathcal{P}_{x_n} \Delta \mathcal{P}_x} \alpha_k \leq \sum_{k=K}^{\infty} \alpha_k < \varepsilon,$$

and $x_n \to x$ in (D, d). $\square$

LEMMA 4.5: Suppose $(D, \sqsubseteq)$ is a domain, and μ is a constructable valuation, then the induced metric space (D, d) is totally bounded.

Proof: For any $\varepsilon > 0$, there exists $N > 1$ such that $\sum_{n=N}^{\infty} \alpha_n < \varepsilon$. We define

$$A = \{\bigsqcup Q \mid Q \subseteq \{q_n \mid 1 \leq n < N\},\ Q \text{ consistent}\} \bigcup \{\bot\}.$$

Then for any $x \in D$, there must exist $a \in A$ such that for all $1 \leq n < N$, $q_n \sqsubseteq a$ if, and only if $q_n \sqsubseteq x$, so we have

$$d(a, x) = \sum_{q_n \in \mathcal{P}_a \Delta \mathcal{P}_x} \alpha_n \leq \sum_{n=N}^{\infty} \alpha_n < \varepsilon.$$

So A is a finite ε-net and (D, d) is totally bounded. ❑

COROLLARY 4.6: Suppose $(D, \sqsubseteq)$ is a domain and μ is a constructable valuation, then the Scott and pmetric topologies coincide ($\sigma_D = \mathcal{T}[p]$) as do the Lawson and metric topologies ($\lambda_D = \mathcal{T}[d]$).

Proof: By Lemmas 4.4 and 4.5, $\mathcal{T}[d]$ is compact, so $(D, \sqsubseteq, \mu)$ is a compact valuation space, and by Lemmas 3.8 and 3.6 $\sigma_D = \mathcal{T}[p]$. By Corollary 3.10, $\mathcal{T}[d]$ is the patch topology, but $(D, \sqsubseteq)$ is continuous (since prime algebraic), so the Lawson topology is also the patch topology (see, e.g., [4]), and we have $\lambda_D = \mathcal{T}[d]$. ❑

5. CONCLUSIONS AND FURTHER WORK

There are two conclusions that we wish to draw from this paper. The first is the applicability of partial metric spaces to domain theory. However, before we can fully justify this, we must further develop a general theory of partial metric spaces, as well as explore in more depth the connections with domain theory. An important initial goal would be to give suitable partial metrics to more than just the domains in ωp-ALG. We are confident that this next step can be taken.

Our second conclusion will be harder to justify. We have seen how to unify (by which we mean; derive from a single function) the Scott and Lawson topologies for the domains of ωp-ALG, where we also have a natural metric for the Lawson topology. Now we claim that this can lead to a new approach for generalizing the domain theoretic and metric approaches to semantics. Since we are unifying the topologies on a domain, rather than developing a general theory which can specialize in either direction, we further claim that this is distinct from existing generalizations such as the enriched categories of [10] or the generalized ultrametrics of [7]. In an attempt to clarify this, we could term our approach an "Internal Unification of Topologies" as opposed to an "External Unification of Theories."

REFERENCES

1. ABRAMSKY, S. & A. JUNG. 1992. Domain Theory. Handbook of Logic in Computer Science, Vol. 3. Oxford Science Publications.
2. BIRKHOFF, G. 1967. Lattice Theory. AMS Colloquium Publication, 3rd edit.
3. EDALAT, A. 1994. Domain Theory and Integration. LICS '94, IEEE Computer Society Press.
4. LAWSON, J.D. 1988. The Versatile Continuous Order. Mathematical Foundations of Programming Language Semantics. Lecture Notes in Computer Science, **298**. Springer-Verlag. New York.

5. MATTHEWS, S.G. 1992. Partial Metric Topology. *In:* Proceedings of the 8th Summer Conference on Topology and Its Applications (S. Andima *et al.*, Eds.). Annals of the New York Academy of Sciences. **728:** 183–197. New York.
6. NIELSEN, M., G. PLOTKIN, & G. WINSKEL. 1981. Petri Nets, Event Structures, and Domains, Part 1. Theoretical Computer Science, Vol. 13.
7. RUTTEN, J.J.M.M. 1995. Elements of Generalized Ultrametric Domain Theory. Report CS-R9507. Computer Science. CWI.
8. SMYTH, M.B. 1992. Topology, Handbook of Logic in Computer Science, Vol. 1. Oxford Science Publications.
9. SUTHERLAND, W.A. 1975. Introduction to Metric and Topological Spaces. Oxford University Press. New York.
10. WAGNER, K.R. 1994. Solving Recursive Domain Equations with Enriched Categories. PhD. Thesis, School of Computer Science, Carneige Mellon University.

Examples of Recurrence

JAN PELANT[a,c] AND SCOTT W. WILLIAMS[b,d]

[a]*Department of Mathematics*
Czech Academy of Sciences
Prague 1
115 67 Czech Republic
and
[b]*Department of Mathematics*
State University of New York at Buffalo
Buffalo, New York 14214

ABSTRACT: We present five examples illustrating that results well known for discrete dynamical systems on (locally) compact metric spaces and continuous functions can fail if either (local) compactness or metrizable is removed. Some samples are: A homeomorphism h from the space $\mathbb{P}$ of irrationals to itself such that $[\mathbb{P}; h]$ has each point recurrent, but $[\mathbb{P}; h]$ has no minimal sets. Commuting homeomorphisms f and g from a compact space X to itself such that $[X; f]$ and $[X; g]$ have no common recurrent points.

INTRODUCTION

The fundamental object considered here is a pair $[X;f]$ called a discrete dynamical system, where X is a Tychonov space and $f: X \to X$ is a continuous function. All the definitions below are standard (see [6], [10], and [17]).

Fix a system $[X;f]$, let $f^1 = f$, and for each $n \in \mathbb{N}$ (= the set of positive integers) let $f^{n+1} = ff^n$. The *orbit* of a point $x \in X$ is the set $\{f^n(x): n \in \mathbb{N}\}$, denoted by $Ob_f(x)$ (or $Ob(x)$ when no confusion results). The *orbit closure* of a point $x \in X$ is the set $\mathrm{cl}(Ob(x))$, denoted by $OC_f(x)$ (or $OC(x)$ when no confusion results). A point $x \in X$ is said to be *recurrent* in $[X; f]$ provided $x \in OC(x)$. In the case f is a homeomorphism, $x \in X$ is said to be *stable* (for Poisson stable) in $[X; f]$ provided it is recurrent in both $[X; f]$ and $[X; f -1]$.

A point $x \in X$ is said to be *almost periodic* (also known as *uniformly recurrent*) in $[X; f]$ provided that for each neighborhood U of x, $\{n \in \mathbb{N} : f^n x \in U\} \neq \varnothing$ is relatively dense in $\mathbb{N}$; i.e., $\exists k = k(U) \in \mathbb{N}$ such that $\forall m \in \mathbb{N}$, $\{n \in \mathbb{N}: km \leq n < km + k, f^n x \in U\} \neq \varnothing$. A set $M \subseteq X$ is said to be *minimal* in $[X; f]$

Mathematics Subject Classification: 54H20, 54D40.
Key words and phrases: dynamical system, recurrence, almost-periodic, irrationals, Stone-Cech compactification.

[c]Work partially conducted while visiting the University of Toronto, and partially supported by the grant GACR 201/94/0069 and the grant of Czech Academy of Science 119401

[d]Work partially supported by N.S.F. grant R118239633, and partially conducted while Fulbright Lecturer at Charles University in Prague.

provided it is a minimal element in the partially ordered, by $\subseteq$, set of all non-empty closed sets $A \subseteq X$ such that $f(A) \subseteq A$. If X is minimal in $[X; f]$, then $[X; f]$ is said to be a *minimal system.*

Let $\mathbb{P}$ denote the subspace of irrationals on the real line, and let $\mathbb{C}$ denote the Cantor set.

In this paper we present six examples exhibiting restrictions of results (some of which are folklore) in topological dynamics:

EXAMPLE 4.1: There is an isometry $f: X \to X$ of a separable complete metric space X such that $[X; f]$ is a system with each point recurrent but no almost-periodic points.

EXAMPLE 4.2: There is a homeomorphism $f: \mathbb{P} \to \mathbb{P}$ such that $[\mathbb{P}; f]$ is a minimal system with no almost-periodic points.

The following result answers the question (rumored due to Gottschalk in the 1940s), open for 40 years: Is there a system $[X; f]$ on a completely metrizable space such that each point of X is recurrent but X contains no minimal sets?

EXAMPLE 4.3: There is a homeomorphism $f: \mathbb{P} \to \mathbb{P}$ such that $[\mathbb{P}; f]$ is a system with each point recurrent but no minimal sets.

Our last three examples concern restrictions to extensions of the Multiple Birkhoff Recurrence Theorem [8], and we consider *multiple systems* $[X; \mathcal{F}]$, where $\mathcal{F}$ is a commuting family of continuous functions (i.e., $\forall f, g \in \mathcal{F}$, $fg = gf$) of functions $f: X \to X$. Given a multiple system $[X; f]$, let $\langle \mathcal{F} \rangle = \{ f_1 f_2 \dots f_k : \langle f_1, f_2, \dots, f_k \rangle$ is a finite sequence in $\mathcal{F}\}$. For $x \in X$, we set $OC_{\mathcal{F}}(x) = \mathrm{cl}\{fx : f \in \langle \mathcal{F} \rangle\}$. A *minimal* set in a multiple system $[X; \mathcal{F}]$ is minimal with respect to the condition: All nonempty closed sets $A \subseteq X$ such that $\forall f \in \mathcal{F}$, $fA \subseteq A$.

Generalizing "stable," $x \in X$ is said to be *jointly recurrent* in the multiple system $[X; \mathcal{F}]$ provided $\forall f \in \mathcal{F}$, x is recurrent in $[X; \mathcal{F}]$. In the case $\mathcal{F}$ is finite, $x \in X$ is said to be *multiply recurrent* in $[X; f]$ provided that for each neighborhood U of x there is $n \in \mathbb{N}$ such that $\forall f \in \mathcal{F}$, $f^n x \in U$.

EXAMPLE 5.4: There is a compact first countable space X and commuting homeomorphisms $f, g: X \to X$ such that no $x \in X$ is jointly recurrent in $[X; \{f, g\}]$.

EXAMPLE 5.5: There are commuting homeomorphisms $f, g: \mathbb{P} \to \mathbb{P}$ such that $[\mathbb{P}, \{f, g\}]$ is minimal, $\forall p \in \mathbb{P}$, $OC_f(p) \cap OC_g(p) = \{p\}$.

βX denotes the Cech-Stone compactification of a space X. So when $\mathbb{N}$ is given the discrete topology, we have $\beta\mathbb{N}$.

EXAMPLE 5.6: There are commuting homeomorphisms $f, g: \beta\mathbb{N} \to \beta\mathbb{N}$ such that $\forall x \in \beta\mathbb{N}$, $OC_f(x) \cap OC_g(x) = \varnothing$.

NOTATION:

$\mathbb{C}$, $\mathbb{N}$, $\mathbb{P}$, and $\beta\mathbb{N}$ are all as above; $\mathbb{Q}$, $\mathbb{R}$, and $\mathbb{Z}$ denote, respectively, the space of rationals, the reals, and the integers. We also use ω to denote the nonnegative integers and S^1 to denote the unit circle. *card*(X) denotes the cardinality of a set X.

Given a space X and a set I, we use Logic's notation ${}^I X$ to denote all functions from I to X. $\Pi^I X$ denotes the Tychonov product of I many copies of X.

cl and int denote, respectively, the closure and interior operators in a space X. Closed and open intervals in $\mathbb{C}$, $\mathbb{N}$, $\mathbb{P}$, $\mathbb{Q}$, $\mathbb{R}$, and $\mathbb{Z}$ are denoted by $[a,b]$ and (a,b).

A *zero-dimensional* space is a space with a base of *clopen* ($\equiv$ simultaneously open and closed) sets.

The domain and range of a function f are denoted, respectively, by dom and rng. In general, we use fx instead of $f(x)$ for the image of a point unless some confusion would result.

1. SOME FOLKLORE

In most texts on topological dynamics (see [1] or [6], proof of 2.5) there is a standard result: Suppose X is a locally compact space and $f: X \to X$ is a homeomorphism. Then $M \subseteq X$ is minimal in $[X; f]$ iff $M = OC_f(x)$, where x is an almost-periodic point in $[X; f]$. This result can and has been extended in several ways — most of which have been folklore the past 40 years. Below we state, for the benefit of the reader, a few of the extensions. Remember, spaces are assumed only to be Tychonov. The first proposition says that minimal systems consist entirely of recurrent points.

PROPOSITION 1.1: A subset $M \subseteq X$ is a minimal set in the multiple system $[X; \mathcal{F}]$ iff $\forall x \in M$ $OC_{\mathcal{F}}(x) = M$. Further, if $h \in \mathcal{F}$ is a homeomorphism, then M is minimal in $[X; f]$ iff it is minimal in $[X; \{h^{-1}\} \cup \mathcal{F} \backslash \{h\}]$.

PROPOSITION 1.2: Suppose x is an almost-periodic point in a system $[X; f]$. Then $OC(x)$ is minimal and each $y \in OC(x)$ is almost-periodic.

Proof: First extend f to F: $\beta X \to \beta X$, whence the proposition is well known to hold as the Birkhoff Recurrence Theorem [8]. Now notice that $OC_f(x) = X \cap OC_F(x)$. ❑

Recall [7] the generalization of homeomorphism that calls a map $f: X \to X$ *perfect* provided it is a continuous, closed surjection, with compact point inverses. It is standard that under perfect maps, the inverse of a compact set is compact. The next result and its proof was essentially communicated to us by B. Balcar. It shows that many of the examples we seek must be nowhere locally compact; i.e., each compact set has empty interior.

LEMMA 1.3: Suppose $[X; \mathcal{F}]$ is a multiple system such that $\mathcal{F}$ is finite, and each f is a perfect map. If $C \subseteq X$ is a compact set with nonempty interior, and if $C \subseteq \bigcup_{f \in \langle \mathcal{F} \rangle} f^{-1}\text{int}(C)$. Then $[X; \mathcal{F}]$ contains a compact minimal set [and hence an almost-periodic point in the case $\mathcal{F} = \{f\}$].

Proof: Let $\mathcal{F} = \{f_1, \ldots, f_k\}$. We know $\{f^{-n}\text{int}(C) : f \in \langle \mathcal{F} \rangle\}$ is an open cover of C; applying compactness, there is $m \in \mathbb{N}$ such that $C \subseteq K = \bigcup\{f_1^{-n_1} \ldots f_k^{-n_k}(C); 0 < \min\{n_1, \ldots, n_k\} \leq \max\{n_1, \ldots, n_k\} \leq m\}$. Now f is perfect, so K is compact. Since we also have each $f[K] \subseteq K$ (remember $\mathcal{F}$ is commuting), we may apply Zorn's lemma, to see K contains a minimal set. ❑

Lemma 1.3 has a number of applications. For instance, it suggests we should probably consider the so-called nowhere locally compact spaces (= each compact set has empty interior) for examples. The application most interesting to us in this section is the following result whose proof is left to the reader:

PROPOSITION 1.4: Suppose $[X; \mathcal{F}]$ is a multiple system such that $\mathcal{F}$ is finite, and each f is a perfect map. Further, suppose X contains a compact open nonempty set C. Then the following is true

(1) If, $\forall x \in C, x \in OC_{\mathcal{F}}(x)$ [in particular, if $\mathcal{F} = \{f\}$ and each point of C is recurrent], then $[X; \mathcal{F}]$ contains a compact minimal set.
(2) If $[X; \mathcal{F}]$ is a minimal multiple system, then X is compact.

LEMMA 1.5: Suppose that $[X; f]$ is a system on a metric space X, and $f: X \to X$ is an isometry. Then $\exists x \in X$ such that $OC_{\mathcal{F}}(x) = X$ then $[X; f]$ is minimal. Further, if some point is almost-periodic in $[X; f]$, then $\forall x \in X$, x is recurrent (almost periodic) in $[X; f]$.

2. SYSTEMS ON THE SPACE OF IRRATIONALS

In this section we develop machinery for use in the next three sections. In [7], it is proved that the space of irrationals is homeomorphic to $\Pi = \Pi^{\mathbb{Z}}\mathbb{Z}$ or the Baire space $\Pi^{\mathbb{N}}\mathbb{N}$, the Tychonov product of countably many copies of the integers. We need a folklore characterization of $\mathbb{P}$.

LEMMA 2.1: A space X is homeomorphic to $\mathbb{P}$ iff it is a separable zero-dimensional completely metrizable nowhere locally compact space.

Proof: *Only if.* $\mathbb{P}$ is well known to be a separable zero-dimensional completely metrizable nowhere locally compact space.

If. Assume X is a separable zero-dimensional nowhere locally compact complete metric space. Since X is a Lindelöf noncompact, it is covered by an open family $\mathcal{R} = \{R_n : n \in \mathbb{N}\}$ such that $\forall R \in \mathcal{R}$, $\mathcal{R}\setminus\{R\}$ doesn't cover X. As X is zero-dimensional, we can assume each $R \in \mathcal{R}$ is clopen and has diameter at most 1. Let $\mathcal{C}$ be the set of all nonempty clopen subsets of X. Thus, $\mathcal{C}_1 = \{R_n \setminus \bigcup_{i<n} R_i : n \in \mathbb{N}\}$ is a pairwise disjoint cover of X by nonempty members of $\mathcal{C}$ with diameter at most 1.

Let $\mathcal{C}_1 = \{C_n : n \in \mathbb{N}\}$. Suppose $m \in \mathbb{N}$ and $\forall k \leq m$ $\forall s \in {}^k\mathbb{N}$, we have defined sets $C_s \in \mathcal{C}$ such that $\{C_s : s \in {}^k\mathbb{N},\ s|k-1 = r\}$ is a pairwise disjoint family with union C_r for $r \in {}^{k-1}\mathbb{N}$ by nonempty members of $\mathcal{C}$ with diameter at most 2^{-k}. Since for each $s \in {}^m\mathbb{N}$, C_s is clopen, it is a separable nowhere locally compact complete subset of X. So the previous paragraph gives us a recipe for defining a pairwise disjoint family $\{C_t : t \in {}^{m+1}\mathbb{N},\ t|m = s\}$, with union C_s, by nonempty members of $\mathcal{C}$ with diameter at most $2^{-(m+1)}$.

For $f \in {}^{\mathbb{N}}\mathbb{N}$, let $A_f = \bigcap_{n \in \mathbb{N}} C_{f|n}$. Since each $C_{f|n}$ is a closed set if diameter at most 2^{-n} and $C_{f|n+1} \subseteq C_{f|n}$, A_f consists of exactly one point, call it x_f. As each C_s is the union of its immediate succesors, it should be clear that for each $x \in X$, there is one and exactly one $f \in {}^{\mathbb{N}}\mathbb{N}$ such that $x = x_f$. Further, $\{C_{f|n} : n \in \mathbb{N}\}$ is a nhbd base at x.

Given another separable zero-dimensional nowhere locally compact complete metric space Y, we can define for the points y_f in the same manner. Clearly, $\Phi x_f = y_f$ defines a homeomorphism from $\Phi : X \to Y$. ❑

LEMMA 2.2: A closed subset A of Π is compact iff it is bounded in the pointwise product partial order.

Proof: If. Suppose A is a bounded closed subset A of Π; i.e, there are $f, g \in \Pi$ such that $A \subseteq K = \Pi_{n \in \mathbb{Z}}[f(n), g(n)]$. As K is the product of finite sets, K is compact. Since A is closed, it is compact.

Only if. Suppose A is a compact set. For $m \in \mathbb{N}$, A projects onto the m-th coordinate as a compact, and hence, a finite set with minimum $f(n)$ and maximum $g(n)$. As $A \subseteq \Pi_{n \in \mathbb{Z}}[f(n), g(n)]$, A is bounded. ❑

THEOREM 2.3: A nonempty closed subspace X of Π is homeomorphic to $\mathbb{P}$ iff each nonempty open set of X is unbounded in the pointwise product partial order.

Proof: Since, from 2.1, X is nowhere-locally compact, 2.2 immediately implies the properties above. Conversely, suppose $X \subseteq \Pi$ has the properties above. Clearly, each subspace of Π is zero-dimensional separable metric, so X is. 2.2 shows X is nowhere locally compact iff each of its nonempty open subsets are unbounded. Therefore, 2.1 applies to prove the result. ❑

We will consider subsystems of $[\Pi; \sigma]$, where σ is the (product-) *shift homeomorphism* $\sigma x(n) = x(n+1)$. Various compact subsystems of $[\Pi; \sigma]$ are well-known in the study of Symbolic Dynamics especially $\Pi^{\mathbb{Z}}2$ and $\Pi^{\mathbb{N}}2$, see [8], [12], and [13]. In each of our examples on $[\Pi; \sigma]$, we study closed subsystems in which every point is recurrent. However, whenever a space is embeddable as a G_δ-set in a compact Hausdorff space (e.g., when X is completely metrizable), and $f: X \to X$ is a homeomorphism for which X has a dense set of recurrent points of $[X; f]$, then X has a dense (completely metrizable, in the metric case) set of recurrent points of $[X; f^{-1}]$ ([18], and [10] for the metric case). So if all points are recurrent, then there is a dense set of stable points. Thus, it will be simpler to consider stable points in $[\Pi; \sigma]$.

As a "basic" nhbd of $x \in \Pi$ is determined by the finite sequences $x|I$ for some interval I in $\mathbb{Z}$. Thus, the following lemma follows directly from the definition.

LEMMA 2.4: For $x \in \Pi$, the following are true:

(1) x is stable in $[\Pi; \sigma]$ iff $\forall m \in \mathbb{N}$, $\exists a, b > m$ such that $\sigma^{-a}(x)|[-m, m] = x|[-m, m] = \sigma^{b}(x)|[-m, m]$.

(2) x is jointly almost-periodic in the system $[\Pi; \{\sigma, \sigma^{-1}\}]$ iff $\forall m \in \mathbb{N}$, $\exists k > 2m \in \mathbb{N}$ such that if $a, b \in \mathbb{Z}$ and if $b - a = k$, then $\exists i$, $0 \le i \le k - 2m$ such that $x|[-m, m] = \sigma^{a+i}(x)|[-m, m]$.

Thus, to build and discuss points in Π, we need some special notions about finite sequences in $\mathbb{Z}$. A *block* will be a finite function with domain dom B an (a possibly empty) interval in $\mathbb{Z}$, and range rng B a subset of $\mathbb{Z}$. All blocks will be assumed to have as domain an initial segment of $\mathbb{N}$ unlesss otherwise stated. If the block B has (non) empty domain, we write $B = \varnothing$ ($B \neq \varnothing$). The length $\lambda B = card(\text{dom } B)$. If $B \neq \varnothing$ is a block, if $a \le b \in \mathbb{Z}$, and if dom $B = [a, b]$, then by a *tail* (*head*) of B, we mean any block of the form $B|[c, b]$ (respectively, $B|[a, c]$), where $a \le c \le b$.

A partial order on blocks: Suppose A and B are blocks. We will say A is a copy of B and write $A \equiv B$ provided the following two conditions are satisfied:

(i) $\lambda A = \lambda B$ and

(ii) $\exists z \in \mathbb{Z}$ such that $\forall n \in \text{dom} A$, $A(n) = B(n + z)$.

So $\equiv$ is an equivalence relation on the set of all blocks. We will also need a partial order on the equivalence classes: Let us agree that A is in B and write $A \le B$ provided there is an interval $I \subseteq \text{dom} B$ such that $A \equiv B|I$. Notice that we can restate 2.4 as:

LEMMA 2.4: (alternate) For $x \in \Pi$, the following are true:

(1) x is stable in $[\Pi; \sigma]$ iff $\forall m \in \mathbb{N}$, $\exists a, b > m$ such that $x|[-m, m] \equiv x|[-a-m, -a+m] \equiv x|[b-m, b+m]$.

(2) x is jointly almost-periodic in the system $[\Pi; \{\sigma, \sigma^{-1}\}]$ iff $\forall m \in \mathbb{N}$, $\exists k > 2m \in \mathbb{N}$ such that if $a, b \in \mathbb{Z}$ and if $b - a = k$, then $x|[-m, m] \le x|[a, b]$.

A finitary operation on classes of blocks: Now suppose I and J are finite intervals in $\mathbb{Z}$, f is a function with dom$f = I$ and rng $f = J$. Further, suppose $(B_z : z \in J)$ is a sequence of blocks. Let us agree that $f(B_z : z \in I)$ or $(B_{fz} : z \in$

J), will denote the unique (up to the equivalence $\equiv$) block obtained by allowing $\forall z$, $\min I \le z < \max I$, B_{fz} to be immediately followed by $B_{f(z+1)}$. When f: $[a,b] \to [a,b]$ is the identity function, we also let $B_a \wedge B_{a+1} \wedge \dots \wedge B_b$ denote $f(B_z\colon z \in [1,n])$.

LEMMA 2.5: For $x \in \Pi$, $OC(x)$ is homeomorphic to $\mathbb{P}$ provided that for each block A in x, there is an $m \in \mathbb{N}$ such that:

$$L = \{\text{rng } C\colon \lambda C = m \text{ and } A \wedge C \text{ is a block in } x\} \text{ is infinite.}$$

Given $a < b \in \mathbb{Z}$ let $G_{a,b} = \{y \in Ob(x)\colon y|[0, b-a] = x|[a,b]\}$. Then $G_{a,b}$ is open in $OC(x)$ and each open set in $OC(x)$ contains a $G_{a,b}$ for suitable $a < b \in \mathbb{Z}$.

Let $C = \{\text{blocks } C\colon \lambda C = m \text{ and } A \wedge C \text{ is a block in } x\}$. Since L is infinite, there is a first $k \le m$ such that $\{C(k)\colon C \in C\}$ is unbounded. As $\forall C \in C$, $\lambda C = m$, $\{C|[1,k)\colon C \in C\}$ is finite. Without loss of generality, we may assume that $card(\{C|[1,k)\colon C \in C\}) = 1$ and $\forall C \in C$, $C = C|[1,k]$. Suppose $C \in C$, Let $A = x|[u,v]$ and $w = v + k$. Then $G_{u,w} \subseteq G_{u,v}$. But if $A = x|[a,b]$, then $G_{u,v} = G_{a,b}$ and $\{y(b-a+k)\colon y \in G_{a,b}\}$ is unbounded. Hence, $G_{a,b}$ is unbounded. According to 2.3, $OC(x)$ is homeomorphic to Π.

A *street* is a sequence $S = (B_z : z \in \mathbb{Z})$ of blocks indexed by $\mathbb{Z}$ and such that $\forall z \in \mathbb{Z}$, $B_z \ne \varnothing$. An infinitary operation on streets: Suppose $x \in \Pi$ and $S = (B_z : z \in \mathbb{Z})$ is a street. Then xS denotes the unique element y of $\mathbb{P}$ such that if $m > 0$, $f = x|[0,m)$, $g = x|(-m,-1]$, if

$$s = \sum_{z=0}^{m-1} \lambda B_{fz}\text{, and if } t = \sum_{z=-m}^{-1} \lambda B_{fz}\ ,$$

then

$$y|[0,s] = f(B_z : z \in [0,m)) \text{ and } y|[-t,-1] = g(B_z : z \in (-m,-1]).$$

Most fundamental to our constructions is the unique element in Π, denoted by α, satisfying

2.6(1). $\alpha(0) = 0$,
2.6(2). $\forall m \ge 0$, $\alpha(-m) = -\alpha(m)$,
2.6(3). $\forall m \ge 0$, $\alpha(3^m) = m + 1$, and
2.6(4). $\forall m \ge 0$, $\forall z \in (-3^m, 3^m)$, $\alpha(2.3^m + z) = \alpha(z)$.

The next lemma has a straightforward proof by induction.

LEMMA 2.7: Suppose $m,z \in \mathbb{Z}$ and $m > 0$. Then the following hold:
(1) $\alpha|(-3^m, 3^m) \le \alpha|(z, z + 4.3^m - 1)$.
(2) If $z \le i < j \le z + 2|m|$, then $\min(|\alpha(i)|\ ,\ |\alpha(j)|) \le 1 + \log_3|m|$.

Notice that 2.7(1) shows α is jointly almost periodic in the system $[\Pi; \{\sigma, \sigma^{-1}\}]$, while 2.5 shows $OC(\alpha)$ is homeomorphic to $\mathbb{P}$.

REMARK 2.8: There is a much simpler version of α when one considers just the shift $\sigma: \Pi^{\mathbb{N}}\mathbb{N} \rightarrow \Pi^{\mathbb{N}}\mathbb{N}$ for which most of the results of the next two sections hold with "stable" replaced by "recurrent." This is the point defined by algebra's 2-adic evaluation plus 1. Define $\alpha(n) = k + 1$ if k is the largest integer such that 2^k divides n.

3. STABLE POINTS IN $[\Pi; \sigma]$

THEOREM 3.1: Suppose $x \in \Pi$. x is stable in $[\Pi; \sigma]$ iff there is a street S such that $x = \alpha S$.

Proof: *If.* Given the street $S = (B_z : z \in \mathbb{Z})$, let $x = \alpha(B_z : z \in \mathbb{Z})$. Then $\forall m \in \mathbb{N}$, $x|[-m, m]$ is a block in $A = \alpha(B_z : |z| \leq m)$. Applying 2.7(1) shows $A \leq \alpha(B_z : -m - 3^{m+1} \leq z \leq -m)$ and $A \leq \alpha(B_z : m \leq z \leq m + 3^{m+1})$. So $\exists a, b \in \mathbb{N}$ such that $x|[-m, m] \equiv x|[-a - m, -a + m] \equiv x|[b - m, b + m]$. From the alternate 2.4, x is stable in $[\Pi; \sigma]$.

Only if. Conversely, suppose x is stable in $[\Pi; \sigma]$. By induction, we construct a street $(B_z : z \in \mathbb{Z})$. Let $B_0 = x|\{0\}$, and suppose $m \geq 0$ is such that $\forall z$, $|z| \leq m$, then the following hold:

(a) B_z is defined (and hence, $\forall z \in (-3^m, 3^m)$, $B_{\alpha z}$ is defined).

(b) $x|[r, s] \equiv f(B_z : z \in (-3^m, 3^m))$, where

$$r = \sum_{z = 1 - 3^m}^{-1} \lambda B_{\alpha z}, \quad s = \sum_{z=1}^{3^m - 1} \lambda B_{\alpha z}, \text{ and } f = \alpha|(-3^m, 3^m).$$

Since x is stable, we may find a largest $a < r$ and a smallest $b > s$ such that $x|[r, s] \equiv x|[a - (r + s), a - 1]$ and $x|[r, s] \equiv x|[b + 1, b + r + s]$. Now define $B_{-m-1} = x|[a, r - 1]$ and $B_{m+1} = x|[s + 1, b]$. As it is clear that we can proceed, in the above fashion, defining B_z and exhausting x, condition (b) shows $x = \alpha(B_z : z \in \mathbb{Z})$. ❑

THEOREM 3.2: Suppose $(B_z : z \in \mathbb{Z})$ is a street with $x = \alpha S$.

(1) If $\{\lambda B_z : z \in \mathbb{Z}\}$ is bounded, then x is jointly almost-periodic in $[\Pi; \{\sigma, \sigma^{-1}\}]$.

(2) If x is jointly almost-periodic in $[\Pi; \{\sigma, \sigma^{-1}\}]$ and if $\exists m \in \mathbb{N}$ such that $I = \{z \in \mathbb{Z}: \alpha(B_w: w \in (-3^{m+1}, 3^{m+1}))$ is not a block in $B_z\}$ is infinite, then $L = \{\lambda B_z : z \in I\}$ is bounded.

Proof: (1) Suppose $u \in \mathbb{N}$ is an upper bound for $\{\lambda B_z : z \in \mathbb{Z}\}$. Then $2u \cdot 3^m$ is an upper bound for $t = \lambda\alpha(B_z : z \in (-3^m, 3^m))$. Applying 2.7(1) shows that $\forall w \in \mathbb{Z}$, $\alpha(B_z : z \in (-3^m, 3^m)) \leq x|(w, w + 2t)$. Hence, x is a jointly almost-periodic point in $[\Pi; \{\sigma, \sigma^{-1}\}]$.

(2) Suppose L is unbounded and $k \in \mathbb{N}$. Since L is unbounded, we may choose $z \in I$ such that $\lambda B_z \geq k + \lambda\alpha(B_w: w \in (-3^{|m|+1}, 3^{|m|+1}))$. But then B_z is a block in x of length greater than k failing to contain a copy of $\alpha(B_w: w \in (-3^{m+1}, 3^{m+1}))$. From the alternate 2.4(2), x is not jointly almost-periodic.

THEOREM 3.3: Suppose $S = (B_z : z \in \mathbb{Z})$ is a street, $f \in {}^{\mathbb{N}}\mathbb{N}$ is an increasing function, and $k \in \mathbb{N}$ has $I = \{z \in \mathbb{Z}: k \leq \lambda B_z\ ,\ |B_z(k)| \geq f(|z|)\}$ infinite. Then $OC(\alpha S)$ is homeomorphic to $\mathbb{P}$.

Proof: Suppose A is a block in $x = \alpha S$. Let $i \in I$ be such that $A \leq \alpha(B_w: w \in (-3^{|i|}, 3^{|i|}))$, say $A = x|[a,b]$ for $-3^{|i|} < a \leq b < 3^{|i|}$. $\forall z \in I$ with $|z| \geq |i|$, let $C_z = x|[3^{|i|} - b, 3^{|i|}] \wedge B_z|[1,k]$. From 2.6(4), $L = \{\text{rng } C_z: z \in I, A \wedge C_z \text{ is a block in } x\}$ is infinite. For $m = k + 2 \cdot 3^{|i|} - b$, the hypothesis of 2.5 is satisfied. Therefore, $OC(\alpha S)$ is homeomorphic to $\mathbb{P}$. ❑

The next result is the key to our primary examples on $\mathbb{P}$.

LEMMA 3.4: Suppose that $S = (B_z : z \in \mathbb{Z})$ is a street, $y \in OC(\alpha S)$, and suppose $\exists\lambda \in \mathbb{N}$ such that for each block F with $\lambda F = \lambda$, $\{z \in \mathbb{Z}: F \text{ is a head or tail of } B_z\}$ is finite. If y is not stable in $[\Pi;\sigma]$ or if $OC(y) \neq OC(\alpha S)$, then $\exists h \in \Pi$ such that $\forall z \in \mathbb{Z}$, $y|[-|z|, |z|] \leq B_{hz}$.

Proof: Let $x = \alpha S$. For simplicity, we use α when we restrict its domain. The definition of h for negative values is analogous to (but not the same as) its definition for positive values, so we only define h for positive values. So let us suppose $m > 0$, we define $h(m)$.

Since $y \in OC(x)$, $\forall m \in \mathbb{N}$, $y|[-m,m] = \sigma^k x|[-m,m]$, so we may choose functions $a,b: \mathbb{N} \to \mathbb{Z}$ such that

(1) $a(m) \leq b(m)$, and

(2) $\forall m \in \mathbb{N}$, $b(m) - a(m)$ is minimal with respect to $(y|[-m,m]) \leq a(B_z: a(m) \leq z \leq b(m))$.

Now, $\forall m \in \mathbb{N}$, find $c(m) \in \mathbb{Z}$ and (possibly empty) blocks $H(m)$ and $T(m)$ such that

(3) $H(m) \wedge (y|[-m,0])$ is a head of $\alpha(B_z: a(m) \leq z \leq c(m))$,

(4) $(y|[0,m]) \wedge T(m)$ is a tail of $\alpha(B_z: c(m) \leq z \leq b(m))$, and

(5) $\alpha(m) \leq c(m) \leq b(m)$ and $H(m) \wedge (y|[-m,m]) \wedge T(m) \equiv \alpha(B_z: a(m) \leq z \leq b(m))$.

Notice that the conditions (2) to (5) imply two more statements

(6) $\alpha(B_z: a(m) < z < b(m)) \leq (y|[-m,m])$, and

(7) $\alpha(B_z: a(m) < z < c(m)) \leq (y|[-m,0])$ and $\alpha(B_z: c(m) < z < b(m)) \leq (y|[0,m])$.

Applying (6) and (7), $\exists f,g: \mathbb{N} \to \mathbb{Z}$ such that $\forall m \in \mathbb{N}$, $f(m) \leq 0 \leq g(m)$ and $(y|[f(m), g(m)]) \equiv B_{\alpha c(m)}$. From the hypothesis of this lemma (concerning heads and tails of B_z), we have one of two possibilities:

(8) either there is an infinite $N \subseteq \mathbb{N}$ such that both $f|N$ and $g|N$ are monotone, or

(9) there is an infinite $N \subseteq \mathbb{N}$ such that $(\alpha c)|N$ is constant.

Of course if (8) holds, we are done — just set $h(m) = \alpha c(i_m)$, where the natural indexing of N is $\{i_m: m \in \mathbb{N}\}$. So we assume (9) is true. We will consider two possibilities for the functions $c - a$ and $b - c$.

Case 1: For each infinite $I \subseteq N$, both $\text{rng}(c - a)|I$ and $\text{rng}(b - c)|I$ are unbounded.

Case 2: There is an infinite $I \subseteq N$, such that at least one of $\text{rng}(c - a)|I$ and $\text{rng}(b - c)|I$ is bounded, while the other is unbounded.

Assume Case 1 holds, and suppose $k \in \mathbb{N}$ is arbitrary. As $y \in OC(x)$, $\exists p \in \mathbb{N}$ such that $\sigma^p x|[-k,k] = y|[-k,k]$. Thus, we may find $k_1 \in \mathbb{N}$ such that:

(10) $x|[-k,k] \leq a(B_z: |z| \leq k_1)$ and $y|[-k,k] \leq \alpha(B_z: |z| \leq k_1)$.

Let $I = \{m \in \mathbb{N}: 3^{k_1+1} + 1 < c(m) - a(m)\}$. Since Case 1 holds, I is infinite. Again, applying Case 1 shows $J = \{m \in I: 3^{k_1+1} + 1 < b(m) - c(m)\}$ is infinite. Now choose $j \in J$. Using (7), 2.7(1) shows:

(11) $\alpha(B_z: |z| \leq k_1) \leq y|[-j,0]$ and $\alpha(B_z: |z| \leq k_1) \leq y|[0,j]$.

Applying (7) to (11), we see $x|[-k,k]$ and $y|[-k,k]$ are in each of $y|[-j,0]$ and $y|[0,j]$. It is now clear that $\exists q,r,s \in \mathbb{N}$ such that each of the following hold: $\sigma^q y|[-k,k] = x|[-k,k]$, $\sigma^r y|[-k,k] = y|[-k,k]$, and $\sigma^{-s} y|[-k,k] = y|[-k,k]$. As $k \in \mathbb{N}$ is arbitrary, we see, respectively, that $OC(y) = OC(x)$ and y is stable.

Assume Case 2 is true — say $\text{rng}(c - a)|I$ is bounded. Then, without loss of generality, we may assume:

(12) $(c - a)|I$ is constant, $\exists c_0 \in \mathbb{N}$ such that $\forall m \in I\ c(m) = a(m) + c_0$.

From (9), (12), and 2.7, $\exists j \in [0, c_0]$ such that $\forall i \in [0, c_0) \setminus \{j\}$,

$$a(c + i)|(I \cap [k,\infty)) \text{ is constant for } k = 1 + \log_3\left(\sum_{z=-c_0}^{c_0} \lambda B_z\right).$$

From (3), $j = 0$. So $\exists a_0, b_0 \in \mathbb{Z}$, $a_0 \leq 0 \leq b_0$ such that $\forall m > k$, $m \in I$, $y|[a_0, b_0] \equiv \alpha(B_z: a(m) < z \leq c(m))$. Therefore, there is an increasing $d: I \to \mathbb{N}$ and an infinite valued $e: I \to \mathbb{Z}$ such that $(y|[-d(m), a_0))$ is a tail of $B_{\alpha ce(m)}$ — a contradiction. The proof, in case $\text{rng}(b - c)|I$ is bounded, is similar. ❑

4. SYSTEMS WITH ALL POINTS STABLE

All of the systems discussed here will have every point recurrent, a condition identified in [16] and continued, among other places in [2] and in [9]. We discovered our first example in 1987. Recently and independently, another very different construction has been found by E. Akin, J. Auslander, and K. Berg, (unpublished) also exhibiting 4.1.

EXAMPLE 4.1: There is an isometry $f\colon X \to X$ of a separable complete metric space X such that $[X; f]$ is a system with each point recurrent but no almost-periodic points.

Proof: Define $k_0 = 1$ and for each $n \in \mathbb{N}$, we define, $k_n = 4^{n(n+1)/2}$; i.e., 4 raised to the sum of the first n integers. We define a partial function δ on $\mathbb{Z}^2$ by $\delta(x,y) = 0$ if $x = y$ and $\delta(x,y) = 2^{-n}$, if $|x - y| = k_n$. Using the partial function δ define a function d: $\mathbb{Z}^2 \to \mathbb{R}$ by $d(x,y)$ = infimum of all

$$\sum_{i=1}^{t-1} \delta(a_i, a_{i+1})$$

satisfying:

(1) $\{a_0, \ldots, a_t\} \subseteq \mathbb{Z}$, $a_0 = \min\{x,y\}$, $a_t = \max\{x,y\}$, $\forall i\ (0 < i \leq t-1)$, and $\exists n(i)$, $|a_i - a_{i+1}| = k_n(i)$.

[Remark: $d(0,4096) = d(4096,8192) = 1/8$, and $d(0,8192) = 1/4$. Also observe that $d(0,4095) = 9/8$ can not be realized by an increasing sequence of $\{a_0, \ldots, a_t\}$.]

Claim 1: If $x \neq y$ and if $d(x,y) < 2^{-n}$, then k_n divides $x - y$. Consider $\{a_0, \ldots, a_t\}$ as in (1) such that

$$\sum_{i=1}^{t-1} \delta(a_i, a_{i+1}) < 2^{-n}.$$

Then $\forall i$, $0 < i \leq t-1$, $d(a_i - a_{i+1}) \leq 2^{-(n+1)}$. Hence, $\forall i\ (0 < i \leq t-1)$, $n(\mathrm{i}) \geq n+1$. So $\forall i\ (0 < i \leq t-1)$, k_n divides $a_i - a_{i+1}$. But one of $x - y$ and $y - x$ equals $a_t - a_0 = (a_1 - a_0) + \ldots + (a_t - a_t - 1)$. Thus, the claim holds.

Both parts of the next claim follow immediately from Claim 1.

Claim 2: (a) If $|x - y| = k_n$, then $d(x,y) = \delta(x,y) = 2^{-n}$.

(b) If $x \neq y$ then $d(x,y) > 0$.

Clearly, d satisfies the symmetric law and the triangle inequality, while Claim 2(b) shows d is a *metric on* $\mathbb{Z}$.

Claim 3: Suppose $n \geq 3$ and p satisfies

(2) $(3^{n-1} + 1) \cdot k_{n-1} < p < (4^n - 3^{n-1}) \cdot k_{n-1}$. Then $d(0,p) \geq 1$.

Find $\{a_i\colon 0 \leq i \leq t\}$ such that in (1):

$$d(0,p) = \sum_{i=1}^{t-1} \delta(a_i, a_{i+1}) .$$

Let $T = \{0, 1, \ldots, t-1\}$, $V = \{i \in T\colon \delta(a_i, a_{i+1}) \leq 2^{-n}\}$, and let

$$v = \sum_{i \in V} a_{i+1} - a_i .$$

Then

$$p - v = \sum_{i \in T\backslash V} a_{i+1} - a_i . \ \forall i \in T\backslash V,\ \delta(a_i, a_{i+1}) \geq 2^{-n+1},$$

so Claim 3 follows from

(3) $card(T\backslash V) \geq 2^{n-1}$. So let us assume

(4) $card(T\backslash V) < 2^{n-1}$. $\forall i \in T\backslash V$, $\delta(a_i, a_{i+1}) \geq 2^{-n+1}$, so from Claim 2(a), $|a_i - a_{i+1}| \leq k_{n-1}$. So (4) shows $|p - v| < 2^{n-1} \cdot k_{n-1}$.

(5) $|p - v| < 2^{n-1} \cdot k_{n-1}$.

As $p > 3^{n-1} \cdot k_{n-1}$, (5) proves $V \neq \varnothing$. Claim 1 shows $\forall i \in V$, k_n divides $a_{i+1} - a_i$. Hence, $\exists m \in \mathbb{Z}\backslash\{0\}$ such that $v = m \cdot k_n$. From (2), $0 < p < (4^n - 3^{n-1}) \cdot k_{n-1} < k_n \leq |m \cdot k_n|$. So $|v - p| \geq k_n - p$. Again using (2), $p + 3^{n-1} \cdot k_{n-1} < k_n$. Thus, $k_n - p \geq 3^{n-1} \cdot k_{n-1}$. So $|v - p| \geq 3^{n-1} \cdot k_{n-1}$ — which contradicts (5). Therefore, (3) holds and Claim 3 is proved.

It should be clear that $\sigma(n) = n + 1$ defines an isometry of the countable metric space $(\mathbb{Z}, d)$ on to itself. We define X to be the completion of $(\mathbb{Z}, d)$ and f the extension of σ to X. Hence, f is also an isometry.

According to 1.5, we need only show that 0 is recurrent and not almost-periodic. But $\forall n \in \mathbb{N}$, $\sigma^{k_n}(0) = k_n$, thus $d(0, \sigma^{k_n}(0)) = 2{-}n$. So 0 is recurrent in $[X; f]$. On the other hand, Claim 3 shows that $\forall m \in \mathbb{N}$, $\exists x, y \in \mathbb{N}$, $y - x > m$ such that $\forall n \in [x, y]$ if $d(0, \sigma^n(0)) > 1$. Therefore, 0 is not almost-periodic in $[X; f]$. ❑

Our next example was discovered before we understood α and the results in Section 3. Please contrast it with 1.2.

EXAMPLE 4.2: There is a homeomorphism $f: \mathbb{P} \to \mathbb{P}$ such that $[\mathbb{P}; f]$ is a minimal system, and no point of $\mathbb{P}$ is almost-periodic in $[\mathbb{P}; f]$.

Proof: Let $\{I_z : z \in \mathbb{Z}\}$ be such that $\{\text{rng} I_z : z \in \mathbb{Z}\}$ is a pairwise-disjoint family of finite subsets of $\mathbb{N}$(= the positive integers) such that $\forall z \in \mathbb{Z}$, $|z| \leq \min I_z < card(I_z)$ [e.g., $\forall z < 0$, define $I_z = (6^{|z|}, 3 \cdot 6^{|z|})$, and $\forall z \geq 0$, define $I_z = (3 \cdot 6^{|z|}, 6^{|z|})$]. Now, for each $z \in \mathbb{Z}$, let $B_z : I_z \to I_z$ be the identity function. Let $x = \alpha(B_z : z \in \mathbb{Z})$. According to 3.3, $OC(x)$ is homemorphic to $\mathbb{P}$.

To see that $[OC(x); \sigma|OC(x)]$ is minimal, suppose $y \in OC(x)$. Since the rng I_z's are pairwise-disjoint, the hypothesis of 3.4 is satisfied. So if $OC(x) \neq OC(y)$, then 3.4 finds an $h \in \Pi$ such that $\forall m \in \mathbb{N}$, $y|[-m, m] \leq B_{h(m)}$. Clearly, $h|\mathbb{N}$ takes at most one value; i.e., $\exists m \in \mathbb{N}$ such that $\lambda B_m < m < card(I_m) = \lambda B_m$ — ridiculous. So $OC(x) = OC(y)$. Therefore, $[OC(x); \sigma|OC(x)]$ is minimal.

According to 1.2 and the minimality of $OC(x)$, in order to show that no point of $OC(x)$ is almost-periodic, it suffices to show that x is not almost-periodic. $U = \{y \in \Pi : y(0) = x(0)\}$ is a neighborhood of x. Of course, $\sigma^n x \in U$ iff $n \in I_0$. Because the rng I_z's are pairwise-disjoint, $\sigma^n x \in U$ implies $\forall z \neq$

$0, n \in I_z$. As each of the I_z's occur infinitely often in x, and because $|z| <$ card $(I_z) = \lambda B_z$, 2.4 (alternate) shows x cannot be almost-periodic. ❑

REMARK: The reader should observe that 4.2 can be made to yield a new explicit proof, as opposed to an (easy) existence proof, of the following example: There is an homeomorphism $h: \mathbb{C} \to \mathbb{C}$ such that in $[\mathbb{C}; h]$ the set of points with dense orbit is second category and the union of the minimal sets is first category. Sketch of proof: Let K be the one-point compactification of $\mathbb{Z}$, then for the point x, in 4.2, let $X = OC(x)$ and $Y = \text{cl}(OC(x))$, as a subspace of $\Pi^{\mathbb{Z}}K$. Then Y is homeomorphic to $\mathbb{C}$. Let h be the extension of σ to Y.

The next example should be contrasted with 1.3.

EXAMPLE 4.3: There is a homeomorphism $f: \mathbb{P} \to \mathbb{P}$ such that each point in $\mathbb{P}$ is stable in $[\mathbb{P}; f]$, but $[\mathbb{P}; f]$ has no minimal sets.

Proof: Let $\{I_n: n \geq 0\}$ be a pairwise-disjoint family of infinite subsets of $\mathbb{N}$ such that $\forall n \in \mathbb{N}$, $n < \min(I_n)$ [for example, define $I_n = \{2^n(2m+1): m \geq 0\}$]. $\forall n \geq 0$, let $\{a(n,z): z \in \mathbb{Z}\}$ be an enumeration of I_n with $a(n,0) = \min(I_n)$. During the remainder of this proof, we consider blocks as either functions or ordered sequences.

$\forall n \geq 0$, let $B(n,0)$ denote the one-element sequence $\langle a(n,0)\rangle$, and define, inductively

(1) $\forall n > 0$, $\forall m \in \mathbb{Z}$, $m \neq 0$, $B_{(n,m)} = \langle a(n,m)\rangle \wedge (B_{(n+1,\alpha z)}: -|m| < z < |m|) \wedge \langle a(n,m)\rangle$.

We claim the following is true:

(2) $\forall n \geq 0$, $\forall m \in \mathbb{Z}$, $\text{rng} B_{(n,m)} \subseteq \bigcup\{I_k: n \leq k \leq n + |m|\}$.

Certainly (2) is true when $m = 0$. Fix $p > 0$ and suppose (2) is true $\forall m, n, m \in \mathbb{Z}$, $n < p$. If $|z| < |m|$, then 2.7(2) shows $|\alpha z| \leq \log_3|m| < |m|$. Hence, $\text{rng} B_{(p+1,\alpha z)} \subseteq \bigcup\{I_k: p+1 \leq k \leq p + |m|\}$. Because $a(p,m) \in I_p$, $\text{rng} B_{(p,m)} \subseteq \bigcup\{I_k: p \leq k \leq p + |m|\}$. Therefore, (2) is true.

Suppose $n \geq 0$ and $m, z \in \mathbb{Z}$, $z < m$. If $k \neq z$, then $a(n,z) \notin I_k$. So (2) shows, $a(n,z) \notin \text{rng} B_{(n,m)}$. As $\langle a(n,z)\rangle$ is both a head and a tail of $B_{(n,z)}$, the following is true:

(3) $\forall n \geq 0$, $\forall m, z \in \mathbb{Z}$, $m \neq z$, no head or tail of $B_{(n,m)}$ is a head or tail of $B_{(n,z)}$.

$\forall n \geq 0$ define $x_n = (B_{(n,\alpha z)}: z \in \mathbb{Z})$. From 3.1, x_n is a stable point in $[\mathbb{P}; \sigma]$. Since $\forall m \in \mathbb{Z}$, $n < a(n,0) \leq a(n,m) = B_{(n,m)}(0)$, 3.3 proves:

(4) $\forall n \geq 0$, $OC(x_n)$ is homeomorphic to $\mathbb{P}$.

$$\forall n, p, n \geq 0, p > 0, \text{ let } G_p = \{y \in \Pi: y|(-p,p) = (B_{(n+1,\alpha z)}: -|p| < z < |p|)\}.$$

Then $\{G_p: p > 0\}$ forms a neighborhood base at x_{n+1}. But (1) shows that $\forall p > 0$, $\exists m \in \mathbb{Z}$ such that $(B_{(n+1,\alpha z)}: -|p| < z < |p|) \leq B_{(n,m)} = x_n|I$ for some interval I of $\mathbb{Z}$. Hence, $\exists q \in \mathbb{N}$ with $\sigma^q x_n \in G_p \cap Ob(x_n)$. Thus, $x_{n+1} \in OC(x_n)$.

Similarly, $\forall r \in \mathbb{N}$, $\sigma^r x_{n+1} \in OC(x_n)$. Further, since $x_n(0) = a(n,0) \notin \bigcup\{I_k : k > n\}$, (2) shows $x \notin OC(x_{n+1})$. So we have:

(5) $\forall n \geq 0$, $OC(x_n + 1) \subsetneq OC(x_n)$.

Since $\forall n \in \mathbb{N}$, $n < \min(I_n)$. (2) implies that $\forall n \in \mathbb{N}$, $n < \min \mathrm{rng} x_n$. Hence, the following holds:

(6) $\forall k \geq 0$, $\bigcap_{n \geq k} OC(x_n) = \varnothing$.

Now fix $n \geq 0$. Suppose that $y \in OC(x_n)$ is not stable or satisfies $OC(y) \neq OC(x_n)$. According to (3), 3.4 finds an $h \in \Pi$ such that $\forall z \in \mathbb{Z}$, $y|[-|z|,|z|] \leq B_{(n,h(\alpha z))}$. So $\forall m \in \mathbb{N}$, $y|[-(m+1), m+1] \leq B_{(n,h(\alpha(m)))}$. Applying (1), $y|[-m,m] \leq (B_{(n+1,\alpha h(\alpha(m)))} : -|h(\alpha(m))| < z < |h(a(m))|)$. Hence, $y \in OC(x_{n+1})$. According to (6), we have:

(7) $\forall y \in OC(x_0)$, y is stable, and $\exists n \geq 0$ such that $OC(y) = OC(x_n)$.

Finally notice that (5), (7), and 1.1 prove $OC(x_0)$ contains no minimal sets. Since (4) shows $OC(x_0)$ is homeomorphic to $\mathbb{P}$, $OC(x_0)$ satisfies are requirements. ❑

5. MULTIPLE RECURRENCE

In this section we are concerned with multiple systems — systems comprised of more than one map from a space to itself. One of the earliest results on multiple systems is due to P. Erdös and A. Stone:

THEOREM 5.1: x is recurrent (almost-periodic) in a system $[X; f]$ iff x is jointly recurrent (almost-periodic) in the system $[X; \{f^n : n \in \mathbb{N}\}]$.

We say that a space X has the MBR provided that for each finite family $\mathcal{F}$ of commuting maps on X, $[X; \mathcal{F}]$ has a multiply recurrent point. The Power Theorem is partially extended by The Multiple Birkhoff Recurrence Theorem:

THEOREM 5.2: [8] (Also, see [4]). Each compact metric space has the MBR.

REMARK 5.3: Notice that 1.4 and 5.2 show that when X is a metric space containing a compact set with nonempty interior and if $[X, \mathcal{F}]$ is minimal, then X has the MBR.

The next result is a modification of an example appearing in [4] and strengthens the result that compact spaces of weight of the continuum need not have the MBR.

EXAMPLE 5.4: There is a compact separable first countable space X and a commuting pair of homeomorphisms $f, g : X \to X$ such that $[X; \{f, g\}]$ has no jointly recurrent points.

Proof: Identify the unit circle S^1 with the group $(\mathbb{R}/\mathbb{Z}, +)$, and let $Y = S^1 \times \{0,1\}$ have the topology induced by sets of form either:

$$I_0(r;\varepsilon) = \{(r,0)\} \cup (r - \varepsilon, r) \times \{0,1\} \text{ or } I_1(r;n) = \{(r,1)\} \cup (r, r + \varepsilon) \times \{0,1\},$$

where $\forall n \in \mathbb{N}$, $(r - \varepsilon, r + \varepsilon)$ is an "open interval" in S^1 centered at r with radius ε. $\{I_0(r;2^{-n}): n \in \mathbb{N}\}$ forms a countable neighborhood base at (r,0) while $\{I_1(r;2^{-n}): n \in \mathbb{N}\}$ forms a countable neighborhood base at $(r,1)$. Fix an irrational θ in $\mathbb{R}/\mathbb{Z}$, and define a "rotation" $h: Y \to Y$ by $h(r,i) = h(r+\theta, i)$. Then, as shown in [4], h is a homeomorphism, Y is separable first countable, and, as a two-to-one closed continuous preimage of S^1, is compact as well.

Since each point of S^1 is recurrent under the usual rotation by θ, each point of Y is jointly recurrent in the system $[Y; \{h, h^{-1}\}]$. Further, it is easy to see [4] the following is true:

$$(\#) \ \forall y \in Y \ \forall i,j \in \{0,1\} \ \forall n \in \mathbb{N},\ h^n(r,i) \in I_j(r;n) \text{ iff } h^{-n}(r,i) \in I_{1-j}(r;n).$$

Define $X = Y^2$ have the product topology, and we consider the product maps $f = h \times h$ and $g = h \times h^{-1}$. Let $A = \{((r,i), (s,i)) \in X: i \in \{0,1\}\}$ and let $B = X\backslash A$. Then (#) immediately implies A is the set of points recurrent in $[X; f]$, while B is the set of points recurrent in $[X; g]$. Clearly, f and g commute. ❑

REMARK: The reader should observe that 5.4 can be made to yield an explicit proof, as opposed to an (easy) existence proof, of the following example: There are commuting homeomorphisms F,G: $\mathbb{C} \to \mathbb{C}$ such that $[\mathbb{C}; \{F,G\}]$ is a minimal system with a dense set of nonjointly recurrent points.

Sketch of proof: Let X, Y, f, and g be as defined in 5.4. Define

$$Q = \{h^z(t,i) \in X: z \in \mathbb{Z},\ t \in \mathbb{Q}/\mathbb{Z},\ i \in \{0,1\}\}.$$

Let $\equiv$ denote the equivalence relation on X which identifies each quadruple $((r,i), (s,j)) \in X\backslash Q^2$, with $i,j \in \{0,1\}$, with a single point $(r,s) \in (S^1)^2$ and leaves Q^2 alone. Then the quotient space $C = X/\equiv$ is homeomorphic to $\mathbb{C}$. Let F and G denote the maps induced on C by $\equiv$ acting on $\{f, g\}$.

EXAMPLE 5.5: There are commuting homeomorphisms $f,g: \mathbb{P} \to \mathbb{P}$ such that $[\mathbb{P}, \{f,g\}]$ is minimal, $\forall p \in \mathbb{P}$, $OC_f(p) \cap OC_g(p) = \{p\}$.

Proof: Let P be $OC(\alpha)$ in Π, where α is defined in 2.6. 2.5 shows $\mathbb{P}$ is homeomorphic to $\mathbb{P}$. From 2.1, P^2 is homeomorphic to $\mathbb{P}$. Let $f = \sigma \times id$ and $g = id \times \sigma$. Then f and g commute. From 2.7(a) and 1.2, $[P;\sigma]$ is minimal; hence, $[P^2;\{f,g\}]$ is a minimal multiple system. Clearly, $\forall (x,y) \in P^2$, $OC_f((x,y)) = P \times \{y\}$ and $OC_g((x,y)) = \{x\} \times P$. So $OC_f((x,y)) \cap OC_g((x,y)) = \{(x,y)\}$. ❑

Our next result was inspired by [3] and methods of N. Hindman (see, e.g., [14] and [15]).

EXAMPLE 5.6: There is a compact space X and commuting homeomorphisms $f,g: X \to X$ such that $\forall x \in X$, $OC_f(x) \cap OC_g(x) = \emptyset$.

Proof: Consider $\beta(\mathbb{Z}^2)$ as the space of all ultrafilters on $\mathbb{Z}^2$, and let βh and βv be, respectively, the extension to $\beta(\mathbb{Z}^2)$ of the "horizontal" and "vertical" shifts $(n, m) \mapsto (n+1, m)$ and $(n,m) \mapsto (n, m+1)$ on $\mathbb{Z}^2$. We define $X = \beta(\mathbb{Z}^2)\backslash\mathbb{Z}^2$, $f = \beta h|X$, and $g = \beta v|X$. Clearly, X is compact and, f and g are commuting homeomorphisms on X.

Fix an ultrafilter $x \in X$, then x contains exactly one element of $\{\mathbb{N}^2, (-\mathbb{N}) \times \mathbb{N}, \mathbb{N} \times (-\mathbb{N}), (-\mathbb{N})^2\}$, where $-\mathbb{N}$ denotes the set of negative integers. As the four proofs are analogous, we assume $\mathbb{N}^2 \in x$. Since X is an F-space, disjoint cozero sets of X have disjoint closures. Thus, to prove our result, it is sufficient to find a disjoint pair of csets G and H, each the union of countably many clopen sets in X, and containing, respectively, $Or_f(x)$ and $Or_g(x)$.

$\forall A \subseteq \mathbb{N}^2$, $\forall (n,m) \in \mathbb{N}^2$, we define $A + (n,m) = \{(n+a, m+b) : (a,b) \in A\}$.

We find $\forall (n,m) \in \mathbb{N}^2$ a pair of sets $I_{(n,m)}, J_{(n,m)} \in x$ such that the following holds:

$$(\#)\quad h^n I_{(n,m)} \cap v^m J_{(n,m)} = (I_{(n,m)} + (n,0)) \cap (J_{(n,m)} + (0,m)) = \emptyset.$$

Given $k \in \mathbb{N}$, $\mathbb{N}^2 = \bigcup\{\{(kr+i, ks+j) : r,s \in \mathbb{N}\} : i,j \in (-k, k)\}$. As x is an ultrafilter, $\exists i(k), j(k) \in [1,k)$ with $M_k = \{(kr+i(k), ks+j(k)) : r,s \in \mathbb{N}\} \in x$. $\forall n \in \mathbb{N}$, let $U_n = \bigcap_{k \leq n} M_k$. Then $U_n \in x$.

$\forall (n,m) \in \mathbb{N}^2$, we show that $I(n,m) = U_{2n}$ and $J_{(n,m)} = U_{2m}$ satisfy (#). In addition, assume $n \leq m$ (the proof for $m \leq n$ is similar) so that $J_{(n,m)} \subseteq I_{(n,m)}$. Suppose $\exists (a,b) \in (I + (n,0)) \cap (I + (0,m))$. Then $\exists r,s \in \mathbb{N}$ such that $a = 2nr + i(2n) + n = 2ns + i(2n)$. Thus, $2r + 1 = 2s$ — a contradiction. Thus, $(I_{(n,m)} + (n,0)) \cap (J_{(n,m)} + (0,m)) = \emptyset$.

But $cl_{\beta(\mathbb{Z}^2)}(I_{(n,m)}) \cap X$ and $cl_{\beta(\mathbb{Z}^2)}(J_{(n,m)}) \cap X$ are clopen in X. Thus, $G = \bigcup_{t \in \mathbb{Z}^2} I_t$ and $H = \bigcup_{t \in \mathbb{Z}^2} J_t$ prove the theorem. ❑

ACKNOWLEDGMENTS

Our collaboration on this paper was begun in 1985. J.P. wishes to thank the University of Toronto in Canada, his host while some of the results in this paper were discovered. S.W.W. wishes to thank Charles University in Prague, his host while some of the results in this paper were discovered. We also acknowledge valuable conversations with B. Balcar, M. Husek, P. Simon, and J. deVries.

Finally, we are most thankful for the referee who found an errors in the statement of 1.5, and in the original proof of 4.1, and who told us about [1], [2], [9], and [16].

REFERENCES

1. AKIN, E. 1993. The general Topology of Dynamical Systems. Graduate Studies in Mathematics 1. American Mathematical Society. Providence, Rhode Island.
2. AKIN, E., J. AUSLANDER, & K. BERG. When in a transitive map chaotic? To appear.
3. BALCAR, B. & P. KALÁSEK. 1989. Nonexistence of multiple recurrence points in ultrafilter dynamical systems. Bull. Pol. Acad., Math. **37:** 525–529.
4. BALCAR, B., P. KALÁSEK, & S.W. WILLIAMS. 1987. Multiple Recurrence in dynamical systems. Comment. Math. Univ. Carolina. **28:** 607–612.
5. BLASS, A. 1993. Ultrafilters: where topological dynamics = algebra = combinatorics. Topology Proc. **19:** 33–56.
6. ELLIS, R. 1969. Lectures on Topological Dynamics. W.A. Benjamin, Inc. New York.
7. ENGELKING, R. 1977. General Topology, Polish Scientific Publishers.
8. FURSTENBERG, H. 1981. Recurrence in Ergodic Theory and Combinatorial Number Theory. Princeton Univ. Press. Princeton. New Jersey.
9. GLASNER, S. & D. MAON. 1989. Rigidity in topological dynamics. Ergodic Theory and Dynamical Systems. **9:** 177–188.
10. GOTTSCHALK, W.H. & G.A. HEDLUND. 1955. Topological Dynamics. American Mathematical Society Colloquium Pub. **36**.
11. GOTTSCHALK, W.H. 1944. Orbit-closure decompositions and almost periodic properties. Bull. Amer. Math. Soc. **50:** 915–919.
12. HEDLUND, G.A. 1966. Transformations commuting with the shift. *In:* Topological Dynamics. W.A. Benjamin. New York. 259–290.
13. HEDLUND, G.A. 1969. Endomorphisms and automorphisms of the shift dynamical system. Math. Systems Theory. **3:** 320–375.
14. HINDMAN, N. 1980. Partitions and sums and products in $\beta\mathbb{N}$ and products of integers, Trans. Am. Math. Soc. **247:** 227–245.
15. HINDMAN, N. 1993. The topological-algebraic system $(\beta\mathbb{N}, +, \cdot)$. Annals of New York Acad. Sci. **70**: 155–163.
16. KATZNELSON, Y. & B. WEISS. 1979/80. When all points are recurrent/generic. *In:* Ergodic Theory and Dynamical Systems I Proceedings, Special Year Maryland.
17. PETERSEN, K. 1983. Ergodic Theory, Cambridge University Press (1983).
18. WILLIAMS, S.W. 1988. Special points arising from self-maps. General Topology and Relations to Modern Analysis. **5:** 629–638.

Bi-neighborhood Posets

FRANK P. PROKOP

Department of Mathematics
The University of Wollongong
Wollongong, N.S.W., Australia

ABSTRACT: Bi-neighborhood (bi-nbhd) lattices are a generaliization of bitopological spaces which use a natural duality that associates nbhd filters with dual nbhd ideals to view a lattice as a "top down" structure, in which points are unnecessary and which uses ideals to determine dual nbhds. This leads to a definition of dual continuous functions between dual neighborhood lattices. The link between dual nbhd continuity and topological (top) continuity is established by proving that if $f: X \rightarrow Y$ is a one-to-one and onto function between top spaces X and Y, then f is top continuous if and only if the direct image function is a dual nbhd continuous function mapping $P(X)$, the power set of X, onto $P(Y)$. In addition, bi-nbhd lattices are used to generate examples of Urysohn collections and quasiproximities.

1. INTRODUCTION

Bi-topological spaces have been used by Kopperman [10], Kelly [8], and others to examine the interaction between two topologies on the same set in order to explain the role that duality plays in topological spaces. Bi-topological spaces allow one to view a set, as it were, from two viewpoints and to examine how the imposed topological structures and the corresponding continuous functions interrelate. Bi-neighborhood (bi-nbhd) posets give one the same latitude of having available differing viewpoints of the same poset but the basic difference is that bi-nbhd posets use a natural duality that associates nbhd filters with dual nbhd ideals to give two differing structural descriptions of the same poset — one as a "bottom up" structure, which may have points like a topological space and which uses filters to determine nbhds, and one as a "top down" structure, in which points are unnecessary, and which uses ideals to determine dual nbhds. Now, not only do each of the nbhd and dual nbhd structures have associated definitions of continuous functions, but also continuity can be defined in each direction between nbhd and dual nbhd posets. Thus, we have up to four types of continuity associated with each bi-nbhd poset. As a consequence of the "richness" of the bi-nbhd structure we have the appropriate framework in which to ask the following question: "Do we live in a 'bottom up' universe in which topology,

Mathematics Subject Classification: 54A05 (06A99, 06B99, 54C60).

Key words and phrases: neighborhood lattices, neighborhood continuity, dual neighborhood lattices, dual continuity, bi-topological space, bi-neighborhood lattice, bi-dual lattice, preneighborhood Urysohn collection, quasiproximity, neighborhood quasiproximity.

which is based on points or "smallest" particles, is the appropriate tool for describing some aspects of "reality" or do we live in a "top down" universe without "smallest" building blocks which requires an alternate description or do we live in a universe which exhibits aspects of both theories?"

To illustrate how we can generate a bi-nbhd structure without points, we start with a topological space (X, T) and $(P(X), \cup, \cap)$, the power set lattice of X. We extend the neighborhoods of points to neighborhoods of sets, while identifying a nbhd of x with a nbhd of $\{x\}$, then we have $T \subseteq P(X)$, $\eta(\{x\}) = \{N \mid (\exists t \in T)(\{x\} \subseteq t \subseteq N)\} \subseteq P(X)$, and $\eta(\{x\})$ is a filter in $P(X)$. Now all of the essential ingredients of a topology are "embedded" in $P(X)$, a lattice. Utilizing this viewpoint of a topological space, it is a "natural" step to define an analogous nbhd structure in an arbitrary lattice ($\wedge$-semilattice) along with a corresponding definition of a continuous function. At this stage, we have an appropriate structure to define two nbhd systems on the same poset, i.e., we are at a stage of development similar to that of bi-topological spaces. However, if the nbhd structure on a lattice is dualized, a dual nbhd structure, based on ideals, is developed which is used to define "closed" elements in a lattice independently of "points" or complementation. In this paper we begin with the development of the theory of dually continuous functions and then show that two additional types of continuity can be defined for functions between nbhd lattices and dual nbhd lattices, so that there are four types of continuous relationships between nbhd and dual nbhd lattices. These nbhd and dual nbhd structures along with the corresponding continuous functions are used to define bi-nbhd posets.

We begin by listing the notation and convention that are used in this paper. If L is an orthocomplemented lattice, then $'$ is used to denote complementation; I is used to denote an arbitrary indexing set; if P is a poset, $\vee$ and $\wedge$ represent the operations of sup and inf; while $\cup$ and $\cap$ are used to denote the set theoretic operations of union and intersection, $\bigvee_{\alpha \in I} x_\alpha$ is used to represent $\vee\{x_\alpha \colon \alpha \in I\}$, with similar abbreviations used for $\bigwedge_{\alpha \in I} x_\alpha$, $\bigcup_{\alpha \in I} A_\alpha$, and $\bigcap_{\alpha \in I} A_\alpha$; 1 represents the greatest element of P and ϕ represents the least element of P; $F(P) = \text{Filt } P = \{F \colon F \text{ is a filter of } P\}$; $I(P) = Id\ P = \{I \colon I \text{ is an ideal of } P\}$; if $x \in P$, then $\uparrow(\{x\}) = \uparrow x = \{y \colon y \in P \text{ and } y \geq x\}$ and $\downarrow(\{x\}) = \downarrow x = \{y \colon y \in P \text{ and } y \leq x\}$. In addition, the following definitions, which are discussed fully in [16], are used. If P is an $\wedge$-semilattice, then a function η: $P \to F(P)$ is a *filter mapping*. If η is a filter mapping on P, then $g \in P$ is *nbhd open* (or simply open) if $\eta(g) = \uparrow g$. Thus, the open elements in P are those elements for which the associated filter is a principal filter. Further, if η is a filter mapping on P, we let $G = \{g \colon g \in P \text{ and } \eta(g) = \uparrow g\}$. A filter mapping η is a *pre-nbhd mapping* if $(\forall x, t \in P)(t \in \eta(x) \leftrightarrow (\exists g \in G)(x \leq g \leq t))$. For a filter map to be a pre-nbhd mapping, the filters associated with each element must be determined by the set G of open elements which satisfies (i) $(\forall x \in P)\ (\exists g \in G)(g \geq x)$ and (ii) G is an $\wedge$-subsemilattice of P. If L is a lattice, a filter mapping η is a *nbhd mapping* if (i) G is a $\vee$-semi-complete sublattice of L, and (ii) $\phi \in L \to \eta(\phi) = \uparrow\phi$. If (X, T) is a topological space, then $\eta: P(X) \to F(P(X))$ defined by $\eta(A) = \{N: (\exists g \in T)(A \subseteq g \subseteq N)\}$ is the *induced nbhd mapping* on $P(X)$ and the pair $(P(X), \eta)$ is the *induced nbhd lattice* of (X, T). Finally, if X and Y are sets and $f: X \to Y$ is a function, then the *direct image function*, $f_*: P(X) \to P(Y)$ is given by $f_*(A) = \{f(a): a \in A\}$, and the *inverse image function*, $f^*: P(Y) \to P(X)$ is given by $f^*(B) = \{x: x \in X \text{ and } f(x) \in B\}$.

In [16], the definitions of pre-nbhd and nbhd $\wedge$-semilattices are dualized, leading to a parallel theory of "closed" elements in a $\vee$-semilattice based on ideals, pre-dual nbhd mappings, and dual nbhd mappings. This dual theory establishes that there is a "natural" lattice theoretic duality between the definitions of open and closed, which is not dependent on complementation. However, if L is an orthocomplemented lattice, then an element is closed in the dual theory if and only if it is the complement of an open element.

A brief summary of the dual definitions: Let P be a $\vee$-semilattice. A function $\gamma: P \to I(P)$ is called an *ideal mapping*. $h \in P$ is said to be γ-*closed* (or simply closed) if $\gamma(h) = \downarrow h$. We let $H = \{h: h \in P \text{ and } \gamma(h) = \downarrow h\}$. Similarly, a *pre-dual nbhd mapping* (or more simply a *pre-dual mapping*) is defined on P. For a lattice L, a pre-dual mapping $\gamma: L \to I(L)$ is a dual nbhd mapping if (i) H is a $\wedge$-semicomplete sublattice of L and (ii) $1 \in L \Rightarrow \gamma(1) = \downarrow\{1\}$ and the pair (L, γ) is called a dual nbhd lattice. If (X, T) is a topological space, then $\gamma: P(X) \to I(P(X))$ defined by $\gamma(A) = \{t: (\exists g \in T)(t \subseteq g' \subseteq A)\}$ is the *induced dual nbhd mapping* on $P(X)$ and the pair $(P(X), \gamma)$ is the *induced dual nbhd lattice* of (X, T).

The proofs of those lemmas that are straightforward computations and which follow directly from the corresponding definitions are omitted.

2. DUAL NEIGHBORHOOD CONTINUITY

Since we are interested in dualizing the definition of a nbhd continuous function, we begin this section with a brief introduction to nbhd continuity. As discussed in [18], nbhd continuity came about after an examination of the equivalent conditions for a function to be topologically continuous (globally) showed that in the statement: "If (X, T_1) and (Y, T_2) are topological spaces, $(P(X), \eta_1)$ and $(P(Y), \eta_2)$ are the induced nbhd lattices and $f: X \to Y$ is a function, then f is continuous if and only if $(\forall A \in P(X)(\forall B \in \eta_2 f_*(A))(\exists Z \in \eta_1(A))(f_*(Z) \subseteq B)$," we need only replace the elements of $P(X)$ with the elements of a pre-nbhd $\wedge$-semilattice P and the nbhds of a set in $P(X)$ by the corresponding nbhds of an element in P to have an appropriate definition of nbhd continuity.

DEFINITION 1: Let (P_1, η_1) and (P_2, η_2) be pre-nbhd $\wedge$-semilattices. A function f: $P_1 \rightarrow P_2$ is said to be *nbhd continuous* (*η-continuous*) *at* $a \in P_1$ if $(\forall y \in \eta_2 f(a))(\exists z \in \eta_1(a))(f(a) \leq f(z) \leq y)$. Further, f: $P_1 \rightarrow P_2$ is said to be *nbhd continuous* (*η-continuous or continuous*) (on P_1) if f is nbhd continuous at each element of P_1.

The idea expressed by Definition 1 can be visualized in Diagram 1.

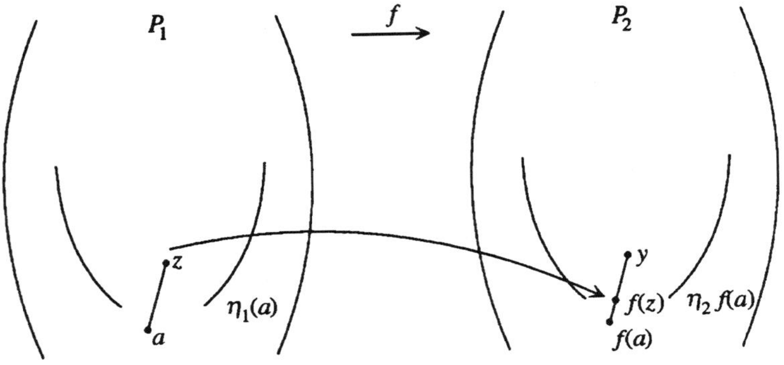

DIAGRAM 1

That Definition 1 is a reasonable generalization of topological continuity is established in [16], where it is proved that if (X, T_1) and (Y, T_2) are topological spaces and f: $X \rightarrow Y$ is a function, then f is a topologically continuous function from X to Y if and only if $f_*: P(X) \rightarrow P(Y)$ is a nbhd continuous function from the induced nbhd lattice $(P(X), \eta_1)$ to the induced nbhd lattice $(P(Y), \eta_2)$.

If Diagram 1 is interpreted in $P(X)$ and $P(Y)$ for X and Y topological spaces, then if a function f moves a continuously to $f(a)$, the continuity of f is determined by what f does to points "outside" of a. From a lattice theoretic point of view continuity seems to give special significance to elements above a given element in the partial order. To try to find out why nbhd filters are so much more "important" than ideals which are their duals, we define "dual" continuity.

DEFINITION 2: Let (P_1, γ_1) and (P_2, γ_2) be pre-dual nbhd $\vee$-semilattices. A function f: $P_1 \rightarrow P_2$ is said to be *pre-dual nbhd continuous* (*dual continuous or γ-continuous*) *at* $a \in P_1$ if $(\forall y \in \gamma_2 f(a))(\exists z \in \gamma_1(a))(y \leq f(z) \leq f(a))$. Further, f: $P_1 \rightarrow P_2$ is said to be *pre-dual nbhd continuous* (*dual continuous or γ-continuous*) (on P_1) if f is pre-dual nbhd continuous at each element of P_1.

The idea expressed by Definition 2 can be visualized in Diagram 2.

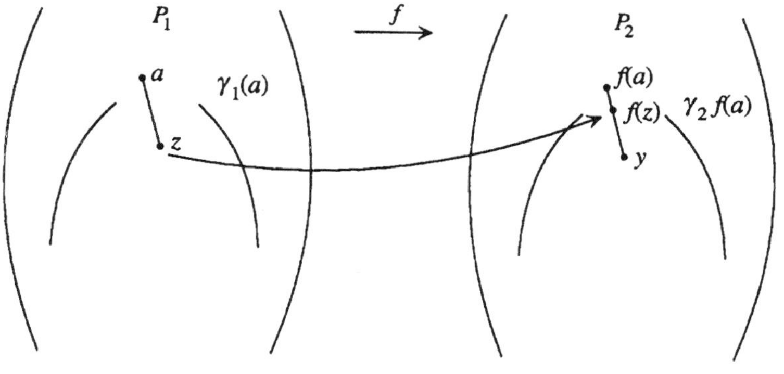

DIAGRAM 2

If Diagram 2 is interpreted in $P(X)$ and $P(Y)$ for X and Y topological spaces, then if a function f moves a dual continuously to $f(a)$, the dual continuity of f is determined by what f does to points "inside" of a. We could say that when a moves dual continuously to $f(a)$, all the elements "inside" of a must be mapped to elements inside of $f(a)$. This seems to be a reasonable interpretation of the movement of objects in reality. Since a function must map a "point" to a "point," it is clear that if (X, T_1) and (Y, T_2) are T_1 topological spaces, $(P(x), \gamma_1)$ and $(P(Y), \gamma_2)$ are the induced dual nbhd lattices, $f: X \to Y$ is any function, then $f_*: P(X) \to P(Y)$ is dual continuous at each atom of $P(X)$.

Since the closed elements determine the dual nbhds of elements, the dual continuity of a function at a is also determined by the closed elements in the dual nbhd of $f(a)$.

LEMMA 3: Let (P_1, γ_1) and (P_2, γ_2) be pre-dual nbhd $\vee$-semilattices.

(i) A function $f: P_1 \to P_2$ is pre-dual nbhd continuous at $a \in P_1$ if and only if $(\forall h \in \gamma_2 f(a) \cap H_2)(\exists z \in \gamma_1(a))(h \leq f(z) \leq f(a))$.

(ii) Let $f: P_1 \to P_2$ be isotone. f is pre-dual nbhd continuous at $x \in P_1$ if and only if $((h \in \gamma_2\ f(x) \cap H_2) \to (\exists h_1 \in \gamma_1(x) \cap H_1)(h \leq f(h_1) \leq f(x)))$.

Proof: (i) Follows from $z \in \gamma_2\ f(x) \leftrightarrow (\exists h \in \gamma_2 f(x) \cap H_2)(z \leq h \leq f(x))$.

(ii) Let f be pre-dual nbhd continuous at x and $h \in \gamma_2\ f(x) \cap H_2$. $(\exists z \in \gamma_1(x))(h \leq f(z) \leq f(x))$. $(z \in \gamma_1(x)) \to (\exists h_1 \in H_1)(z \leq h_1 \leq x)$. Since f is isotone, we have $f(z) \leq f(h_1) \leq f(x)$. The proof of the converse is clear. ❑

We note that if (L, γ) is a dual neighborhood lattice and a $\in L$, then $f_a: L \to L$ given by $f_a(x) = x \wedge a$. is dual continuous, as are $f_1: L \to L$ and $f_2: L \to L$ given by $f_1(x) = a$ and $f_2(x) = x$.

PROPOSITION 4: Let (L_1, η_1) and (L_2, η_2) be orthocomplemented nbhd lattices. Let γ_1 and γ_2 be the induced dual nbhd mappings given, for $i = 1, 2$, by $t \in \gamma_i(x) \leftrightarrow t' \in \eta_i(x)$. If $f: L_1 \rightarrow L_2$ is a function such that $(\forall x \in L_1)(f(x') = (f(x))'$, then f is η-continuous at a if and only if f is γ-continuous at a'.

Proof: The proof follows from Definition 1, Definition 2, Lemma 3, and the following equivalences: $(g \in \eta_i f(x) \cap G_i \leftrightarrow g' \in \gamma_i f(a) \cap H_i)$; $(z \in \eta_i(x) \leftrightarrow z' \in \gamma_i(x))$; and $(g \leq f(z) \leq f(x) \leftrightarrow f(x)' \leq f(z)' \leq g'$. ❑

COROLLARY 4.1: Let (X, T_1) and (Y, T_2) be topological spaces, let $(P(X), \gamma_1)$ and $(P(Y), \gamma_2)$ be the induced dual nbhd lattices, and let $f: X \rightarrow Y$ be a one-to-one and onto function. f is a topologically continuous function on X if and only if f_* is dual continuous on $P(X)$.

Proof: f is topologically continuous on X if and only if $f_*$is η-continuous on $P(X)$, but by Proposition 4, f_* is γ-continuous on $P(X)$. ❑

It follows from Proposition 4, that a Boolean homomorphism $f: X \rightarrow Y$ between Boolean nbhd lattices is continuous on X if and only if f is dual continuous on X.

COROLLARY 4.2: Let (L_1, η_1) and (L_2, η_2) be orthocomplemented nbhd lattices. Let γ_1 and γ_2 be the induced dual nbhd mappings. If $f: L_1 \rightarrow L_2$ is an onto function such that $(\forall x, y \in L_1)(f(x \vee y) = f(x) \vee f(y))$, then f is η-continuous at $a \leftrightarrow f$ is γ-continuous at a'.

Proof: Clearly, f is isotone (monotone). $f(x') = f(x)'$ follows from $f(x \vee x') = f(1) = f(x) \vee f(x')$ and $(\forall x)(f(x) \leq f(1))$. If $w \in L_2$, then $(\exists t \in L_1)(f(t) = w)$. Hence, $w \leq f(1)$. Thus, $f(1) = 1'$ and $f(x') = f(x)'$. The result now follows from Proposition 4. ❑

Lemma 5 indicates what must happen if the dual continuous image of an element is closed.

LEMMA 5: Let (P_1, γ_1) and (P_2, γ_2) be pre-dual nbhd $\vee$-semilattices. Let $f: P_1 \rightarrow P_2$ be dual continuous at x and $f(x)$ be closed in P_2. If x is not closed in P_1, then there exists $z \in \gamma_1(x)$ such that $z < x$ and $f(z) = f(x)$.

Proof: Let f be dual continuous at x, and $f(x)$ be closed in P_2. $f(x) \in \gamma_2 f(x) \rightarrow$ there exists $z \in \gamma_1(x)$ such that $f(z) = f(x)$. But $z \in \gamma_1(x) \rightarrow$ there exists h closed in P_1 such that $z \leq h \leq x$. Now, $x \notin \eta_1(x) \rightarrow x \neq h \rightarrow z \leq h < x$. ❑

COROLLARY 5.1: Let (P_1, γ_1) and (P_2, γ_2) be pre-dual nbhd $\vee$-semilattices and let $f: P_1 \rightarrow P_2$ be an isotone mapping. Further, let f be dual continuous at x and let $f(x)$ be closed in P_2. If x is not closed in P_1, then there exists h closed in P_1 such that $h < x$ and $f(x) = f(h)$.

COROLLARY 5.2: Let (X, T_1) and (Y, T_2) be topological spaces, let $(P(X), \gamma_1)$ and $(P(Y), \gamma_2)$ be the induced dual nbhd lattices, and let $f_*: P(X) \to P(Y)$ be dual continuous. If $f_*(A)$ is closed in Y, and A is not closed in X, then there exists h closed in X such that $f_*(A) = f_*(h)$.

COROLLARY 5.3: Let (P_1, γ_1) and (P_2, γ_2) be pre-dual nbhd $\vee$-semilattices. If $f: P_1 \to P_2$ is dual continuous, one-to-one, and onto, then $\forall h$ closed in P_2, $f^{-1}(h)$ is closed in P_1.

COROLLARY 5.4: Let (L_1, γ_1) and (L_2, γ_2) be pre-nbhd lattices and let $f: L_1 \to L_2$ be a dual continuous function such that $f(\bigwedge_{\alpha \in I} x_\alpha) = \bigwedge_{\alpha \in I} f(x_a)$. If h is closed in L_2, and $A = f^*(h)$, then $\bigwedge_{a \in A} a$ (if it exists) is closed in L_1.

Proof: Let h be closed in L_2, $A = f^*(h)$ and $z = \bigwedge_{a \in A} a$. We first note that $f(z) = f(\bigwedge_{a \in A}) a = \bigwedge_{a \in A} f(a) = h$. If z is not closed in L_1, then, by Lemma 5, there exists $w \in \gamma_1(z)$ such that $w < z$ and $f(w) = f(z) = h$. Thus, $w \in A$ and $z \leq w$. A contradiction. ❑

Corollaries 5.3 and 5.4 show that if f is dual continuous, 1–1, and onto, then the inverse of a closed element is closed, and that if f is dual continuous and preserves infima, then the infimum of all elements that map onto a closed element is closed.

EXAMPLE 6: Let (X, T_1) and (Y, T_2) be compact topological spaces and let $(P(X), \gamma_1)$ and $(P(Y), \gamma_2)$ be the induced dual nbhd lattices. Since the union of compact sets is compact, we can let $H_1 = \{A : A \in P(X) \text{ and } A \text{ is compact}\}$ and $H_2 = \{B : B \in P(Y) \text{ and } B \text{ is compact}\}$ and let γ_i^c be the corresponding pre-dual nbhd mapping determined by H_i. $f_*: (P(X), \gamma_1^c) \to (P(Y), \gamma_2^c)$ is γ continuous at A if and only if $(\forall h \in \gamma_2^c f_*(A) \cap H_2)(\exists w^{\text{compact}} \in \gamma(A))(h \subseteq f_*(w) \subseteq f_*(A)$. Let f_* be γ continuous at A and consider the case when $f_*(A)$ is compact but A isn't known to be compact. By Corollary 5.2, $(\exists h^{\text{compact}})(f_*(h) = f_*(A))$. So, $f_*(A)$ is the image of a compact set.

Proposition 7 gives a partial answer to the question: "When will the inverse of a closed element being closed ensure the dual continuity of the function?"

PROPOSITION 7: Let (P_1, γ_1) and (P_2, γ_2) be pre-dual nbhd $\vee$-semilattices, let $f: P_1 \to P_2$ be one-to-one, onto, and, further, let f^{-1} be isotone. If $\forall h$ closed in L_2, $f^{-1}(h)$ is closed in L_1, then f is dual continuous.

Proof: Let $x \in P_1$, and let h be closed in L_2 such that $h \in \gamma_2 f(x)$. $h \leq f(x) \to f^{-1}(h) \leq f^{-1}(f(x)) = x$. If we let $z = f^{-1}(h)$, then $h \leq f(z) \leq x$. Thus, f is dual continuous at x.

PROPOSITION 8: Let (X, T_1) and (Y, T_2) be topological spaces, let $(P(X), \gamma_1)$ and $(P(Y), \gamma_2)$ be the induced dual nbhd lattices, and let $f: X \to Y$ be a function mapping X onto Y. The following statements are equivalent.

(i) f^* is γ-continuous.
(ii) f_* is a closed mapping.
(iii) $(\forall A \in P(X))(f_*(A)^- \subseteq f_*(A^-))$.

Proof: (i) ↔ (ii). Let f^* be γ-continuous, let H be closed in X and let $B = f^*(H)$. Now, $H \subseteq f^*(f_*(H)) = f^*(B) \to (\exists W^{\text{closed}} \in \gamma_1(B))(H \subseteq f^*(W) \subseteq f^*(B)) \to f_*(H) \subseteq f_*(f^*(W)) = W \subseteq B = f_*(H)$. Thus, $W = f_*(H)$ is closed. Let f_* be a closed mapping and let H (closed) $\in \gamma_1(f^*(B))$. $H \subseteq f^*(B) \to f_*(H) \subseteq f_*(f^*(B)) = B \to f_*(H) \in \gamma_1(B)$ and $H \subseteq f^*(f_*(H)) \subseteq f^*(B)$.

(ii) ↔ (iii). Let f_* be a closed mapping and let $A \in P(X)$. $A \subseteq A^- \to f_*(A) \subseteq f_*(A^-) = f_*(A^-)^- \to f_*(A)^- \subseteq f_*(A^-)$. Let $A \in P(X)$ and $A = A^-$. But, $f_*(A)^- \subseteq f_*(A^-) = f_*(A) \to f_*(A) = f_*(A)^-$.

DEFINITION 9. Let (L_1, η_1) be a pre-nbhd lattice and (L_2, γ_2) be a pre-dual nbhd lattice. A function f: $L_1 \to L_2$ is said to be ηγ *continuous at* $a \in L_1$ if $(\forall y \in \gamma_2 f(a)((\exists z \in \eta_1(a))(y \leq f(z) \leq f(a))$. Further, f: $L_1 \to L_2$ is said to be ηγ *continuous* (on L_1) if f is pre-nbhd continuous at each element of L_1. In a similar way we define a function f: $L_1 \to L_2$ to be γη *continuous at* $a \in L_1$ and γη *continuous* (on L_1).

The idea of ηγ continuity can be visualized in Diagram 3.

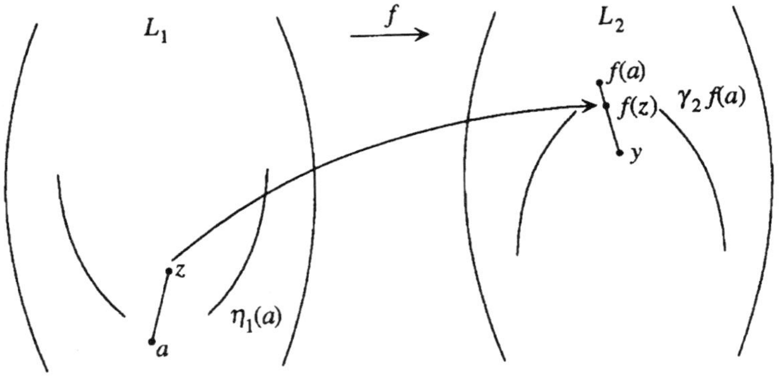

DIAGRAM 3

In Section 3 we show that if (L, η, γ) is both a pre-nbhd and a pre-dual lattice, then ηγ and γη continuous functions from L to L arise in a natural way. For example, if (L, η, γ) is an orthocomplemented pre-nbhd lattice and γ is the induced pre-dual mapping, then $'$: $L \to L$ is both ηγ and γη continuous.

3. BI-NEIGHBORHOOD POSETS

We begin this section by noting that if (L, η, γ) is both a pre-nbhd and a pre-dual nbhd lattice then L has both a bottom up and a top down structure as well as a distinguished collection of clopen elements. Further, if $f: L \rightarrow L$ is both η and γ continuous, then f can "approximate" a value of $f(a)$ from above using the images of open elements and from below using the images of closed elements. In particular, as we show in Example 12, if $f: L \rightarrow L$ is both η and γ continuous and the set of open elements is "dense," then $f(a)$ can be "approximated" from both above and below by the images of open elements.

Since (X, T_1, T_2) is a bitopological space if T_1 and T_2 are each topologies on X, it is clear that by examining the relationship between two structures on the same lattice (poset), we are adopting a viewpoint of nbhd lattices which is consistent with the perspective of topology displayed by bitopological spaces. Thus, we now define bi-neighborhood (bi-nbhd) posets as a generalization of bitopological spaces.

DEFINITION 10: If P is a pre-nbhd $\wedge$-semilattice with respect to both η_1, η_2, then (P, η_1, η_2), is called a *bi-pre-neighborhood* (*bi-pre-nbhd*) *poset*. Similarly, we define a *bi-nbhd poset*, a *bi-dual nbhd poset*, a *bi-*η γ*-poset* and a *bi-*γ η*-poset*.

To see that Definition 10 is an appropriate generalization of bitopological spaces we simply rephrase, for bitopological spaces, results proved in this paper and in [16]: (X, T_1, T_2) is a bitopological space if and only if $(P(X), \eta_1, \eta_2)$ is a bi-nbhd lattice if and only if $(P(X), \gamma_1, \gamma_2)$ is a bi-dual nbhd lattice. Further, $f: (X, T_1) \rightarrow (X, T_2)$ is (top) continuous if and only if the direct image function, $f_*: (P(X), \eta_1) \rightarrow (P(X), \eta_2)$ is (nbhd) continuous. So, bi-nbhd posets generalize both bitopological spaces and the corresponding continuous functions. Finally, it follows from Proposition 4, that for one-to-one functions from X onto X, the γ-continuity and the η-continuity of f_* agree with the continuity of f. Thus, the bi-nbhd and bi-dual nbhd homeomorphisms are exactly the topological homeomorphisms.

DEFINITION 11: If (L, η_1, η_2) is a pre-nbhd lattice then $f: L \rightarrow L$ is *pairwise continuous* if f is both η_1 and η_2 continuous. Similarly, (L, η, γ) is a bi-$\eta\gamma$ lattice, then $f: L \rightarrow L$ is said to be *pairwise* $\eta\gamma$*-continuous* if f is both η pre-nbhd continuous and γ pre-dual nbhd continuous. In the same manner we have *pairwise dual continuous and pairwise* $\gamma\eta$*-continuous*.

EXAMPLE 12: Consider the chain of real numbers $\mathbb{R}$. Define a pre-nbhd map η and a pre-dual nbhd map γ on $\mathbb{R}$ by:

$$\eta(x) = \begin{cases} [x, \infty), & x \in \mathbb{Q}, \\ (x, \infty), & x \in \mathbb{R} - \mathbb{Q}; \end{cases}$$

$$\gamma(x) = \begin{cases} (-\infty, x], & x \in \mathbb{Q}, \\ (-\infty, x), & x \in \mathbb{R} - \mathbb{Q}. \end{cases}$$

The elements of $\mathbb{Q}$ are both open and closed

(a) Let f: $\mathbb{R} \to \mathbb{R}$ be isotone. f is continuous in the usual topology on $\mathbb{R}$ if and only if f: $(\mathbb{R}, \eta, \gamma) \to (\mathbb{R}, \eta, \gamma)$ is pairwise $\eta\gamma$-continuous, i.e., f is both continuous and dual continuous.

Proof: f: $\mathbb{R} \to \mathbb{R}$ is both η and γ continuous at $a \in \mathbb{R}$ iff $(\forall g \in \eta\, f(a))(\forall h \in \gamma\, f(a))(\exists z \in \eta(a))(\exists w \in \gamma(a))(h \le f(w) \le f(a) \le f(z) \le g)$. f: $\mathbb{R} \to \mathbb{R}$ is continuous in the usual topology iff $(\forall \varepsilon > 0)(\exists \delta > 0)(a - \delta < x < a + \delta \to f(a) - \varepsilon < f(x) < f(a) + \varepsilon)$. Let f: $\mathbb{R} \to \mathbb{R}$ be isotone and continuous in the usual topology. Let $h, g \in \mathbb{Q}$, such that $h < f(a) < g$ and let $\varepsilon < \min\{|f(a) - h|, |g - f(a)|\}$. $(\exists \delta > 0)(\exists w, z \in \mathbb{Q})(a - \delta < w < a < z < a + \delta$ and $h \le f(w) \le f(a) \le f(z) \le g)$. Let f: $\mathbb{R} \to \mathbb{R}$ be isotone and both η and γ continuous at $a \in \mathbb{R}$, $\varepsilon > 0$, let $h \in \mathbb{Q}$ such that $f(a) - \varepsilon < h < f(a)$ and let $g \in \mathbb{Q}$ such that $f(a) < g < f(a) + \varepsilon$. $(\exists z \in \eta(a) \cap \mathbb{Q})(\exists w \in \gamma(a) \cap \mathbb{Q})(f(a) - \varepsilon < h \le f(w) \le f(a) \le f(z) \le g < f(a) + \varepsilon) \to (\exists z \in \eta(a) \cap \mathbb{Q})(\exists w \in \gamma(a) \cap \mathbb{Q})(w \le a \le z)(f(a) - \varepsilon < h \le f(w) \le f(a) \le f(z) \le g < f(a) + \varepsilon)$.($\alpha$) Now, since f is isotone, $\forall t \in (w, z)$, $f(w) \le f(t) \le f(z)$. Thus, $\forall t \in (w, z)$, $f(a) - \varepsilon < f(t) < f(a) + \varepsilon$. So, let $\delta = \min\{|a - w|, |z - a|\}$.

(b) Let f: $(\mathbb{R}, \eta, \gamma) \to (\mathbb{R}, \eta, \gamma)$ be an isotone function. The function f is said to be *approximated by rationals* if and only if $(\forall a \in \mathbb{R})(\forall \varepsilon > 0)(\exists q \in \mathbb{Q})(|f(a) - f(q)| < \varepsilon)$. If one were only interested in approximating the values of f using rational numbers, then it follows from (α) that if f is either η continuous or γ continuous, f can be approximated by rationals.

EXAMPLE 13: Let (X, T) be a compact Hausdorff space and let $C = \{A : A$ is compact$\}$. Since the intersection of compact sets is compact, we can let $H = G = C$ be the set of clopen elements and $(P(X), \eta, \gamma)$ be the corresponding bi-pre-$\eta\gamma$ nbhd lattice. $f_* : P(X) \to P(X)$ is pairwise $\eta\gamma$ continuous if and only if $(\forall A \in P(X))(\forall g \in \eta\, f_*(A) \cap C)(\forall h \in \gamma\, f_*(A) \cap C)(\exists z \in \eta(A) \cap C)(\exists w \in \gamma(A) \cap C)(h \subseteq f_*(w) \subseteq f_*(A) \subseteq f_*(z) \subseteq g)$.

(a) $f_* : P(X) \to P(X)$ is $\eta\gamma$ continuous at A if and only if $(\exists z \in \eta(A) \cap C)(f_*(z) = f_*(A))$.

(b) Let f: $X \to X$ be a continuous function mapping (X, T) onto (X, T). $f^* : (P(X), \eta, \gamma) \to (P(X), \eta, \gamma)$ is $\eta\gamma$-pairwise continuous.

Proof: (a) $\eta\gamma$ continuity at $A \to (\forall h \in \gamma\, f_*(A) \cap C)(\exists z \in \eta(A) \cap C)(h \le f_*(z) \le f_*(A))$. But, $A \subseteq z \to f_*(w) = f_*(A)$. The converse is clearly true.

(b) f_* is both a closed and open mapping. Thus, by Proposition 8 and its dual, f^* is $\eta\gamma$-pairwise continuous. ❑

In the next section, we show that pre-nbhd lattices are a useful tool for creating examples of Urhsohn collections and bitopological spaces. The definitions and notation are essentially that found in [10].

DEFINITION 14: A *Urysohn collection* is a set P of pairs of subsets of X such that:

(i) $(A, B) \in P \rightarrow A \subseteq B$;

(ii) $(A, B) \in P \rightarrow (\exists E \in P(X))((A, E), (E, B) \in P)$;

(iii) $(A, B) \in P$ and $E \subseteq A$, $B \subseteq F \rightarrow (E, F) \in P$.

The *dual of a Urysohn collection P* is $P^* = \{(B', A'): (A, B) \in P\}$. Let $T_p = \{G: x \in G \rightarrow (\exists F$ a finite subset of $P)(x \in \cap \operatorname{Dom} F$ and $\cap \operatorname{Range} F \subseteq G)\}$. If T_p is a topology on X, then T_p is called the *topology arising from P* and the *bitopological space arising from P* is $\chi_P = (X, T_p, T_{P^*})$.

EXAMPLE 15: (a) If we consider $P = \{(\phi, B): B \in P(X)\}$, then P is a Urysohn collection of sets and if $T_P = \{G: x \in G \rightarrow (\exists F$ a finite subset of $P)$ $(x \in \cap \operatorname{Dom} F$ and $\cap \operatorname{Range} F \subseteq G)\} = \{\phi\}$ then T_p not a topology on X. However, $P^* = \{(A, X): A \in P(X)\}$ and $T_{P^*} = \{\phi, X\}$ is a toplogy on X.

(b) If we consider $P = \{(\phi, B): B \in P(X)\} \cup \{(\{1\}, B): B \in \uparrow\{1\}\}$, then P is a Urysohn collection of sets and $T_p = \{G: x \in G \rightarrow (\exists F$ a finite subset of $P)(x \in \cap \operatorname{Dom} F$ and $\cap \operatorname{Range} F \subseteq G)\} = \{\phi, \{1\}\}$ and T_p is not a topology on $X \neq \{1\}$. $P^* = \{(A, X): A \in P(X)\} \cup \{(A, X - \{1\}): A \in X - \downarrow\{1\}\}$ and $T_{P^*} = \{\phi, X - \{1\}, X\}$ is a topology on X.

LEMMA 16: Let P be a Urysohn collection on a set X and let $T_p = \{G: x \in G \rightarrow (\exists F$ a finite subset of $P)(x \in \cap \operatorname{Dom} F$ *and* $\cap \operatorname{Range} F \subseteq G)\}$.

(a) $(A, B) \in P \rightarrow \downarrow A \times \uparrow B \subseteq P$. In particular, $(A, B) \in P \rightarrow (\forall a \in A)((\{a\}, B) \in P)$.

(b) T_p is a topology on X if and only if $X \in T_p$.

(c) $X \in T_p$ if and only if $(\forall x \in X)(\exists B \in P(x))((\{x\}, B) \in P)$.

(d) $G \in T_p$ if and only if $(\forall x \in G)((\exists n \in \mathbb{R})(\exists(\{x\}, B_1)\ (\{x\}, B_2), \ldots, (\{x\}, B_n) \in P)(\bigcap_i B_i \subseteq G)$.

Proof: (a) and (b) follow from the definitions.

(c) $X \in T_p \rightarrow (\forall x \in X)(\exists F$ a finite subset of P, $x \in \cap \operatorname{Dom} F$ and $\cap \operatorname{Range} F \subseteq X) \rightarrow (\exists n \in \mathbb{N})(\exists(A_1, B_1)(A_2, B_2), \ldots, ((A_n, B_n) \in P)(x \in \bigcap_i A$ and $\bigcap_i B_i \subseteq X) \rightarrow x \in A_1 \rightarrow (\{x\}, B_1) \in P$. Conversely, $(\forall x \in X)(\exists B \in P(x))((\{x\}, B) \in P) \rightarrow (\forall x \in X)(\exists(\{x\}, B) \in P)(x \in \{x\}$ and $B \subseteq X) \rightarrow X \in T_p$.

(d) Similar to (c). ❑

Lemma 17 shows how pre-nbhd and nbhd lattices can be used to construct Urysohn collections and T_P topologies.

LEMMA 17: Let $(P(X), \eta)$ be a pre-nbhd lattice. Define P on X by $(A, B) \in P \leftrightarrow B \in \eta(A) \leftrightarrow (\exists G^{\text{open}})(A \subseteq G \subseteq B)$, i.e., $P = \bigcup_{A \in P(X)} (A \times \eta(A))$.

(a) P is a Urysohn collection of sets and G is η open if and only if $(G, G) \in P$.

(b) $(A, B) \in P^*$ if and only if $B \in \gamma(A)$, i.e., the dual of P is $P^* = \bigcup_{B \in P(X)} (\gamma(B) \times B)$.

Proof: The proof follows from the definition of a pre-nbhd lattice. ❑

DEFINITION 18: Let $(P(X), \eta)$ be a pre-nbhd lattice and $P = \bigcup_{A \in P(X)} (A \times \eta(A))$. P is said to be a *pre-nbhd Urysohn collection on X*.

PROPOSITION 19: Let $(P(X), \eta)$ be a pre-nbhd lattice and P be the corresponding pre-nbhd Uryson collection on X. $B = \{G: G \text{ is } \eta \text{ open}\}$ is a base for the topology T_P on X.

Proof: Let $T \in T_p$. $x \in T \rightarrow (\exists n \in \mathbb{R})(\exists(\{x\}, B_1)(\{x\}, B_2), \ldots, (\{x\}, B_n) \in P)(\bigcap_i B_i \subseteq T) \rightarrow (\forall i \leq n)(\exists G_i^{x \text{ open}})(\{x\} \subseteq \bigcap_I G_i^x \subseteq \bigcap_I B_i \subseteq T) \rightarrow x \in (\bigcap_i G_i^x) = G^x$. Thus, $T \subseteq \bigcup_{x \in T} G^x \subseteq T$. Hence, $T = \bigcup_{x \in T} G^x$. Clearly, if $T = \bigcup_{\alpha \in i} G_\alpha$, then $T \in T_p$.

It follows from Proposition 19 that if $(P(X), \eta)$ is a pre-nbhd lattice and P the corresponding pre-nbhd Uryson collection on X, then every η-open element is T_p open. Further, if (X, T) is a topological space and $(P(X), \eta)$ is the induced nbhd lattice and P be the corresponding nbhd Uryson collection on X, then $T = T_P$. Hence, we have proved Corolloary 19.1. ❑

COROLLARY 19.1: Every topology is T_P for some nbhd Urysohn collection P.

COROLLARY 19.2: Let $(P(X), \eta)$ be a pre-nbhd lattice and P be the corresponding pre-nbhd Uryson collection on X. If $f_*: P(X) \rightarrow P(X)$ is η continuous then $f: X \rightarrow X$ is T_p continuous.

It follows from Proposition 19 and its proof, that if we begin with a topological space (X, T) and construct its nbhd collection of sets and corresponding T_p, then $T_p = T$ and $\chi_P = (X, T, T_{P*})$. Further, if we start with a Urysohn collection of sets to construct a topology T_p, then use T_p to construct its nbhd Urysohn collection of sets and a "new" topology T_P^1, then $Tp = T_P^1$.

DEFINITION 20: A *quasiproximity* is a Urysohn collection P of pairs of subsets of X such that:

(i) $(\phi, \phi) \in P$ and $(X, X) \in P$:
(ii) $(A, B) \in P$ and $(A, C) \in P \rightarrow (A, B \cap C) \in P$;
(iii) $(A, B) \in P$ and $(C, B) \in P \rightarrow (A \cup C, B) \in P$.

A quasiproximity is called a *nbhd quasiproximity* if the collection P of pairs of subsets of X is a nbhd Urysohn collection.

It should be noted that if P is a quasiproximity then so is P^*. Further, it follows from Lemma 16(d) and Definition 20(ii) that for a quasiproximity P on X, $G \in T_p$ iff $(\forall x \in G)((\{x\}, B) \in P)(B \subseteq G)$(also see [10]). Finally, it follows immediately from Definition 14(iii) that for a quasiproximity P on X, $G \in T_p$ if and only if $(\forall x \in G)((\{x\}, G) \in P)$.

LEMMA 21: Every nbhd Urysohn collection is a quasiproximity.

Proof: (i) and (ii) follow immediately from $(A, B) \in P \leftrightarrow B \in \eta(A)$. (iii) follows from the fact that for a nbhd mapping η on $P(X)$, $\eta(A \cup C) = \eta(A) \cap \eta(C)$. ❑

COROLLARY 21.1: Every topology is T_P for some nbhd quasiproximity P.

COROLLARY 21.2: Let $(P(X), \eta)$ be a pre-nbhd lattice and P and P^* be the corresponding pre-nbhd and pre-dual quasiproximities on X. $B = \{H: H$ is γ closed$\}$ is a base for the topology T_{P^*} on X.

Proof: $G^* \in T_{P^*} \leftrightarrow (\forall x \in G^*)((\{x\}, G^*) \in P^*) \leftrightarrow (\forall x \in G^*)((\{x\} \in \gamma(G^*)) \leftrightarrow (\forall x \in G^*)\,(\exists H_x{}^{\text{closed}})(\{x\} \subseteq H_x \subseteq G^*)$. Thus, $G^* \subseteq \bigcup_{x \in G^*} H_x \subseteq G^*$. Hence, $G^* = \bigcup_{x \in G^*} H_x$. Finally, if $G^* = \bigcup_{\alpha \in \iota} H_\alpha$, then $G^* \in T_{P^*}$. ❑

COROLLARY 21.3: Let $(P(X), \eta)$ be a T_1 pre-nbhd lattice and P and P^* be the corresponding pre-nbhd and pre-dual quasiproximities on X. T_{P^*} is the discrete topology on X.

Proof: In [17] it was proved that $(P(X), \eta)$ is a T_1 pre-nbhd lattice $\leftrightarrow (\forall x \in X)(\{x\}$ is closed).

DEFINITION 22: Let P and P_1 be Urysohn collections on X and Y, respectively, and let $f: X \to Y$. f is an *Urysohn map* if $(A_1, B_1) \in P_1 \to (f^*(A_1), f^*(B_1)) \in P$.

PROPOSITION 23: Let $(P(X), \eta_1)$ and $(P(Y), \eta_2)$ be pre-nbhd lattices and let P and P_1 be the pre-nbhd Urysohn collections on X and Y, respectively. $f: X \to Y$ is an Urysohn map if and only if $f_*: (P(X), \eta_1) \to (P(Y), \eta_2)$ is a pre-nbhd continuous function.

Proof: Let $f: X \to Y$ be an Urysohn map and let G be η_2-open. $(G, G) \in P_1 \to (f^*(G), f^*(G)) \in P \to f^*(G) \in \eta_1(f^*(G)) \to f^*(G)$ is open $\to f_*$ is a pre-nbhd continuous function. Conversely, let $f_*: (P(X), \eta_1) \to (P(Y), \eta_2)$ be a pre-nbhd continuous function. $(A_1, B_1) \in P_1 \to B_1 \in \eta(A_1) \to (\exists G_1{}^{\text{open}})(A_1 \subseteq G_1 \subseteq B_1) \to (\exists G_1{}^{\text{open}})(f^*(A_1) \subseteq f^*(G_1) \subseteq (f^*(B_1))$. But $f^*(G_1)$ is η_1-open by the continuity of f. Thus, $f^*(B_1) \in \eta_1\, f^*(B_1) \to (f^*(A_1), f^*(B_1)) \in P \to f$ is an Urysohn map.

COROLLARY 23.1: If (X, T_1) and (Y, T_2) are topological spaces and $(P(X), \eta_1)$ and $(P(Y), \eta_2)$ are the induced nbhd lattices and P and P_1 are the nbhd quasiproximities on X and Y, respectively, then $f: X \to Y$ is an Urysohn map if and only if $f: X \to Y$ is a continuous function.

COROLLARY 23.2: Let (X, T_1) and (Y, T_2) be topological spaces and let $(P(X), \eta_1, \gamma_1)$ and $(P(Y), \eta_2, \gamma_2)$ be the induced bi-$\eta\gamma$ lattices and let P_1 and P_2 be the pre-nbhd Urysohn collections on X and Y, respectively, and P_{1*} and P_{2*} be their duals. $f: X \to$ Y is an Urysohn map with respect to P_{1*} and P_{2*} if and only if $f: X \to Y$ is a continuous function.

Proof: Let $f: X \rightarrow Y$ be an Urysohn map and let H be closed ($\gamma_2{}^{\text{closed}}$). $(H, H) \in P_{2*} \rightarrow (f^*(H), f^*(H)) \in P_{1*} \rightarrow f^*(H) \in \gamma_1(f^*(H)) \rightarrow f^*(H)$ is closed $\rightarrow f$ is a continuous function. Conversely, let $f: X \rightarrow Y$ be a continuous function. $(A_1, B_1) \in P_2 \rightarrow B_1 \in \gamma(A_1) \rightarrow (\exists H_1{}^{\text{closed}})(B_1 \subseteq H_1 \subseteq A_1) \rightarrow (\exists H_1{}^{\text{closed}}))(f^*(B_1) \subseteq f^*(H_1) \subseteq (f^*(A_1))$. But $f^*(H_1)$ is closed by the continuity of f. Thus, $f^*(B_1) \in \gamma_1 f^*(B_1) \rightarrow (f^*(A_1), f^*(B_1)) \in P_{1*} \rightarrow f$ is an Urysohn map. ❑

We conclude this paper by raising a question about the structure of "reality" where there is a significant difference between a "top down" model and a "bottom up" or topological model. A topological model of reality presumes the existence of that elusive "smallest" particle, the "point." Although it was shown in [18] that most of the concepts of topology could be defined in a lattice without points, the dual theory suggests a structure which begins "at the top" may be appropriate. Thus, we say an $\wedge$-semilattice P with 1 is *dually atomistic* if each element in P is the infimum of dual atoms. A *universal poset* is a dually atomistic pre-dual $\wedge$-semilattice. The universal poset structure along with dual continuous functions provides a framework for an examination of the merits of a "top down" model of some aspects of reality which uses dual atoms as "large" building blocks.

REFERENCES

1. BIRKHOFF, G. 1967. Lattice Theory. Amer. Math.Soc. Providence, Rhode Island
2. ČECH, E., 1966. Topological Spaces, Vols. I & II. Interscience Publishers. London-New York -Sydney
3. DUGUNDJI, J. 1972. Topology. Allyn and Bacon, Inc. Boston, Massachusetts.
4. EILENBERG, S. 1941. Ordered Topological Spaces. Amer. J. Math. **63:** 39–45.
5. GRATZER, G. 1971. Lattice Theory: First Concepts and Distributive Lattices. Freeman. San Francisco, California.
6. ISBELL, J.R. 1972. Atomless parts of spaces. Math. Scand. **31:** 5–32.
7. JOHNSTONE, P.T. 1983. The Point of Pointless Topology. Bull. Amer. Math. Soc. **8:** (1) 41–53.
8. KELLY, W.C. 1963. Bitopological Spaces. Proc. London Math. Soc. **13:** 71–89.
9. KELLEY, J.L. 1955. General Topology. D. Van Nostrand Co., Inc. Princeton, New Jersey.
10. KOPPERMAN, R. 1995. Asymmetry and duality in topology. *In:* Topology and its Application. **66:** 1–39.
11. KOPPERMAN, R. 1988. All topologies come from generalized metrics. Amer. Math. J. Monthly. **95:** 89–97.
12. MACLANE, S., & BIRKHOFF, G. 1967. Algebra. The Macmillan Co. New York.
13. MCKINSEY, J.C.C. & TARSKI, A. 1944. The algebra of topology. Ann. of Math.**45:** (2) 141–191.
14. NACHBIN, L. 1965. Topology and Order, D. Van Nostrand Co, Inc. Princeton, New Jersey.
15. NAIMPALLY, S.A. & WARRACK, B.D. 1970. Proximity Spaces. The University Press. Cambridge.

16. PROKOP, F.P., 1989. Neighborhood Lattices-A Poset Approach to Topological Spaces. Bull. Aust. Math. Soc. **39:** 31–49.
17. PROKOP, F.P. 1989. Convergence in Neighborhood Lattices. Bull. Aust. Math. Soc. **40:** 129–147.
18. PROKOP, F.P. 1994. A step beyond topology. *In:* Papers on General Topology and Applications. Annals N.Y. Academy of Sci. **728:** 96–113.
19. SIKORSKY, R. 1964. Boolean Algebras. Springer-Verlag. Berlin.

On Upper Weightable Spaces

M. SCHELLEKENS

Imperial College
Department of Computing
180 Queen's Gate
London SW7 2BZ UK

ABSTRACT: The weightable quasi-pseudo-metric spaces have been introduced by Matthews as part of the study of the denotational semantics of dataflow networks (e.g., [6] and [7]). The study of these spaces has been continued in the context of Nonsymmetric Topology by Künzi and Vajner [4], [5]. We introduce and motivate the class of upper weightable quasi-pseudo-metric spaces. The relationship with the development of a topological foundation for the complexity analysis of programs [10] is discussed, which leads to the study of the weightable optimal (quasi-pseudo-metric) join semilattices.

1. INTRODUCTION

Weightable quasi-pseudo-metric spaces have been introduced by Matthews in the context of the study of the Denotational Semantics of dataflow networks (e.g., [3], [6], [7], and [14]).

We recall that the weightable quasi-pseudo-metrics originally have been introduced as "partial metrics," which are a new kind of generalized metrics for which the "reflexivity-axiom," $\forall x.\ d(x, x) = 0$, is not required to hold. Since the partial metric spaces have been shown to be equivalent to the weightable quasi-pseudo-metric spaces [6], their topological study can be carried out in the context of Nonsymmetric Topology.

The study of the weightable quasi-pseudo-metric spaces has been the subject of [4] and also of the survey paper "Nonsymmetric Topology" [5], where several open characterization problems on weightable spaces have been stated[1]: "Characterize those quasi-uniformities having a countable base which are induced by a weighted quasi-pseudo-metric" (Problem 7), "Which topological spaces admit weightable quasi-pseudo-metrics?" (Problem 8) and "Develop a concept of a weighted quasi-uniformity" (Problem 10).

For the open problems stated above, some partial results in connection to Problem 7 and Problem 8 are known [4], [5]. The results are limited to restricted classes as for instance the class of the Alexandroff topologies in con-

Mathematics Subject Classification: 06, 54.
Keywords and phrases: (weightable) quasi-pseudo-metrics, complexity, directed partial orders, join semilattices.

[1]In [4] the problems are actually stated in terms of "quasi-metrics," which correspond to the "quasi-pseudo-metrics" as originally defined in [2].

nection to a partial solution of Problem 8 [4]. In [5] the remark is made in connection to Problem 7 that any totally bounded quasi-uniform space with a countable base can be induced by a weighted quasi-pseudo-metric.

The paper introduces and motivates a subclass of the weightable spaces, the upper weightable spaces. A characterization of the upper weightable spaces has been obtained in [11], which is briefly discussed in the paper (no results of the present paper are based on this characterization). The relevance of these spaces with respect to the development of a topological foundation for the complexity analysis of programs [10] is discussed, which leads to the introduction of the class of the weightable optimal (quasi-pseudo-metric) join semilattices.

These upper weightable quasi-pseudo-metric spaces posses by definition an associated partial order which is a join semilattice for which the join operation is "optimal" with respect to the distance function (in a sense explained below).

The paper presents two characterization results for this class of spaces.

2. QUASI-PSEUDO-METRIC SPACES

Quasi-pseudo-metrics are generalized metrics which do not necessarily satisfy the axiom of symmetry. Their study belongs to the field of Nonsymmetric Topology (e.g., [1], [2], [4], and [5]) and these generalized metrics have for instance been applied in the context of Denotational Semantics (e.g., [3], [6], [7], [12], and [14]) as well as in Complexity Theory (e.g., [10] and [13]).

We use the following notation. $\mathcal{R}$ denotes the set of real numbers and $\mathcal{N}$ denotes the set of natural numbers. We define

$$\mathcal{R}^+ = (0, \infty) \text{ and } \mathcal{R}_0^+ = [0, \infty), \text{ while } \overline{\mathcal{R}^+} = \mathcal{R}^+ \cup \{\infty\} \text{ and } \overline{\mathcal{R}_0^+} = \mathcal{R}_0^+ \cup \{\infty\}.$$

A function $d: X \times X \to \mathcal{R}_0^+$ is a *quasi-pseudo-metric* iff

(1) $\forall x,y,z.\ d(x, y) + d(y, z) \geq d(x, z)$
(2) $\forall x.\ d(x, x) = 0$.

A quasi-pseudo-metric d is a *quasi-metric* iff

(3) $\forall x, y.\ d(x, y) = 0 \Rightarrow x = y$.

We give a few examples of quasi-pseudo-metric spaces which will be frequently referred to later on.

EXAMPLES: The function $d_1 : \mathcal{R}^2 \to \mathcal{R}_0^+$, defined by $d_1(x, y) = y - x$ when $x < y$ and $d_1(x, y) = 0$ otherwise, and its conjugate are quasi-pseudo-metrics. We refer to d_1 as the "left distance" and to its conjugate as the "right distance" (e.g., [5]). These quasi-pseudo-metrics correspond to the nonsymmetric versions of the standard metric m on the reals, where $\forall x, y \in \mathcal{R}.\ m(x, y) = |x - y|$.

The left distance induces the topology with a base consisting of the intervals $\{(-\infty, a) | a \in \mathcal{R}\}$ while the right distance induces the topology with a base consisting of the intervals $\{(a, \infty) | a \in \mathcal{R}\}$.

Note that the right distance has the usual order on the reals as associated order, that is, $\forall x, y \in \mathcal{R}.\ x \leq_{d_1^{-1}} y \Leftrightarrow x \leq y$, while for the left distance we have $\forall x, y \in \mathcal{R}.\ x \leq_{d_1} y \Leftrightarrow x \geq y$.

The function d_2: $(\mathcal{R} - \{0\})^2 \rightarrow \mathcal{R}_0^+$, defined by $d_2(x, y) = \frac{1}{y} - \frac{1}{x}$ when $y < x$ and 0 otherwise, and its conjugate are quasi-pseudo-metrics.

Each quasi-pseudo-metric d induces a topology $\mathcal{T}_d$ generated by the base $\{B_\epsilon[x] | \epsilon > 0, x \in X\}$, where $\forall \epsilon > 0\ \forall x \in X.\ B_\epsilon[x] = \{y | d(x, y) < \epsilon\}$.

Let $\Delta = \{(x, x) | x \in X\}$. The topology $\mathcal{T}_d$ is T_0 iff $\leq_d$ is a partial order and is T_1 iff $\leq_d$ is a discrete order (e.g., [8]). We assume all quasi-pseudo-metric spaces to satisfy the T_0 separation axiom in what follows.

The *conjugate* d^{-1} of a quasi-pseudo-metric d is defined to be the function $d^{-1}(x, y) = d(y, x)$, which is again a quasi-pseudo-metric (e.g., [2]). The *conjugate* of a quasi-pseudo-metric space (X, d) is the quasi-pseudo-metric space (X, d^{-1}). The *metric* d^* induced by a quasi-pseudo-metric d is defined by $d^*(x, y) = \max\{d(x, y), d(y, x)\}$.

The *associated preorder* $\leq_d$ of a quasi-pseudo-metric d is defined by $x \leq_d y$ iff $d(x, y) = 0$.

A quasi-pseudo-metric space (X, d) *has a maximum* (*minimum*) iff its associated partial order $(X, \leq_d)$ has a maximum (minimum). (We recall that all spaces are assumed to be T_0.)

A function f: $(X, d) \rightarrow \mathcal{R}$ is decreasing (increasing) iff $\forall x, y \in X.\ x \leq_d y \Rightarrow f(x) \geq f(y)$ ($f(x) \leq f(y)$) and a function f: $(X, d) \rightarrow \mathcal{R}$ is strictly decreasing (strictly increasing) iff $\forall x, y \in X.\ x <_d y \Rightarrow f(x) > f(y)$ ($f(x) < f(y)$).

A partial order $(X, \leq)$ is *directed* when for any two elements x, y of X there exists an element $z \in X$ such that $x, y \leq z$. A quasi-pseudo-metric space is *directed* when its associated order is directed. A *join semilattice* is a partially ordered set $(X, \leq)$ such that any two points $x, y \in X$ have a supremum $x \sqcup y$ in X. A *lattice* is a partially ordered set $(X, \leq)$ such that any two elements $x, y \in X$ have a supremum $x \sqcup y$ and an infimum $x \sqcap y$ in X.

Given a quasi-pseudo-metric space (X, d) then for any point $x \in X$, we define $x\uparrow = \{(x_n)_{n \geq 1} | x_1 \geq_d x \text{ and } \forall n \geq 1.\ x_n \leq_d x_{n+1}\}$. Note that $\forall x \in X.\ x\uparrow \neq \emptyset$, since $x\uparrow$ contains the sequence with constant value x.

For any function f: $A \rightarrow B$ and for any set $X \subseteq A$, $f|X$ indicates the restriction of f to the set X. A *subspace* of a quasi-pseudo-metric space (X, d) is a pair $(Y, d|Y^2)$, where $Y \subseteq X$. An *extension* of a quasi-pseudo-metric space (X, d) is a quasi-pseudo-metric space (Y, d), such that (X, d) is a subspace of (Y, d).

A quasi-pseudo-metric space (X, d) satisfies a given property *hereditarily* when every subspace of the space (X, d) satisfies this property.

For any set X, we define $X^{\omega} = \{f | f: \mathcal{N} \to X\}$. We will consider function spaces which are obtained from a given quasi-pseudo-metric space (X, d), of the kind $(\mathcal{F}, d^{\omega})$, where $\mathcal{F} \subseteq X^{\omega}$ and where the distance d^{ω} is defined by $\forall f, g \in \mathcal{F}.\ d^{\omega}(f, g) = \sum_n d(f(n), g(n)) \frac{1}{2^n}$, on condition that the sum converges on the elements of $\mathcal{F}$. In what follows, whenever a function space is considered, we implicitly assume that this condition is satisfied.

For instance in case d is bounded, that is $\exists K \in \mathcal{R}_0^+\ \forall x, y \in X.\ d(x, y) \leq K$, the distance d^{ω} is defined on the entire set X^{ω}.

It is easy to verify that a function space $(\mathcal{F}, d^{\omega})$ has the pointwise order generated from $\leq_d$ as associated order, that is $\forall f, g \in \mathcal{F}.\ f \leq_{d^{\omega}} g \Leftrightarrow \forall n.\ f(n) \leq_d g(n)$.

For any set $A \subset \mathcal{R}^+$, a function $f \in A^{\omega}$ is *bounded from below* (*above*) iff $\exists c > 0\ \forall n \in \mathcal{N}.\ f(n) \geq c\ (f(n) \leq c)$.

3. WEIGHTABLE SPACES

We recall the definition of a weightable quasi-pseudo-metric space.

DEFINITION 1: A quasi-pseudo-metric space (X,d) is *weightable* iff there exists a function $w: X \to \mathcal{R}_0^+$ such that $\forall x, y \in X.\ d(x, y) + w(x) = d(y, x) + w(y)$. This equality is referred to as the *weighting equality*. The function w is called a *weighting function*, $w(x)$ is the *weight* of x and the quasi-pseudo-metric d is *weightable by the function* w. A *weighted* space is a triple (X, d, w) where (X, d) is a quasi-pseudo-metric space weightable by the function w.

EXAMPLES: The quasi-pseudo-metric space $(\mathcal{R}_0^+, d_1)$ is weightable by the identity function, $w_1(x) = x$ and the space $(\overline{\mathcal{R}^+}, d_2)$ is weightable by the function $w_2(x) = \frac{1}{x}$.

We recall the example of the weightable Baire quasi-pseudo-metric discussed in [6] and [7] (cf. also [3] and [14]).

Let $S^{\leq \omega}$ be the set of countably infinite and finite sequences of elements from a given set S. Given a sequence $s \in S^{\leq \omega}$, say of length $L \geq 1$, then for any natural number n such that $1 \leq n \leq L$, $s(n)$ denotes the n-th element of the sequence. Define the function $p: S^{\leq \omega} \times S^{\leq \omega} \to \mathcal{R}_0^+$ as follows:

$$\forall x, y \in S^{\leq \omega}.\ p(x, y) = 2^{-\alpha}, \text{ where } \alpha = \max \{n | x(n) = y(n)\}$$

when the sequences x and y have a common nonempty initial segment and $\alpha = 0$ otherwise.

The function p is a "partial metric," the "Baire partial metric," also referred to as the "Kahn partial metric" (e.g., [3] and [7]). The weightable quasi-pseudo-metric corresponding to the Baire partial metric, is the Baire quasi-pseudo-metric b, defined by $\forall x, y \in X.\ b(x, y) = p(x, y) - p(x, x)$, with weighting function $w_b(x) = p(x, x)$ (cf. [7]).

We remark that the conjugate quasi-pseudo-metric space $(\mathcal{R}_0^+, d_1^{-1})$ is not weightable.

We sketch the argument. Assume by way of contradiction that the space $(\mathcal{R}_0^+, d_1^{-1})$ is weightable by a function w. For $x \geq y \geq 0$ we obtain, by weightedness of d_1^{-1} with respect to w, that $d_1^{-1}(x, y) = x - y = w(y) - w(x)$, which implies that w is unbounded. However, for $y = 0$, we obtain that $\forall x \geq 0.\ x = w(0) - w(x)$ and thus $\forall x \geq 0.\ w(x) \leq w(0)$, which contradicts the unboundedness of w.

Since the paper focuses on weightable quasi-pseudo-metric spaces, we will, in what follows, solely refer to the weightable space $(\mathcal{R}_0^+, d_1)$ and not to its conjugate.

For more information on conjugates of weightable spaces we refer the reader to [5]. We recall that a topological space induced by a quasi-pseudo-metric whose conjugate is weightable, need not admit any weightable quasi-pseudo-metric [5].

We show the following fact on weightable quasi-pseudo-metric spaces.

LEMMA 2: The weighting functions of a weightable quasi-pseudo-metric space are descending.

Proof: Let (X,d) be a weightable quasi-pseudo-metric space and let w be a weighting function for (X,d). For any two points $x, y \in X$ such that $x \leq_d y$, we have by the weighting equality that $w(x) = d(y,x) + w(y) \geq w(y)$. ❑

We introduce the "descending path condition" in connection to weightable spaces.

We use the following terminology from [5]. Given a set X, then a *path* in X is a finite sequence $(x_1, \ldots, x_n)$ of points in X. For any two points x, y of a given set X, a path from x to y is a path $p = (x_1, \ldots, x_n)$ in X such that $x_1 = x$, $x_n = y$ and $n \geq 2$. A path $p = (x_1, \ldots, x_n)$ is *descending* iff $\forall i: 1 \ldots n-1.\ x_i \geq_d x_{i+1}$.

DEFINITION 3: A quasi-pseudo-metric space satisfies the descending path condition (DPC) iff $\forall x, y, z \in X.\ x \geq_d y \geq_d z \Rightarrow d(x, y) + d(y, z) = d(x, z)$.

The descending path condition intuitively expresses that the shortest distance between two points is obtained by following descending paths.

We show that weightable quasi-pseudo-metric spaces satisfy the descending path condition. An example of a nonweightable quasi-pseudo-metric space which satisfies the condition is given by the space $(\mathcal{R}_0^+, d_1^{-1})$.

LEMMA 4: Weightable quasi-pseudo-metric spaces satisfy the descending path condition.

Proof: If (X, d) is a weightable quasi-pseudo-metric space and w is a weighting function for this space, then we obtain by the weighting equality that $\forall x, y \in X.\ x \geq_d y \Rightarrow d(x, y) = w(y) - w(x)$. So for any three points $x, y, z \in X$ such that $x \geq_d y \geq_d z$ we have $d(x, y) + d(y, z) = d(x, z)$ since $(w(y) - w(x)) + (w(z) - w(y)) = w(z) - w(x)$. ❑

We will use the following notation for the function which represents the distance from a given point. If (X, d) is a quasi-pseudo-metric space then for any point $x_0 \in X$, the function $f_{x_0}: (X, d) \rightarrow \mathcal{R}_0^+$ is defined by $\forall x \in X.\ f_{x_0}(x) = d(x_0, x)$.

We show that these functions are decreasing.

LEMMA 5: If (X, d) is a quasi-pseudo-metric space then $\forall x, y, z \in X.\ (x' \leq_d x$ and $y' \geq_d y) \Rightarrow d(x', y') \leq d(x, y)$.

Proof: If (X, d) is a quasi-pseudo-metric space and $x, y, z \in X$ such that $x' \leq_d x$ and $y' \geq_d y$, then $d(x', y') \leq d(x', x) + d(x, y) + d(y, y') = d(x, y)$. ❑

LEMMA 6: If (X, d) is a quasi-pseudo-metric space then for any point $x_0 \in X$ the function f_{x_0} is decreasing.

Proof: Assume that (X, d) is a quasi-pseudo-metric space and let x_0 be a point of X. If x and y are two points of X such that $x \leq_d y$ then, by Lemma 5, $f_{x_0}(y) = d(x_0, y) \leq d(x_0, x) = f_{x_0}(x)$. ❑

LEMMA 7: If (X, d) is a quasi-pseudo-metric space which satisfies the descending path condition and which has a maximum x_0 then the function f_{x_0} is strictly decreasing.

Proof: Assume that (X, d) satisfies the descending path condition and has a maximum x_0. If x and y are points of X such that $x <_d y$ then, since f_{x_0} is decreasing (Lemma 6), we have $f_{x_0}(x) \geq f_{x_0}(y)$.

Assume by way of contradiction that $f_{x_0}(x) = f_{x_0}(y)$. Since $x_0 \geq_d y >_d x$, the descending path condition implies that $f_{x_0}(y) + d(y, x) = f_{x_0}(x)$. So we obtain that $d(y, x) = 0$ and thus, since the spaces are assumed to be T_0, we obtain that $x = y$, which contradicts the hypothesis $x <_d y$. ❑

We show that weightable quasi-pseudo-metric spaces with a maximum essentially posses a unique weighting function.

LEMMA 8: If (X, d) is a weightable quasi-pseudo-metric space with a maximum x_0 then its weighting functions are exactly the functions $f_{x_0} + c$ where $c \geq 0$.

Proof: Let (X, d) be a weightable space with a maximum x_0 and let w be a weighting function for the space (X, d). Then by the weighting equality we obtain that $\forall x \in X.\ d(x, x_0) + w(x) = d(x_0, x) + w(x_0)$. So we have that $\forall x \in X.\ w(x) = d(x_0, x) + w(x_0) = f_{x_0}(x) + w(x_0)$. Hence, every weighting function is of the form $f_{x_0} + c$ for some constant $c \geq 0$. Note that if, for some $c \geq 0$, $f_{x_0} + c$ is a weighting function for the space (X, d), then by substracting the constant c from the weighting equality we obtain that f_{x_0} is a weighting function for (X, d). This implies that for any constant $c \geq 0$ the function $f_{x_0} + c$ is a weighting function for this space. ❑

4. UPPER WEIGHTABLE SPACES

DEFINITION 9: A quasi-pseudo-metric d is functionally bounded iff there exists a function $f: X \to \mathcal{R}_0^+$ such that $\forall x, y \in X.\ d(x, y) \leq f(y)$. A quasi-pseudo-metric is functionally bounded with respect to a point $x_0 \in X$ iff the quasi-pseudo-metric is functionally bounded by the function f_{x_0}. A quasi-pseudo-metric space (X, d) is functionally bounded when d is functionally bounded.

COMMENT: There is no need to introduce notions of functionally boundedness "with respect to the second argument" ("$d(x, y) \leq f(y)$") and of functionally boundedness "with respect to the first argument ("$d(x, y) \leq f(x)$") since the last property can be expressed by the statement that d^{-1} is functionally bounded.

EXAMPLES: Any bounded quasi-pseudo-metric space is functionally bounded. This is, for instance the case for the Baire quasi-pseudo-metric spaces $(S^{\leq \omega}, b)$ and for the space $(\overline{\mathcal{R}^+}, d_2)$, which are bounded by 1. The space $(\mathcal{R}_0^+, d_1)$ is functionally bounded by the function w_1. The spaces $(\mathcal{R}_0^+, d_1)$ and $(\overline{\mathcal{R}^+}, d_2)$ are functionally bounded with respect to 0 and ∞, respectively.

DEFINITION 10: A quasi-pseudo-metric space (X, d) is upper weightable iff there exists a weighting function w for (X, d) such that the space (X, d) is functionally bounded with respect to w. In that case the space (X, d, w) is called upper weighted. The space (X, d) is upper weightable with respect to a point $x_0 \in X$ iff (X, d) is functionally bounded with respect to the point x_0, such that the function f_{x_0} is a weighting function for (X, d). In that case the space (X, d, w) is called upper weighted with respect to the point x_0.

EXAMPLES: The spaces $(\mathcal{R}^+, d_1, w_1)$ and $(\mathcal{R}^+, d_2, w_2)$ are upper weighted, while the space $(\mathcal{R}_0^+, d_1, w_1)$ is upper weighted with respect to 0 and the space $(\overline{\mathcal{R}^+}, d_2, w_2)$ is upper weighted with respect to ∞.

The notion of an upper weightable space has originally been motivated by the study of the complexity spaces. These spaces have been introduced in [10] as part of the development of a topological foundation for the complexity analysis of programs. The development of this foundation has been continued in [11] based on a function space construction for upper weightable spaces which we briefly discuss below.

DEFINITION 11: If (X, d, w) is an upper weighted space then we define the upper weighted function space generated from (X, d, w) to be the space $(\tilde{X}, d^{\omega}, w^{\omega})$, where

$$\tilde{X} = \{f\colon \mathcal{N} \to X \mid w \circ f \text{ is bounded from above}\},$$

$$d^{\omega}(f, g) = \textstyle\sum_{n \geq 0} d(f(n), g(n))\frac{1}{2^n}$$

and

$$w^{\omega}f = \textstyle\sum_{n \geq 0} w(f(n))\frac{1}{2^n}.$$

To verify that the definition is sound, we note that it is easy to show that the upper weightedness of the space (X, d, w) implies that the sum used in the definition of d^{ω} converges and also that $(\tilde{X}, d^{\omega}, w^{\omega})$ is upper weighted.

EXAMPLE 1: The upper weighted function space

$$(\tilde{\overline{\mathcal{R}^+}}, d_2{}^{\omega}, w_2{}^{\omega})$$

generated from the upper weighted space $(\overline{\mathcal{R}^+}, d_2, w_2)$ coincides with the complexity space (C, d, w) originally introduced in [10]. We recall that the set C consists of all functions from $\overline{\mathcal{R}^+}^{\omega}$ which are bounded from below, which coincides with the set

$$\tilde{\overline{\mathcal{R}^+}}.$$

The quasi-pseudo-metric $d_2{}^{\omega}$ on C, referred to as "the complexity distance" in [11], is defined by:

$$\forall f, g \in C.\ d_2{}^{\omega}(f, g) = \textstyle\sum_{n \geq 0}\{(\frac{1}{g(n)} - \frac{1}{f(n)})\frac{1}{2^n} \mid f(n) > g(n)\}.$$

It is easy to verify that the space $(C, d_2{}^{\omega}, w_2{}^{\omega})$ is upper weighted with respect to the function $\top$, where $\top \in C$ is the function with constant value ∞. This function intuitively corresponds to the complexity function of a program which is undefined on all inputs (cf. [10]).

EXAMPLE 2: The upper weighted function space generated from the upper weighted quasi-pseudo-metric space $(\mathcal{R}_0^+, d_1, w_1)$ consists of the functions which are bounded from above.

A second motivation for the study of the upper weightable spaces lies in the fact that these spaces are tightly related to the directed weightable quasi-pseudo-metric spaces. In fact, the following characterization is obtained in [11]: "*A weightable space is upper weightable iff it has a directed weightable extension.*"

In connection to this result, we remark that Lemma 8 implies that any weightable space with a maximum is upper weightable. Indeed, if (X, d) is a weightable space with a maximum, say x_0, then from Lemma 8 we obtain that the function f_{x_0} is a weighting function for (X, d). So $\forall x, y \in X.\ d(x, y) \leq d(x, x_0) + d(x_0, y) = f_{x_0}(y)$.

We show the more general result that any directed weightable space is upper weightable.

Assume that the quasi-pseudo-metric space (X, d) is directed and weightable via a function w. Then we have that $\forall x, y \in X\ \exists z \geq_d x, y$ and thus $w(y) - w(z) = d(z, y) \geq d(x, y)$, by weightedness and by Lemma 5, which implies that $d(x, y) \leq w(y)$.

This implies in particular that any weightable space with a directed weightable extension is upper weightable. The converse is shown in [11].

The upper weightable spaces can be interpreted to form a class of spaces which in a sense is "orthogonal" to the class of the metric spaces inside the class of the weightable spaces.

We remark that any weighted quasi-pseudo-metric space (X, d, w) satisfies the following property: $\forall x, y \in X.\ d(x, y) \leq w(y) \Leftrightarrow d(y, x) \leq w(x)$, or equivalently: $\forall x, y \in X.\ d(x, y) \geq w(y) \Leftrightarrow d(y, x) \geq w(x)$. These properties are immediate consequences of the weighting equality. The spaces which arise as "extreme cases" with respect to these equivalent properties are the upper weightable spaces, satisfying the inequality $\forall x, y \in X.\ d(x, y) \leq w(y)$ and the lower weightable spaces, satisfying the inequality $\forall x, y \in X.\ d(x, y) \geq w(y)$.

The class of lower weightable spaces is easily characterized, since the class coincides with the class of the metric spaces. Indeed, if (X, d) is lower weightable then $\forall x \in X.\ w(x) \leq d(x, x) = 0$ and thus $\forall x \in X.\ w(x) = 0$, which implies that (X, d) is a metric space. The converse holds since any metric space is lower weightable by the function with constant value 0.

So we will restrict our attention to the upper weightable spaces in what follows.

Of course, not every weightable space falls under one of these extreme cases. The Baire quasi-pseudo-metric space $(\mathcal{N}^{\leq\omega}, b, w_b)$ provides an example of a space which is neither lower nor upper weightable. Note that the space would be upper weightable iff $\forall x, y \in \mathcal{N}^{\leq\omega}.\ b(x, y) \leq w_b(x) = p(x, x)$, that is $\forall x, y \in \mathcal{N}^{\leq\omega}.\ p(x, y) \leq 2p(x, x)$. This condition is violated for the case where x is an infinite sequence and y is a finite initial segment of x. Since the space is not a metric space, the space is not lower weightable (we leave the verifications to the reader).

We focus in what follows on a characterization of a subclass of the upper weightable spaces: the weightable optimal join semilattices.

We recall that one of our motivations to consider upper weightable spaces is to continue the development of the topological foundation for Complexity Analysis discussed in [11]. Since, as indicated above, upper weightable function spaces play a central role in this study, we aim at providing a characterization of a class of upper weightable spaces which is sufficiently large to include such function spaces. This leads to the study of the weightable optimal join semilattices, which include the examples discussed above on upper weightable spaces and the function spaces they generate.

The following section discusses the class of functionally bounded directed quasi-pseudo-metric spaces which will be useful in the characterization of the weightable optimal join semilattices.

5. FUNCTIONALLY BOUNDED DIRECTED SPACES

We show that any functionally bounded directed space has an extension which is functionally bounded with respect to a point.

LEMMA 12: A quasi-pseudo-metric space has a maximum x_0 iff the space is functionally bounded with respect to the point x_0.

Proof: If a quasi-pseudo-metric space (X, d) has a maximum x_0 then $\forall x, y \in X.\ d(x, y) \leq d(x, x_0) + d(x_0, y) = d(x_0, y)$ and thus the space is functionally bounded with respect to the point x_0. To show the converse, note that when a quasi-pseudo-metric space (X, d) is functionally bounded with respect to a point, say x_0, then $\forall x \in X.\ d(x, x_0) \leq d(x_0, x_0) = 0$ and thus $\forall x \in X.\ x \leq_d x_0$ or equivalently, x_0 is the maximum of the space (X, d). ❑

THEOREM 13: Every functionally bounded directed quasi-pseudo-metric space has an extension which is functionally bounded with respect to a point.

Proof: Assume that (X, d) is a directed quasi-pseudo-metric space, which is functionally bounded by the function f.

If the quasi-pseudo-metric space (X, d) has a maximum x_0 then the result holds by Lemma 12.

In case the space (X, d) does not posses a maximum, let $X_0 = X \cup \{x_0\}$ for some point $x_0 \notin X$ and define the function d_0 as follows:

$$\forall x, y \in X.\ d_0(x, y) = d(x, y),$$

$$\forall x \in X_0.\ d_0(x, x_0) = 0$$

and

$$\forall x \in X.\ d_0(x_0, x) = \sup\{\lim_{n\to\infty} d(x_n, x) \mid (x_n)_n \in x\uparrow\}.$$

We verify that the function d_0 is well-defined.

Note that for any $x \in X$ and for any sequence $(x_n)_n \in x\uparrow$, the sequence $(d_0(x_n, x))_n$ is increasing by Lemma 5 and thus the sequence $(d_0(x_n, x))_n$ converges since it is bounded by $f(x)$. The fact that for each sequence $(x_n)_n$ in $x\uparrow$, the limit of the sequence $(d_0(x_n, x))_n$ is bounded by $f(x)$ guarantees that the supremum of these limits is finite.

We show that for any $x \in X$ there exists a sequence $(y_n)_n \in x\uparrow$ such that $d_0(x_0, x) = \lim_{n\to\infty} d(y_n, x)$. Indeed, since $d_0(x_0, x) = \sup\{\lim_{n\to\infty} d(x_n, x) \mid (x_n)_n \in x\uparrow\}$, there exists a sequence of sequences in $x\uparrow$, say $[(x_n^k)_n]_k$, such that $d(x_0, x) = \sup_k\{\lim_{n\to\infty} d_0(x_n^k, x)\}$. Since the quasi-pseudo-metric space (X, d) is directed, we can define a sequence $(y_n)_n$ where $y_1 = x^1{}_1$ and $\forall n \geq 1$. y_{n+1} is an element such that $y_{n+1} \geq_d y_n, x^1{}_n, \ldots, x^n{}_n$. Note that $(y_n)_n \in x\uparrow$ and that $\forall$k. $\lim_{n\to\infty} d(x_n^k, x) \leq \lim_{n\to\infty} d(y_n, x)$. So $d_0(x_0, x) = \lim_{n\to\infty} d(y_n, x)$.

Next we verify that d_0 is a quasi-pseudo-metric.

The fact that $\forall x \in X_0.\ d_0(x, x) = 0$ follows since d_0 coincides with d on X and since $d_0(x_0, x_0) = 0$.

In order to verify the triangle inequality, $\forall x, y, z \in X_0.\ d_0(x, y) + d_0(y, z) \geq d_0(x, z)$, we distinguish cases.

Case 1: $x, y, z \in X$.
The result follows by the fact that d is a quasi-pseudo-metric.

Case 2: $z = x_0$.
The inequality reduces to the inequality $\forall x, y \in X.\ d_0(x, y) \geq 0$, which obviously holds. So we can assume that $z \neq x_0$ in what follows.

Case 3: $x = x_0$.
If $y = x_0$ then the triangle inequality reduces to the trivial inequality $\forall z \in X.\ d_0\,(x_0, z) \geq d_0(x_0, z)$.

We verify the case where $y \neq x_0$. Let $(y_n)_n$ be a sequence in $y\uparrow$ such that $d_0(x_0, y) = \lim_{n\to\infty} d(y_n, y)$ and similarly let $(z_n)_n$ be a sequence in $z\uparrow$ such that $d_0(x_0, z) = \lim_{n\to\infty} d(z_n, z)$. Since the quasi-pseudo-metric space (X, d) is directed, we can construct a sequence $(u_n)_n$ of $y\uparrow \cap z\uparrow$ where $u_1 \geq_{d_0} y_1, z_1$ and where $\forall n \geq 1.\ u_{n+1} \geq_d u_n, y_{n+1}, z_{n+1}$. Then we have that $d_0(x_0, y) = \lim_{n\to\infty} d(u_n, y)$ and $d_0(x_0, z) = \lim_{n\to\infty} d(u_n, z)$. Since $\forall n \geq 1.\ d(u_n, y) + d(y, z) \geq d(u_n, z)$, the result follows by taking the limit of both sides.

Case 4: $x \neq x_0$ and $y = x_0$.
We need to verify that $d_0(x_0, z) \geq d_0(x, z)$; that is whether $d_0(x_0, z) \geq d(x, z)$. Let $(x_n)_n$ be a sequence in $x\uparrow$ such that $d_0(x_0, z) = \lim_{n\to\infty} d(x_n, z)$. Then since $\forall n \geq 1.\ x_n \geq_d x$, we have that $d_0(x_0, z) = \lim_{n\to\infty} d(x_n, z) \geq d(x, z)$.

Finally we verify that the quasi-pseudo-metric d_0 is functionally bounded with respect to x_0. If x and y are two points of X, then, for the case where x

$\neq x_0$, let $(x_n)_n$ be a sequence in $x\uparrow$ such that $d_0(x_0, y) = \lim_{n\to\infty} d(x_n, y)$. Then, by directedness, we can construct a sequence $(y_n)_n$ such that $\forall n \geq 1.\ y_n \geq_d x, x_n$. So we have, as above, that $\lim_{n\to\infty} d(y_n, y) = d(x_0, y)$. Then we have that $\forall n \geq 1.\ d(x, y) \leq d(y_n, y)$ and thus $d(x, y) \leq \lim_{n\to\infty} d(y_n, y) = d_0(x_0, y)$. The verification for the case where $x = x_0$ is trivial. ❑

6. OPTIMALITY

We consider in what follows quasi-pseudo-metric spaces for which the associated order is a join semilattice.

By Lemma 5 we obtain that for any quasi-pseudo-metric space (X,d) the following holds: $\forall x,y,z \in X.\ z \geq_d x \Rightarrow d(z, y) \geq d(x, y)$. In particular, any quasi-pseudo-metric space (X, d) for which the associated order is a join semilattice, satisfies the inequality $\forall x,y \in X.\ d(x \sqcup y, y) \geq d(x, y)$. Similarly we obtain that any quasi-pseudo-metric space (X, d) for which the associated order is a lattice satisfies the inequality: $\forall x, y \in X.\ d(x, x \sqcap y) \geq d(x, y)$, in addition to the previous one.

We will consider the "optimal" cases where $d(x \sqcup y, y) = d(x, y)$ and $d(x, x \sqcap y) = d(x, y)$.

DEFINITION 14: A quasi-pseudo-metric (join semi)lattice is a quasi-pseudo-metric space for which the associated order is a (join semi)lattice. A quasi-pseudo-metric join semilattice (X, d) is optimal iff $\forall x,y \in X.\ d(x \sqcup y, y) = d(x, y)$. The notion of an optimal quasi-pseudo-metric lattice is defined in a similar way. For the sake of brevity, we refer to optimal quasi-pseudo-metric (join semi)lattices in what follows as optimal (join semi)lattices, in which case the quasi-pseudo-metric space is referred to as the "underlying quasi-pseudo-metric space."

EXAMPLES: Any quasi-pseudo-metric space for which the associated partial order is linear is an optimal lattice.

Any upper weighted function space generated from a given optimal upper weighted (join semi)lattice is optimal, where the operations meet and join are defined by pointwise extension of the operations on the original space. We leave the straightforward verifications to the reader. Some specific examples are the complexity space (C, d) and the upper weighted function space generated from the upper weighted space $(\mathcal{R}_0^+, d_1, w_1)$.

Of course, not every quasi-pseudo-metric join semilattice (X, d) is optimal. A counterexample is given by the space (X, d), where $X = \{(0, 1), (1, 3), (4, 1)\}$ and where $\forall (x_1, y_1), (x_2, y_2) \in X.\ d((x_1, y_1), (x_2, y_2)) = d_1(x_1, x_2) + d_1(y_1, y_2)$.

This space is upper weightable via the function $w((x_1, y_1)) = x_1 + y_1$. So the condition of upper weightability does not imply optimality.

THEOREM 15: An optimal join semilattice is weightable iff the underlying quasi-pseudo-metric space is functionally bounded and satisfies the descending path condition.

Proof: Let (X, d) be an optimal join semilattice.

If (X, d) is weightable then, since the space is directed, it is upper weightable and thus in particular functionally bounded. By Lemma 4, (X, d) satisfies the descending path condition.

To show the converse, assume that (X, d) is functionally bounded and satisfies the descending path condition.

By Theorem 13, there exists an extension (X_0, d_0) of (X, d) which is functionally bounded with respect to a point x_0, where $X_0 = X \cup \{x_0\}$ in case the space (X, d) does not have a maximum and $X_0 = X$ otherwise (we use the notation introduced in the proof of Theorem 13).

Since weightability is a hereditary property, in order to show that the space (X, d) is weightable it suffices to show that the space (X_0, d_0) is weightable. We will verify that the space (X_0, d_0) is weightable by the weighting function w_0, defined by $\forall x \in X_0.\ w_0(x) = d_0(x_0, x)$.

We show that $\forall x, y \in X_0.\ x \geq_{d_0} y \Rightarrow d_0(x_0, x) + d_0(x, y) = d_0(x_0, y)$. Equivalently we need to show that $\forall x, y \in X_0.\ x \geq_{d_0} y \Rightarrow d_0(x, y) = w_0(y) - w_0(x)$. We distinguish cases.

The verifications of the cases where $x = x_0$ or $y = x_0$ are straightforward.

If $x, y \in X$ such that $x \geq_{d_0} y$, then, since d_0 and d coincide on X, we have $x \geq_d y$ and thus for some sequence $(x_n)_n \in x\uparrow$ we have that $d_0(x_0, x) + d_0(x, y) = \lim_{n\to\infty} d(x_n, x) + d_0(x, y) = \lim_{n\to\infty} (d(x_n, x) + d(x, y)) = \lim_{n\to\infty} d(x_n, y)$, where the last equality follows by the fact that (X,d) satisfies the descending path condition. So to obtain the result it suffices to show that $\lim_{n\to\infty} d(x_n, y) = d_0(x_0, y)$. Since $(x_n)_n \in x\uparrow$ and $x \geq_d y$, we have that $(x_n)_n \in y\uparrow$ and thus the inequality $\lim_{n\to\infty} d(x_n, y) \leq d_0(x_0, y)$ holds.

We assume by way of contradiction that $\lim_{n\to\infty} d(x_n, y) < d_0(x_0, y)$. Then there exists a sequence $(y_n)_n \in y\uparrow$ such that $\lim_{n\to\infty} d(y_n, y) > \lim_{n\to\infty} d(x_n, y)$. However, for the sequence $(y_n \sqcup x)_n \in x\uparrow$, we obtain that $\lim_{n\to\infty} d(y_n \sqcup x, y) > \lim_{n\to\infty} d(x_n, y)$. Since $d(y_n \sqcup x, y) = d(y_n \sqcup x, x) + d(x, y)$ and $d(x_n,y) = d(x_n,x) + d(x, y)$ by the descending path condition, we obtain that $\lim_{n\to\infty} d(y_n \sqcup x, x) > \lim_{n\to\infty} d(x_n, x) = d_0(x_0, x)$, which yields a contradiction.

It is straightforward to verify that (X_0, d_0) is still an optimal join semilattice. So we have that $\forall x, y \in X_0.\ d_0(x, y) = d_0(x \sqcup y, x)$ and $d_0(y, x) = d_0(y \sqcup x, x)$. Since we have shown that $\forall x, y \in X_0.\ x \geq_{d_0} y \Rightarrow d_0(x, y) = w_0(y) - w_0(x)$, we obtain that $\forall x, y \in X_0.\ d_0(x, y) - d_0(y, x) = d_0(x \sqcup y, y) - d_0(x \sqcup y, x) = (w_0(y) - w_0(x \sqcup y)) - (w_0(x) - w_0(x \sqcup y)) = w_0(y) - w_0(x)$. So (X_0, d_0) is weighted with respect to w_0 and hence (X, d) is weighted with respect to the function $w_0|X$. ❑

The following result provides some information on weighted optimal lattices.

PROPOSITION 16: A weighted optimal lattice $(X, d, w, \sqcup, \sqcap)$ satisfies the following laws:

$$\forall x, y \in X.\ w(x \sqcap y) + w(x \sqcup y) = w(x) + w(y) \text{ and}$$

$$\forall x, y \in X.\ w(x \sqcap y) - w(x \sqcup y) = d(x, y) + d(y, x).$$

Proof: By weightedness we have that $\forall x, y \in X.\ d(x, y) - d(y, x) = w(y) - w(x)$, which by the optimality of the quasi-pseudo-metric lattice is equivalent to $\forall x,y \in X.\ d(x \sqcup y, y) - d(y, x \sqcap y) = w(y) - w(x)$. Since $\forall x,y \in X.\ x \sqcup y \geq_d y$ and $y \geq_d x \sqcap y$ we obtain, again by weightedness, that $d(x \sqcup y, y) = w(y) - w(x \sqcup y)$ and $d(y, x \sqcap y) = w(x \sqcap y) - w(y)$. Hence $\forall x, y \in X.\ (w(y) - w(x \sqcup y)) - (w(x \sqcap y) - w(y)) = w(y) - w(x)$ and thus $\forall x, y \in X.\ w(x) + w(y) = w(x \sqcup y) + w(x \sqcap y)$. To verify the second equality, note that $\forall x, y \in X.\ d(x, y) + d(y, x) = d(x \sqcup y, y) + d(y, x \sqcap y) = w(x \sqcap y) - w(x \sqcup y)$. ❑

We conclude the paper with an alternative characterization of weightable optimal join semilattices.

Note that for optimal weightable join semilattices (X, d), say with a weighting function w, we have that $\forall x, y \in X.\ w(x \sqcup y) = w(y) - d(x, y)$. Since these spaces are directed they are in particular upper weightable.

Conversely, an upper weightable space allows one to define the following equation with unknown z in X: $\forall x,y \in X.\ w(z) = w(y) - d(x, y)$, based on the fact that the expression $w(y) - d(x, y)$ is guaranteed to be positive.

The solution intuitively corresponds to an "optimal supremum" of the points x and y. This intuition lies at the basis of the following definition and theorem.

DEFINITION 17: A weighting function is upper solvable iff $\forall x,y \in X\ \exists z \in X.\ z \geq_d x,y$ and $w(z) = w(y) - d(x, y)$. A weightable space is upper solvable iff it has an upper solvable weighting function.

We remark that any upper solvable weightable space is upper weightable.

LEMMA 18: If (X, d, w) is a weighted quasi-pseudo-metric space then

$$(*)\ \forall x, y, z \in X.\ z \geq_d x, y \Rightarrow w(z) \leq_d w(y) - d(x, y).$$

Proof: Note that for $x, y, z \in X$ such that $z \geq_d x, y$ we have that $d(x, y) \leq d(z, y)$ (by Lemma 5). By the fact that $z \geq_d y$ and by the weighting equality, we have that $d(z, y) = w(y) - w(z)$ and thus $d(x, y) \leq w(y) - w(z)$, which is equivalent to the desired inequality. ❑

THEOREM 19: A weightable space (X, d, w) is an optimal join semilattice iff the space is upper solvable and satisfies the following condition:

$$\forall x,y,z_1,z_2 \in X.\ (z_1 \geq_d x,y \text{ and } z_2 \geq_d x,y) \Rightarrow (\exists z \in X.\ z_1,\ z_2 \geq_d z \geq_d x,y).$$

Proof: Assume that (X, d) is a weightable quasi-pseudo-metric space which has an upper solvable weighting function w and which satisfies $\forall x,y,z_1,z_2 \in X.\ (z_1 \geq_d x,y$ and $z_2 \geq_d x,y) \Rightarrow (\exists z \in X.\ z_1, z_2 \geq_d z \geq_d x,y)$.

Next we show that any equation of the form $w(z) = w(y) - d(x, y)$ where $x,y \in X$ and $z \geq_d x,y$ has a unique solution. Let $x,y \in X$ and assume that z_1 and z_2 are points in X such that $z_1 \geq_d x,y$ and $z_2 \geq_d x,y$, where $w(z_1) = w(z_2) = w(y) - d(x, y)$.

Note that by hypothesis there exists an element z such that $z_1, z_2 \geq_d z \geq_d x,y$. By (*) (Lemma 18) and by the fact that $z \geq_d x, y$, we obtain that $w(z) \leq w(y) - d(x, y)$.

Since $z_1, z_2 \geq_d z$ we also have by Lemma 2 that $w(y) - d(x, y) = w(z_1) = w(z_2) \leq w(z)$. Thus $w(z) = w(y) - d(x, y)$ and thus $w(z_1) = w(z_2) = w(z)$.

Since z_1, $z_2 \geq_d z$, by the weighting equality we have that $d(z_1, z) = w(z) - w(z_1)$ and $d(z_2, z) = w(z) - w(z_2)$. Thus $d(z_1, z) = d(z_2, z) = 0$, that is $z_1 \leq_d z$ and $z_2 \leq_d z$, and thus $z_1 = z_2 = z$.

Given two points $x,y \in X$, let z_0 be the unique solution to the equation $w(z) = w(y) - d(x, y)$, where $z_0 \geq_d x,y$. We show that z_0 is the supremum of x and y with respect to the associated order. Assume that $z_1 \in X$ is such that $z_1 \geq_d x,y$. Then we have that $\exists z \in X.\ z_0,\ z_1 \geq_d z \geq_d x, y$. Since $z \leq_d z_0$ we have that $w(z_0) \leq w(z)$. However since $z \geq x,y$, by (*) we obtain that $w(z) \leq w(y) - d(x, y)$ and thus $w(z) = w(y) - d(x, y)$. Since $z \geq_d x,y$, we obtain by uniqueness of the solution that $z = z_0$ and thus, since $z \leq_d z_1$, we have that $z_0 \leq_d z_1$.

We leave the straightforward verifications of the converse to the reader. ❑

Finally we remark that Problem 10 of [4] "Develop a notion of a weighted quasi-uniform space" cannot be solved via an axiomatization in terms of the elements of the quasi-uniform space which would guarantee the weightability of all quasi-pseudo-metrics which induce the quasi-uniformity (for the case of quasi-uniformities with a countable base). This is easily shown by the following counterexample.[2]

Consider the weightable quasi-pseudo-metric space $(\mathcal{R}_0^+, d_1)$. The quasi-pseudo-metric d' defined by $d'(x, y) = \min(d_1(x, y),1)$ induces the same quasi-uniformity as d_1 but does not satisfy the descending path condition and hence is not weightable.

Since the counterexample provides a space which belongs to the weightable optimal join semilattices, the same remark holds for this class as well as for the class of the upper weightable spaces.

[2]After an example suggested by W. Lindgren.

ACKNOWLEDGMENT

The author is grateful to the referee for helpful comments on the paper.

REFERENCES

1. FLAGG, R.C. & R.D. KOPPERMAN. 1993. The asymmetric topology of Computer Science. Mathematical Foundations of Programming Semantics.
2. FLETCHER, P. & W. LINDGREN. 1982. Quasi-uniform spaces, Marcel Dekker, Inc. New York.
3. KAHN, G. 1974. The semantics of a simple language for parallel processing. Proc. IFIP Conf.
4. KÜNZI, H.P. & V. VAJNER. 1993. Weighted quasi-metrics. Proc. Summer Conf. Queens College. Papers on General Topology and Applications. **728:** 64–77. Annals of the New York Academy of Sciences. New York.
5. KÜNZI, H.P. 1993. Nonsymmetric topology. Proceedings Szekszard Conference.
6. MATTHEWS, S.G. 1994. Partial metric spaces. Papers on General Topology and Applications. **728:** 183–197. Annals of the New York Academy of Sciences. New York.
7. MATTHEWS, S.G. 1992. The topology of partial metric spaces. Papers on General Topology and Applications. Annals of the New York Academy of Sciences. New York.
8. MURDESHWAR, M.G. & S.A. NAIMPALLY. 1966. Quasi-uniform topological spaces. Noordhoff.
9. NACHBIN, L. 1965. Topology and order. New York Mathematical Studies, Vol. 4. Princeton, NJ.
10. SCHELLEKENS, M. 1995. The Smyth completion: a common foundation for denotational semantics and complexity analysis. Electronic Notes in Theoretical Computer Science, Vol. I. Proc. 11th Conf. on the Mathematical Foundations of Programming Semantics. Elsevier. New York. 211–232.
11. SCHELLEKENS, M. Weightable directed spaces. Manuscript.
12. SMYTH, M. 1987. Quasi-uniformities: Reconciling domains with metric spaces. Lecture Notes in Computer Science **298**. Springer Verlag. New York.
13. SMYTH, M. 1991. Totally bounded spaces and compact ordered spaces as domains of computation. *In:* Topology and Category Theory in Computer Science. 207-229. Oxford University Press. Oxford.
14. WADGE, W.W. 1981. An extensional treatment of dataflow deadlock. Theoretical Computer Science. **13**.

Direct Topologies from Discrete Rings

NIEL SHELL

Department of Mathematics
The City College of New York
Convent Avenue at 138th Street
New York, New York 10031

ABSTRACT: An exposition of a portion of Zobel's results on direct topologies is given. A construction by Zobel of direct topologies on the rational field is generalized to produce direct topologies on certain other discretely valued fields.

In [8] we constructed a topology on the rational field, making use of the integers. In [11] we showed that a discretely topologized order in an absolute valued field could play the role of the integers in the construction in [8]. In [13] we will describe similar generalizations of two constructions of Mutylin (see [5] and [6]) and some results in [4] can also similarly be generalized. The constructions in [3], [4], [5] and [6] are special cases of direct topologies introduced by Zobel in [15], and we consider here our generalization from the point of view of direct topologies. We give an exposition of a portion of the unpublished work in [15], somewhat different from the original, as background for our generalization. See [14, pp.101–106] for an outline of the work in [15] more closely matching the original exposition.

We use $\mathbf{Z}$, $\mathbf{Z}_{>0}$, $\mathbf{Q}$, and $\boldsymbol{R}$ to denote the sets of integers, positive integers, rational numbers and real numbers, respectively.

In what follows G will denote an Abelian group (written additively), and A^* will denote the nonzero elements of a subset A of G, and $n \times A$ will denote the sum of n copies $A + \cdots + A$ of a subset A of G. Similarly, for a subset A of a ring S, $A^{\times n}$ denotes the set of all products of n elements from A, and $S^{\times}$ will denote the set of invertible elements of S. If a group topology (i.e., a topology with respect to which addition and negation are continuous) is being considered on G, the closure in the completion of G of a subset A of G will be denoted by $\hat{A}$. With this notation $\hat{G}$ is the completion of G. The neighborhood filter at 0 in a topology $\mathcal{T}$ on G will be denoted by $\mathcal{B}(\mathcal{T})$.

Recall that a topology is *ultraregular* if it has a base of clopen sets. Also recall (see, e.g., [9, Definition 3.4.1]) that a family of topologies on a set X is *independent* if $\cap U_i = \varnothing$ for $U_i \in \mathcal{T}_i$, where the $\mathcal{T}_i$ are distinct members of a finite subfamily of the family, implies one of the sets U_i must be empty.

Mathematics Subject Classification: Primary 54H13; Secondary 54H11.
Key words and phrases: direct system, direct topology, discrete ring.

A nonzero element x of a ring is called *algebraically nilpotent* if $x^n = 0$ for some positive integer n. A nonzero element x of a topological ring is called *topologically nilpotent* if the sequence $\{x^n\}$ converges to 0. By a valuation we mean either a non-Archimedean valuation or a power of the usual absolute value on a subfield of the complex numbers. By an absolute value on a field we mean either a real valued (i.e., rank one) non-Archimedean valuation or a power of the usual absolute value on a subfield of the complex numbers. We denote by $\mathcal{T}_{|\,|}$ the topology induced by an absolute value $|\ |$.

A ring topology is a group topology on a ring with respect to which multiplication is continuous. A subset of a topological ring is said to be *bounded away from 0* if it has empty intersection with at least one neighborhood of 0. We say that inversion is *maximally discontinuous* on a topological ring S with identity if each net $\{x_i\}$ in $S^\times$ such that $\{x_i\}$ converges to 1 and $\{x_i^{-1}\}$ converges to 1 is eventually equal to 1. Let $\mathcal{T}$ be a topology induced by a nontrivial valuation on a field, and let $\mathcal{S}$ be a ring topology on the field such that, for each set E that is bounded with respect to the valuation inducing $\mathcal{T}$, there is an $\mathcal{S}$-neighborhood U of 0 such that $E \cap U^* = \varnothing$. Then inversion is maximally discontinuous with respect to $\mathcal{S}$ (see [10]).

A collection of filters on a set will be considered partially ordered by containment. (In [15] the opposite order, reverse containment, is used.) As usual, we say a filter $\mathcal{A}$ is finer (strictly finer) than a filter $\mathcal{F}$ if $\mathcal{F}$ is a subset (proper subset) of $\mathcal{A}$. A collection of topologies on a set will also be considered partially ordered by containment.

We use $a \vee b$ ($a \wedge b$) to denote the least upper bound (respectively, the greatest lower bound) of a and b, where a and b come from a lattice (such as the reals with the usual order or a lattice of group topologies); $\vee A$ ($\wedge A$) denotes the least upper bound (greatest lower bound) of an arbitrary subset A in a lattice.

1. DIRECT TOPOLOGIES ON GROUPS

DEFINITION 1: A sequence of subsets $\{M_{in}\}_{i,n\in\mathbf{Z}_{>0}}$ of an Abelian group G will be called a *direct system* in G if:

(D1) $0 \in M_{in}$ for all i and n and $M_{in} = \{0\}$ if $i < n$;

(D2) $-M_{in} = M_{in}$ for all i and n;

(D3) $M_{i\,n+1} + M_{i\,n+1} \subset M_{in}$ for all i and n; and

(D4) $[\sum_{i=1}^{k-1}(M_{i1} + M_{i1} + M_{i1})] \cap (M_{k1} + M_{k1} + M_{k1}) = \{0\}$ for all $k > 1$.

For each positive integer n, $U_n = \bigcup_{k=1}^{\infty} \sum_{i=1}^{k} M_{in}$. The phrase "$\sum x_i$ is a *canonical representation* of $x \in U_n$" will mean $x_i \in M_{in}$ for all i, $x_i = 0$ for all but finitely many i and x is the sum of the nonzero elements x_i. (Define the empty sum to be 0.) The sequence $\{U_n\}$ will be called the neighborhood base associated with the direct system. The symbols M with a double subscript and U with a single subscript will be used exclusively for sets in a direct system and their associated neighborhoods, respectively.

From (D1) and (D3) it follows that the sets M_{in} decrease as (i is fixed and) n increases. Thus, $M_{in} \subset M_{i1}$ for all i and n, from which it follows that, for all n, $U_n \subset U_1$ and a canonical representation of an element in U_n is a canonical representation when viewed as an element in U_1. Condition (D4) readily implies that canonical representations of elements in U_1 are unique: If $\sum x_i$ and $\sum y_i$ were distinct canonical representations of an element of U_1, with m being the largest index for which the finitely nonzero sequences $\{x_i\}$ and $\{y_i\}$ are different, then the equality $x_m - y_m = \sum_{i=1}^{m-1} y_i - x_i$ would violate (D4). The function $\varphi_i\colon U_1 \longrightarrow M_{i1}$ (which we call the i-th coordinate function) which assigns to x the element x_i of its canonical representation is well-defined. By the *height* of $x \in U_1{}^*$, which we write as hgt(x), we mean the largest integer i such that $\varphi_i(x) \neq 0$. We define hgt(0) = $-\infty$. If $x, y, x + y \in U_1$, condition (D4) implies $\varphi_i(x + y) = \varphi_i(x) + \varphi_i(y)$ for all i: If the equality did not hold for all i, and if m were the largest index of a nonzero term in the finitely nonzero sequence $\varphi_i(x + y) - \varphi_i(x) - \varphi_i(y)$, then rearranging the equality $\sum_{i \leq m} \varphi_i(x + y) = \sum_{i \leq m} \varphi_i(x) + \sum_{i \leq m} \varphi_i(y)$ so that the three index m terms (and nothing else) are one side of the equality would contradict (D4). In particular, $\varphi_i|_{U_2}$ is a "homomorphism."

For $x, y \in U_1$ such that $x - y \in U_n$, one easily sees that $\varphi_i(x) = \varphi_i(y)$ for $i < n$.

We denote by π_n (the truncation to the first n coordinates) the function $\sum_{i=1}^{n} \varphi_i$. $\varphi_k(x)$ is called the *leading* (nonzero) *term* of x if $\varphi_k(x) \neq 0$ and $\varphi_i(x) = 0$ for $i < k$.

The following superficially different definition for a direct system is sometimes more convenient to use than the given one. Call $\{M_{in}\}_{i,j \geq -1}$ a direct system if it satisfies conditions (D1)–(D3) and:

(D4′) $[\sum_{i=-1}^{k-1} M_{i(-1)}] \bigcap M_{k(-1)} = \{0\}$ for all $k > -1$.

If the sets $M_{i(-1)}$, M_{i0}, $M_{(-1)n}$, and M_{0n} are removed from the newly defined direct system, the remaining sets satisfy the original definition of direct system. Conversely, each direct system in the original sense is a direct system in the new sense (provided the sets M_{in} are relabelled so that the indexing begins at (−1)).

THEOREM 1: [15, Theorems 2.1 and 2.3] If $\{M_{in}\}$ is a direct system in G, then the associated neighborhood base is a neighborhood base at 0 for a Hausdorff group topology on G (which will be referred to as the topology associated with — or induced by — the direct system). The sets U_n, $n \geq 2$, are clopen in this topology.

Proof: Conditions (D1)–(D3) immediately imply that, for all n, $0 \in U_n$, $U_n = -U_n$ and $U_{n+1} + U_{n+1} \subset U_n$. Thus, the sets U_n form a base at zero for a group topology. If $x \in U_n$, then $\varphi_i(x) = 0$ for $i < n$. Hence, $\bigcap_{n=1}^{\infty} U_n = \{0\}$ and the topology is Hausdorff.

Note that $\overline{U_n} = \bigcap_{t=1}^{\infty}(U_n + U_{n+t})$. For $n \geq 2$, choose $x \in \overline{U_n}$ and $t \in \mathbf{Z}_{>0}$. The first equality in this paragraph implies there are canonical representations $\sum y_i$ and $\sum z_i$ of elements in U_n and U_{n+t}, respectively, such that

$$\begin{aligned} x &= \sum y_i + \sum z_i \\ &= \sum_{i=1}^{n+t-1} y_i + \sum_{i \geq n+t} y_i + z_i . \end{aligned}$$

Therefore, $\varphi_i(x) = y_i \in M_{in}$ for $i < n + t$. Since t is an arbitrary positive integer, $x \in U_n$. (If $n = 1$, the elements $y_i + z_i$ may not be terms of a canonical representation.)

Finally, we observe that U_n is open for each positive integer n: If $x \in U_n$, then $x + U_k \subset U_n$ for $k > n \vee \operatorname{hgt}(x)$. ❑

A topology induced by some direct system will be called a *direct topology*. A group topology on a group H will be called an *extended direct topology* if H contains a dense directed subgroup; or, equivalently, if there exists a direct topology on a group G such that $G \subset H \subset \hat{G}$, and the topology of H is the subspace topology from $\hat{G}$.

The following two lemmas facilitate the description of the completion of a group with a direct topology.

LEMMA 2.1: (cf. [15, Theorem 2.7]) Let $\{M_{in}\}$ be a direct system on the group G. If $x_i \in M_{i1}$ for each positive integer i and the series $\sum_{i=1}^{\infty} x_i$ (is convergent and) has sum 0 in the topology associated with the direct system, then $x_i = 0$ for each i.

Proof: Given $k > 0$, we have $\sum_{t=1}^{n} x_t \in U_k$ for some $n \geq k$ (in fact for all sufficiently large n). Therefore, $x_i = \varphi_i(\sum_{t=1}^{n} x_t) = 0$ for $i < k$. Since k was arbitrary, we have $x_i = 0$ for all i. ❑

LEMMA 2.2: Let $\{M_{in}\}$ be a direct system on the group G. If $\{y_k\}$ is a Cauchy sequence in U_2 (with respect to the topology on G associated with $\{M_{in}\}$), then there exists a unique sequence $\{x_i\}$ such that $x_i \in M_{i2}$ for all i and $y_k - \sum_{t=1}^{k} x_t$ converges to 0. (Consequently, the sequence $\{\sum_{t=1}^{k} x_t\}$ is Cauchy.)

Proof: (cf. [3, Lemma 9]) Let $\{N(m)\}$ be a strictly increasing sequence such that $y_i - y_j \in U_{m+1}$ whenever $i, j \geq N(m)$. Then $N(m) \geq m$ for all m. Let $x_m = \varphi_m(y_{N(m)})$. For $j \geq N(m)$ and $i \leq m$,

$$y_j - y_{N(i)} \in U_{i+1};$$

$$\varphi_i(y_j) = \varphi_i(y_{N(i)}) = x_i.$$

Since $\{y_i\}$ is in U_2, $x_i \in M_{i\,2}$, for all i, and $\sum_{t=1}^{k} x_t \in U_2$ for all k. Thus, $y_k - \sum_{t=1}^{k} x_t \in U_1$.

We show that $y_k - \sum_{t=1}^{k} x_t$ converges to zero. For a given value of m, we choose $k \geq N(m) \vee \mathrm{hgt}(y_{N(m)})$. Write

$$y_k - \sum_{t=1}^{k} x_t = (y_k - y_{N(m)}) + (y_{N(m)} - \sum_{t=1}^{k} x_t).$$

Since $k \geq N(m)$, $y_k - y_{N(m)} \in U_{m+1}$. Since $N(k) \geq k \geq N(m)$, $y_{N(m)} - y_{N(k)} \in U_{m+1}$. Hence, for $i \leq k$,

$$\begin{aligned} \varphi_i(y_{N(m)} - \textstyle\sum_{t=1}^{k} x_t) &= \varphi_i(y_{N(m)}) - \varphi_i(\textstyle\sum_{t=1}^{k} x_t) \\ &= \varphi_i(y_{N(m)}) - x_i \\ &= \varphi_i(y_{N(m)}) - \varphi_i(y_{N(k)}) \\ &= \varphi_i(y_{N(m)} - y_{N(k)}) \in M_{i\,m+1}; \end{aligned}$$

and, for $i > k$,

$$\begin{aligned} \varphi_i(y_{N(m)} - \textstyle\sum_{t=1}^{k} x_t) &= \varphi_i(y_{N(m)}) - \varphi_i(\textstyle\sum_{t=1}^{k} x_t) \\ &= 0 - 0 \in M_{i\,m+1}. \end{aligned}$$

Suppose $y_k - \sum_{t=1}^{k} w_t$ also converges to 0, where $w_i \in M_{i\,2}$ for all i. For any integer m, there is an integer $k \geq N(m)$ ($\geq m$) such that $y_k - \sum_{t=1}^{k} w_t \in U_{m+1}$. Then $w_m = \varphi_m(\sum_{t=1}^{k} w_t) = \varphi_m(y_k) = x_m$. ❑

Recall that $\{\hat{U}_n\}$ is a neighborhood base at 0 in the completion, $\hat{G}$, of G (see, e.g., [9, Theorem A.3.5]), and $\hat{U}_n = -\hat{U}_n$ and $\hat{U}_{n+1} + \hat{U}_{n+1} \subset \hat{U}_n$. A series $\sum_{i=1}^{\infty} x_i$ is called a canonical representation of $x \in \hat{U}_n$ if $x_i \in M_{i\,n}$ for all i and x is the sum of the series.

THEOREM 2: (cf. [15, Theorems 2.12 and 2.13]) Let $\hat{U}_n$ be the closure of U_n in the completion $\hat{G}$ of the group G with respect to the topology associated with the direct system $\{M_{i\,n}\}$.

(1) For $n \geq 2$,

$$\hat{U}_n = \{\textstyle\sum_{i=1}^{\infty} x_i : x_i \in M_{i\,n}\ \forall i,$$

and given m $\exists I(m)$ such that $x_i \in M_{i\,m} \forall i \geq I(m)\}$.

(2) Canonical representations in $\hat{U}_2$ are unique. Thus, the definition

$$\varphi_i(\textstyle\sum_{t=1}^{\infty} x_t) = x_i \text{ extends } \varphi_i \text{ to } U_1 \cup \hat{U}_2;$$

and, when $x, y, x + y \in \hat{U}_2$, $\varphi_i(x + y) = \varphi_i(x) + \varphi_i(y)$. In particular, $\varphi_i|_{\hat{U}_3}$ is a "homomorphism."

(3) If $\sum_{i=1}^{\infty} x_i$ is a canonical representation of an element $x \in \hat{U}_2$, then $x \in G$ if and only if $\{x_i\}$ is finitely nonzero, and $x \in \hat{U}_n$ if and only if $x_i \in M_{in}$ for all i.

Hence, a nondiscrete direct topology is not complete.

(4) For $n \geq 2$, the sets $\hat{U}_n$ are clopen.

Thus, each group with an extended direct topology has an ultraregular completion.

Proof: (1) Let S denote the set of all sums described above. Certainly $S \subset \hat{U}_n$. Conversely, if $x \in \hat{U}_n$, then there exists a Cauchy sequence $\{y_k\}$ in U_n converging to x. The sum of the sequence $\{x_i\}$ described in Lemma 2.2 is in S. Since x is the sum of the series, the proof is complete.

(2) The uniqueness of the series representation follows from Lemma 2.1. If $x, y, x+y \in \hat{U}_2$ have canonical representations $\sum x_i$, $\sum y_i$, and $\sum z_i$, respectively, then, for any $m > 1$, there exists $k > m$ such that $\sum_{t=1}^{k} z_t - (x_t + y_t) \in \hat{U}_m \cap G = U_m$. Let $\sum_{t=1}^{r} w_i$, where $r \geq k$, be the canonical representation of this element in U_m. Let $u_i = z_i - (x_i + y_i) - w_i$ for $i \leq k$ and $u_i = -w_i$ for $i > k$. Then $u_i \in M_{i1} + M_{i1} + M_{i1}$ for all i. Condition (D4) implies $u_i = 0$ for all i; and $\sum w_i \in U_m$ implies $w_i = 0$ for $i < m$. Hence, $z_i = x_i + y_i$ for $i < m$. Since m is arbitrary, the proof is complete.

(4) By definition, the sets $\hat{U}_n$ are closed. If $n \geq 2, x \in \hat{U}_n$ and $\varphi_i(x) \in M_{i\,n+1}$ for $i \geq k$, then $x + \hat{U}_m \subset \hat{U}_n$ for $m \geq k \vee (n+1)$. ❑

Define $\mathrm{hgt}(x) = \infty$ for $x \in \hat{U}_2 \backslash U_2$.

Note that Theorem 2 implies that the usual topology on the set of rational numbers (even though it is ultraregular) is not an extended direct topology, since the completion is connected.

Theorem 3 below (for whose proof we refer the reader to [15]) can be interpreted as saying that the set of direct topologies is "fairly thickly spread out" among the set of first countable topologies in the lattice of all group topologies on a group. Recall (see, e.g., [15, Theorem 1.2] or [1, p.49]) that the infimum in the lattice of group topologies of two group topologies S and $\mathcal{T}$ on a group G has a neighborhood base at 0 consisting of sets of the form $U + V$, where U and V vary over bases of neighborhoods at 0 in S and $\mathcal{T}$, respectively. This is finer than $S \cap \mathcal{T}$, the infimum of S and $\mathcal{T}$ in the lattice of *all* topologies on G.

THEOREM 3: ([15, Theorems 2.14 and 2.15 and Remark 2.15.1]; cf. Theorem 8 below) Let $\mathcal{T}$ be a nondiscrete first countable Hausdorff group topology on the Abelian group G, and let $\mathcal{A}$ be a filter, strictly finer than $\mathcal{B}(\mathcal{T})$, such that 0 belongs to each set in $\mathcal{A}$. Then there exists a direct topology $\mathcal{T}_{\mathcal{A}}$ such that $\mathcal{B}(\mathcal{T}) \leq \mathcal{B}(\mathcal{T}_{\mathcal{A}})$ and $\mathcal{A} \not\leq \mathcal{B}(\mathcal{T}_{\mathcal{A}})$.

Consequently, each nondiscrete first countable Hausdorff group topology on an Abelian group G is the intersection of a family of direct topologies.

2. DIRECT TOPOLOGIES ON RINGS

DEFINITION 2: A *direct system* on a ring S is a sequence $\{M_{in}\}$ which is a direct system on the additive group of S and which satisfies:
(D5) $\sum_{j \vee k = i} M_{j\,n+1} M_{k\,n+1} \subset M_{in}$ for all i and n; and
(D6) for all $a \in S$ and $n > 0$, there exists $k \geq 0$ such that, for all i, $aM_{i\,n+k} \subset M_{in}$ and $M_{i\,n+k}\, a \subset M_{in}$.

The left side of the containment in (D5) can be written

$$\left(\sum_{j=1}^{i-1} M_{j\,n+1} \right) M_{i\,n+1} + M_{i\,n+1} \left(\sum_{j=1}^{i-1} M_{j\,n+1} \right) + M_{i\,n+1} M_{i\,n+1}.$$

THEOREM 4:
(1) The topology associated with a direct system on a ring is a ring topology.
(2) $U_{n+1}\, U_{n+1} \subset U_n$ and $\hat{U}_{n+1}\, \hat{U}_{n+1} \subset \hat{U}_n$ for all n.
(3) $\varphi_i(xy) = \sum_{j \vee k = i} \varphi_j(x)\varphi_k(y)$ for $x,y \in \hat{U}_2$.
(4) $\pi_n(xy) = \pi_n(x)\, \pi_n(y)$ for $x,y \in \hat{U}_2$.
(5) $\mathrm{hgt}(xy) \leq \mathrm{hgt}(x) \vee \mathrm{hgt}(y)$ for $x,y \in \hat{U}_2$.

Proof: (2) and (3) follow from (D5), and (1) follows from (2) and (D6) (which has as an immediate conequence that, given $a \in S$ and $n \geq 1$, there exists $k \geq 0$ such that $aU_{n+k} \subset U_n$ and $U_{n+k}a \subset U_n$).

(4) is obtained by grouping together all terms $\varphi_j(x)\varphi_k(y)$ on the right side of the equality for which $j \vee k$ has a fixed value. For $x,y \in U_2$, (5) follows from (3). ❑

LEMMA 5.1: If a ring with a nondiscrete ring topology has a topologically nilpotent element, then every neighborhood of 0 contains all positive integral powers of some topologically nilpotent element.

Proof: If x is topologically nilpotent and U is a neighborhood of 0, then x^n is topologically nilpotent and, for sufficiently large n, $(x^n)^k \in U$ for all k. ❑

The following analogue of Lemma 5.1 would be convenient to prove Theorem 5(2) below: If S is a nondiscrete topological ring with divisors of zero and U is a neighborhood of 0, then there exist $u,v \in U^*$ such that $uv = 0$. However, only a weaker form of the statement is correct: Suppose $xy = 0$, $x \neq 0$, $y \neq 0$ and U is a neighborhood of zero. Choose a neighborhood V of 0 such that $Vx \subset U$ and $V \subset U$. For $v \in V^*$, $vxy = 0$. If $vx \neq 0$, then let $u = vx$ (so $uy = 0$). Otherwise, let $u = v$ (so $ux = 0$). Thus, if $xy = 0$, then there exists $u \in U^*$ such that $ux = 0$ or $uy = 0$.

To see that the original statement is not true, note that if a product S of two copies of almost any topological integral domain (say $\mathbf{Q}$ with the usual topology for a concrete example) is given coordinatewise multiplication and addition and the product of the given topology on the first factor and the discrete topology on the second factor, then $\mathbf{Q} \times \{0\}$ is a neighborhood of 0 which does not contain a pair of nonzero elements whose product is 0.

THEOREM 5: [15, Theorems 3.6 and 3.7] Let S be a ring with a direct topology.

(1) If S has no algebraically nilpotent elements, then the completion of S has no topologically nilpotent elements.

(2) If S is an integral domain (or, in particular, a field), then the completion of S is an integral domain and has no topologically nilpotent elements.

Proof: (1) If $\hat{S}$ has topologically nilpotent elements, then (Lemma 5.1) there exists a topologically nilpotent element x such that $x^k \in \hat{U}_3$ for all k. If x_i is the leading term of x, then $x_i^{\,k} = \varphi_i(x^k)$. If k is sufficiently large, then $x^k \in \hat{U}_{i+1}$ and $x_i^{\,k} = 0$; thus x_i is an algebraically nilpotent element in S.

(2) Suppose S has no divisors of zero. [It is easy to see that $uv \neq 0$ for $u,v \in \hat{U}_3{}^*$: If u and v have canonical representations $\sum_{i=m}^{\infty} u_i$ and $\sum_{i=n}^{\infty} v_i$ with $u_m, v_n \neq 0$ and, say, $m \le n$, then the canonical sum $u_m + \cdots + u_n \neq 0$. Therefore, $\varphi_n(uv) = (u_m + \cdots + u_n)v_n \neq 0$. However, the discussion above shows this is not sufficient.] Suppose $x,y \in \hat{S}^*$ and $xy = 0$. Choose $x_0, y_0 \in S$ such that $x - x_0$, $y - y_0 \in \hat{U}_4$; let $\sum_{i=1}^{\infty} x_i$ and $\sum_{i=1}^{\infty} y_i$ be the canonical representations of $x - x_0$ and $y - y_0$, respectively. For $k \ge 0$, let $w_k = \sum_{i \vee j = k} x_i y_j$. Then $\sum_{k=0}^{n} w_k = (\sum_{i=0}^{n} x_i)(\sum_{i=0}^{n} y_i)$, and, since the factors on the right converge to x and y, the series on the left converges to ($xy =$)0. According to Theorem 2(1) and condition (D6), there exists m such that $x_0 y_i, x_i y_0 \in M_{i\,4}$ for all $i \ge m$. For $k \ge m$,

$$w_k = x_0\, y_k + x_k\, y_0 + (x_1 + \cdots + x_{k-1})y_k + x_k(y_1 + \cdots + y_{k-1}) + x_k\, y_k$$
$$\in M_{k4} + M_{k\,4} + M_{k\,3} \subset M_{k\,3} + M_{k\,3} \subset M_{k\,2}.$$

Therefore, $\sum_{k=m}^{\infty} w_k$ is the canonical representation of an element in $\hat{U}_2$. Since $\sum_{k=m}^{\infty} w_k = -\sum_{k=0}^{m-1} w_k \in S$, there exists an integer n' such that $w_k = 0$ for $k \ge n'$, and for $n \ge n'$,

$$0 = \sum_{k=0}^{n} w_k = (\sum_{i=0}^{n} x_i)(\sum_{i=0}^{n} y_i).$$

Therefore, one of the two factors (both in S) on the right side of the display is 0 for infinitely many n. Hence, the limit, x or y, of one of the factors is 0.

The second assertion in (2) follows from (1). ❑

COROLLARY 5.1:

(1) A nondiscrete ring topology on a field with respect to which there is a topological nilpotent (such as a ring topology induced by a norm

or, more specially, an absolute value) is not an extended direct topology.

(2) A supremum of a family of at least two independent ring topologies on a field is not an extended direct topology.

(3) A locally bounded topology on **Q** is not a direct topology.

Proof: (2) The completion of a supremum of independent topologies is the Cartesian product of the completions of the individual topologies, and, hence, has divisors of zero.

(3) Locally bounded topologies on **Q** are either induced by absolute values or have completions having divisors of 0. (See [7] or [14, Section 33].) ❑

In [15, p.38] it is shown that the p-adic topologies are not even direct topologies on the additive group of **Q**.

3. DIRECT TOPOLOGIES ON FIELDS WITH DISCRETE RINGS

Throughout this section, $(K, |\ |)$ will be a field with a nontrivial absolute value, and R will be a unique factorization domain which is a discrete order of K (i.e., a subring of K with identity which has quotient K and which is discrete in the relative topology from K — see [12] for a discussion of discrete orders).

Our purpose here is to show that there is a nondiscrete direct ring topology on K. The results below were obtained in [15] for the special case in which $R = \mathbf{Z}$ and $K = \mathbf{Q}$, with the usual topology, and conditions (Q2) or (Q4) (defined below) are satisfied.

In a discrete ring, nonzero elements have absolute value at least one, and invertible elements have absolute value equal to one. From this follows the frequently used fact that, for $a,b \in R^*$, we have $|a| \le |ab|$, with equality holding if b is invertible.

An element $a \in R^*$ is called a *numerator* (*denominator*) of an element $x \in K^*$ if $x = a/b$ (b/a, respectively), where a and b are relatively prime elements in R^*; and a is a numerator (denominator) of a set E in K if it is a numerator (denominator) of some element of E; $s(E) := \vee \{|a|: a$ is a denominator of $E\}$ (and $s(\{0\}) = s(\varnothing) = 0$).

DEFINITION 3: Let s be a real number greater than or equal to 1. A subset E of R is called s-denominator complete if $a/b \in E$, where a and b are relatively prime elements in R^*, implies (1) $|b| \le s$, and (2) $a/c \in E$ for all $c \in R^*$ such that $|c| \le s$. A set E is called *denominator complete* if it is s-denominator complete for some real number s. If there is a smallest s-denominator complete set containing a subset E of K, it will be denoted by $D_s(E)$ and called the *s-complete hull of E*.

An s-denominator complete set E is symmetric (i.e., $E = -E$). A union or an intersection (of a nonempty family) of s-denominator complete sets is again s-denominator, and $\bigcup_{1 \le |b| \le s} R/b$ is the largest s-denominator complete subset in K. (The discreteness of R guarantees that, when an element in the union above is reduced, the reduced denominator will have absolute value no bigger than the original denominator.) The s-complete hull of a set E exists if and only if $s(E) \le s$, and E is contained in some denominator complete set if and only if $s(E)$ is finite; and $0 \in D_s(E)$ if and only if $0 \in E$, and $D_1(\{0, \pm 1\}) = \{0\} \cup \{a/b : a,b \in R^*, a^{-1} \in R, |b| = 1\}$.

THEOREM 6: [15, Theorem 4.1] If $s(E) \le s$, then $D_s(E)^*$ equals the set of all fractions a/b satisfying the following conditions:

(1) a and b are relatively prime elements in R^*;
(2) $|b| \le s$;
(3) there exists a numerator c of E such that a divides c and each prime factor of c/a has absolute value not exceeding s.

Proof: If a/b satisfies the conditions and c/a is the product $p_1 \cdots p_n$ of (not necessarily distinct) primes with absolute value not exceeding s, then $c/p_1 \in D_s(E)$. But $(c/p_1)/1$ is a reduced representation, so $(c/p_1)/p_2 \in D_s(E)$. Continuing in this fashion we see that $a = c/(p_1 \cdots p_n) \in D_s(E)$. Hence $a/b \in D_s(E)^*$. (It is also easy to verify that $a/b \in D_s(E)^*$ if c/a is invertible.)

Conversely, the set T of all fractions satisfying (1)–(3) is easily seen to be an s-denominator complete set containing E^*. Thus, $D_s(E)^* = D_s(E^*) \subset T$. ❑

COROLLARY 6.1: If $E \subset K$ and $s(E) \le s$, then $\vee\, |D_s(E)| \le (\vee\, |E|)s$.

Proof: With notation as in Theorem 6, $|a/b| \le |a| \le |c| \le (\vee\, |E|)s$. ❑

LEMMA 7.1: Suppose that A_1 and A_2 are denominator complete, and $0 \in A_1 \subset A_2$. Suppose $d \in R^*$ and $|d|s(A_1) \le s(A_2)$. If $x_i \in K$ for $1 \le i \le l$ and $k \le l$ and

$$N_1 = \textstyle\sum_{i=1}^{k} x_i A_1, \quad \text{and} \quad N_2 = \sum_{i=1}^{l} x_i A_2,$$

then $(1/d)N_1 \subset N_2$.

Proof: If a/b is a reduced fraction in A_1, then $a/b \in A_2$ and, since $|db| \le s(A_2)$, $a/db \in A_2$; i.e., $(1/d)A_1 \subset A_2$. The desired containment now follows readily. ❑

Fix elements $x_1, \ldots, x_n \in K$, and let $L(A) = \sum_{i=1}^{n} x_i A$. The easily verified facts that $L(A + B) = L(A) + L(B)$ and $AL(B) \subset L(AB)$ for all $A, B \subset K$ are used below.

LEMMA 7.2: Suppose $A, A_1, \ldots, A_n \subset K$; $a_1, \ldots, a_n, b_1, \ldots, b_n \in R^*$; and $E \subset R^{\times}$. Then:

(1) $s(\bigcup A_i) = \vee\, s(A_i)$.

(2) $s(\sum A_i) \le \prod s(A_i)$.
(3) $s(A_1 \cdots A_n) \le \prod s(A_i)$.
(4) $s(\sum(a_i/b_i) A_i) \le \prod|b_i|s(A_i)$.
(5) $s(EA) = s(A)$.
(6) $\wedge |A^*| \ge 1/s(A^*)$. Thus, $s(A) < \infty$ implies A^* is bounded away from 0.

The set $\{3/2, 5/3, 7/5, \ldots\}$ of all quotients of consecutive primes (with the larger prime in the numerator) shows that $s(A)$ need not be finite when A is bounded away from 0.

We are now ready to construct a direct system on K. The sequence M_{in} is constructed inductively. We define $M_{11} = D_1(\{0, \pm1\})$ and $M_{1n} = \{0\}$ for $n > 1$. Given that the sets of i-th coordinates of all sets U_n have been defined for all $i < j$ (i.e., all sets M_{in} with $i < j$ and $n \in \mathbf{Z}_{>0}$), we define the sets M_{jn} for all n. If $n > j$, then $M_{jn} = \{0\}$. We will define the sets M_{jn} for $n = 1, \ldots, j$ in three steps:

(1) Inductively define sets $M^+_{jj}, M^+_{jj-1}, \ldots, M^+_{j1}$ in that order.
(2) Choose an element $q_j \in K$.
(3) Define $M_{jn} = \sum_{t=1}^{2^{j-n}} q_j{}^t M^+_{jn}$, for $1 \le n \le j$.

Step (1): Let $\{B_n\}$ be an increasing sequence of bounded sets whose union is R (e.g., we could let $B_n = \{x \in R : |x| \le n\}$). Given that the set M^+_{jn+1} has been defined, we define M^+_{jn} to be an s-complete hull of a set consisting of sums and products of elements from B_n, M^+_{jn+1}, and the previously constructed sets M_{in} as follows: $M^+_{jn} = D_{s(j,n)}(M^{++}_{jn})$, where

$$M^{++}_{jj} = \{0, \pm1\}; \qquad s(j, j) = 1,$$

and, for $n < j$,

$$M^{++}_{jn} = (M^+_{jn+1} + M^+_{jn+1}) \cup (B_n M^+_{jn+1}) \cup$$

$$\left\{2 \times \left[\left(\sum_{t=1}^{j-1} M_{t\,n+1}\right) M^+_{jn+1}\right] + 2^{j-n} \times [M^+_{jn+1} M^+_{jn+1}]\right\};$$

$$s(j, n) = s(M^{++}_{jn}) \vee 2s(j, n+1).$$

Given that $s(M_{in}) < \infty$ for all $i < j$ and $n > 0$, Lemma 7.2 implies that $s(j,n)$ is finite for all n, so that the denominator complete set M^+_{jn} is well defined. Corollary 6.1 and Lemma 7.2(4) then imply that $s(M_{jn}) < \infty$ for all n.

Step (2): The element q_j will be chosen to satisfy any of the four conditions (Q1)–(Q4) below that will guarantee that the set M_{j1} defined in Step (3) will satisfy (D4). Namely, we will choose q_j so that it is not a solution to any polynomial equation of the form

$$m_0 = \sum_{i=1}^{2^{j-1}} m_i q_j^i, \quad m_0 \in 3 \times \sum_{i=1}^{j-1} M_{i1}, \quad m_i \in 3 \times M_{j1}^{+}$$

for $i \geq 1$, except when $m_0 = 0$.

Lemma 7.2 implies that the set of all possible nonzero coefficients of the polynomials in q_j is bounded away from 0; say $|m_i| > \gamma_j > 0$ for all possible choices of m_i, $0 \leq i \leq 2^{j-1}$. Observe that, given that all sets M_{in} are bounded when $i < j$ and all sets M_{jk}^{+} are bounded for $k \geq n+1$, it follows that M_{jn}^{+} is bounded by virtue of Corollary 6.1 and the fact that a finite union, a finite sum and a finite product of bounded sets is again bounded. Finally, the sets M_{jn} are all bounded because each one is a finite sum of bounded sets. Therefore, the set of all absolute values of possible coefficients has a finite upper bound, say, β_j.

(Q1) If $|q_j| > \alpha_j$, where $\alpha_j = 2^j\beta_j/\gamma_j (\geq 1)$, then (D4) is satisfied: Let k be the largest index such that $m_k \neq 0$. If $|q_j| > \alpha_j$, we see that $|m_k q_j^k| > (k+1)\beta_j|q_j|^{k-1}$; but an upper bound for the absolute value of the sum of the k terms $m_0 - \sum_{i=1}^{k-1} m_i q_j^i$ is $k\beta_j|q_j|^{k-1}$.

(Q2) If $0 < |q_j| < \delta_j$, where $\delta_j = \gamma_j/2\beta_j$ ($\leq 1/2$), then (D4) is satisfied:

$$|m_0| > \gamma_j \text{ and } \textstyle\sum_{i=1}^{k} |m_i q_j^i| \leq \sum_{i=1}^{k} \beta_j(\gamma_j/2\beta_j)(1/2)^{i-1} < \gamma_j.$$

Let p be any prime in R such that $|p| > 1$. Since the set M of possible coefficients of the polynomial considered above is bounded and $s(M) < \infty$, there is a largest integer e_j such that p^{e_j} divides at least one numerator of M; $s(M) < \infty$ implies there is a largest integer f_j such that p^{f_j} divides at least one denominator of M. Let $|\ |_p$ denote the p-adic absolute value on K. If $q_j = a/(p^{n_j}b)$, where $a,b \in R^*$ and a and p are relatively prime, and $n_j > 2(e_j + f_j)$, then each nonzero term of the polynomial above has distinct p-adic absolute value, which leads to the third condition.

(Q3) If $p \in R$ is prime and $|p| > 1$ and $|q_j|_p > \eta_j$, where $\eta_j = |p|_p^{-2(e_j+f_j)}$, then (D4) is satisfied.

Finally, we may choose $q_j = p^{n_j}(a/b)$, where $a,b \in R^*$ and b and p are relatively prime and $n_j > e_j + f_j$ to obtain the fourth condition.

(Q4) If $p \in R$ is prime and $|p| > 1$ and $0 < |q_j|_p < \varepsilon_j$, where $\varepsilon_j = |p|_p^{e_j+f_j}$, then (D4) is satisfied.

Step (3) has already been described completely.

R must always contain a prime p such that $|p| > 1$ (since otherwise R would be bounded and could not have K as its quotient), so that it is always possible to use (Q3) and (Q4) to choose q_j. The ring $R = \mathbf{Z}[x]$ in $\mathbf{Q}(x)$ with the $1/x$-adic absolute value shows that some primes may have absolute value one (viz., those primes in **Z**). On the other hand, in a maximal discrete ring (see [12]), one easily sees that the set of elements with absolute value one is exactly the set of invertible elements.

Note that, in the statement preceding (Q4), if we choose $b = 1$ and increase our lower bound on n_j sufficiently, then (Q1) and (Q4) are both satisfied. Similarly, in (Q2), if we also choose $|q_j|_p$ to be sufficiently large, then (Q3) is also satisfied.

THEOREM 7: The sequence $\{M_{in}\}$ defined above is a direct system on the field K. In the associated topology, $\{q_n\}$ converges to 0. Consequently, the associated topology is not discrete.

If each element q_j satisfies (Q1) or each element q_j satisfies (Q3), then inversion is maximally discontinuous in the associated topology.

Proof: It is clear from the construction that (D1) and (D2) are satisfied. We verify that (D3) is satisfied. If we let $x_i = q_j{}^i$, then $M_{j\,n} = L(M^+_{j\,n})$ (for L as defined after Lemma 7.1), and

$$\begin{aligned} M_{j\,n+1} + M_{j\,n+1} &= L(M^+_{j\,n+1}) + L(M^+_{j\,n+1}) \\ &= L(M^+_{j\,n+1} + M^+_{j\,n+1}) \subset L(M^+_{j\,n}) = M_{j\,n}. \end{aligned}$$

We already observed that (D4) was satisfied because of our choice of q_j.

(D5) is verified as follows:

$$2 \times \Big[\Big(\sum_{t=1}^{j-1} M_{t\,n+1}\Big) M_{j\,n+1}\Big] + M_{j\,n+1} M_{j\,n+1}$$

$$= 2 \times \Big[\Big(\sum_{t=1}^{j-1} M_{t\,n+1}\Big)\Big(\sum_{k=1}^{2^{j-n-1}} q_j{}^k M^+_{j\,n+1}\Big)\Big]$$

$$+ \Big(\sum_{i=1}^{2^{j-n-1}} q_j{}^i M^+_{j\,n+1}\Big)\Big(\sum_{l=1}^{2^{j-n-1}} q_j{}^l M^+_{j\,n+1}\Big)$$

$$\subset \sum_{k=1}^{2^{j-n-1}} q_j{}^k \Big\{2 \times \Big[\Big(\sum_{t=1}^{j-1} M_{t\,n+1}\Big) M^+_{j\,n+1}\Big]\Big\}$$

$$+ \sum_{k=2}^{2 \cdot 2^{j-n-1}} q_j{}^k [(k-1) \times (M^+_{j\,n+1} M^+_{j\,n+1})]$$

$$\subset \sum_{k=1}^{2^{j-n}} q_j{}^k \Big\{2 \times \Big[\Big(\sum_{t=1}^{j-1} M_{t\,n+1}\Big) M^+_{j\,n+1}\Big]\Big\} + \sum_{k=1}^{2^{j-n}} q_j{}^k [2^{j-n} \times (M^+_{j\,n+1} M^+_{j\,n+1})]$$

$$= \sum_{k=1}^{2^{j-n}} q_j^k \left\{2 \times \left[\left(\sum_{t=1}^{j-1} M_{t\,n+1}\right) M_{j\,n+1}^+\right] + 2^{j-n} \times (M_{j\,n+1}^+ M_{j\,n+1}^+)\right\}$$

$$\subset \sum_{k=1}^{2^{j-n}} q_j^k M_{j\,n}^+ = M_{j\,n}.$$

We verify that (D6) is satisfied. Observe that, for $k \le n$,

$$B_k M_{j\,n+1} \subset B_n M_{j\,n+1} = B_n L(M_{j\,n+1}^+) \subset L(B_n M_{j\,n+1}^+) \subset L(M_{j\,n}^+) = M_{j\,n}.$$

Given a positive integer n and an element $x \in K$, we choose $a,b \in R$ such that $x = a/b$; we choose $k > 1$ such that $a \in B_k$; and we choose a positive integer m such that $m \ge k$ and $|b| \le 2^m$. Then

$$\frac{a}{b} M_{j\,n+m+1} = \frac{1}{b}(aM_{j\,n+m+1}) \subset \frac{1}{b}(B_k M_{j\,n+m+1}) \subset \frac{1}{b}(M_{j\,n+m}) \subset M_{j\,n},$$

where the last containment follows from Lemma 7.1 and the fact that $s(j, n) \ge 2^m s(j, n + m)$.

Since $q_n \in U_n$, $\{q_n\}$ converges to 0.

To show that inversion is maximally discontinuous if q_j satisfies (Q1), we use the result quoted in the introduction, stating that it suffices to find, given a real number α, an integer n such that $|x| \ge \alpha$ for all $x \in U_n^*$. Suppose x is any element in U_1 and $j \ge 2$ is the height of x, then, according to (and using the notation in) our discussion, in Step (2), of the choice of the elements q_j, we have (that $\sum_{i=1}^{j-1} \varphi_i(x) = m_0$ and)

$$|x| \ge 2^j \beta_j |q_j|^{k-1} - 2^{j-1} \beta_j |q_j|^{k-1} > \beta_j.$$

Since, for $j \ge 2$, $\beta_{j+1} \ge |q_j|^2 > \beta_j^2$, it follows that the quantity β_j goes to infinity as j goes to infinity; and, since $j \ge n$ for $x \in U_n^*$, the desired result follows.

If q_j satisfies (Q3), then the terms of x as a polynomial in q_j (whose degree k is greater than or equal to one) all have distinct values, so

$$|x|_p = |m_k q_j^k|_p \ge \left|\frac{1}{p}\right|_p^{n_j - e_j}.$$

Now $n_j - e_j \ge e_j + 2f_j$ and $f_{j+1} > n_j \ge 2f_j$. Therefore, $\lim(n_j - e_j) \ge \lim f_j = \infty$, and again our claim is established. ❑

Although fields with a direct topology do not have topologically nilpotent elements, we can choose the elements q_j all to be powers $x^{n(j)}$ of any element x with absolute value greater than one and obtain an example of a direct topology in which a subsequence of x^n converges to 0. (This was observed in [10] for the special case $R = \mathbf{Z}$ and K is $\mathbf{Q}$ with the usual topology.)

When the completion of K is locally compact, the construction above produces sufficiently many direct topologies that any nondiscrete first countable Hausdorff ring topology is the infimum of a family of direct ring topologies. The remainder of this paper is devoted to establishing this.

Suppose that the completion of K is locally compact. It is easy to verify that each bounded subset of R is finite. [*Proof:* R is closed (as was observed in [12]), and the intersection of a closed discrete subset of K and a compact set (such as $\{x \in K: |x| \le c\}$) is finite.] Since the invertible elements in R have absolute value one, the set of invertible elements is finite. Also, each prime has absolute value greater than one, since, otherwise, the set of all positive integral powers of the prime (which all must be distinct) would be an infinite bounded subset of R.

The following result is easily established:

LEMMA 8.1: [15, Lemma 4.3] Let $(X,\mathcal{T})$ be a T_1 topological space, and let $\mathcal{N}_x$ be the neighborhood filter of a point $x \in X$. If x belongs to each set in a filter $\mathcal{A}$ on X, and if $\mathcal{N}_x$ is not a finer filter than $\mathcal{A}$, then there exists $A \in \mathcal{A}$ such that $U\backslash A$ is an infinite set for each $U \in \mathcal{N}_x$.

For $E \subset K$, let prnum(E) (prden(E)) be the set of all numbers p^n such that p is prime, n is a positive integer, and, for some numerator (respectively, denominator) a of E, a is divisible by p^n but not by p^{n+1}.

LEMMA 8.2: (cf. [15, Lemma 4.4]) Let R be a discrete order of the nontrivial absolute valued field $(K, |\ |)$. Suppose the completion of $(K, |\ |)$ is locally compact. Let $\mathcal{T}$ be a T_1 topology on K, and let $\mathcal{A}$ be a filter such that 0 belongs to each set in $\mathcal{A}$.

If $\mathcal{B}(\mathcal{T}) \vee \mathcal{B}(\mathcal{T}_{|\ |}) = \mathcal{A} \vee \mathcal{B}(\mathcal{T}_{|\ |})$ and $\mathcal{B}(\mathcal{T})$ is not a finer filter than $\mathcal{A}$, then there exists $A \in \mathcal{A}$ such that prnum$(U\backslash A)$ is an infinite set for each $U \in \mathcal{B}(\mathcal{T})$.

Proof: Choose $A \in \mathcal{A}$ such that $U\backslash A$ is infinite for each $U \in \mathcal{B}(\mathcal{T})$; and choose $V \in \mathcal{B}(\mathcal{T})$ and $n \in \mathbf{Z}_{>0}$ such that $V \cap \{x \in K: |x| \le 1/n\} \subset A$. For $W \in \mathcal{B}(\mathcal{T})$ such that $W \subset V$, suppose that prnum$(W\backslash A)$ is finite. If also prden$(W\backslash A)$ is finite, then $W\backslash A$ is finite, a contradiction. However, if prden$(W\backslash A)$ is infinite, then the set of all denominators of $W\backslash A$ is unbounded; but the set of all numerators of $W\backslash A$ is bounded. Thus, $W\backslash A$ has elements with absolute value less than $1/n$, which, by the choice of V and n, is again a contradiction. Since any $\mathcal{T}$-neighborhood contains a $\mathcal{T}$-neighborhood contained in V, the lemma is proved. ❑

LEMMA 8.3: If V is a neighborhood of 0 in an additive topological group, then there exists a sequence $\{V_i\}$ of neighborhoods of 0 such that $V_1 + \cdots + V_n \subset V$ for all n.

Proof: Choose a neighborhood V_1 of 0 such that $V_1 + V_1 \subset V$; and given neighborhoods $V_1, \ldots, V_n$ such that $(V_1 + \cdots + V_n) + V_n \subset V$, choose a neighborhood V_{n+1} of 0 such that $V_{n+1} + V_{n+1} \subset V_n$. ❑

LEMMA 8.4:

(1) Let $\mathcal{T}_i$, $i \in I$, be a family of topologies on a set X. Suppose $\mathcal{T}$ is a topology on X such that, for each $x \in X$, the $\mathcal{T}$-neighborhood filter at x is the intersection over $i \in I$ of the $\mathcal{T}_i$-neighborhood filters at x. Then $\mathcal{T} = \bigcap_{i \in I} \mathcal{T}_i$.

(2) Let $\mathcal{T}_i$, $i \in I$, be a family of group topologies on an additive group G. Suppose $\mathcal{T}$ is a group topology on G such that the $\mathcal{T}$-neighborhood filter at 0 is the intersection over $i \in I$ of the $\mathcal{T}_i$-neighborhood filters at 0. Then $\mathcal{T} = \bigcap_{i \in I} \mathcal{T}_i$.

THEOREM 8: (cf. [15, Theorem 4.5]) Let R be a discrete order of the nontrivial absolute valued field $(K, |\ |)$. Suppose the completion of $(K, |\ |)$ is locally compact. Let $\mathcal{T}$ be a nondiscrete first countable Hausdorff ring topology on K, and let $\mathcal{A}$ be a filter, strictly finer than $\mathcal{B}(\mathcal{T})$, such that 0 belongs to each set in $\mathcal{A}$. Then there exists a direct ring topology $\mathcal{T}_{\mathcal{A}}$ such that $\mathcal{B}(\mathcal{T}) \leq \mathcal{B}(\mathcal{T}_{\mathcal{A}})$ and $\mathcal{A} \nleq \mathcal{B}(\mathcal{T}_{\mathcal{A}})$.

Consequently, each nondiscrete first countable Hausdorff ring topology on K is an intersection of a family of direct ring topologies.

Proof: Case 1: $\mathcal{B}(\mathcal{T}) \vee \mathcal{B}(\mathcal{T}_{|\ |}) < \mathcal{A} \vee \mathcal{B}(\mathcal{T}_{|\ |})$. In this case we have $\mathcal{B}(\mathcal{T}) \vee \mathcal{B}(\mathcal{T}_{|\ |})$ is not finer than $\mathcal{A}$. Otherwise, we would have the contradiction

$$\mathcal{B}(\mathcal{T}) \vee \mathcal{B}(\mathcal{T}_{|\ |}) < \mathcal{A} \vee \mathcal{B}(\mathcal{T}_{|\ |}) \leq (\mathcal{B}(\mathcal{T}) \vee \mathcal{B}(\mathcal{T}_{|\ |})) \vee \mathcal{B}(\mathcal{T}_{|\ |}) = \mathcal{B}(\mathcal{T}) \vee \mathcal{B}(\mathcal{T}_{|\ |}).$$

As in Lemma 8.1, choose $A \in \mathcal{A}$ such that $V \cap \{x \in K: |x| < 1/r\} \backslash A$ is infinite for all $V \in \mathcal{B}(\mathcal{T})$ and all positive integers r. Let $\{V_n\}$ be a $\mathcal{T}$-neighborhood base at 0 such that $V_1 = K$; and, for each n, choose a sequence $\{V_{in}\}$ of $\mathcal{T}$-neighborhoods of 0 such that $V_{11} = K$ and such that $V_{1n} + \cdots + V_{kn} \subset V_n$ for all k and n. Define the sequence $\{M_{in}\}$ and a sequence $\{W_{in}\}$ of $\mathcal{T}$-neighborhoods of 0 inductively as follows. Let $M_{11} = D_1(\{0, \pm 1\})$, let $M_{1n} = \{0\}$ for $n > 1$, let $W_{11} = V_{11}$, and let $W_{jn} = V_{jn}$ for $j < n$. Suppose that the sets M_{in} have been chosen, as described before Theorem 7, for $i < j$ and for all n, and suppose also that $M_{in} \subset V_{in}$. Define the sets M_{jn}^+ as before.

Note that the set of denominators of any set M_{jn}^+ is finite, since the set of denominators is a $\mathcal{T}_{|\ |}$-bounded subset of R. Since M_{jn}^+ is bounded and $s(M_{jn}^+) < \infty$, the set of numerators of M_{jn}^+ is bounded, and, hence, also finite. Thus, the set M_{jn}^+ is finite. Therefore, the set M_{jn}^+ is $\mathcal{T}$-bounded, and we may choose a $\mathcal{T}$-neighborhood W_{jn} such that

$$\sum_{t=1}^{2^{j-n}} M_{j\,n}^{+} W_{j\,n}^{\times t} \subset V_{j\,n}.$$

Since $(\bigcap_{n=1}^{j} W_{j\,n}) \cap \{x \in K: |x| < 1/r\} \setminus A$ is infinite for all r, we may choose q_j from $\bigcap_{n=1}^{j} W_{j\,n}$ which satisfies (Q2) and define $M_{j\,n}$ as before. Then $M_{j\,n} \subset V_{j\,n}$, which implies the standard neighborhood U_n associated with $\{M_{i\,n}\}$ is contained in V_n. Since $q_j \notin A$ for $n > 1$, the proof is complete in this case.

Case 2: $\mathcal{B}(\mathcal{T}) \vee \mathcal{B}(\mathcal{T}_{|\,|}) = \mathcal{A} \vee \mathcal{B}(\mathcal{T}_{|\,|})$. Using Lemma 8.2, we choose $A \in \mathcal{A}$ such that prnum$(V \setminus A)$ is infinite for each $V \in \mathcal{B}(\mathcal{T})$. Choose the sets V_n, $V_{i\,n}$, and $M_{1\,n}$ as in Case 1. Also, define inductively the sets $M_{i\,n}$ and $W_{i\,n}$ as before. If, for some prime p, there are infinitely many exponents k such that $p^k \in$ prnum$([\bigcap_{n=1}^{j} W_{j\,n}] \setminus A)$, then we may choose $k > e_j + f_j$ and choose $q_j \in [\bigcap_{n=1}^{j} W_{j\,n}] \setminus A$ such that p^k is a factor of a numerator of q_j. Otherwise, the set of primes, some power of which belongs to prnum$([\bigcap_{n=1}^{j} W_{j\,n}] \setminus A)$, is an infinite subset of R and, hence, unbounded. By choosing a sufficiently large prime p from this unbounded set, we have $e_j = f_j = 0$, and we can choose q_j to be any element in $[\bigcap_{n=1}^{j} W_{j\,n}] \setminus A$ such that p divides the numerator of q_j.

The proof of the theorem is completed by applying Lemma 8.4, with the index set I consisting of all filters $\mathcal{A}$ finer than $\mathcal{B}(\mathcal{T})$ and verifying that $\mathcal{B}(\mathcal{T}) = \bigcap_{\mathcal{A}} \mathcal{B}(\mathcal{T}_{\mathcal{A}})$. ❑

Let **T** be a collection of ring topologies on a ring S. Let $\wedge_t$ **T**, $\wedge_g$ **T**, and $\wedge_r$ **T** denote the finest topology, group topology and ring topology, respectively, which is weaker than or equal to each topology in **T**. (Thus $\wedge_t$ **T** $= \bigcap$ **T**.) Certainly, $\wedge_r$ **T** $\leq \wedge_g$ **T** $\leq \wedge_t$ **T**; and it is possible that both inequalities are strict. Suppose K is as in Theorem 8; and we use the notation of Theorem 8; and **T** consists of all topologies of the form $\mathcal{T}_{\mathcal{A}}$, where $\mathcal{A}$ is a filter, strictly finer than $\mathcal{B}(\mathcal{T})$, such that 0 belongs to each set in $\mathcal{A}$. Then, from the proof of Theorem 8, it follows that $\wedge_t$ **T** is a ring topology (viz. $\mathcal{T}$). An immediate consequence of this is that, for this particular collection **T**, $\wedge_t$ **T** $= \wedge_g$ **T** $= \wedge_r$ **T**.

REFERENCES

1. DIKRANJAN, D.N., I.R. PRODANOV, & L.Z. STOYANOV. 1989. Topological Groups, Vol. 130. Dekker. New York. **MR 91e:22001**.
2. KELLEY, J.L. 1955. General Topology. van Nostrand. Princeton, New Jersey. **MR 16 1136**.
3. MARCOS, J.E. 1996. An example of a nontrivial ring topology on the algebraic closure of a finite field. J. Pure Appl. Algebra. **108:** 265–278.
4. MARCOS, J.E. The field of rational numbers is algebraically closed in some of its completions. J. Pure Appl. Algebra. To appear.
5. MUTYLIN, A.F. 1968. An example of a nontrivial topologization of the field of rational numbers. Complete locally bounded fields. Izv. Akad. Nauk. SSSR ser. Mat. (Amer. Math. Soc. Transl.). **30:** (73) 873–890 (159–179) [see correction **30:** (2) 245–246 (237–238)]. **MR 33 #4052 and 36 #5110**.

6. MUTYLIN, A.F. 1968. Connected complete locally bounded fields. Complete not locally bounded fields. Mat. Sbornik (N.S.) (Math. USSR Sbornik). **76:** (5) 454–472 (433–449). **MR 37 #6265.**
7. SHANKS, M.E. & S. WARNER 1973. Topologies on the rational field. Bull. Amer. Math. Soc. **79:** 1281–1282. **MR 51 #5568.**
8. SHELL, N. 1978. Maximal and minimal ring topologies. Proc. Amer. Math. Soc. **68:** 23–26. **MR 57 #299.**
9. SHELL, N. 1990. Topological Fields and Near Valuations, Vol. 135. Dekker, New York, 1990. **MR 91m:12012.**
10. SHELL, N. 1991. Big sets. Conference on General Topology and Applications, Brookville, New York, 1990. Lecture Notes in Pure and Applied Mathematics. Dekker, New York. **134:** 172–175
11. SHELL, N. 1994. Residue class topologies. Conference on General Topology and Applications, Flushing, New York, 1992. Annals of the New York Academy of Sciences. New York. **Vol. 728:** 339–345.
12. SHELL, N. 1996. Discrete rings. J. Pure Appl. Algebra. **111:** 277–284.
13. SHELL, N. Direct topologies from discrete rings, II. To appear.
14. WARNER, S. 1989. Topological Fields. North-Holland. Amsterdam. **MR 90L:12012.**
15. ZOBEL, R. 1973. Direkte Gruppen- und Ringtopologien. Dissertation. Tech. Univ. Braunschweig.

Metrizability of Asymmetric Spaces

S.D. SHORE[a] AND SALVADOR ROMAGUERA[b,c]

[a]Department of Mathematics
University of New Hampshire
Durham, New Hampshire 03824
and
[b]Escuela de Caminos
Universidad Politécnica de Valencia
46071 Valencia, Spain

ABSTRACT: Recent research has spawned an evolving application of non-Hausdorff topologies [22] generated by *nonsymmetric* distance functions to areas within mathematics (e.g., posets and continuous lattices [6]) as well as to allied areas that might seem remote (e.g., computer science ([24],[4],[16]) and biology [25]).

The study of nonsymmetric distance can be traced to the 1910 thesis of T.H. Hildebrandt [11], written at the University of Chicago under the direction of E.H. Moore. In a spirit paralleling that of Hildebrandt's, this paper undertakes a study of nonsymmetric distance for which an equivalent symmetric distance can be constructed. This illuminates situations in which an associated *metric* exists.

Of particular interest in our study are distances that are either locally symmetric or locally satisfy the triangle inequality. We recall Niemytzki's classical result [19] that a topological space is metrizable if, and only if, there is a semimetric for the space that *locally* satisfies the triangle inequality. We investigate two forms in which a nonsymmetric distance might locally satisfy the triangle inequality. In either case, we show that such spaces are metrizable, when the distance is *locally* symmetric. Although many of our results are known, the approach is particularly straightforward in providing an explicit construction of a distance with the desired properties.

1. INTRODUCTION

A nonnegative real-valued function $d\colon X \times X \to \mathbf{R}$ is a *distance function*, or simply a *distance*, for X iff $d(p, p) = 0$ for every $p \in X$. For any distance function d, if $S_d(p, \epsilon) = \{x \in X \mid d(p, x) < \epsilon\}$, the sphere centered at p of radius ϵ, then

$$\mathcal{T}_d = \{A \subseteq X \mid \text{when } a \in A, \text{ then } S_d(a, \epsilon) \subseteq A \text{ for some } \epsilon > 0\}$$

Mathematics Subject Classification: 54E35, 54E25, 54D55.

Key words and phrases: metrizable, asymmetric, symmetric, quasimetric, semimetric, Nagata space, γ-space.

[c]The second author acknowledges with gratitude the support of the DGICYT grant GV-2223/94.

is a topology for X. In general, $S_d(p, \epsilon)$ need not be a neighborhood for p in $(X, \mathcal{T}_d)$. However, if $\{S_d(p, \epsilon) | \epsilon > 0\}$ is a neighborhood base in $(X, \mathcal{T})$ for every p, then $\mathcal{T} = \mathcal{T}_d$

A distance function d is an *asymmetric*, or an *o-metric* [18], iff

$$\text{for every } p, q \in X,\ d(p, q) = 0 \text{ iff } p = q.$$

A topological space $(X, \mathcal{T})$ is *asemimetrizable* iff there is an asymmetric d for X such that for every $p \in X$, $\{S_d(p, \epsilon) | \epsilon > 0\}$ is a neighborhood base for p in $(X, \mathcal{T})$; in this case we say that d is an *asemimetric for* $(X, \mathcal{T})$. A topological space $(X, \mathcal{T})$ is *asymmetrizable* iff there is an asymmetric d for X such that $\mathcal{T} = \mathcal{T}_d$; in this case we say that d is an *asymmetric for* $(X, \mathcal{T})$.

An asymmetric d for X is *a symmetric* [19], [1], or *an écart* [5], iff

$$d(p, q) = d(q, p) \text{ for every } p, q \in X.$$

An asymmetric d is *locally symmetric* [3], is *a local symmetric* or *has property* (s_1) [18], iff for every $p \in X$ and every $\epsilon > 0$ there is $\delta > 0$ such that,

$$\text{when } d(p, x) < \delta, \text{ then } d(x, p) < \epsilon;$$

or equivalently, for every $p \in X$,

$$\text{when } d(p, x_n) \to 0, \text{ then } d(x_n, p) \to 0.$$

An asymmetric d for X *satisfies the triangle inequality*, is a *quasimetric* [27] or is a *Δ-metric* [20], iff for every $p, q \in X$

$$d(p, q) \leq d(p, x) + d(x, q) \text{ for every } x \in X.$$

In considering the triangle inequality locally we note two alternatives. An asymmetric d for X is a *γ-asymmetric*, has the *local axiom of the triangle* [19], or is a *γ-metric* [15], iff for any $p \in X$ and every $\epsilon > 0$ there is $\delta > 0$ such that,

$$\text{when } d(p, y) < \delta \text{ and } d(y, x) < \delta, \text{ then } d(p, x) < \epsilon.$$

Equivalently, d is a γ-asymmetric for X iff for any $p \in X$,

$$\text{when } d(p, y_n) \to 0 \text{ and } d(y_n, x_n) \to 0, \text{ then } d(p, x_n) \to 0;$$

An asymmetric d for X is *coherent* [21], or *has property* (Π_1) [18], iff for any $p \in X$,

$$\text{when } d(p, y_n) \to 0 \text{ and } d(x_n, y_n) \to 0, \text{ then } d(p, x_n) \to 0.$$

Obviously, for symmetrics, metrics are identical with quasimetrics and γ-asymmetrics are identical with coherent asymmetrics. Note also that any quasimetric is a γ-asymmetric.

A topological space $(X, \mathcal{T})$ is *locally symmetrizable* iff there is a locally symmetric asymmetric for $(X, \mathcal{T})$; it is *locally semimetrizable* iff there is a locally symmetric asemimetric for $(X, \mathcal{T})$; it is *coherently symmetrizable* iff there is a coherent symmetric for $(X, \mathcal{T})$. Similarly, we define topological spaces determined by any variety of distance.

An asymmetric *d has unique limits* iff for any $p, q \in X$,

$$\text{when } d(p, x_n) \to 0 \text{ and } d(q, x_n) \to 0, \text{ then } p = q;$$

a topological space $(X, \mathcal{T})$ *has unique (sequential) limits* iff for any $p, q \in X$,

$$\text{when } x_n \to p \text{ and } x_n \to q, \text{ then } p = q.$$

Note that coherent asymmetrics have unique limits. It is well known that this need not be the case for quasimetrics and, hence, for γ-asymmetrics. In fact, if d is a quasimetric with unique limits, then $\mathcal{T}_d$ is Hausdorff. The existence of non-Hausdorff quasimetrizable spaces is well known; see also Example 3.5 below.

For any distance function d for X and $A, B \subseteq X$:

$S_d[A, \epsilon) = \bigcup\{S_d(a, \epsilon) | a \in A\}$ is the sphere of radius ϵ about A;

$d[A, B] = \inf\{d(a, b) | a \in A, b \in B\}$ is the distance from A to B.

As foundation for our work we note the following lemmas, which highlight essential properties relating distance functions to the topologies they generate.

LEMMA 1.1: For any asymmetric d, the following are equivalent:
(i) $(X, \mathcal{T}_d)$ has unique limits;
(ii) countably compact subsets of $(X, \mathcal{T}_d)$ are closed;
(iii) if $p \neq q$, then $S_d(p, \epsilon) \cap S_d(q, \epsilon) = \varnothing$ for some $\epsilon > 0$;
(iv) d has unique limits.

LEMMA 1.2: If d is an asymmetric for X with unique limits, then

$$d(p, x_n) \to 0 \Leftrightarrow x_n \to p \text{ (in } \mathcal{T}_d).$$

LEMMA 1.3: For any distance function d the following are equivalent:
(i) For every p, $\{S_d(p, \epsilon) | \epsilon > 0\}$ is a neighborhood base for p in $(X, \mathcal{T})$;
(ii) for every $\epsilon > 0$ and every $p \in X$

$$\{x \in X | \text{ there is } \alpha > 0 \text{ such that } S_d(x, \alpha) \subseteq S_d(p, \epsilon)\} \in \mathcal{T}.$$

(iii) $(X, \mathcal{T})$ is first countable and $d(p, x_n) \to 0 \Leftrightarrow x_n \to p$ (in $\mathcal{T}$) and $\mathcal{T} \subseteq \mathcal{T}_d$;
(iv) $(X, \mathcal{T})$ is Fréchet and $d(p, x_n) \to 0 \Leftrightarrow x_n \to p$ (in $\mathcal{T}$);
(v) for every nonempty set $A \subseteq X$, $\mathrm{cl}_{\mathcal{T}}(A) = \{x \in X | d(x, A] = 0\}$;
(vi) for any $A \subseteq X$, $\mathcal{T}|A = \mathcal{T}_{d|(A \times A)}$.

2. SPACES GENERATED BY COHERENT ASYMMETRICS

Coherent symmetrics are called *coherent distance functions* by Chittenden and Pitcher [21] although distances with the defining property had been considered earlier by Chittenden [2]. An expository study of coherently symmetrizable spaces is given in [23].

We begin our investigation of asymmetry with some fundamental observations.

LEMMA 2.1: For any coherent asymmetric d,

(a) d has unique limits;

(b) $d(p, x_n) \to 0 \Leftrightarrow x_n \to p$ (in $\mathcal{T}_d$).

(c) If $d(x_n, p) \to 0$, then $d(p, x_n) \to 0$; hence,

(d) the distance function d_* for X such that $d_*(p, q) = \min\{d(p, q), d(q, p)\}$ is a symmetric for $(X, \mathcal{T}_d)$.

Our next result extends to coherent asymmetrics some important characterizations that have arisen periodically in studies of coherent semimetrics; see [23] and the historical remarks following Corollary 2.5.

LEMMA 2.2: For any asymmetric d, the following are equivalent:

(i) d is coherent;

(ii) $d[F, K] > 0$ for any closed set F and disjoint compact set K (in $\mathcal{T}_d$);

(iii) for any closed set F (in $\mathcal{T}_d$),

$$\text{if } p \notin F, \text{ then there is } \epsilon > 0 \text{ such that } S_d(p, \epsilon) \cap S_d[F, \epsilon) = \varnothing.$$

Proof: (i) $\Rightarrow$ (iii). Suppose that d is coherent and that p is not in the closed set F. There is $\epsilon > 0$ such that $S_d(p, \epsilon) \cap F = \varnothing$; hence, from coherency there is $\delta > 0$ such that $d(p, y) < \delta$, $d(x, y) < \delta$ implies that $d(p, x) < \epsilon$ and, thus, $S_d(p, \delta) \cap S_d[F, \delta) = \varnothing$.

(iii) $\Rightarrow$ (ii). Suppose that d satisfies (iii); equivalently, for a_n and b_n,

$$\text{if } d(p, a_n) \to 0 \text{ and } d(b_n, a_n) \to 0 \text{ with } b_n \in F, \text{ then } p \in F.$$

Now, suppose K is compact with $d[F, K] = 0$. Choose $a_n \in K$ and $b_n \in F$ such that $d(b_n, a_n) \to 0$. From the compactness of K there is $p \in X$ and a subsequence $\langle a_{k(n)} \rangle$ of $\langle a_n \rangle$ such that $d(p, a_{k(n)}) \to 0$. (Otherwise, let $F_1 = \{a_k \mid k \geq 1\}$. For each $p \notin F_1$ there $\epsilon > 0$ such that $S_d(p, \epsilon) \cap F_1 = \varnothing$; hence, F_1 is closed. Similarly $F_n = \{a_k \mid k \geq n\}$ is closed. Thus, $\{(X \backslash F_k) \mid k \geq 1\}$ is an open cover of K with no finite subcover which contradicts the compactness of K.) From our assumptions $p \in F$. But, distinct points are contained in disjoint spheres; hence, countably compact sets are closed (Lemma 1.1) so that also $p \in K$, which establishes the desired result.

(ii) $\Rightarrow$ (i). First, note that if d satisfies (ii), then $x_n \to p \Leftrightarrow d(p, x_n) \to 0$. Otherwise, there is $p \in X$, $\epsilon > 0$ and $\langle x_n \rangle$ such that $x_n \to p$, while $d(p, x_n) \geq \epsilon$.

Thus, if $A = \{x_n \mid n \in \mathbf{N}\}$, then A is closed without containing its limit point p which is a contradiction.

Finally, d is coherent. Otherwise, there are $p \in X$, x_n and y_n such that

$$d(p, y_n) \to 0,\ d(x_n, y_n) \to 0, \text{ while } d(p, x_n) \not\to 0 \text{ (i.e., } x_n \not\to p).$$

Thus, there is an open neighborhood G of p such that $x_n \notin G$, for infinitely many n. We conclude that $F = X \backslash G$ is closed, while $K = \{y_n | y_n \in G\} \cup \{p\}$ is a disjoint compact set with $d[F, K] = 0$, which contradicts that d satisfies (ii). ❑

Since coherent asymmetrics have unique limits, we have from Lemmas 1.2 and 1.3 an immediate corollary

THEOREM 2.3: $(X, \mathcal{T})$ is coherently asemimetrizable iff it is coherently asymmetrizable and first countable.

Coherent symmetrics were introduced in order to ensure that the derived set of any set is closed. Such distances always have property (v) of Lemma 1.3 and, hence, being coherently symmetrizable is equivalent to being coherently semimetrizable. This is not the case for asymmetrics.

EXAMPLE 2.4: A symmetrizable space $(X, \mathcal{T})$ can be coherently asymmetizable without being coherently asemimetrizable.

Let $X = [0, 1]$ and $A = \{1/3^n | \ n \in \mathbf{N}\} \cup \{0\}$. The asymmetric d for X such that

$$d(x, y) = \begin{cases} 1, & \text{if } x = 0 \text{ and } y \notin A; \\ 1, & \text{if } x \notin A \text{ and } y \in A; \\ |x - y|, & \text{otherwise,} \end{cases}$$

is coherent. Since $(X, \mathcal{T}_d)$ is not first countable at 0, there can be no coherent asemimetric for $(X, \mathcal{T}_d)$. Note also that the distance function d_* for X such that

$$\mathrm{d}_*(p, q) = \min\{d(p, q), d(q, p)\}$$

is a symmetric for $(X, \mathcal{T})$.

HISTORICAL NOTE: This example appears in a somewhat different form as Arhangel'skiĭ's Example 2.2 in [1] and in many different guises throughout the literature; see, for example, [9, Example 1]. However, the essential properties of this example are observed by Fréchet [5, pp.55–56] and Hahn [8] in showing that in limit spaces the derived set of a set might not be closed.

Of prime importance for our study is the 1927 theorem of Niemytzki and its immediate corollary from Lemma 2.2.

NIEMYTZKI'S THEOREM: A $(X, \mathcal{T})$ is metrizable iff it is coherently symmetrizable.

COROLLARY 2.5: For any topological space $(X, \mathcal{T})$ the following are equivalent:

(i) $(X, \mathcal{T})$ is metrizable;

(ii) there is a symmetric d for $(X, \mathcal{T})$ such that

$$d[F, K] > 0, \text{ for any closed set } F \text{ and any disjoint compact set } K;$$

(iii) there is a symmetric d for $(X, \mathcal{T})$ such that for any closed set F, if $p \notin F$, then there is $\epsilon > 0$ such that

$$S_d(p, \epsilon) \cap S_d[F, \epsilon) = \varnothing.$$

HISTORICAL NOTES: Arhangel'skiĭ [1] showed (i) is equivalent (ii) for Hausdorff spaces, while Kenton [14] and Martin [17] independently showed that Hausdorff could be omitted. Kenton [14] and Harley and Faulkner [9] independently showed (ii) is equivalent (iii).

We generalize Niemytzki's Theorem by showing that the symmetric property need only hold *locally*.

THEOREM 2.6: $(X, \mathcal{T})$ is metrizable iff there is a coherent *local* symmetric for $(X, \mathcal{T})$.

Proof: Suppose that d is a locally symmetric, coherent asymmetric for $(X, \mathcal{T})$. Then the distance function d^* for X such that

$$d^*(p, q) = \max\{d(p, q), d(q, p)\}$$

is a symmetric for X.

Since $d \leq d^*$, $\mathcal{T} = \mathcal{T}_d \subseteq \mathcal{T}_{d^*}$. Now, suppose that $A \in \mathcal{T}_{d^*}$; we claim $A \in \mathcal{T}_d$. Otherwise, there is $p \in A$ and $\alpha > 0$ such that $S_{d^*}(p, \alpha) \subseteq A$ and, for every n, $S_d(p, 2^{-n}) \nsubseteq A$. Thus, for every n there is $x_n \notin A$, even though $d(p, x_n) \to 0$. But, d is locally symmetric and, therefore, $d(x_n, p) \to 0$ and, hence, $d^*(p, x_n) \to 0$. Eventually $x_n \in S_{d^*}(p, \alpha)$ which contradicts that $x_n \notin A$ for all n. That is, d^* is a symmetric for $(X, \mathcal{T})$.

Finally, d^* is coherent. Suppose that $d^*(p, y_n) \to 0$ and $d^*(x_n, y_n) \to 0$. Since $d \leq d^*$ and d is coherent, $d(p, x_n) \to 0$; by local symmetry $d^*(p, x_n) \to 0$.

Metrizability of $(X, \mathcal{T})$ now follows from Niemytzki's Theorem. ❑

Note that Example 2.4 is symmetrizable and coherently asymmetrizable without being coherently symmetrizable. Thus, the distance properties of Theorem 2.6 do not "factor" metrizability. That is, to ensure metrizability of a space the defining distance must be simultaneously locally symmetric and coherent.

COROLLARY 2.7: For any topological space $(X, \mathcal{T})$ the following are equivalent:

(i) $(X, \mathcal{T})$ is metrizable;

(ii) there is a local symmetric d for $(X, \mathcal{T})$ such that $d[F, K] > 0$ for any closed set F and any disjoint compact set K;

(iii) there is a local symmetric d for $(X, \mathcal{T})$ such that for any closed set F, if $p \notin F$, then there is $\epsilon > 0$ such that

$$S_d(p, \epsilon) \cap S_d[F, \epsilon) = \varnothing.$$

3. SPACES GENERATED BY γ-ASYMMETRICS

Chittenden [2] and Niemytzki [19] investigated γ-symmetrics; however, Kofner [15] specifically isolated the γ-distance function property for non-symmetric distances. Quasimetrics were studied much earlier by Wilson [27] and Niemytzki [20].

The open spheres generated by a quasimetric are always topologically open. This need not be the case for γ-distances or even for γ-symmetrics. However, if d is a γ-distance, then for any sphere $S_d(p, \epsilon)$ there is $\delta > 0$ such that, when $x \in S_d(p, \delta)$, then there is $\alpha > 0$ such that $S_d(x, \alpha) \subseteq S_d(p, \epsilon)$ (see Lemma 1.3(ii)); that is, the open spheres are neighborhoods of their center. Thus, we note the following properties of γ-distances.

LEMMA 3.1: For every γ-distance d,

(a) for every p, $\{S_d(p, \epsilon) | \epsilon > 0\}$ is a *neighborhood base* for p in $(X, \mathcal{T}_d)$;

(b) hence, $(X, \mathcal{T}_d)$ is first countable and

(c) d is a γ-asymmetric for $(X, \mathcal{T})$ iff d is a γ-asemimetric for $(X, \mathcal{T})$.

(d) d need not have unique limits.

The sequence of results which follow parallel our study of coherent distance functions. Since the proofs are similar, we omit the details.

LEMMA 3.2: An asymmetric d is a γ-asymmetric iff

(i) for any $p \in X$, $d(p, x_n) \to 0 \Leftrightarrow x_n \to p$ (in $\mathcal{T}_d$) and

(ii) $d[K, F] > 0$ for any closed set F and any disjoint compact set K.

In this context Niemytzki's Theorem may be stated as:

NIEMYTZKI'S THEOREM: A topological space is metrizable iff it is γ-symmetrizable.

As in Theorem 2.6, symmetry need only hold locally. This theorem occurs as Theorem 5 in [3] and can be derived from Corollary 2.4 in [13]. Our proof is subsumed in our proof of Lemma 3.3.

From Theorem 3.3 and Lemma 1.2, one concludes immediately that a topological space $(X, \mathcal{T})$ is metrizable iff there is a local symmetric *with unique limits* for $(X, \mathcal{T})$ such that $d[K, F] > 0$ for every closed set F and disjoint compact set K. Actually the uniqueness of limits is implicit in the assumed distance properties.

LEMMA 3.3: For any topological space $(X, \mathcal{T})$ the following are equivalent:

(i) $(X, \mathcal{T})$ is metrizable;

(ii) there is a locally symmetric quasimetric for $(X, \mathcal{T})$;

(iii) there is a local symmetric d for $(X, \mathcal{T})$ such that

$$d[K, F] > 0 \text{ for every closed set } F \text{ and disjoint compact set } K.$$

Proof: Clearly, (i) implies (ii); (ii) implies (iii) from Lemma 3.2.

Suppose that d is a local symmetric with the property of condition (iii). Then, the distance d^* for X such that

$$d^*(p, q) = \max\{d(p, q), d(q, p)\}$$

is a symmetric such that $d^*[K, F] > 0$ for every closed set F and every disjoint compact set K. Metrizability now follows from Corollary 2.5. ❑

THEOREM 3.4: $(X, \mathcal{T})$ is metrizable iff there is a γ-local symmetric for $(X, \mathcal{T})$.

Example 2.4 shows that a coherently asymmetrizable space need not be γ-asymmetrizable. The next example shows that a γ-asymmetrizable (or even quasimetrizable) space need not be coherently asymmetrizable.

EXAMPLE 3.5: A quasimetrizable and semimetrizable compact space that is not coherently asymmetrizable.

Let $X = \{1/3^n \mid n = 0, 1, 2, \ldots\}$ and d be the asymmetric for X such that for $p \neq q$:

$$d(p, q) = \begin{cases} q, & \text{if } q < p; \\ 1, & \text{if } q > p. \end{cases}$$

Then, d is a quasimetric for $(X, \mathcal{T})$, where $\mathcal{T}$ is the finite-complement topology.

Thus, quasimetrics and γ-asymmetrics need not have unique limits. Since coherent asymmetrics have unique limits, there is no coherent asymmetric for $(X, \mathcal{T})$. Note also that the distance function d_* for X such that

$$d_*(p, q) = \min\{d(p, q), d(q, p)\}$$

is a semimetric for $(X, \mathcal{T})$.

The examples given in 2.4 and 3.5 are, of course, not metrizable, since both coherently asymmetrizable and γ-asymmetrizable are necessary properties of metrizable spaces. Unexpectedly, the two properties "factor" metrizability.

THEOREM 3.6: A topological space is metrizable iff it is γ-asymmetrizable and coherently asymmetrizable.

Proof: Let d_1 be a coherent asymmetric for $(X, \mathcal{T})$ and d_2 be a γ-asymmetric for $(X, \mathcal{T})$. The distance function d^* for X such that

$$d^*(p,q) = \max\{d_1(p, q), d_2(p, q)\}$$

is a coherent, γ-asymmetric for $(X, \mathcal{T})$; hence, the distance d_{**} for X such that

$$d_{**}(p, q) = \min\{d^*(p, q), d^*(q, p)\}$$

is a coherent symmetric for $(X, \mathcal{T})$, which from Niemytzki's Theorem concludes that $(X, \mathcal{T})$ is metrizable.

4. NEIGHBORHOOD CHARACTERIZATIONS

In the study of spaces defined by distance functions one frequently finds connections to spaces defined by sequences of open covers; see, for example, [10], [12] and [7]. We say that $\{U_n(p) | n \in \mathbf{N}, p \in X\}$ is a *neighborhood assignment* in $(X, \mathcal{T})$ iff

$$p \in U_n(p) \in \mathcal{T} \text{ for every } p \in X \text{ and every } n \in \mathbf{N}.$$

A topological space $(X, \mathcal{T})$ is a *Nagata space* [12] iff there is a neighborhood assignment, $\{U_n(p) | n \in \mathbf{N}, p \in X\}$, in $(X, \mathcal{T})$ such that

$$\text{for every } p, \text{ if } U_n(p) \cap U_n(x_n) \neq \varnothing \text{ for every } n, \text{ then } x_n \to p.$$

In this case we say the $\{U_n(p) | n \in \mathbf{N}, p \in X\}$ is a *Nagata* neighborhood assignment in $(X, \mathcal{T})$; note that $\{U_n(p) | n \in \mathbf{N}\}$ is, in fact, a local base for p.

A topological space $(X, \mathcal{T})$ is a γ-*space* [12] iff there is a neighborhood assignment, $\{U_n(p) | n \in \mathbf{N}, p \in X\}$, in $(X, \mathcal{T})$ such that

$$\text{for every } p, \text{ if } x_n \in U_n(y_n) \text{ and } y_n \in U_n(p) \text{ for every } n, \text{ then } x_n \to p.$$

In this case we say the $\{U_n(p) | n \in \mathbf{N}, p \in X\}$ is a γ-neighborhood assignment in $(X, \mathcal{T})$; again in this case $\{U_n(p) | n \in \mathbf{N}\}$ is a local base for p.

THEOREM 4.1: A T_0-space $(X, \mathcal{T})$ is a Nagata space iff it is coherently asemimetrizable.

Proof: If d is a coherent asemimetric for $(X, \mathcal{T})$, then $(X, \mathcal{T})$ is a T_1-space and, if $U_n(p) = \text{int}_{\mathcal{T}}\{S_d(p, 1/2^n)\}$, then $\{U_n(p) | n \in \mathbf{N}, p \in X\}$ is a Nagata neighborhood assignment for $(X, \mathcal{T})$. Conversely, if $\{U_n(p) | n \in \mathbf{N}, p \in X\}$ is a Nagata neighborhood assignment for $(X, \mathcal{T})$, then we use Hildebrandt's construction [11] to obtain the distance d for X such that

$$d(p, q) = 1/2^n, \text{ where } n = \min\{k | q \notin U_k(p)\}.$$

Then $U_n(p) = S_d(p, 1/2^n)$ and d is an asemimetric for $(X, \mathcal{T})$ such that $d(p, x_n) \to 0 \Leftrightarrow x_n \to p$ (in $\mathcal{T}$). That d is coherent is immediate from the Nagata property of the neighborhood assignment. ❑

Similarly, we obtain

THEOREM 4.2: A T_1-space is a γ-space iff it is γ-asymmetrizable.

From Theorem 3.6 we get an easy proof of Hodel's theorem [12].

COROLLARY 4.3: $(X, \mathcal{T})$ is pseudometrizable iff it is a Nagata γ-space.

Note that coherently asymmetrizable spaces are symmetrizable (Lemma 2.1(d)) and being a $w\Delta$-space is sufficient for developability in regular Hausdorff symmetrizable space [7, Theorem 9.13]. Since developable Nagata spaces are metrizable [12, Theorem 6.1], we have the following result.

COROLLARY 4.4: A regular Hausdorff coherently asymmetrizable space $(X, \mathcal{T})$ is metrizable iff it is a $w\Delta$-space.

REFERENCES

1. ARHANGEL'SKIĬ, A.V. 1966. Mappings and spaces. Russian Math. Surveys. **21:** 115–162.
2. CHITTENDEN, E.W. 1915. On the existence of continuous functions. Bull. Amer. Math. Soc. **21:** 440.
3. COLLINS, P.J. & A.W. ROSCOE. 1984. Criteria for metrisability. Proc. Amer. Math. Soc. **90:** 631–640.
4. FLAGG, R.C. & R.D. KOPPERMAN. 1994. The asymmetric topology of computer science. *In:* Mathematical Foundations of Programming Semantics. Lecture Notes in Computer Science,Vol. 802. Springer-Verlag. New York. 544–553.
5. FRÉCHET, M. 1918. Relations entre les notions de limite et de distance. Trans. Amer. Math. Soc. **19:** 53–65.
6. GIERZ, G., K.H. HOFFMANN, K. KEIMEL, J.D. LAWSON, M. MISLOVE, & D.S. SCOTT. 1980. A Compendium of Continuous Lattices. Springer. Berlin.
7. GRUENHAGE, G. 1984. Generalized metric spaces. *In:* Handbook of Set–Theoretic Topology (K. Kunen & J.E. Vaughan, Eds.). North–Holland. Amsterdam. 423–501.
8. HAHN, H. 1908. Bemerkungen zu den Untersuchungen des Herrn M. Fréchet: Sur quelques points du calcul fonctionnel. Monatshefte für Mathematik und Physik. **19:** 247–257.
9. HARLEY, III, P.W. & G.D. FAULKNER. 1975. Metrization of symmetric spaces. Canadian J. Math. **27:** 986–990.

10. HEATH, R.W. 1962. Arc-wise connectedness in semi-metric spaces. Pacific J. Math. **12:** 1301–1319.
11. HILDEBRANDT, T.H. 1912. A contribution to the foundations of Fréchet's calcul fonctionnel. Amer. J. Math. **34:** 237–290.
12. HODEL, R.E. 1972. Spaces defined by sequences of open covers which guarantee that certain sequences have cluster points. Duke Math. J. **39:** 253–263.
13. HUNG, H.H. 1977. A contribution to the theory of metrization. Canadian J. Math. **29:** 1145–1151.
14. KENTON, S.A. 1971. Neometrizable Spaces. Dissertation. University of New Hampshire.
15. KOFNER, J. 1980. On quasi-metrizability. Topology Proceedings. **5:** 111–138.
16. KOPPERMAN, R.D. 1995. Asymmetry and duality in topology. Topology and Its Applications. **66:** 1–39..
17. MARTIN, H.W. 1972. Metrization of symmetric spaces and regular maps. Proc. Amer. Math. Soc. **35:** 269–274.
18. NEDEV, S.Ĭ. 1971. *o*–metrizable spaces. Trans. Moscow Math. Soc. **24:** 213–247.
19. NIEMYTZKI, V.W. 1927. On the "third axiom of metric space." Trans. Amer. Math. Soc. **29:** 507–513.
20. NIEMYTZKI, V.W. 1931. Über die Axiome des metrischen Raumes. Math. Ann. **104:** 666–671.
21. PITCHER, A.D. & E.W. CHITTENDEN. 1918. On the foundations of the calcul fonctionnel of Fréchet. Trans. Amer. Math. Soc. **19:** 66–78.
22. REILLY, I.L. 1992. On non–Hausdorff spaces. Topology and Its Applications. **44:** 331–340.
23. SHORE, S.D. 1981. Coherent distance functions. Topology Proceedings. **6:** 405–422.
24. SMYTH, M.B. 1991. Totally bounded spaces and compact ordered spaces as domains of computation. *In:* Topology and Category Theory in Computer Science (G.M. Reed, A.W. Roscoe, & R.F. Wachter, Eds.). Clarendon Press. Oxford. 207–229.
25. WATERMAN, M.S., T.F. SMITH, & W.A. BEYER. 1976. Some biological sequence metrics. Advances in Math. **20:** 367–387.
26. WILSON, W.A. 1931. On semi–metric spaces. Amer. J. Math. **53:** 361–373.
27. WILSON, W.A. 1931. On quasi-metric spaces, Amer. J. Math. **53:** 675–684.

Net Spaces in Categorical Topology[a]

JOSEF ŠLAPAL[b]

Department of Mathematics
Technical University of Brno
Technická 2
616 69 Brno, Czech Republic

ABSTRACT: In the paper we introduce the notion of S-nets, S a nonempty construct, as a generalization of the usual nets. S-net spaces and continuous maps between them are then defined in a natural way. For S-net spaces we introduce a number of axioms and study the categories obtained. We find some topological and Cartesian closed categories of S-net spaces and describe their relations to certain known topological categories.

0. INTRODUCTION

Cartesian closed topological categories belong to the most important objects studied in categorical topology. The usefulness of these categories for applications to many classical branches of mathematics is well known. And the interest in Cartesian closed categories became even stronger after discovering the important role they play in computer science. When looking for new Cartesian closed topological categories, categorical topologists usually start from the category of topological spaces and generalize or modify the considered axioms of topology. These axioms depend on the chosen access to topology (closure operations, neighborhood structures, systems of open sets, etc.). In this connection the most often used access is the one which views topologies as convergence structures and uses filters for describing the convergence. This filter access gave rise to a rich hierarchy of useful categories and an intensive study is being performed in the area at present.

In this paper, we also view topologies as convergence structures, but we use nets instead of filters for expressing the convergence. This access is closely related to the one using filters, and some connections between certain types of categories of convergence filter spaces and those of convergence net spaces are described in [14]. Of course, from the classical point of

Mathematics Subject Classification: 54A05, 54A20, 54B30, 18B30, 18D15.
Key words and phrases: S-net, subnet, S-net space, convergence S-net space, limit S-net space, topological category, Cartesian closed category, topological universe, diagonality axiom, condition of iterated limits, preclosure space, closure space, pseudo-topological space, topological space, R_0-space, T_0-space, T_1-space.
[a]This research was supported by Grant Agency of the Czech Republic grant no. 201/95/0468.
[b]E-mail: slapal@kinf.fme.vutbr.cz.

view, filters are more suitable for investigations in topology than nets which have, essentially, an order-theoretic nature. However, in our considerations we do not deal with the usual nets, but with a certain type of generalized nets which have lost their order-theoretic nature in the benefit of a category-theoretic one. Our use of the generalized nets, called $\mathcal{S}$-nets (where $\mathcal{S}$ is a construct), hardly could be duplicated by using filters.

For $\mathcal{S}$-net spaces introduced, we consider a number of axioms including, at the first place, the axioms that are extensions of the usual net axioms. This way we obtain a number of full subcategories of the category of $\mathcal{S}$-net spaces (with extended continuous maps as morphisms) which are then studied. We especially deal with the question of Cartesian closedness of these subcategories, and with their relations to some known topological categories (subcategories and supercategories of *Top*). We discover a number of Cartesian closed topological categories of $\mathcal{S}$-net spaces some of which are trivial, some are isomorphic to certain well-known categories, but many are new and useful for applications to various branches of mathematics. We also obtain some known results [5], [7] as trivial consequences of certain more general facts.

1. BASIC DEFINITIONS

Throughout the paper, $\mathcal{S}$ will denote a nonempty construct (i.e., concrete category over *Set*) whose objects have nonempty underlying sets. As usual, we will not distinguish notationally between $\mathcal{S}$-objects (respectively, $\mathcal{S}$-morphisms) and their underlying sets (respectively, maps). If X, Y are sets and $y \in Y$ is an element, then the constant map $f: X \to Y$ with $f(x) = y$ for each $x \in X$ will be called the *y-constant map*.

DEFINITION 1.1:

(1) Let X be a set. By an *$\mathcal{S}$-net* in X we understand any pair (S, f) where $S \in \mathcal{S}$ is an object and $f: S \to X$ is a map. The class of all $\mathcal{S}$-nets in X will be denoted by $\langle X \rangle_{\mathcal{S}}$.

(2) If $(S, f), (T, g) \in \langle X \rangle_{\mathcal{S}}$, then (S, f) is called a subnet of (T, g) if there is an $\mathcal{S}$-morphism $\varphi: S \to T$ such that $f = g \circ \varphi$.

EXAMPLE 1.2:

(1) Let *Dir* be the category of directed sets whose morphisms are the cofinal maps, i.e., maps $\varphi: (S_1, \leq_1) \to (S_2, \leq_2)$ having the property that for any $s_2 \in S_2$ there is $s_1 \in S_1$ such that $s_2 \leq_2 \varphi(s)$ whenever $s \in S_1$, $s_1 \leq_1$ s. If $\mathcal{S} = Dir$, then the notions of an $\mathcal{S}$-net and a subnet (of an $\mathcal{S}$-net) coincide with the usual notions of a net and a subnet [7], respectively.

(2) Let $\alpha > 0$ be an ordinal and let $\underline{\alpha}$ be the category whose only object is

α and whose morphisms are isotone injections of α into itself. Clearly, $\underline{\alpha}$ is a subcategory of *Dir*. For $S = \underline{\alpha}$ the notions of an S-net and a subnet coincide with the notions of a sequence of type α and a subsequence, respectively. Sequences of type ω are usually called sequences.

(3) Let Set^+ be the category of nonempty sets. If $S = Set^+$ and (S, f), $(T, g) \in \langle X \rangle_S$, then, clearly, (S, f) is a subnet of (T, g) iff $f(S) \subseteq g(T)$.

Denotation 1.3: If X is a set and $\pi: \langle X \rangle_S \to \exp X$ a map, then we will usually write $(S, f) \xrightarrow{\pi} x$ instead of $x \in \pi(S, f)$, and $(S, f) \not\xrightarrow{\pi} x$ instead of $x \in X - \pi(S, f)$. The set $\pi(S, f)$ can be viewed as the set of all limits of the net (S, f) with respect to π and therefore, we will sometimes say that (S, f) converges to x (with respect to π) if $(S, f) \xrightarrow{\pi} x$.

DEFINITION 1.4:

(1) A pair (X, π) where X is a set and $\pi: \langle X \rangle_S \to \exp X$ a map is called an *S-net space* provided that $(S, f) \xrightarrow{\pi} x$ whenever f is the x-constant map.

(2) An S-net space (X, π) is called a *convergence S-net space* provided that from $(S, f) \xrightarrow{\pi} x$ it follows that $(T, g) \xrightarrow{\pi} x$ for each subnet (T, g) of (S, f).

(3) A convergence S-net space (X, π) is called a *limit S-net space* provided that, whenever $(S, f) \not\xrightarrow{\pi} x$, there is a subnet (T, g) of (S, f) such that $(U, h) \not\xrightarrow{\pi} x$ for any subnet (U, h) of (T, g).

EXAMPLE 1.5:

(1) Let $S = Dir$ or $S = \underline{\alpha}$, where $\alpha > 0$ is an ordinal. Let $(X, \mathcal{O})$ be a topological space (given by the system $\mathcal{O}$ of open sets). Let (X, π) be the S-net space associated with $(X, \mathcal{O})$, i.e., for any $(S, f) \in \langle X \rangle_S$, $(S, f) \xrightarrow{\pi} x$ iff for each $A \in \mathcal{O}$ with $x \in A$ there exists $s_A \in S$ such that $f(s) \in A$ for every $s \in S$, $s \geq s_A$. Then, obviously, (X, π) is a limit S-net space.

(2) The $\mathcal{B}$-convergence structures studied in [15] for special subcategories $\mathcal{B}$ of *Dir* are nothing else than the maps $\pi: \langle X \rangle_{\mathcal{B}} \to \exp X$ for which (X, π) are convergence $\mathcal{B}$-net spaces.

(3) Let $S = \underline{\omega}$. Then for convergence S-net spaces (X, π) the maps π are precisely the multivalued convergences on X from [10]. The convergence or limit S-net spaces (X, π) fulfilling card $\pi(S, f) \leq 1$ for each $(S, f) \in \langle X \rangle_S$ are known as the (Fréchet) $\mathcal{L}$-spaces or $\mathcal{L}^*$-spaces, respectively [4].

DEFINITION 1.6: Let (X, π), (Y, ρ) be a pair of S-net spaces. A map $F: X \to Y$ is called a *continuous map* of (X, π) into (Y, ρ) provided that $(S, F \circ f) \xrightarrow{\rho} F(x)$ whenever $(S, f) \xrightarrow{\pi} x$. We denote by Net_S the category of S-net spaces and continuous maps, and by $Conv_S$ and Lim_S its full subcategories of convergence S-net spaces and limit S-net spaces, respectively. The set of all continuous maps of (X, π) into (Y, ρ) will be denoted by $C((X, \pi), (Y, \rho))$.

EXAMPLE 1.7:

(1) Let $(X, \pi) \in Net_S$ be an object and let π': $\langle X\rangle_{Set^+} \to \exp X$ be the map given by $(S, f) \xrightarrow{\pi'} x$ iff f is the x-constant map or there exists $(S, f) \in \langle X\rangle_S$ such that $(S, f) \xrightarrow{\pi} x$. Then the assignment $(X, \pi) \mapsto (X, \pi')$ determines a concrete embedding of Net_S into Net_{Set^+}.

(2) For $S = Dir$ the category Lim_S is studied in [5] where its objects are called L*-spaces.

2. PROPERTIES OF Net_S, $Conv_S$, and Lim_S

The category Net_S is not fibre-small in general (it is fibre-small whenever S is small). Therefore, we do not suppose fibre-smallness when working with initially structured or topological categories. More precisely, by an *initially structured category* we understand a construct which has initial structures for all mono-sources and in which each constant map between its objects is a morphism. An initially structured category which is initially complete (i.e., has initial structures for all sources) is said to be a *topological category*. Also the notion of *topological universe* here is meant in the generalized sense obtained by omitting the requirement of fibre-smallness [1]. Thus, a topological universe is a Cartesian closed topological category $\mathcal{K}$ with the property that every $\mathcal{K}$-object C has a one-point extension, i.e., a $\mathcal{K}$-object $\overline{C} = C \cup \{\infty_C\}$ where $\infty_C \notin C$ such that C is a subobject of $\overline{C}$ and for any $\mathcal{K}$-object A, any subobject B of A and any $\mathcal{K}$-morphism φ: $B \to C$ the map $\bar{\varphi}: A \to \overline{C}$, given by $\bar{\varphi}(a) = \varphi(a)$ whenever $a \in B$ and $\bar{\varphi}(a) = \infty_C$ whenever $a \notin B$, is a $\mathcal{K}$-morphism. Let us remark here that by a subobject of an object we always understand an initial subobject with regard to an inclusion map.

A topological universe is said to be *strong* provided that its quotient maps are productive [12]. Throughout the paper, products in S are meant to be products over nonempty index sets.

THEOREM 2.1: Net_S is a strong topological universe.

Proof: Clearly, each constant map between objects of Net_S is continuous. Let $F_i: X \to (X_i, \pi_i)$, $i \in I$, be a source in Net_S and let π: $\langle X\rangle_S \to \exp X$ be given as follows: $(S, f) \xrightarrow{\pi} x$ iff $(S, F_i \circ f) \xrightarrow{\pi_i} F_i(x)$ for each $i \in I$. Clearly, $(X, \pi) \in Net_S$. Let $(Y, \rho) \in Net_S$ be an object, and $G: Y \to X$ a map such that $F_i \circ G$: $(Y, \rho) \to (X_i, \pi_i)$ is continuous for each $i \in I$. Let $(T, g) \in \langle Y\rangle_S$, $(T, g) \xrightarrow{\rho} y$. Then $(T, F_i \circ G \circ g) \xrightarrow{\pi_i} F_i \circ G(y)$ for each $i \in I$. Hence, $(T, G \circ g) \xrightarrow{\pi} G(y)$, i.e., G: $(Y, \rho) \to (X, \pi)$ is a continuous map. Therefore, π is the initial structure of the source F_i, $i \in I$, and it is unique because Net_S is transportable. Consequently, Net_S is initially complete. We have shown that Net_S is a topological category.

Let $(X, \pi), (Y, \rho) \in Net_S$ be objects. Let $Z = C((X, \pi), (Y, \rho))$ and let σ: $\langle Z\rangle_S \rightarrow \exp Z$ be given by $z \in \sigma(S, g)$ iff from $(S, f) \xrightarrow{\pi} x$ it follows that $(S, g^f) \xrightarrow{\rho} z(x)$ where g^f: $S \rightarrow Y$ is defined by $g^f(s) = g(s)(f(s))$. Then $(Z, \sigma) \in \mathrm{Net}_S$ and we show that (Z, σ) is a function space of (X, π) and (Y, ρ) in Net_S. To this end, let e: $(X, \pi) \times (Z, \sigma) \rightarrow (Y, \rho)$ be the evaluation map, i.e., the map with $e(x, z) = z(x)$ whenever $(x, z) \in X \times Z$ ($\times$ denotes the Cartesian products in both Set and Net_S; in Net_S their existence follows from the initial completeness of Net_S). Put $(V, \tau) = (X, \pi) \times (Z, \sigma)$ and let $(S, f) \in \langle V\rangle_S$, $(S, f) \xrightarrow{\tau} (x, z)$. Then $(S, pr_X \circ f) \xrightarrow{\pi} x$ and $(S, pr_Z \circ f) \xrightarrow{\sigma} z$. Hence, $(S, h^p) \xrightarrow{\rho} z(x)$ where $h = pr_Z \circ f$ and $p = pr_X \circ f$. We have $h^p(s) = h(s)(p(s)) = e(p(s), h(s)) = e(pr_X(f(s)), pr_Z(f(s))) = e(f(s))$ for every $s \in S$. Therefore, $h^p = e \circ f$ which yields $(S, e \circ f) \xrightarrow{\rho} e(x, z)$. Consequently, e: $(X, \pi) \times (Z, \sigma) \rightarrow (Y, \rho)$ is continuous. Now, let F: $(X, \pi) \times (W, \upsilon) \rightarrow (Y, \rho)$ be a continuous map and let F^*: $W \rightarrow Z$ be the map given by $F^*(w)(x) = F(x, w)$. Let $(S, g) \in \langle W\rangle_S$, $(S, g) \xrightarrow{\upsilon} w$, and $(S, f) \in \langle X\rangle_S$, $(S, f) \xrightarrow{\pi} x$. Then $(S, (f, g))$ converges to (x, w) in the space $(X, \pi) \times (W, \upsilon)$. Hence, $(S, F \circ (f, g)) \xrightarrow{\rho} F(x, w)$. This results in $(S, F^* \circ g) \xrightarrow{\sigma} F^*(w)$ because $F((f, g)(s)) = F^*(g(s))(f(s)) = (F^* \circ g)^f(s)$ whenever $s \in S$. Consequently, F^*: $(W, \upsilon) \rightarrow (Z, \sigma)$ is a continuous map. Since clearly $e \circ (id_X \times F^*) = F$, we have shown that (Z, σ) is a function space of (X, π) and (Y, ρ) in Net_S. Therefore, Net_S is Cartesian closed.

Let $(X, \pi) \in Net_S$ be an object and ∞_X a point with $\infty_X \notin X$. Let $(X \cup \{\infty_X\}, \pi') \in Net_S$ be defined as follows: if $(S, f) \in \langle X \cup \{\infty_X\}\rangle_S$, then $\pi'(S, f) = \pi(S, f) \cup \{\infty_X\}$ whenever $\infty_X \notin f(S)$ and $\pi'(S, f) = X \cup \{\infty_X\}$ whenever $\infty_X \in f(S)$. Clearly, (X, π) is a subspace (i.e., a subobject) of $(X \cup \{\infty_X\}, \pi')$. Let $(Y, \rho) \in Net_S$, let (Z, σ) be a subspace of (Y, ρ) and let F: $(Z, \sigma) \rightarrow (X, \pi)$ be a continuous map. Let $\overline{F}$: $Y \rightarrow X \cup \{\infty_X\}$ be defined by $\overline{F}(y) = F(y)$ whenever $y \in Z$ and $\overline{F}(y) = \infty_X$ whenever $y \in Y - Z$. It easily can be seen that $\overline{F}$: $(Y, \rho) \rightarrow (X \cup \{\infty_X\}, \pi')$ is a continuous map. Consequently, Net_S is a topological universe. We will show that this universe is strong.

Let $(X, \pi) \in Net_S$ be an object, Y a set and $F: X \rightarrow Y$ a surjection. Evidently, an object $(Y, \rho) \in Net_S$ is the quotient object of (X, π) with regard to F if and only if the following condition is fulfilled: $(S, g) \xrightarrow{\rho} y$ iff there is $(S, f) \in \langle X\rangle_S$, $(S, f) \xrightarrow{\pi} x$, such that $F \circ f = g$ and $F(x) = y$. Let $F_i: (X_i, \pi_i) \rightarrow (Y_i, \rho_i)$, $i \in I$, be a family of quotient maps in Net_S. Let $(X, \pi) = \prod_{i \in I} (X_i, \pi_i)$, $(Y, \rho) = \prod_{i \in I} (Y_i, \rho_i)$ be the (Cartesian) products and let p_i: $(X, \pi) \rightarrow (X_i, \pi_i)$, q_i: $(Y, \rho) \rightarrow (Y_i, \rho_i)$, $i \in I$, be the corresponding projections. Finally, let F: $(X, \pi) \rightarrow (Y, \rho)$ be the unique continuous map with $q_i \circ F = F_i \circ p_i$ for all $i \in I$. Obviously, F is a surjection. Let $(S, g) \in \langle Y\rangle_S$, $(S, g) \xrightarrow{\rho} y$. Then $(S, q_i \circ g) \xrightarrow{\rho_i} q_i(y)$ for each $i \in I$. Thus, for each $i \in I$ there exists $(S, f_i) \in \langle X_i\rangle_S$ such that $(S, f_i) \xrightarrow{\pi_i} x_i$, $F_i \circ f_i = q_i \circ g$ and $F_i(x_i) = q_i(y)$. Let f: $S \rightarrow X$ be the map with $p_i \circ f = f_i$ for each $i \in I$, and let $x \in X$ be the element with $p_i(x) = x_i$.

Then $(S, f) \xrightarrow{\pi} x$ and we have $F \circ f = g$ because $q_i \circ F \circ f = F_i \circ p_i \circ f = F_i \circ f_i = q_i \circ g$ for each $i \in I$. Next, we have $F(x) = y$ because $q_i \circ F(x) = F_i \circ p_i(x) = F_i(x_i) = q_i(y)$ for each $i \in I$. Conversely, let there be $(S, f) \in \langle X \rangle_S$, $(S, f) \xrightarrow{\pi} x$, $F \circ f = g$ and $F(x) = y$. Then $(S, p_i \circ f) \xrightarrow{\pi_i} p_i(x)$ and, for each $i \in I$, we have $F_i \circ p_i \circ f = q_i \circ F \circ f = q_i \circ g$ and $F_i(p_i(x)) = q_i(F(x)) = q_i(y)$. Consequently, $(S, q_i \circ g) \xrightarrow{\rho_i} q_i(y)$ for each $i \in I$, i.e., $(S, g) \xrightarrow{\rho} y$. We have shown that (Y, ρ) is a quotient object of (X, π) with regard to F. Hence, quotient maps are productive in Net_S. Thus, Net_S is a strong topological universe. The proof is complete. ❑

$Conv_S$ and Lim_S are obviously initially closed in Net_S and so they are topological. Consequently, $Conv_S$ is a bireflective subcategory of Net_S and Lim_S is a bireflective subcategory of $Conv_S$. The bireflection of an object $(X, \pi) \in Net_S$ in $Conv_S$ is given by the identity map id_X: $(X, \pi) \to (X, \hat{\pi})$ where $\hat{\pi}$: $\langle X \rangle_S \to \exp X$ is defined by $(T, g) \xrightarrow{\hat{\pi}} x$ iff there is $(S, f) \in \langle X \rangle_S$ with $(S, f) \xrightarrow{\pi} x$ such that (T, g) is a subnet of (S, f). The bireflection of an object $(X, \pi) \in Conv_S$ in Lim_S is given by the identity map id_X: $(X, \pi) \to (X, \hat{\pi})$ where $\hat{\pi}$: $\langle X \rangle_S \to \exp X$ is defined by $(T, g) \xrightarrow{\hat{\pi}} x$ iff for each subnet (S, f) of (T, g) there is a subnet (U, h) of (S, f) with $(U, h) \xrightarrow{\pi} x$.

THEOREM 2.2: *$Conv_S$ is a bicoreflective subcategory of Net_S.*

Proof: Let $(X, \pi) \in Net_S$ and let π^*: $\langle X \rangle_S \to \exp X$ be the map given by $(T, g) \xrightarrow{\pi^*} x$ iff, whenever (S, f) is a subnet of (T, g), $(S, f) \xrightarrow{\pi} x$. Clearly, $(X, \pi^*) \in Conv_S$ and id_X: $(X, \pi^*) \to (X, \pi)$ is a bimorphism. Let $(Y, \rho) \in Conv_S$ be an object and F: $(Y, \rho) \to (X, \pi)$ a continuous map. Let $(T, g) \in \langle Y \rangle_S$, $(T, g) \xrightarrow{\rho} y$. Then each subnet (U, h) of (T, g) has the property that $(U, h) \xrightarrow{\rho} y$. Let (S, f) be a subnet of $(T, F \circ g)$. Then there is an S-morphism φ: $S \to T$ with $f = F \circ g \circ \varphi$. Since $(S, g \circ \varphi)$ is a subnet of (T, g), we have $(S, g \circ \varphi) \xrightarrow{\rho} y$. Consequently, $(S, f) = (S, F \circ g \circ \varphi) \xrightarrow{\pi} F(y)$. Hence, $(T, F \circ g) \xrightarrow{\pi^*} F(y)$, i.e., F: $(Y, \rho) \to (X, \pi^*)$ is continuous. Hence, id_X: $(X, \pi^*) \to (X, \pi)$ is a bicoreflection of (X, π) in $Conv_S$ and the proof is complete. ❑

Because $Conv_S$ is a full isomorphism closed subcategory of Net_S which is closed under formation of products in Net_S, Theorems 2.1 and 2.2 immediately result in

COROLLARY 2.3: *$Conv_S$ is a Cartesian closed category.*

THEOREM 2.4: *If S has finite Cartesian products, then Lim_S is Cartesian closed.*

Proof: Let (X, π), $(Y, \rho) \in Lim_S$. Let $Z = C((X, \pi), (Y, \rho))$ and let σ: $\langle Z \rangle_S \to \exp Z$ be the map defined by $(T, g) \xrightarrow{\sigma} z$ iff from $(S, f) \xrightarrow{\pi} x$ it follows that $(S \times T, g^f) \xrightarrow{\rho} z(x)$ where g^f: $S \times T \to Y$ is the map given by $g^f(s, t) = g(t)(f(s))$. Let $z \in Z$ and let $(T, g) \in \langle Z \rangle_S$ where g is the z-constant map. Let $(S, f) \in \langle X \rangle_S$, $(S, f) \xrightarrow{\pi} x$. Then $g^f(s, t) = g(t)(f(s)) = z(f(s))$. Hence, $g^f = z \circ f \circ pr_S$.

As z is continuous, $(S, z \circ f) \xrightarrow{\pi} z(x)$. Therefore, $(S \times T, g^f) \xrightarrow{\rho} z(x)$, i.e., $(T, g) \xrightarrow{\sigma} z$.

Let $(T, g) \in \langle Z \rangle_S$, $(T, g) \xrightarrow{\sigma} z$, and let $(S, f) \xrightarrow{\pi} x$. Let (U, h) be a subnet of (T, g). Then there is an S-morphism $\varphi: U \to T$ such that $h = g \circ \varphi$. We have $(S \times T, g^f) \xrightarrow{\rho} z(x)$. As the map $\psi: S \times U \to S \times T$ given by $\psi(s, u) = (s, \varphi(u))$ is obviously an S-morphism, $(S \times U, g^f \circ \psi)$ is a subnet of $(S \times T, g^f)$ and thus $(S \times U, g^f \circ \psi) \xrightarrow{\sigma} z(x)$. But $g^f(\psi(s, u)) = g^f(s, \varphi(u)) = g(\varphi(u))(f(s)) = h(u)(f(s)) = h^f(s, u)$ for any $(s, u) \in S \times U$ and consequently $(U, h) \xrightarrow{\sigma} z$.

Let $(T, g) \in \langle Z \rangle_S$, $(T, g) \not\xrightarrow{\sigma} z$. Then there exists $(S, f) \in \langle X \rangle_S$ such that $(S, f) \xrightarrow{\pi} x$ but $(S \times T, g^f) \not\xrightarrow{\rho} z(x)$. Thus, there exists a subnet (U, h) of $(S \times T, g^f)$ such that $(V, p) \not\xrightarrow{\rho} z(x)$ for any subnet (V, p) of (U, h). Let $\varphi: U \to S \times T$ be the S-morphism with $h = g^f \circ \varphi$. As $pr_T \circ \varphi: U \to T$ is an S-morphism, $(U, g \circ pr_T \circ \varphi)$ is a subnet of (T, g). Let (W, r) be a subnet of $(U, g \circ pr_T \circ \varphi)$. Then there is an S-morphism $\psi: W \to U$ such that $r = g \circ pr_T \circ \varphi \circ \psi$. Let $\chi: W \to S \times W$ be the map given by $\chi(w) = (pr_S(\varphi(\psi(w))), w)$. Then χ is an S-morphism and we have

$$g^f(\varphi(\psi(w))) = g^f(pr_S(\varphi(\psi(w))), pr_T(\varphi(\psi(w))))$$

$$= g(pr_T(\varphi(\psi(w))))(f(pr_S(\varphi(\psi(w)))))$$

$$= r(w)(f(pr_S(\chi(w)))) = r^f(pr_S(\chi(w)), w) = r^f(\chi(w))$$

for every $w \in W$. Hence, $(W, g^f \circ \varphi \circ \psi)$ is a subnet of $(S \times W, r^f)$. Since $(W, g^f \circ \varphi \circ \psi)$ is a subnet of (U, h), we have $(W, g^f \circ \varphi \circ \psi) \not\xrightarrow{\rho} z(x)$ and consequently $(S \times W, r^f) \not\xrightarrow{\rho} z(x)$. Hence, $(W, r) \not\xrightarrow{\sigma} z$ which means that no subnet of $(U, g \circ pr_T \circ \varphi)$ converges to z. We have shown that $(Z, \sigma) \in Lim_S$.

Let $e: (X, \pi) \times (Z, \sigma) \to (Y, \rho)$ be the evaluation map. Denote $(V, \tau) = (X, \pi) \times (Z, \sigma)$ and let $(S, f) \in \langle X \rangle_S$, $(S, f) \xrightarrow{\tau} (x, z)$. Then $(S, pr_X \circ f) \xrightarrow{\pi} x$ and $(S, pr_Z \circ f) \xrightarrow{\sigma} z$. Hence, $(S \times S, h^g) \xrightarrow{\rho} z(x)$ where $g = pr_X \circ f$ and $h = pr_Z \circ f$. Let $\varphi: S \to S \times S$ be the map given by $\varphi(s) = (s, s)$. Then φ is an S-morphism and consequently $(S, h^g \circ \varphi)$ is a subnet of $(S \times S, h^g)$. Thus $(S, h^g \circ \varphi) \xrightarrow{\rho} z(x)$. For any $s \in S$ we have $h^g(\varphi(s)) = h^g(s, s) = pr_Z(f(s))(pr_X(f(s))) = e(pr_X(f(s)), pr_Z(f(s))) = e(f(s))$. Therefore, $(S, e \circ f) \xrightarrow{\rho} e(x, z)$ and we have shown that e is continuous.

Let $(W, \upsilon) \in Lim_S$ and let $F: (X, \pi) \times (W, \upsilon) \to (Y, \rho)$ be a continuous map. Let $F^*: W \to Z$ be the map given by $F^*(w)(x) = F(x, w)$. Let $(S, f) \in \langle S \rangle_S$, $(S, f) \xrightarrow{\pi} x$, and $(T, g) \in \langle W \rangle_S$, (T, g) by $(T, g) \xrightarrow{\upsilon} w$. Then the S-net $(S \times T, (f \circ pr_S, g \circ pr_T)) = (S \times T, f \times g)$ converges to (x, w) in $(X, \pi) \times (W, \upsilon)$ and thus $(S \times T, F \circ (f \times g)) \xrightarrow{\rho} F(x, w)$. For any $(s, t) \in S \times T$ we have $F((f \times g)(s, t)) = F(f(s), g(t)) = F^*(g(t))(f(s)) = (F^* \circ g)^f(s, t)$. Hence, $(S \times T, (F^* \circ g)^f) \xrightarrow{\rho} F^*(w)(x)$ which results in $(S, F^* \circ g) \xrightarrow{\sigma} F^*(w)$. Thus, $F^*: (W, \upsilon) \to (Z, \sigma)$ is a continuous map. As the equality $e \circ (id_X \times F^*) = F$ is clearly valid, we have shown that (Z, σ) is the function space of (X, π) and (Y, σ) in Lim_S. The proof is complete. ❑

EXAMPLE 2.5: The Cartesian closedness of Lim_S for $S = Dir$, which is an immediate consequence of Theorem 2.4, is also proved in [5].

3. DIAGONAL S-NET SPACES

In the previous paragraph it has been shown that $Conv_S$ is a Cartesian closed topological subcategory of the topological universe Net_S but the function spaces in $Conv_S$ are not inherited from Net_S. The same is also valid for Lim_S provided that S has finite Cartesian products. In this paragraph, we find such Cartesian closed topological subcategories of $Conv_S$ and Lim_S whose function spaces are inherited from Net_S. These subcategories are obtained as intersections of $Conv_S$ and Lim_S with certain full subcategory of Net_S. This subcategory itself will be topological and Cartesian closed with function spaces inherited from Net_S and, moreover, its function spaces will have the pointwise structure.

DEFINITION 3.1: An S-net space (X, π) is called *diagonal* provided that it satisfies the following condition (called the *diagonality axiom*):

Let $\{(S, f_i); i \in S\}$ and $\{(S, g_j); j \in S\}$ be families of S-nets in X such that $f_i(j) = g_j(i)$ whenever $i, j \in S$ and let $(S, f_i) \xrightarrow{\pi} x_i$ whenever $i \in S$ and $(S, g_j) \xrightarrow{\pi} y_j$ whenever $j \in S$. Let $f, g, h: S \to X$ be the maps given by $f(i) = x_i$, $g(j) = y_j$ and $h(i) = f_i(i)$. Then from $(S, f) \xrightarrow{\pi} x$ and $(S, g) \xrightarrow{\pi} x$ it follows that $(S, h) \xrightarrow{\pi} x$.

We denote by $DNet_S$ the full subcategory of Net_S whose objects are precisely the diagonal S-net spaces.

EXAMPLE 3.2:

(1) If $S = \underline{\omega}$, then the diagonality of S-net spaces coincides with the diagonality defined in [13].

(2) Let $S = \underline{1}$. Then $DNet_S$ is isomorphic with the category of quasi-discrete topological spaces [3] (i.e., topological spaces with completely additive closures).

LEMMA 3.3: Let $(X, \pi) \in Net_S$ and $(Y, \rho) \in DNet_S$ be objects, let $Z = C((X, \pi), (Y, \rho))$, and let (Z, σ) be the function space of (X, π) and (Y, ρ) in Net_S. Then the space (Z, σ) has pointwise structure, i.e., whenever $(S, g) \in \langle Z \rangle_S$, $(S, g) \xrightarrow{\sigma} z$ iff $(S, g^x) \xrightarrow{\rho} z(x)$ for each $x \in X$ where $g^x: S \to Y$ is the map given by $g^x(i) = g(i)(x)$.

Proof: Recall that $(S, g) \xrightarrow{\sigma} z$ iff $(S, f) \xrightarrow{\pi} x$ implies $(S, g^f) \to z(x)$ where $g^f: S \to Y$ is the map given by $g^f(i) = g(i)(f(i))$ — see the proof of Theorem 2.1. Let $\tau: \langle Z \rangle_S \to \exp Z$ be the map given by $(S, g) \xrightarrow{\tau} z$ iff $(S, g^x) \xrightarrow{\rho} z(x)$ for each $x \in X$. Let $(S, g) \in \langle Z \rangle_S$, $(S, g) \xrightarrow{\tau} z$, and let $(S, f) \in \langle X \rangle_S$, $(S, f) \xrightarrow{\pi} x$. Then for each $i \in S$ we have $(S, g(i) \circ f) \xrightarrow{\rho} g(i)(x)$. Put $u_i = g(i) \circ f$ for each

$i \in S$ and $v_j = g^{f(j)}$ for each $j \in S$. Then $u_i(j) = g(i)(f(j)) = g^{f(j)}(i) = v_j(i)$ whenever $i, j \in S$. Obviously, $(S, u_i) \xrightarrow{\rho} g(i)(x)$ for each $i \in S$ and $(S, v_j) \xrightarrow{\rho} z(f(j))$ for each $j \in S$. Further, we have $(S, g^x) \xrightarrow{\rho} z(x)$ and $(S, z \circ f) \xrightarrow{\rho} z(x)$. As $g^f(i) = g(i)(f(i)) = u_i(i)$ for each $i \in S$, we have $(S, g^f) \xrightarrow{\rho} z(x)$, and consequently $(S, g) \xrightarrow{\rho} z$. Thus, $(S, g) \xrightarrow{\tau} z$ implies $(S, g) \xrightarrow{\sigma} z$. The inverse implication is obvious: For any $x \in X$ we have $(S, f) \xrightarrow{\pi} x$ whenever $f: S \to X$ is the x-constant map, thus from $(S, g) \xrightarrow{\sigma} z$ it follows that $(S, g^f) \xrightarrow{\rho} z(x)$ where $g^f(i) = g(i)(f(i)) = g(i)(x) = g^x(i)$. Hence, $(S, g) \xrightarrow{\tau} z$. Therefore, $\sigma = \tau$ which proves the statement. ❑

Let us denote $DConv_S = DNet_S \cap Conv_S$ and $DLim_S = DNet_S \cap Lim_S$. Then we have

THEOREM 3.4: *$DNet_S$, $DConv_S$ and $DLim_S$ are Cartesian closed topological categories. Their function spaces are inherited from Net_S and have pointwise structure.*

Proof: Let $F_k: X \to (X_k, \pi_k)$, $k \in K$, be a source in $DNet_S$ and let π be its initial structure in Net_S, i.e., $(S, f) \xrightarrow{\pi} x$ iff $(S, F_k \circ f) \xrightarrow{\pi_k} F_k(x)$ for each $k \in K$ (see the proof of Theorem 2.1). Let $\{(S, f_i); i \in S\}$ and $\{(S, g_j); j \in S\}$ be families of S-nets in X such that $f_i(j) = g_j(i)$ whenever $i, j \in S$ and let $(S, f_i) \xrightarrow{\pi} x_i$ whenever $i \in S$ and $(S, g_j) \xrightarrow{\pi} y_j$ whenever $j \in S$. Further, let $f, g, h: S \to X$ be the maps given by $f(i) = x_i$, $g(j) = y_j$ and $h(i) = f_i(i)$, and suppose that $(S, f) \xrightarrow{\pi} x$, $(S, g) \xrightarrow{\pi} x$. Then, for each $k \in K$, $\{(S, F_k \circ f_i); i \in S\}$ and $\{(S, F_k \circ g_j), j \in S\}$ are families of S-nets in X_k such that: $(F_k \circ f_i)(j) = (F_k \circ g_j)(i)$, whenever $i, j \in S$; $(S, F_k \circ f_i) \xrightarrow{\pi_k} F_k(x_i)$, whenever $i \in S$; and $(S, F_k \circ g_j) \xrightarrow{\pi_k} F_k(y_j)$, whenever $j \in S$. Next, for each $k \in K$, the map $F_k \circ f: S \to X_k$ fulfills $(F_k \circ f)(i) = F_k(x_i)$ for each $i \in S$, the map $F_k \circ g: S \to X_k$ fulfills $(F_k \circ g)(j) = F_k(y_j)$ for each $j \in S$, and the map $F_k \circ h: S \to X_k$ fulfills $(F_k \circ h)(i) = F_k(f_i(i))$ for each $i \in S$. As $(S, F_k \circ f) \xrightarrow{\pi_k} F_k(x)$ and $(S, F_k \circ g) \xrightarrow{\pi_k} F_k(x)$ for each $k \in K$, $(S, F_k \circ h) \xrightarrow{\pi_k} F_k(x)$ for each $k \in K$ too. Consequently, $(S, h) \xrightarrow{\pi} x$. Hence, $(X, \pi) \in DNet_S$ and we have shown that $DNet_S$ is initially closed in Net_S. As $Conv_S$ and Lim_S are initially closed in Net_S, so are also $DConv_S$ and $DLim_S$. Hence $DNet_S$, $DConv_S$ and $DLim_S$ are topological.

Let (X, π), $(Y, \rho) \in DNet_S$ be objects and let (Z, σ) be the function space of (X, π) and (Y, ρ) in Net_S. By Lemma 3.3 we have $(S, p) \xrightarrow{\sigma} z$ iff $(S, p^x) \xrightarrow{\sigma} z(x)$ for each $x \in X$ where $p^x: S \to Y$ is the map defined by $p^x(s) = p(s)(x)$. Let $\{(S, f_i); i \in S\}$ and $\{(S, g_j); j \in S\}$ be families of S-nets in Z such that $f_i(j) = g_j(i)$ whenever $i, j \in S$ and let $(S, f_i) \xrightarrow{\sigma} y_i$ whenever $i \in S$ and $(S, g_j) \xrightarrow{\sigma} z_j$ whenever $j \in S$. Let $f, g, h: S \to Z$ be the maps given by $f(i) = y_i$, $g(j) = z_j$, $h(i) = f_i(i)$, and let $(S, f) \xrightarrow{\sigma} z$, $(S, g) \xrightarrow{\sigma} z$. Then for any $x \in X$, $\{(S, f_i^x); i \in S\}$ and $\{(S, g_j^x); j \in S\}$ are families of S-nets in Y such that $f_i^x(j) = f_i(j)(x) = g_j(i)(x) = g_j^x(i)$ whenever $i, j \in S$, $(S, f_i^x) \xrightarrow{\rho} y_i(x)$ whenever $i \in S$, and $(S, g_j^x) \xrightarrow{\rho} z_j(x)$ whenever $j \in S$. Further, for any $x \in X$ we have $f^x(i) = y_i(x)$ whenever $i \in S$, $g^x(j) = z_j(x)$ whenever $j \in S$, $h^x(i) = f_i^x(i)$ whenever $i \in S$,

$(S, f^x) \xrightarrow{\rho} z(x)$ and $(S, g^x) \xrightarrow{\rho} z(x)$. Therefore, $(S, h^x) \xrightarrow{\rho} z(x)$ for each $x \in X$. This results in $(S, h) \xrightarrow{\sigma} z$, i.e., $(Z, \sigma) \in DNet_S$. Consequently, $DNet_S$ is Cartesian closed, and its function spaces are inherited from Net_S and have pointwise structure.

Let (X, π), $(Y, \rho) \in DConv_S$ be objects and let (Z, σ) be their function space in Net_S. Let $(S, p) \xrightarrow{\sigma} z$ and let $x \in X$. Let (T, q) be a subnet of (S, p). Then there is an S-morphism $\varphi: T \to S$ with $q = p \circ \varphi$, and $(T, p^x \circ \varphi)$ is a subnet of (S, p^x). Therefore, $(T, p^x \circ \varphi) \xrightarrow{\rho} z(x)$. But, for every $t \in T$ we have $p^x(\varphi(t)) = p(\varphi(t))(x) = q(t)(x) = q^x(t)$, i.e., $p^x \circ \varphi = q^x$. Hence $(T, q^x) \xrightarrow{\rho} z(x)$. We have shown that $(T, q) \xrightarrow{\sigma} z$. Consequently, $(Z, \sigma) \in Conv_S$ and regarding the previous part of the proof, $(Z, \sigma) \in DConv_S$. Hence $DConv_S$ is Cartesian closed, and its function spaces are inherited from Net_S and have pointwise structure.

Finally, let (X, π), $(Y, \rho) \in DLim_S$ be objects and let (Z, σ) be their function space in Net_S. Let $(S, f) \in \langle Z \rangle_S$ have the property that each of its subnet has a subnet converging to z (with respect to σ). Let $x \in X$ and let (T, g) be a subnet of (S, f^x). Then there is an S-morphism $\varphi: T \to S$ with $g = f^x \circ \varphi$ and we have $(f^x \circ \varphi)(t) = f(\varphi(t))(x) = (f \circ \varphi)^x(t)$ for each $t \in T$, i.e., $g = (f \circ \varphi)^x$. As $(T, f \circ \varphi)$ is a subnet of (S, f), there is a subnet (U, g) of $(T, f \circ \varphi)$ such that $(U, g) \xrightarrow{\sigma} z$. Hence $(U, g^x) \xrightarrow{\rho} z(x)$. Let $\psi: U \to T$ be the S-morphism with $g = f \circ \varphi \circ \psi$. As $g^x(u) = g(u)(x) = f(\varphi(\psi(u)))(x) = (f \circ \varphi)^x \psi(u) = g(\psi(u))$ for each $u \in U$, (U, g^x) is a subnet of (T, g). Consequently, $(S, f^x) \xrightarrow{\rho} z(x)$ and hence $(S, f) \xrightarrow{\sigma} z$. Thus, with regard to the previous part of the proof, $(Z, \sigma) \in DLim_S$. Therefore, $DLim_S$ is Cartesian closed, and its function spaces are inherited from Net_S and have pointwise structure. The statement is proved. ❑

4. RELATIONS OF *Net*$_S$, *Conv*$_S$ AND *Lim*$_S$ TO SOME SUPERCATEGORIES OF *Top*

We denote by *Pcl* the category of preclosure spaces (i.e., pairs (X, u) where X is a set and u: exp$X \to$ exp$X$ is a map fulfilling $u\varnothing = \varnothing$, $A \subseteq X \Rightarrow A \subseteq uA$ and $A \subseteq B \subseteq X \Rightarrow uA \subseteq uB$) and continuous maps (i.e., maps $F: (X, u) \to (Y, v)$ with $f(uA) \subseteq vf(A)$ whenever $A \subseteq X$). Further, we denote by *Clo* and *Wet* the full subcategories of *Pcl* whose objects are precisely the closure spaces (i.e., the preclosure spaces (X, u) fulfilling $u(A \cup B) = uA \cup uB$ whenever $A, B \subseteq X$) and the weakly topological spaces (i.e., the preclosure spaces (X, u) fulfilling $uuA = uA$ whenever $A \subseteq X$), respectively. Finally, we denote by *Top* the category of topological spaces, i.e., *Top* = *Clo* ∩ *Wet*.

It is well known that for $S = Dir$ the category *Top* can be fully concretely embedded into Lim_S [7]. It is also known that for $S = \underline{\omega}$ the full subcategory of Lim_S given by the limit S-net spaces in which each sequence converges to at most one point can be fully concretely embedded into *Top* [4]. We will find some other subcategories of Net_S and Lim_S that can be, for special categories S, fully concretely embedded into *Pcl*, *Clo*, *Wet* or *Top*.

For any S-net space (X, π) let u_π: exp$X \to$ exp$X$ be the map given by $A \subseteq X \Rightarrow u_\pi A = \{x \in X$; there exists $(S, f) \in \langle X \rangle_S$ such that $f(S) \subseteq A$ and $(S, f) \xrightarrow{\pi} x\}$. The following assertion is evident:

PROPOSITION 4.1: The assignment $\pi \mapsto u_\pi$ determines a concrete functor from Net_S into *Pcl*.

The concrete functor from Proposition 4.1, and also its restrictions, will be denoted by $\mathcal{F}_S$.

EXAMPLE 4.2: For $S = \underline{1}$ the functor $\mathcal{F}_S$ is a full concrete embedding of Net_S into *Pcl*. Namely, in this case, $\mathcal{F}_S$ is an isomorphism of Net_S onto the full subcategory of *Clo* whose objects are precisely the quasi-discrete closure spaces [3], i.e., the closure spaces with completely additive closures.

DEFINITION 4.3: The category S is called:

(1) an *additively hereditary category* if for any S-object S and any pair of sets T, U with $S = T \cup U$ there is a subobject of S whose underlying set is T or U,

(2) a *fork category* if for any pair of S-objects T, U and any pair of elements $t \in T$, $u \in U$ there exists an S-object S such that both the t-constant map of S into T and the u-constant map of S into U are S-morphisms.

EXAMPLE 4.4:

(1) If $S = Dir$, then S is additively hereditary.

(2) If in S all constant maps are S-morphisms, then S is a fork category.

(3) If S is initially complete, then S is both additively hereditary and fork category.

PROPOSITION 4.5: If S is additively hereditary, then $\mathcal{F}_S$ is a concrete functor from $Conv_S$ into *Clo*.

Proof: Let $(X, \pi) \in Net_S$, $A, B \subseteq X$ and $x \in u_\pi(A \cup B)$. Then there is $(S, f) \in \langle X \rangle_S$ with $(S, f) \xrightarrow{\pi} x$ such that $f(S) \subseteq A \cup B$. Put $C = f(S) \cap A$ and $D = f(S) \cap B$. Then $f(S) = C \cup D$, i.e., $S = f^{-1}(C) \cup f^{-1}(D)$. Thus, there is a subobject T of S whose underlying set is $f^{-1}(C)$ or $f^{-1}(D)$. Without loss of generality, we can suppose that $f^{-1}(C)$ is the underlying set of T. Then $(f^{-1}(C), f \restriction f^{-1}(C))$ is a subnet of (S, f). Hence $(f^{-1}(C), f \restriction f^{-1}(C)) \xrightarrow{\pi} x$ and, since $(f \restriction f^{-1}(C))(f^{-1}(C)) = C \subseteq A$, we have $x \in u_\pi A \subseteq u_\pi A \cup u_\pi B$. Therefore, $u_\pi(A \cup B) \subseteq u_\pi A \cup u_\pi B$. As the converse inclusion is obvious, $(X, u_\pi) \in Clo$. Now the statement follows from Proposition 4.1. ❑

DEFINITION 4.6: We denote by $\widetilde{Net}_S$, $\widetilde{Conv}_S$, and $\widetilde{Lim}_S$ the full subcategories of Net_S, $Conv_S$ and Lim_S, respectively, given by those objects (X, π) for which the following condition is satisfied:

If $(S, f) \xrightarrow{\pi} x$ and $(T_s, g_s) \xrightarrow{\pi} f(s)$ for each $s \in S$, then there is $(U, h) \in \langle X \rangle_S$ such that $(U, h) \xrightarrow{\pi} x$ and $h(U) \subseteq \bigcup_{s \in S} g_s(T_s)$.

PROPOSITION 4.7: $\mathcal{F}_S$ is a concrete functor from $\widetilde{Net}_S$ into *Wet*.

Proof: Let $(X, \pi) \in \widetilde{Net}_S$, $A \subseteq X$, and $x \in u_\pi u_\pi A$. Then there is a net $(S, f) \in \langle X \rangle_S$ with $(S, f) \xrightarrow{\pi} x$ such that $f(S) \subseteq u_\pi A$. Further, for each $s \in S$ there is a net $(T_s, g_s) \in \langle X \rangle_S$ with $(T_s, g_s) \xrightarrow{\pi} f(s)$ such that $g_s(T_s) \subseteq A$. Thus, there is a net $(U, h) \in \langle X \rangle_S$ with $(U, h) \xrightarrow{\pi} x$ and $h(U) \subseteq \bigcup_{s \in S} g_s(T_s) \subseteq A$. Therefore, $x \in u_\pi A$ and we have proved the inclusion $u_\pi u_\pi A \subseteq u_\pi A$. As the converse inclusion is obvious, $(X, u_\pi) \in Wet$ and the statement follows from Proposition 4.1. ❑

COROLLARY 4.8: If S is additively hereditary, then $\mathcal{F}_S$ is a concrete functor from $\widetilde{Conv}_S$ into *Top*.

LEMMA 4.9: Let S be a fork category, X a set and (S, f), $(T, g) \in \langle X \rangle_S$. If $f(S) \subseteq g(T)$, then there exists a subnet of (S, f) which is also a subnet of (T, g).

Proof: Let $x \in f(S)$ and let $s \in S$, $t \in T$ be arbitrary elements with $f(s) = g(t) = x$. Then there is an S-object U such that both the s-constant map φ: $U \to S$ and the t-constant map ψ: $U \to T$ are S-morphisms. Let $h: U \to X$ be the x-constant map. Then clearly (U, h) is a subnet of both (S, f) and (T, g). ❑

LEMMA 4.10: Let S be a fork category and $(X, \pi) \in Lim_S$ an object. Then for any $(S, f) \in \langle X \rangle_S$ and any $x \in X$ we have $(S, f) \xrightarrow{\pi} x$ iff $x \in u_\pi g(T)$ for each subnet (T, g) of (S, f).

Proof: Suppose that $(S, f) \not\xrightarrow{\pi} x$. Then there is a subnet (T, g) of (S, f) such that for any subnet (U, h) of (T, g) there holds $(U, h) \not\xrightarrow{\pi} x$. Let $(V, p) \in \langle X \rangle_S, p(V) \subseteq g(T)$. By Lemma 4.9, there exists a subnet (W, q) of (V, p) which is also a subnet of (T, g). But then $(W, q) \not\xrightarrow{\pi} x$, hence $(V, p) \not\xrightarrow{\pi} x$. Therefore, $x \notin u_\pi g(T)$. We have proved that from $x \in u_\pi g(T)$ for each subnet (T, g) of (S, f) it follows that $(S, f) \xrightarrow{\pi} x$. As the converse implication is obvious, the proof is complete. ❑

THEOREM 4.11: If S is a fork category, then $\mathcal{F}_S$ is a full concrete embedding of Lim_S into *Pcl*.

Proof: Let (X, π), $(X, \rho) \in Lim_S$ be objects, $\pi \neq \rho$. Without loss of generality we can suppose that there are $(S, f) \in \langle X \rangle_S$ and $x \in X$ such that $(S, f) \xrightarrow{\pi} x$ but $(S, f) \not\xrightarrow{\rho} x$. Thus, by Lemma 4.10, $x \in u_\pi g(T)$ for each subnet (T, g)

of (S, f) and there is a subnet (T_0, g_0) of (S, f) such that $x \notin u_\rho g_0(T_0)$. Hence, $u_\pi \neq u_\rho$, i.e., $\mathcal{F}_S$ is injective on objects.

Let $(X, \pi), (Y, \rho) \in Lim_S$ and let $F: (X, u_\pi) \to (Y, u_\rho)$ be a continuous map. Let $(S, f) \xrightarrow{\pi} x$. Then, by Lemma 4.9, for each subnet (T, g) of (S, f) we have $x \in u_\pi g(T)$ and consequently $F(x) \in u_\rho F(g(T))$. But every subnet of $(S, F \circ f)$ has the form $(T, F \circ g)$ where (T, g) is a subnet of (S, f). Thus, by Lemma 4.10, $(S, F \circ f) \xrightarrow{\rho} F(x)$. Therefore, $\mathcal{F}_S$ is full. The assertion now results from Proposition 4.1. ❑

COROLLARY 4.12: If S is an additively hereditary and fork category, then $\mathcal{F}_S$ is a full concrete embedding of Lim_S into *Clo*.

Proof: It follows from Proposition 4.5 and Theorem 4.11. ❑

COROLLARY 4.13:

(a) If S is a fork category, then $\mathcal{F}_S$ is a full concrete embedding of $\widetilde{Lim}_S$ into *Wet*.

(b) If S is an additively hereditary and fork category, then $\mathcal{F}_S$ is a full concrete embedding of $\widetilde{Lim}_S$ into *Top*.

Proof: It is a consequence of Proposition 4.7, Corollary 4.8, and Theorem 4.11. ❑

EXAMPLE 4.14: Let $S = \underline{\omega}$. It is well known that the full subcategory of Lim_S whose objects are precisely the $\mathcal{L}^*$-spaces (see 1.5, Example 3) is embeddable by $\mathcal{F}_S$ into *Clo*. The $\mathcal{L}^*$-spaces that are objects of $\widetilde{Lim}_S$ are called S^*-spaces [6]. It is known that $\mathcal{F}_S$ is a full concrete embedding of the category of S^*-spaces into *Top* (more precisely, $\mathcal{F}_S$ is an isomorphism of the category of S^*-spaces onto the category of those Fréchet topological spaces in which every sequence has at most one limit).

5. THE CONDITION OF ITERATED LIMITS

The condition introduced in Definition 4.6 has several lacks. It is known that for $S = Dir$ the functor $\mathcal{F}_S: \widetilde{Lim}_S \to Top$ is surjective on objects but it is not an isomorphism. Next, $\widetilde{Net}_S$ and $\widetilde{Lim}_S$ are evidently not closed under formation of finite products in Net_S, and the same is valid for the full subcategories of $\widetilde{Net}_S$ and $\widetilde{Lim}_S$ given by their diagonal objects. Consequently, these subcategories are not closed under formation of function spaces in Net_S. We will therefore replace the condition from Definition 4.6 with a more convenient condition, the condition of iterated limits, which is an extension of the known condition introduced by J.L. Kelley for the case $S = Dir$ [7].

DEFINITION 5.1: Let $\mathcal{S}$ have Cartesian products and let (X, π) be a limit $\mathcal{S}$-net space. We say that (X, π) fulfills the *condition of iterated limits* if the following is valid: Let $S \in \mathcal{S}$ be an object and let $(T_s, g_s) \in \langle X \rangle_{\mathcal{S}}$ for each $s \in S$. Let $U = S \times \prod_{s \in S} T_s$ and let $h: U \to X$ be the map given by $h(s, t) = g_s(t(s))$. Then from $(T_s, g_s) \xrightarrow{\pi} x_s$ for each $s \in S$ and $(S, f) \xrightarrow{\pi} x$ where $f: S \to X$ is given by $f(s) = x_s$ it follows that $(U, h) \xrightarrow{\pi} x$.

We denote by $Cil_{\mathcal{S}}$ the full subcategory of $Lim_{\mathcal{S}}$ whose objects are precisely the limit $\mathcal{S}$-net spaces fulfilling the condition of iterated limits. Clearly, $Cil_{\mathcal{S}}$ is a full subcategory of $\widetilde{Lim}_{\mathcal{S}}$ and thus $\mathcal{F}_{\mathcal{S}}$ is a concrete functor from $Cil_{\mathcal{S}}$ into *Wet*.

DEFINITION 5.2: Let $\mathcal{S}$ have Cartesian products. Then $\mathcal{S}$ is called *fine* if for each $\mathcal{S}$-object S there is a map $i \mapsto S^{(i)}$ of S into the class of all subobjects of S such that:

(1) $\varphi: S \to T$ is an $\mathcal{S}$-morphism iff for any $j \in T$ there exists $i \in S$ such that $\varphi(S^{(i)}) \subseteq T^{(j)}$,
(2) $j \in S^{(i)} \Rightarrow S^{(j)} \subseteq S^{(i)}$,
(3) $(\prod_{k \in K} S_k)^{(\nu)} \subseteq \prod_{k \in K} S_k^{(\nu(k))}$.

EXAMPLE 5.3:
(1) The category *Dir* is fine. For any directed set $S = (S, \leq)$ and any $i \in S$ the directed subset $S^{(i)}$ of S is the set $S^{(i)} = \{s \in S;\ s \geq i\}$ directed by $\leq$.
(2) The category Set^+ is fine. For any nonempty set S and any $i \in S$ the subset $S^{(i)}$ of S is given by $S^{(i)} = S$.

LEMMA 5.4: Let $\mathcal{S}$ be fine and let $(X, \pi) \in Cil_{\mathcal{S}}$ be an object. Then for any $(S, f) \in \langle X \rangle_{\mathcal{S}}$ and any $x \in X$ we have $(S, f) \xrightarrow{\pi} x$ iff $x \in u_\pi g(T)$ for each subnet (T, g) of (S, f).

Proof: Let $x \in u_\pi g(T)$ for each subnet (T, g) of (S, f). Let (T, g) be a subnet of (S, f). Then, for each $i \in T$, $(T^{(i)}, g \restriction T^{(i)})$ is a subnet of (S, f) and hence $x \in u_\pi g(T^{(i)})$. Consequently, for each $i \in T$ there is a net $(U_i, h_i) \in \langle X \rangle_{\mathcal{S}}$ with $(U_i, h_i) \xrightarrow{\pi} x$ and $h_i(U_i) \subseteq g(T^{(i)})$. Put $W = T \times \prod_{i \in T} U_i$ and let $p: W \to X$ be the map given by $p(i, \nu) = h_i(\nu(i))$. Now, since $(T, q) \xrightarrow{\pi} x$ whenever $q: T \to X$ is the x-constant map, we have $(W, p) \xrightarrow{\pi} x$. We will show that (W, p) is a subnet of (T, g). For each $i \in T$ let $\varphi_i: U_i \to T^{(i)}$ be an arbitrary map with the property $h_i = g \circ \varphi_i$. Now we can define a map $\varphi: W \to T$ by putting $\varphi(i, \nu) = \varphi_i(\nu(i))$. Let $j \in T$ be an arbitrary element. Then for any $\nu \in \prod_{i \in T} U_i$ and any $(k, w) \in W^{(j, \nu)}$ we have $(k, w) \in (T \times \prod_{i \in T} U_i)^{(j, \nu)} \subseteq T^{(j)} \times \prod_{i \in T} U_i^{(\nu(i))}$. Hence, $k \in T^{(j)}$ and $w(i) \in U_i^{(\nu(i))}$ for each $i \in T$. Consequently, $\varphi(k, w) = \varphi_k(w(k)) \in T^{(k)} \subseteq T^{(j)}$. Thus $\varphi(k, w) \in T^{(j)}$ and therefore, $\varphi: W \to T$ is an $\mathcal{S}$-morphism. For any $(i, \nu) \in W$ we have $g(\varphi(i, \nu)) = g(\varphi_i(\nu(i))) = h_i\ (\nu(i)) =$

$p(i, v)$, hence $g \circ \varphi = p$. We have shown that (W, p) is a subnet of (T, g). But (T, g) has been chosen arbitrarily and so it follows that $(S, f) \xrightarrow{\pi} x$. As the converse inclusion is obvious, the proof is complete. ❑

THEOREM 5.5: If S is fine, then $\mathcal{F}_S$ is a full concrete embedding of Cil_S into Wet.

Proof: Using Lemma 5.4 we can prove Theorem 5.5 in an analogous way as we proved Theorem 4.11 by using Lemma 4.10. ❑

From Proposition 4.5 and Theorem 5.5 we immediately get

COROLLARY 5.6: If S is fine and additively hereditary, then $\mathcal{F}_S$ is a full concrete embedding of Cil_S into Top.

EXAMPLE 5.7: It is well known that for $S = Dir$ the functor $\mathcal{F}_S$ is a concrete isomorphism of Cil_S onto Top [7]. Thus, if S is fine and additively hereditary, then Cil_S can be fully concretely embedded into Cil_{Dir}.

If S has Cartesian products, then we denote by $DCil_S$ the full subcategory of Cil_S determined by those of its objects that are diagonal.

THEOREM 5.8: Let S have Cartesian products. Then $DCil_S$ is a Cartesian closed topological category with function spaces inherited from Net_S.

Proof: Let $F_k: X \to (X_k, \pi_k)$, $k \in K$, be a source in $DCil_S$ and let π be its initial structure in Net_S, i.e., let $(S, f) \xrightarrow{\pi} x$ iff $(S, F_k \circ f) \xrightarrow{\pi_k} F_k(x)$ for each $k \in K$ (see the proof of Theorem 2.1). Let $S \in S$ be an object, let $(T_s, g_s) \in \langle X\rangle_S$, $(T_s, g_s) \xrightarrow{\pi} x_s$ for each $s \in S$, and let $(S, f) \xrightarrow{\pi} x$ where $f: S \to X$ is the map given by $f(s) = x_s$. Further, let $U = S \times \prod_{s\in S} T_s$ and let $h: U \to X$ be the map given by $h(s, t) = g_s(t(s))$. Then for each $k \in K$ we have $(T_s, F_k \circ g_s) \in \langle X_k\rangle_S$, $(T_s, F_k \circ g_s) \xrightarrow{\pi_k} F_k(x_s)$ for each $s \in S$, and $(S, F_k \circ f) \xrightarrow{\pi_k} F_k(x)$ where $F_k \circ f(s) = F_k(x_s)$. Next, we have $F_k \circ h: U \to X_k$, $(F_k \circ h)(s, t) = (F_k \circ g_s)(t(s))$. Therefore, $(U, F_k \circ h) \xrightarrow{\pi_k} F_k(x)$ and consequently $(U, h) \xrightarrow{\pi} x$. Hence, the object (X, π) fulfills the condition of iterated limits. Now, because $DLim_S$ is initially closed in Net_S, so is also $DCil_S$. Therefore, $DCil_S$ is a topological category.

Let (X, π), $(Y, \rho) \in DCil_S$ be objects and let (Z, σ) be their function space in Net_S, i.e., let $Z = C((X, \pi), (Y, \rho))$ and let $(T, g) \xrightarrow{\sigma} z$ iff $(V, g^x) \xrightarrow{\rho} z(x)$ for each $x \in X$ where $g^x: V \to Y$ is the map defined by $g^x(s) = g(s)(x)$ (see Lemma 3.3). Let $S \in S$ be an object, let $(T_s, g_s) \in \langle Z\rangle_S$, $(T_s, g_s) \xrightarrow{\sigma} z_s$ for each $s \in S$, and let $(S, f) \xrightarrow{\pi} z$ where $f: S \to Z$ is the map given by $f(s) = z_s$. Further, let $U = S \times \prod_{s\in S} T_s$ and let $h: U \to Z$ be the map given by $h(s, t) = g_s(t(s))$. Then for each $x \in X$ we have $(T_s, g_s^x) \in \langle Y\rangle_S$, $(T_s, g_s^x) \xrightarrow{\rho} z_s(x)$ for each $s \in S$, and $(S, f^x) \xrightarrow{\rho} z(x)$ where $f^x: S \to Y$ fulfills $f^x(s) = f(s)(x) = z_s(x)$ for each $s \in S$. Next, we have $h^x: U \to Y$ and $h^x(s, t) = h(s, t)(x) = g_s(t(s))(x) = g_s^x(t(s))$. Therefore, $(U, h^x) \xrightarrow{\rho} z(x)$ and consequently $(U, h) \xrightarrow{\sigma} z$. We have shown that the function space (Z, σ) fulfills the condition of iterated limits. Because $DLim_S$ has function spaces inherited from Net_S, so has $DCil_S$. ❑

EXAMPLE 5.9: If $S = Dir$, then from Corollary 5.6 and Theorem 5.8 it follows that $\mathcal{F}_S(DCil_S)$ is a Cartesian closed topological subcategory of *Top*. But this subcategory is precisely the category of indiscrete topological spaces.

REMARK 5.10: Obviously, the condition of iterated limits can be extended to convergence S-net spaces and also to S-net spaces in general. Then, by Theorem 3.4 and Corollary 4.8, the full subcategories of $DConv_S$ and $DNet_S$ given by those of their objects that fulfil the condition of iterated limits are Cartesian closed topological categories with function spaces inherited from Net_S.

6. S-NET SPACES AND SEPARATION AXIOMS

In what follows, we deal with the question under which conditions on an S-net space (X, π) the preclosure space (X, u_π) fulfills the lower separation axioms R_0, T_0 or T_1, respectively. For preclosure spaces these axioms are defined analogously to topological spaces. A preclosure space (X, u) is said to be:

an R_0-*space* if $x \in u\{y\} \Rightarrow y \in u\{x\}$ whenever $x, y \in X$;

a T_0-*space* if $x \in u\{y\}, y \in u\{x\} \Rightarrow x = y$ whenever $x, y \in X$;

a T_1-*space* if $u\{x\} = \{x\}$ whenever $x \in X$.

In [8], R_0-spaces are called B^*-spaces, and in [2] T_0-spaces and T_1-spaces are called K-spaces and B-spaces, respectively. We denote by R_0-*Pcl*, T_0-*Pcl*, and T_1-*Pcl* the full subcategories of *Pcl* whose objects are precisely the R_0-spaces, T_0-spaces, and T_1-spaces, respectively. Clearly, a preclosure space is a T_1-space iff it is both R_0-space and T_0-space, i.e., $T_1\text{-}Pcl = R_0\text{-}Pcl \cap T_0\text{-}Pcl$.
The following assertion is obvious.

PROPOSITION 6.1: If (X, π) is an S-net space, then:
(1) $(X, u_\pi) \in R_0\text{-}Pcl$ iff for any pair $x, y \in X$ and any $(S, f) \in \langle X\rangle_S$ such that f is the x-constant map and $(S, f) \xrightarrow{\pi} y$ there exists $(T, g) \in \langle X\rangle_S$ such that g is the y-constant map and $(T, g) \xrightarrow{\pi} x$,
(2) $(X, u_\pi) \in T_0 - Pcl$ iff for any $x, y \in X$ and any pair $(S, f), (T, g) \in \langle X\rangle_S$ such that f is the x-constant map and g is the y-constant map from $(S, f) \xrightarrow{\pi} y$ and $(T, g) \xrightarrow{\pi} x$ it follows that $x = y$,
(3) $(X, u_\pi) \in T_1\text{-}Pcl$ iff card $\pi(S, f) = 1$, whenever $(S, f) \in \langle X\rangle_S$ and f is a constant map.

The full subcategory of Net_S whose objects are the S-net spaces (X, π)

fulfilling $(X, u_\pi) \in R_0\text{-}Pcl$ is a bicoreflective subcategory of Net_S but, in general, it is not closed under formation of finite products in Net_S and hence is not closed under formation of function spaces in Net_S. Neither is the full subcategory of Net_S whose objects are the S-net spaces (X, π) fulfilling $(X, u_\pi) \in T_0\text{-}Pcl$ closed under formation of function spaces in Net_S in general, despite it is closed under formation of initial structures for mono-sources in Net_S (and hence, initially structured). In each of these two categories we will therefore find a full Cartesian closed subcategory with function spaces inherited from Net_S.

DEFINITION 6.2: We denote by $R_0\text{-}Net_S$ the full subcategory of Net_S whose objects are precisely the S-net spaces (X, π) fulfilling the following condition:

Let $\{(S, f_i);\ i \in S\}$ be a family of S-nets in X such that $f_i(i) = x$ ($x \in X$ an element) for each $i \in S$. Let $(S, f_i) \xrightarrow{\pi} x_i$ for each $i \in I$ and let $(S, f) \in \langle X \rangle_S$, $(S, f) \xrightarrow{\pi} x$. For every $j \in S$ let $g_j\colon S \to X$ be the map given by $g_j(i) = f_i(j)$ and let $g\colon S \to X$ be the map given by $g(i) = x_i$. Then $(S, g_j) \xrightarrow{\pi} f(j)$ for each $j \in S$ and $(S, g) \xrightarrow{\pi} x$.

$R_0\text{-}Net_S$ is clearly initially closed in Net_S, i.e., $R_0\text{-}Net_S$ is a topological category.

As an immediate consequence of Proposition 4.1 and Definition 6.2 we get:

COROLLARY 6.3: $\mathcal{F}_S$ is a concrete functor from $R_0\text{-}Net_S$ into $R_0\text{-}Pcl$.

LEMMA 6.4: Let $(X, \pi) \in R_0\text{-}Net_S$, $(S, f) \in \langle X \rangle_S$, and $(S, f) \xrightarrow{\pi} x$.

(1) If $j, k \in S$ and $f_{j,k}\colon S \to X$ is the map given by $f_{j,k}(i) = f(j)$ whenever $i \neq j$ and $f_{j,k}(j) = f(k)$, then $(S, f_{j,k}) \xrightarrow{\pi} f(j)$.
(2) If $j \in S$ and $g_j\colon S \to X$ is the map given by $g_j(i) = f(j)$ whenever $i \neq j$ and $g_j(j) = x$, then $(S, g_j) \xrightarrow{\pi} f(j)$.
(3) If $g\colon S \to X$ is the x-constant map, then $(S, g) \xrightarrow{\pi} f(j)$ for each $j \in S$.

Proof: Let the assumption of (1) be fulfilled and for each $i \in S$ let f_i: $S \to X$ be the map given by $f_i(m) = f(j)$ whenever $i \neq j$ and $f_j(m) = f(m)$. Then $(S, f_i) \in \langle X \rangle_S$ and $f_i(i) = f(j)$ for each $i \in S$. Next, we have $(S, f_i) \xrightarrow{\pi} f(j)$ whenever $i \neq j$ and $(S, f_j) \xrightarrow{\pi} x$. Let $h\colon S \to X$ be the $f(j)$-constant map. Then $(S, h) \xrightarrow{\pi} f(j)$. Finally, let $g_k\colon S \to X$ be the map defined by $g_k(i) = f_i(k)$. Then, by the definition of $R_0\text{-}Net_S$, $(S, g_k) \xrightarrow{\pi} h(k) = f(j)$. But it is evident that $g_k = f_{j,k}$.

Let the assumption of (2) be fulfilled and let f_i and h be defined in the same way as in the previous part of the proof. Let $g\colon S \to X$ be the map given by $g(i) = f(j)$ whenever $i \neq j$ and $g(j) = x$. Then, by the definition of $R_0\text{-}Net_S$, $(S, g) \xrightarrow{\pi} f(j)$. But it is evident that $g = g_j$.

Finally, let the assumption of (3) be fulfilled and for each $i \in S$ let $f_i\colon S \to X$ be the x-constant map. Then, for each $i \in S$, $f_i(i) = x$ and $(S, f_i) \xrightarrow{\pi} x$. Let

$j \in S$ and let $g_j: S \to X$ be the x-constant map. Then $g_j(i) = f_i(j)$ for each $i \in S$ and, because $(S, f) \xrightarrow{\pi} x$, by the definition of $R_0\text{-}Net_S$ we get $(S, g_j) \xrightarrow{\pi} f(j)$. But $g_j = g$. The proof is complete. ☐

THEOREM 6.5: $R_0\text{-}Net_S$ is a Cartesian closed topological category with function spaces inherited from Net_S.

Proof: Clearly, $R_0\text{-}Net_S$ is initially closed in Net_S. Let (X, π), $(Y, \rho) \in R_0\text{-}Net_S$ and let (Z, σ) be the function space of (X, π) and (Y, ρ) in Net_S. Let $\{(S, h_i); i \in S\}$ be a family of S-nets in Z such that $h_i(i) = z$ ($z \in Z$) for each $i \in S$. Let $(S, h_i) \xrightarrow{\sigma} z_i$ for each $i \in S$ and let $(S, h) \in \langle Z \rangle_S$, $(S, h) \xrightarrow{\sigma} z$. Let $j \in S$ and let $q_j: S \to Z$ be the map given by $q_j(i) = h_i(j)$. Let $(S, f) \in \langle X \rangle_S$, $(S, f) \xrightarrow{\pi} x$. For any $k \in S$ let $f_{j,k}: S \to X$ be the map defined in Lemma 6.4. Then $(S, f_{j,k}) \xrightarrow{\pi} f(j)$ and, obviously, $f_{j,k}(k) = f(j)$ for each $k \in K$. Let $g_j: S \to X$ be the map defined in Lemma 6.4. Then $(S, g_j) \xrightarrow{\pi} f(j)$. Thus, for each $i \in S$ we have $(S, h_i{}^p) \xrightarrow{\rho} z_i(f(j))$ where $p = f_{j,i}$, and $h_i{}^p(i) = z(f(j))$ for each $i \in S$. Next, we have $(S, h^{g_j}) \xrightarrow{\rho} z(f(j))$. Consequently, if $r: S \to Y$ is the map given by $r(i) = h_i{}^p(j)$, then $(S, r) \xrightarrow{\rho} h^{g_j}(j)$. But $r(i) = h_i{}^p(j) = h_i(j)(p(j)) = h_i(j)(f_{j,i}(j)) = h_i(j)(f(i)) = q_j(i)(f(i)) = q_j{}^f(i)$ for each $i \in S$ and $h^{g_j}(j) = h(j)(g_j(j)) = h(j)(x)$. Therefore, $(S, q_j) \xrightarrow{\sigma} h(j)$. Now, let $q: S \to Z$ be the map given by $q(i) = z_i$. For any $i \in S$ let $f_i: S \to X$ be the x-constant map. Then $f_i(i) = x$ and by Lemma 6.4 we have $(S, f_i) \xrightarrow{\pi} f(i)$. Consequently, $(S, h_i{}^{f_i}) \xrightarrow{\rho} z_i(f(i))$ and $h_i{}^{f_i}(i) = z(x)$. Next, we also have $(S, h^f) \xrightarrow{\rho} z(x)$. Thus, if $g: S \to Y$ is the map given by $g(i) = z_i(f(i))$, then $(S, g) \xrightarrow{\rho} z(x)$. But $g(i) = z_i(f(i)) = q(i)(f(i)) = q^f(i)$ and therefore $(S, q) \xrightarrow{\sigma} z$. We have proved that $(Z, \sigma) \in R_0\text{-}Net_S$. This completes the proof. ☐

EXAMPLE 6.6:   If $S = \underline{1}$, then $\mathcal{F}_S$ is an isomorphism of $R_0\text{-}Net_S$ onto the category of quasi-discrete closure R_0-spaces. In [3] closure R_0-spaces are called semi-uniformizable spaces.

DEFINITION 6.7:   We denote by $T_0\text{-}Net_S$ the full subcategory of Net_S whose objects are precisely the S-net spaces (X, π) fulfilling the following condition:
    Let $\{(S, f_i); i \in S\}$ and $\{(S, g_j); j \in S\}$ be families of S-nets in X such that $f_i(j) = g_j(i)$ for each $i, j \in S$ and $f_i(i) = x$ ($x \in X$ an element) for each $i \in S$. Let $(S, f_i) \xrightarrow{\pi} x_i$ for each $i \in S$ and $(S, g_j) \xrightarrow{\pi} y_j$ for each $j \in S$. Let $f, g: S \to X$ be the maps given by $f(i) = x_i$, $g(j) = y_j$, and let $(S, f) \xrightarrow{\pi} x$ and $(S, g) \xrightarrow{\pi} x$. Then f_i is x-constant for each $i \in S$ and also both f and g are x-constant.

Clearly, $T_0\text{-}Net_S$ is closed under formation of initial structures for mono-sources in Net_S, hence $T_0\text{-}Net_S$ is initially structured. The following statement is an immediate consequence of Proposition 4.1 and Definition 6.7:

COROLLARY 6.8: $\mathcal{F}_S$ is a concrete functor from $T_0\text{-}Net_S$ into $T_0\text{-}Pcl$.

THEOREM 6.9: T_0-Net_S is a Cartesian closed initially structured category with function spaces inherited from Net_S.

Proof: Clearly, T_0-Net_S is closed under formation of initial structures for mono-sources in Net_S. Let (X, π), $(Y, \rho) \in T_0$-Net_S and let (Z, σ) be the function space of (X, π) and (Y, ρ) in Net_S. Let $\{(S, p_i); i \in S\}$ and $\{(S, q_j); j \in S\}$ be families of S-nets in Z such that $p_i(j) = q_j(i)$ for each $i, j \in S$ and $p_i(i) = z$ for each $i \in S$. Let $(S, p_i) \xrightarrow{\sigma} z_i$ for each $i \in S$ and $(S, q_j) \xrightarrow{\sigma} t_j$ for each $j \in S$. Let $p, q: S \to Z$ be the maps given by $p(i) = z_i$, $q(j) = t_j$ and let $(S, p) \to z$, $(S, q) \to z$. Let $x \in X$ be an element and $f: S \to X$ the x-constant map. Then $p_i^f(j) = p_i(j)(f(j)) = q_j(i)(f(i)) = q_j^f(i)$ for each $i, j \in S$ and $p_i^f(i) = z(x)$ for each $i \in S$. Next, as $(S, f) \xrightarrow{\pi} x$, we have $(S, p_i^f) \xrightarrow{\rho} z_i(x)$ and $(S, q_j^f) \xrightarrow{\rho} t_j(x)$. Let $g, h: S \to Y$ be the maps given by $g(i) = z_i(x)$ and $h(j) = t_j(x)$. Then $(S, g) \xrightarrow{\rho} z(x)$ and $(S, h) \xrightarrow{\rho} z(x)$ because $g(i) = z_i(x) = p(i)(f(i)) = p^f(i)$ for each $i \in S$ and $h(j) = t_j(x) = q(j)(f(j)) = q^f(j)$ for each $j \in S$. Consequently, p_i^f is the $z(x)$-constant map and both g and h are the $z(x)$-constant maps. Thus, for each $i \in S$ we have $p_i(i)(x) = p_i^f(i) = z(x)$, hence, p_i is the z-constant map. Similarly, we have $p(i)(x) = z_i(x) = g(i) = z(x)$ for each $i \in S$ and $q(j)(x) = t_j(x) = h(j) = z(x)$ for each $j \in S$. Hence, p and q are also the z-constant maps. This completes the proof. ❑

EXAMPLE 6.10: Let $S = \underline{1}$. Then $\mathcal{F}_S$ is an isomorphism of T_0-Net_S onto the category of quasi-discrete closure T_0-spaces. In [3] closure T_0-spaces are called feebly semiseparated spaces (and closure T_1-spaces are called semiseparated spaces).

REMARK 6.11: R_0-$Net_S \cap T_0$-Net_S is precisely the category of discrete S-net spaces.

REFERENCES

1. ADÁMEK J., H. HERRLICH, & G.E. STRECKER. 1990. Abstract and Concrete Categories. Wiley-Interscience. New York.
2. ČECH, E. 1968. Topological papers of E. Čech. Chap. 28. Academia. Prague. 436–472
3. ČECH, E. 1966. Topological Spaces. (Revised by Z.Frolík & M.Katětov). Academia. Prague.
4. DUDLEY, R.M. 1964. On sequential convergence. Trans. Amer. Math. Soc. **112:** 483–507.
5. EDGAR, G.A. 1976. A Cartesian closed category for topology. Gen. Top. Appl. **6:** 65–72.
6. ENGELKING, R. 1977. General Topology. Polish Scientific Publishers. Warszawa.
7. KELLEY, J.L. 1963. General Topology. Van Nostrand. Princeton, New Jersey.
8. KOUTSKÝ K. & M. SEKANINA. 1960. On the $\mathcal{R}$-modification and several other modifications of a topology. Publ. Fac. Sci. Univ. Brno 410. 45–64.
9. KURATOWSKI, K. 1966. Topology I. Polish Scientific Publishers. Warszawa.
10. NOVÁK, J.. 1964. On some problems concerning multivalued convergence. Czech. Math. J. **14:** 548–561.

11. PREUSS, G. 1988. Theory of Topological Structures. D. Reidel Publ. Comp. Dordrecht.
12. PREUSS, G. 1993. Cauchy spaces and generalizations. Math. Japonica. **38:** 803–812
13. ŠLAPAL, J. 1990. A Cartesian closed topological category of sequential spaces. Periodica Math. Hungar. **21:** 109–112
14. TAYLOR W.. 1970. Convergence in relational structures. Math. Ann. **186:** 215–227
15. WICHTERLE K. 1968. On $\mathcal{B}$-convergence spaces. Czech. Math. J. **18:** 569–588.

Lines as Topological Graphs[a]

M.B. SMYTH[b]
Department of Computing
Imperial College
London SW7 2BZ, United Kingdom

ABSTRACT: We present a unified theory of digital and continuous lines, arcs, etc., in terms of a (relatively) new notion of *topological graph.*

0. INTRODUCTION

Topological graphs are structures which embody a topology as well as a binary relation (which need not be transitive), thus constituting a generalization of ordered topological spaces. One of the principal theses of [11] was that digital ("discrete") topology can be unified with standard ("continuous") topology by means of the theory of topological graphs (and certain closely related structures). The aim of the present paper is to carry out this program in some detail for the case of linear structures (continua or extended continua): arcs, unbounded lines, etc. The unified theory we have in mind has two main aspects: the framework of definitions and basic results that applies equally to the finitary digital structures and to the continuous structures, and the approximation (via inverse limits) of the classical continuous structures by finitary structures. In the present paper, the first of these aspects will receive somewhat more attention than the second.

Among previous studies in digital topology, the nearest antecedent to the present work is the theory of COTS (connected ordered topological spaces) presented in [5]. Two main ranges of examples of interest for us are not captured by the theory of COTS. The first is that of the symmetric linear graphs (in the finite case: the graph having nodes 0, 1, ..., $n - 1$, in which each node is related to its neighbors in the sequence as well as to itself). The second, properly topological, range of examples is provided by certain linearly ordered spaces which arise as inverse limits of finite COTS. Among these is the space we call sI, which has a special rôle in our theory since arcs in general can be characterized as the images, via suitable maps, of sI (Theorem 4.3). We find that the second, as well as the first, type of example can be

Mathematics Subject Classification: 54F15, 54H99.
Key words and phrases: digital topology, continuum, linear structure.
[a]The work was partially supported by the EPSRC grant "Foundational Structures for Computer Science."
[b]E-mail: mbs@doc.ic.ac.uk.

readily accommodated by admitting a graph structure to the spaces in addition to the topology.

Continua and their basic properties are considered in Section 1. Various formulations of the linear case are given in Section 2. The final section deals with the characterization of arcs, arcwise connectivity and Peano continua. The treatment of the topics given here generalizes the standard theory of them: the latter can be recovered by restricting the graph component of the structures to be the identity relation. A possibly unusual feature of the present study is that (Section 2) we relate the material to the characterization of lines in terms of a "betweenness" relation adopted in studies of the foundations of (Euclidean) geometry since the late nineteenth century. Also, the rôle of connectedness-reflecting maps is perhaps noteworthy: these are needed for inverse limits (Section 3), and also for the characterization of arcs (as images of sI: Section 4).

Independently of the present author's work, Latecki and Prokop [7] have advocated the use of closure spaces and related structures (semiproximities) in digital topology. It is interesting to note that maps which reflect connectedness also play a key rôle in their work.

1. TOPOLOGICAL GRAPHS. CONTINUA

DEFINITION 1.1: (*restricted form*) A *topological graph* is a triple $(X, \mathcal{T}, R)$, where $\mathcal{T}$ is a Hausdorff topology on the set X, and $R \subseteq X \times X$ is a closed (with respect to the product topology), reflexive relation.

Essentially this notion (under the name *closed graph*) has been used previously by Krawczyk and Steprans [6] in their study of continuous graph colourings (they restrict attention to the compact zero-dimensional case). A less restrictive definition (as in [11]) can dispense with both the requirements that $\mathcal{T}$ be Hausdorff and that R be reflexive, but the simple condition that R be closed may then require modification as well.

On taking as morphisms the continuous R-preserving maps, we have the category **TopGr** of topological graphs. We need very few properties of this category, but we note that inverse limits are readily constructed by calculating separately the limit topology and the limit graph. (In checking that the limit relation is closed, and elsewhere, it is convenient to use the following criterion for a relation R to be closed: if $\neg\, xRy$, then there exist neighborhoods U, V of x, y such that $R(U) \cap V = \varnothing$, where $R(U) = \{v \mid \exists u \in U.\ uRv\}$.)

EXAMPLE 1.2: (A) Let $(L, \leq)$ be a linearly ordered set, and $\mathcal{T}$ its interval topology, that is, the topology having the $\leq$-closed intervals as a subbase for the closed sets. (We count the closed rays $[x, \infty)$, $(-\infty, x]$, which we also sometimes write as $\uparrow x$, $\downarrow x$, among the closed intervals. Likewise for the open rays $(x, \infty) = \{y \mid x < y\}$, $(-\infty, x) = \{y \mid y < x\}$, vis-a-vis the open intervals. We occasionally write $\Uparrow x$, $\Downarrow x$ for these open rays.) We note the following alternative descriptions of the interval topology: the closed rays are a subbase for the closed sets; the open intervals, or again just the open rays, are a subbase of $\mathcal{T}$. Clearly, $(L, \mathcal{T}, \leq)$ is a topological graph. However, we are more interested in the *adjacency* (*topological*) *graph* $(L, \mathcal{T}, A)$ derived from L, where A is the adjacency relation of $\leq$, defined by:

$$xAy \equiv xLAy \vee xRAy$$

where

$$xLAy \equiv x \leq y \wedge \neg \exists w.\, x < w < y \quad (x \text{ is } \textit{left-adjacent} \text{ to } y),$$

$$xRAy \equiv y \leq x \wedge \neg \exists w.\, y < w < x \quad (x \text{ is } \textit{right-adjacent} \text{ to } y).$$

Now let K_n, $n = 0, 1, \ldots$, be the linear order with 2^n elements, which we identify with the set $\{0, 1\}^n$ of n-element binary strings taken in lexicographic order. We take each K_n with its adjacency graph structure. The interval topology is, of course, discrete. Next, we consider the inverse system $\Sigma = (K_n, f_n)$, where $f_n\colon K_{n+1} \to K_n$ is the truncation map:

$$f_n(a_1 \ldots a_n a_{n+1}) = a_1 \ldots a_n.$$

Let $K_\infty = Lim_{\leftarrow}\Sigma$. As a topological graph, K_∞ is $(2^\omega, \mathcal{C}, A)$, where 2^ω is the set of binary sequences, $\mathcal{C}$ is the Cantor topology, and A is the reflexive symmetric closure of $\{\langle \sigma 01^\omega, \sigma 10^\omega \rangle \mid \sigma \in \{0, 1\}^*\}$. Notice that K_∞ is indeed the adjacency graph of the lexicographically ordered set 2^ω. (This is an instance of a general situation, Proposition 1.14). Moreover, A is an equivalence relation; and the quotient of $(2^\omega, \mathcal{C})$ with respect to A is the usual unit interval I.

(B) We now consider a slightly different procedure for approximating the unit interval, namely by successive subdivision into (open) subintervals together with dyadic points of division. That is, we let $H_0 = \{\{0\}, (0, 1), \{1\}\}$, $H_1 = \{\{0\}, (0, 2^{-1}), \{2^{-1}\}, (2^{-1}, 1), \{1\}\}$, and so on. For the relation R_n on H_n we take now, not adjacency, but a subrelation of adjacency: each division point $k.\, 2^{-n} (n = 0, \ldots, 2^n)$ is taken to be related to each of the subintervals adjacent to it, but not conversely. This makes R_n a partial order, in which all maximal chains have length 2. (This choice of R_n makes H_n, in its Alexandroff topology, a "Khalimsky space" which is a quotient of I; see, e.g., [5]). We assign labels to the elements of the H_n as follows. The interval (0, 1) is labelled Λ (null string), while the end-points 0, 1 are labelled L, R. An interval σ in H_n splits into "left" and "right" subintervals $\sigma-$, $\sigma+$, and the mid-point (division point) $\sigma 0$, in H_{n+1}. A dyadic point labelled π in H_n receives the label $\pi 0$ in H_{n+1}. As in the previous example, we have an inverse system (H_n, g_n), with the connecting maps g_n given by truncation (of labels). In the inverse limit $(H_\infty, \mathcal{T}, \sqsubseteq)$ we have the dyadic points $\sigma 0^\omega$ $(\sigma \in \{-, +\}^*)$

and, with each dyadic d, its neighbors (in the lexicographic order) $d_- = \sigma{-}{+}^\omega$ and $d_+ = \sigma{+}{-}^\omega$. Between these there hold the relations

$$d \sqsubseteq d_-, \quad d \sqsubseteq d_+.$$

Of course, each of the end-points $L0^\omega$, $R0^\omega$ has only one neighbor of this kind.) The remaining elements of H_∞, with labels in $\{-, +\}^\omega$, stand in an obvious bijection with the nondyadic reals in I.

In viewing these structures as topological graphs, the H_n are taken with discrete *topology*. The inverse limit topology of H_∞ will receive further attention later in the section.

The *uniformity* of the compact space H_∞ will also be of interest subsequently (in the proof of Proposition 4.2). In fact the uniformity of H_∞, as indeed the induced uniformity of any subset (in particular, of the set of dyadics) of H_∞ can be conveniently described as follows: it has as base the set of relations U_n, $n = 0, 1, \ldots$, where

$$x U_n y \Leftrightarrow x \restriction n = y \restriction n.$$

Here, $x \restriction n$ is the initial substring of x of length n.

(C) Generalizing (B), every compact ordered space is clearly a topological graph. This means that any Scott domain D can, if desired, be viewed as a topological graph $(D, \mathcal{L}, \sqsubseteq_D)$, where $\mathcal{L}$ is the Lawson topology of D. (For Scott domain/Lawson topology see, e.g., Lawson [8], Abramsky and Jung [1].)

In our previous work on this topic [11], topological graphs have been introduced in the context of Čech closure spaces. Indeed, a graph $(X, \mathcal{T}, R)$ induces a Čech space via:

$$x \in Int(S) \Leftrightarrow \exists U, V \in \mathcal{T}.\, x \in U \wedge R(U) \subseteq V \subseteq S.$$

The *Int* so defined is only a Čech interior operator rather than a topological interior since it lacks idempotence (in general). In some of the more significant cases of interest to us (notably, the compact ordered spaces) the induced closure space is topological, and indeed coincides with the *saturated* topology of $(X, \mathcal{T}, R)$: the collection of open sets U such that $R(U) \subseteq U$. In this paper we avoid explicit consideration of closure spaces; we occasionally make incidental use of the saturated topology.

EXAMPLE 1.3: When a Scott domain is viewed as a graph $(D, \mathcal{L}, \sqsubseteq)$ (Example 1.2(C)), the Scott topology is recovered as the saturated topology of D (equally, as the induced closure space). More generally, the saturated topology of a compact ordered space $(X, \mathcal{T}, \leq)$ consists of the open upper sets. Applied to the graphs considered in Example 1.2(B), the H_n yield in this way (as already noted) finite Khalimsky spaces; while H_∞ yields the spectral space which we denoted as sI in [10] (a "spectralization" of I). We mostly use the notation sI for this structure, even when it is viewed as a graph rather than a spectral space.

The definition of connectedness for closure spaces (Čech [3]) reduces to the following in the case of graphs.

DEFINITION 1.4: A graph $(X, \mathcal{T}, R)$ is *connected* if, for every partition $\{U, V\}$ of X into $\mathcal{T}$-open sets, there exist points x, y in U, V, respectively, such that x, y are R-related (that is, such that xRy or yRx).

Evidently the image of a connected graph under a morphism of **TopGr** is connected. Notice also that a graph is connected if and only if it is, in the ordinary topological sense, connected with respect to the saturated topology.

Definition 1.4 illustrates the pattern which will be followed in the rest of the paper, with regard to topological terminology. That is to say, the majority of topological terms, such as "open" and "compact," in relation to a graph $(X, \mathcal{T}, R)$, are taken in the sense of $\mathcal{T}$; the exception is "connected," which is in effect taken in the sense of the (coarser) saturated topology.

A simple set-theoretic notion, which can be helpful in formulating definitions and results in a way that facilitates comparison with their classical counterparts, is that of R-disjointness: given a reflexive relation R defined on a set X, we say that $S, S' \subseteq X$ are R-*disjoint* if $R(S) \cap S' = R(S') \cap S = \varnothing$. Thus, a graph $(X, \mathcal{T}, R)$ is connected if and only if X has no partition into R-disjoint open sets.

As a further illustration of R-disjointness, we may cite the following proposition, proved the same way as its standard prototype (bearing in mind that the relation R is closed):

PROPOSITION 1.5: Any two R-disjoint compact subsets of a graph $(X, \mathcal{T}, R)$ have R-disjoint neighborhoods.

Turning now to continua, we have:

DEFINITION 1.6: A *continuum* is a compact connected graph.

The presence of the relation gives this definition a far greater scope than its classical counterpart. For example, every Scott domain, viewed (as above) as a graph $(D, \mathcal{L}, \sqsubseteq)$, is evidently a continuum, when taken with its Lawson topology along with its order $\sqsubseteq_D$. More significant than this, for our present study, is that we can have nontrivial finite continua: every finite connected reflexive graph (in the ordinary sense), with discrete topology, is a continuum. The structures K_∞, H_∞ considered in Example 1.2, along with their finite approximations, are continua. (That K_∞ and H_∞ are connected as topological graphs is easy to check directly, but is also a consequence of general results (see Theorem 3.5).)

It is a fairly straightforward exercise to check that the basic theory of continua, as presented, e.g., in Hocking and Young [4], can be developed on the basis of Definition 1.6, with, at the most, minor adjustment of the standard proofs. Taking a *subspace* of a graph $(X, \mathcal{T}, R)$ to be a subset endowed with the subspace topology and the subgraph relation (so that the meaning of *subcontinuum* is clear) we have in particular the following counterparts of Lemma 2.8 and Theorem 2.11 of [4]. The proofs are as usual, except that at the appropriate points one works with R-disjointness (and Proposition 1.5) in place of simple disjointness.

PROPOSITION 1.7: In any topological graph, the intersection of a filtered collection of continua is a continuum.

PROPOSITION 1.8: If A is any subset of a continuum K, then K contains a subcontinuum irreducible about A.

Defining *cut-point* as usual, we have the following simple observation:

PROPOSITION 1.9: If p is a cut-point of the connected graph X, with $X - p = A \cup B$ (A, B separated), then A and B are open subsets of X.

In the usual manner we can prove:

PROPOSITION 1.9: Every nondegenerate continuum has at least two non-cut-points. Moreover, every continuum is irreducible about the set of its non-cut-points.

Finally we note that the definition and basic theory of local connectedness read as usual (although having, of course, a wider range of application).

DEFINITION 1.11: The graph $(X, \mathcal{T}, R)$ is *locally connected* at $x \in X$ if every neighborhood of x contains a connected open neighborhood of x.

PROPOSITION 1.12: Let X be a (topological) graph. The following are equivalent:
(1) X is locally connected (at every point);
(2) the topology of X has a base of connected (open) sets;
(3) for any open set U of X, each component of U is open.

We conclude the section with some considerations on interval topologies. A (partially) ordered set P will be said to be *complete* if every nonempty subset of P that is bounded above has a least upper bound (equivalently, if every nonempty subset that is bounded below has a greatest lower bound). In case P is totally ordered, this notion coincides with Dedekind completeness (see Definition 2.8). We note that, if P is complete, then the intersection of any collection of order-closed intervals is a closed interval, or P, or $\varnothing$, and conversely. As a slight extension of this observation we obtain:

PROPOSITION 1.13: A linear order $(P, \leq)$ is complete if and only if every interval of P which is a closed set in the interval topology is a $\leq$-closed interval.

Thus, in the complete case, "closed interval" has an unambiguous meaning.

We note that, if $f: X \to Y$ is a continuous, isotone (= monotone increasing) map of complete linear orders, then f preserves existing sups (and infs) of nonempty sets. Indeed, suppose that $S \subseteq X$ is nonempty, and $x = \sup S$, while $x \notin S$. Then $f(x)$ is an upper bound of $f(S)$ by monotonicity. If $y < f(x)$ is also an upper bound of $f(S)$, then the inverse image of the closed set (ray) $\downarrow y$ is the nonclosed ray $\downarrow S$, contradicting the continuity of f.

Next, a proposition to the effect that, in the case of complete linear orders, passage to the interval topology commutes with inverse limit:

PROPOSITION 1.14: Let $(X_i, \mathcal{T}_i, \leq_i)$, $i = 0, 1, \ldots$, be a sequence of complete totally ordered sets, each endowed with the interval topology. Let Σ be the inverse sequence given by the continuous, isotone (= monotone increasing = monotonic) maps $f_i : X_{i+1} \to X_i$. Let $Lim_{\leftarrow}\Sigma$ be $(X, \mathcal{T}, \leq)$, with projections $\pi_i : X \to X_i$. Then $(X, \leq)$ is complete and $\mathcal{T}$ is the interval topology of $\leq$. Suppose, further, that the maps f_i preserve adjacency. Then, for any $x = (x_i)$, $y = (y_i)$ in X,

$$xAy \Leftrightarrow \forall i.\, x_i A_i y_i,$$

where A_i, A are the adjacency relations of X_i, X.

Proof: Let S be a nonempty bounded subset of X. Then each projection $\pi_i(S)$ is nonempty and bounded, with sup (say) x_i. By the preceding remarks, $f_i(x_{i+1}) = x_i$. Thus the sequence (x_i) is an element of X, which is readily seen to be the sup of S. This shows that X is complete.

Next, let k be chosen arbitrarily, and let $J = [x_k, y_k]$ be a closed interval in X_k. Then, for each $l > k$, $f_l^{-1}(J)$ is a closed interval $[x_l, y_l]$, and from this we have that $\pi_k^{-1}(J)$ is the order-closed interval $[(x_l), (y_l)]$ (with slight abuse of notation). Thus, each subbasic $\mathcal{T}$-closed set is order-closed. On the other hand, an arbitrary closed interval $[(x_i), (y_i)]$ in X is clearly the intersection of the $\mathcal{T}$-closed sets $\pi_i^{-1}([x_i, y_i])$. Hence, $\mathcal{T}$ coincides with the interval topology of $(X, \leq)$.

We omit the verification of the adjacency property (see also the proof of Theorem 3.5). ❑

REMARK: The proposition could as well have been formulated for general inverse systems as for sequences.

It follows that topological graphs defined via inverse limits in the manner of K_∞ and H_∞ can be presented simply as linear orderings: the order fixes the topology (as the interval topology) and the relation (as the adjacency). The ordering of H_∞ in particular is that of the unit interval, with the insertion, for each dyadic $d \in I$, of "copies" d_- and d_+ adjacent to d on the left and on the right. We note that in the (interval) topology of H_∞ the dyadics are isolated points, which at the same time constitute a dense subset of H_∞.

2. LINEAR STRUCTURES

Given three points x, y, z of a connected (topological) graph $(X, \mathcal{T}, R)$, we say that y *separates* $\{x, z\}$ if $X - \{y\}$ is disconnected, and x, z belong to distinct members of a separation of $X - \{y\}$, that is, of a partition (U, V) of $X - \{y\}$ into R-disjoint open subsets.

DEFINITION 2.1: A linear (*topological*) graph is a connected locally connected graph satisfying the following ("COTS") condition:
For any three distinct points, one of them separates the other two.

Apart from the use of topological graphs, the main difference between our Definition 2.1 and the original definition of COTS ("connected ordered topological spaces") in [5] is the requirement of local connectivity. The main effect of this is to rule out examples in which there are "irrelevant" open sets, that is, open sets not related to the intended linear order. Thus, if we enlarge the (Euclidean) topology of the unit interval I by admitting $I \cap \mathbb{Q}$ as an open set, the result is still a COTS, but (taking the graph relation as the identity) not a linear graph.

A further slight difference between our formulation and that of [5] lies in the definition of "separates." In [5], x is said to separate $\{y, z\}$ if y, z lie in different *components* of $X - \{x\}$. The two definitions are easily seen to be equivalent in the presence of local connectivity, but we have not been able to establish their equivalence in the absence of this condition. For expository reasons we prefer not to appeal to local connectivity until later in the argument (see 2.6 and 2.12), and that being so it is the (apparently) stronger notion of "separates" that we need for our development.

It is important to observe that the structures K_∞, H_∞ of Example 1.2 are linear when viewed as topological graphs, but not otherwise. The interval topologies of these structures are of course not connected, while the saturated topologies do not satisfy the COTS condition. In particular, the points d_-, d_+ for d dyadic are cut-points of sI when this is viewed as a topological graph, but not in its saturated topology.

Given three points x, y, z of a linear graph X, we shall say that y, z are *on the same side of* x if x does not separate $\{y, z\}$. This amounts to saying that, for any partition $X = U \cup \{x\} \cup V$, with U, V open, y and z belong to the same one of U, V. (Note: it is a fact that, for any given $x \in X$, there is at most one nontrivial partition of this type, but we prefer not to demonstrate or use this fact just yet.)

The next proposition is best understood in relation to studies of the foundations of Euclidean geometry which, following Pasch, Hilbert, and others, characterize lines in terms of a "betweenness" relation. We follow the version in Borsuk and Szmielew [2], including the numbering of the properties of the relation; omitted are those properties which assert that a line is dense, unbounded, and (Dedekind) complete.

PROPOSITION 2.2: Let X be a linear graph. Write $B(x, y, z)$ for "y separates $\{x, z\}$." Then the following assertions hold (for any points a, b, c, d of X):

(O2) $B(a, b, c) \Rightarrow B(c, b, a)$
(O3) $B(a, b, c) \Rightarrow \neg B(b, a, c)$
(O4) a, b, c mutually distinct $\Rightarrow B(a, b, c) \vee B(b, c, a) \vee B(c, a, b)$
(O7) $B(a, b, c) \wedge B(b, c, d) \Rightarrow B(a, b, d)$
(O8) $B(a, b, d) \wedge B(b, c, d) \Rightarrow B(a, b, c)$.

Proof: (O3): Suppose $B(a, b, c)$; that is, we have a partition (U_b, b, V_b) of X with $a \in U_b$, $c \in V_b$. Suppose, if possible, that we also have a partition (U_a, a, V_a) with $b \in U_a$, $c \in V_a$. Then it can be checked that $(U_a \cup U_b, V_a \cap V_b)$ is a partition of X into R-disjoint open sets. But this is impossible since X is connected. Hence, $\neg B(b, a, c)$

(O7): Assume $B(a, b, c)$ and $B(b, c, d)$. By (O3), $\neg B(d, b, c)$. That is, c, d are on the same side of b. But c, a are on opposite sides of b (that is, they are not on the same side), and therefore, so are d, a. Thus $B(a, b, d)$.

(O8): Similar to O7. ❑

Starting from a characterization of the line in terms of a betweenness relation, geometers customarily derive a total ordering of the points of the line. We can adapt their procedure to yield a canonical (up to reversal) total order of a linear graph. For example, one may follow the development in [3], checking that only the properties listed in Proposition 2.2 (that is, omitting density, unboundedness, and completeness) are needed for the derivation. Alternatively, one may fill in the details of the relatively quick and direct development we now sketch.

Let L be a set equipped with a ternary relation B satisfying the properties (O2)–(O8) listed in Proposition 2.2. Given an arrangement (ordering) (a, b, c, d) of four distinct elements of L, let us say that the arrangement *yields* an assertion $B(x, y, z)$ if y is indeed between x, z, in the ordinary sense, in the arrangement.

PROPOSITION 2.3: Let $A \subseteq L$ have cardinality 4. Then there is a unique (up to reversal) arrangement of A which yields exactly the true assertions of the form $B(x, y, z)$ $(x, y, z \in A)$.

Proof: (*sketch*) Choose three elements of A, and let $B(a, b, c)$ be an assertion true for them (cf. (O4)). This will fix the order of a, b, c within the arrangement. Let d be the fourth element of A. If either $B(b, c, d)$ or $B(b, d, c)$, the whole arrangement is fixed. Suppose, however, that $B(d, b, c)$. Consider $\{a, b, d\}$. We see that $B(a, b, d)$ is not possible, by considering in turn $\{a, c, d\}$. Indeed, each assertion about the last set leads to a contradiction. For example, suppose $B(a, c, d)$. By (O2) and (O8) we have

$$B(a, b, c), B(a, c, d) \Rightarrow B(b, c, d),$$

so that we get a contradiction with $B(d, b, c)$ (cf. (O3)). Thus we have either $B(d, a, b)$ or $B(a, d, b)$, fixing the arrangement of A in either case.

Lastly, we must show that the arrangement thus determined yields exactly the true assertions. There are four (very similar) cases to check. The first case involves checking that $B(a, b, c) \wedge B(b, c, d)$ implies an "assertion" P if the arrangement (a, b, c, d) yields P, and implies $\neg P$ otherwise. There are three further cases corresponding to the other three possible ways in which the position of d in the arrangement was fixed. ❑

With the aid of this Proposition it is straightforward to define the canonical ordering(s) of L. Assume that $|L| > 2$, and fix two elements a, b of L. For any distinct $x, y \in L$, write $x < y$ if, in the unique arrangement of $\{a, b, x, y\}$ $(= A)$ yielding the correct B-assertions about A, and in which a precedes b, x also precedes y. (We have to allow for the possibility of overlap between $\{a, b\}$ and $\{x, y\}$. But it is trivial that, if A has only two or three elements, it has a unique arrangement satisfying the stated constraints.) It is readily verified that the relation $<$ is a (strict) total ordering of L. We shall denote this order, and its nonstrict version, by $<_B$, $\leq_B$.

In terms of this notation, we have:

THEOREM 2.4: The relation $<_B$ is, up to inversion, the unique ordering $<$ of L such that

$$x < y < z \Leftrightarrow B(x, y, z).$$

Proof: That $<_B$ satisfies the condition is clear. On the other hand, two total orderings $<$, $<'$ of a set X coincide up to inversion if and only if, for no $a, b, c \in X$, do we have $a < b < c$ while neither $a <' b <' c$ nor $c <' b <' a$; that is, if and only if the order-open intervals with respect to the two orders coincide. This establishes uniqueness. ❑

Starting with a linear graph $(L, \mathcal{T}, R)$ we have defined its "betweenness" relation B, and thence the linear order $<_B$. We now aim to show that the topology $\mathcal{T}$ can be recovered from the order $<_B$ as its interval topology. Denote this interval topology by $\mathcal{T}_B$. A subbase for $\mathcal{T}_B$ is given by the sets of the form $\downarrow x$, $\uparrow x$.

PROPOSITION 2.5: For any $x \in L$, the subspace $L - \{x\}$ is connected if x is an endpoint (that is, a least or greatest element with respect to $<_B$), and otherwise has exactly two components. In the latter case the components are $\downarrow x$ (that is, $\{y \mid y < x\}$) and $\uparrow x$.

Proof: Suppose that the subspace $L - \{x\}$ has two or more components. Then, by the definition of B and Theorem 2.4, we have points u, v such that $u < x < v$. By contraposition, if x is an end-point, then $L - \{x\}$ is connected.

Suppose now that x is not an end point, so that we have u, v with $u < x < v$. Let (U_x, V_x) be a partition of $L - \{x\}$ into saturated open sets such that $u \in U_x$, $v \in V_x$. We claim that $U_x = \downarrow x$ and $V_x = \uparrow x$. For suppose this is not so, so that, say, U_x meets $\uparrow x$. If $w \in U_x \cap \uparrow x$ we have (by the definition of B) $B(w, x, v)$. But this is impossible, since both $x < w$ and $x < v$. ❑

COROLLARY: $\mathcal{T}_B \subseteq \mathcal{T}$.

Notice that local connectivity has played no part in the argument until now. In fact, it is only so that we can prove the converse of the corollary just stated, that we require linear graphs to be locally connected:

PROPOSITION 2.6: With the notation of Proposition 2.5 we have:

$$\mathcal{T} \subseteq \mathcal{T}_B .$$

Proof: Let V be a $\mathcal{T}$-neighborhood of the point $x \in L$. By local connectivity, V contains a connected set, that is to say an interval, I, such that I is a $\mathcal{T}$-open neighborhood of x. If x is an (order-) interior point of I, then V is a $\mathcal{T}'$-neighborhood of x, and we are done. The only case in which this may fail to hold is that x is an end-point of I, say a left end-point, while there is no point adjacent to x on the left (although $(-\infty, x)$ is nonempty). But this case cannot arise, since, in view of the already established fact that order-open sets are $\mathcal{T}$-open, we then would have $\{(-\infty, x), [x, \infty)\}$ as a partition of L into R-disjoint $\mathcal{T}$-open sets. ❑

We have seen that, starting with a linear topological graph $(X, \mathcal{T}, R)$, we can define a ternary relation B on X satisfying (O2)–(O8), and thence a total order $<_B$ on X, from which the original topology can be recovered as its interval topology. Until now, the relation R has been largely ignored in this development. In fact, it is clear that the following is satisfied:

(P) $\exists y.\, B(x, y, z) \Rightarrow \neg R(x, z)$.

(For, if we had $B(a, b, c)$ while $R(a, c)$, b would fail to separate $\{a, c\}$.) As a partial converse, we note that if condition (P) is satisfied, then the induced interval topology, taken together with R, gives a topological graph. Indeed, suppose that $\neg R(a, c)$. If $\exists x.\, B(a, x, c)$, let b be some point between a and c, and put $U = \downarrow b$, $V = \uparrow b$; otherwise put $U = \downarrow c$, $V = \uparrow a$. In either case we have that $R(U)$ is disjoint from V.

In view of these considerations, we adopt the following definition:

DEFINITION 2.7: A (*linear*) *B-structure* is a triple (X, B, R) satisfying conditions (O2)–(O8) and (P).

In terms of the associated order $<_B$, condition (P) states that the relation R holds between points x, y only if the points are adjacent (which we take to include the case that $x = y$). A criterion for the connectedness of a B-structure, or its associated linear order, is easily given in terms of (the lack of) "gaps." To be precise:

DEFINITION 2.8: Let $(X, \leq)$ be a total order, and R a relation satisfying (P). A partition (A, B) of X, where A, B are (nonempty) lower and upper sets, respectively, is a *gap* if one of the following holds:

(1) A has no greatest element and B has no least element;
(2) A has a greatest element a and B has a least element b, but a, b are not R-related.

Notice that X has no gap of type (1) (that is, X is Dedekind-complete) if and only if it is complete in the sense studied in Section 1.

PROPOSITION 2.9: A B-structure X is connected (that is, induces a connected topological graph, via the interval topology) if and only if X, viewed as a total order, has no gaps.

Since a connected B-structure is automatically locally connected, it is evident that the various notions of (connected) linear structure that we have formulated are essentially equivalent. A slight technical difficulty here is that the choice of total order ($<_B$) is unique only up to inversion. This could be accommodated by taking a linear graph, or B-structure, with a specific *orientation*. We prefer, however, to modify the notion of an ordered set by treating the order and its inverse as being on an equal footing. Specifically:

DEFINITION 2.10: A *linear order with adjacency* is a triple $(X, \{\leq, \leq^{-1}\}, R)$, where $\leq$ is a total order on X, and R is a restricted adjacency relation with respect to $\leq$, in the sense that

$$x(R \cup R^{-1})y \Leftrightarrow xA\,y,$$

where A is the adjacency relation (Example 1.2).

The above findings can be summarized:

THEOREM 2.11: In terms of the preceding constructions, the following notions are equivalent:

(1) Linear topological graph;
(2) Connected B-structure;
(3) Complete linear order with adjacency.

Since closed intervals of a Dedekind-complete total order are compact in the interval topology, a linear graph is necessarily locally compact. Let us consider further the compact case. Suppose that we have a compact connected topological graph $(X, \mathcal{T}, R)$ (not assumed to be lc) satisfying the COTS condition. Observe that the proof of Proposition 2.5 does not make use of local connectivity of the structure, and so this proposition and its corollary hold for X. Now recall that the compact Hausdorff topologies on a given set S are incomparable (in the lattice of all topologies on S): specifically, if $\mathcal{T}'$ is a compact Hausdorff topology on S, then any topology strictly coarser than $\mathcal{T}'$ is non-Hausdorff, while any topology strictly finer than $\mathcal{T}'$ fails to be compact. Since an interval topology is necessarily Hausdorff, the corollary to Proposition 2.5 implies, in the present instance, that $\mathcal{T}_B = \mathcal{T}$. It of course follows that $\mathcal{T}$ is locally connected. If an "arc" is understood to be the compact metrizable (2nd-countable) case of a "line," we thus arrive at the following definition:

DEFINITION 2.12: An *arc* is a 2nd-countable compact connected topological graph satisfying the COTS condition.

In terms of the betweenness or order version of a linear structure, compactness simply means that the structure has two end-points. Indeed, it is evident that a nondegenerate arc in the sense of Definition 2.12 has exactly two end-points (non-cut-points). For a converse, let M be a 2nd-countable continuum having exactly two non-cut-points. Then one may proceed as in the usual theory (Hocking and Young, Chapter 2), although using Propositions 1.7 to 1.10, to develop the canonical (up to inversion) linear ordering of M. As a (partial) alternative to the usual development, more in line with what we have done above, one may formulate the notion of a *bounded* B-structure: a B-structure in which there are two points a, b such that for every x distinct from a and from b, $B(a, x, b)$. (The axiomatization of a bounded B-structure can be significantly simpler than that for a general B-structure.) One readily deduces from Proposition 1.10 that a continuum having exactly two non-cut-points is a bounded B-structure.

In summary we have:

PROPOSITION 2.13: Let X be a nondegenerate 2nd-countable continuum. Then X is an arc iff it has exactly two non-cut-points.

3. MAPPINGS. INVERSE LIMITS

We seek a notion of morphism according to which inverse systems of linear structures have limits, and, in addition, images of arcs are again arcs.

DEFINITION 3.1: A morphism $f: X \to Y$ of topological graphs X, Y is said to be *connectedness-reflecting* (abbreviated as: *c-morphism*) if, for every connected subset S of $f(X)$, $f^{-1}(S)$ is connected.

It can be checked, either directly or as a consequence of Proposition 3.3, that in the case of linear graphs, the composition of two c-morphisms is a c-morphism.

The above definition has a somewhat provisional character. In particular, there are some indications that it would be advantageous to consider morphisms which reflect compactness as well as connectedness. There is evidently some connection with the established topological notion of a monotone map (a continuous map such that the inverse image of each point is a compact connected set). In the case of a continuous function between ordinary compact spaces, the two notions coincide (see Whyburn [12, Chapter 8(2.2)]). For topological graphs this is no longer true: for example, an injective graph morphism $f: L \to G$, where L is discrete linear, is trivially monotone but, clearly, need not be a c-morphism. We leave the further investigation of these matters to another occasion.

LEMMA 3.2: Let $f: X \to Y$ be a mapping of totally ordered sets. Then f is either isotone or antitone if and only if the following holds:

$$\forall x, y, z \in X.\, y \in [x, z] \Rightarrow f(y) \in [f(x), f(z)].$$

It is convenient to use the notation $B_=(x, y, z)$ for $B(x, y, z) \vee x = y \vee y = z$. Thus, Lemma 3.2 says that, under the stated condition, f is isotone or antitone iff f preserves $B_=$. We should mention that we view $[u, v]$ as the interval with end-points u, v, regardless of whether $u \leq v$ or $v \leq u$.

If X, Y are B-structures (or linear orders with adjacency), a map $f: X \to Y$ will be called *monotone* if it is isotone for some choice of the orders of X, Y; equivalently if, the total order of each of X, Y having been fixed, f is either isotone or antitone.

PROPOSITION 3.3: Let (X, B, R), (X', B', R') be connected B-structures, and $f: X \to X'$. Then the following assertions about f are equivalent:

(1) f is continuous (with respect to the interval topologies $\mathcal{T}_B$, $\mathcal{T}_{B'}$) and monotone;

(2) f is continuous, and $B_=$-preserving (that is, $B_=(x, y, z) \Rightarrow B_='(f(x), f(y), f(z))$);

(3) the inverse image of any closed interval of Y is a closed interval of X.

Moreover, the conjunction of any of (1)–(3) with the assertion that f is R-preserving is equivalent to:

(4) f is a c-morphism.

Proof: The equivalence of (1) and (2) is given by Lemma 3.2 together with Theorem 2.4.

(3) ⇒ (2): Condition (3) implies continuity of f since the closed intervals are a subbase for the closed sets. But it also implies preservation of $B_=$(by f); for if, for some triple a, b, c we have $B_=(a, b, c)$ while $\neg B_='(f(a), f(b), f(c))$, then the inverse image of the closed interval $[f(a), f(c)]$ contains the points a and c but not the point b which lies between them (thus, this inverse image is not an interval).

(2) ⇒ (3): That f is $B_=$-preserving clearly implies that the inverse image of a (closed) interval C is an interval; continuity of f gives that $f^{-1}(C)$ is closed (recalling Proposition 1.13).

Finally, since intervals are the same as connected subsets, we have the equivalence of (2) (or (3)), conjoined with R-preservation, with (4). ❑

Some variations of condition (3) in the preceding proposition are evidently possible. For example, open intervals may be substituted for closed intervals. And, of course, it suffices to consider, in X', some restricted set of intervals that still forms a subbase for the open (or the closed) sets. In this connection, the following proposition, giving a necessary and sufficient condition that a given set of points yield a set of rays that is a subbase, will prove useful (see Proposition 4.2).

PROPOSITION 3.4: Let (X, B, R) be a B-structure, and P a subset of X. The following are equivalent:

(1) The open rays from points of P are a subbase for $\mathcal{T}_B$.
(2) The closed rays from points of P are a subbase for the closed sets.
(3) P is dense, and for any $x \neq y$ such that x, y are adjacent, both x and y belong to P.

Proof: The equivalence of (1) and (2) is trivial, since the open rays are the complements of the closed rays.

(1) ⇒ (3): Assume that the open rays from points of P (P-rays, for short) form a subbase. Let $x \neq y$ be adjacent. Clearly, only a ray from x can contain y but not x. Hence, $x \in P$. Next, suppose (u, v) is a nonempty open interval. If (u, v) contains only isolated points, then by what we have seen, these belong to P. If (u, v) contains a nonisolated point z, then some points "close to" z must belong to P.

(3) ⇒ (1): Suppose that (3) is satisfied. Consider an arbitrary ray, say $E = (x, \infty)$. If there is a point adjacent to x on the right, then $x \in P$; if not, then, since P is dense, E is the union of (open) P-rays. ❑

THEOREM 3.5: Let Σ be an inverse system $(X_i, \mathcal{T}_i, R_i)_{i \in I}$ of linear graphs, with c-morphisms $f_{ik}: X_i \to X_k$ as bonding maps. Let $(X, \mathcal{T}, R)$ be the inverse limit of Σ in the category of topological graphs, with projections π_i: $X \to X_i$. Then X is a linear graph, and the π_i are c-morphisms.

Proof: We shall view the X_i as connected linear orders with adjacency, via Theorem 2.11 and Proposition 3.3. Let distinct points u, v of X be chosen (the case that X has fewer than two points being trivial). We can without loss of generality assume that $\pi_i(u) \neq \pi_i(v)$ for all $i \in I$; for the set of indices i for which this inequality holds is clearly cofinal with I, so that X_i in which $\pi_i(u) = \pi_i(v)$ can be discarded without affecting the inverse limit obtained. Choose for each i the ordering of X_i for which $\pi_i(u) < \pi_i(v)$. The bonding maps are monotonic with respect to these (linear) orderings, and so a linear order $\leq$ is induced on X (with $u < v$). Since the bonding maps are also continuous, we can conclude that the order $\leq$ has no gaps of type (1) (cf. remarks following Definition 2.8). Moreover (Proposition 1.14), the interval topology induced by $\leq$ coincides with $\mathcal{T}$. It remains to investigate R.

We note that only adjacent points in X can be R-related; for if $x < y < z$ while $xR \cup R^{-1}z$, the corresponding situation would (by monotonicity of the π_i) have to obtain in some X_i. Next, suppose that $x = (x_i)$ and $y = (y_i)$ are adjacent in X with $x < y$. We claim that each pair x_i, y_i is adjacent (in X_i). For suppose we have $x_i < t_i < y_i$ for some i. Given any $k \geq i$, the interval $[x_k, y_k]$ is mapped onto $[x_i, y_i]$ by f_{ki}, since graph morphisms preserve connectedness; thus there exists t_k, lying between x_k and y_k, with $f_{ki}(t_k) = t_i$. In this way we would obtain $t = (t_i)$ with $x < t < y$. Thus the (x_i, y_i) are R_i-related, in a coherent fashion: say $x_i R_i y_i$ for all $i \in I$. Then we have xRy. We have shown that X is a connected linear order with adjacency. ❑

4. ARCS

We first consider images of arcs under c-morphisms.

PROPOSITION 4.1: Let C be an arc, $(X, \mathcal{T}, R)$ a topological graph, and $f: C \rightarrow X$ a c-morphism. Then $f(C)$, as a subspace of X, is an arc.

Proof: The image $f(C)$ is compact and connected. Let $x, y, z \in f(C)$ be distinct. Then $f^{-1}(x)$, $f^{-1}(y)$, $f^{-1}(z)$ are disjoint closed intervals of C and one of them, say $f^{-1}(y)$, separates the other two. We claim that y separates x, z in $f(C)$. Indeed, let $\{U, V\}$ be the partition of $C - \{f^{-1}(y)\}$ into open sets, with $f^{-1}(x) \subseteq U$, $f^{-1}(z) \subseteq V$. Then the sets $f(U)$, $f(V)$ are open in $f(C)$, and form a partition of $f(C) - \{y\}$ such that $x \in f(U)$, $z \in f(V)$. Further, these sets are R-disjoint: for if we had $u \in f(U)$, $v \in f(V)$ with $u(R \cup R^{-1})v$, then $\{U, V\}$ would be a connected set although its inverse image under f is not connected. Thus $f(C)$ satisfies the COTS condition. ❑

PROPOSITION 4.2: Every arc is the image of sI under a c-morphism.

Proof: Let $(C, \mathcal{T}, R)$ be an arc, with end-points c_0, c_1 (assumed distinct). Let $p_0, p_1, \ldots,$ be an enumeration of a dense subset P of C such that the intervals (p_i, p_j) form a base of open sets for the space (c_0, c_1) of non-end-points of C. We now proceed (roughly) by repeatedly subdividing the interval (c_0, c_1), assigning the dyadic points of sI to the points of division of the resulting intervals in a systematic fashion.

In more detail, begin by assigning the end-points L, R of sI to c_0, c_1. As regards internal points: subintervals of (c_0, c_1) arising in the construction will be labelled with strings $\in \{+, -\}^*$, with (c_0, c_1) itself receiving the label Λ. Suppose that, at a given stage, we have the (sub)interval (p, q), labelled σ. We have two cases to consider.

(A) The interval (p, q) is empty. We note that the following assertions hold:

(a) $p \neq q$

(b) $p, q \in P$ (by construction of P)

(c) p, q are R-related (since C is connected).

If it happens that pRq, then all the dyadic points $\tau 0^\omega$, for extensions τ of σ, are assigned to q; otherwise they are assigned to p.

(B) The interval (p, q) is nonempty. In this case we let r be the first term p_i in the enumeration of P such that $p_i \in (p, q)$; assign (map) the dyadic point $\sigma 0^\omega$ to r; and construct the new subintervals (p, r), (r, q), labelled $\sigma-$, $\sigma+$, respectively.

By this construction, every dyadic point (of sI) is mapped to some element of P, and conversely. The mapping is easily seen to be monotonic. Indeed, we have the following Fact, easily proved by induction on the generation (by subdivision) of intervals (p, q) along with their labels.

FACT: Let (p, q) be an interval of C, with label σ arising in the above construction. Suppose that the dyadics x, y are mapped to p, q, respectively. Then every dyadic which lies strictly between p and q is of the form $\tau 0^\omega$, where τ extends σ, and moreover, every such dyadic is mapped into the interval $[p, q]$.

Let $f\colon D \to C$ be the mapping from dyadics to points of P defined in the above construction. We claim that f is uniformly continuous, with respect to the (unique) uniformity of (C, R) and the uniformity of D induced from that of sI (that is, H_∞: Example 1.2(B)). Indeed, given any uniform (open) cover $\mathcal{C}$ of C, we may choose a (finite) uniform cover $\mathcal{C}'$ refining $\mathcal{C}$, such that the members of $\mathcal{C}'$ are (closed) intervals whose end-points belong to $\{c_0, c_1, p_0, p_1, \ldots\}$. Moreover, we may assume that the (interiors of the) members of $\mathcal{C}'$ are intervals which have received labels in the above labelling process, the maximum length of such labels being n. In view of the Fact, the inverse image of $\mathcal{C}'$ under f is a uniform cover of D (corresponding to some relation $U_{k(n)}$: Example 1.2(B)). Let $\overline{f}$ be the continuous extension of f over sI. Then $\overline{f}$ is monotonic.

It remains to show that $\overline{f}$ is R-preserving. Let $d, d' \in sI$ with $d \neq d'$, $d \sqsubseteq d'$. This means that d is dyadic, and d' is adjacent to d; suppose without loss of generality that $d \leq d'$ (so that d' is d_+). We deduce that $\overline{f}(d)$, $\overline{f}(d')$ are adjacent in C; for if not, then there exists some point p of P lying strictly between $\overline{f}(d)$ and $\overline{f}(d')$, and this point p must be the image of some dyadic point e that (by monotonicity of $\overline{f}$) lies between d and d', which is impossible. Thus $\overline{f}(d) = p$, $\overline{f}(d') = r$ are R-related elements of P, and (q, r) is an empty interval that arise in the above construction with a label, say σ.

Suppose, if possible, that $\neg qRr$. Then rRq. Let e be a dyadic such that $f(e) = r$, and e' a dyadic, lying strictly between d and e; clearly, $d < d' < e' < e$. According to the Fact, e' is of the form $\tau 0^{\omega}$, where τ extends σ; thus, by the construction of f , e' is mapped to q. But now we have a contradiction, since, by monotonicity of $\overline{f}$, e' is mapped to r. Thus qRr, and we conclude that $\overline{f}$ is a c-morphism. ❑

Combining the preceding results, we obtain:

THEOREM 4.3: A subspace C of a topological graph X is an arc if and only if C is the image of sI under a c-morphism $f\colon sI \to X$.

Our last topic is that of the relation between connectivity and arcwise connectivity. The treatment will be brief; our main observation is that the standard theory can be adapted, with minor adjustments, to the more general context considered here. We follow, broadly, the exposition of Hocking and Young [4], as being the most accessible. However, as pointed out by the referee, that exposition contains a significant error. The source of the error lies in the definition of "goes straight through" (applied to chains of open sets). With the appropriate definition of this concept, as in Schurle [9], the argument proceeds without difficulties.

DEFINITION 4.4: Given two points a, b of a space (graph) $(X, \mathcal{T}, R)$, a collection $\{A_1, \ldots, A_n\}$ of subsets of X is a *simple chain from a to b* provided that A_1 (and only A_1) contains a, A_n (and only A_n) contains b, and A_i is R-disjoint from A_j if and only if $|i - j| > 1$.

PROPOSITION 4.5: If a and b are two points of a connected graph $(X, \mathcal{T}, R)$, and $\{U_\alpha\}$ is a collection of open sets covering X, then there is a simple chain of elements of $\{U_\alpha\}$ from a to b.

Proof: Exactly as in [4], or [9], one shows that the set Y of points y, such that there is a simple chain of U_α from a to y is clopen. Then, we claim that Y is also R-disjoint from $X - Y$. Indeed, suppose that we have $y \in Y$ and $x \in X - Y$ with $xR \cup R^{-1}y$. Let $K = \{U_1, \ldots, U_n\}$ be a simple chain from a to y; and let U_α be a member of the given open cover such that $x \in U_\alpha$. Then by taking $\{U_1, \ldots, U_m\}$, where U_m is the first term of K that is not R-disjoint from U_α, together with U_α, we have a simple chain from a to x. Since X is given to be connected, we conclude that $Y = X$. ❑

Given chains $K = \{U_1, \ldots, U_m\}$ and $L = \{V_0, \ldots, V_n\}$ from a to b, in a graph X, we say, as in [9], that L *goes straight through* K if there are integers $t_0, t_1, \ldots, t_m$ such that $0 = t_0 < t_1 < \ldots < t_m = n$, $V_i \subseteq U_j$, whenever $t_{j-1} \leq i < t_j$, and $V_n \subseteq U_m$.

We now have the counterpart of Theorem 3.14 of [4], or 4.2.4 of [9]:

PROPOSITION 4.6: Let X be a graph, and let $K = \{U_1, \ldots, U_n\}$ be a simple chain of connected open sets from a to b, where $a, b \in X$. Suppose that $\mathcal{V}$ is a collection of open sets such that each U_i is a join of members of $\mathcal{V}$. Then there is a simple chain of members of $\mathcal{V}$, from a to b, that goes straight through K.

Proof: (*outline*) The proof is as in [4], [9] except that we select from each U_i *two* (not necessarily distinct) points x_i, x_i', with $a = x_1$, $b = x_n'$, and each x_i' R-related to x_{i+1} ($i = 1, \ldots, n-1$). Then we take, within each U_i, a simple $\mathcal{V}$-chain from x_i to x_i'; putting these chains together, we get (after, if necessary, some pruning) a simple $\mathcal{V}$-chain from a to b, that goes straight through K. ❑

A compact, connected, locally connected graph $(X, \mathcal{T}, R)$, such that $\mathcal{T}$ is metrizable, may be termed a *Peano graph*. Using Proposition 4.6, and bearing in mind Proposition 1.7 as well as the characterization of arcs given by Proposition 2.13, we can now prove, in exactly the style of [4, Theorem 3.15]:

THEOREM 4.7: Every Peano graph is arcwise connected.

ACKNOWLEDGMENT

Thanks are due to Julian Webster and an anonymous referee for several corrections and improvements to the text. Helpful comments on the material were also provided by Peter Collins.

REFERENCES

1. ABRAMSKY, S & A. JUNG. 1992. Domain Theory. *In:* Handbook of Logic in Computer Science, Vol. 3 (Abramsky, Gabbay, & Maibaum, Eds.). Oxford University Press. New York.
2. BORSUK, K, & W. SZMIELEW. 1960. Foundations of Geometry. North-Holland. Amsterdam.
3. E. ČECH. 1966. Topological Spaces. Wiley-Interscience. New York.
4. HOCKING, G. & J. YOUNG. 1961. Topology. Addison-Wesley. Reading, Massachusetts.

5. KHALIMSKY, E., R. KOPPERMAN, & P. MEYER. 1990. Computer graphics and connected topologies on finite ordered sets. Topology and its Applications. **36:** 1–17.
6. KRAWCZYK, A. & J. STEPRANS. 1993. Continuous colourings of closed graphs. Topology and its Applications. **51:**
7. LATECKI, L. & F. PROKOP. 1995. Semi-proximity continuous functions in digital images. Preprint.
8. LAWSON, J. 1987. The versatile continuous order. *In:* Mathematical Foundations of Programming Language Semantics. Lecture Notes in Computer Science, 298. Springer-Verlag. New York.
9. SCHURLE, A. 1979. Topics in Topology. North-Holland. Amsterdam.
10. SMYTH, M. 1992. Stable compactification I. J. Lond. Math. Soc. **45:** (2) 321–340.
11. SMYTH, M. 1995. Semi-Metrics, closure spaces and digital topology. Theoret. Comp. Sci. **151:** 257–276.
12. WHYBURN, G. 1963. Analytic Topology. Amer. Math. Soc. Coll. Publications, **Vol. 28.**

Versions of Monotone Paracompactness

IAN S. STARES

Department of Mathematics
University of North Carolina at Greensboro
Greensboro, North Carolina 27412

ABSTRACT: We consider the notion of monotone paracompactness introduced by Gartside and Moody. We show, in particular, that not all "possible" definitions for monotone paracompactness coincide, and we provide a new characterization of proto-metrizability.

1. INTRODUCTION

In [2] Gartside and Moody introduced the following notion.

DEFINITION 1.1: A space X is said to be *monotonically paracompact* if there is a map $m: C \to C$, where C is the set of open covers of X, such that,
(a) $m(\mathcal{U})$ star refines $\mathcal{U}$ for each $\mathcal{U} \in C$,
(b) $m(\mathcal{U})$ refines $m(\mathcal{V})$ whenever $\mathcal{U}$ refines $\mathcal{V}$.

Essentially, this is the idea behind monotone normality applied to paracompact spaces. Since then, others have applied this idea to other covering properties and we now have the definitions of monotone orthocompactness [3], monotone compactness [6], and monotone Lindelöfness [7].

The motivation for the definition of monotone paracompactness was a study of the important class of generalized metric spaces; the proto-metrizable spaces. Although originally defined in terms of special bases, the following characterization of proto-metrizability due to Nyikos [10] is very striking. The class of proto-metrizable spaces is the class of spaces which can be constructed as follows. Given a metrizable space X_0 and a subspace Y of X_0, then, for each $y \in Y$, we replace y by a clopen copy of some metrizable space (possibly a different one for each y). This process is continued transfinitely, taking some subspace of the inverse limit at limit ordinals and stopping at some particular stage.

Gartside and Moody proved the following elegant theorem.

THEOREM 1.2: [2] A space X is proto-metrizable if and only if X is monotonically paracompact.

Mathematics Subject Classification: Primary 54D20; Secondary 54E20.
Key words and phrases: Collins-Roscoe mechanism, McAuley's bow tie space, monotone paracompactness, proto-metrizable.

As Gartside and Moody pointed out, there is a question of notation here. Should the property above actually be called monotone paracompactness? It would perhaps be more correct to call the above property monotone full normality and reserve the term monotone paracompactness for the following property: for a space X there is a map $m\colon C \to C$ such that,

(a) $m(\mathcal{U})$ is a locally finite refinement of $\mathcal{U}$ for each $\mathcal{U} \in C$,

(b) $m(\mathcal{U})$ refines $m(\mathcal{V})$ whenever $\mathcal{U}$ refines $\mathcal{V}$.

Moreover, are these two properties equivalent? More generally, if we "monotonize" a given characterization of paracompactness in a similar fashion to the above, do we get a property equivalent to monotone paracompactness as defined by Gartside and Moody? In this paper we show that in general we do not. We take two characterizations of paracompactness and "monotonize" them both in exactly the same manner, yet produce two distinct classes of spaces. Along the way, we provide new characterizations of proto-metrizable spaces in the spirit of Theorem 1.2 above.

Before we proceed, we first recall the two characterizations of paracompactness which we intend to monotonize.

THEOREM 1.3: [9] X is paracompact iff for each open cover $\mathcal{U}$ of X there is a semineighborhood D of the diagonal in X^2 such that $\{D[x]: x \in X\}$ refines $\mathcal{U}$.

THEOREM 1.4: [4, pp.156–161] X is paracompact iff for each open cover $\mathcal{U}$ of X there is a neighborhood D of the diagonal in X^2 such that $\{D[x]: x \in X\}$ refines $\mathcal{U}$.

We now "monotonize" these two conditions. For the remainder of the paper, C denotes the collection of open covers on the space X, S denotes the collection of semineighborhoods of the diagonal in X^2, and $\mathcal{N}$ denotes the collection of neighborhoods of the diagonal in X^2.

DEFINITION 1.5: A space X is said to be *monotonically semineighborhood refining* (MSNR) if there is a map $m\colon C \to S$ such that,

(a) $\{m(\mathcal{U})[x]: x \in X\}$ refines $\mathcal{U}$ for each $\mathcal{U} \in C$,

(b) $m(\mathcal{U}) \subseteq m(\mathcal{V})$ whenever $\mathcal{U}$ refines $\mathcal{V}$.

DEFINITION 1.6: A space X is said to be *monotonically neighborhood refining* (MNR) if there is a map $m\colon C \to \mathcal{N}$ such that,

(a) $\{m(\mathcal{U})[x]: x \in X\}$ refines $\mathcal{U}$ for each $\mathcal{U} \in C$,

(b) $m(\mathcal{U}) \subseteq m(\mathcal{V})$ whenever $\mathcal{U}$ refines $\mathcal{V}$.

Clearly, all MNR spaces are MSNR. We shall see that although, without the monotonicity conditions (b), these two properties both characterize paracompactness, they actually give distinct classes of spaces. We shall see, however, that MNR spaces are proto-metrizable and hence, Definition 1.6 is a valid alternative to the original definition of monotone paracompactness.

As mentioned above, other authors have considered adding monotonicity conditions to covering properties. Almost 2-full normality is a considerable weakening of paracompactness first considered by Mansfield [5]. We give the following definition.

DEFINITION 1.7: A space X is said to be *monotonically almost 2-fully normal* if there is a map $m: C \to C$ such that,

(a) for each $x \in X$, if $y, z \in St(x, m(\mathcal{U}))$, then $y, z \in U$ for some $U \in \mathcal{U}$,

(b) $m(\mathcal{U})$ refines $m(\mathcal{V})$ whenever $\mathcal{U}$ refines $\mathcal{V}$.

Before we proceed, we recall one definition from [1]. A space X is said to satisfy well-ordered open (neighborhood) (F) if for each $x \in X$ there is a family $\mathcal{W}(x)$ of open subsets containing x (neighborhoods of x) with each $\mathcal{W}(x)$ well-ordered by reverse inclusion and such that, for each open U containing a point x there is an open set $V(x, U)$ containing x such that if $y \in V(x, U)$ then $x \in W \subseteq U$ for some $W \in \mathcal{W}(y)$. It is an open question, due to Gartside and Moody, whether spaces satisfying well-ordered open (F) are proto-metrizable. The converse is immediate from Theorem 1 of [2].

2. A THEOREM

THEOREM 2.1: The following are equivalent for space X:

(i) X is proto-metrizable

(ii) X is MNR

(iii) X is paracompact and monotonically almost 2-fully normal.

Proof: (i) $\Rightarrow$ (ii) Assume m is the operator given in the definition of monotone paracompactness. For an open cover $\mathcal{U}$ of X define $n(\mathcal{U}) := \bigcup\{O \times O : O \in m(\mathcal{U})\}$. Since $m(\mathcal{U})$ is an open cover, $n(\mathcal{U})$ is a neighborhood of the diagonal in X^2. If $y \in n(\mathcal{U})[x]$ then $y \in St(x, m(\mathcal{U})) \subseteq U_x$ for some $U_x \in \mathcal{U}$. Hence, $\{n(\mathcal{U})[x] : x \in X\}$ refines $\mathcal{U}$. Finally, if $\mathcal{U}$ refines $\mathcal{V}$, then $m(\mathcal{U})$ refines $m(\mathcal{V})$ and hence, $n(\mathcal{U}) \subseteq n(\mathcal{V})$. Thus, X is MNR.

(ii) $\Rightarrow$ (iii) Clearly by Theorem 1.4, MNR spaces are paracompact. Assume that m is the operator given in the definition of MNR. For an open cover $\mathcal{U}$, define $n(\mathcal{U}) := \{O \subseteq X : O \text{ open and } O \times O \subseteq m(\mathcal{U})\}$. Since $m(\mathcal{U})$ is a neighborhood of the diagonal, $n(\mathcal{U})$ is an open cover of X and clearly, if $\mathcal{U}$ refines $\mathcal{V}$ then $m(\mathcal{U}) \subseteq m(\mathcal{V})$ and hence, $n(\mathcal{U})$ refines $n(\mathcal{V})$.

If $y, z \in St(x, n(\mathcal{U}))$ then there are open sets O and P such that $x, y \in O$ and $O \times O \subseteq m(\mathcal{U})$ and $x, z \in P$ and $P \times P \subseteq m(\mathcal{U})$. Therefore, $O, P \subseteq m(\mathcal{U})[x] \subseteq U_0$ for some $U_0 \in \mathcal{U}$ and hence $y, z \in U_0$. We have therefore shown that X is monotonically almost 2-fully normal.

(iii) $\Rightarrow$ (i) This is proved in a similar fashion to the proof of (2) implies (3) implies (4) implies (1) in Theorem 1 in [2], with only the first step requiring any modification. ❑

One obvious question arising from this theorem is can we omit the assumption of paracompactness in (iii)? That is, are monotonically almost 2-fully normal spaces proto-metrizable?

3. A COUNTEREXAMPLE

THEOREM 3.1: If X satisfies well-ordered neighborhood (F) then X is MSNR.

Proof: Let $\mathcal{U}$ be an open cover of X. For each $x \in X$ define

$$\mathcal{F}(x,\mathcal{U}) = \{W : W \in \mathcal{W}(x),\ W \subseteq U \text{ for some } U \in \mathcal{U}\}.$$

Since each $\mathcal{W}(x)$ is well ordered, $\mathcal{F}(x,\mathcal{U})$ has a least element $W(x,\mathcal{U})$ (largest, with respect to set inclusion).

Define $m(\mathcal{U})$ by $m(\mathcal{U}) := \bigcup \{\{x\} \times W(x,\mathcal{U}) : x \in X\}$. Clearly, $m(\mathcal{U})[x] = W(x,\mathcal{U}) \subseteq U$ for some $U \in \mathcal{U}$, hence $\{m(\mathcal{U})[x] : x \in X\}$ refines $\mathcal{U}$. We claim $m(\mathcal{U})$ is a semineighborhood of Δ. Obviously $\Delta \subseteq m(\mathcal{U})$ and $m(\mathcal{U})[x]$ is a neighborhood of x. We need to show that $m(\mathcal{U})^{-1}[x]$ is also a neighborhood of x. We claim

$$x \in V(x, W(x, \mathcal{U})^\circ) \subseteq m(\mathcal{U})^{-1}[x].$$

Take $y \in V(x, W(x, \mathcal{U})^\circ)$. This implies there is $W \in \mathcal{W}(y)$ such that $x \in W \subseteq W(x, \mathcal{U})^\circ$. Now $W(x, \mathcal{U}) \subseteq U \in \mathcal{U}$, therefore, $W \in \mathcal{F}(y,\mathcal{U})$ and hence, $W \subseteq W(y, \mathcal{U})$. Consequently, $x \in W(y,\mathcal{U})$ which implies $(y, x) \in \{y\} \times W(y, \mathcal{U})$, i.e., $y \in m(\mathcal{U})^{-1}[x]$. Condition (a) is therefore satisfied. As for condition (b), if $\mathcal{U}$ refines $\mathcal{V}$, then $\mathcal{F}(x,\mathcal{U}) \subseteq \mathcal{F}(x,\mathcal{V})$ for each $x \in X$ and hence, $W(x,\mathcal{U}) \subseteq W(y,\mathcal{V})$ (by minimality) and therefore, $m(\mathcal{U}) \subseteq m(\mathcal{V})$. ❑

EXAMPLE 3.2: McAuley's bow tie space is MSNR but not proto-metrizable (hence, not MNR).

Proof: Let X denote McAuley's bow tie space [8]. It is known that X is stratifiable but not metrizable. Hence, X cannot be proto-metrizable since spaces which are both stratifiable and proto-metrizable are metrizable [10]. It is known that X satisfies well-ordered neighborhood (F) [1] and therefore, by the above theorem, X is MSNR. ❑

We have shown that one must be careful in choosing which definition of paracompactness to "monotonize." Gartside and Moody's question remains as to whether the locally finite version of paracompactness, when monotonized, gives monotone paracompactness. More generally, the same question may be asked of the plethora of characterizations of paracompactness! Which of them coincide when monotonized and, in particular, which characterizations of paracompactness when monotonized provide characterizations of proto-metrizability?

ACKNOWLEDGMENT

The author would like to thank the referee for helpful comments concerning this paper.

REFERENCES

1. COLLINS, P.J. & A.W. ROSCOE. 1984. Criteria for metrisability. Proc. Amer. Math. Soc. **90:** 631–640.
2. GARTSIDE, P.M. & P.J. MOODY. 1993. A note on proto-metrisable spaces. Topology Appl. **52:** 1–9.
3. JUNNILA, H.J.K. & H.P.A. KÜNZI. 1993. Ortho-bases and monotonic properties. Proc. Amer. Math. Soc. **119:** 1335–1345.
4. KELLEY, J.L. 1955. General Topology. Van Nostrand, New York.
5. MANSFIELD, M.J. 1957. Some generalizations of full normality. Trans. Amer. Math. Soc. **86:** 489–505.
6. MATVEEV, M. Monotone compactness. Preprint.
7. MATVEEV, M. Monotone Lindelöfness. Preprint.
8. MCAULEY, L.F. 1956. A relation between perfect separability, completeness and normality in semi-metric spaces. Pacific J. Math. **6:** 315–326.
9. MICHAEL, E. 1959. Yet another note on paracompact spaces. Proc. Amer. Math. Soc. **10:** 309–314.
10. NYIKOS, P.J. 1975. Some surprising base properties in topology. *In:* Studies in topology (N.H. Stavrakas & K.R. Allen, Eds.). Academic Press. New York. 427–450.

Covering Dimension from Large Sets

A.H. STONE

Department of Mathematics
Northeastern University
Boston, Massachusetts 02115

ABSTRACT: The notion of "covering dimension" involves coverings by "small enough" sets; how small is "enough" is investigated for the n-cube and spherical n-ball. The results are applied to give a partial answer to a question of J. Tišer about the existence of σ-discrete open covers of metric spaces (X, ρ) in which the n-th discrete system consists of sets of small diameters.

The dimension of a finite-dimensional compact metric space (X, ρ) is the largest integer k such that every finite cover of X by small enough closed subsets has order at least k — that is, some point of X (assumed nonempty) belongs to more than k of them (cf. [2, p. 67].) How small is "small enough?" Of course this depends on X and ρ. We show that, when (X, ρ) is the unit cube Q^d in Euclidean d-space $\mathbf{R}^d$, "of diameters less than 1" suffices; clearly 1 is "best possible" here.

We deduce this from a slightly more specific statement. A subset of Q^d will be called "transverse" if it meets both of an opposite pair of $(d-1)$-dimensional faces of Q^d. We show (Theorem 1) that it is enough to assume that none of the sets in the cover is transverse.[1]

Another natural example for consideration is the unit spherical ball B^d in $\mathbf{R}^d$. Here we obtain a criterion (Theorem 2) but conjecture a better one.

NOTATION: Throughout, d is a fixed positive integer. A typical point of $\mathbf{R}^d$ is $\mathbf{x} = (x_1, x_2, \ldots, x_d)$. The "integer lattice" consists of the points $\mathbf{c} = (c_1, c_2, \ldots, c_d)$ of $\mathbf{Z}^d$. The unit cubes having these lattice points as vertices are typified by

$$Q^d(\mathbf{c}) = \{x \in \mathbf{R}^d : c_i \le x_i \le c_i + 1 \quad \text{for } i = 1, 2, \ldots, d\}$$

where $\mathbf{c} \in \mathbf{Z}^d$. They form a covering of $\mathbf{R}^d$ by closed sets with pairwise disjoint interiors, but of course there are common boundary points in the union of their $(d-1)$-dimensional faces. It will be convenient to use the term "face" as an abbreviation for "$(d-1)$-dimensional face" in what follows. Thus, each $Q^d(\mathbf{c})$ has just $2d$ "faces."

Mathematics Subject Classification: 54E35, 54F45.
Key words and phrases: covering dimension, n-cube, metric space, σ-discrete covering.

[1] I am grateful to Dr. J. Vermeer for pointing out to me that Theorem 1 is known; see [5, p. 78]. The present proof is believed to be new.

We prove the following:

THEOREM 1: If $\mathcal{F}$ is a finite cover of the unit cube $Q^d(\mathbf{0})$ in $\mathbf{R}^d$ by n closed sets, none of which is transverse, then $\mathcal{F}$ has order at least d. In particular, $n \geq d + 1$.[1]

Proof: Suppose $\mathcal{F} = \{F_1, F_2, \ldots, F_j, \ldots, F_n\}$. Our object is to show that some point of $Q^d(\mathbf{0})$ is in at least $d + 1$ of these sets. Without loss of generality, we can assume that each F_j has nonempty interior (by expanding the F's slightly to open sets relative to $Q^d(\mathbf{0})$ and taking their closures).

We extend $\mathcal{F}$ to an infinite closed cover $\mathcal{F}^*$ of $\mathbf{R}^d$ by the following "kaleidoscope" process. Imagine the $2d$ faces of $Q^d(\mathbf{0})$ to be mirrors. The (iterated) reflections map $Q^d(\mathbf{0})$ onto each $Q^d(\mathbf{c})$ according to the rule $f_\mathbf{c}(\mathbf{x}) = \mathbf{y}$ where (for $i = 1, 2, \ldots, d$ and $0 \leq x_i \leq 1$)

$$\begin{aligned} y_i &= c_i + x_i \quad \text{if } c_i \text{ is even} \\ &= c_i + 1 - x_i \quad \text{if } c_i \text{ is odd.} \end{aligned}$$

And $f_\mathbf{c}(\mathcal{F})$ (meaning $\{f_\mathbf{c}(F_j)\colon 1 \leq j \leq n\}$) will be a cover of $Q^d(\mathbf{c})$ congruent to $\mathcal{F}$. Together, $\bigcup\{f_\mathbf{c}(\mathcal{F})\colon \mathbf{c} \in \mathbf{Z}^d\}$ is a cover, denoted by $\mathcal{F}^*$, of $\mathbf{R}^d$ by closed sets with nonempty interiors.

We next group the sets of $\mathcal{F}^*$ into "clumps" as follows. Fix j $(1 \leq j \leq n)$ for the moment. For each i $(1 \leq i \leq d)$ pick one of the faces $x_i = 0$, $x_i = 1$ of $Q^d(\mathbf{0})$ that F_j does *not* meet. (There is at least one such face, by hypothesis; if there are two, choose one at random.) Suppose the selected faces are specified by $x_i = 0$ when $i \in G$, $x_i = 1$ when $i \in H$ (where G, H are disjoint sets with union $\{1, 2, \ldots, d\}$). Since F_j is disjoint from these faces, F_j is contained in the "box" B_j defined by

$$B_j = \{\mathbf{y} \in \mathbf{R}^d\colon \varepsilon \leq y_i \leq 1 \quad \text{for } i \in G, 0 \leq y_i \leq 1 - \varepsilon \quad \text{for } i \in H\}$$

for some (small enough) positive ε. Thus B_j contains a unique integer-lattice point $\mathbf{p}_j$, with i-th coordinate 1 if $i \in G$, 0 if $i \in H$.

To F_j we adjoint the $2^d - 1$ reflections of F_j in the neighboring unit cubes (those having $\mathbf{p}_j$ as a vertex), forming the "primary clump" $K(\mathbf{0})_j$ of F_j. The analogously defined "primary clump" $C(\mathbf{0})_j$ of B_j is the set

$$\{x \in \mathbf{R}^d\colon -(1 - \varepsilon) \leq x_i \leq 1 - \varepsilon \quad \text{if } i \in G, \quad \varepsilon \leq x_i \leq 2 - \varepsilon \text{ if } i \in H\},$$

a cube of side $2(1 - \varepsilon)$.

Now the *clumps* $K(2\mathbf{c})_j$ of F_j(where $\mathbf{c} \in \mathbf{Z}^d$) are obtained by translating $K(\mathbf{0})_j$ by the *even* translations

$$\mathbf{x} \mapsto \mathbf{x} + 2\mathbf{c}.$$

The clump $C(2\mathbf{c})_j$ of B_j is defined in the same way. We see that different clumps $C(2\mathbf{c})_j$, $C(2\mathbf{c}')_j$ (where $\mathbf{c} \neq \mathbf{c}'$) of B_j are pairwise disjoint and that each

$C(2\mathbf{c})_j$ is of diameter less than $2\sqrt{d}$. Hence, the same is true of the clumps $K(2\mathbf{c})_j$ of F_j (for fixed j), since $K(2\mathbf{c})_j \subset C(2\mathbf{c})_j$. Each $Q^d(\mathbf{c})$ overlaps (and is largely contained in) a unique even translate of the box-clump $C(\mathbf{0})_j$, which carries a clump of F_j whose intersection with Q^d ($\mathbf{c}$) will be the corresponding reflection $f_{\mathbf{c}}(F_j)$ of F_j.

Now let j vary. From the foregoing, the clumps $K(2\mathbf{c})_j$ (where $1 \le j \le n$ and $\mathbf{c} \in \mathbf{Z}^d$) together form a closed cover (say) $\mathcal{K}$ of $\mathbf{R}^d$ whose trace on each lattice cube $Q^d(\mathbf{c})$ is congruent to $\mathcal{F}$. (The point here is that the reflection process ensures that no extra intersections can arise on the boundary of $Q^d(\mathbf{c})$ from copies of $\mathcal{F}$ in neighboring cubes. Some verbal awkwardness has been avoided here by our having eliminated the possibility of some F_j being contained in the boundary of a cube.)

There is some $\eta > 0$ such that, for every finite cover of $Q^d(\mathbf{0})$ by closed sets of diameters less than η, some point belongs to $d + 1$ (or more) of them. Take an integer $N > 2\sqrt{d}/\eta$ and consider the trace of $\mathcal{K}$ on the cube $[0, N]^d$. The similarity transformation $\mathbf{x} \mapsto (1/N)\mathbf{x}$ takes this trace to a cover of $Q^d(\mathbf{0})$ by closed sets of diameter less than η, so at least one point of $[0, 1]^d$ belongs to $d + 1$ (or more) clumps $K(2\mathbf{c})_j$ — and, of course, to some $Q^d(\mathbf{c}')$. Hence, by conguence, some point of $Q^d(\mathbf{0})$ belongs to the intersection of $Q^d(\mathbf{0})$ with $d + 1$ different clumps — i.e., to $d + 1$ of the sets F_j; and the theorem is proved. ❑

COROLLARY: Every finite closed cover of the unit cube $Q^d(\mathbf{0})$ in $\mathbf{R}^d$ by sets of diameter less than 1 has order at least d.

REMARK: It is easy to see that the condition "less than 1" in this corollary cannot be replaced by "less than b" if $b > 1$.

QUESTION: But can it be replaced by "at most 1?" (No, if $d = 1$, but assume $d > 1$.)

THEOREM 2: Every finite closed cover of the spherical ball B^d of unit radius in $\mathbf{R}^d$, by sets of diameters less than $2/\sqrt{d}$, has order at least d.

Proof: Consider the cube $2Q^d$ of side 2 in $\mathbf{R}^d$, defined as $\{\mathbf{x} \in \mathbf{R}^d: |x_1| \le 1, |x_2| \le 1, \ldots, |x_d| \le 1\}$, circumscribing B^d. Let r denote the radial retraction of $2Q^d$ onto B^d; that is,

$$r(\mathbf{x}) = \frac{1}{|\mathbf{x}|}\mathbf{x} \quad \text{if } |\mathbf{x}| \ge 1, \qquad r(\mathbf{x}) = \mathbf{x} \text{ if } |\mathbf{x}| \le 1.$$

Let $\mathcal{F} = \{F_1, F_2, \ldots, F_n\}$ be a closed cover of B^d by sets of diameter less than $2/\sqrt{d}$. The sets $r^{-1}(F_j)$, $1 \le j \le n$, form a closed cover of $2Q^d$. We note that these sets are not transverse in $2Q^d$. For if (say) $r^{-1}(F_1)$ contains points $(1, x_2, \ldots, x_d)$ and $(-1, y_2, \ldots, y_d)$ (where each $|x_j|$ and $|y_j|$ is at most 1), put $a = (1 + x_2^2 + \cdots + x_d^2)^{1/2}$ and $b = (1 + y_2^2 + \cdots + y_d^2)^{1/2}$, and note that F_1 contains the points $\frac{1}{a}(1, x_2, ..., x_d)$ and $\frac{1}{b}(-1, y_2, ..., y_d)$. Thus the diameter of F_1 is at least $\frac{1}{a} + \frac{1}{b} \geq 2/\sqrt{d}$, contrary to hypothesis.

From Theorem 1, some point $\mathbf{x} \in 2Q^d$ is in $d + 1$ or more of the sets $r^{-1}(F_j)$; and then $r(\mathbf{x})$ is in $d + 1$ or more of the sets F_j, completing the proof. ❑

REMARK: Here "less than $2/\sqrt{d}$" is unlikely to be "best possible." I conjecture that the "right" condition in Theorem 2 is that the diameters (of the sets covering B^d) be less than 2 — i.e., that none of these sets contains an antipodal pair of points of the boundary $(d - 1)$-sphere. If true, this would be an elegant generalization of the classical theorem (of Lusternik and Schnirelman [3] and Borsuk [1]), which is essentially the case $n = d$.[2] In my talk at the conference I sketched an argument that I believed proved the above conjecture, but it did not survive close checking.

MOTIVATION: This paper arose from the following question of J. Tišer (asked orally, in 1991). Given a metric space (X, ρ) and a decreasing sequence $a_1 \geq a_2 \geq \cdots \geq a_n \geq \cdots$ of positive real numbers converging to 0, does there exist a σ-discrete open cover of X, say $\mathcal{U} = \bigcup \{\mathcal{U}_n : n \in \mathbf{N}\}$ such that, for each n, $\mathcal{U}_n$ is discrete and each $U \in \mathcal{U}_n$ has diameter less than a_n?[3]

THEOREM 3: If the (covering) dimension $\dim X$ is finite, the answer is "yes."

Proof: Suppose $\dim X = d\ (< \infty)$, and take a locally finite open cover $\mathcal{V}$ of X, of order d, by sets of diameter less than a_{d+1} (cf., e.g., [4, p. 132, Cor. 4.4]). Index $\mathcal{V}$ as $\{V_\lambda : \lambda \in \Lambda\}$ and shrink it to a closed cover $\mathcal{F} = \{F_\lambda : \lambda \in \Lambda\}$ with each F_λ closed and contained in V_λ. For each $\lambda \in \Lambda$, take open sets $G_\lambda{}^k$ and closed sets $F_\lambda{}^k$ $(k = 1, 2, ..., d + 1)$ such that

$$F_\lambda = F_\lambda{}^0 \subset G_\lambda{}^1 \subset F_\lambda{}^1 \subset \cdots \subset G_\lambda{}^{d+1} \subset F_\lambda{}^{d+1} \subset V_\lambda.$$

Let $\Lambda^{(k)}$ denote the family of all subsets of Λ with exactly k (distinct) elements. For each $S \in \Lambda^{(k)}$, say $S = \{\lambda_1, \lambda_2, ..., \lambda_k\}$, we use the notation

$$F_S{}^i = F_{\lambda_1}{}^i \cap F_{\lambda_2}{}^i \cap \cdots \cap F_{\lambda_k}{}^i, \text{ and similarly}$$

$$G_S{}^i = G_{\lambda_1}{}^i \cap G_{\lambda_2}{}^i \cap \cdots \cap G_{\lambda_k}{}^i$$

(so that the sets $F_S{}^i$, $G_S{}^i$, with $|S| > d + 1$ are empty).

Now define (for $k = 1, 2, ..., d + 1$ and $S \in \Lambda^{(k)}$)

$$A_S = F_S{}^{k-1} \backslash \bigcap \{G_T{}^k : T \in \Lambda^{(k+1)}\}.$$

[2]I have not seen [3]; it may have been motivated by the calculus of variations (cf. [6, pp. 250–254]). The relevant theorem in [1] is that if $A_0, A_1, ..., A_n$ are closed sets none of which contains two antipodal points of the n-sphere S_n, their union cannot contain S_n. The paper [1] has led to many extensions and generalizations; see [6, pp. 1117–1125].

[3]The question occurred to him as an interesting one with no application in mind.

The sets $A_S (S \in \Lambda^{(k)})$ are, for fixed k, closed and pairwise disjoint; hence (from the local finiteness of $\mathcal{V}$) they form a discrete system. Their union, as k ranges from 1 to $d+1$, is X; and their diameters are less than a_{d+1}, *a fortiori* less than a_k for $k \leq d$. All that remains is to expand them slightly to open sets without losing the properties just established — a routine construction via collectionwise normality. ❑

Thus, Tišer's question is of interest mainly for infinite-dimensional spaces. We investigate the Hilbert cube $H = I^{\mathbf{N}}$ as perhaps the simplest example, taking H to be the product of closed intervals in which the n-th has length $1/n$. More precisely, we regard H as the set of all real sequences $\mathbf{x} = \langle x_1, x_2, \ldots, x_n, \ldots \rangle$ with $0 \leq x_n \leq 1$ for all $n \in \mathbf{N}$, and with "Euclidean" metric ρ given by

$$\rho(\mathbf{x}, \mathbf{y}) = \left(\sum_{n=1}^{\infty} \frac{(x_n - y_n)^2}{n^2} \right)^{\frac{1}{2}}.$$

Since every open cover of H has a finite subcover, we interpret Tišer's question as asking: given an arbitrary sequence $a_1 \geq a_2 \geq \cdots \geq a_n \geq \cdots$ of real numbers converging to 0, whether there exists, for arbitrarily large n, an open cover $\mathcal{U}$ of H, consisting of n discrete subsystems $\mathcal{U}_1, \mathcal{U}_2, \ldots, \mathcal{U}_n$, the sets in each $\mathcal{U}_i$ having diameters less than a_i (for $i = 1, 2, \ldots, n$). If this is so we say that the sequence $\langle a_1, a_2, \ldots \rangle$ has the *Tišer property* for (H, ρ). Our next result gives a sufficient condition for this. First define

$$b_n = \left(\sum_{k \geq n} \frac{1}{k^2} \right) \left(= n^{-1/2} + \mathcal{O}\left(n^{-3/2}\right)\right).$$

THEOREM 4: If $a_n > b_n$ for arbitrarily large n, the sequence $\langle a_1, a_2, \ldots, a_n, \ldots \rangle$ has the Tišer property for (H, ρ).

Proof: Fixing an arbitrarily large n for which $a_n > b_n$, write $n - 1 = k$ and regard H as $I^k \times J_k$ where $I^k = \prod_{i=1}^{k} [0, 1]_i$ and $J_k = \prod_{i>k} [0, 1]_i$, metrized, respectively, by

$$\sigma\left((x_1, \ldots, x_k), (y_1, \ldots, y_k)\right) = \left(\sum_{i=1}^{\infty} \frac{(x_i - y_i)^2}{i^2} \right)^{\frac{1}{2}} \quad \text{and}$$

$$\tau\left((x_{k+1}, x_{k+2} \ldots), (y_{k+1}, y_{k+2} \ldots)\right) = \left(\sum_{j>k}^{\infty} \frac{(x_j - y_j)^2}{j^2} \right)^{\frac{1}{2}},$$

so that $\rho^2 = \sigma^2 + \tau^2$.

Fixing $\varepsilon > 0$ for the moment, take a finite open cover of I^k consisting of $k + 1$ discrete systems, say consisting of sets $U_q^p (p = 1, 2, \ldots, k + 1, q = 1, 2, \ldots, n_p)$, of σ-diameters less than ε, and discrete for fixed p. The sets $U_q^p \times J^k$ provide an open cover of H consisting of $k + 1$ discrete systems, and each of these sets has ρ-diameter less than $(\varepsilon^2 + b_{k+1}{}^2)^{1/2}$. For small enough ε this is less than a_n, and *a fortiori* less than $a_1, a_2, \ldots, a_{n-1}$ also. ❑

In the reverse direction, we obtain a result only in a rather special case. We say that the decreasing sequence $a_1, a_2, \ldots$ (converging to 0) has the *special Tišer property* for (H, ρ) if the n discrete systems $\mathcal{U}_1, \ldots, \mathcal{U}_n$ in the definition of the Tišer property can be made to consist of sets *all* of which have diameters less than a_n. Note that the coverings produced in the proofs of Theorems 3 and 4 are of this type.

THEOREM 5: If $a_n < b_n$ for all n greater than some n_0, the sequence $\langle a_1, a_2, \ldots \rangle$ does not have the special Tišer property for (H, ρ).

Proof: Given an open cover (which we can assume to be finite), $\{G_1, G_2, \ldots, G_j\}$ of H, we shrink it slightly to a closed cover $\{F_1, F_2, \ldots, F_j\}$ of H with $F_i \subset G_i$, and cover each F_i by a finite number of basic open product subsets of H, say with union U_i, in such a way that the closure $\overline{U}_i$ of U_i is contained in G_i. Each U_i restricts only finitely many coordinates, so we may write $\overline{U}_i = A_i^k \times J_k$ for some closed subset A_i^k of I^k, where k is independent of i and greater than n. With the same notation as before, the ρ-diameter of $\overline{U}_i$ is then $((\sigma\text{-diam } A_i^k)^2 + (\tau\text{-diam } J_k)^2)^{1/2}$. Thus $(a_k)^2 \geq (\sigma\text{-diam } A_i^k)^2 + (b_{k+1})^2$, showing that each A_i^k has σ-diameter less than $1/k$. The sets $A_1^k, A_2^k, \ldots, A_j^k$ thus form a closed cover of the cube I^k by sets none of which is transverse. By Theorem 1, there is a point in $k + 1$ (or more) of these sets, and *a fortiori* in $k + 1$ of the sets G_i. Thus the G_i's cannot be grouped into as few as k discrete (or even pairwise disjoint) systems, and the special Tišer property fails. ❑

REFERENCES

1. BORSUK, K. 1933. Drei Sätze über die *n*-dimensionale euklidische Sphäre. Fundamenta Math. **20:** 177–190.
2. HUREWICZ, W. & H. WALLMAN. 1941. Dimension Theory. Princeton Univ. Press. Princeton, New Jersey.
3. LUSTERNIK, L. & L. SCHNIRELMAN. 1930. Méthodes topologiques dans les problèmes variationels. Isseldowatelskij Inst. Mat. i Mechanik, pri IMGO. Moscow.
4. PEARS, A.R. 1975. Dimension Theory of General Spaces. Cambridge Univ. Press. Cambridge, UK.
5. ENGELKING, R. 1978. Dimension Theory. North-Holland. Amsterdam.
6. STEENROD, N.E. (classifier). 1968. Reviews of Papers in Algebraic and Differential Topology, Topological Groups, and Homological Algebra. American Mathematics Society. Providence, Rhode Island.

The Krull Dimension of Alexandroff T_0-spaces

PETRA WIEDERHOLD[a] AND RICHARD G. WILSON[b]

Universidad Autónoma Metropolitana
Unidad Iztapalapa, Departamento de Matemáticas
México 09340 D.F., México

ABSTRACT: Alexandroff T_0-spaces have been studied as topological models of the supports of digital images and as discrete models of continuous spaces in theoretical physics. In [17] we have defined (topologically) the *Alexandroff dimension* for arbitrary Alexandroff spaces. In this paper we will prove that the Alexandroff dimension of a T_0-space is equal to its *Krull dimension*, which is defined in terms of the lattice of closed sets of the space and which was first studied in [16]. Since the category of Alexandroff T_0-spaces is known to be isomorphic to the category of posets (see [3, Theorem 3.5]), the results could be formulated in this latter category as well.

1. INTRODUCTION

In digital image processing and computer graphics, it is necessary to describe topological properties of n-dimensional digital image arrays, hence the search for models of the supports of such images. Recently, topological models have been constructed. Based on the Khalimsky topology τ on the integers, given by the subbase $\{\{2n - 1, 2n, 2n + 1\}: n \in \mathbb{Z}\}$, a *digital n-space* was defined as a product of n copies of $(\mathbb{Z}, \tau)$ and Jordan curve (surface) theorems were proved for digital 2-space [8] and digital 3-space [10]. A more general construction of a digital space was proposed in [12] where a collection of locally finite disjoint open subsets of $\mathbb{R}^n$, whose union is dense in $\mathbb{R}^n$, is used to define a partition of $\mathbb{R}^n$ giving a digital image which is a locally finite T_0-space, that is, a space in which each point x has a finite, and hence also a minimal, neighborhood which we will denote by $U(x)$. Spaces in which every point has a minimal neighborhood are called *Alexandroff*; such spaces were first considered by Alexandroff [1] under the name of *discrete spaces*. Other approaches to digital topology, which lead to locally finite spaces are, for example, the models based on cellular complexes developed by Kovalevsky [11] and Lee and Rosenfeld [13] or the model of molecular

Mathematics Subject Classification: Primary 54F45; Secondary 54D10, 06B10.
Key words and phrases: Alexandroff space, partially ordered set, Alexandroff dimension, Krull dimension, lattice of closed sets, prime filter, locally finite space.
[a]Permanent address: Centro de Investigación y Estudios Avanzados del IPN, Depto. de Ingeniería Eléctrica, Secc. de Control Automático, Apdo. Postal 14-740, México 07000 D.F., México.
[b]Research supported by Consejo Nacional de Ciencia y Tecnología, Grant No. 4874E-9406

spaces developed by Ivashchenko [5], [6]. These latter papers apply discrete topological models in theoretical physics as well.

A problem on which research has been focused recently, is that of the dimension of a digital space. A digital image which is obtained by "discretization" of an image defined on $\mathbb{R}^n$, should be modeled by an n-dimensional topological space. A dimension function for Alexandroff spaces called *Alexandroff dimension*, which is essentially the small inductive dimension of [14], was studied in [17] and later by Ivashchenko *et al.* in [6]. An Alexandroff T_0-space (X, τ) is completely determined by a poset $(X, \leq)$, where $\leq$ is defined by $x \leq y \Leftrightarrow x \in cl(\{y\}) \Leftrightarrow y \in U(x)$. The partial order $\leq$ was first introduced in [1] and has been called the *specialization order* by [7], [9], and others. The *partial order dimension* or *poset dimension* of a poset $(X, \leq)$ is defined as the supremum of all lengths of chains in $(X, \leq)$ (see [2]). It was shown in [17] that for any Alexandroff T_0-space, its Alexandroff dimension equals its poset dimension. This implies immediately that digital n-space has Alexandroff dimension n. In another context the same poset dimension was defined in [6] for transitive graphs, and the same equality was proved. In [6] it was also proved, that both dimensions coincide with a third dimension function, which is defined inductively for directed graphs, and which was studied previously by Ivashchenko [5] in relation to his model of molecular space.

In this note we study another dimension function, the *Krull dimension*, first applied to topological spaces in [16], but previously known in algebra as a dimension function for rings and lattices. In general spaces, this dimension is difficult to calculate, but it is interesting to note that the Krull dimension agrees with other dimension functions in some classes of spaces. In particular, in [16] it was proved, that the Krull dimension is equal to the small inductive and to the covering dimensions for separable metrizable spaces. We prove in this work that for any Alexandroff T_0-space, its Krull dimension coincides with its Alexandroff dimension (and hence equals its poset dimension). Our results generalize those of [4] and [15] for finite spaces, but we note that in [4], Krull dimension is defined differently in terms of chains of irreducible closed sets and coincides with the definition used here in the case of finite spaces.

2. PRELIMINARIES

Recall that if $A(X)$ is the set of all closed sets of a topological space X, then $(A(X), \cap, \cup)$ is a distributive lattice with greatest and least elements X and $\varnothing$, respectively. It is clear that in the case of a discrete space this lattice is a Boolean algebra.

DEFINITION 1: Let $(L, \wedge, \vee)$ be a lattice; a nonempty subset $F \subset L$ is said to be a *filter* if $a, b \in F$, $c \in L$ implies that $a \wedge b$, $a \vee c \in F$ and $F \neq L$. A filter F is said to be *prime* if whenever $a, b \in L$ and $a \vee b \in F$ then $a \in F$ or $b \in F$. An *ultrafilter* is a maximal filter, and a *set-theoretic ultrafilter* is an ultrafilter (or, equivalently a prime filter) in the lattice of all subsets of X.

It is clear that if F is a filter in $A(X)$ then $X \in F$ and $\varnothing \notin F$.

DEFINITION 2: The *Krull dimension* of a nonempty lattice $(L, \wedge, \vee)$ is defined as

$$KdimL = \sup\{n \in \omega: \exists \text{ chain } F_0 \subset F_1 \subset \ldots \subset F_n \text{ of distinct prime filters in } L\}.$$

It is known that a lattice with a greatest and a least element has Krull dimension zero if and only if it is a Boolean algebra [16]. Consequently, $KdimA(X) = 0$ if X is a discrete space. In general, the function *Kdim* is not monotone with respect to sublattices, as the following example shows.

EXAMPLE 1: Let X be an infinite discrete space. Clearly, $A(X)$ itself is a sublattice of $A(X)$, and $KdimA(X) = 0$. Now let $\mathcal{B}$ be the sublattice generated by $\{X \backslash \{x\}, x \in X\} \cup \{\varnothing, X\}$. Then

$$\mathcal{B} = \{X \backslash F, F \subset X, F \text{ finite or empty}\} \cup \{\varnothing\}.$$

Since $\mathcal{B} \backslash \{\varnothing\}$ is closed under finite intersections, it is a prime filter in $\mathcal{B}$. Moreover, for any $x \in X$,

$$F_x = \{M \in \mathcal{B}: x \in M\} = \{X \backslash F,\ F \subset X,\ F \text{ finite or empty},\ x \notin F\}$$

is a prime filter in $\mathcal{B}$, and $F_x \subset (\mathcal{B} \backslash \{\varnothing\})$ implies $Kdim\mathcal{B} \geq 1$.

Definition 2 was first applied to the lattice of closed sets of a topological space by Vinokurov [16]; the same definition was used in [15], although the term *ideal* was used instead of filter.

The *Krull dimension* of a space Y, will be defined in terms of the Krull dimensions of bases of the lattice $(A(Y), \cap, \cup)$, where again, $A(Y)$ is the set of all closed subsets of Y. For completeness, we first summarize some definitions and results from [17], and in the next section we apply these concepts to Alexandroff T_0-spaces.

DEFINITION 3: The *Alexandroff dimension DIM* of any Alexandroff space (X, τ) is defined inductively in terms of a local dimension *DIL* determined by the minimal neighborhoods:

(i) $DIM(X) = -1 \Leftrightarrow X = \varnothing$.
(ii) If $X \neq \varnothing$ then define

$$DIM(X) = \sup\{DIL(x), x \in X\}$$

where, for $x \in X$ and $n \in \omega$, we define

$$DIL(x) \leq n \Leftrightarrow \text{DIM}\,(FrU(x)) \leq n-1,$$

$$DIL(x) = n \Leftrightarrow DIL(x) \leq n, \text{ and } DIL(x) \not\leq n-1,$$

$$DIL(x) = \infty \Leftrightarrow DIL(x) \leq n \text{ is false } \forall n.$$

For full details, we refer the reader to [17]. It is easy to see that a discrete space has Alexandroff dimension 0, and for an Alexandroff T_0-space the converse is also true. As mentioned in the introduction, the function *DIM* is essentially the small inductive dimension *ind* of [14].

DEFINITION 4: The *poset dimension ODIM* of a poset $(X, \leq)$ is defined in terms of the local poset dimension *ODIL* as follows:

$$ODIL(x) = \sup\{l\colon \exists\, a_1, a_2, \ldots, a_l \in X \text{ such that } a_l < a_{l-1} < \cdots < a_0 = x\}$$

$$ODIM(X) = \sup\{ODIL(x),\ x \in X\}$$

Again, full details may be found in [17] where the following was proved:

PROPOSITION 1: If (X,τ) is an Alexandroff T_0-space with specialization order $\leq$, then

$$DIM(X,\tau) = ODIM(X, \leq). \quad \square$$

Suppose now that (Y, τ) is a topological space. In order to define the Krull dimension of the space (Y, τ), we will need the concept of a base for the lattice $A(Y)$:

DEFINITION 5: A subset $\mathcal{B}$ of $A(Y)$ is said to be a *lattice base* of $A(Y)$ if it is a (topological) base for the closed sets, $A(Y)$ and additionally, is a sublattice of $(A(Y), \cap, \cup)$ which contains $\varnothing$ and Y.

Clearly, $A(Y)$ itself is a lattice base of $A(Y)$ and since a lattice base of $A(Y)$ is closed under finite intersections, $A(Y)$ is the unique lattice base in the case that Y is a finite set.

DEFINITION 6: The *Krull dimension* of a topological space (Y, τ) is defined by

$$KDIM(Y) = \min\{Kdim\,\mathcal{B}\colon \mathcal{B} \textit{ is a lattice base of } A(Y)\}.$$

It is a consequence of the remarks following Definition 2, that if Y is a discrete space, then $KDIM(Y) = 0$, since we may take $A(Y) = \mathcal{P}(Y)$ which is a Boolean algebra. In [15] and [16], many properties of *KDIM* were obtained; specifically, it was shown in [16] that $KDIM(Y) = \text{ind}\,Y = \dim Y$ for any separable metrizable space Y (where ind, dim are the small inductive and the covering dimension, respectively). However, in general spaces, *KDIM* is frequently difficult to calculate.

LEMMA 1: If L is an arbitrary sublattice of $A(Y)$ such that $\cap L = \emptyset$, then

$$F_{y,L} = \{M \in L: cl(\{y\}) \subset M\}$$

is a prime filter in L.

Proof: We leave the routine verifications to the reader and note only that if $\cap L = \emptyset$, then $F_{y,L} \neq L$. ❑

3. THE KRULL DIMENSION OF AN ALEXANDROFF T_0-SPACE

The aim of this section is to prove that if (X, τ) is an Alexandroff T_0-space, then the Krull dimension of X coincides with the Alexandroff dimension of X, thus generalizing a result of [15] for finite spaces.

Throughout, $\leq$ will denote the specialization order of the space (X, τ). Note that $ODIL(x) = 0$ if and only if $\{x\}$ is closed, and x is maximal with respect to $\leq$ if and only if $\{x\}$ is open.

The main theorem will be a consequence of the following two propositions.

PROPOSITION 2: If (X, τ) is an Alexandroff T_0-space, and L is a topological base for the closed sets which is also a sublattice of $(A(X), \cap, \cup)$, *then* $Kdim L \geq DIM(X)$.

Proof: From Proposition 1, it suffices to show that $Kdim L \geq ODIM(X)$. Since L is a base for the closed sets, it follows that $\cap L = \emptyset$. Hence, by Lemma 1 we have that

$$F_a = \{M \in L: cl(\{a\}) \subset M\}$$

is a prime filter in L for any $a \in X$.

We claim that if $a_1 < a_2$ for $a_1, a_2 \in X$, then $F_{a_2} \subset F_{a_1}$, and $F_{a_2} \neq F_{a_1}$. For, if $M \in F_{a_2}$ then $cl(\{a_2\}) \subset M$; but clearly, $cl(\{a_1\}) \subset cl(\{a_2\}) \subset M$, and so $M \in F_{a_1}$. Now, since (X, τ) is a T_0-space, $cl(\{a_1\}) \neq cl(\{a_2\})$ and so, since L is a base of $A(X)$, there is some closed set $M \in L$ such that $cl(\{a_1\}) \subset M$ and $a_1 \notin M$. Thus $M \in F_{a_1} \backslash F_{a_2}$ and the claim is proved.

Consequently, for any $a \in X$ with $ODIL(a) = k$ there exists a chain of prime filters $F_{a_k} \subset F_{a_{k-1}} \subset \cdots \subset F_{a_0}$ (with $a_k = a$). Therefore, if $DIM(X) = ODIM(X) \geq n$ then there is $x \in X$ with $ODIL(x) \geq n$, implying $Kdim L \geq n$. The result follows. ❑

COROLLARY 1: If (X, τ) is an Alexandroff T_0-space, then $KDIM(X) \geq DIM(X)$. ❑

PROPOSITION 3: If (X, τ) is an Alexandroff T_0-space, then $Kdim A(X) = DIM(X)$.

Proof: From Proposition 2, we have $KdimA(X) \geq DIM(X)$; hence, if $DIM(X)$ is infinite, then so is $KdimA(X)$ and we are done. Thus, in the sequel we suppose that $DIM(X) = ODIM(X) = n$ is finite and we prove that $KdimA(X) \leq n$.

Since $ODIM(X) = n$, we can write $X = \cup\{X_i : 0 \leq i \leq n\}$ where $X_i = \{x \in X: ODIL(x) = i\}$; thus X_0 consists precisely of the closed points of X, while all points of X_n are open. We note that the relative topology on each X_i is discrete.

Now let F be a prime filter in the lattice base $A(X)$. For each $m \leq n$, we denote by $\mathcal{TR}_m(F)$ the *trace of* F on X_m, that is, $\mathcal{TR}_m(F) = \{A \cap X_m: A \in F\}$. Note that $\mathcal{TR}_m$ may contain the empty set, but if $\varnothing \notin \mathcal{TR}_m(F)$, then $\varnothing \notin \mathcal{TR}_k(F)$ for each $k \leq m$.

Now let $\mathcal{F}_X$ be the set of prime filters in $A(X)$. We define a function h: $\mathcal{F}_X \to \omega$ by

$$h(F) = \max\{m: \varnothing \notin \mathcal{TR}_m(F)\}.$$

Clearly, for each $F \in \mathcal{F}_X$, we have $h(F) \leq ODIM(X)$ and so the proof of Proposition 3 will follow from the following observations which show that h is injective when restricted to a chain in $\mathcal{F}_X$.

Claim (*i*): If $h(F) = m$ and $A \in \mathcal{TR}_m(F)$, then $A \cup \cup \{X_i : 0 \leq i \leq m-1\} \in F$.

First note that if $h(F) = m$, then $\cup\{X_i : 0 \leq i \leq m\} \in F$. Now if $J \in F$ such that $J \cap X_m = A$, then $J \cap \cup\{X_i : 0 \leq i \leq m\} \in F$. But this latter set is a subset of $A \cup \cup \{X_i : 0 \leq i \leq m-1\}$ and the claim is proved.

Claim (*ii*): If $h(F) = m$, then $\mathcal{TR}_m(F)$ is a (set-theoretic) ultrafilter on X_m.

Suppose that $A,B \subset X_m$ are such that $A \cup B \in \mathcal{TR}_m(F)$. From (*i*) it follows that $(A \cup B)\cup\cup\{X_i : 0 \leq i \leq m-1\} \in F$ and hence that $(A \cup \cup \{X_i : 0 \leq i \leq m - 1\}) \cup (B \cup \cup \{X_i : 0 \leq i \leq m-1\}) \in F$. Since these sets are closed and F is prime, it follows that one of them lies in F and hence either A or B lies in $\mathcal{TR}_m(F)$. Thus $\mathcal{TR}_m(F)$ is a prime filter, and hence an ultrafilter, on X_m.

Claim (*iii*): If F,G are distinct prime filters in the lattice base $A(X)$ and $F \subset G$, then $h(F) > h(G)$.

That $h(F) \geq h(G)$ is clear; hence, we need only show that equality does not occur. To this end, suppose to the contrary that $h(F) = h(G) = m$. If $\mathcal{TR}_m(F) \neq \mathcal{TR}_m(G)$, then by (*ii*), since these filters are ultrafilters on X_m, they have disjoint elements A_F and A_G, respectively. Hence, by (*i*), $A_F \cup \cup\{X_i: 0 \leq i \leq m-1\} \in F \subset G$ and $A_G \cup \cup\{X_i : 0 \leq i \leq m-1\} \in G$. Thus, the intersection of these sets is an element of G which misses X_m, giving a contradiction.

Hence, we can assume that $\mathcal{TR}_m(F) = \mathcal{TR}_m(G) = \mathcal{K}$. We claim that both F and G are generated by the filter base $\{cl(K): K \in \mathcal{K}\}$. To prove this, we first note that since each element $A \in F$ is closed and $A \cap X_m \in \mathcal{K}$, each element of F contains $cl(K)$ for some $K \in \mathcal{K}$. It now suffices to show that $cl(K) \in F$ for each $K \in \mathcal{K}$. To this end, we assume to the contrary that there is some $K \in \mathcal{K}$ such that $cl(K) \notin F$. Since $K \in \mathcal{K}$, there is some $A \in F$ such that $A \cap X_m = K$ and hence $A \backslash cl(K) \subset \cup\{X_i : 0 \leq i \leq m-1\}$. Now if we let $C = cl(A \backslash cl(K))$, then $C \notin F$ since $C \subset \cup \{X_i : 0 \leq i \leq m-1\} \notin F$. However, $C \cup cl(K) = A \in F$, which contradicts the fact that F is prime. Thus $\{cl(K): K \in \mathcal{K}\}$ is a filter base for F and G which then must be equal.

Thus a chain of prime ideals in $A(X)$ is of length less than or equal to $ODIM(X, \leq)$ and the proof is complete. ❑

Combining this result with Corollary 1, we obtain:

THEOREM 1: If X is an Alexandroff T_0-space, then

$$KDIM(X) = DIM(X) = ODIM(X). \quad ❑$$

It is well known that the category of Alexandroff T_0-spaces is isomorphic to the category of partially ordered sets under the functor which takes the space (X, τ) to the poset $(X, \leq)$, where $\leq$ is the specialization order of (X, τ) (see [3]). Thus if we define the Krull dimension of a poset to be the Krull dimension of the corresponding Alexandroff space, Theorem 1 can be stated in the following order-theoretic form:

THEOREM 2: If $(X, \leq)$ is a poset, then

$$KDIM(X, \leq) = ODIM(X, \leq). \quad ❑$$

4. THE KRULL DIMENSION OF A LOCALLY FINITE T_0-SPACE

In Section 3 we have seen that the the Krull dimension of an Alexandroff T_0-space X equals the Krull dimension of the lattice $A(X)$ (which is a lattice base of itself). Our aim now is to show that in the case of a locally finite T_0-space (X, τ), there is a minimal sublattice of $A(X)$ which is a (topological) base for the closed sets in X and which has the same Krull dimension as the whole lattice $A(X)$. This base for the closed sets does not necessarily contain $\varnothing$, and so in general it is not a lattice base.

Recall that for any $x \in X$, $U(x)$ denotes its minimal open neighborhood. Clearly, $\{U(x), x \in X\}$ is a topological base of τ. This base is minimal among all topological bases of τ, in the sense that it is contained in any other base of τ. Consequently, the sublattice generated by the complements of the sets $U(x)$ is minimal with respect to being a sublattice of $(A(X), \cap, \cup)$ and being a base for the closed sets of (X, τ). This sublattice is a lattice base for $A(X)$ if it contains $\varnothing$ and X. We summarize these remarks in the following lemma:

LEMMA 2: If $\mathcal{B}_{\min}$ is the sublattice of $A(X)$ generated by $\{X\backslash U(x): x\in X\}$, then $\mathcal{B}_{\min}$ is minimal with respect to being a (topological) base and a sublattice of $(A(X), \cap, \cup)$. ❑

We call $\mathcal{B}_{\min}$ defined in Lemma 2 the *canonical base* of the lattice of closed sets $A(X)$ of the Alexandroff space X.

LEMMA 3: If $\mathcal{B}_{\min}$ is the canonical base of $A(X)$, and if $F_a = \{M\in\mathcal{B}_{\min}: cl(\{a\})\subset M\}$, then $\cap F_a = cl(\{a\})$.

Proof: The statement is trivially true if $X = \{a\}$,, so now suppose that X has at least two elements. Clearly $cl(\{a\})\subset \cap F_a$ and if there is some $b\in\cap F_a$ such that $b\notin cl(\{a\})$, then $cl(\{a\})\subset U(b)^c$. But $U(b)^c\in\mathcal{B}_{\min}$, which implies that $U(b)^c\in F_a$, a contradiction. ❑

Example 1 shows that the Krull dimension applied to sublattices of $A(X)$ is not monotone in the sense that there may exist sublattices $\mathcal{B}$, C, $\mathcal{B}\subset C$ such that $Kdim\,\mathcal{B} > Kdim\,C$. Nevertheless, in the class of locally finite (and hence, Alexandroff) T_0-spaces the canonical base $\mathcal{B}_{\min}$ has the same Krull dimension as $A(X)$; that is to say, it has minimal dimension among all sublattices:

THEOREM 3: If (X, τ) is a locally finite T_0-space, and $\mathcal{B}_{\min}$ is the canonical base of the lattice $(A(X), \cap, \cup)$, then $Kdim\,\mathcal{B}_{\min} = DIM(X)$.

Proof: From Proposition 2 we have $Kdim\,\mathcal{B}_{\min} \geq DIM(X)$, and for any $a\in X$ with $ODIL(a) = k$ there exists a chain of prime filters $F_{a_k}\subset F_{a_{k-1}}\subset\cdots\subset F_{a_0}$ (with $a_k = a$) in $\mathcal{B}_{\min}$, where

$$F_a = \{M\in\mathcal{B}_{\min}: cl(\{a\})\subset M\}.$$

It follows that if $DIM(X)$ is infinite, then so is $Kdim\,\mathcal{B}_{\min}$ and so $DIM(X) = Kdim\,\mathcal{B}_{\min}$. Hence, it is sufficient to suppose that $DIM(X) = ODIM(X) = n$ is finite and to prove that $Kdim\,\mathcal{B}_{\min}\leq n$.

First note that any filter F of $\mathcal{B}_{\min}$ satisfies $\cap F\neq\varnothing$. For, suppose $\cap F=\varnothing$, then for each $x\in X$ there is $J_x\in F$ such that $x\notin J_x$. Hence, $J_x\subset U(x)^c$ and thus $U(x)^c\in F$ for each $x\in X$. It follows that $F = \mathcal{B}_{\min}$.

We now claim that if a is a closed point of X then F_a is a maximal filter in $\mathcal{B}_{\min}$. To show this, note first that if $\mathcal{B}_{\min}\backslash F_a = \{\varnothing\}$, then clearly F_a is maximal; so suppose that $N\in\mathcal{B}_{\min}\backslash F_a$ is nonempty, and let F be the filter generated by $F_a\cup\{N\}$. By Lemma 3, $\cap F_a = \{a\}$, which implies that $\cap F = \cap\{M\cap N: M\in F_a\} = (\cap F_a)\cap N = \{a\}\cap N = \varnothing$, which now implies that F is not a filter in $\mathcal{B}_{\min}$. That the filters F_a are the only prime filters of $\mathcal{B}_{\min}$, will follow from the following observations:

Let F be a prime filter in $\mathcal{B}_{\min}$.

(i) We have shown that $\cap F\neq\varnothing$. So, if we choose $x\in\cap F$, then $F\subset F_x$.

(ii) By (i) and the fact that all chains in $(X, \leq)$ are of length at most n, we can find $b \in X$ which is a maximal element in $(\cap F, \leq)$ and $F \subset F_b$. Let S denote the set of (strict) successors of b in $(X, \leq)$ (note that if b is maximal in $(X, \leq)$, that is if $\{b\}$ is open, then $S = \varnothing$). From the maximality of b in $\cap F$, it follows that $F \not\subset F_s$ for each $s \in S$ and hence, for each $s \in S$, there is some $A_s \in F$ such that $s \notin A_s$.

(iii) $\cap F = cl(\{b\})$: Since $F \subset F_b$ implies that $\cap F_b \subset \cap F$, it follows from Lemma 3 that $cl(\{b\}) \subset \cap F$. For the converse, suppose that there exists $x \in \cap F$ such that $x \notin cl(\{b\})$. We have also $b \notin cl(\{x\})$, because, if $b \in cl(\{x\})$ $(x \neq b)$, then $x \in S$ and it follows from (ii) that $x \notin \cap F$. Thus b and x are incomparable points in $(X, \leq)$, which implies that $U(x)^c \cup U(b)^c \supset \{b, x\}$. Now let $A \in F$; then if $z \in U(b) \cap U(x)$, $b < z$ and so $z \notin \cap F$; hence there exists $A_z \in F$ such that $z \notin A_z$. Since $U(b)$ is finite, $B = \cap\{A_z: z \in U(b) \cap U(x)\} \cap A \in F$. Hence, $B \subset (U(x)^c \cup U(b)^c)$, implying that

$$B = (B \cap U(x)^c) \cup (B \cap U(b)^c).$$

But neither of these latter sets lies in F (because $F \subset F_b$, $F \subset F_x$), which contradicts the fact that F is prime.

(iv) We claim that $F = F_b$: From (iii) it follows that for each $z \notin cl(\{b\})$, there exists $A_z \in F$ such that $z \notin A_z$. Hence, $A_z \subset U(z)^c$ which in turn implies that $U(z)^c \in F$ since F is filter. Consequently $\{U(z)^c: z \notin cl(\{b\})\} \subset F$. However, any such $U(z)^c$ belongs to $\mathcal{B}_{\min}$ and contains b, which implies that the family $\{U(z)^c: z \notin cl(\{b\})\}$ generates F_b. Thus, $F_b \subset F$ and so the filters F_b are the only prime filters in $\mathcal{B}_{\min}$ and the proof is complete. ❑

We note that in the proof of (iii) above, we require only that $U(b) \cap U(x)$ be finite; hence, it suffices that either $U(x)$ *or* $U(b)$ be finite. Thus, the condition of local finiteness in the statement of Theorem 3 can be weakened to "locally finite at all but at most one point." However, if the space is not locally finite at two points, then $Kdim\,\mathcal{B}_{\min} > DIM(X)$ can occur as the following example shows:

EXAMPLE 2: X will denote the poset $(\{0, 1\} \times \{0\}) \cup (\omega \times \{1\})$ ordered in such a way that each point with second coordinate 1 is greater than each point with second coordinate 0 and with no other relations. It is easy to see that $ODIM\,(X, \leq) = 1$, and it is straightforward to verify that if F denotes the filter of all cofinite closed sets, then F is prime in $\mathcal{B}_{\min}$ and $F_{(0,1)} \subset F \subset F_{(0,0)}$; hence $Kdim\,\mathcal{B}_{\min} \geq 2$.

REFERENCES

1. ALEXANDROFF, P. 1937. Diskrete Räume. Mat. Sbornik. **2:** (44) 501–519.
2. EISENREICH, G. 1989. Lexikon der Algebra. Akademie-Verlag. Berlin.

3. ERNÉ, M. 1977. The ABC of order and topology. *In:* Category Theory at Work (H. Herrlich & H.-E. Porst, Eds.). Helderman Verlag. Berlin. 57–83.
4. ISBELL, J. 1985. Graduation and dimension in locales. *In:* Aspects of Topology (I.M. James & E.H. Kronheimer, Eds.). London Math. Soc. Lecture Notes, Series 93. Cambridge University Press, Cambridge, 1985.
5. IVASHCHENKO, A.V. 1985. Dimension of molecular spaces. VINITI. Moscow. **6422-84**: 3–11.
6. EVAKO (IVASHCHENKO), A.V., R. KOPPERMAN, & Y.V. MUKHIN. 1996. Dimensional properties of graphs and digital spaces. J. Math. Imaging and Vision. **6:** 109–119..
7. JOHNSTONE, P.T. 1982. Stone Spaces. Cambridge University Press. Cambridge.
8. KHALIMSKY, E., R. KOPPERMAN, & P.R. MEYER. 1990. Computer graphics and connected topologies on finite ordered sets. Topology and its Applications. **36:** 1–17.
9. KONG, T.Y., R. KOPPERMAN, & P.R. MEYER. 1992. A topological approach to digital topology. Amer. Math. Monthly. **98:** 901–917.
10. KOPPERMAN, R., P.R. MEYER, & R.G. WILSON. 1991. A Jordan surface theorem for three-dimensional digital surfaces. Discrete and Comput. Geometry. **6:** 155–161.
11. KOVALEVSKY, V.A. 1989. Finite topology as applied to image analysis. CVGIP. **46:** 141–161.
12. KRONHEIMER, E.H. 1992. The topology of digital images. Topology and its Applications. **46:** 279–303.
13. LEE, C.N. & A. ROSENFELD. 1986. Connectivity issues in 2D and 3D images. Proceedings of the Intern. Conf. on Computer Vision and Pattern Recognition, Miami Beach, Florida (CH 2290-5/86, IEEE Comp. Soc. Press).
14. PEARS, A.R. 1975. Dimension Theory of General Spaces. Cambridge University Press. Cambridge.
15. SANCHO DE SALAS, J.B. & M.T. SANCHO DE SALAS. 1991. Dimension of distributive lattices and universal spaces. Topology and its Applications. **42:** 25–36.
16. VINOKUROV, V.G. 1966. A lattice method of defining dimension. Soviet Math. Dokl. **168:** (3) 663–666.
17. WIEDERHOLD, P. & R.G. WILSON. 1993. Dimension for Alexandroff spaces. *In:* Vision Geometry (R.A. Melter & A.Y. Wu, Eds.). Proceedings of the Society of Photo-Optical Instrumentation Engineers (SPIE). **1832:** 13–22.

Maximally Almost Periodic Groups and a Theorem of Glicksberg[a]

TA SUN WU[b] AND LEMUEL RIGGINS

Department of Mathematics
Case Western Reserve University
Cleveland, Ohio 44106

ABSTRACT: It is well known that if G is a locally compact Abelian group (LCA group) with Bohr compactification $(\beta(G), \sigma)$ then $\sigma(G)$ is normal in $\beta(G)$ and, by a beautiful theorem of Glicksberg, we have that $A \subset \sigma(G)$ is compact if and only if $\sigma^{-1}(A) \subset G$ is compact. The aim of this paper is to study maximally almost periodic (MAP) groups which have these properties and the results obtained are as follows. (1) If G is a σ-compact locally compact MAP group with Bohr compactification $(\beta(G), \sigma)$ and $\sigma(G)$ is normal in $\beta(G)$, then for each $g \in \beta(G)$, the automorphism induced by σ and conjugation by g is actually a topological isomorphism. (2) A finite extension of a *L*CA group is a MAP group and it has the property that $A \subset \sigma(G)$ is compact if and only if $\sigma^{-1}(A) \subset G$ is compact, and (3) A discrete MAP group G with Bohr compactification $(\beta(G), \sigma)$ satisfying both of the properties being considered must be Abelian by finite, i.e., a finite extension of an Abelian group.

Let G be a (Hausdorff) topological group. G is called maximally almost periodic (MAP for short) if there is a continuous injective homomorphism from G into a compact (Hausdorff) group. If G is a MAP group, then there exists a compact group denoted by $\beta(G)$ (called the Bohr compactification of G) and an injective continuous homomorphism $\sigma: G \rightarrow \beta(G)$ such that for any continuous homomorphism θ from G into a compact group H there exists a unique continuous homomorphism $\tilde{\theta}: \beta(G) \rightarrow H$ such that $\theta = \tilde{\theta} \circ \sigma$, i.e., the following diagram commutes:

$$\begin{array}{ccc} & & \beta G \\ & \sigma \nearrow & \\ G & & \downarrow \theta^{\sim} \\ & \theta \searrow & \\ & & H \end{array}$$

Loomis [10] or Weil [15].

Mathematics Subject Classification: 22A05.
Key words and phrases: MAP groups, $\overline{FC}$ groups.
[a]This material is based upon work supported under a National Science Foundation Graduate Fellowship.
[b]E-mail: txw3@po.cwru.edu.

We recall the amazing result first proved by Freudenthal for groups with countable open bases and the general case due to Weil [15, pp. 126–130]. A connected locally compact MAP group has the form $R^n \times G_0$ where R^n is a n-dimensional vector group (over the real number field $\mathbf{R}$) and G_0 is a compact group. We shall sketch a proof of this theorem which is somewhat different from the original proof (it does not mean to be more elegant or shorter) for the purpose of introducing some notions which we need for later discussions. First we state a theorem.

THEOREM 1.1: (Goto [5]) Let G be a locally compact connected and locally connected group. Let θ be a continuous injective homomorphism from G into a topological group F such that $\theta(G)$ is dense in F. Then $\theta(G)$ is a normal subgroup of F. Now define a homomorphism $\alpha: F \rightarrow \operatorname{Aut} G$ by $\alpha(a)(x) = \theta^{-1}(a\theta(x)a^{-1})$ for $a \in F$ and $x \in G$. Then α is a continuous homomorphism from F into the group of automorphisms of G. (Note: G is locally compact, connected, and locally connected, so $\operatorname{Aut} G$ is a topological group when it is topologized by the compact-open topology.)

Throughout the rest of this paper, let G denote a locally compact MAP group and Γ denote a MAP group with respect to the discrete topology on Γ. Also, all topological groups are assumed to be Hausdorff.

DEFINITION: Let $x \in G$. x is a bounded element if $\mathcal{O}_x := \{gxg^{-1}: g \in G\}$ is relatively compact. Let $B(G)$ denote the set of all bounded elements in G. Then $B(G)$ is a characteristic subgroup of G, i.e., if ψ is a topological isomorphism of G, then we must have that $\psi(B(G)) = (B(G))$. (For some properties of $B(G)$, see [16]).

Sketch of the proof of Freudenthal Theorem: Let G be a locally compact connected MAP group. So there exists a continuous isomorphism σ from G into a compact group F such that $\sigma(G)$ is dense in F. Because G is connected and locally compact, G contains a unique maximal compact normal subgroup M such that G/M is an analytic group (thus, it is locally connected). Then $\sigma(M)$ is a compact normal subgroup of F. This implies we have the induced isomorphism σ' from G/M into $F/\sigma(M)$. Now we may apply Goto's Theorem to conclude that G is an $\overline{FC}$ group, i.e., every element of G is a bounded element $G = B(G)$. It is known that G contains a maximal normal compact subgroup N such that G/N is isomorphic with a vector group [16] and since vector groups are analytic, we must have that G/M is a vector group. Now by a theorem of Iwasawa [17] we have that $G = C_G(M)M$, where $C_G(M)$ is the centralizer of M in G, and our next goal is to show that $G = C_G(M)_0M$, where $C_G(M)_0$ is the connected component of $C_G(M)$.

Observe that (1) $C_G(M) \cap M$ equals the center of M so $C_G(M) \cap M$ is Abelian, (2) by uniqueness of M, we have that M is a characteristic subgroup of G and this implies that $C_G(M)$ is a characteristic subgroup of G. Also, this gives us that $C_G(M)_0$ is a characteristic subgroup of G and hence, any characteristic subgroup of M, $C_G(M)$ or $C_G(M)_0$ is a normal subgroup of G. Next, we notice that because G is connected (hence, σ-compact) and $C_G(M)$ is closed in G (hence, σ-compact), we have that G/M is topologically isomorphic to $C_G(M)/(C_G(M) \cap M)$ so that the latter group is a vector group, hence it is connected. Now $C_G(M)/(C_G(M)_0(C_G(M) \cap M))$ is a totally disconnected group and it is also a continuous homomorphic image of $C_G(M)/C_G(M) \cap M$ so we must have that $C_G(M) = C_G(M)_0(C_G(M) \cap M)$. From this fact, it follows easily that $G = C_G(M)_0M$.

With this information in mind, we see that G/M is topologically isomorphic with $C_G(M)_0/(C_G(M)_0 \cap M)$ and because G/M is a vector group and $C_G(M)_0 \cap M$ is Abelian, we have that $C_G(M)_0$ is a connected solvable characteristic subgroup of G. Now it is well known that a compact connected solvable group is Abelian and since $\overline{\theta(C_G(M)_0)}$ is a closed connected solvable subgroup of the compact group F, we must have that $\overline{\theta(C_G(M)_0)}$ is compact and hence Abelian. But θ is 1–1 implies $C_G(M)_0$ is a connected *LCA* group so it has the form $V \times C$ where V is a vector group and C is a compact connected Abelian group. We also have that V and C are characteristic subgroups of $C_G(M)_0$, hence, V and C are normal in G. This gives us the fact that $G = VM$ where V is a normal vector subgroup of G because M is the unique maximal compact normal subgroup of G such that G/M is analytic. But a vector group has no nontrivial compact subgroups so $V \cap M$ is trivial. This implies that G is a direct product of a vector group and a compact group, hence, the theorem is proved. ❑

REMARK: In general, an $\overline{FC}$ group is not a direct product of a vector group and a compact group. For example: Let G be the Heisenberg group; i.e., the group consisting of all the upper triangular matrices with real number entries.

$$\left\{\begin{pmatrix} 1 & x & z \\ 0 & 1 & y \\ 0 & 0 & 1 \end{pmatrix} : x, y, z \in \mathbf{R}\right\}.$$

Let Z be a discrete central subgroup of G. For instance

$$\left\{\begin{pmatrix} 1 & 0 & n \\ 0 & 1 & 0 \\ 0 & 0 & 1 \end{pmatrix} : n \in \mathbf{Z}\,(\text{integers})\right\}.$$

Then G/Z is an $\overline{FC}$ group. G/Z is not faithfully representable by linear groups. (See Hochschild [9, Chapter 18] for further discussion).

DEFINITION: Let Γ be a (discrete) group. Γ is an *FC* group if every conjugacy class of Γ is finite, i.e., for all $x \in \Gamma$, we have that $\mathcal{O}_x := \{yxy^{-1} : y \in \Gamma\}$ is finite. For properties of such groups, see Robinson [11].

Assume G is a σ-compact LCH (locally compact Hausdorff) MAP group. Then G is a Lindelöf space and every closed subset of G is a Lindelöf space (when given the relative topology).

THEOREM 1.2: If $A \in \beta(G)$ is such that $A\sigma(G)A^{-1} = \sigma(G)$, then the map from G to G induced by A is a continuous automorphism, in fact it is a homeomorphism.

Proof: Let $C_A : \beta(G) \to \beta(G)$ be conjugation by A then the algebraic automorphism induced by A on G is given by $L_A = \sigma^{-1} \cdot C_A \cdot \sigma$. Since $A\sigma(G)A^{-1} = \sigma(G)$ clearly implies that $A^{-1}\sigma(G)A = \sigma(G)$, we have that $L_{A^{-1}}$ is well-defined and $L_{A^{-1}} = (L_A)^{-1}$. So if we show that L_A is an open map for any $A \in \beta(G)$ s.t. $A\,\sigma(G)A^{-1} = \sigma(G)$, then it follows that L_A and $L_{A^{-1}} = (L_A)^{-1}$ are both open which implies that L_A is a homeomorphism. So we must show that L_A is open.

Claim: If $K \subseteq G$ is compact, then $L_A(K)$ is closed. This follows because if K is compact, then $\sigma(K)$ is compact because σ is continuous, but this implies that $C_A(\sigma(K)) \subset \beta(G)$ is compact and hence closed, because $\beta(G)$ is Hausdorff. But this implies that $L_A(K) = \sigma^{-1}(C_A(\sigma(K)))$ is closed in G because σ is continuous, hence the claim is true. Now let U be a neighborhood of 1_G, then since G is LCH, there exists V an open neighborhood of 1_G such that $\overline{V}$(= closure of V) is compact and $\overline{V}^{-1}\overline{V} \subseteq U$. Now G is Lindelöf so we have that countably many translates of $\overline{V}$ cover G which implies that countably many translates of $L_A(\overline{V})$ cover G since L_A is an algebraic automorphism. Now by the Baire category theorem we must have that $L_A(\overline{V})$ is not nowhere dense which implies that there exists a nontrivial open subset of G call it W s.t. $W \subset (\overline{L_A(\overline{V})})^0$ (= the interior of $\overline{L_A(\overline{V})}$))} but $\overline{V}$ is compact so by our claims, we have that $L_A(\overline{V}) = \overline{L_A(\overline{V})}$ which implies that $W \subset L_A(\overline{V})$. But then $W^{-1}W$ is an open neighborhood of 1_G, and $L_A(U) \supset L_A(\overline{V}^{-1}\overline{V}) \supset L_A(\overline{V}^{-1})L_A(\overline{V}) \supset W^{-1}W$ by L_A is a homomorphism which implies that $L_A(U)$ is a neighborhood of 1_G for any U such that U is a neighborhood of 1_G which implies that L_A is an open map. Hence, the proof is complete. ❑

In view of the above result, we clearly must have that if G is a σ-compact locally compact MAP group with Bohr compactification $(\beta(G), \sigma)$ and $\sigma(G)$ is normal in $\beta(G)$, then for each $g \in \beta(G)$, the automorphism induced by σ and conjugation by g is actually a topological isomorphism.

DEFINITION: M is called a conjugacy class of a group G if there exists $x \in G$ such that $M = \mathcal{O}_x = \{gxg^{-1} : g \in G\}$.

COROLLARY 1.2(a): If $\sigma(G)$ is normal in $\beta(G)$ and $M \subset \sigma(G)$ is a conjugacy class of $\beta(G)$, then $\sigma^{-1}(M)$ is compact in G.

Proof: Let $C = \sigma^{-1}(M)$, then $\sigma(C) = M$ by $M \subseteq \sigma(G)$. Now $M \subset \sigma(G)$ and M is a conjugacy class of $\beta(G)$ implies that there exists a $x \in G$ such that M

$= \mathcal{O}_{\sigma(x)} = \{g\sigma(x)g^{-1}: g \in \beta(G)\}$. Now because $\beta(G)$ is compact Hausdorff we have that M is compact, hence closed, which implies that $C = \sigma^{-1}(M)$ is closed because σ is continuous. Now if V is a compact neighborhood of 1_G, then $\{yV \cap C: y \in C\}$ is a cover for C when C is given the relative topology and since G is Lindelöf and $C \subset G$ is closed, we have that C is Lindelöf which implies that the above cover has a countable subcover $\{y_nV \cap C\}_{n=1}^{\infty}$ where $y_n \in C$, $\forall n \in \mathbf{N}$. So we have that $\{\sigma(y_nV \cap C)\}_{n=1}^{\infty}$ cover $\sigma(C) = M$, but M is compact Hausdorff so by the Baire category theorem there exists $k \in \mathbf{N}$ such that $\sigma(y_kV \cap C)$ has nonempty interior with respect to the relative topology on M (Note: C closed and V compact $\Rightarrow yV \cap C$ compact which implies that $\sigma(yV \cap C)$ is closed compact). But M is a conjugacy class so that $\{g(\sigma(y_kV \cap C))^0g^{-1}: g \in \beta(G)\}$ (where if A is a set, A^0 denotes the interior of A and the interior of $\sigma(y_kV \cap C)$ is with respect to the topology of M) is an open cover for M which implies it has a finite subcover, i.e., $M = \bigcup_{i=1}^{n} g_i\sigma(y_kV \cap C)g_i^{-1}$ which implies that $C = \sigma^{-1}(M) = \bigcup_{i=1}^{n} \sigma^{-1}(g_i(\sigma(y_kV \cup C))g_i^{-1}) = \bigcup_{i=1}^{n} L_{g_i}(y_kV \cap C)$. But by the previous work, we have that L_{g_i} is continuous for all $i \in P_n$ which shows that C is a finite union of compact sets which implies that C is compact, hence the proof is complete. ❑

COROLLARY 1.2(b): If G is a σ-compact locally compact MAP group such that $\sigma(G)$ is normal in $\beta(G)$, then G is an $\overline{FC}$ group.

Proof: Given $x \in G$, we have that $\mathcal{O}_{\sigma(x)}$ is a conjugacy class in $\beta(G)$ and $\mathcal{O}_{\sigma(x)} \subset \sigma(G)$ because $\sigma(G)$ is normal in $\beta(G)$. So by the previous work in Corollary 1.2, $\sigma^{-1}(\mathcal{O}_{\sigma(x)})$ is compact in G, but $\mathcal{O}_x$ is contained in $\sigma^{-1}(\mathcal{O}_{\sigma(x)})$ which implies that $\overline{0}_x$ is compact, hence, corollary is true. ❑

Note: Clearly, an $\overline{FC}$ MAP group is normal in its Bohr compactification.

2

It would be very interesting to generalize Freudenthal's theorem to more general groups, or to ask: Are there other interesting conditions which can be used to characterize MAP groups. In general, this is easily seen to be a hopeless task, as one simply takes any dense subgroup of a compact group and give it the discrete topology. It is well known that every locally compact Abelian group (*LCA* group) is a MAP group [8]. On the other hand, recall the following class of groups.

DEFINITION: Γ is a residually finite group if for every element x, distinct from the identity 1 of Γ, there is a normal subgroup Γ_x of finite index in Γ such that $x \notin \Gamma_x$.

REMARK: If Γ is a residually finite group, then Γ is a MAP group. (Simply consider the map $\pi: \Gamma \rightarrow \prod_{x \neq 1} \Gamma/\Gamma_x$ defined by $\pi(Y) = \{Y\Gamma_x\}_{x \neq 1}$ for all

$Y \in \Gamma$). We note that a free group is a residually finite group. Therefore, every free group is a MAP group.

Now, we give a few simple observations on *LCH* MAP groups.

(a) $G = \prod G_i$ is the Cartesian product of the groups G_i. Then G is MAP if and only if each G_i is a MAP group.

(b) A finite extension of a MAP group is MAP. (If K is a subgroup of a *LCH* group G such that G/K is finite, then G is a finite extension of K.)

(c) A continuous homomorphic image of a MAP group may not be a MAP group. (As an example, let G be a discrete group which is not a MAP group (an infinite simple torsion group will do). Then since every group is a homomorphic image of a free group, we have a counter example. However, if G' is a compact normal subgroup of a MAP group G, then G/G' is a MAP group.)

The following theorem is useful when we are dealing with discrete groups.

THEOREM 2.1: (Hall [11]) Let Γ be a finitely generated group which is an extension of an Abelian group by a nilpotent group. Then Γ is residually finite. Therefore, Γ is MAP.

In view of the above discussion, the class of locally compact MAP groups contains many diverse groups, and there does not seem to be any uniform structural characterization of MAP groups. Therefore, we need to narrow down to some subclass. We notice that the class of locally compact Abelian groups is already a huge class. And given any LCA group G, there are some special relations between G and its Bohr compactification $\beta(G)$. Here, we name a few.

(1) G when considered as a subgroup of $\beta(G)$ is a normal subgroup.

(2) Let H be a closed subgroup of G. Then $\beta(H)$ is isomorphic with $\overline{\sigma(H)}$. (Observe, every character of H extends to a character of G).

(3) Let H be a closed subgroup of G. Then $\overline{\sigma(H)} \cap \sigma(G) = \sigma(H)$. (This follows from the fact that if $x \notin H$, then there is a character χ of G such that $\chi(x) \neq 1$ and $\chi(H) = \{1\}$.

(4) Let A be a compact subset contained in $\sigma(G)$. Then $\sigma^{-1}(A)$ is a compact subset of G [2].

As stated at the beginning of this paper, we wish to consider MAP groups with properties (1) and (4), and since we have already studied groups with property (1), we shall now consider property (4). Given any locally compact Abelian group G, let A be any compact subset of $\sigma(G)$. Then $\sigma^{-1}(A)$ is also compact. Let us first consider the case: G is discrete. Then the above property says that a subset of $\sigma(G)$ is compact if and only if it is finite. This indicates that $\sigma(G)$ is, roughly speaking, sparsely distributed inside $\beta(G)$. The reason for this seems to come from (2) and (3), and these observations seem to say that $\beta(G)$ is very big in comparison to $\sigma(G)$. Topologically, if we compare the cardinality of the topology of G with that of $\beta(G)$, we see that this indicates that this is the case. Furthermore, every closed subgroup H of G also has the same property. This means that H not only has "large" compactification, but also this compactification extends $\beta(G)$. For Abelian compact groups, we know that the weight of G equals the cardinality of its character group [8, (24.14)]. Counting the cardinality of the character group shows this fact clearly in particular when G is a discrete group.

The question now begging to be asked is: Is property (4) something special only for Abelian groups?

THEOREM 2.2: Let F be a finite extension of a locally compact Abelian group G. Then F is MAP and if $A \subset \psi(F)$ is compact, then $\psi^{-1}(A) \subset F$ is compact, where $(\beta(F), \psi)$ is the Bohr compactification of F.

Proof: First we show how to construct a compactification of F from a Bohr compactification of G. Let $a \in F$. Then the conjugation defined by a induces an automorphism I_a of G. Using the universal property of $\beta(G)$, I_a induces an automorphism of $\beta(G)$ for which we use the same notation I_a. Then F acts on $\beta(G)$ as a group of automorphisms of $\beta(G)$. Let $(\beta(G) \ltimes F)$ be the semidirect product defined by the above action. Let $\triangle = \{(\sigma(g), g) \in (\beta(G) \ltimes F), g \in G\}$. Then $\triangle$ is a closed normal subgroup of $(\beta(G) \ltimes F)$ (note $\triangle$ is closed in $(\beta(G) \ltimes F)$ because F is a finite extension of G implies that G is closed in F, σ is continuous and $(\beta(G) \ltimes F)$ has the product topology), so we have the quotient group $(\beta(G) \ltimes F)/\triangle$. Now, since F is a finite extension of G, we have that $(\beta(G) \ltimes F)/\triangle$ is compact. Let $\sigma' : F \to (\beta(G) \ltimes F)/\triangle$ be given by $\sigma'(a) = (1, a)\triangle$. Then we have that $((\beta(G) \ltimes F)/\triangle, \sigma')$ is a compactification of F. Let A be a compact subset contained in $\sigma'(F)$. Now because $\sigma'(G)$ has finite index in $\sigma'(F)$; i.e., $\sigma'(F)$ is a finite disjoint union of cosets of $\sigma'(G)$, and $\sigma'(G)$ has property (4) above, we therefore get that $\sigma'^{-1}(A)$ is compact in F. Since property (4) holds for some compactification of F, it must also hold for $(\beta(F), \psi)$ (use universal property of $\beta(F)$ and ψ, σ' with 1–1). Hence, the proof is complete. ❑

This shows that G may be non-Abelian, noncompact, and yet still poseses property (4). On the other hand, from the work [2], it is not necessary to use the Bohr compactification of G, some weaker compactification of G may have the same property. So we define the following class of locally compact groups.

DEFINITION: Let G be a locally compact MAP group. We say G is a (g) group if there exists an injective continuous homomorphism θ from G into a compact group H such that $\theta(G)$ is dense in H and for every compact subset A of $\theta(G)$, $\theta^{-1}(A)$ is also compact. It is clear that if G is a (g) group, we may use $(\beta(G), \sigma)$ for (H, θ).

DEFINITION: Let G be a locally compact MAP group and $(\beta(G), \sigma)$ be its Bohr compactification. Then G is said to be β-invariant if $\sigma(G)$ is normal in $\beta(G)$.

We wish to characterize (g) groups and we start by looking at discrete (g) groups. In fact, we shall consider infinite discrete MAP groups. Let Γ be such a group. Now if Γ is a (g) group and Γ' is a subgroup of Γ, then $(\sigma(\Gamma'), \sigma|_{\Gamma'})$ has property (4) which implies that Γ' is also a (g) group. Now given an infinite discrete MAP group Γ, let Γ' be an infinite subgroup of Γ. If $\sigma(\Gamma')$ is metrizable, then Γ is not a (g) group because $\sigma(\Gamma')$ metrizable implies that either $\sigma(\Gamma')$ is discrete or that $\sigma(\Gamma')$ contains a sequence of distinct elements which converges to the identity of Γ. Now if $\sigma(\Gamma')$ is discrete in $\beta(\Gamma)$, then since $\beta(G)$ is Hausdorff, we have that $\sigma(\Gamma')$ is closed which implies that $\sigma(\Gamma')$ is compact since $\beta(\Gamma)$ is compact, but Γ' is infinite and σ is 1–1, so Γ cannot be a (g) group. Now if there exists a sequence $\{\sigma(x_n)\}_{n=1}^{\infty}$ as described above, then clearly $\{\{\sigma(x_n)\}_{n=1}^{\infty}, 1\}$ is a compact subset of $\sigma(\Gamma)$ and since $x_n = x_m$ if and only if $n = m$, we have that the above set is infinite which implies that Γ cannot be a (g) group. It turns out that this observation is very useful. In fact we are going to show that certain MAP groups are not (g) groups by using this observation.

EXAMPLE: Let $\mathbf{Z}_3$ be the cyclic group of order three. The automorphism group of $\mathbf{Z}_3$ is a cyclic group of order 2, and it is isomorphic with $\mathbf{Z}_2$. We can form the semidirect product $\mathbf{Z}_3 \ltimes \mathbf{Z}_2 \simeq S_3$. Now, let $B_i \ltimes A_i$ be a copy of $\mathbf{Z}_3 \ltimes \mathbf{Z}_2$, $i \in \mathbf{N}$. Let $\Gamma = \sum_{i=1}^{\infty} B_i \ltimes A_i$ with the discrete topology. Then Γ is a MAP group since it is residually finite. Since the exponent of Γ is 6 we have that any compactification of Γ is totally disconnected. Let $B = \sum_{i=1}^{\infty} B_i \simeq \sum_{i=1}^{\infty} (\mathbf{Z}_3)_i$. Then any cofinite normal subgroup of B has the form $\sum_{i \in K} B_i$, where K is a subset of the natural numbers which misses only at most finitely many numbers. Thus, $\overline{\sigma(B)}$ is metrizable where σ is the map from Γ into $\beta(\Gamma)$. Hence, Γ is not a (g) group.

Note: Γ is a solvable 2-step FC group, i.e., the commutator of Γ is Abelian.

GOAL: To show that if Γ is a discrete β-invariant (g) group, then Γ is Abelian by finite.

Let G be a LCH MAP group, and Γ denote a MAP group with the discrete topology.

PRELIMINARY 1: If G is a β-invariant (g) group, then G is $\overline{FC}$.

Proof: Let $(\beta(G), \sigma)$ be the Bohr compactification of G and $x \in G$, then $\mathcal{O}_{\sigma(x)} = \{g\sigma(x)g^{-1} : g \in \beta(G)\}$ is compact in $\beta(G)$ since $\beta(G)$ is compact. But $\sigma(G)$ is normal in $\beta(G)$, so we have that $\mathcal{O}_{\sigma(x)} \subset \sigma(G)$ which implies that $\sigma^{-1}(\mathcal{O}_{\sigma(x)})$ is compact in G because G is a (g) group. But $\mathcal{O}_x$ is contained in $\sigma^{-1}(\mathcal{O}_{\sigma(x)})$ which is closed and compact since G is Hausdorff. This implies that $\mathcal{O}_x$ is compact in G, so Preliminary 1 is true. In particular, if G is discrete, then G is FC and also $\sigma(G)$ is an FC subgroup of $\beta(G)$ since $\mathcal{O}_{\sigma(x)}$ must be finite (actually, in this case $\sigma(\mathcal{O}_x) = \mathcal{O}_{\sigma(x)}$). ❑

PRELIMINARY 2: If L is a LCH group with a dense FC subgroup K, then L_0, the connected component of the identity of L, is central.

Proof: Given $x \in K$, we have a continuous map $\pi_x: L_0 \to L$ given by $\pi_x(j) = jxj^{-1}$, $\forall j \in L_0$, but $\pi_x(L_0) \subset \mathcal{O}_x$ which is finite because K is a FC subgroup of L (i.e., $x \in K \Rightarrow \mathcal{O}_x = \{LxL^{-1}: L \in L\}$ is finite) and since L is Hausdorff, the relative topology $\mathcal{O}_x$ inherits from L is discrete which implies $\mathcal{O}_x$ is totally disconnected. But L_0 is connected and since L_0 contains the identity in L, we must have that $\pi_x(L_0) = \{x\}$ which implies that $jxj^{-1} = x$, $\forall j \in L_0$ and $\forall x \in K$ which implies that L_0 is central in L since K is dense in L. Hence, Preliminary 2 is true. ❑

In particular, if L is a compact Lie group with a dense FC subgroup, then L is central by finite because by our previous work, L_0 is central and since L is a compact Lie group we must have that L/L_0 is finite.

PRELIMINARY 3: Let T be any group, and $F \subseteq T$ be a finite subset of FC elements of T. Let A be the normal subgroup of T generated by F. Then A is finitely generated and $C_T(A)$, the centralizer of A in T has finite index in T and is normal in T.

Proof: Since F is a set of FC elements of T and F is finite, we clearly have that A is finitely generated. This is because if $F = \{f_i\}_{i=1}^n$, then $\bigcup_{i=1}^n \mathcal{O}_{f_i}$ is a finite set because each $\mathcal{O}_{f_i}$ is finite, but $A = \langle \bigcup_{i=1}^n \mathcal{O}_{f_i} \rangle$ so A is finitely generated, also, we clearly have that $C_T(A) = \bigcap_{x \in \bigcup_{i=1}^n \mathcal{O}_{f_i}} C_T(x)$ which is a finite intersection of subgroups of finite index in T, hence $C_T(A)$ has finite index in T. (Note: $C_T(x)$ has finite index in T, because we have a natural bijection between $T/C_T(x)$ and $\mathcal{O}_x$ given by $gC_T(x) \to gxg^{-1}$ and $\mathcal{O}_x$ is finite because x is an FC element). Finally, since A is normal in T, we have that $C_T(A)$ is normal in T because given $t \in T$, $z \in C_T(A)$ and $a \in A$, we have that:

$$
\begin{aligned}
tzt^{-1}a(tzt^{-1})^{-1} &= tzt^{-1}atz^{-1}t^{-1} && \text{by } (xy)^{-1} = y^{-1}x^{-1} \\
&= tz(t^{-1}at)\,z^{-1}t^{-1} && \text{associativity} \\
&= t(t^{-1}at)t^{-1} && \text{by } t^{-1}at \in A \quad \text{by } A \text{ is normal and } z \in C_T(A) \\
&= a && \text{by } tt^{-1} = 1_T
\end{aligned}
$$

and since $a \in A$ is arbitrary, we see that $tzt^{-1} \in C_T(A)$. Hence, $C_T(A)$ is normal in T because $t \in T$ and $z \in C_T(A)$ are arbitrary. ❑

In particular, if T is a compact topological group, then $C_T(A)$ is an open subgroup of T.

THEOREM 2.3: If Γ is a discrete β-invariant (g) group, then Γ is Abelian by finite.

Let Γ be such a group and $(\beta(\Gamma), \sigma)$ be its Bohr compactification. Then by Preliminary 1, Γ is an *FC* group, and $\sigma(\Gamma)$ is an *FC* subgroup of $\beta(\Gamma)$.

Claim: If Γ is not Abelian by finite, then Γ contains a distinct sequence $\{\Gamma_i\}_{i=1}^{\infty}$ of normal non-Abelian subgroups such that:

(1) For each $k \in \mathbf{N}$, there exists $x_k, y_k \in \Gamma_k$ s.t. Γ_k equals the normal subgroup generated by x_k, y_k, and $c_k = [x_k, y_k] = x_k y_k x_k^{-1} y_k^{-1}$ does not equal the identity in Γ.
(2) $a_1 \in \Gamma_i$ and $a_j \in \Gamma_j$, $i \neq j$ implies that $a_i a_j = a_j a_i$, and
(3) $c_i = c_j$ if and only $i = j$.

Proof: We will identify Γ with $\sigma(\Gamma) \subseteq \beta(\Gamma)$ since σ is 1–1. Now since Γ is not Abelian by finite, there exist $x_1, y_1 \in \Gamma$ such that $c_1 = [x_1, y_1]$ is not equal to the identity in Γ. Now let Γ_1 be the normal subgroup of $\beta(\Gamma)$ generated by X_1, Y_1. Now assume that $\{\Gamma_i\}_{i=1}^{n}$ have been chosen and satisfy the properties (1), (2), and (3). Then we must show that we can pick Γ_{n+1} because then by induction, we have that the claim is true. Now by properties (1) and (2), and the fact that $n \in \mathbf{N}$, we have that $\prod_{i=1}^{n} \Gamma_i$ is the normal subgroup of $\beta(\Gamma)$ generated by $\{x_1, y_1, \ldots, x_n, y_n\}$ so by Preliminary 3, we have that $C_{\beta(\Gamma)}(\prod_{i=1}^{n} \Gamma_i)$ is open in $\beta(\Gamma)$ which implies that there exists V a symmetric neighborhood of 1 in $\beta(\Gamma)$ such that $V^4 \subset C_{\beta(\Gamma)}(\prod_{i=1}^{n} \Gamma_i)$ and x_i, y_i, c_i are not elements of V^4 for all $i \in P_n = \{1, 2, \ldots, n\}$. Now if $\sigma(\Gamma) \cap V$ is a commutative set, then since $\sigma(\Gamma)$ is dense in $\beta(\Gamma)$, we have that V is an open commutative subset of $\beta(\Gamma)$ which implies that $\langle V \rangle$ is an open Abelian subgroup of $\beta(\Gamma)$ and since $\beta(\Gamma)$ is compact, this implies that $\beta(\Gamma)$ is Abelian by finite which implies that Γ is Abelian by finite which is false, hence there exists $x_{n+1}, y_{n+1} \in V$ such that $c_{n+1} = [x_{n+1}, y_{n+1}]$ is not the identity in $\beta(\Gamma)$. Now let Γ_{n+1} = the normal subgroup of $\beta(\Gamma)$ generated by x_{n+1} and y_{n+1}. Then since $V^4 \subset C_{\beta(\Gamma)}(\prod_{i=1}^{n} \Gamma_i)$ which is normal by Preliminary 3 and $x_{n+1}, y_{n+1} \in V$, we clearly have that property (2) is satisfied. Also, property (3) is satisfied because $[x_{n+1}, y_{n+1}] \in V^4$ which does not contain c_i $\forall_i \in P_n$. Now property (1) is satisfied because for $i \in P_n$, we have that x_i does not centralize Γ_i which implies that $x_i \notin \Gamma_{n+1}$, hence by induction such a sequence exists.

Now, if we show that this contradicts Γ being a (g) group, then we must have that Γ is Abelian by finite. We claim that $\{c_i\}_{i=1}^{\infty}$ converges to 1 as n goes to infinity. If we show this, then since $\{\{c_i\}_{i=1}^{\infty}, 1\}$ is contained in $\sigma(\Gamma)$ and is compact, then this set with the discrete topology must be compact, but this contradicts $c_i = c_j$ if and only if $i = j$ and we will have found our contradiction. What's left to show is that $\{c_i\}_{i=1}^{\infty}$ converges to 1. Let V be an arbitrary neighborhood of 1 in $\beta(\Gamma)$, then since $B(\Gamma)$ is compact, there exists a N which is a normal subgroup of $\beta(\Gamma)$ such that $N \subseteq V$ and $\beta(\Gamma)/N$ is a compact Lie group. Now $\sigma(\Gamma)$ is a dense *FC* subgroup of $\beta(\Gamma)$ implies that $\beta(\Gamma)/N$ is a compact Lie group with a dense *FC* subgroup which implies (via Preliminary 2) that $(\beta(G)/N)_0 = C/N$ is central in $\beta(\Gamma)/N$ and $\beta(\Gamma)/C$ is finite, so that $[\beta(\Gamma), C] \subset N$ (where C is the inverse image in $\beta(\Gamma)$ of $\beta(\Gamma)/N)_0$ under the canonical map from $\beta(\Gamma)$ to $\beta(\Gamma)/N)$. Now let $A = \langle\{\Gamma_i\}_{i=1}^{\infty}\rangle$, then $A \cap C$ is a cofinite normal subgroup of A which implies there exists $n \in \mathbf{N}$ such that $A = (\prod_{i=1}^{n} \Gamma_i)(A \cap C)$. So if $k > n$, then $y_k = ac$ where $a \in \prod_{i=1}^{n} \Gamma_i$ and $c \in A \cap C$, but $k > n$ implies that $ay_k = y_k a$ which implies that $ac = ca$. Also, because $k > n$, we have that $ax_k = x_k a$, but then $c_k = x_k y_k x_k^{-1} y_k^{-1} = x_k a c x_k^{-1} c^{-1} a^{-1} = x_k c x_k^{-1} c^{-1} a a^{-1} = x_k c x_k^{-1} c^{-1} \in N \subset V$ by $c \in C$ so we must have that $c_k \in V$ for all $k > n$ which implies that c_k converges to 1 as k goes to infinity. Hence, if Γ is a discrete β-invariant (g) group, then Γ is Abelian by finite. ❑

REFERENCES

1. COMFORT, W.W. & F.J. TRIGOS-ARRIETA. 1991. Remarks on a Theorem of Glicksberg, in General Topology and Applications. Lecture Notes in Pure and Appl. Math. **134:** 25–33. Dekker, New York.
2. COMFORT, W.W., F. JAVIER TRIGOS-ARRIETA, & TA-SUN WU. 1993. The Bohr compactification, modulo and metrizable subgroup. Fundamenta Math. **143:** 119–136.
3. DIXON, J.D. 1972. Free Subgroups of Linear Groups. Conference on Group Theory (Univ. of Wisconsin-Parkside). Springer-Verlag Lecture Note, No. 319: 45–56.
4. EPSTEIN, D.B.A. 1971. Almost all subgroups of Lie groups are free. J.Algebra. **19:** 261–262.
5. GLICKSBERG, I. 1962. Uniform Boundedness for Groups. Canad. J. Math. **14:** 269–276.
6. GOTO, M. 1980. Immersion of Lie groups. J. Math. Soc. Japan. **32:** 727–749.
7. HALL, M. 1950. A topology for free groups and related groups. Annals of Math. **52:** 127–139.
8. HEWITT, E. & K. ROSS. 1963. Abstract Harmonic Analysis, Vol. 1. Springer-Verlag. New York.
9. HOCHSCHILD, G. 1965. Structure of Lie Groups. Holden-Day. San Francisco, California.
10. LOOMIS, L. 1963. An Introduction to Abstract Harmonic Analysis. van Nostrand. New York.
11. ROBINSON, D.J.S. 1982. A Course in the Theory of Groups. Springer-Verlag. New York.
12. SEGAL, D. 1989 Residually finite groups. *In:* Groups-Canberra. Springer-Verlag Lecture Notes, No. 1456: 85–95.
13. TITS, J. 1972. Free Subgroups in linear groups. J. Algebra. **20:** 250–270.
14. WANG, S.P. 1974. On Jordan's Theorem of Torsion Groups. J. Algebra. **31:** 514–416.
15. WEIL, A. 1941/51. L' integration dans les Groupes Topologiques et ses Applications, Actualits Sci et Ind. Hermann & Cie. Paris.
16. WU, T.S. 1986. Closures of Lie subgroups and almost periodic groups. Bull. Academia Sinica. **14:** 325–347.
17. IWASAWA, K. 1948. On some types of topological groups. Ann. of Math. **3:** 507–558.

Convergence Axioms for Topology

OSWALD WYLER

Department of Mathematics
Carnegie Mellon University
Pittsburgh, Pennsylvania 15213-3890

ABSTRACT: We obtain axioms for a topology using filter convergence as the basic construct, and two very simple axioms for a topology in terms of ultrafilter convergence.

INTRODUCTION

Point set topology was created to deal with continuity and convergence. However, topologies are usually defined in terms of open sets or closed sets or closures or neighborhoods, and hardly ever in terms of convergence or limits.

There is of course a good reason for this state of affairs. Topologies had to be defined before convergence and limits could be fully understood. In fact, one of the first attempts to define topologies [5] used convergence of sequences as the basic notion, but it soon became clear that convergent sequences need not suffice to characterize a topology. This forced topologists to use other basic notions until Moore and Smith [10] defined generalized sequences which they called nets. Eventually, Kelley [6] characterized topologies in terms of net convergence.

Convergence of nets has several drawbacks. Convergent nets on a nonempty topological space always form a proper class; even subnets of a net form a proper class. The definition of subnets is somewhat involved, and Kelley's iterated limit axiom is quite messy. These complications disappear if nets are replaced by the filters of H. Cartan [2], [3].

We begin by comparing nets and filters, and we use this to translate Kelley's axioms from nets to filters, replacing his diagonal net construction by a natural diagonalization due to Kowalsky [8]. This leads to a simple system of two axioms for convergence of ultrafilters in topology, and as a bonus to a simple proof of the theorem that compact Hausdorff spaces are the algebras for the ultrafilter monad.

The results of this paper have been around for some time, but the results of Sections 3 and 4 have never been published.

Mathematics Subject Classification: 54A20.
Key words and phrases: filter convergence, net convergence, ultrafilter convergence, diagonal net, compression (of filter of filters).

Some remarks about notations: Set inclusion $A \subset S$ includes the case $A = S$; there is no need for "proper" subsets. Parentheses around arguments of functions are sometimes omitted. Direct and inverse images for a mapping $f: S \to T$ are denoted by $f^{\to}A$ and $f^{\leftarrow}B$, for $A \subset S$ and $B \subset T$. For a subset A of a preordered set D, the increasing subset of D generated by A is denoted by $\uparrow A$, and the decreasing set generated by A is $\downarrow A$. These notations become $\uparrow i$ and $\downarrow i$ for $A = \{i\}$.

1. NETS AND FILTERS

1.1. Nets

A *net* in a set S is a function $s: D \to S$ from a directed set D to S. Sequences are special nets, with D the set $\mathbb{N}$ of natural numbers, or the set $\mathbb{N}\backslash\{0\}$ of positive integers. More generally, a *directed set* is a set D preordered by a relation $\geqslant$, in which every finite subset A, including $\varnothing$, has an upper bound, an element u of D such that $u \geqslant x$ for all x in A. As for sequences, entries $s(i)$ of a net s are often denoted s_i.

1.2. Filters and Filterbases

A *filter* on a set S is an increasing set $\mathcal{F}$ of subsets of S which is closed under finite intersections, so that $S \in \mathcal{F}$, and $A \cap B \in \mathcal{F}$ for any two subsets A, B in $\mathcal{F}$.

Many authors also require that $\varnothing \notin \mathcal{F}$, but it is better for many purposes not to require this. If $\mathcal{F}$ is increasing in the powerset $\mathsf{P}S$ and $\varnothing \in \mathcal{F}$, then $\mathcal{F} = \mathsf{P}S$. We call this filter the *nullfilter* on S. All other filters on S are said to be *proper*. Filter convergence is usually restricted to proper filters.

A *filterbase* on a set S is a set $\mathcal{B}$ of subsets of S which is directed downwards, i.e., directed by inclusion $\subset$, so that the set $\uparrow \mathcal{B}$ of subsets $F \supset B$ of S, for some B in $\mathcal{B}$, is a filter on S. This filter is said to be *generated* by $\mathcal{B}$.

For $A \subset S$, the singleton $\{A\}$ is a filterbase, generating the filter $\uparrow A$ on S. Filters of this kind are called *principal*. In particular, filters $\uparrow \{x\}$ are denoted by $\dot{x}$ and called *point filters*.

1.3. Filter Functors

We denote by $\mathsf{F}S$ the set of all filters on a set S, and by F_0S the set of all proper filters on S. For a mapping $f: S \to T$ and a filter $\mathcal{F}$ on S, we denote by $(\mathsf{F}f)\mathcal{F}$ the filter on T generated by the direct images $f^{\to}A$ of the sets A in $\mathcal{F}$, or in a filterbase of $\mathcal{F}$. This defines a mapping $\mathsf{F}f$: $\mathsf{F}S \to \mathsf{F}T$. We note that

$$B \in (\mathsf{F}f)\mathcal{F} \Leftrightarrow f^{\leftarrow}B \in \mathcal{F},$$

for a subset B of T and its inverse image $f^{\leftarrow}B$ by f. We note that $\mathsf{F}f$ always maps proper filters to proper filters, and a nullfilter to a nullfilter.

It is easily seen that the sets and mappings of filters just defined define a *filter functor* F and by restrictions a *proper filter functor* F_0 on sets.

1.4. Cofinal Mappings and Subnets

We say that a mapping $f\colon E \to D$ of directed sets is *cofinal* if for every i in D, there is j in E such that $f^{\to}(\uparrow j) \subset \uparrow i$, i.e., $f(j') \geqslant i$ for all $j' \geqslant j$ in E. We do not require that f preserves preorders.

We say that a net $t\colon E \to S$ is a *subnet* of a net $s\colon D \to S$ if $t = s \cdot f$ for a cofinal mapping $f\colon E \to D$. The subnet relation is clearly reflexive and transitive, and subnets of a net always form a proper class.

Subsequences are special subnets, but subnets of a sequence need not be subsequences.

1.5. Finer Filters

We say that a filterbase $\mathcal{C}$ on a set S is *finer* than a filterbase $\mathcal{B}$ on S, and we put $\mathcal{C} \leqslant \mathcal{B}$, if for every set $B \in \mathcal{B}$ there is a subset C of B in $\mathcal{C}$. Note that $\mathcal{C} \leqslant \mathcal{B}$ iff $\uparrow\mathcal{C} \leqslant \uparrow\mathcal{B}$ for the induced filters.

This clearly defines a preorder for filterbases on S, with $\mathcal{B}$ and $\mathcal{C}$ equivalent iff they generate the same filter. For filters $\mathcal{F}$ and $\mathcal{G}$ on S, $\mathcal{G} \leqslant \mathcal{F}$ means that $\mathcal{F} \subset \mathcal{G}$ as subsets of PS.

With the order just defined, FS is a complete lattice, with the nullfilter as least element and $\{S\}$ as greatest element. For filters $\mathcal{F}$ and $\mathcal{G}$ on S, the meet $\mathcal{F} \wedge \mathcal{G}$ consists of all $A \cap B$, and the join $\mathcal{F} \vee \mathcal{G}$ of all $A \cup B$, with $A \in \mathcal{F}$ and $B \in \mathcal{G}$. The supremum $\sup \mathcal{F}_i$ of a family of filters on S consists of all sets $\bigcup A_i$ with $A_i \in \mathcal{F}_i$ for every $\mathcal{F}_i$. It follows easily that every map Ff preserves suprema. We note that $\sup \mathcal{F}_i$ is the intersection of the sets $\mathcal{F}_i$.

Using $\leqslant$ for "finer" avoids a lot of order reversals and considerably simplifies filter algebra; see [8] or [11].

1.6. Induced Filters

For a net $s\colon D \to S$, the sets $s^{\to}\uparrow i = \{s_j \mid j \geqslant i\}$, for $i \in D$, form a filterbase. The filter on S with this base is called the *induced filter* of s; Bourbaki calls it the *Fréchet filter* of s.

If we denote by $\mathcal{I}_D$ the induced filter of the net id_D in D, with the sets $\uparrow i$ as base, then the induced filter of a net $s\colon D \to S$ is $(\mathsf{F}s)\mathcal{I}_D$.

A mapping $f\colon E \to D$ of directed sets is cofinal iff $(\mathsf{F}f)\mathcal{I}_E \leqslant \mathcal{I}_D$; this follows immediately from the definitions. It follows from this that the induced filter $(\mathsf{F}s)(\mathsf{F}f)\mathcal{I}_E$ of a subnet $s \cdot f$, with $f\colon E \to D$, is always finer than the induced filter of the net s. We shall need the following well-known results.

PROPOSITION:

(1) Every proper filter $\mathcal{F}$ on S is induced by a net u such that every net in S with induced filter $\mathcal{F}$ is a subnet of u.

(2) If $\mathcal{F} \wedge \mathcal{G}$ is proper for nets $s\colon D \to S$ and $t\colon E \to S$ with induced fil-

ters $\mathcal{F}$ and $\mathcal{G}$, and in particular if $\mathcal{G} \leqslant \mathcal{F}$, then s and t have a common subnet with induced filter $\mathcal{F} \wedge \mathcal{G}$.

Proof: For (1), consider the set of all pairs (x,A) with $x \in A \in \mathcal{F}$, with $(x', A') \geqslant (x, A)$ iff $A' \subset A$. Then $u\colon (x, A) \mapsto x$ defines u, with $s = u \cdot f$ for a net s with induced filter $\mathcal{F}$ and $f\colon i \mapsto (s_i, s^{\rightarrow}{\uparrow}i)$.

For (2), let P be the set of all (i,j) in $D \times E$ such that $s_i = t_j$, with the order induced by the product order of $D \times E$. This set is directed, the projections $D \xleftarrow{p} P \xrightarrow{q} E$ are cofinal, and $s \cdot p = t \cdot q$ is the desired subnet.

From here on, everything is straightforward; we omit the details. ❑

2. FILTER AND NET CONVERGENCE

2.1. Convergence

We consider in this paper spaces consisting of a set X with a relation q from filters or nets on X to X, called *convergence*, with suitable axioms. For such spaces X and Y, a function $f\colon X \to Y$ is called *continuous* if it preserves convergence, i.e., whenever a filter $\mathcal{F}$ or a net s converges to x in X, then the filter $(\mathsf{F}f)\mathcal{F}$ or the net $f \cdot s$ converges to $f(x)$ in Y.

For a topological space X, we say that a filter $\mathcal{F}$ on X converges to x in X if $\mathcal{F}$ is finer than the neighborhood filter of x, i.e., every neighborhood of x in X is in $\mathcal{F}$, and we say that a net s in X converges to x in X if the induced filter of s converges to x, i.e., s is eventually in V for every neighborhood V of x in X.

We say here that a net $s\colon D \to X$ is *eventually* in a subset A of X if there is $i \in D$ such that $s_j \in A$ for all $j \geqslant i$. This is the case iff A is in the induced filter of s.

2.2. Convergence Axioms

Convergence of filters in a topological space X is a relation q from $\mathsf{F}X$ to X which clearly satisfies the following axioms, for proper filters on X and points of X.

FQ$_1$. $\dot{x}\,q\,x$.
FQ$_2$. If $\mathcal{F} q x$ and $\mathcal{G} \leqslant \mathcal{F}$, then $\mathcal{G} q x$.
FQ$_3$. If every filter finer than $\mathcal{F}$ admits a finer filter which converges to x, then $\mathcal{F}$ converges to x.

Convergence of nets in a topological space X is a relation q from nets in X to X which satisfies the following axioms, for nets in X and points of X.

NQ$_1$. If $s\colon D \to X$ is constant, with $s_i = x$ for all $i \in D$, then $s q x$.

NQ_2. If sqx, then every subnet of s converges to x.
NQ_3. If every subnet of a net s has a subnet which converges to x, then s converges to x.

FQ_3 and NQ_3 are known as *Urysohn conditions*.

Note that the axioms listed above do not suffice to characterize topological spaces by convergence of filters or nets; this will be taken up in Section 4.4.

2.3. Comparison

Every convergence relation for filters induces a convergence relation for nets, with a net s converging to a point x iff the filter induced by s converges to x. In this situation, NQ_1 clearly is equivalent to FQ_1, NQ_2 to FQ_2, and NQ_3 to FQ_3, and continuity for filter convergences is equivalent to continuity for the induced net convergences.

Not every net convergence is induced by a filter convergence. Discrete net convergence, with sqx allowed only if s is constant, with $s_i = x$ for all elements of its domain, is an example. A net s induces a point filter iff s is eventually constant, and if X has more than one point, then an eventually constant net need not be constant.

The following result deals with this situation.

PROPOSITION: A convergence relation for nets which satisfies NQ_2 and NQ_3 is always induced by a convergence relation for filters.

Proof: Assume that a net $s: D \to X$ converges to a point x, and let r be a net in X with the same induced filter. For every subnet t of r, the nets s and t have a common subnet by 1.6, which converges to x by NQ_2. But then r converges to x by NQ_3. ❑

2.4. Ultrafilters

For a proper filter $\mathcal{F}$ on a set S, the following three properties are equivalent.

(i) The only filters finer than $\mathcal{F}$ are $\mathcal{F}$ and the nullfilter.

(ii) For every subset A of S, either $A \in \mathcal{F}$ or $S \backslash A \in \mathcal{F}$, but not both.

(iii) For subets A, B of S with $A \cup B \in \mathcal{F}$, we always have $A \in \mathcal{F}$ or $B \in \mathcal{F}$. A proper filter with these properties is called an *ultrafilter*. We denote by $\mathsf{U}S$ the set of all ultrafilters on S.

For a mapping $f: S \to T$, the mapping $\mathsf{F}f: \mathsf{F}S \to \mathsf{F}T$ preserves ultrafilters, since the inverse image mapping $f^{\leftarrow}$ preserves set unions and complements. We denote by $\mathsf{U}f: \mathsf{U}S \to \mathsf{U}T$ the restriction of $\mathsf{F}f$ to ultrafilters. This defines an *ultrafilter functor* U on sets.

We state the main property of ultrafilters without proof.

PROPOSITION: Every filter $\mathcal{F}$ on a set S is a supremum of ultrafilters finer than $\mathcal{F}$. In particular, every proper filter admits a finer ultrafilter.

2.5. Pseudotopological Spaces

A set X with a convergence relation for filters which satisfies FQ_1 and FQ_2 is called a *convergence space*. We recall from 2.1 that a map $f: X \to Y$ of convergence spaces is called *continuous* if f preserves filter convergence, i.e., if $\mathcal{F}$ converges to x for X, then $(\mathsf{F}f)\ \mathcal{F}$ must converge to $f(x)$ for Y.

In view of 2.4, FQ_3 can also be formulated as follows.

FQ_3. If every ultrafilter on X finer than a filter $\mathcal{F}$ on X converges to x in X, then $\mathcal{F}$ converges to x.

Convergence spaces with this property were introduced by Choquet [4]; they are called *pseudotopological spaces* or *Choquet spaces*. Since every filter is a supremum of ultrafilters, and every map $\mathsf{F}f$ preserves suprema, pseudotopological spaces and their continuous maps are characterized by the restrictions of their convergence relations to ultrafilters, with only one axiom:

$UQ_1 = FQ_1$. $\dot{x}\, qx$, for every $x \in X$.

Pseudotopological spaces define a full subcategory of convergence spaces, and topological spaces a full subcategory of pseudotopological spaces.

3. THE ITERATED LIMIT CONDITION

3.1. The Iterated Limit Condition for Nets

For a directed set D, and for a net of nets $s_m: E_m \to X$, one for every $m \in D$, we define a *diagonal net*

$$\Delta: D \times \textstyle\prod_{m \in D} E_m \to X: (m, f) \mapsto s_m(f_m)$$

with the product set domain directed by the product preorder. With this notation, Kelley's iterated limit axiom is

NQ_4. For a net $t: D \to X$ and nets $s_m: E_m \to X$, one for every $m \in D$, such that $s_m q t_m$ for every $m \in D$, if qx in X then also Δqx for the diagonal net Δ just defined.

Our task in this section is to reformulate NQ_4 for filter convergence, replacing the diagonal net construction by a natural transformation called compression. We postpone to 4.4 the proof that NQ_4 characterizes topological spaces among pseudotopological spaces.

3.2. Filter Compression

For a set S, we assign to every subset A of S a subset A^* of $\mathsf{F}S$ by putting

$$\mathcal{F} \in A^* \Leftrightarrow A \in \mathcal{F}$$

for a filter $\mathcal{F}$ on S, and we define κ_S: $\mathsf{FF}S \to \mathsf{F}S$ by putting

$$A \in \kappa_S(\Phi) \Leftrightarrow A^* \in \Phi,$$

for $A \subset S$ and a filter Φ on $\mathsf{F}S$.

This mapping was first defined by Kowalsky [8] in a different way and called *compression* The simpler definition just given was obtained in [11].

We need compression for proper filters, with the null filter $\uparrow\varnothing$ removed. This also removes $\varnothing^* = \{\uparrow\varnothing\}$. Thus, the formulas displayed above also define compression from $\mathsf{F}_0\mathsf{F}_0\, S$ to $\mathsf{F}_0 S$ (which we also denote by κ_S).

LEMMA 3.3: In 3.1, let Φ be the induced filter of the net $u: D \to \mathsf{F}_0 X$ with u_m the induced filter of s_m for $m \in D$. Then $\kappa_X(\Phi)$ is the induced filter of the diagonal net Δ.

Proof: If $A \subset X$, then $A \in \kappa_X(\Phi)$ iff $A^* \in \Phi$, and this means that there is $m \in D$ with $u_i \in A^*$, i.e., $A \in u_i$, for all $i \geqslant m$ in D. Thus $A \in \kappa_X(\Phi)$ iff for some $m \in D$ and each $i \geqslant m$ there is $f_i \in E_i$ such that $s_i^{\to}\uparrow f_i \subset A$. The f_i form the restriction to $\uparrow m$ of a family f in $\prod E_i$, and we have $h \geqslant f_i$ in E_i iff there is $g \geqslant f$ in $\prod E_i$ with $h = g_i$. Thus we require that there is $m \in D$ and $f \in \prod E_i$ with $\Delta^{\to}\uparrow(m, f) \subset A$, and this completes the proof. ❑

3.4

If we replace the filter I_D on the directed set D by a proper filter $\mathcal{F}$ on a set S, then NQ_4 becomes:

FQ_4. If t: $S \to X$ and u: $S \to \mathsf{F}X$ are mappings such that $u(s)$ converges to $t(s)$ for every $s \in S$, and if $(\mathsf{F}\,t)\mathcal{F}$ converges to x in X, then the filter $\kappa_X((\mathsf{F}\,u)\mathcal{F})$ also converges to x in X.

This only seems more general than NQ_4 for a filter convergence. The filter $\mathcal{F}$ is induced by a net $r: D \to S$; thus we just have replaced the nets t and u in NQ_4 and 3.3 by the nets $t \cdot r$ and $u \cdot r$.

If we define the *graph* of q as the set

$$\Gamma_q = \{(\mathcal{F}, x) \in \mathsf{F}X \times X \mid \mathcal{F} q x\},$$

with projections p_1, p_2 to $\mathsf{F}X$ and X, then u and t in FQ$_4$ satisfy $u = p_1 \cdot v$ and $t = p_2 \cdot v$ for a mapping $v: S \to \Gamma_q$. Thus it suffices to require FQ$_4$ only for the projections of Γ_q.

4. AXIOMS FOR ULTRAFILTER CONVERGENCE

4.1. Compression for Ultrafilters

Restriction to the set $\mathsf{U}S$ of ultrafilters on a set S defines a bijection between the set of filters on $\mathsf{F}S$ which contain $\mathsf{U}S$, and the set of filters on $\mathsf{U}S$. This replaces the sets A^* of 3.2 by the subsets $A^\#$ of $\mathsf{U}S$ characterized by

$$\varphi \in A^\# \Leftrightarrow A \in \varphi,$$

for ultrafilters φ on S.

With this notation, the formal properties of ultrafilter become:

$$(A \cap B)^\# = A^\# \cap B^\#, \qquad (A \cup B)^\# = A^\# \cup B^\#, \qquad (S\backslash A)^\# = \mathsf{U}S\backslash A^\#,$$

for subsets A, B of S. We also have $\varnothing^\# = \varnothing$, and $S^\# = \mathsf{U}S$.

It follows immediately that compression of an ultrafilter on $\mathsf{U}S$, or of its extension to an ultrafilter on $\mathsf{F}S$, is an ultrafilter on S. Thus compression restricted to ultrafilters defines a mapping μ_X: $\mathsf{UU}S \to \mathsf{U}S$, characterized by

$$A \in \mu_X\Phi \Leftrightarrow A^\# \in \Phi,$$

for $\Phi \in \mathsf{UU}S$ and $A \subset S$.

Restriction of FQ$_4$ to ultrafilter convergence now becomes

UQ$_4$. If t: $S \to X$ and u: $S \to \mathsf{U}X$ are mappings such that $u(s)$ converges to $t(s)$ for every $s \in S$, and if $(\mathsf{U}t)\varphi$ converges to x in X for an ultrafilter φ on S, then the ultrafilter $\mu_X((\mathsf{U}u)\varphi)$ also converges to x in X.

As in 3.4, it suffices to require UQ$_4$ only for the case that u and t are the projections of the graph Γ_q of q, restricted to ultrafilters.

4.2. Closures

Ultrafilter convergence in a convergence space X allows us to define the *closure* $\overline{A}$ of every set $A \subset X$ as the set of all limits of ultrafilters φ with $A \in \varphi$, i.e.,

$$\overline{A} = t^{\to}\, u^{\leftarrow}\, A^\#$$

for the projections u and t of Γ_q, restricted to ultrafilter convergence. We call A *closed* if $\overline{A} = A$.

This closure always satisfies three of the four Kuratowski axioms:

$$\overline{\varnothing} = \varnothing, \qquad A \subset \overline{A}, \qquad \overline{A \cup B} = \overline{A} \cup \overline{B}.$$

The first two axioms say that ultrafilters are proper, and that $\dot{x} q x$ for every x in X, and the third one follows from the fact that $(A \cup B)^{\#} = A^{\#} \cup B^{\#}$.

4.3

The following Lemma will be useful.

LEMMA [7]: For an ultrafilter φ on a convergence space X and a subset A of X with $\overline{A} \in \varphi$, there is an ultrafilter ψ on X with $A \in \psi$, and with $\overline{B} \in \varphi$ for every set $B \in \psi$.

Proof: Assign to every $x \in \overline{A}$ an ultrafilter σ_x converging to x with $A \in \sigma_x$, putting $\sigma_x = \dot{x}$ for $x \in A$. For $B \subset A$, let B' be the set of all x in $\overline{A}$ with $B \in \sigma_x$. Then $\varnothing' = \varnothing$ and $A' = \overline{A}$, and

$$B \subset B' \subset \overline{B}, \qquad (B \cap C)' = B' \cap C', \qquad (B \cup C)' = B' \cup C',$$

for subsets B, C of A. It follows that the sets B with $B' \in \varphi$ form an ultrafilter on A, extending to an ultrafilter ψ on X such that $\overline{B} \in \varphi$ for $B \subset A$ in ψ, and hence for all B in ψ. ❑

4.4

We now state our main result.

THEOREM: For a pseudotopological space X, the following are equivalent.

(1) X satisfies NQ_4.
(2) X satisfies FQ_4.
(3) X satisfies UQ_4.
(4) If φ and ψ are ultrafilters on X such that $\overline{B} \in \varphi$ for every B in ψ and $\varphi q x$, then $\psi q x$.
(5) Closures in X define a topology, and ultrafilter convergence in X is ultrafilter convergence for this topology.

Proof: We have seen in 3.4 and 4.1 that (1) $\Leftrightarrow$ (2) and (2) $\Rightarrow$ (3).

Now assume UQ_4 with u and t the projections of the graph Γ_q of q, restricted to ultrafilters. For ultrafilters φ and ψ with $\overline{B} \in \varphi$ for B in ψ, the sets $A \cap \overline{B}$ with $A \in \varphi$ and $B \in \psi$ are not empty. Thus there is always a pair (ρ, x) in Γ_q with $B \in \rho$ and $x \in A$, hence $\rho \in B^{\#}$ and

$$(\rho, x) \in u^{\leftarrow}(B^{\#}) \cap t^{\leftarrow}(A).$$

The sets $u^{\leftarrow}(B^{\#}) \cap t^{\leftarrow}(A)$ clearly form a filterbase on Γ_q; thus, there is an ultrafilter Φ on Γ_q with these sets in Φ. The sets $B^{\#}$ then are in $(\mathrm{U}u)\Phi$ and the sets B in $\mu_X((\mathsf{U}u)\Phi)$, and the sets A are in $(\mathsf{U}t)\Phi$. As these filters are ultrafilters, we have

$$(\mathsf{U}t)\Phi = \varphi \quad \text{and} \quad \mu_X((\mathsf{U}u)\Phi) = \psi.$$

If UQ_4 holds and $\varphi q x$, then $\psi q x$ follows.

Now assume (4). For x in the closure of $\overline{A}$, there is an ultrafilter φ on X with $\overline{A} \in \varphi$ and $\varphi q x$. By 4.3, there is an ultrafilter ψ on X with $A \in \psi$, and $\overline{B} \in \varphi$ for all $B \in \psi$. But then $\psi\ q\ x$ by (4). Thus $\overline{A}$ is closed, and the closure for q topological.

If $\psi\ q\ x$, then $x \in \overline{B}$ for all $B \in \psi$; thus ψ converges to x for the topology given by the closures. Conversely, if $x \in \overline{B}$ for all $B \in \psi$, then $\overline{B} \in \dot{x}$ for all $B \in \psi$. Since $\dot{x}\, q x$, we get $\psi q x$ from (4), and the given convergence is convergence for the topology.

Assume finally that X is a topological space. For an open neighborhood U of x in FQ_4, there is a set A in $\mathcal{F}$ with $t^{\rightarrow}A \subset U$. Since $u(s)$ converges to $t(s)$ and U is open, it follows that $U \in u(s)$, hence $u(s) \in U^*$ for every $s \in A$. But then $u^{\rightarrow}A \subset U^*$, so that U^* is in $(\mathsf{F}u)\mathcal{F}$, and U in $(\kappa_X \cdot \mathsf{F}u)\mathcal{F}$. Thus this filter converges to x as claimed. ❑

4.5. Discussion

The equivalence of (2) and (3) with (5) in Theorem 4.4 goes back to Kowalski [8] and Barr [1], but their equivalence with (4) is new. Conditions (1) through (3) can be considered as algebraic, but (4) is definitely topological. Condition (4) with the last part turned around to $\psi q x \Rightarrow \varphi q x$ would characterize regularity for pseudotopological spaces; thus being topological is in a sense a converse of regularity.

Compact Hausdorff spaces are characterized by the requirement that convergence of ultrafilters is a mapping $q\colon \mathsf{U}X \to X$. If we replace S in UQ_4 with $\mathsf{U}X$, with $t = q$ and $u = \mathrm{id}$, then UQ_1 of 2.5 and UQ_4 become

$$q \cdot \eta_X = \mathrm{id}\, X, \qquad \text{and} \qquad q \cdot \mu_X = q \cdot \mathsf{U}q,$$

with $\eta_X\colon x \mapsto \dot{x} : X \to \mathsf{U}X$. Thus, specializing 4.4 to compact Hausdorff spaces provides a simple proof of the theorem of Manes [9] that compact Hausdorff spaces are the algebras for the ultrafilter monad on sets.

REFERENCES

1. BARR, M. 1970. Relational algebras. Reports of the Midwest Category Seminar IV. Lecture Notes in Math. **137:** 39–55.
2. CARTAN, H. 1937. Théorie des filtres. C.R. Acad. Sci. Paris. **205:** 595–598.
3. CARTAN, H. 1937. Filtres et ultrafiltres. C.R. Acad. Sci. Paris. **205:** 777–779.
4. CHOQUET, G. 1948. Convergences. Ann. Univ. Grenoble. Sect. Sci. Math. Phys. (N.S.). **23:** 57–112.

5. FRÉCHET, M. 1906. Sur quelques points du calcul fonctionnel. Rendic. Circ. Mat. Palermo. **22:** 1–74.
6. KELLEY, J.L. 1950. Convergence in topology. Duke Math. Jour. **17:** 277–283.
7. KELLEY, J.L. 1955. General Topology. Van Nostrand. New York.
8. KENT, D.C. & G.R. RICHARDSON. 1973. The decomposition series of a convergence space. Czech. Math. Journal. **23:** (98) 437–446.
9. KOWALSKY, H.-J. 1954. Beiträge zur topologischen Algebra. Math. Nachrichten. **11:** 143–186.
10. MANES, E.G. 1967. A Triple Miscellany. Some aspects of the theory of algebras over a triple. Ph.D. Dissertation. Wesleyan University. Middletown, Connecticut.
11. MOORE, E.H. & H.L. SMITH. 1922. A general theory of limits. Amer. Jour. Math. **44:** 102–121.
12. WYLER, O. 1974. Filter space monads, regularity, completions. TOPO 72 — General Topo logy and Its Applications. Lecture Notes in Math. **378:** 591–637.

Characterization of the Club Forcing

JINDRICH ZAPLETAL[a]

Mathematical Sciences Research Institute
1000 Centennial Drive
Berkeley, California 94720

ABSTRACT: We provide an external characterization of the algebra $RO(R)$, where R is Baumgartner's poset for adding a club with finite conditions. As a corollary, we show that many complete subalgebras of $RO(R)$ must be isomorphic to either $\mathbb{C}_{\aleph_1}$ or $RO(R)$.

0. INTRODUCTION

It appears that there are not many qualitatively different forcings of size $\aleph_1$ in ZFC. Among these, $C_{\aleph_1}$ or adding $\aleph_1$ Cohen reals occupies a very special place. Koppelberg in [4] gave an external characterization of the complete Boolean algebra $\mathbb{C}_{\aleph_1} = RO(C_{\aleph_1})$:

THEOREM 1: Let P be a poset of uniform density $\aleph_1$ such that:
(1) the set $\{Q \in [P]^{\aleph_0}: Q$ is a regular subposet of $P\}$ contains a club.
Then $RO(P) \equiv \mathbb{C}_{\aleph_1}$.

Since the criterion in question is hereditary to regular subposets, it follows that $\mathbb{C}_{\aleph_1}$ is a minimal algebra of uniform density $\aleph_1$ in the quasiorder $\lessdot$, the complete embeddability. This fact opened the possibility of proving [6] that under the Proper Forcing Axiom, $\mathbb{C}_{\aleph_1}$ is indeed the smallest algebra in the given quasiorder among the complete Boolean algebras of uniform density $\aleph_1$.

There is another very simple forcing notion R of size $\aleph_1$, introduced by Baumgartner [1] — "adding a club with finite conditions," for a definition see below. The poset R differs from $C_{\aleph_1}$ basically only by the fact that it is nowhere c.c.c. (countable chain condition). Our thesis is that R occupies a similar position to that of $C_{\aleph_1}$ among the *nowhere c.c.c.* forcings. To support this thesis, we prove the following external characterization of $RO(R)$:

THEOREM 2: Let P be a forcing of size $\aleph_1$, $P = \{p_\alpha: \alpha \in \omega_1\}$, such that:
(1) P is proper;
(2) P is nowhere c.c.c.;

Mathematics Subject Classification: 03E40, 06E10.
Key words and phrases: club forcing.
[a]Research at MSRI supported in part by NSF grant #DMS 9022140.
E-mail: jindra@msri.org.

(3) $P \Vdash$ "there is a club $C \subset \omega_1$ such that for any $\alpha \in C$, the filter $\dot{G} \cap \{p_\beta : \beta \in \alpha\}$ is a V-generic subset of $\{p_\beta : \beta \in \alpha\}$."

Then $RO(P) \equiv RO(R)$.

Upon a moment's thought, the criterion (3) is quite parallel to that of (1) of Theorem 1. It easily follows that $RO(R)$ is minimal among complete nowhere c.c.c. algebras in the quasiorder $\lessdot_p$, where $\mathbb{A} \lessdot_p \mathbb{B}$ if the algebra $\mathbb{A}$ is completely embeddable into $\mathbb{B}$ so that $\mathbb{A} \Vdash$ "$\mathbb{B}/\mathbb{A}$ is proper" — Corollary 5(2). The minimality of $RO(R)$ with respect to $\lessdot$ in the above class remains an open problem, as well as whether under the Proper Forcing Axiom, $RO(R)$ is the smallest complete nowhere c.c.c. algebra of weight $\aleph_1$ with respect to $\lessdot$.

Our notation follows the set-theoretic standard as set forth in [3]. The words "algebra, embedding, embeddable" always stand for "complete Boolean algebra, complete embedding, completely embeddable," respectively. For partially ordered sets $Q \subset P$ we write $Q \lessdot P$ to mean that Q is a regular subposet of P, i.e., $\forall p \in P \, \exists q \in Q \, \forall r \in Q \; r \leq q \rightarrow r$ is compatible with p. A witness $q \in Q$ for $p \in P$ in the above formula is called a pseudoprojection of p to Q. If $\mathbb{A} \lessdot \mathbb{B}$ are complete algebras and $b \in \mathbb{B}$ then $\mathrm{proj}_{\mathbb{A}}\, b$ denotes the projection of b into $\mathbb{A}$. If $0 \neq b \in \mathbb{B}$ then $\mathbb{B} \restriction b$ is the Boolean algebra with relativized operations on the set $\{c \in \mathbb{B} : c \leq b\}$. The relation $\equiv$ is used to indicate isomorphism between algebras. The symbol $\dot{G}$ always denotes the canonical name for a generic filter on the relevant partial order or algebra. For a statement ϕ of the $\mathbb{A}$-forcing language, $\|\phi\|_{\mathbb{A}}$ stands for the Boolean value of ϕ in $\mathbb{A}$. κ denotes a large regular cardinal, say $(2^{\aleph_1})^+$, and H_κ is the collection of all sets hereditarily of size $< \kappa$. The relation $\prec$ is that of elementary embeddings between models.

1. THE PROOFS

In [1], Baumgartner introduced a version of the following forcing:

DEFINITION 3: The poset R is the set $\{\langle a, b\rangle : a \in [\omega_1]^{<\aleph_0}$, b is a finite set of clopen intervals of countable ordinals and $a \cap \bigcup b = 0\}$ ordered by $\langle a_0, b_0\rangle \leq \langle a_1, b_1\rangle$ just in case $a_1 \subset a_0$ and $\bigcup b_1 \subset \bigcup b_0$.

If a set $S \subset \omega_1$ is stationary then R_S is defined exactly as R only with the restriction that $a \in [S]^{<\aleph_0}$.

Thus, R_S shoots a closed unbounded set through S. The forcing R has a number of useful properties, of which we list just a few. No part of the following lemma is due to us, and we refer the reader to [1] for detailed proofs.

LEMMA 4:

(1) Let $G \subset R$ be a generic filter. Then G can be decoded from the club

set $D = \bigcup\{a : \langle a, 0\rangle \in G\} \subset \omega_1$.

(2) R is proper and nowhere c.c.c.

(3) Let $R = \{r_\alpha : \alpha \in \omega_1\}$. Then $R \Vdash$ "there is a club $\dot{C} \subset \omega_1$ such that for any $\alpha \in \dot{C}$ the filter $\dot{G} \cap \{r_\beta : \beta \in \alpha\} \subset \{r_\beta : \beta \in \alpha\}$ is V-generic."

Proof:

(1) Obviously, $G = \{\langle a, b\rangle \in R : a \subset D \text{ and } \bigcup b \cap D = 0\}$. That D is closed unbounded in ω_1 follows from the usual genericity arguments.

(2) R is proper: let $M \prec H_\kappa$ be a countable elementary submodel of a large enough structure, $r_0 = \langle a, b\rangle \in R \cap M$ and $\delta = M \cap \omega_1$. Then $r_0 > r_1 = \langle a \cup \{\delta\}, b\rangle$ is a master condition for M; indeed $r_1 \Vdash$ "$\dot{G} \cap M \subset \check{R} \cap M$ is a V-generic filter." R is nowhere c.c.c. since D is a new club subset of ω_1 which is forced by 1_R not to have an infinite subset in the ground model.

Let $\langle M_i : i \in \omega_1\rangle$ be a continuous $\in$-tower of countable elementary submodels of some large H_κ with the enumeration of R in question an element of M_0. Let $\dot{C}$ be the R-name defined by $\dot{C} = \dot{D} \cap \{M_i \cap \omega_1 : i \in \omega_1\}$. Then $\dot{C}$ is forced to be a club and by the observation in (2) above, $R \Vdash$ "for all $\alpha \in \dot{C}$, the filter $\dot{G} \cap \{r_\beta : \beta \in \alpha\} \subset \{r_\beta : \beta \in \alpha\}$ is V-generic." ❑

The set $D \subset \omega_1$ from (1) is referred to as the R-generic club. In this paper, we show that conditions (2) and (3) above fully characterize the forcing R or rather its completion $RO(R)$.

THEOREM 2: Let P be a forcing of size $\aleph_1$, $P = \{p_\alpha : \alpha \in \omega_1\}$, such that:

(1) P is proper;

(2) P is nowhere c.c.c.;

(3) $P \Vdash$ "there is a club $\dot{C} \subset \omega_1$ such that for any $\alpha \in \dot{C}$, the filter $\dot{G} \cap \{p_\beta : \beta \in \alpha\}$ is a V-generic subset of $\{p_\beta : \beta \in \alpha\}$."

Then $RO(P) \equiv RO(R)$.

It should be noted that requirement (3) above is really a property of $RO(P)$ whose validity does not depend on the particular choice of a dense subset of $RO(P)$ and its enumeration.

REMARK: None of the requirements can be dropped. If properness is dropped then R_S would be a counterexample for a stationary costationary set $S \subset \omega_1$. The poset R_S, unlike R, collapses stationarity of $\omega_1 \setminus S$ and so $RO(R)$ and $RO(R_S)$ cannot be isomorphic. If (2) is replaced with "P is c.c.c." then the conclusion is changed to $RO(P) \equiv \mathbb{C}_{\aleph_1}$ according to Theorem 1. Note that for a c.c.c. forcing P the club set $\dot{C}$ can be taken from the ground model, giving a club of countable regular subposets of P. Also requirement (3) cannot be weakened to "$\dot{C}$ stationary." Here, a counterexample is provided by $P = R \times Q$, with Q any pseudo-Cohen not Cohen forcing of size $\aleph_1$, see [2]. An isomorphism of $RO(P)$ and $RO(R)$ would contradict Corollary 5(1).

Theorem 2 results in a minimality of sorts of the algebra $RO(R)$. Let $\lessdot_p$, "properly complete embeddability" be the quasiorder defined in the introduction. The relation $\lessdot_p$ is quite natural on the class of proper algebras and by virtue of Theorem 2, $RO(R)$ has a distinguished place in this quasiorder.

COROLLARY 5: Let $\mathbb{A} \lessdot_p RO(R)$ be an algebra of uniform density $\aleph_1$ which is uniform in chain condition. Then there are only two cases:

(1) $\mathbb{A}$ is c.c.c. Then $\mathbb{A} \equiv \mathbb{C}_{\aleph_1}$.

(2) $\mathbb{A}$ is nowhere c.c.c. Then $\mathbb{A} \equiv RO(R)$.

Thus $RO(R)$ is minimal in the class of nowhere c.c.c. algebras with respect to $\lessdot_p$.

REMARK: It is impossible to drop the subscript p in Corollary 5. To show that, we must define a new c.c.c. forcing of size $\aleph_1$. Let us first choose a ladder $\langle l_\alpha: \alpha < \omega_1 \text{ limit}\rangle$, i.e., every $l_\alpha \subset \alpha$ is a cofinal set of ordertype ω. For a costationary set $S \subset \omega_1$, we define a poset P_S [7], [2]: $P_S = \{\langle a, b\rangle: a$ is a finite set of countable limit ordinals and f is a finite function from ω_1 to $2\}$ ordered by $\langle a, f\rangle \leq \langle b, g\rangle$ just in case $b \subset a$, $g \subset f$ and $\forall\alpha \in b\ \forall\beta \in \mathrm{dom}(f\backslash g)$ $\beta \in l_\alpha \rightarrow f(\beta) = 1$. This poset is easily seen to be c.c.c. and moreover, it has the following property. Let $P_S{}^\alpha \subset P_S$ be the set $\{\langle a, f\rangle \in P_S: a \subset \alpha, \mathrm{dom}(f) \subset \alpha\}$; then for limit $\alpha \in \omega_1$, $\alpha \notin S$ iff $P_S{}^\alpha \lessdot P_S$. Now we can define a c.c.c. complete subalgebra $\mathbb{B}$ of $RO(R)$ with $\mathbb{B} \not\equiv \mathbb{C}_{\aleph_1}$. Just let $\mathbb{B} = RO(C_{\omega_1} * P_{\dot{S}})$, where $\dot{S} = \{\alpha \in \omega_1: \dot{c}(\alpha) = 1\}$ and $\dot{c}: \omega_1 \rightarrow 2$ is the C_{ω_1}-generic function. $\mathbb{B} \not\equiv \mathbb{C}_{\aleph_1}$ by Theorem 1, since it has very few regular subalgebras — note that $Q_\alpha = C_\alpha * P_{\dot{S}}{}^\alpha \not\lessdot C_{\omega_1} * P_{\dot{S}}$ for any $\alpha \in \omega_1$, since the element $\langle\{\langle\alpha, 1\rangle\}, 0\rangle \in C_{\omega_1} * P_{\dot{S}}$ does not have a pseudoprojection into Q_α. The reader can check that $RO(R) \equiv RO(C_{\omega_1} * P_{\dot{S}} * R_{\omega_1\backslash\dot{S}})$ by Theorem 2; so $\mathbb{B} \lessdot RO(R)$ is the desired counterexample.

To define a nowhere c.c.c. complete subalgebra $\mathbb{A}$ of $RO(R)$ with $\mathbb{A} \not\equiv RO(R)$ is only a minor variation on the above scheme: we shall put $\mathbb{A} = RO(C_{\omega_1} * P_{\dot{S}} * \check{R}) \equiv \mathbb{B} \times RO(R)$. $\mathbb{A}$ is not isomorphic to $RO(R)$ since $\mathbb{B} \lessdot_p \mathbb{A}$ would contradict Corollary 5(1) and $\mathbb{A}$ can be embedded into $RO(R)$ in the same way $\mathbb{B}$ can: by Theorem 2, $RO(R) = \mathbb{A} * RO(R_{\omega_1\backslash\dot{S}})$.

Proof of Corollary 5: It is enough to show that the property (3) from Theorem 2 is inherited by $\lessdot_p$ smaller algebras. To this end let $\mathbb{A} \lessdot_p \mathbb{B}$ be such that $Q = \{q_\alpha: \alpha \in \omega_1\}$ and $P = \{p_\alpha: \alpha \in \omega_1\}$ are dense subsets of $\mathbb{A}, \mathbb{B}$ respectively, with $Q \subset P$ and let P satisfy (3) of the Theorem, with a P-name $\dot{C}$. We claim that the set $\dot{D} \in V^{\mathbb{A}}$ defined by $\alpha \in \dot{D}$ iff $\dot{G} \cap \{q_\beta: \beta \in \alpha\}$ is a V-generic filter over $\{q_\beta: \beta \in \alpha\}$ contains a club in $V^{\mathbb{A}}$. (Here, $\dot{G}$ is the $\mathbb{A}$-name for the generic filter on $\mathbb{A}$.) To prove this, note that there is a club $E \subset \omega_1$ such that for every $\alpha \in E$ we have $\{q_\beta: \beta \in \alpha\} \lessdot \{p_\beta: \beta \in \alpha\}$. This is because $Q \lessdot P$ and $\lessdot$ is a first-order relation on posets. Thus, $\mathbb{B} \Vdash$ "$\dot{C} \cap \check{E}$

$\subset \dot{D}$" and in $V^{\mathbb{B}}$, the set $\dot{D} \subset \omega_1$ contains a club. But since $\mathbb{A} \Vdash$ "$\mathbb{B}/\mathbb{A}$ is proper," the set $\dot{D}$ must contain a club already in $V^{\mathbb{A}}$. ❑

Proof of Theorem 2: Let us fix a poset $P = \{p_\alpha : \alpha \in \omega_1\}$ satisfying the requirements (1), (2), (3) of Theorem 2 with a P-name $\dot{C}$ witnessing the condition (3). Let $\langle M_i : i \in \omega_1 \rangle$ be a continuous $\in$-tower of countable elementary submodels of some large H_κ such that both $P = \{p_\alpha : \alpha \in \omega_1\}$ and $\dot{C}$ belong to M_0. Let $\delta_i = M_i \cap \omega_1$ for $i \in \omega_1$. The following lemma is crucial.

LEMMA 6: $P \Vdash$ "$\dot{D} = \{i \in \omega_1 : \delta_i \in \dot{C}\}$ is an R-generic club."

Of course, the set $\dot{D}$ does not code the whole P-generic filter, but it does provide us with the most important information.

Proof of Lemma 6: We shall prove the lemma by computing a pseudoprojection of an arbitrary condition in P into R associated with the name $\dot{D}$ for an R-generic club. More precisely, for each $p \in P$ an element $\langle a, b \rangle \in R$ will be produced such that for every condition $\langle \bar{a}, \bar{b} \rangle \leq \langle a, b \rangle$ in R there is $\bar{p} \leq p$ in P forcing that "$\bar{a} \subset \dot{D}$, $\bigcup \bar{b} \cap \dot{D} = 0$." By a standard argument, this proves that $\dot{D}$ is forced to be R-generic. So fix an arbitrary $p \in P$.

First, a piece of notation. If $q \Vdash \check{\delta}_i \in \dot{C}$" then there is a condition $\bar{q} \in M_i \cap P = \{p_\alpha : \alpha \in \delta_i\}$ so that for every $\bar{\bar{q}} \leq \bar{q}$, if $\bar{\bar{q}} \in M_i$ then $\bar{\bar{q}}$ is compatible with q. (Otherwise, the set $X = \{r \in M_i \cap P : r$ is incompatible with $q\}$ would be dense in $M_i \cap P$ and $q \Vdash$ "$\check{X} \cap \dot{G} = 0$" contrary to the assumption that $q \Vdash$ "$\dot{G} \cap M_i \subset P \cap M_i$ is a V-generic filter.") Any such condition $\bar{q}$ will be called an i — *pseudoprojection* of q. By induction construct sequences $p = q_0, q_1, \ldots, q_{2n}, q_{2n+1}, \ldots$ of conditions in P and $i_0, i_1, \ldots, i_{2n}, i_{2n+1}, \ldots$, of ordinals so that:

(1) $q_{2n} \in P \cap M_{i_{2n}}$ and i_{2n} is the least such ordinal. In particular, i_{2n} is a successor ordinal.

(2) $q_{2n+1} \leq q_{2n}$, $q_{2n+1} \in P \cap M_{i_{2n}}$ and $q_{2n+1} \Vdash$ "i_{2n+1} = the index of $\max(\dot{C} \cap \{\delta_i : i < i_{2n}\})$." Note that both sets in the intersection are (forced to be) closed (i_{2n} is a successor ordinal!), their names are in $M_{i_{2n}}$ and $\{\delta_i : i < i_{2n}\}$ is bounded in $\delta_{i_{2n}}$. If the intersection is forced to be empty, stop the induction.

(3) q_{2n+2} is any i_{2n+1} — pseudoprojection of q_{2n+1}.

This is readily done, and since $i_0 > i_1 \geq i_2 > i_3 \geq \ldots$ there will be some $\bar{n} \in \omega$ such that $q_{2\bar{n}+1} \Vdash$ "$\dot{C} \cap \{\delta_i : i < i_{2\bar{n}}\} = 0$." We set

$$a = \{i_1, i_3, \ldots, i_{2\bar{n}-1}\} \text{ listed in the decreasing order, and}$$

$$b = \{(i_1, i_0), (i_3, i_2), \ldots\} \text{ listed in the decreasing order.}$$

Here, the last interval in b will be either $(i_{2\bar{n}-1}, i_{2\bar{n}-2})$ or $[0, i_{2\bar{n}})$ depending on whether $i_{2\bar{n}-1} = 0$ or not.

We claim that $\langle a, b\rangle \in R$ has the desired properties of a pseudoprojection of the condition $p \in P$ into R. To show this, choose a condition $r = \langle \bar{a}, \bar{b}\rangle \leq \langle a, b\rangle$ in R. We shall produce a condition $\bar{p} \leq p$ in P such that $\bar{p} \Vdash$ "$\bar{a} \subset \dot{D}$, $\bigcup \bar{b} \cap \dot{D} = 0$, or in other words, $\langle \bar{a}, \bar{b}\rangle$ belongs to the filter on $\check{R}$ generated by $\dot{D}$." This will complete the proof of the Lemma.

Let us split the condition r into $\bar{n} + 1$ pieces $r_{\bar{n}}, \ldots, r_0 \in R$ so that each r_n includes exactly the ordinals and intervals mentioned in r that fall between i_{2n+1} inclusive and i_{2n-1}, where $i_{2n+1} = 0$ and $i_{-1} = \omega_1$. Obviously $r = \inf_{n \in \bar{n}+1} r_n$ in R. We now work from below, constructing a decreasing chain $1_P = z_{\bar{n}+1} \geq \ldots \geq z_0$ of conditions in P so that:

(1) $z_n \in P \cap M_{i_{2n-1}}$;

(2) $z_n \leq z_{n+1}$ and $z_n \leq q_{2n+1}$;

(3) $z_n \Vdash$ "$\check{r}_n$ belongs to the filter on $\check{R}$ generated by $\dot{D}$."

The conjunction in (2) is possible to realize in P since $z_{n+1} \in M_{i_{2n+1}}$ is less than q_{2n+2} which is an i_{2n+1} — pseudoprojection of q_{2n+1}. Then $\bar{p} = z_0 \leq q_1 \leq q_0 = p$ will be the desired condition. We must show how to perform task (3) above, and we do it by means of an example, which illustrates all possible substantial difficulties.

So assume that we are given $z_{n+1} \in M_{i_{2n+1}}$ and that our r_n is $\langle\{i_{2n+1}, i_{2n}\}, \{(i_{2n+1}, i_{2n}), (j, k+1)\}\rangle$, where $i_{2n} \leq j < k + 1 \leq i_{2n+1}$. Note that the ordinal i_{2n+1} and the interval (i_{2n+1}, i_{2n}) were mentioned already in the original condition $\langle a, b\rangle$. Now z_{n+1} is compatible with q_{2n+1} and both of these conditions are in $M_{i_{2n}}$, so this model also contains one of their common lower bounds, say x_0. We take care of r_n from below, ordinal by ordinal, interval by interval.

First, $q_{2n+1} \Vdash$ "$\check{\delta}_{i_{2n+1}} \in \dot{C}$ and $\dot{C} \cap \{\delta_l : i_{2n+1} < l < 2n\} = 0$." So $x_0 \leq q_{2n+1}$ forces that $i_{2n+1} \in \dot{D}$ and $(i_{2n+1}, i_{2n}) \cap \dot{D} = 0$.

Second, we take care of the ordinal $i_{2n} \in \bar{a}$. Here, the properness of the forcing P is used. We just choose any condition $x_1 \leq x_0$ in the model M_{j+1} which is master for $M_{i_{2n}}$. Then $x_1 \Vdash$ "$\dot{C}$ is unbounded in $M_{i_{2n}} \cap \omega_1$, so $\check{\delta}_{i_{2n}} \in \dot{C}$, so $i_{2n} \in \dot{D}$."

Last, we take care of the interval $(j, k+1) \in \bar{b}$. Here, the nowhere c.c.c. of the forcing P is used. Since $x_1 \in M_{j+1}$, in this model there is a maximal antichain $A \subset P$ together with its one-to-one enumeration $\{a_\alpha : \alpha \in \omega_1\}$ such that there are uncountably many elements of A below x_1. Now, both x_1 and A are also in M_{k+1}, and thus by elementarity there is in this model an ordinal α, $\delta_k < \alpha < \delta_{k+1}$ such that $x_2 = a_\alpha < x_1$. Then $x_2 \Vdash$ "$\dot{C} \cap \{\delta_l : j + 1 \leq l \leq k\} = 0$, since $A \cap M_l$ is a maximal antichain of $P \cap M_l$ which is not met by the generic filter for $j + 1 \leq l \leq k$." Therefore, $x_2 \Vdash$ "$(j, k+1) \cap \dot{D} = 0$."

Thus, we have constructed a chain $x_0 \geq x_1 \geq x_2$ taking care of all the tasks one-by-one and $x_2 \in M_{k+1} \subset M_{i_{2n-1}}$. Setting $z_n = x_2$, the job is finished. ❑

An inspection of the above proof makes it obvious that it really gives more:

LEMMA 7: If $p_0 \in P$ forces "$\delta_i \in \dot{C}$" then $p_0 \Vdash$ "$\dot{G} \cap M_i$ and $\dot{D} \setminus (i+1)$ represent mutually generic filters on $(P \cap M_i)$ and R."

Proof: Essentially a repetition of the argument for Lemma 6. For each $p \leq p_0$ in P a pseudoprojection of p into $(P \cap M_i) \times R \upharpoonright \langle 0, \{[0, i+1)\}\rangle$ will be computed, that is, conditions $q \in P \cap M_i$ and $\langle a, b\rangle \in R$, $[0, i+1) \subset \bigcup b$, will be found such that for every $\bar{q} \leq q$ and $\langle \bar{a}, \bar{b}\rangle \leq \langle a, b\rangle$ in the respective posets there is a condition $\bar{p} \leq p$ such that $\bar{p} \leq \bar{q}$ (so $\bar{p} \Vdash \bar{q} \in \dot{G} \cap M_i$) and $\bar{p} \Vdash$ "$\bar{a} \subset \dot{D}$, $\bigcup \bar{b} \cap \dot{D} \subset (i+1)$." That completes the proof.

Fix $p \in P$ and repeat the argument from the previous proof. Stop the first induction on an integer $\bar{n}$ such that either (*) $i_{2\bar{n}+1} = i$ or (**) $i_{2\bar{n}} \leq i$. Define $q = q_{2\bar{n}+2}$ or $q = q_{2\bar{n}}$ depending on whether (*) or (**) holds, and define $\langle a, b\rangle \in R$ as in that proof with the change that the last interval in b is either $[0, i_{2n})$ or $[0, i+1)$ again depending on whether (*) or (**) holds. Then q, $\langle a, b\rangle$ have the desired properties of a pseudoprojection of p into $(P \cap M_i) \times R \upharpoonright \langle 0, \{[0, i+1)\}\rangle$. ❑

The rest of the proof of Theorem 2 can be heuristically described as follows. Let $\mathbb{B} = RO(P)$ and let $\mathbb{A}$ be the complete subalgebra of $\mathbb{B}$ isomorphic to $RO(R)$ given by the name for an R-generic from Lemma 6. ($\mathbb{A}$ is the complete closure in $\mathbb{B}$ of the set $\{\| i \in \dot{D} \| : i \in \omega_1\}$.) Work in $V^{\mathbb{A}}$. Let $\langle i_\alpha : \alpha \in \omega_1\rangle$ enumerate the $\mathbb{A}$-generic club $\dot{D}$. We simply, step by step, add generic filters for each $M_{i_\alpha} \cap P$ so that they form a continuous chain under inclusion and agree with the generic action done on $\mathbb{A}$. Adding each of those generics is a countable action, and this action is repeated ω_1 times. Moreover, it is not hard to see that $\mathbb{B}/\mathbb{A}$ is forced by $\mathbb{A}$ to have uniform density $\aleph_1$. We shall conclude that $\mathbb{A} \Vdash$ "$\mathbb{B}/\mathbb{A} \equiv \mathbb{C}_{\aleph_1}$" and so $\mathbb{B} = \mathbb{A} * \mathbb{C}_{\aleph_1}$. Thus $RO(P)$ of every poset P satisfying (1), (2), (3) of Theorem 2 is isomorphic to $RO(R) * \mathbb{C}_{\aleph_1}$. Since R satisfies these conditions too, we have $RO(R) \equiv RO(R) * \mathbb{C}_{\aleph_1} \equiv RO(P)$ as needed.

It is not difficult to make the above heuristic precise. Let $\mathbb{A}_\alpha : \alpha \in \omega_1$ be the complete algebras defined as $\mathbb{A}_\alpha$ = the complete closure in $\mathbb{B}$ of $\mathbb{A}$ and $\{p \wedge \| i_\beta = \check{j} \|_{\mathbb{A}} : p \in P \cap M_j, j \in \omega_1, \beta \leq \alpha\}$.

LEMMA 8: For each $\alpha \in \omega_1$, $\mathbb{A} \Vdash$ "$\mathbb{A}_\alpha/\mathbb{A}$ has countable density, so is isomorphic to the Cohen algebra."

LEMMA 9: For each limit $\alpha \in \omega_1$, $\mathbb{A}_\alpha$ is the direct limit of $\mathbb{A}_\beta$, $\beta \in \alpha$.

Note that $\mathbb{B}$ is the direct limit of the $\mathbb{A}_\alpha$'s, since $\bigcup_{\alpha \in \omega_1} \mathbb{A}_\alpha \subset \mathbb{B}$ is dense. It follows that $\mathbb{A} \Vdash$ "$\mathbb{B}/\mathbb{A} \equiv \mathbb{C}_{\aleph_1}$" and $\mathbb{B} \equiv RO(R) * \mathbb{C}_{\aleph_1}$. By the reasoning above, $\mathbb{B} = RO(P) \equiv RO(R)$ and we are finished.

Towards the proof of the lemmas fix ordinals $\alpha, j \in \omega_1$ and restrict everything to $x = \| i_\alpha = j \| \in \mathbb{A} \subset \mathbb{B}$. Set $\mathbb{A}^*$ to be $\mathbb{A} \upharpoonright x$. Thus $\mathbb{A}^* \equiv \mathbb{A}_{\leq j} \times \mathbb{A}_{>j}$, where $\mathbb{A}_{\leq j}$ is the complete closure in $\mathbb{A}^*$ of the set $\{x \wedge \| k \in \dot{D} \| : k \leq j\}$ and $\mathbb{A}_{>j}$ is the complete closure of $\{x \wedge \| k \in \dot{D} \| : k > j\}$. This follows immediately from the definition of R. Moreover, set $\mathbb{B}^* = \mathbb{B} \restriction x$, $\mathbb{A}_\alpha^* = \mathbb{A}_\alpha \restriction x$ and let $\mathbb{C}$ be the complete closure in $\mathbb{B}^*$ of $\{p \wedge x : p \in P \cap M_j\}$. The following hold:

(1) $\mathbb{A}_{\leq j} \subset \mathbb{C}$;

(2) $\mathbb{A}_\alpha^* \equiv \mathbb{C} \times \mathbb{A}_{>j}$;

(3) the set $\{p \wedge x : p \in P \cap M_j\}$ is dense in $\mathbb{C}$.

The easiest way to see (1) is to note that x forces in $\mathbb{B}$ that $\dot{D} \cap j$ depends only on the behavior of the $P \cap M_j$-generic filter $G \cap M_j$. (2) follows from Lemma 7 and (1). Finally, for (3) argue that below x, the poset $P \cap M_j$ is a regular subposet of P.

Proof of Lemma 8: Fix an ordinal $\alpha \in \omega_1$ and let $a \in \mathbb{A}$. It is enough to produce a condition $b \leq a$ in $\mathbb{A}$ which forces the algebra $\mathbb{A}_\alpha/\mathbb{A}$ to have countable density. There is $j \in \omega_1$ such that $b = a \wedge \| i_\alpha = j \| \neq 0$. It easily follows from (1) and (2) above that $b \Vdash$ "$\mathbb{A}_\alpha/\mathbb{A} \equiv \mathbb{A}_\alpha^*/\mathbb{A}^* \equiv \mathbb{C}/\mathbb{A}_{\leq j}$." Since $\mathbb{C}$ has countable density from (3), necessarily $b \Vdash$ "$\mathbb{A}_\alpha/\mathbb{A}$ has countable density." That proves the Lemma. ❑

Proof of Lemma 9: Fix a limit ordinal $\alpha \in \omega_1$. It is obvious that for $\beta \in \alpha$, $\mathbb{A}_\beta \subset \mathbb{A}_\alpha$ holds. It is necessary to show that $\bigcup_{\beta \in \alpha} \mathbb{A}_\beta \subset \mathbb{A}_\alpha$ is dense. To this aim, choose $a \in \mathbb{A}_\alpha$. There is an ordinal $j \in \omega_1$ such that $b = a \wedge \| i_\alpha = j \| \neq 0$. Now by (2) and (3), there is $c \leq b$ such that $c = p \wedge d \wedge \| i_\alpha = j \|$, where $p \in P \cap M_j$ and $d \in \mathbb{A}_{>j}$. Since α is a limit ordinal and so is j, one can conclude that $M_j = \bigcup_{k \in j} M_k$ and there are ordinals $\beta \in \alpha$ and $k \in j$ such that $p \in M_k$ is compatible with $\| i_\beta = k \| \wedge \| i_\alpha = j \|$. Note that $\| i_\beta = k \| \wedge \| i_\alpha = j \| \in \mathbb{C}$ by (1) above — "$i_\beta = k$" is a statement of $\mathbb{A}_{\leq j}$ forcing language. Thus by (2) again, $e = c \wedge \| i_\beta = k \| = p \wedge \| i_\beta = k \| \wedge \| i_\alpha = j \| \wedge d \neq 0$. As $e \in \mathbb{A}_\beta$, $e \leq a$, the proof is complete. ❑

Perusing the proof of Theorem 2 one immediately finds out that a similar external characterization can be proved for the forcings R_S, $S \subset \omega_1$ stationary, only with (1) of the Theorem amended to "P is exactly S-proper," that is $P \Vdash$ "$\dot{S}$ contains a club and the stationarity of every $T \subset S$ from the ground model is preserved." This reduces many facts, which would be otherwise awkward to prove, to trivialities. For example, it follows immediately that $RO(R_S) \equiv RO(R_T)$ iff $S = T$ modulo the nonstationary ideal, and $RO(R_S) \lessdot RO(R_T)$ iff $T \subset S$ modulo the nonstationary ideal. To prove the latter fact,use the above characterization to show that $RO(R_S \times R_T) \equiv RO(R_T)$ whenever $T \subset S$. The following remains open:

QUESTION 10: Is $RO(R)$ $\lessdot$-minimal among complete nowhere c.c.c. algebras?

Confronting the remark after Corollary 5, it still can happen that every nowhere c.c.c. subalgebra of $RO(R)$ contains a copy of $RO(R)$ as a complete subalgebra!

QUESTION 11: Under Proper Forcing Axiom, is $RO(R)$ $\lessdot$-smallest among the nowhere c.c.c. algebras of density $\aleph_1$?

REFERENCES

1. BAUMGARTNER, J. 1984. Applications of the proper forcing axiom. *In:* Handbook of Set-theoretical Topology (K. Kunen & J. E. Vaughan, Eds.). North Holland. Amsterdam. 913–959.
2. BALCAR, B., T. JECH, & J. ZAPLETAL. Generalizations of Cohen algebras. To appear.
3. JECH, T. 1978. Set Theory. Academic Press. New York.
4. KOPPELBERG, S. 1993. Characterization of Cohen algebras. *In:* Papers on General Topology and Applications (S. Andima, R. Kopperman, P.R. Misra, & A.R. Todd, Eds.). Annals of the New York Academy of Sciences. New York. **704:** 227–237.
5. SHELAH, S. 1982. Proper Forcing. Springer-Verlag. New York.
6. SHELAH, S., & J. ZAPLETAL. Embeddings of Cohen algebras. Advances in Mathematics. To appear .
7. TODORCEVIC, S. Classification of transitive relations on ω_1. To appear.

Subject Index

I

J

K

L

M

N

O

P

Q

R

S

T

U

V

W

Y

Z

AMS Classifications Code Index[a]

[a]As given by authors, page numbers in parentheses indicate author's choice of Secondary AMS code.

Index of Contributors